Square: Opposite sides are
parallel, adjacent sides are
perpendicular, and all sides
have the same length.
$P = 4s$
$A = s^2$

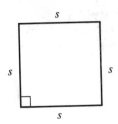

Sphere (ball):
$S = 4\pi r^2$
$V = \frac{4}{3}\pi r^3$

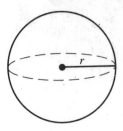

Circle:
$\quad P$ = circumference
$\quad\quad = C = 2\pi r$
or $\quad C = \pi d$
$A = \pi r^2$
$\pi \doteq 3.14 \quad$ or $\quad \pi \doteq \frac{22}{7}$

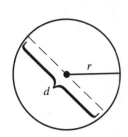

Right circular cylinder (can):
$S = 2\pi r^2 + 2\pi rh$
$V = \pi r^2 h$

Rectangular
parallelepiped (box): surface
area = S
$\quad = 2WL + 2LH + 2WH$
volume = $V = LWH$

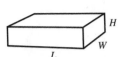

Right circular cone:
$S = \pi rs + \pi r^2$
$V = \frac{1}{3}\pi r^2 h$

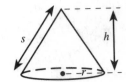

Cube: All edges have the same
length.
$S = 6e^2$
$V = e^3$

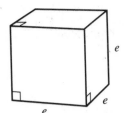

Pyramid: Let n = number of
sides of the base
A = area of base
$S = A + \frac{1}{2}nal$
$V = \frac{1}{3}Ah$

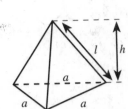

INTERMEDIATE ALGEBRA

INTERMEDIATE ALGEBRA

Fourth Edition

Dennis Weltman
North Harris County College

Gilbert Perez
San Antonio College

Terry Tiballi
Oswego State University

Harcourt College Publishers
Fort Worth   Philadelphia   San Diego   New York   Orlando   Austin   San Antonio
Toronto   Montreal   London   Sydney   Tokyo

Custom Publisher   Felix Frazier
Senior Production Manager   Sue Dunaway

Intermediate Algebra, Fourth Edition

Permissions Department
Harcourt, Inc.
6277 Sea Harbor Drive
Orlando, FL 32887-6777

Portions of this Work were published in previous editions.

Printed in the United States of America

10   9   8   7   6   5   4   3   2   1

0-03-045323-2

To our parents,
who have always encouraged
and supported us

Contents

To the Instructor   xiii

To the Student   xvii

Chapter **1**

Fundamental Concepts   1

1.1   Sets   2

1.2   Natural Numbers and Integers   6

1.3   Rational Numbers   14

1.4   Irrational Numbers   26

1.5   Real Numbers   31

1.6   Order of Operations, Absolute Value, and Algebraic Expressions   40

Applied Algebra   48

Chapter 1 Review   49

Chapter 1 Test   55

Chapter **2**

Linear Equations and Inequalities   57

2.1   Linear Equations   58

2.2   Linear Inequalities   66

2.3   Literal Equations and Applications   76

2.4   Applications   80

2.5 Absolute Value Equations 89
Group Activity 95

2.6 Absolute Value Inequalities 96
Group Activity 101
Applied Algebra 101
Chapter 2 Review 103
Chapter 2 Test 106

Test Your Memory 107

Chapter **3**

Polynomials and Factorable Quadratic Equations 109

3.1 Adding and Subtracting Polynomials 110

3.2 Multiplication of Polynomials 118

3.3 Factoring Polynomials—Greatest Common Factor and Grouping 126

3.4 Factoring Binomials 130

3.5 Factoring Trinomials 135
Group Activity 144

3.6 Factoring Summary 144

3.7 Solving Quadratic Equations by Factoring 146
Group Activity 151

3.8 Applications 151
Applied Algebra 157
Chapter 3 Review 158
Chapter 3 Test 162

Test Your Memory 164

Chapter **4**

Rational Expressions 167

4.1 Integer Exponents 168

4.2 Reducing Rational Expressions to Lowest Terms 177

4.3   Multiplying and Dividing Rational Expressions   **182**

4.4   Adding and Subtracting Rational Expressions   **187**

4.5   Complex Fractions   **194**

4.6   Division of Polynomials   **202**

4.7   Synthetic Division (Optional)   **207**

Group Activity   **212**

4.8   Equations Involving Rational Expressions   **212**

4.9   Applications   **219**

Applied Algebra   **226**

Chapter 4 Review   **227**

Chapter 4 Test   **233**

Test Your Memory   **235**

Chapter   **5**

Exponential and Radical Expressions   **237**

5.1   Rational Exponents   **238**

Internet Connection   **244**

5.2   Radicals   **248**

5.3   Simplifying Radical Expressions   **252**

Group Activity   **260**

5.4   Operations with Radical Expressions   **261**

5.5   More Operations with Radical Expressions   **263**

Group Activity   **267**

5.6   Radical Equations   **268**

5.7   Complex Numbers   **275**

Applied Algebra   **283**

Chapter 5 Review   **284**

Chapter 5 Test   **289**

Test Your Memory   **291**

Chapter **6**   **Relations and Functions**   **293**

6.1   The Cartesian Coordinate System   **294**
Group Activity   **303**

6.2   Relations and Functions   **303**

6.3   Function Notation and Combinations of Functions   **312**
Group Activity   **318**
Internet Connection   **318**

6.4   Linear Functions   **320**
Group Activity   **334**

6.5   Equations of Lines   **334**
Group Activity   **346**
Internet Connection   **347**

6.6   The Distance and Midpoint Formulas   **350**
Group Activity   **360**

6.7   Linear Inequalities in Two Variables   **360**

6.8   Variation   **366**

6.9   Branch Functions (Optional)   **375**
Applied Algebra   **382**
Chapter 6 Review   **383**
Chapter 6 Test   **387**

Test Your Memory   **389**

Chapter **7**   **Quadratic and Higher-Degree Equations and Inequalities**   **391**

7.1   Graphing Quadratic Functions   **392**

7.2   Completing the Square to Find the Vertex   **400**
Group Activity   **413**

7.3   Solving Quadratic Equations by Completing the Square   **413**

7.4   The Quadratic Formula   **418**
Group Activity   **427**

7.5   Quadratic Equations Summary   **427**

7.6   Applications   **429**

7.7   Equations in Quadratic Form and
Higher-Degree Equations   **437**

Group Activity   **444**

7.8   Polynomial and Rational Inequalities   **444**

Group Activity   **458**

Applied Algebra   **458**

Chapter 7 Review   **460**

Chapter 7 Test   **466**

Test Your Memory   **467**

Chapter **8**

Conic Sections   **469**

8.1   Horizontal Parabolas   **470**

8.2   Circles   **478**

Group Activity   **486**

8.3   Ellipses   **487**

8.4   Hyperbolas   **494**

8.5   Graphing Second-Degree Inequalities   **502**

Applied Algebra   **510**

Chapter 8 Review   **512**

Chapter 8 Test   **518**

Test Your Memory   **519**

Chapter **9**

Systems of Equations   **521**

9.1   Linear Systems in Two Variables   **522**

9.2   Linear Systems in Three Variables   **531**

Group Activity   **538**

Internet Connection   **538**

9.3   Applications   **543**

     Group Activity   **550**

9.4   Nonlinear Systems of Equations   **550**

     Group Activity   **560**

     Applied Algebra   **560**

     Chapter 9 Review   **563**

     Chapter 9 Test   **567**

Test Your Memory   **568**

Chapter **10**   **Exponential and Logarithmic Functions**   **571**

10.1   Inverse Functions   **572**

     Group Activity   **586**

10.2   Exponential Functions and Equations   **586**

     Group Activity   **596**

     Internet Connection   **597**

10.3   Logarithmic Functions   **602**

     Group Activity   **608**

10.4   Properties of Logarithms   **608**

     Group Activity   **617**

10.5   Logarithmic and More Exponential Equations   **617**

     Group Activity   **624**

     Applied Algebra   **624**

     Chapter 10 Review   **626**

     Chapter 10 Test   **629**

Test Your Memory   **631**

Chapter **11**   **Sequences and Series**   **633**

11.1   Sequences and Series   **634**

11.2   Arithmetic Sequences and Series   **639**

11.3   Geometric Sequences and Series   **646**
Group Activity   **656**

11.4   The Binomial Theorem and Pascal's Triangle   **656**
Applied Algebra   **661**
Chapter 11 Review   **662**
Chapter 11 Test   **666**

Test Your Memory   **668**

Appendix *A*   **Tables**   **671**

Appendix *B*   **Determinants and Cramer's Rule**   **679**
Group Activity   **689**

Appendix *C*   **Counting, Permutations, Combinations, and Probability**   **691**

Appendix *D*   **Graphing Calculators**   **711**

Answers to Odd-Numbered Problems   **717**

Index   **771**

To the Instructor

Purpose

This text is designed for the student who has completed a course in introductory algebra. The purpose of *Intermediate Algebra* is not only to prepare the student for college algebra but to do it in a nonthreatening and organized way. This edition maintains the proven strengths of the successful third edition—the text is mathematically sound, yet written so that students feel comfortable using it. The new features of the fourth edition support this purpose.

New to the Fourth Edition

New Features and Coverage

1. **Group Activities** provide the student with an opportunity to work with classmates on more challenging problems that extend or synthesize concepts and techniques discussed in class. These activities are often exploratory in nature and encourage students to make conjectures and then verify them. Because students are asked to use their own thoughts and wording, some of these problems do not have answers in the back of the book. Answers are included when one answer is appropriate.

2. **Internet Connections** take the mathematics studied in the classroom into the real world. Real data obtained from the internet related to a particular topic are analyzed using concepts and techniques discussed in class. Students may be asked to perform calculations and interpret results, use a model to make predictions, or create their own model. Group activity is often part of the process, and questions for further exploration are included. These exercises have several purposes:

   ■ To encourage students to explore the internet.

   ■ To introduce the power of the internet at an appropriate level.

■ To relate algebra to real-world data and help students develop algebraic models based on that data.

■ To encourage group work and pattern recognition.

3. **Inclusion of interval notation** provides the instructor and student with an alternative way to express solutions to problems involving intervals of real numbers. Introduced at the end of Section 2.2 in the context of linear inequalities, interval notation is used throughout the text wherever appropriate. Answers are often presented using both interval notation and inequality notation, giving the instructor the flexibility to omit this topic, if desired.

Established Strengths

Features

Our motivation for writing this text was the traditionally high attrition rate in college algebra classes. We have attempted to write a book that will help a student with a weak algebra background bridge the gap betwen intermediate and college algebra. To do this, we have included the following features:

1. **Conversational Writing Style.** We have attempted to address the student in a conversational style to make the material less formal and distant but without sacrificing mathematical integrity. We have also injected some humor to make the subject less dry to a student who typically views math with a combination of fear and boredom.

2. **Examples with Detailed Step-by-Step Solutions.** Examples are thoroughly worked out, including every step that a student should need to follow the solutions.

3. **Incorrect/Correct Boxes.** Boxes containing common student errors are provided to warn the student of mistakes to avoid.

4. **Exercises.** Each exercise set contains a large number of exercises for practice, for a total of over 6,500 exercises. Each exercise set starts with problems that test basic skills. The degree of difficulty gradually increases through the set. Similar exercises are typically paired so that a student has both an even and an odd problem that are similar. This encourages students to check their answers, yet gives them practice doing problems without the added security of the answer. Calculator exercises are marked appropriately: scientific ▉ and graphing ▱ .

5. **Graphing Calculator Exercises.** In several sections, you will find examples and exercises that are appropriate to work on a graphing calculator, clearly marked by a graphing calculator screen ▱ . Their purpose is to introduce the power of technology at an appropriate level. These examples and exercises center heavily on pattern recognition. Appendix D goes into greater detail on the basics of using the TI-81. These problems are optional and are therefore easily skipped without losing continuity.

6. **Write Algebra.** At the end of most exercise sets are problems called "Write Algebra." These writing problems require students to reflect on what they have

learned in the section. We believe that the process of putting ideas and concepts into one's own words can be very beneficial. These problems are at the end of each exercise set, and therefore can be omitted without difficulty if desired. Since students are asked to use their own thoughts and wording, most of these problems do not have answers in the back of the book. Answers are included when one answer is appropriate.

7. **Applications.** Applications, although primarily presented in four sections devoted entirely to word problems, are included wherever appropriate throughout the text (see, for example, the exercises in Sections 3.1, 6.2, 7.2, and 10.3). In addition, applications are used at times to motivate the introduction of a new topic (see, for example, the beginning of Section 8.3). Also, each chapter begins with a discussion of an application from everyday life. These applications, called **Applied Algebra,** incorporate some of the mathematics that is covered in the chapter. At the end of each chapter, the problem is solved and similar problems are given as exercises. These real-world applications drive home the usefulness of mathematics.

8. **Review Sections.** All the important definitions, properties, and formulas are repeated in the chapter reviews to help the student remember them. In addition, the review exercises contain an ample number of problems, including all the different types of problems covered in the chapter. A timed chapter test also helps the student determine if he or she has mastered the concepts of the chapter (see, for example, the Chapter 4 review). Chapters 2 through 11 end with cumulative exercises to help the student continually review topics covered in previous chapters.

9. **Advanced Topics.** Some advanced topics, such as functions and higher-degree equations, are included, but at an introductory level.

Organization

In deciding in what order to present the topics in our book, we were aware of varying needs in intermediate algebra classrooms. Our reasons for choosing the present order of topics include the following:

1. Much of the material in the first three or four chapters will be review; however, we have found that the student can benefit from reviewing some fundamental concepts. Although you may choose not to spend class time on some or all of this material, having it at the beginning of the book helps the student easily find a point he or she may want to back up to, in order to review.

2. **The properties of exponents** are presented in three chapters. The student is given the first treatment of the product laws of exponents in Chapter 3 and then has ample opportunity to work with them before the next discussion in Chapter 4. Here the student receives the remainder of the definitions and laws governing integer exponents, and he or she is given the opportunity to work with them in several sections. Finally, the presentation of rational exponents and radicals in Chapter 5 reinforces the laws of exponents further.

3. **Solving quadratic equations by factoring** was placed in Chapter 3 so that equations leading to quadratic equations could be included in Sections 4.8 (rational equations), 4.9 (applications of rational equations), and 5.6 (radical equations).

4. **Graphing functions and equations in two variables** is introduced early, in Chapter 6, so that these important concepts for college algebra can be reinforced several times in Chapters 7, 8, 9, and 10.

Supplements

A comprehensive package of supplements is available to facilitate teaching and learning. For the instructor, a test bank is available that includes free-response and multiple choice tests for each chapter. For the student, a partial solutions manual/study guide, tutorial software, and a videotape series are available. In addition, online support is provided at http://www.hbcollege.com.

Acknowledgments

The authors would like to express their gratitude to those individuals at Harcourt College Publishing who helped us put this book together, in particular Felix Frazier, Ann Coburn, and Tina Landman.

We would also like to thank the following reviewers for their significant contributions:

Bobby Avila
Mt. San Jacinto College

Judith Cantey
Jefferson State Community College

Robert B. Eicken
Illinois Central College

Robert Forrester
Volunteer State Community College

Odene Forsythe
Westark Community College

Lenore Frank
State University of New York at Stony Brook

Kimberly McFadden
Volunteer State Community College

Martha Pratt
Mississippi State University

Ross Rueger
College of the Sequoias

Carol Russell
Indiana University Southeast

Paul Schell
Ferris State University

Kenneth Schoen
Worcester State College

Michael Simon
Housatonic Community-Technical College

Jan Vandever
South Dakota State University

Ray Watkins
Saddleback College

Cindy Wilson
Mississippi State University

In addition, we would like to thank our colleagues at North Harris County College, San Antonio College, and Oswego State University for their suggestions and encouragement. Most of all, we thank our families and friends, who have offered much support from the beginning.

DENNIS WELTMAN
GILBERT PEREZ
TERRY TIBALLI

To the Student

Have you ever watched a tennis match on television and marveled at how the pros made it look so easy? You might have noticed that your math instructor makes algebra look easy. And yet in both cases, when you try to play tennis, or work some algebra problems, you find it is not so easy. What's the reason? A tennis pro practices for hours each day. He or she played tennis for years before becoming good enough to play professionally. Your math instructor has been working with math for *years*, spending many hours a day in preparation for class. This is not to imply that you should strive for, or expect to achieve, the level of proficiency of your math instructor. However, if you want to do well in this math course, you have to practice. The more you practice, the better you should be. If you don't want to waste your time and money registering for the same algebra course semester after semester, here are some suggestions:

1. *Attend class regularly and ask questions.* Almost no student can learn algebra without going to class. You need to see and hear the instructor's explanations. While you can't expect to understand everything, you should have an idea of how to work the problems. If not, *ask!*

2. *Do your homework soon after class.* You will forget most of what was discussed in class within a few hours unless you do your homework during this time. Waiting until the last minute before the next class meeting to do your homework is the worst thing you can do. Work all problems that are assigned. *Do not* skip over sections or groups of problems that you think are "easy." They may very well be easy for you but you need to practice in order to reduce your chances of making a careless mistake. Remember the tennis pro! If you have time, do problems that were not assigned. Use your notes and text as a reference while doing your homework. However, make it a habit to work some problems during each session without any notes or the textbook. This practice pays off at test time.

3. *Read the book carefully.*

   ■ It is best to read each section before the material is discussed in class. Try to follow the examples and work some of the exercises. Write questions you have in the book itself. Make notes to yourself in the book. Underline sentences that explain key concepts.

■ After class, reread the section carefully. Read the statement of the problem in each example and try to solve the problem before reading the solution given in the example.

■ Exercises that may require a calculator are identified by a small calculator that looks like this: scientific ▨ or graphing ▱ .

■ We have boxed some common mistakes, labelled Incorrect/Correct, that many students make. Make a special note of these when you work the exercises.

■ The answers to the odd-numbered exercises are in the back of the book. You should not look at the answers until you have finished working each problem. Remember, on a quiz or test, you don't have the answers to refer to. When working "Write Algebra" problems, you will find that most will not have answers in the back. Try to work these problems by expressing your own thoughts in your own words. This can be a big help in becoming familiar with concepts.

■ When you have completed a chapter, use the chapter review to go back over the important properties, formulas, and so on, that were introduced in that chapter. Exercises from each section are provided for you to get more practice. When you think you have reviewed the chapter sufficiently, take the timed chapter test provided. The Test Your Memory exercises at the end of each chapter will help to keep "old" topics fresh in your mind.

4. *Try to make the class fun.* Get into a friendly competition with someone from your class. If Group Activities are included during classtime, use this as an opportunity to develop productive relationships with some of your classmates. Study with other people in a small group. Explain concepts to each other when someone in the group does not understand something. You will find that this benefits the tutor as well as the other students.

5. If you need additional help, a student solutions manual/study guide is also available. Ask your instructor if the videotapes and computer software supplements are available at your campus.

Fundamental Concepts

The National Weather Service warns that if the following conditions are expected to prevail for 3 hr or longer, then a blizzard warning must be issued.

1. Sustained wind speeds of 35 mph or greater or frequent gusts to 35 mph or greater, and

2. Considerable falling and/or blowing snow, frequently reducing the visibility to less than $\frac{1}{4}$ mi.

Although there is no set temperature requirement for a blizzard warning, when the temperature falls below 20°F, forecasters should highlight the life-threatening nature of the cold temperature in combination with other hazardous conditions such as the wind. The chilling effect of the wind during cold temperatures is called the **wind-chill factor.** Although the thermometer may indicate the temperature is 20° F, a 35-mph wind will make it seem as if the temperature is much colder. The formula to determine wind chill is an example of an algebraic expression. At the end of this chapter we will be able to compute what temperature we would perceive under the conditions just described.

1.1 Sets

One of the most basic concepts in any mathematics course is that of a **set.** A set is a collection of objects; for example, the set that contains the counting numbers from 1 to 5, inclusive, is written

$$\{1, 2, 3, 4, 5\}$$

The numbers within a set are called **elements** or **members** of the set, and they are set apart by commas and written within braces { }. This method of writing the members of a set is called **set notation.** We often name a set by using a capital letter; for example,

$$A = \{1, 2, 3, 4, 5\}$$

When we mention set A as defined above, we know that we are referring to the set $\{1, 2, 3, 4, 5\}$.

If we want to indicate that 2 is an element of a set A, we write

$$2 \in A$$

If we want to indicate that 8 is not an element of A, we write

$$8 \notin A$$

Consider the set $B = \{2, 4\}$. Each element of set B is also an element of set A, so we say that set B is a **subset** of set A and write

$$B \subseteq A$$

The concept that B is contained within A is sometimes illustrated as follows:

On the other hand, consider set $C = \{1, 3, 5, 7\}$. Since $7 \in C$ but $7 \notin A$, set C is not a subset of set A, which is written

$$C \nsubseteq A$$

Note the difference between the use of $\in$ and $\subseteq$. An element (like 7) precedes $\in$, and a set (like C) precedes $\subseteq$.

EXAMPLE 1.1.1   Consider the sets $F = \{1, 3, 5, 7, 9\}$, $G = \{3, 5\}$, and $J = \{3, 4, 5\}$. Determine whether each of the following statements is true or false.

1. $G \subseteq F$

True. Each element of G is also an element of F.

2. $J \subseteq F$

False. 4 is an element of J, but 4 is not an element of F.

3. $J \subseteq J$

True. Each element of J is also an element of J.

NOTE ▶   *Every set is a subset of itself.* Thus if A is a set, $A \subseteq A$.

An interesting set is the **empty set** or **null set,** which contains no elements. It is denoted by either { } or $\varnothing$. An example of the empty set is the set of counting numbers between 2 and 3. Consider the set $A = \{1, 2\}$. We can say that $\varnothing \subseteq A$ because if $\varnothing$ were not a subset of A, then there would have to be an element of $\varnothing$ that is not in A; but there aren't any elements in $\varnothing$! We can say that *the empty set is a subset of any set.*

Up to now, we have considered only **finite sets,** sets that have a limited number of elements. The set that contains all the counting numbers is written

$$N = \{1, 2, 3, 4, \ldots\}$$

This is an **infinite set** because the number of elements in this set is not limited. We write only the first few numbers, establishing a pattern, and then write the three dots (called an ellipsis) to indicate that this pattern continues forever. Another example of an infinite set is the set of all even counting numbers:

$$E = \{2, 4, 6, 8, \ldots\}$$

We can also use the three dots when we are dealing with a finite set that has many elements. For example, the set of the first 100 counting numbers can be written

$$P = \{1, 2, 3, 4, \ldots, 100\}$$

EXAMPLE 1.1.2   Consider the sets $N = \{1, 2, 3, \ldots\}$, $E = \{2, 4, 6, 8, \ldots\}$, and $A = \{1, 2, 3, 4, 5\}$. Determine whether each of the following statements is true or false.

1. $E \subseteq N$

True. Each element of E is also an element of N.

2. $24 \in E$

True. 24 is even and is therefore an element of E.

3. $N \subseteq A$

False. N contains numbers like 6, 7, 8,... that are not elements of A.

4. $\varnothing \subseteq A$

True. The empty set is a subset of every set.

Two sets are **equal** if they contain exactly the same elements. For example, the sets

$$X = \{2, 7, 9\} \quad \text{and} \quad Y = \{7, 2, 7, 9\}$$

are equal because each contains the same elements, 2, 7, and 9, even though they are listed in different orders and one of the elements is repeated in set Y.

Suppose we are interested in the set of elements that two sets D and G have in common. This set is called the **intersection** of D and G and is denoted by

$$D \cap G$$

Other times we might be interested in putting the elements of sets D and G together into one set. This set is called the **union** of D and G and is denoted by

$$D \cup G$$

For example, suppose you are looking for a job. You might find that you can divide the prospective jobs into two sets. Let's label one set M (for money), which will represent the set of jobs that pay well. Another set might be the set of jobs in a desirable location, which we will label L. When you start looking for a job, you might want one that satisfies both conditions. The set that contains these jobs is the intersection of the two sets,

$$M \cap L$$

To be in this set, a job must be both in set M (pay well) *and* in set L (good location). But suppose, after looking for some time, you cannot find a job that satisfies both requirements. You might be willing to settle for a job that satisfies either one condition *or* the other. In that case, the set of jobs you would consider is the union of two sets M and L,

$$M \cup L$$

We have taken all the jobs in set M and all the jobs in set L and formed one large set.

As another example, let $S = \{1, 2, 3, 5\}$ and $T = \{-2, 0, 2, 4\}$. Then the intersection of S and T is $S \cap T = \{2\}$. The union of S and T is $S \cup T = \{1, 2, 3, 5, -2, 0, 2, 4\} = \{-2, 0, 1, 2, 3, 4, 5\}$. You can think of the intersection as where the two sets meet, like the intersection of two streets (see Figure 1.1.1). The union of sets resembles the union of states in one country (see Figure 1.1.2).

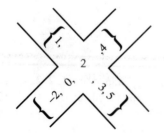

Figure 1.1.1

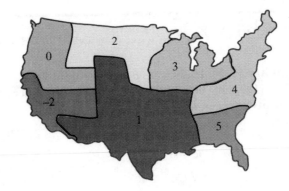

Figure 1.1.2

XAMPLE 1.1.3   Consider the sets $A = \{1, 3, 5, 7, 9\}$, $B = \{1, 2, 3, 4\}$, $C = \{2, 4, 6, 8, \ldots\}$, $D = \{4, 8, 12, 16, \ldots\}$, $F = \{\text{all females}\}$, and $M = \{\text{all males}\}$. Find the following.

1. $F \cup M$

$F \cup M = \{\text{all people}\}$. Every person is either male or female.

2. $F \cap M$

$F \cap M = \varnothing$. These sets have no element in common.

3. $A \cap B$

$A \cap B = \{1, 3\}$, since the elements 1 and 3 appear in both set A *and* set B.

4. $A \cup B$

$A \cup B = \{1, 2, 3, 4, 5, 7, 9\}$, since each element appears in either set A *or* set B.

5. $C \cap D$

$C \cap D = \{4, 8, 12, 16, \ldots\} = D$. Why?

6. $C \cup D$

$C \cup D = \{2, 4, 6, 8, \ldots\} = C$. Why?

Try to avoid these mistakes:

Incorrect	Correct
1. $5 \subseteq \{3, 4, 5\}$	$5 \in \{3, 4, 5\}$
2. $\{4, 5\} \in \{3, 4, 5\}$	$\{4, 5\} \subseteq \{3, 4, 5\}$

Exercises 1.1

Consider the sets $A = \{5, 10, 15, 20, \ldots\}$, $B = \{10, 20, 30, 40, \ldots\}$, $C = \{25, 50, 75, 100\}$, and $D = \{100, 200, 300, \ldots, 1000\}$. Determine whether each of the following statements is true or false.

1. $A \subseteq B$

2. $B \subseteq A$

3. $100 \in A$

4. $100 \in B$

5. $75 \in B$

6. $700 \in B$

7. $A \subseteq A$

8. $\varnothing \subseteq A$

9. $15 \subseteq A$

10. $\{20\} \in B$

11. $C \subseteq A$

12. $C \subseteq B$

13. $D \subseteq A$

14. $D \subseteq B$

15. $700 \in D$

16. $A \cap B = B$

17. $A \cup B = A$

18. $A \cap C = \varnothing$

19. $B \cap D = \varnothing$

20. $\varnothing \cap C = \varnothing$

Consider the sets $M = \{1, 2, 3, \ldots, 10\}$, $N = \{1, 2, 3, 4, \ldots\}$, $D = \{1, 3, 5, 7, \ldots\}$, and $P = \{2, 4, 6, 8, 10\}$. Find the following.

21. $M \cap N$

22. $M \cap D$

23. $M \cap P$

24. $M \cup N$

25. $M \cup P$

26. $M \cup D$

27. $N \cup D$

28. $N \cap D$

29. $D \cap P$

30. $D \cup P$

31. $N \cup P$

32. $N \cap P$

33. $M \cap M$

34. $M \cup M$

35. $M \cup \varnothing$

36. $M \cap \varnothing$

37. $(M \cap D) \cup P$

38. $(M \cap N) \cap D$

39. $(M \cup D) \cap P$

40. $(M \cap P) \cup (D \cap P)$

41. List all the subsets of the set $\{1, 2, 3\}$.

Write Algebra

42. In your own words, describe how to form the union of two sets.

43. In your own words, describe how to find the intersection of two sets.

1.2 Natural Numbers and Integers

In algebra there are instances when we wish to deal with only a specific set of numbers. In this section we will look at two sets of numbers, the natural numbers and the integers. We will consider other specific sets of numbers in the next three sections.

The set of counting numbers $\{1, 2, 3, 4, \ldots\}$ in mathematics is generally called the set of **natural numbers** and is labeled with an N. The set of **whole numbers,** labeled W, includes the natural numbers and zero. $W = \{0, 1, 2, 3, \ldots\}$. The set of **integers** includes the whole numbers and the negatives of the natural numbers. It is usually labeled with an I:

$$I = \{\ldots, -3, -2, -1, 0, 1, 2, 3, \ldots\}$$

Note the relationship between these sets:

$$N \subseteq W \subseteq I$$

In this section we will review the rules covering the arithmetic of integers; however, we first need to introduce the concept of **absolute value.**

| Definition 1.2.1 | The *absolute value* of a number x, denoted by $|x|$, is defined as follows: |

- $|x|$ is x if x is greater than or equal to zero
- $|x|$ is $-x$ if x is less than zero

REMARK

The absolute value of a number that is positive or zero is that number. To evaluate the absolute value of a negative number, change its sign.

For example,

- The absolute value of 3 is 3,
- The absolute value of 0 is 0,
- The absolute value of -2 is 2.

It is assumed that you are able to work with positive numbers. We will review the rules for arithmetic involving integers, but keep in mind that these rules generalize to problems involving numbers that are not integers.

Adding Negative Integers	Examples
To add two or more negative integers, add the absolute values of the numbers and make the answer negative.	$(-27) + (-6) = -33$ $-6 + (-5) + (-9) = -20$

Adding a Positive and a Negative Integer	Examples
To add two integers with opposite signs, subtract the smaller absolute value from the larger absolute value. The answer will have the same sign as the number with the larger absolute value.	$-23 + 15 = -8$ $45 + (-29) = 16$

EXAMPLE 1.2.1   Add the following integers:

$$(-21) + 7 + (-8) + (-13) + 12$$

Let us add the negative numbers first. Combining their absolute values, $21 + 8 + 13 = 42$, so

$$(-21) + (-8) + (-13) = -42$$

Now add the positive numbers

$$7 + 12 = 19$$

Since $42 - 19 = 23$,

$$(-42) + 19 = -23$$

Before subtracting integers, we need to discuss what we mean by the additive inverse of a number.

Additive Inverse	Examples
If the sum of two numbers is zero, then they are **additive inverses.** To find the additive inverse of a number, change the sign in front of the number to its opposite.	The additive inverse of 7 is -7, since $7 + (-7) = 0$. The additive inverse of -5 is 5, since $-5 + 5 = 0$. The additive inverse of 0 is 0, since $0 + 0 = 0$.

Placing a negative sign in front of a number changes it to its additive inverse. Thus, $-(-8)$ is the additive inverse of -8. Since 8 is the additive inverse of -8, then $-(-8) = 8$. In general,

$$-(-a) = a \quad \text{where } a \text{ is any number}$$

Subtracting Integers	Examples
To subtract an integer, add its additive inverse. Hence, if a and b represent integers, then $a - b = a + (-b)$	$7 - (-12) = 7 + 12 = 19$ $-3 - 22 = -3 + (-22) = -25$

If we have a string of integers to add and subtract, we change all the subtractions to additions, using the preceding rule, and then apply the rules for adding integers.

EXAMPLE 1.2.2   Perform the following addition and subtraction operations:

$$-8 - 19 + 29 - (-40) + (-11)$$

Change 19 to -19 and -40 to 40, and change the operations before them to addition:

$$\begin{aligned}
-8 - 19 + 29 - (-40) + (-11) &= -8 + (-19) + 29 + 40 + (-11) \\
&= -8 + (-19) + (-11) + 29 + 40 \\
&= -38 + 69 \\
&= 31
\end{aligned}$$

In addition problems, the numbers we add are called **terms.** For instance, in the problem $7 + (-16) = -9$, the terms are 7 and -16. The answer, -9, is called the **sum.** In multiplication problems, the numbers we multiply are called **factors.** For example, in the problem $5 \cdot 8 = 40$, the factors are 5 and 8. The answer, 40, is called the **product.**

Multiplying Two Integers with the Same Sign	Examples
The product of two integers with the same sign is positive.	$16 \cdot 5 = 80$ $(-3) \cdot (-14) = 42$

Multiplying Two Integers with Opposite Signs	Examples
The product of two integers with opposite signs is negative.	$13 \cdot (-3) = -39$ $(-4) \cdot 16 = -64$

When multiplying more than two integers, we multiply two integers at a time, starting at the left and multiplying the product each time by the next integer until all the integers have been multiplied.

EXAMPLE 1.2.3   Perform the following multiplications.

1. $(-5) \cdot (3) \cdot (-2) \cdot (-4)$

= $(-15) \cdot (-2) \cdot (-4)$

= $(30) \cdot (-4)$

= -120

2. $(-6) \cdot (-1) \cdot (-2) \cdot (7) \cdot (-5)$

= $(6) \cdot (-2) \cdot (7) \cdot (-5)$

= $(-12) \cdot (7) \cdot (-5)$

= $(-84) \cdot (-5)$

= 420

From the preceding examples, we can detect a pattern for predicting the sign of the product.

Multiplying Integers	Examples
If the number of negative factors is even, the product will be positive.	$(-5)(11)(-2) = 110$ $(-2)(3)(-5)(-1)(-5) = 150$ $(-2)(-2)(-2)(-2)(-2)(-2) = 64$
If the number of negative factors is odd, the product will be negative.	$(-12)(9) = -108$ $(1)(-4)(-7)(2)(-1) = -56$ $(-1)(-2)(-3)(-4)(-5) = -120$

Note that the number of positive factors does not affect the sign of the product.

Multiplying by Zero	Examples
The product of zero and any other integer(s) is zero.	$(-45)(0) = 0$ $(0)(-20)(-12)(7) = 0$

When all the factors in a product are the same number, we have a shorter way of writing it, which we call **power notation,** or **exponential notation.** For example, $2 \cdot 2 \cdot 2 \cdot 2 \cdot 2$ can be written as 2^5, where 2, the factor that is being multiplied, is called the **base,** and 5, the number of factors, is called the **power** or **exponent.**

$$\text{base} \to 2^5 \leftarrow \text{exponent}$$

The expression 2^5 is read "two to the fifth power" or "two raised to the fifth power." Other examples are

- $3 \cdot 3 \cdot 3 \cdot 3 = 3^4 = 81$, read "three to the fourth power"
- $5 \cdot 5 \cdot 5 = 5^3 = 125$, read "five to the third power" or "five **cubed**"
- $7 \cdot 7 = 7^2 = 49$, read "seven to the second power" or "seven **squared**"

In general,

E xponential Notation

$$a^n = \underbrace{a \cdot a \cdot a \cdots a}_{n \text{ factors}} \quad \text{where } a \text{ is any number and } n \text{ is a natural number}$$

NOTE ▶ When the base is negative, we have to place parentheses around the base and write the exponent outside of the parentheses. Consider the following examples. In $(-5)^2 = (-5)(-5) = 25$, the base is -5. In $-5^2 = -(5 \cdot 5) = -25$, the exponent 2 applies to only 5, the base, and not to the negative sign in front. We write the negative sign in front of the answer after we have multiplied the product 5^2.

XAMPLE 1.2.4   Multiply the following.

1. $(-13)^2$ $(-13)^2 = (-13)(-13) = 169$

2. -6^3 $-6^3 = -(6 \cdot 6 \cdot 6) = -216$

In a division problem like $6 \div 2 = 3$, 6 is called the **dividend,** 2 is called the **divisor,** and 3 is called the **quotient.** Since the quotient is an integer, we say that 6 is **divisible** by 2. In this section we consider only those cases where one integer is divisible by another.

Dividing Two Integers with the Same Sign	Examples
The quotient of two integers with the same sign is positive.	$48 \div 3 = 16$ $$\frac{-56}{-2} = 28$$

Dividing Two Integers with Opposite Signs	Examples
The quotient of two integers with unlike signs is negative.	$$\frac{-45}{15} = -3$$ $52 \div (-4) = -13$

We can check the answer when dividing by determining whether

$$(\text{dividend}) = (\text{quotient}) \cdot (\text{divisor})$$

For instance, we stated that $52 \div (-4) = -13$. This is true because $52 = (-13)(-4)$ is a true statement.

Division Involving Zero	Examples
1. Zero divided by any *nonzero* integer is zero.	$0 \div 5 = 0$ $$\frac{0}{-24} = 0$$
2. *You cannot divide by zero.*	$7 \div 0$ is undefined. $\dfrac{0}{0}$ is indeterminate.

Let us check some of the preceding examples. Since $(-24) \cdot (0) = 0$, we can say that $\frac{0}{-24} = 0$. On the other hand, if we try to produce an answer for $7 \div 0$, we run into trouble. A common mistake is to say that $7 \div 0 = 0$. But this does not check, since $(0) \cdot (0) \neq 7$. We have a different problem trying to get an answer for $\frac{0}{0}$. Since $(\ \ ) \cdot (0) = 0$ for any number we place in the parentheses, we can give any number as an answer for $\frac{0}{0} = (\ \ )$. Since there is no unique answer, we say $\frac{0}{0}$ is indeterminate.

Try to avoid these mistakes:

Incorrect	Correct
1. "Two negatives make a positive" (an overgeneralization).	The sum of two negative integers is negative: $$(-5) + (-8) = -13$$ The product (or quotient) of two negative integers is positive: $$(-12) \cdot (-8) = 96$$ $$(-54) \div (-3) = 18$$
2. $-3^2 = 9$	$$(-3)^2 = (-3) \cdot (-3) = 9$$ $$-3^2 = -(3 \cdot 3) = -9$$
3. $\dfrac{5}{0} = 0$	$\dfrac{5}{0}$ is undefined. $\dfrac{0}{5} = 0$
4. $\dfrac{0}{0} = 0$	$\dfrac{0}{0}$ is indeterminate.

E xercises 1.2

Perform the indicated operations, if possible.

1. $43 + (-18)$

2. $-25 + (-11)$

3. $14 - 73$

4. $(-20)(-5)$

5. $16(-30)$

6. 6^2

7. $(-12)^2$

8. -8^2

9. $-7 - 16$

10. $-26 - (-19)$

11. $-30 - (-52)$

12. $(-18) \div (-3)$

13. $39 \div (-3)$

14. $\dfrac{36}{-18}$

15. $\dfrac{68}{-2}$

16. $\dfrac{-46}{-23}$

17. $\dfrac{-47}{47}$

18. $(-5)(-8)(-3)$

19. $(-2)(-1)(15)(-4)(-5)$

20. $\frac{0}{14}$

21. $(-6)(-7)$

22. $(-18)(5)$

23. $(17)(0)$

24. $(-61) + 29$

25. $(-13) + (-36)$

26. -5^3

27. $(-2)^3$

28. $(-3)^3$

29. -5^4

30. 2^7

31. 0^5

32. $(-4)^4$

33. (-2^4)

34. $9 - 51 + (-32)$

35. $-57 + 17 - (-23) - 7$

36. $6 + 6 - 6 - (-6) + (-6)$

37. $59 + 41 - (-10) - 9$

38. $(-6)(-1)(-2)(3)(-5)(-1)$

39. $(-7)(2)(10)(-3)$

40. $(-13)(2)(0)(-4)$

41. $\frac{8}{0}$

42. $\frac{0}{-12}$

43. $\frac{0}{0}$

44. $-15 - (-22) - 18$

45. $(-2)(2)(-3)(-4)(8)$

46. $-12 - 21 - (-9) + 11 - 91$

47. $(-10)(100)(1000)$

48. $(-10)^6$

49. $\frac{-1,000,000,000}{1000}$

Write Algebra

50. In a multiplication problem, describe how you determine the sign of the answer.

51. Explain the difference in how the following are evaluated: $(-7)^2$ and -7^2.

1.3 Rational Numbers

The next set of numbers we will consider is the set of **rational numbers,** which we label with a Q. We can describe it as the set of all numbers that can be written as fractions, where the numerator and denominator are integers and the denominator is not zero. Since any integer can be written as a fraction (for example, $2 = \frac{2}{1}$, $-4 = -\frac{8}{2}$, $7 = \frac{-42}{-6}$), the set of integers is a subset of the set of rational numbers. So

$$N \subseteq W \subseteq I \subseteq Q$$

Before we review the arithmetic of rational numbers, we need to state the definition of a prime number.

| Definition 1.3.1 | If a natural number other than one is divisible by only itself and one, we say that number is **prime.** |

If you were to cut a pizza into eight equal pieces and then eat four of them, you would have eaten $\frac{4}{8}$ or $\frac{1}{2}$ of the pizza. Both these fractions represent the same rational number, but $\frac{1}{2}$ is a ratio of smaller integers and is usually simpler to work with. "Simplifying" a fraction from $\frac{4}{8}$ to $\frac{1}{2}$ is called **reducing** the fraction. This is accomplished by finding the prime factors of the numerator and denominator and then dividing out the factors that the numerator and denominator have in common. In our example,

$$\frac{4}{8} = \frac{\overset{1}{\cancel{2}} \cdot \overset{1}{\cancel{2}}}{\underset{1}{\cancel{2}} \cdot \underset{1}{\cancel{2}} \cdot 2} = \frac{1}{2}$$

Each time a common factor is divided out, a factor of one is left. When all the common prime factors have been divided out, the fraction is said to be in **lowest terms.**

EXAMPLE 1.3.1   Reduce the following fractions to lowest terms.

1. $\dfrac{24}{90}$ $\dfrac{24}{90} = \dfrac{\cancel{2} \cdot 2 \cdot 2 \cdot \cancel{3}}{\cancel{2} \cdot \cancel{3} \cdot 3 \cdot 5} = \dfrac{4}{15}$

2. $\dfrac{13}{39}$ $\dfrac{13}{39} = \dfrac{\overset{1}{\cancel{13}}}{3 \cdot \cancel{13}} = \dfrac{1}{3}$ *Don't forget the 1 on top!*

3. $\dfrac{27}{25}$ $\dfrac{3 \cdot 3 \cdot 3}{5 \cdot 5} = \dfrac{27}{25}$ *Since there are no common factors, the fraction is already in lowest terms.*

REMARK   After performing an operation involving fractions, we always want the answer in lowest terms.

Multiplying Fractions		**Example**
To multiply fractions		
1. Write the product of the numerators over the product of the denominators.		$\dfrac{6}{5} \cdot \dfrac{15}{4} = \dfrac{6 \cdot 15}{5 \cdot 4}$
2. Factor the numerators and denominators and reduce the fraction.		$= \dfrac{\cancel{2} \cdot 3 \cdot 3 \cdot \cancel{5}}{\cancel{5} \cdot \cancel{2} \cdot 2}$
3. Multiply any remaining factors in the numerator and denominator.		$= \dfrac{9}{2}$

EXAMPLE 1.3.2 Perform the indicated operations.

1. $\dfrac{11}{9} \cdot \dfrac{3}{22} = \dfrac{11 \cdot 3}{9 \cdot 22} = \dfrac{\cancel{11} \cdot \overset{1}{\cancel{3}}}{\cancel{3} \cdot 3 \cdot 2 \cdot \cancel{11}} = \dfrac{1}{6}$

2. $\dfrac{4}{5} \cdot \dfrac{15}{14} \cdot \dfrac{7}{6} = \dfrac{4 \cdot 15 \cdot 7}{5 \cdot 14 \cdot 6} = \dfrac{2 \cdot \overset{1}{\cancel{2}} \cdot \cancel{3} \cdot \cancel{5} \cdot \cancel{7}}{\cancel{5} \cdot \cancel{2} \cdot \cancel{7} \cdot \cancel{2} \cdot \cancel{3}} = \dfrac{1}{1} = 1$

Since $\frac{-6}{2} = -3$, $\frac{6}{-2} = -3$, and $-\left(\frac{6}{2}\right) = -3$, a negative sign in a fraction can be placed in the numerator, in the denominator, or in front of the division bar. So

$$\frac{-a}{b} = \frac{a}{-b} = -\frac{a}{b}$$

NOTE ▶ When multiplying fractions involving negative signs, we use the multiplication rules from the preceding section and merely count the number of negative factors in the problem. If the number of negative factors is even, the answer is positive, and if the number of negative factors is odd, the answer is negative.

EXAMPLE 1.3.3 Perform the indicated operations.

1. $\left(-\dfrac{7}{8}\right)\left(\dfrac{6}{-35}\right) = \dfrac{7 \cdot 6}{8 \cdot 35}$

$= \dfrac{\cancel{7} \cdot 3 \cdot \cancel{2}}{\cancel{2} \cdot 2 \cdot 2 \cdot 5 \cdot \cancel{7}}$

$= \dfrac{3}{20}$

The answer is positive, since there are two negative factors.

2. $\left(\dfrac{-1}{2}\right)\left(\dfrac{10}{-9}\right)\left(-\dfrac{6}{25}\right) = -\dfrac{1 \cdot 10 \cdot 6}{2 \cdot 9 \cdot 25}$

$= -\dfrac{\cancel{2} \cdot \cancel{5} \cdot 2 \cdot \cancel{3}}{\cancel{2} \cdot \cancel{3} \cdot 3 \cdot 5 \cdot \cancel{5}}$

$= -\dfrac{2}{15}$

The answer is negative, since there are three negative factors.

Dividing Fractions	Example
To divide fractions, invert the divisor and multiply.	$\dfrac{2}{3} \div \dfrac{4}{15} = \dfrac{2}{3} \cdot \dfrac{15}{4} = \dfrac{2 \cdot 15}{3 \cdot 4}$ $= \dfrac{2 \cdot \cancel{3} \cdot 5}{\cancel{3} \cdot \cancel{2} \cdot 2} = \dfrac{5}{2}$

EXAMPLE 1.3.4   Perform the indicated operations.

1. $-\dfrac{6}{5} \div \dfrac{18}{7} = -\dfrac{6}{5} \cdot \dfrac{7}{18} = -\dfrac{6 \cdot 7}{5 \cdot 18} = -\dfrac{\cancel{2} \cdot \cancel{3} \cdot 7}{5 \cdot \cancel{2} \cdot \cancel{3} \cdot 3}$

$= -\dfrac{7}{15}$

2. $5 \div \dfrac{10}{13} = \dfrac{5}{1} \div \dfrac{10}{13}$ *Write 5 as a fraction.*

$= \dfrac{5}{1} \cdot \dfrac{13}{10} = \dfrac{5 \cdot 13}{10} = \dfrac{\cancel{5} \cdot 13}{2 \cdot \cancel{5}} = \dfrac{13}{2}$

3. $\dfrac{\dfrac{2}{11}}{12} = \dfrac{\dfrac{2}{11}}{\dfrac{12}{1}}$ *Write 12 as a fraction.*

$= \dfrac{2}{11} \div \dfrac{12}{1} = \dfrac{2}{11} \cdot \dfrac{1}{12} = \dfrac{2}{11 \cdot 12}$

$= \dfrac{\cancel{2}}{11 \cdot \cancel{2} \cdot 2 \cdot 3} = \dfrac{1}{66}$

Let us return to the pizza that was cut into eight equal pieces. If you eat one piece first and later eat three more pieces, you will have eaten one-eighth of the pizza and then three-eighths or, altogether, four-eighths of the pizza. Written with fractions,

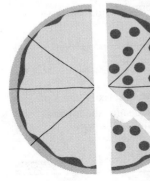

$$\frac{1}{8} + \frac{3}{8} = \frac{4}{8} \quad \text{or} \quad \frac{1}{2}$$

We add the numerators only and use the denominator they have in common. (Of course, we also reduce the answer to lowest terms.)

Adding and Subtracting Fractions with the Same Denominator	Examples
To add (or subtract) fractions that have the same denominator, add (or subtract) the numerators and write this answer over the common denominator.	$\dfrac{2}{7} + \dfrac{3}{7} = \dfrac{2+3}{7} = \dfrac{5}{7}$ $\dfrac{2}{9} + \dfrac{5}{9} - \dfrac{10}{9} = \dfrac{2+5+(-10)}{9}$ $= \dfrac{7+(-10)}{9} = \dfrac{-3}{9}$ $= -\dfrac{\cancel{3}}{\cancel{3}\cdot 3} = -\dfrac{1}{3}$

EXAMPLE 1.3.5

Perform the indicated operations.

1. $\dfrac{2}{5} - \dfrac{9}{5} + \dfrac{7}{5} = \dfrac{2+(-9)+7}{5} = \dfrac{0}{5} = 0$

2. $\dfrac{7}{12} + \dfrac{1}{30}$

These fractions do not have the same denominators. Therefore, we cannot use the preceding rule to add them.

The last example leads to the next type of problem, adding fractions with different denominators. We need to find a common denominator that is divisible by both 12 and 30. One such common denominator is $12 \cdot 30 = 360$. But 12 and 30 also divide into 180 and into 60. We would like to find the smallest or least common denominator. The **least common denominator,** which is abbreviated *LCD,* can be found in the following manner:

How to Find the LCD	Example
1. Factor each of the denominators into products of prime numbers using power notation.	$12 = 2 \cdot 2 \cdot 3 = 2^2 \cdot 3$ $30 = 2 \cdot 3 \cdot 5$
2. Form the LCD by taking the product of all the different factors raised to their highest power in each factorization.	Since we have different factors of 2, 3, and 5, and since the highest power 2 is raised to is the second power, $LCD = 2^2 \cdot 3 \cdot 5 = 60$.

When we have found the LCD, we need to convert each of the fractions to equivalent fractions that have the LCD as their denominators. This conversion is done in each fraction by dividing the denominator into the LCD, and multiplying this answer times the numerator and denominator. For example, $60 \div 12 = 5$, so we multiply $\frac{7}{12}$ on both top and bottom by 5:

$$\frac{7}{12} \cdot \frac{5}{5} = \frac{35}{60}$$

Similarly, $60 \div 30 = 2$, so we multiply $\frac{1}{30}$ on both top and bottom by 2:

$$\frac{1}{30} \cdot \frac{2}{2} = \frac{2}{60}$$

Note that we are multiplying each fraction by another fraction equivalent to one. This multiplication changes the form but not the value of the fraction. Thus,

$$\frac{7}{12} + \frac{1}{30} = \frac{7}{12} \cdot \frac{5}{5} + \frac{1}{30} \cdot \frac{2}{2}$$
$$= \frac{35}{60} + \frac{2}{60} = \frac{37}{60}$$

EXAMPLE 1.3.6   Perform the indicated operations.

1. $\dfrac{3}{10} - \dfrac{-7}{50} = \dfrac{3}{10} + \dfrac{7}{50}$

$$= \frac{3}{10} \cdot \frac{5}{5} + \frac{7}{50}$$

$$= \frac{15}{50} + \frac{7}{50}$$

$$= \frac{22}{50} = \frac{\cancel{2} \cdot 11}{\cancel{2} \cdot 5 \cdot 5} = \frac{11}{25}$$

2. $\dfrac{5}{7} + \dfrac{8}{21} - \dfrac{11}{9} = \dfrac{5}{7} \cdot \dfrac{9}{9} + \dfrac{8}{21} \cdot \dfrac{3}{3} - \dfrac{11}{9} \cdot \dfrac{7}{7}$

$$= \frac{45}{63} + \frac{24}{63} - \frac{77}{63}$$

$$= \frac{69}{63} + \frac{-77}{63}$$

$$= \frac{-8}{63}$$

3. $3\dfrac{1}{2} + \dfrac{7}{8}$

We first convert $3\frac{1}{2}$ from a mixed number (an integer and a fraction) to an improper fraction (a fraction whose numerator is larger than the denominator):

$$3\frac{1}{2} = 3 + \frac{1}{2} = \frac{3}{1}\cdot\frac{2}{2} + \frac{1}{2} = \frac{6}{2} + \frac{1}{2} = \frac{7}{2}$$

A shortcut is to multiply the integer times the denominator and then add the result to the numerator (still over the denominator):

$$3\frac{1}{2} + \frac{7}{8} = \frac{7}{2} + \frac{7}{8}$$
$$= \frac{7}{2}\cdot\frac{4}{4} + \frac{7}{8}$$
$$= \frac{28}{8} + \frac{7}{8} = \frac{35}{8}$$

In general, we will leave answers as reduced improper fractions instead of mixed numbers.

EXAMPLE 1.3.7

A rectangular template that is 5 in. long and 3 in. wide has a square piece removed, as shown in Figure 1.3.1.

a. What is the length of each side of the square?

b. What is the area of the remaining template?

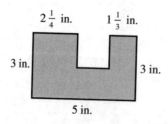

$2\frac{1}{4}$ in. $1\frac{1}{3}$ in.

3 in.   3 in.

5 in.

Figure 1.3.1

SOLUTION

a. Since the original length was 5 in. and the parts that remain have lengths of $2\frac{1}{4}$ in. and $1\frac{1}{3}$ in., the side of the square has a length equal to 5 in. minus the sum of $2\frac{1}{4}$ in. and $1\frac{1}{3}$ in.:

$$2\frac{1}{4} + 1\frac{1}{3} = \frac{9}{4} + \frac{4}{3} = \frac{9}{4} \cdot \frac{3}{3} + \frac{4}{3} \cdot \frac{4}{4} = \frac{27}{12} + \frac{16}{12} = \frac{43}{12}$$

$$5 - \frac{43}{12} = \frac{5}{1} - \frac{43}{12} = \frac{5}{1} \cdot \frac{12}{12} - \frac{43}{12} = \frac{60}{12} - \frac{43}{12} = \frac{17}{12}$$

Therefore, each side of the square is $\frac{17}{12}$ in., or $1\frac{5}{12}$ in.

b. To find the area of the remaining template, we subtract the area of the square from the area of the original rectangle. The area of the square is

$$\left(\frac{17}{12}\right)^2 = \frac{17}{12} \cdot \frac{17}{12} = \frac{289}{144} \text{ sq. in.}$$

The area of the rectangle is

$$3 \cdot 5 = 15 \text{ sq in.}$$

The area of the remaining template is

$$15 - \frac{289}{144} = \frac{15}{1} - \frac{289}{144} = \frac{15}{1} \cdot \frac{144}{144} - \frac{289}{144}$$

$$= \frac{2160}{144} - \frac{289}{144} = \frac{1871}{144} \quad \text{or} \quad 12\frac{143}{144} \text{ sq in.}$$

At the beginning of the section, we described the set of rational numbers as the set of all numbers that can be written as fractions with integral numerators and denominators. But any rational number can also be written as a terminating or repeating decimal by dividing the denominator into the numerator. For example:

$$\frac{1}{4} = 1 \div 4 \qquad \begin{array}{r} 0.25 \\ 4\overline{)1.00} \\ \underline{8} \\ 20 \\ \underline{20} \\ 0 \end{array}$$

Place a decimal point after the 1 and follow it with as many zeros as needed. The decimal point in the answer is above the decimal point in the dividend: $\frac{1}{4} = 0.25$.

$$\frac{7}{3} \qquad \begin{array}{r} 2.333\ldots \\ 3\overline{)7.000} \\ \underline{6} \\ 10 \\ \underline{9} \\ 10 \\ \underline{9} \\ 10 \end{array}$$

This time the decimal does not terminate but repeats 3 over and over. We can write this decimal in two ways: $\frac{7}{3} = 2.333\ldots$ or $\frac{7}{3} = 2.\overline{3}$, with the bar indicating that the number(s) under it is (are) repeated.

Conversely, we can go from terminating decimals to fractions by using the following steps:

$$2.7 = 2 + \frac{7}{10} = \frac{20}{10} + \frac{7}{10} = \frac{27}{10}$$

$$-0.15 = -\frac{15}{100} = -\frac{3}{20}$$

Repeating decimals can also be converted to fractions, but this requires a different method, which we will examine in Chapter 11, Section 11.3 (see also exercise 61 at the end of this section).

The arithmetic of positive and negative decimals can be performed using the rules for positive and negative integers.

EXAMPLE 1.3.8   Perform the indicated operations.

1. $64.59 - 128.6$

First, let's change the subtraction to addition: $64.59 + (-128.6)$. Then subtract the absolute values:

$$\begin{array}{r} 128.6 \\ -\ \ 64.59 \\ \hline 64.01 \end{array}$$

The answer is -64.01.

2. $7.4 - 15.04 + 24$

Rewrite as $7.4 + (-15.04) + 24.0$. Add the positives:

$$\begin{array}{r} 7.4 \\ +24.0 \\ \hline 31.4 \end{array}$$

leaving $31.4 + (-15.04)$:

$$\begin{array}{r} 31.4 \\ -15.04 \\ \hline 16.36 \end{array}$$

The answer is 16.36.

3. $\dfrac{0.1173}{-0.253}$

Moving the decimal points three places to the right gives us the equivalent problem

$$\frac{117.3}{-253}$$

$$
\begin{array}{r}
0.46363\ldots \\
253\overline{)117.30000} \\
101\ 2 \\
\overline{16\ 10} \\
15\ 18 \\
\overline{920} \\
759 \\
\overline{1610} \\
1518 \\
\overline{920}
\end{array}
$$

The division does not terminate but develops a pattern. The answer is $-0.4636363\ldots$, or $-0.4\overline{63}$.

EXAMPLE 1.3.9

Ima Cooke earns \$6.25 an hour at the Moss County Deli. Last week she worked $28\frac{1}{2}$ hr. What was her gross income for last week?

SOLUTION

We need to multiply her hourly wage by the number of hours she worked. First, convert $28\frac{1}{2}$ to a decimal: $28\frac{1}{2} = 28 + \frac{1}{2} = 28 + 0.5 = 28.5$. Now multiply:

$$
\begin{array}{r}
6.25 \\
\times\ \ 28.5 \\
\hline
3\ 125 \\
50\ 00 \\
125\ 0 \\
\hline
178.125
\end{array}
$$

Rounding off the answer to the nearest cent, she earned \$178.13.

Exercises 1.3

Reduce the following fractions to lowest terms.

1. $\dfrac{64}{48}$

2. $\dfrac{75}{175}$

3. $\dfrac{35}{18}$

4. $\dfrac{102}{17}$

5. $\dfrac{56}{63}$

6. $\dfrac{120}{84}$

7. $\dfrac{14}{70}$

8. $\dfrac{462}{154}$

9. $\dfrac{243}{162}$

10. $\dfrac{49}{132}$

Perform the indicated operations. Be sure your answer is in lowest terms.

11. $\dfrac{7}{12} - \dfrac{5}{12}$

12. $\dfrac{2}{9} \div 12$

13. $\dfrac{7}{8} - \dfrac{-11}{20}$

14. $\dfrac{9}{10} + \left(\dfrac{-3}{-5}\right)$

15. $\dfrac{8}{21} \cdot \dfrac{35}{6}$

16. $\left(\dfrac{-4}{9}\right)\left(\dfrac{-33}{-20}\right)$

17. $\dfrac{2}{15} \div \dfrac{10}{27}$

18. $\dfrac{7}{4} + \dfrac{3}{4}$

19. $\dfrac{5}{6} - \dfrac{7}{6} + \dfrac{11}{6}$

20. $\left(\dfrac{5}{-12}\right)\left(\dfrac{-8}{15}\right)\left(\dfrac{9}{2}\right)$

21. $\dfrac{\dfrac{18}{-4}}{9}$

22. $\dfrac{1}{6} + \dfrac{7}{8} - \dfrac{5}{18}$

23. $2 - \left(\dfrac{-11}{39}\right) - \dfrac{25}{26}$

24. $\dfrac{\dfrac{-49}{24}}{\dfrac{21}{-16}}$

25. $\left(3\dfrac{1}{4}\right)\left(\dfrac{48}{-13}\right)\left(\dfrac{-5}{8}\right)$

26. $-\dfrac{85}{48} - \dfrac{23}{18} + \dfrac{11}{16}$

27. $5\dfrac{1}{2} - \left(-1\dfrac{1}{6}\right) - 8\dfrac{3}{4}$

28. $\dfrac{7}{16} - \dfrac{11}{4} + \dfrac{3}{-8} + \dfrac{5}{2}$

29. $\left(\dfrac{25}{-9}\right)(-6)\left(-\dfrac{21}{40}\right)\left(\dfrac{2}{35}\right)$

30. $\dfrac{7}{15} - \dfrac{4}{9} + \dfrac{-3}{10}$

Convert the following fractions to decimals.

31. $\dfrac{3}{8}$

32. $\dfrac{1}{9}$

33. $-\dfrac{2}{9}$

34. $-\dfrac{5}{16}$

35. $\dfrac{1}{7}$

36. $\dfrac{2}{7}$

Convert the following decimals to fractions in lowest terms.

37. 0.65

38. 2.4

39. 3.875

40. 0.005

41. 0.0025

42. 12.08

Perform the indicated operations.

43. $-16.45 - 8.403$

44. $(-6.8)(30.21)$

45. $(-13.356) \div (-6.3)$

46. $228 \div 3.75$

47. $7.81 + 16.095 - 9.96$

48. $2 - 0.02 + (-0.2) + 0.002$

49. $(-2.5)(-40.8)(-7.25)$

50. $(4.6)(-7.02)(3.55)(-26.4)$

51. $14.5 + 8.91 - 3.407$

52. $20.63 - 4.8003 - (-72.004)$

53. $(-152.52) \div (6.15)$

54. $260.1 \div (-6.375)$

Word problems.

55. Find the perimeters of the following figures. (*Hint:* Use the formulas on the inside front cover.)

a.

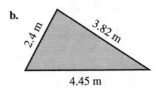

$3\frac{2}{3}$ in.

$1\frac{5}{6}$ in. $1\frac{5}{6}$ in.

$3\frac{2}{3}$ in.

b.

2.4 m   3.82 m

4.45 m

56. Find the areas of the following figures. (*Hint:* Use the formulas on the inside front cover.)

a.

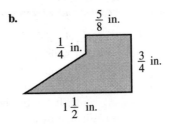

2.3 cm

1.4 cm

5.5 cm

b.

$\frac{5}{8}$ in.

$\frac{1}{4}$ in.

$\frac{3}{4}$ in.

$1\frac{1}{2}$ in.

57. Wanda Tip sacks groceries at the neighborhood supermarket. She is paid $4.50 an hour if she works no more than 40 hr a week. She is paid time and a half for each hour beyond her regular 40-hr week.
  a. One week she worked $56\frac{1}{4}$ hr. What was her gross income for that week?
  b. Another week she worked less than 40 hr but she forgot exactly how long she worked. She remembers that her gross income for that week was $164.25. How many hours did she work that week?

58. Garbonzo's Groceries sells granola in a 14-oz package for $1.19 and in a 22-oz package for $1.76. Which is the better buy; that is, which is cheaper per ounce?

59. Rich N. Young is investigating the price of a share of stock in the Fabulous Fig Company. One day it opened at $16\frac{3}{4}$ and then rose $1\frac{1}{2}$ points within 30 min. Later it rose $\frac{3}{8}$ of a point more. But it plummeted $2\frac{1}{4}$ points from its high of the day. At what price did the stock close?

60. Susie Stretch ran a $4\frac{1}{2}$-mi race in 32:37.5 min, a personal record for her. What was her pace, or time per mile?

61. The repeating decimal 0.4444 ... can be converted to a fraction as follows:

Give it a name: $n = 0.4444\ldots$

Multiply n by 10: $10n = 4.4444\ldots$

Subtract $1n$ from $10n$: $9n = 4$ Subtract 0.444 ... from 4.444 ...

So $n = \dfrac{4}{9}$

Use this method to convert the following decimals to fractions.
  a. 0.55555 ...   **b.** 0.4545 ...

Write Algebra

62. Explain the process for reducing a fraction to lowest terms.

63. Explain the process for converting a fraction to a decimal.

64. Explain the process for converting a terminating decimal to a fraction.

65. Describe the process for determining the LCD of two fractions.

1.4 Irrational Numbers

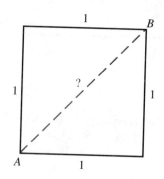

Figure 1.4.1

We will now look at a set of numbers called the **irrational numbers,** which we label with an H. Before their discovery circa 400 B.C., it was thought that all distances could be measured by rational numbers. But then came the realization that the rational numbers were not enough. Consider a square stone block, one unit on a side. It can be shown that the distance from A to B in Figure 1.4.1 cannot be represented by a rational number. We now know this distance to be the irrational number $\sqrt{2}$, the square root of two, which is some number that when squared is 2. Since all rational numbers can be written as terminating or repeating decimals, $\sqrt{2}$ must be a decimal that *does not terminate or repeat:*

$$\sqrt{2} = 1.4142135623\ldots$$

In general, any irrational number can be written as a nonterminating and nonrepeating decimal.

What are some other examples of irrational numbers? Let us consider the square roots of the natural numbers. The following chart is a small portion of Table II in Appendix A. The numbers in this list can also be generated using a calculator.

n	$\sqrt{n}$
*1	1
2	1.414…
3	1.732…
*4	2
5	2.236…
6	2.449…
7	2.646…
8	2.828…
*9	3
10	3.162…

Except for those numbers that are perfect squares—that is, squares of other natural numbers (denoted by asterisks)—all of the square roots of the natural numbers are irrational. Besides square roots, there are many other types of irrational numbers, such as cube roots and fourth roots, which we will study in Chapter 5. Probably the most famous irrational number is π, the ratio of the circumference of a circle to its diameter:

$$\pi = 3.1415926536\ldots$$

If π has a rival in its widespread use in mathematics, it is e, which we will study in Chapter 10. There we will talk about logarithms:

$$e = 2.718281828459045\ldots$$

So far we have seen that $N \subseteq W \subseteq I \subseteq Q$. But H, the set of irrational numbers, has no common elements with Q; that is,

$$H \cap Q = \varnothing$$

EXAMPLE 1.4.1   Determine whether the following are true or false.

1. $6.\overline{25} \in H$

False. $6.\overline{25}$ is a nonterminating *repeating* decimal, so it is a rational number.

2. $\sqrt{16} \in Q$

True, $16 = 4^2$; thus $\sqrt{16}$ is 4, a rational number.

3. $H \cap I = \varnothing$

True. All integers are rational numbers, so none of them is irrational.

Earlier, when we considered the square stone block, the distance we stated to be $\sqrt{2}$ was the **diagonal**. In all squares and most rectangles (see Figures 1.4.2 and 1.4.3), when the sides are natural numbers, the diagonals are irrational numbers. The length of the diagonal can be found using the Pythagorean theorem.

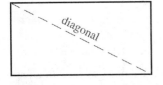

Figure 1.4.2

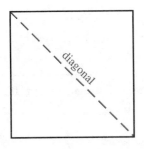

Figure 1.4.3

*T*he Pythagorean Theorem

If a and b are the lengths of the legs of a right triangle and c *is* the length of the hypotenuse, then

$$a^2 + b^2 = c^2$$

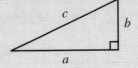

EXAMPLE 1.4.2   Find the lengths of the following diagonals. If the sides are given in metric units, approximate the diagonal to two decimal places. If the sides are given in feet or inches, approximate the diagonal to two decimal places and convert to a fraction.

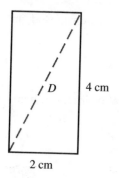

4 cm

2 cm

Figure 1.4.4

1. The diagonal (let's call it D) cuts the rectangle into two right triangles. In Figure 1.4.4 the length, $L = 4$ cm, and the width, $W = 2$ cm, are the legs, while the diagonal D is the hypotenuse. By the Pythagorean theorem,

$$D^2 = L^2 + W^2$$
$$D^2 = 4^2 + 2^2$$
$$D^2 = 16 + 4$$
$$D^2 = 20$$

Thus, D must be $\sqrt{20}$ or $D \doteq 4.47$ cm. (The symbol $\doteq$ means "approximately equal to.")

2. For the square in Figure 1.4.5,

$$D^2 = 7^2 + 7^2$$
$$D^2 = 49 + 49$$
$$D^2 = 98$$

Thus, $D = \sqrt{98} \doteq 9.90$ ft, or $9\frac{9}{10}$ ft.

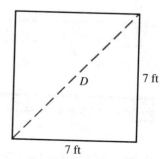

7 ft

7 ft

Figure 1.4.5

When finding the circumference or area of a circle, we must work with an approximation for π. You may remember using $\frac{22}{7}$ for π, *but this is just a rational approximation.* Similarly, you may have used 3.14 as a rational approximation for π in decimal form.

XAMPLE 1.4.3

Find the indicated quantities. If the radius is given as a decimal, use 3.14 for π. If the radius is given as a fraction, use $\frac{22}{7}$ for π.

1.

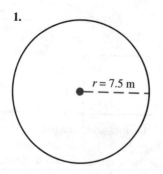

$r = 7.5$ m

$C = ?$

$$C = 2\pi r$$
$$C \doteq 2(3.14)(7.5)$$
$$C \doteq (6.28)(7.5)$$
$$C \doteq 47.1 \text{ m}$$

2.

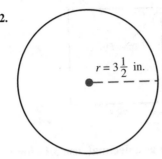

$A = ?$ $A = \pi r^2$

$$A \doteq \frac{22}{7}\left(3\frac{1}{2}\right)^2$$

$$A \doteq \frac{22}{7}\left(\frac{7}{2}\right)^2$$

$$A \doteq \frac{\cancel{2} \cdot 11}{\cancel{7}} \cdot \frac{\cancel{7}}{\cancel{2}} \cdot \frac{7}{2}$$

$$A \doteq \frac{77}{2} = 38\frac{1}{2} \text{ sq in.}$$

At this point we will not go into the arithmetic of irrational numbers, which we will leave until we discuss radicals in Chapter 5.

Exercises 1.4

Determine whether the following are true or false.

1. $\sqrt{12} \in H$

2. $\sqrt{121} \in H$

3. $\sqrt{84} \in Q$

4. $\sqrt{64} \in Q$

5. $1.010010001 \ldots \in H$

6. $3.1415926536 \in H$

7. $2.317317 \ldots \in Q$

8. $0.2626626662 \ldots \in Q$

9. $\pi = \frac{22}{7}$

10. $\pi = 3.14$

11. $H \cap N = N$

12. $H \cup N = H$

Find the lengths of the following diagonals. If the sides are given in metric units, approximate the diagonal to two decimal places. If the sides are given in feet or inches, approximate the diagonal to two decimal places and convert to a fraction.

13.

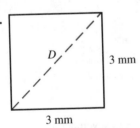

14.

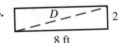

15.

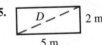

16.

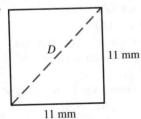

17.

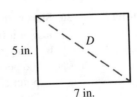

18.

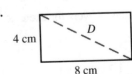

19.

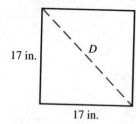

17 in.

17 in.

20.

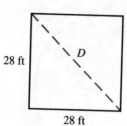

28 ft

28 ft

Find the indicated quantities. If the radius is given as a decimal, use 3.14 for π. If the radius is given as a fraction, use $\frac{22}{7}$ for π.

21.

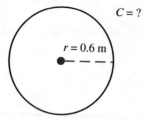

$C = ?$

$r = 0.6$ m

22.

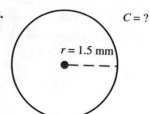

$C = ?$

$r = 1.5$ mm

23.

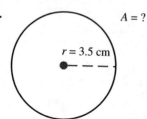

$A = ?$

$r = 3.5$ cm

24.

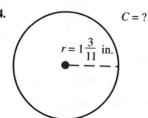

$C = ?$

$r = 1\frac{3}{11}$ in.

25.

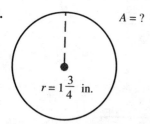

$A = ?$

$r = 1\frac{3}{4}$ in.

26.

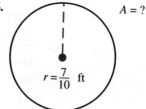

$A = ?$

$r = \frac{7}{10}$ ft

 The following problems involve the *Golden Ratio*, $\Phi = \dfrac{1 + \sqrt{5}}{2}$, which was used in ancient Greece and during the Renaissance in architecture and other art forms.

27. Find an approximation for Φ to seven decimal places.

28. Find an approximation for $\frac{1}{\Phi}$ to seven decimal places.

29. How do the answers for exercises 27 and 28 differ?

2 in.

Figure 1.4.6

30. Start with a square 2 in. on a side.
  a. Cut it in half (see Figure 1.4.6).
  b. Using a protractor, put the point at point A and the pencil at point B and draw an arc down to point C (see Figure 1.4.7).
  c. Draw a line straight up from point C and straight across from point B (see Figure 1.4.8).

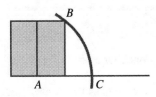

Figure 1.4.7

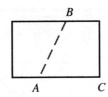

Figure 1.4.8

d. What are the dimensions of this rectangle? (*Hint:* Find the distance from A to B. How does this compare with the distance from A to C?)

e. Where does the Golden Ratio appear in this rectangle?

Most calculators with a square root key show an approximation to seven decimal places for any square root of a number less than 100. But the calculator may actually have a few more "hidden" digits stored in its memory.

31. **a.** Use a calculator to find $\sqrt{2}$.
   b. Multiply by 1000.
   c. Subtract away the number formed by the first four digits.
   d. The "new" digits that now appear at the end of this number should be attached to the end of your answer from part a.
   e. Compare your result with the value for $\sqrt{2}$ found at the beginning of this section.

32. Use a calculator to find a decimal approximation to 10 decimal places for the following.
   a. $\sqrt{8}$ **b.** $\sqrt{12}$

Write Algebra

33. Explain the difference between an irrational number and a rational number.

34. Can the product of two irrational numbers be rational? Explain your answer.

1.5 *Real Numbers*

When you measure something, whether distance, weight, time interval, or temperature, you use numbers that are either rational or irrational. Therefore, we combine these two sets into one larger set, which we call the set of **real numbers.** This set, which we label with an R, is the union of two sets:

$$R = Q \cup H$$

Let us briefly review the sets of numbers we have discussed and their relation to one another.

Real Numbers, *R*

Rational numbers, *Q*	Irrational numbers, *H*
1. Can be written as fractions with integral numerators and denominators	**1.** Cannot be written as fractions with integral numerators and denominators
2. Can be written as terminating or repeating decimals	**2.** Can be written as decimals that are both nonterminating and nonrepeating

> *Integers, I*
> $\dots, -3, -2, -1, 0, 1, 2, 3, \dots$
>
> > *Whole numbers, W*
> > $0, 1, 2, 3, \dots$
> >
> > > *Natural numbers, N*
> > > $1, 2, 3, \dots$

EXAMPLE 1.5.1   Determine whether the following are true or false.

1. There are real numbers that are neither rational nor irrational.

False. The real numbers consist of *only* the rational and irrational numbers.

2. $\sqrt{5} \in R$

True. $\sqrt{5} \in H$ and $H \subseteq R$, so $\sqrt{5} \in R$.

An important concept in mathematics is that of the **real number line.** A ruler is a common example of part of a number line, in that numbers are assigned to specific points on the ruler. In a similar fashion, we can assign every real number to a point on a line, and every point on the line to a real number. We commonly draw the real number line with hash marks on it, like a ruler, labeling the location of some of the integers.

When we locate the point assigned to a specific number and draw a dot there, we say that we are **graphing** that number. To **graph a set** means to graph each of the numbers in the set.

E XAMPLE 1.5.2   Graph the following sets.

1. $\left\{ -\dfrac{1}{2}, \dfrac{5}{2}, \dfrac{11}{3}, 5.2, 7 \right\}$

$-\frac{1}{2}$ is one-half of the way from 0 to -1. $\frac{5}{2} = 2\frac{1}{2}$ is one-half of the way from 2 to 3. $\frac{11}{3} = 3\frac{2}{3}$ is two-thirds of the way from 3 to 4. $5.2 = 5\frac{2}{10}$ is two-tenths of the way from 5 to 6.

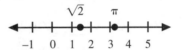

2. $\{\sqrt{2}, \pi\}$

Although in theory any irrational number is assigned to an exact point on the real number line, in practice we use an approximation to graph any irrational number. For $\sqrt{2}$, use 1.4, and for π, use 3.1.

The real number line gives us a pictorial way of understanding the concept of the **order of the real numbers.** Given two distinct real numbers, one is smaller than the other. This relationship is noted with one of the **inequality symbols** given here.

Symbol	Stands for	Examples
$<$	is less than	$2 < 5, \; -8 < -3$
$\leq$	is less than or equal to	$-4 \leq 1, 3 \leq 3$
$>$	is greater than	$5 > 1, 0 > -2$
$\geq$	is greater than or equal to	$6 \geq 0, \dfrac{6}{10} \geq \dfrac{3}{5}$

NOTE ▶   1. $a < b$ means that there is a positive number p such that $a + p = b$.

2. On the real number line the numbers get larger to the right and smaller to the left. Therefore, given two real numbers, the graph of the smaller one lies to the left of the graph of the larger one.

3. A number is positive if it is greater than zero, and it is negative if it is less than zero. Zero is neither positive nor negative. A nonnegative number is either positive or zero.

4. The two symbols $\leq$ and $\geq$ state that one of two possible relationships holds. For example, $-4 \leq 1$ says that either -4 is less than 1 or $-4 = 1$.

5. The inequality symbols $<$ and $>$ always point toward the smaller number.

EXAMPLE 1.5.3   Place the appropriate inequality symbol, $<$ or $>$, between each pair of numbers.

1. $\dfrac{4}{3}$ _____ 1

$\dfrac{4}{3} = 1\dfrac{1}{3}$, so $\dfrac{4}{3} > 1$.

2. -5 _____ -7

The graph of -5 lies to the right of the graph of -7; hence, $-5 > -7$.

3. 3 _____ $\sqrt{10}$

$\sqrt{10} \doteq 3.162$; hence, $3 < \sqrt{10}$.

For the remainder of this section, we will consider some properties that will be useful in our study of algebra.

Properties of Equality	
Let a, b, and c represent real numbers.	
1. Reflexive property	$a = a$
2. Symmetric property	If $a = b$, then $b = a$.
3. Transitive property	If $a = b$ and $b = c$, then $a = c$.
4. Substitution property	If $a = b$, then any expression involving a retains the same value if a is replaced by b.

Properties of Inequalities

Let a, b, and c represent real numbers.

1. Transitive property	If $a < b$ and $b < c$, then $a < c$.
2. Trichotomy property	One and only one of the following statements is true: $a < b$, $a = b$, or $a > b$.

EXAMPLE 1.5.4   State the property illustrated by each of the following statements.

1. $-4 < 0$ and $0 < 3$, so $-4 < 3$ *Transitive property of inequality*

2. If $y = 3$, then $3 = y$.   *Symmetric property of equality*

3. $g = 2$ and $P = g^2$, so $P = 2^2$ *Substitution property of equality*

Properties of Real Numbers If a, b, and c represent real numbers, the following properties are true.		
		Operation
Name	Addition	Multiplication
1. Commutative	$a + b = b + a$	$a \cdot b = b \cdot a$
2. Associative	$(a + b) + c = a + (b + c)$	$(a \cdot b) \cdot c = a \cdot (b \cdot c)$
3. Identity	$a + 0 = a = 0 + a$	$a \cdot 1 = a = 1 \cdot a$
4. Inverse	$a + (-a) = 0$ $= -a + a$	$a \cdot \dfrac{1}{a} = 1 = \dfrac{1}{a} \cdot a$, if $a \neq 0$
5. Distributive	$a \cdot (b + c) = a \cdot b + a \cdot c$	

NOTE ▶   The commutative and associative properties do not hold for the operations of subtraction and division. For example,

$$4 \div 8 \neq 8 \div 4$$
$$10 - (7 - 3) \neq (10 - 7) - 3$$

We discussed additive inverses in Section 1.2. Now let us look at **multiplicative inverses,** or **reciprocals.** For any nonzero number x, the multiplicative inverse is $\frac{1}{x}$.

Number	Multiplicative Inverse	Product of a Number and Its Multiplicative Inverse
6	$\dfrac{1}{6}$	$6 \cdot \dfrac{1}{6} = 1$
-4	$-\dfrac{1}{4}$	$(-4)\left(-\dfrac{1}{4}\right) = 1$
$\dfrac{2}{3}$	$\dfrac{1}{\frac{2}{3}}$ or $\dfrac{3}{2}$	$\left(\dfrac{2}{3}\right)\left(\dfrac{3}{2}\right) = 1$
$-\dfrac{7}{3}$	$-\dfrac{3}{7}$	$\left(-\dfrac{7}{3}\right)\left(-\dfrac{3}{7}\right) = 1$
$\dfrac{1}{5}$	5	$\dfrac{1}{5} \cdot 5 = 1$

Note that the multiplicative inverse agrees in sign with the original number. Also the product of a number and its multiplicative inverse is always positive one.

EXAMPLE 1.5.5   Find the additive and multiplicative inverses of each number.

1. 0.7 Additive inverse: -0.7 (opposite sign); Multiplicative inverse:

$$\frac{1}{0.7} \text{ or } \frac{1}{\frac{7}{10}} = \frac{10}{7} \quad \text{(same sign)}$$

2. $-\dfrac{4}{11}$ Additive inverse: $\dfrac{4}{11}$ (opposite sign);   Multiplicative inverse: $-\dfrac{11}{4}$ (same sign)

3. $\sqrt{3}$ Additive inverse: $-\sqrt{3}$ (opposite sign): Multiplicative inverse: $\dfrac{1}{\sqrt{3}}$

(same sign)

The distributive property is one of the most important properties we will encounter. It is used in two equally important ways, as illustrated here.

EXAMPLE 1.5.6   Use the distributive property to rewrite each of the following.

1. $3(x + a + 5)$

It is understood that 3 is being multiplied times $(x + a + 5)$:

$$3(x + a + 5) = 3 \cdot x + 3 \cdot a + 3 \cdot 5$$
$$= 3x + 3a + 15$$

Note that the factor 3 is "distributed" across all the terms contained in the parentheses.

2. $4(7 - 2) = 4[7 + (-2)]$
$$= 4 \cdot 7 + 4(-2)$$
$$= 28 + (-8)$$
$$= 20$$

3. $2x + 2y = 2(x + y)$

Here the distributive property is used to write a sum as a product.

4. $7x + 5x = (7 + 5)x$
$$= 12x$$

EXAMPLE 1.5.7   Name the property that justifies each step in the following.

1. $3 + (x + 5) = 3 + (5 + x)$ *Commutative property of addition*
$$= (3 + 5) + x$$ *Associative property of addition*
$$= 8 + x$$

2. $\dfrac{1}{3}(3x) = \left(\dfrac{1}{3} \cdot 3\right)x$ *Associative property of multiplication*
$$= 1x$$ *Inverse property of multiplication*
$$= x$$ *Identity property of multiplication*

3. $-6x + 6x = (-6 + 6)x$ *Distributive property*

$\qquad\qquad = 0x$ *Inverse property of addition*

$\qquad\qquad = 0$ *(The product of zero and any other number(s) is zero)*

Try to avoid these mistakes:

Incorrect	Correct
1. $\left(\dfrac{1}{2}\right)(-2x) = x$	$\left(\dfrac{1}{2}\right)(-2x) = \left[\left(\dfrac{1}{2}\right)(-2)\right]x$
	$= -1x = -x$
2. $6(3 \cdot 2) = 6 \cdot 3 \cdot 6 \cdot 2$ Multiplication does not "distribute" over multiplication.	$6(3 \cdot 2) = (6 \cdot 3)2$ Associative property
	$6(3 + 2) = 6 \cdot 3 + 6 \cdot 2$ Distributive property
3. $4(y + 3) = 4y + 3$	$4(y + 3) = 4y + 4 \cdot 3$
4. $2x + 2y = 4(x + y)$	$2x + 2y = 2(x + y)$

Exercises 1.5

Determine whether the following are true or false.

1. $N \subseteq R$

2. $N \subseteq H$

3. $I \cup H = R$

4. $H \cap R = H$

5. $\pi \in R$

6. $0 \in R$

7. $-6 > -2$

8. $\sqrt{3} \leq 3$

9. Zero is positive.

10. Zero is negative.

11. $4 < 4$

12. $4 \leq 4$

13. $-\dfrac{2}{3} + \dfrac{3}{2} = 0$

14. $\left(-\dfrac{2}{3}\right)\left(\dfrac{3}{2}\right) = 1$

15. $6 \cdot x = x \cdot 6$

16. $a(b + c) = (b + c)a$

17. $2x + 7x = 9x^2$

18. $5x + 5y = 10(x + y)$

Graph the following sets.

19. $\left\{-4, \dfrac{4}{3}, \dfrac{9}{10}\right\}$

20. $\left\{-\dfrac{5}{4}, 2, \dfrac{4}{5}\right\}$

21. $\left\{1.3, \dfrac{7}{2}, -2.6\right\}$

22. $\left\{-1.5, -\dfrac{15}{4}, 4.4\right\}$

23. $\{\sqrt{3}, \sqrt{4}, \sqrt{5}\}$

24. $\{\sqrt{8}, \sqrt{9}, \sqrt{10}\}$

Place the appropriate inequality symbol, $<$ or $>$, between each pair of numbers.

25. $\dfrac{5}{4}$ _____ 1.4

26. 2.1 _____ $\dfrac{5}{2}$

27. -7 _____ 0

28. 7 _____ 0

29. -6 _____ -2

30. -4 _____ -9

31. $\sqrt{6}$ _____ 2

32. 4 _____ $\sqrt{17}$

33. π _____ 3

34. e _____ 3

35. $2.\overline{2}$ _____ 2.22

36. $1.50\overline{4}$ _____ $1.5\overline{04}$

State the property illustrated by each of the following statements.

37. If $x = 5$, then $5 = x$.

38. If $p = a$ and $p = 4$, then $a = 4$.

39. $x = x$

40. If $ab = 0$, then $0 = ab$.

41. If $-8 < 0$ and $0 < 4$, then $-8 < 4$.

42. If $2 \leq \sqrt{5}$ and $\sqrt{5} \leq \pi$, then $2 \leq \pi$.

43. If $x = 3$ and $y = x^2$, then $y = 3^2$.

44. If $y = 4$ and $x = \sqrt{y}$, then $x = \sqrt{4}$.

45. $x = 1x$

46. $2 \cdot (3 \cdot 4) = (2 \cdot 3) \cdot 4$

47. $3 + 5 = 5 + 3$

48. $x \cdot y = y \cdot x$

49. $2x + 0 = 2x$

50. $0 + 1 = 1$

51. If $x \nleq 4$, then $x > 4$.

52. If $y \nless a$, then $y \geq a$.

Find the additive and multiplicative inverses of each number.

53. -6

54. $\dfrac{1}{2}$

55. $-\dfrac{3}{7}$

56. $\dfrac{5}{8}$

57. $-\sqrt{8}$

58. $\sqrt{5}$

59. -0.375

60. 0.25

61. $0.333\ldots$

62. $-0.666\ldots$

Use the distributive property to rewrite each of the following.

63. $2(5 + 7)$

64. $-3(4 + 8)$

65. $6(x + y + 4)$

66. $2(x + b + 7)$

67. $5(x - 3)$

68. $4(y - 5)$

69. $5a + 5b$

70. $4x + 4y + 4z$

71. $7x + 10x + 5x$

72. $6y - 3y$

73. $x(y + 1)$

74. $ay + a$

Name the property that justifies each step in the following.

75. $6(3x) = (6 \cdot 3)x$
$= 18x$

76. $-5(2y) = (-5 \cdot 2)y$
$= -10y$

77. $4 + (a + 7) = 4 + (7 + a)$
$= (4 + 7) + a$
$= 11 + a$

78. $-5 + (y + 5) = -5 + (5 + y)$
$= (-5 + 5) + y$
$= 0 + y$
$= y$

79. $-3 + (x + 3) = -3 + (3 + x)$

$= (-3 + 3) + x$

$= 0 + x$
$= x$

80. $(8x) \cdot \dfrac{1}{8} = \dfrac{1}{8} \cdot (8x)$

$= \left(\dfrac{1}{8} \cdot 8\right)x$

$= 1 \cdot x$

$= x$

81. $(-4y) \cdot \left(-\dfrac{1}{4}\right) = \left(-\dfrac{1}{4}\right) \cdot (-4y)$

$= \left[\left(-\dfrac{1}{4}\right) \cdot (-4)\right]y$

$= 1 \cdot y$
$= y$

82. $5a + 9a = (5 + 9)a$
$= 14a$

83. $-8y + 8y = (-8 + 8)y$
$$= 0y$$
$$= 0$$

84. $(3x + 7) + (4x + 9) = 3x + [7 + (4x + 9)]$
$$= 3x + [(4x + 9) + 7]$$
$$= 3x + [4x + (9 + 7)]$$
$$= 3x + [4x + 16]$$
$$= [3x + 4x] + 16$$
$$= [3 + 4]x + 16$$
$$= 7x + 16$$

Write Algebra

85. According to the trichotomy property, if $x \neq 2$, then what can we say about the numbers that x can represent?

86. For each of the following expressions, explain (when appropriate) how the distributive property can be used.
 a. $a(b + c)$
 b. $(a + b) - c$
 c. $(a + b)c$
 d. $a(bc)$
 e. $a(b - c)$

1.6 Order of Operations, Absolute Value, and Algebraic Expressions

Some arithmetic problems would be ambiguous if we did not have rules to interpret them. For example, the expression $3 + 2 \cdot 4$ could be interpreted two ways.

Incorrect	Correct
$\underbrace{3 + 2} \cdot 4$	$3 + \underbrace{2 \cdot 4}$
$= \quad 5 \quad \cdot 4$	$= 3 + \quad 8$
$= 20$	$= 11$
Adding first	Multiplying first

Because of problems like this one, we have the following rules to determine the order in which we perform operations.

Order of Operations	
Rule	**Example**
1. Perform all operations within grouping symbols first.	$6(15 - 8) \div 14$ $= 6 \cdot (7) \quad\quad \div 14$ $= 42 \quad\quad\quad \div 14$ $= 3$

2. If there are no grouping symbols—that is, parentheses or brackets—perform the operations in the following order:
  a. Powers and roots
  b. Multiplications and divisions in order from left to right
  c. Additions and subtractions in order from left to right
	$6 + 5 \cdot 12 \div 2^2 - 1$
	$= 6 + 5 \cdot 12 \div 4 \ - 1$
	$= 6 + 60 \quad\quad \div 4 \ - 1$
	$= 6 + 15 \quad\quad\quad\quad - 1$
	$= 21 \quad\quad\quad\quad\quad\quad - 1$
	$= 20$

3. When grouping symbols are included within other grouping symbols, start with the innermost set and work outward.	$3\left[2^2 - \dfrac{1}{2} + \left(\dfrac{1}{2}\right)(1 - 7) \right]$ $= 3\left[2^2 - \dfrac{1}{2} + \left(\dfrac{1}{2}\right)(-6) \right]$ $= 3\left[4 - \dfrac{1}{2} + \left(\dfrac{1}{2}\right)(-6) \right]$ $= 3\left[4 - \dfrac{1}{2} + (-3) \right]$ $= 3\left[\dfrac{7}{2} \quad\quad + (-3) \right]$ $= 3\left[\dfrac{1}{2} \right]$ $= \dfrac{3}{2}$

E X A M P L E   1 . 6 . 1   Perform the indicated operations following the rules governing the order of operations.

$$\frac{18 \div (0.9) \cdot 2 + 5.6}{1 + \sqrt{7 + 2}}$$

The division line acts as parentheses, indicating that the operations in the numerator and denominator must be performed before the division.

$$\frac{18 \div (0.9) \cdot 2 + 5.6}{1 + \sqrt{7 + 2}} = \frac{20 \cdot 2 + 5.6}{1 + \sqrt{9}} \qquad \textit{The square root symbol acts as parentheses.}$$

$$= \frac{40 + 5.6}{1 + 3} = \frac{45.6}{4} = 11.4$$

Try to avoid these mistakes:

Incorrect	Correct
1. $7 - 5 + 2 = 7 - 7$ $= 0$	$7 - 5 + 2 = 2 + 2$ $= 4$ Work from left to right.
2. $16 \div 2 \cdot 4 = 16 \div 8$ $= 2$	$16 \div 2 \cdot 4 = 8 \cdot 4$ $= 32$ Work from left to right.
3. $3 \cdot 4^2 = 12^2$ $= 144$	$3 \cdot 4^2 = 3 \cdot 16$ $= 48$ Evaluate the exponent first.
4. $6 - 2(5 + 1) = 4(5 + 1)$ $= 4(6)$ $= 24$	$6 - 2(5 + 1) = 6 - 2(6)$ $= 6 - 12$ $= -6$ Outside of the parentheses, multiply before you subtract.

In the preceding section we used letters to represent numbers. We usually call the letters **variables.** A variable is a character used to represent any number from a specified set of numbers. On the other hand, when we have a character that represents a particular number, we call it a **constant.** In the expression

$$3x + \frac{1}{2}y + \pi$$

the variables are x and y, whereas the constants are 3, $\frac{1}{2}$, and π. We use a variable to define a set of numbers when working with **set builder notation.** An example of this is

$$\{x \quad : \quad x \le 3\}$$

$\{x$	:	$x \le 3\}$
the set of all real numbers x	such that	x is less than or equal to 3

Note that this is an infinite set of real numbers satisfying a condition that is indicated after the colon. We have graphed finite sets before, but we can also graph infinite sets like this one:

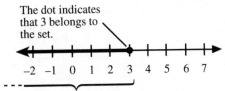

The dot indicates that 3 belongs to the set.

The heavy line indicates that all the numbers to the left of 3 belong to the set.

XAMPLE 1.6.2   Graph the following sets.

1. $\left\{x : x > \dfrac{3}{2}\right\}$

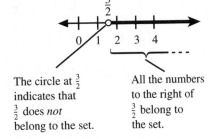

The circle at $\frac{3}{2}$ indicates that $\frac{3}{2}$ does *not* belong to the set.

All the numbers to the right of $\frac{3}{2}$ belong to the set.

2. $\{x : x < -1\} \cup \{x : x \geq 2\}$

The union of these two sets gives a larger set that can be written as $\{x : x < -1$ or $x \geq 2\}$.

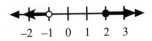

A very important concept in mathematics is that of *absolute value*. We discussed absolute value in Section 1.2. Now let us define it in terms of distance on the real number line.

Absolute Value

*The **absolute value** of a number is the distance from zero to that number on the real number line.* The absolute value of a number x is denoted by $|x|$.

EXAMPLE 1.6.3   Evaluate each of the following.

1. $|5|$

Since 5 is five units to the right of zero, $|5| = 5$.

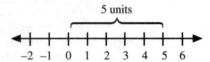

2. $\left|-\dfrac{4}{3}\right|$

Since $-\frac{4}{3}$ is four-thirds units to the left of zero, $\left|-\frac{4}{3}\right| = \frac{4}{3}$.

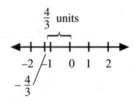

3. $|0|$

Since 0 is zero units from zero, $|0| = 0$.

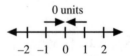

Note that the absolute value of a number is never negative, since distance is never negative. It is always positive or zero.

EXAMPLE 1.6.4   Place the appropriate symbol, $<$, $>$, or $=$, between the following pairs of numbers.

1. $|-4|$ _____ 2

Since $|-4| = 4$, $|-4| > 2$.

2. $|6|$ _____ $|-6|$

Since $|6| = 6 = |-6|$, $|6| = |-6|$.

The last topic we will consider in this chapter is **algebraic expressions.** An algebraic expression is any combination of constants and variables that involves powers, roots, or the four operations of addition, subtraction, multiplication, and division. The following are examples of algebraic expressions, some of which we have encountered earlier in this chapter:

$$a^2 + b^2, \qquad 2\pi r, \qquad \frac{1}{2}ax^3 + \frac{\sqrt{2}}{y}, \qquad \sqrt{x^2 - 4}$$

Many times we want to substitute numbers for the variables in an algebraic expression and then simplify the resulting expression using the rules for order of operations. This process is called **evaluating an algebraic expression** for the given value(s) of the variable(s).

EXAMPLE 1.6.5   Evaluate the following algebraic expressions for the given values of the variables.

1. $3x^2 - x + 8$ for $x = 2$

Replacing each x with 2, we get

$$
\begin{aligned}
3 \cdot 2^2 - 2 + 8 &= 3 \cdot 4 - 2 + 8 \\
&= 12 - 2 + 8 \\
&= 10 + 8 \\
&= 18
\end{aligned}
$$

2. $\dfrac{2x^2 + 3xy - 20y^2}{8x - 20y}$ for $x = -2$ and $y = 3$

$$
\begin{aligned}
\frac{2(-2)^2 + 3(-2)(3) - 20 \cdot 3^2}{8(-2) - 20 \cdot 3} &= \frac{2(4) + 3(-2)(3) - 20 \cdot 9}{8(-2) - 20 \cdot 3} \\
&= \frac{8 + (-18) - 180}{-16 - 60} \\
&= \frac{-10 - 180}{-76} \\
&= \frac{-190}{-76} = \frac{5}{2}
\end{aligned}
$$

3. $|4 - xy^2| - \sqrt{x + 4}$ for $x = 5$ and $y = 2$

NOTE ▶   The absolute value signs, $|\quad|$, and the square root symbol, $\sqrt{\quad}$, both act as parentheses, telling us to perform the operations within them before we subtract.

$$\begin{aligned}
|4 - 5 \cdot 2^2| - \sqrt{5 + 4} &= |4 - 5 \cdot 4| - \sqrt{5 + 4} \\
&= |4 - 20| - \sqrt{5 + 4} \\
&= |-16| - \sqrt{9} \\
&= 16 - 3 = 13
\end{aligned}$$

Exercises 1.6

Perform the indicated operations, following the rules governing the order of operations.

1. $17 - 8 + 7 + (-6)$

2. $-5 + 4 - 9 - (-3)$

3. $24 \div 4 \cdot 3$

4. $36 \div 9 \cdot 2$

5. $5 \cdot 3^2$

6. $(-3)^2 - 25 \div 4 + \dfrac{3}{2}$

7. $-8 + 30 \div 2^2 - \dfrac{1}{4}$

8. $2.5 \div (2.5)^2 - 0.27$

9. $1.8 \div (1.5)^2 + 0.2$

10. $7(3 - 8) \div 14$

11. $11(7 - 15) \div 66$

12. $8 - 3(4 \cdot 2 - 5) + 21$

13. $12 - 2(6 + 3 \cdot 4) + 15$

14. $-3 + \sqrt{5 \cdot 3 + 21}$

15. $-3 - \sqrt{6 \cdot 2 - 8}$

16. $6 \cdot 2^2$

17. $4 \cdot 7 \cdot \sqrt{16}$

18. $8 \cdot 3 \cdot \sqrt{25}$

19. $7 + 8(-11) \div 4$

20. $16 - (-3)(-12) \div 9$

21. $\dfrac{2.1 + 5(0.7)}{(0.2)^2 + 0.1}$

22. $\dfrac{7.33 - 2(0.05)}{0.12 - (0.3)^2}$

23. $4 \cdot \left| 6 - 4 \div \dfrac{1}{2} \right|$

24. $9 \cdot \left| 3 - 6 \div \dfrac{1}{3} \right|$

25. $[7 + 3(2^3 - 1)] \div 21$

26. $[11 - 4(2 - 3^3)] \div 37$

27. $3 - [32 \div (2 \cdot 7 - 6)]$

28. $13 - [28 \div (5 \cdot 3 - 1)]$

Graph the following sets.

29. $\{x : x < -2\}$

30. $\left\{ x : x < \dfrac{1}{2} \right\}$

31. $\{x : x \geq 2.3\}$

32. $\{x : x \geq 1.8\}$

33. $\left\{ x : x > -\dfrac{2}{3} \right\}$

34. $\{x : x > -3\}$

35. $\left\{ x : x \leq \dfrac{5}{2} \right\}$

36. $\left\{ x : x \leq -\dfrac{1}{4} \right\}$

37. $\{x : x \leq -4\} \cup \{x : x \geq -1\}$

38. $\{x : x < 0\} \cup \left\{ x : x > \dfrac{1}{3} \right\}$

39. $\{x : x \leq 2\} \cap \{x : x \leq 5\}$

40. $\{x : x \geq -2\} \cap \left\{ x : x \geq \dfrac{3}{4} \right\}$

Evaluate the following.

41. $|7|$

42. $\left|-\dfrac{2}{5}\right|$

43. $|0|$

44. $5|3|$

45. $-|-4|$

46. $6\left|-\dfrac{4}{3}\right|$

47. $-2|\pi|$

48. $-3|-3|$

Place the appropriate symbol, $<$, $>$, or $=$, between the following pairs of numbers.

49. 3 _____ $|-5|$

50. $|-8|$ _____ 6

51. $|-4|$ _____ $|-7|$

52. $|-9|$ _____ $|-3|$

53. $|-5|$ _____ $|5|$

54. $\left|\dfrac{2}{3}\right|$ _____ $\left|-\dfrac{2}{3}\right|$

55. $|-2|$ _____ $|0|$

56. $|0|$ _____ $\left|-\dfrac{3}{4}\right|$

Evaluate the following algebraic expressions for the given values of the variables.

57. $x^3 + 4x^2 - x + 3$ for $x = -2$

58. $x^4 - 5x^2 - 2x + 1$ for $x = -1$

59. $-2(x + 3)^2 - 4$ for $x = 1$

60. $-3(x + 1)^2 + 2$ for $x = 2$

61. $\dfrac{1}{2}(x - 1)^2 + 3$ for $x = -3$

62. $\dfrac{1}{3}(x - 2)^2 - 5$ for $x = -1$

63. $\dfrac{2}{3}\sqrt{13 - x^2}$ for $x = 2$

64. $\dfrac{1}{2}\sqrt{7 + x^2}$ for $x = 3$

65. $|2x + 3| - 4$ for $x = -4$

66. $|5x + 1| - 7$ for $x = -2$

67. $4x^2 - xy + 3y^2$ for $x = 3, y = -1$

68. $2x^2 + xy - y^2$ for $x = 4, y = 2$

69. $\dfrac{2x + 1}{x}$ for $x = \dfrac{1}{4}$

70. $\dfrac{3x}{x + 2}$ for $x = \dfrac{1}{2}$

71. $\dfrac{x}{y} + 3y$ for $x = \dfrac{1}{3}, y = \dfrac{1}{2}$

72. $\dfrac{y}{2x} - y$ for $x = \dfrac{1}{3}, y = \dfrac{3}{2}$

73. $\dfrac{10x^2 + 11xy - 6y^2}{5x - 2y}$ for $x = \dfrac{1}{2}, y = 2$

74. $\dfrac{12x^2 - 5xy - 2y^2}{4x + y}$ for $x = -\dfrac{2}{3}, y = -1$

75. $2x[x^2 + y(3x - z^2)]$ for $x = 5, y = -2, z = 3$

76. $y[z^2 - 3x(1 - y^2)] + 5x$ for $x = -2, y = 4, z = -3$

77. Find values of x, y, and z to illustrate the following statements.
   a. $|x + y| \le |x| + |y|$
   b. $\sqrt{x + y} \ne \sqrt{x} + \sqrt{y}$
   c. $(x + y)^2 \ne x^2 + y^2$
   d. $x \cdot (y \cdot z) \ne x \cdot y \cdot x \cdot z$

Applied Algebra

When the temperature outdoors is cold and the wind is blowing, it seems colder than what the thermometer indicates. This effect is called the **wind-chill factor.** The formula to compute the wind-chill temperature was calculated by the National Weather Service in 1984 to be

$$T_{WC} = 0.0817(3.71\sqrt{v} + 5.81 - 0.25v)(T - 91.4) + 91.4$$

where T_{WC} = wind-chill temperature in degrees Fahrenheit

T = actual temperature in degrees Fahrenheit

v = wind velocity in miles per hour

Under blizzard conditions, the wind blows at 35 mph. If the thermometer reads 20° F, how cold will it feel like it really is?

SOLUTION

We need to find T_{WC} when $v = 35$ and $T = 20$.

$T_{WC} = 0.0817[3.71\sqrt{35} + 5.81 - 0.25(35)](20 - 91.4) + 91.4$

$\doteq 0.0817[3.71(5.916) + 5.81 - 0.25(35)](20 - 91.4) + 91.4$ *Calculate the square root first.*

$\doteq 0.0817(21.95 + 5.81 - 8.75)(20 - 91.4) + 91.4$ *Multiply inside the parentheses.*

$\doteq 0.0817(19.01)(-71.4) + 91.4$ *Add and subtract inside the parentheses.*

$\doteq -110.9 + 91.4$ *Multiply before you add.*

$\doteq -19.5$

Under these conditions, 20° F would feel like 19.5° *below* zero.

Your Turn

Find the wind-chill temperature under the following conditions:

1. $v = 25$ mph, $T = 32°$

2. $v = 20$ mph, $T = 10°$

Chapter 1 Review

Terms to Remember

Set	**p. 2**	Quotient	**p. 12**
Element or member	**p. 2**	Divisible	**p. 12**
Set notation	**p. 2**	Rational numbers	**p. 14**
Subset	**p. 2**	Prime number	**p. 14**
Empty or null set	**p. 3**	Reducing	**p. 14**
Finite set	**p. 3**	Lowest terms	**p. 15**
Infinite set	**p. 3**	Least common denominator	**p. 18**
Equal sets	**p. 4**	Irrational numbers	**p. 26**
Intersection of sets	**p. 4**	Diagonal	**p. 27**
Union of sets	**p. 4**	Real numbers	**p. 31**
Natural numbers	**p. 6**	Real number line	**p. 32**
Whole numbers	**p. 6**	Graphing a number	**p. 32**
Integers	**p. 6**	Graphing a set	**p. 32**
Additive inverse	**p. 8**	Order of the real numbers	**p. 33**
Terms	**p. 9**	Inequality symbols	**p. 33**
Sum	**p. 9**	Multiplicative inverse	**p. 36**
Factors	**p. 9**	Reciprocal	**p. 36**
Product	**p. 9**	Variables	**p. 42**
Power or exponential notation	**p. 11**	Constant	**p. 42**
Base	**p. 11**	Set builder notation	**p. 42**
Power or exponent	**p. 11**	Absolute value	**p. 7, 43**
Cubed	**p. 11**	Algebraic expression	**p. 45**
Squared	**p. 11**	Evaluating an algebraic	**p. 45**
Dividend	**p. 12**	expression	
Divisor	**p. 12**		

Notation

Set notation	$\{\ldots\}$
Element of	$\in$
Subset of	$\subseteq$
Empty set	$\emptyset$ or $\{\ \ \}$
Intersection	$\cap$
Union	$\cup$
Natural numbers	N
Whole numbers	W
Integers	I
Power notation	a^n
Rational numbers	Q
Irrational numbers	H
Square root	$\sqrt{\ \ }$
Approximately equal to	$\doteq$
Real numbers	R
Less than	$<$
Less than or equal to	$\leq$

Greater than	$>$
Greater than or equal to	$\geq$
Set builder notation	$\{x : \text{property that } x \text{ must satisfy}\}$
Absolute value	$\mid \ \mid$

Sets of Numbers

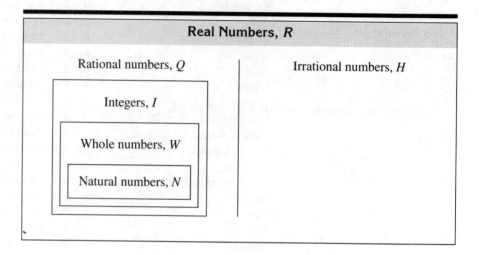

Properties

If a, b, and c represent real numbers, then the following properties are true.

- Reflexive: $a = a$

- Symmetric: If $a = b$, then $b = a$.

- Transitive: If $a = b$ and $b = c$, then $a = c$.

- Substitution: If $a = b$, then any expression involving a retains the same value if a is replaced by b.

- Transitive (inequality): If $a < b$ and $b < c$, then $a < c$.

- Trichotomy: One and only one of the following statements is true: $a < b$, $a = b$, or $a > b$.

- Commutative: Addition: $a + b = b + a$. Multiplication: $a \cdot b = b \cdot a$.

- Associative: Addition: $(a + b) + c = a + (b + c)$. Multiplication: $(a \cdot b) \cdot c = a \cdot (b \cdot c)$.

- Identity: Addition: $a + 0 = 0 + a = a$. Multiplication: $a \cdot 1 = 1 \cdot a = a$.

- Inverse: Addition: $a + (-a) = (-a) + a = 0$. Multiplication: $a \cdot \dfrac{1}{a} = \dfrac{1}{a} \cdot a = 1$, if $a \neq 0$.

- Distributive: $a \cdot (b + c) = a \cdot b + a \cdot c$; $a \cdot b + a \cdot c = a \cdot (b + c)$.

Review Exercises

1.1 Consider the sets $W = \{2, 4, 6, 8\}$, $X = \{4, 8, 12\}$, $Y = \{4, 8, 12, \ldots\}$, and $Z = \{2, 4, 6, 8, \ldots, 100\}$. Determine whether each of the following statements is true or false.

1. $X \subseteq Z$ **2.** $Y \subseteq Z$ **3.** $X = Y$ **4.** $W = Z$ **5.** $16 \in Y$

6. $20 \in Z$ **7.** $\varnothing \subseteq W$ **8.** $X \subseteq X$ **9.** $48 \in X$ **10.** $2468 \in W$

Using the same sets W, X, Y, and Z, find the following.

11. $W \cap X$ **12.** $X \cap Y$ **13.** $W \cap Z$ **14.** $X \cap Z$ **15.** $Y \cap Z$

16. $W \cup X$ **17.** $X \cup Z$ **18.** $X \cup Y$ **19.** $(W \cap Y) \cup X$ **20.** $(Y \cup Z) \cap W$

1.2 Perform the indicated operations, if possible.

21. $-13 + (-24)$ **22.** $-5 - (-12)$ **23.** $-14 - (-2)$ **24.** $5 + (-5)$

25. $72 + (-85)$ **26.** $45 - 65$ **27.** $-19 - 33$ **28.** $24 - (-31)$

29. $17 - (-5) - 14$ **30.** $-16 + (-53) + (-20)$ **31.** $22(-6)$

32. $(-3)(6)(-5)$ **33.** $(-4)(2)(-5)(-3)$ **34.** $5(21)(0)(-13)$

35. $(-19)(-5)$ **36.** -7^2 **37.** $(-11)^2$ **38.** $\dfrac{16}{0}$ **39.** $0 \div 7$

40. $-\dfrac{44}{11}$ **41.** $\dfrac{-35}{-7}$ **42.** $\dfrac{120}{-40}$ **43.** $0 \div 0$ **44.** $10^8 \div 10^5$

1.3 Reduce the following fractions to lowest terms.

45. $\dfrac{54}{81}$ **46.** $\dfrac{90}{72}$ **47.** $\dfrac{19}{38}$

48. $\dfrac{9}{63}$ **49.** $\dfrac{60}{35}$ **50.** $\dfrac{52}{105}$

Perform the indicated operations. Be sure your answer is in lowest terms.

51. $\dfrac{3}{4} \cdot \dfrac{5}{6}$ **52.** $\dfrac{3}{4} \div \dfrac{5}{6}$ **53.** $\dfrac{3}{4} + \dfrac{5}{6}$ **54.** $\dfrac{3}{4} - \dfrac{5}{6}$

55. $\dfrac{7}{8} \div 14$ **56.** $\dfrac{5}{16} + \dfrac{7}{20}$ **57.** $\left(\dfrac{-21}{40}\right)\left(\dfrac{-88}{-49}\right)$ **58.** $\dfrac{\frac{24}{9}}{-4}$

59. $\dfrac{2}{15} - \dfrac{4}{9}$ **60.** $\dfrac{23}{18} + \dfrac{7}{30}$ **61.** $\left(-\dfrac{7}{8}\right)\left(\dfrac{-4}{21}\right)(9)$

62. $\left(\dfrac{15}{-22}\right)(-4)\left(\dfrac{-33}{10}\right)$ **63.** $\dfrac{5}{12} - \dfrac{11}{42} + \dfrac{3}{7}$ **64.** $\dfrac{4}{25} - \dfrac{3}{10} + \dfrac{7}{20}$

65. $-\dfrac{28}{45} \div \left(\dfrac{-77}{-60}\right)$ **66.** $\left(-\dfrac{-27}{8}\right) \div \left(\dfrac{21}{-20}\right)$ **67.** $4\dfrac{1}{2} + 3\dfrac{1}{4} - 2\dfrac{5}{8}$

68. $2\dfrac{2}{3} + 1\dfrac{1}{2} - 5\dfrac{5}{6}$ **69.** $\left(2\dfrac{1}{2}\right)\left(-\dfrac{14}{15}\right)\left(\dfrac{-8}{-7}\right)\left(\dfrac{-1}{4}\right)$ **70.** $(-6)\left(3\dfrac{1}{3}\right)\left(-\dfrac{5}{-12}\right)\left(-\dfrac{3}{25}\right)$

Convert the following fractions to decimals.

71. $\frac{5}{16}$

72. $\frac{5}{32}$

73. $\frac{4}{11}$

74. $\frac{5}{22}$

Convert the following decimals to fractions in lowest terms.

75. 0.24

76. 0.35

77. 2.6

78. 3.8

Perform the indicated operations.

79. $-27.81 + 9.638$

80. $(-14.75)(243.8)$

81. $17.172 \div 6.48$

82. $6.933 - 18.4$

83. $-43.1 - 8.905 + 17.46$

84. $(-214.42) \div 35.5$

85. $(7.08)(-5.5)(-31.25)$

86. $13.78 + (-6.4) - 0.043$

87. Find the areas of the following figures.

a.
$1\frac{3}{4}$ in.

$1\frac{3}{4}$ in.

b.
1.8 cm

4.15 cm

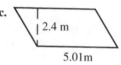

c.
2.4 m

5.01m

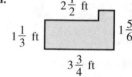

d.
$2\frac{1}{2}$ ft

$1\frac{1}{3}$ ft

$1\frac{5}{6}$

$3\frac{3}{4}$ ft

88. Stan Bayou Mann makes $2.45 for each of his C & W record albums that is sold. Each album sells for $8.95. If Stan's royalties from the album were $45,266.20 last month, how much did the record company gross last month?

89. Rico's Deli sells matzo ball soup two ways. For an appetizer, you can buy an 8-oz bowl for $1.95. For a meal, you can buy an 18-oz pot for $4.45. Which is the better buy?

1.4   **Determine whether the following are true or false.**

90. $\sqrt{49} \in H$

91. $\sqrt{60} \in H$

92. $\sqrt{27} \in Q$

93. $2.4141\ldots \in Q$

94. $5.2121121112\ldots \in H$

95. $16.49783 \in H$

Find the lengths of the following diagonals. If the sides are given in metric units, approximate the diagonal to two decimal places. If the sides are given in feet or inches, approximate the diagonal to two decimal places and convert to a fraction.

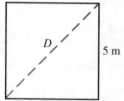

96.
D

5 m

5 m

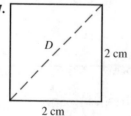

97.
D

2 cm

2 cm

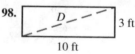

98.
D
3 ft

10 ft

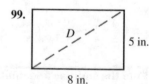

99.
D

5 in.

8 in.

Find the indicated quantities. If the radius is given as a decimal, use 3.14 for π.
If the radius is given as a fraction, use $\frac{22}{7}$ for π.

100.

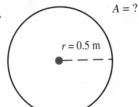

$A = ?$

$r = 0.5$ m

101.

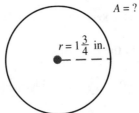

$A = ?$

$r = 1\frac{3}{4}$ in.

102.

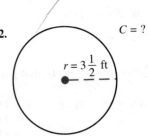

$C = ?$

$r = 3\frac{1}{2}$ ft

103.

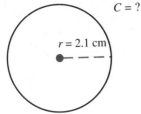

$C = ?$

$r = 2.1$ cm

1.5 Determine whether the following are true or false.

104. $I \cap R = I$

105. $\sqrt{9} \in R$

106. $\sqrt{7} \in R$

107. $5 \geq -5$

108. $0 < -4$

109. $I \cup Q = Q$

110. $3 \cdot (5x) = 15x$

111. $y = 1y$

112. $2(x + 3) = 2x + 3$

113. $2x + 2y = 4(x + y)$

Graph the following sets.

114. $\left\{ -\frac{5}{3}, \frac{7}{4}, \frac{22}{9}, -\frac{7}{8} \right\}$

115. $\left\{ 4.3, -3.7, -\frac{16}{5}, \frac{9}{2} \right\}$

116. $\{\sqrt{0}, \sqrt{1}, \sqrt{2}, \sqrt{3}\}$

Place the appropriate inequality symbol, $<$ or $>$, between each pair of
numbers.

117. -11 _____ -5

118. $-\frac{1}{2}$ _____ -4.1

119. $\sqrt{10}$ _____ 3

120. 5 _____ $\sqrt{24}$

121. 0 _____ -8

122. 0 _____ 5

123. $\frac{7}{5}$ _____ 1.3

124. 2.6 _____ $\frac{8}{3}$

State the property illustrated by each of the following statements.

125. If $z = 3$, then $3 = z$.

126. $9 = 9$

127. If $8 > 4$ and $4 > 0$, then $8 > 0$.

128. If $r = 4$ and $A = \pi r^2$, then $A = \pi 4^2$.

129. $x(x + 3) = x^2 + 3x$

130. $-7 + 7 = 0$

131. $-8 \cdot 1 = -8$

132. $x + 2y = 2y + x$

133. $(x + 4) + 7 = x + (4 + 7)$

134. $5 \cdot \frac{1}{5} = 1$

Find the additive and multiplicative inverses of each number.

135. $-\frac{11}{2}$

136. $\sqrt{21}$

137. 26

138. 1.5

Use the distributive property to rewrite each of the following.

139. $5(x + 7)$

140. $8(4 - 1)$

141. $7(a + b + 3)$

142. $6m + 6n$

143. $8x + 12x$

144. $2y + xy - y$

Name the property that justifies each step in the following.

145. $(-8 + 17) + (-25) = [17 + (-8)] + (-25)$
$= 17 + [-8 + (-25)]$
$= 17 + (-33) = -16$

146. $\left(\frac{1}{2}\right)(2x + 6) = \left(\frac{1}{2}\right) \cdot (2x) + \left(\frac{1}{2}\right) \cdot 6$
$= \left[\left(\frac{1}{2}\right) \cdot 2\right]x + 3$
$= 1 \cdot x + 3$
$= x + 3$

147. $-6x + 6x = (-6 + 6)x$
$= 0x$
$= 0$

1.6   **Perform the indicated operations, following the rules governing the order of operations.**

148. $45 \div 3 \cdot \sqrt{25}$

149. $8 - 24 + \sqrt{16}$

150. $4^2 - 7 \cdot 8 + 2$

151. $5 \cdot 3^2 + 6 \cdot 3 - 17$

152. $\left(\frac{3}{4} - \frac{4}{3} + \frac{2}{5}\right) \div 11$

153. $3 - 2\left(\frac{1}{2} + 4\right) \div 6$

154. $4 + \sqrt{15 - 2 \cdot 3}$

155. $4 - \sqrt{15 - 2 \cdot 7}$

156. $\frac{7}{6} - 5\left(\frac{3}{10}\right) \div 2$

157. $\frac{35.6 - 3(5.01)}{2.6 - 6}$

158. $\frac{(0.04)(2.5) - 3.52}{(0.4)^2 - 0.01}$

159. $2 - |7 \cdot 5 - 46|$

160. $6 - 2[7 + 3(-8 + 5)]$

161. $8 - 5[-4 - 6(4 \cdot 3 - 7)]$

Graph the following sets.

162. $\left\{x : x < \frac{5}{3}\right\}$

163. $\{x : x > 2.8\}$

164. $\left\{x : x \le -\frac{3}{4}\right\}$

165. $\{x : x \ge -5\} \cup \{x : x \ge 1\}$

Place the appropriate symbol, $<$, $>$, or $=$, between the following pairs of numbers.

166. $|-14|$ _____ 12

167. $|-18|$ _____ $|0|$

168. $|-10|$ _____ $|-3|$

169. $\left|-\frac{3}{2}\right|$ _____ $\frac{4}{3}$

170. $|4|$ _____ $|-4|$

171. $|0|$ _____ $|9|$

Evaluate the following algebraic expressions for the given values of the variables.

172. $2x^3 - 7x^2 + x - 5$ for $x = -3$

173. $x^5 - 8x^3 - x + 2$ for $x = -2$

174. $\left(-\frac{1}{2}\right)(x + 2)^2 + 5$ for $x = 2$

175. $\left(-\frac{1}{5}\right)(x - 3)^2 - 2$ for $x = -2$

176. $5\sqrt{x^2 + 4y^2}$ for $x = 8$, $y = 3$

177. $x - |x^2 + 3y|$ for $x = 5$, $y = -9$

178. $\dfrac{x+3}{2x+1}$ for $x = \dfrac{1}{2}$

179. $\dfrac{3x}{x^2+1}$ for $x = \dfrac{2}{3}$

180. $\dfrac{3x-7y}{3x^2-xy-14y^2}$ for $x = 2, y = 3$

181. $\dfrac{x}{2a} + y^2 - \dfrac{2}{3}$ for $x = -3, a = 3, y = \dfrac{1}{2}$

182. $x[y^2 + 3(x - 4z)]$ for $x = 7, y = -2, z = 3$

Chapter 1 Test

(You should be able to complete this test in 60 minutes.)

I. Use the sets $A = \{1, 2, 3, 4\}, B = \{2, 4, 6\}, E = \{2, 4, 6, ...\}, N, I, Q, H$, and R for problems 1–12. Are the following true or false?

1. $0 \in N$

2. $12 \in E$

3. $A \subseteq N$

4. $E \subseteq N$

5. $B = E$

6. $\sqrt{8} \in H$

7. $4.012012... \in Q$

8. $\varnothing \subseteq Q$

Find the following.

9. $A \cap B$

10. $A \cup B$

11. $E \cap N$

12. $A \cup E$

II. Place the appropriate symbol, $<, >$, or $=$, between each pair of numbers.

13. -23 _____ -18

14. $|-7|$ _____ -3

15. $\sqrt{35}$ _____ 6

16. $|14|$ _____ $|-14|$

III. Graph the following sets.

17. $\left\{ -\dfrac{11}{4}, 2, \dfrac{8}{9} \right\}$

18. $\{x : x < -2.1\}$

IV. Find the additive and multiplicative inverses of each number.

19. $\dfrac{5}{7}$

20. -3.5

V. Use the distributive property to rewrite each of the following.

21. $10(8 - 2 + 4)$

22. $13y + 17y$

VI. State the property illustrated by each of the following statements.

23. Either $2 < |-2|, 2 = |-2|$, or $2 > |-2|$.

24. If $x = 16$ and $y = \sqrt{x}$, then $y = \sqrt{16}$.

VII. Name the property that justifies each step in the following.

25. $\dfrac{2}{3} \cdot \left(\dfrac{3}{2} \cdot x \right) = \left(\dfrac{2}{3} \cdot \dfrac{3}{2} \right) x$

$= 1 \cdot x$

$= x$

VIII. Perform the indicated operations, if possible.

26. $(-3)(-2)(-8)$

27. $-14 - 8 - (-21) - 5$

28. $-40 + (-35)$

29. -8^2

30. $\dfrac{60}{-12}$

31. $\dfrac{\frac{-8}{15}}{20}$

32. $\dfrac{5}{0}$

33. $16.7 + (-8.94) - 4.506$

34. $(-0.75)(8.64)(-200)$

35. $\left(\dfrac{2}{3} + 2 \cdot 3\right) \div \dfrac{5}{6} - 7$

36. $1.2768 \div 0.15$

37. $\dfrac{(-2)^4 - 3^2}{(-7 - 9) - (3 - 5)}$

38. $|7 - 16|$

39. $\left(\dfrac{-5}{16}\right)(-8)\left(\dfrac{-26}{65}\right)$

40. $\dfrac{7}{12} + \dfrac{11}{30} - 2$

41. $96 \div 2^2 \cdot 8$

42. $\sqrt{5 \cdot 7 + 7^2 - 3}$

43. $2|4^2 - 3|$

44. $-\dfrac{2}{3} - \dfrac{1}{5} \div \dfrac{3}{10} + 5\left(-\dfrac{2}{15}\right)$

IX. Make the following conversions.

45. Fraction to decimal: $\frac{3}{25}$

46. Decimal to fraction in lowest terms: 0.36

X. Evaluate the following algebraic expressions for the given values of the variables.

47. $2x^4 - x^2 + 5x$ for $x = 2$

48. $\sqrt{25 - x^2}$ for $x = -3$

49. $x - |2x - 9|$ for $x = 3$

50. $(x^2 + 4y^2) - 3xy$ for $x = 5, y = -1$

Linear Equations and Inequalities

When constructing highways and bridges, engineers sometimes install *expansion joints*. Expansion joints are used to control the changes in the length of the highway due to changes in temperature. The expansion joints allow the highway to expand in hot weather. If there were no expansion joints, the highway might buckle or even break on a hot summer day. The problem facing the highway engineer is to determine how much a particular length of roadway will expand.

At the end of this chapter we will give a formula that determines the amount of expansion of a particular length of roadway. In working with this formula, we will use properties of linear equations, which will be developed in this chapter.

2.1 Linear Equations

An equation is an algebraic way of stating that two quantities are equal. Some examples of equations are $C = 2\pi r$, $P = 2l + 2w$, $4x - 3 = 0$, $x^2 - 25 = 0$, and $x^3 + y^3 = z^3$.

Before we begin our study of linear equations, two concepts need to be mentioned. The first is that of a coefficient. Consider the algebraic expression $7x - 3x + y$. The terms of this expression are $7x$, $-3x$, and y, since $7x - 3x + y = 7x + (-3x) + y$. (Remember that terms are quantities that are being added.) The numerical coefficient, or just **coefficient,** of a term is the numerical factor of that term. Thus in the algebraic expression $7x - 3x + y$, the coefficient of the first term is 7, the coefficient of the second term is -3, and the coefficient of the last term is 1, since $y = 1 \cdot y$.

The second concept is like terms. Terms of an algebraic expression whose variable factors are *exactly* the same are called **like terms.** In the algebraic expression $7x - 3x + y$, the terms $7x$ and $-3x$ are like terms. Like terms can be combined by applying the distributive property in the following manner:

$$7x - 3x + y = 7x + (-3x) + y$$
$$= [7 + (-3)]x + y$$
$$= 4x + y$$

REMARK

After we identify like terms, algebraic expressions can be simplified by adding or subtracting the coefficients of the like terms.

EXAMPLE 2.1.1   Simplify the following algebraic expressions by combining like terms.

1. $5x - 8x + 3y + 7y = -3x + 10y$

2. $6x - 4 - x - 2 = 6x - x - 4 - 2$
    $$= 5x - 6$$

In this section we will investigate equations in *one variable*. An example of an equation in one variable is $2x + 1 = 11$. The solution of this equation is $x = 5$. 5 is called the solution because when we replace x with 5 in

$$2x + 1 = 11$$

we obtain

$$2(5) + 1 = 11$$
$$10 + 1 = 11 \qquad \textit{A true statement}$$

| Definition 2.1.1 | The **solutions** of an equation are those numbers that when substituted into the equation for the variable, yield a true statement. |

REMARK When we want to find the solution(s) of an equation, we say we want to **solve the equation** or, if the variable is x, we want to **solve for x.**

EXAMPLE 2.1.2   Find the solutions of the following equations.

1. $4x = 28$

By observation the solution is $x = 7$ because $4(7) = 28$ is a true statement.

REMARK Set notation is sometimes used to denote the set of all solutions of an equation. Using set notation, the **solution set** is indicated as $\{7\}$.

2. $\dfrac{1}{2}x = 40$

By observation the solution is $x = 80$ because $(\frac{1}{2}) \cdot (80) = 40$ is a true statement. The solution set is $\{80\}$.

| Definition 2.1.2 | Two equations are said to be **equivalent** if they have the same solution set. |

REMARK The equations $2x = 10$ and $8x = 40$ are equivalent, since they both have the solution set $\{5\}$.

   The following property will help us to find solutions of equations.

Multiplicative Property of Equality

Let a, b, and c denote real numbers, where $c \neq 0$.

If $a = b$, then $ac = bc$.

REMARK This property states that we can multiply both sides of an equation by any *nonzero number* and obtain a new equation that is equivalent to the original one. We use this property to make the coefficient of the variable equal to 1.

EXAMPLE 2.1.3   Use the multiplicative property of equality to solve the following equations.

1. $4x = 28$

$$\frac{1}{4} \cdot 4x = \frac{1}{4} \cdot 28 \qquad \textit{The coefficient is 4, and we multiply by its reciprocal, } \tfrac{1}{4}.$$

$$1 \cdot x = \frac{28}{4}$$

$$x = 7$$

REMARK   Multiplying by the reciprocal of the coefficient of x changed the left side of the equation from $4x$ to x.

2. $\dfrac{1}{2}x = 40$

$$2 \cdot \frac{1}{2}x = 2 \cdot 40 \qquad \textit{The coefficient is } \tfrac{1}{2}, \textit{ and we multiply by its reciprocal, 2.}$$

$$x = 80$$

Suppose we had an equation of the form $x - 8 = 2$. It should be clear that $x = 10$ is the solution of this equation, since $10 - 8 = 2$ is a true statement. This problem illustrates another property that can be used to find solutions of equations.

Additive Property of Equality

> Let a, b, and c denote real numbers.
>
> If $a = b$, then $a + c = b + c$.

REMARK   This property states that we can add any number to both sides of an equation and obtain a new equation that is equivalent to the original equation. We use this property to isolate the variable term.

EXAMPLE 2.1.4   Find the solutions of the following equations.

1. $x - 3 = 2$

$$x - 3 + 3 = 2 + 3 \qquad \textit{Using the additive property of equality, we add}$$
$$x + 0 = 5 \qquad\qquad \textit{3 to both sides and thereby isolate x.}$$
$$x = 5$$

Checking the solution:

$$5 - 3 \overset{?}{=} 2$$

$$2 = 2 \qquad \textit{So x = 5 is the solution.}$$

NOTE ▶   The symbol $\overset{?}{=}$ indicates that we are checking a potential solution.

2. $x + 6 = -7$

$\quad x + 6 - 6 = -7 - 6 \qquad$ *Here we need to add -6 (or subtract 6) on both sides.*

$\qquad\qquad x = -13$

The check is left to the reader.

Definition 2.1.3	Any equation that can be written in the form $ax + b = 0$, where a and b are real numbers with $a \neq 0$, is called a **linear equation** *in the variable x.*

The following steps can be used to solve any linear equation.

Steps for Solving Any Linear Equation

1. Use the distributive property to eliminate any grouping symbols and then combine all like terms on each side of the equation.

2. (Optional) Eliminate any fractions by using the multiplicative property of equality.

3. Eliminate the variable on one side of the equation by using the additive property of equality.

4. Eliminate the constant term that is being added to the variable term using the additive property of equality.

5. Using the multiplicative property of equality, multiply through by the reciprocal of the coefficient of the variable.

6. (Optional) Check the solution.

EXAMPLE 2.1.5   Find the solutions of the following equations and check your solutions.

1. $2x + 1 = 11$

$\quad 2x + 1 - 1 = 11 - 1 \qquad$ *Eliminate the constant term that is being added to 2x.*

$\qquad\qquad 2x = 10$

$\qquad \dfrac{1}{2} \cdot 2x = \dfrac{1}{2} \cdot 10 \qquad$ *Multiply by the reciprocal of 2.*

$\qquad\qquad x = 5$

Checking the solution:

$$2(5) + 1 \overset{?}{=} 11$$ *Substitute 5 for x in the original equation.*

$$10 + 1 \overset{?}{=} 11$$

$$11 = 11$$ *So x = 5 is the solution.*

2. $$4x + 7 = x - 2$$

$$4x - x + 7 = x - x - 2$$ *Eliminate the variable on the right side of the equation.*

$$3x + 7 = -2$$

$$3x + 7 - 7 = -2 - 7$$ *Eliminate the constant term on the left side.*

$$3x = -9$$

$$\frac{1}{3} \cdot 3x = \frac{1}{3} \cdot (-9)$$ *Multiply by the reciprocal of 3.*

$$x = -3$$

Checking the solution:

$$4(-3) + 7 \overset{?}{=} -3 - 2$$ *Substitute −3 for x in the original equation.*

$$-12 + 7 \overset{?}{=} -5$$

$$-5 = -5$$ *So x = −3 is the solution.*

REMARK

a. The operations in steps 3 and 4 leave the variable term on the left-hand side of the equation and the constant term on the right-hand side.

b. The difference between factors and terms is important here. In steps 3 and 4 we isolate *terms* on each side of the equation by *adding* (or subtracting). In step 5 we make the coefficient, which is a numerical *factor,* equal to 1 by *multiplying.*

3. $$4(x + 2) - 10 = 3(2x - 1)$$

$$4x + 8 - 10 = 6x - 3$$ *Eliminate the grouping symbols.*

$$4x - 2 = 6x - 3$$ *Combine like terms.*

$$4x - 6x - 2 = 6x - 6x - 3$$ *Eliminate the variable on the right side.*

$$-2x - 2 = -3$$

$$-2x - 2 + 2 = -3 + 2$$ *Eliminate the constant term on the left side.*

$$-2x = -1$$

$$-\frac{1}{2} \cdot (-2x) = -\frac{1}{2} \cdot (-1)$$ *Multiply both sides by $-\frac{1}{2}$.*

$$x = \frac{1}{2}$$

The check of the solution is left to the reader.

4. $3(2x + 1) - 2(x - 4) = 7$

$\qquad 6x + 3 - 2x + 8 = 7$ *Eliminate the grouping symbols.*

$\qquad\qquad\quad 4x + 11 = 7$

$\qquad 4x + 11 - 11 = 7 - 11$ *Subtract 11 from both sides.*

$\qquad\qquad\qquad 4x = -4$

$\qquad \dfrac{1}{4} \cdot 4x = -4 \cdot \dfrac{1}{4}$ *Multiply both sides by $\frac{1}{4}$.*

$\qquad\qquad\qquad\quad x = -1$

Again, the check is left to the reader.

Try to avoid these mistakes:

Incorrect	Correct
1. $4(x + 2) - 10 = 3(2x - 1)$ $\quad 4x + 2 - 10 = 6x - 1$ 2 was not multiplied by 4; -1 was not multiplied by 3.	$4(x + 2) - 10 = 3(2x - 1)$ $4x + 8 - 10 = 6x - 3$ Use the distributive property.
2. $3(2x + 1) - 2(x - 4) = 7$ $\quad 6x + 3 - 2x - 8 = 7$ -4 was not multiplied by *negative* 2.	$3(2x + 1) - 2(x - 4) = 7$ $6x + 3 - 2x + 8 = 7$

5. $\qquad 3x - 7 = 5(x + 2) - 2x - 17$

$\qquad 3x - 7 = 5x + 10 - 2x - 17$ *Eliminate grouping symbols.*

$\qquad 3x - 7 = 3x - 7$ *Combine like terms.*

$3x - 3x - 7 = 3x - 3x - 7$ *Subtract 3x from both sides.*

$\qquad\qquad -7 = -7$

Thus, x is any real number.

When all the variables drop out, leaving a true statement, the solution set is the set of *all real numbers* because any value for x will result in a true statement.

6. $\qquad 2x + 4 = 3(x - 1) - x$

$\qquad 2x + 4 = 3x - 3 - x$ *Eliminate grouping symbols.*

$\qquad 2x + 4 = 2x - 3$ *Combine like terms.*

$2x - 2x + 4 = 2x - 2x - 3$ *Subtract 2x from both sides.*

$\qquad\qquad\quad 4 = -3$

Thus, there is no solution.

When all the variables drop out, leaving a false statement, the solution set is the *empty set* because any value for x will result in a false statement.

7.

$$\frac{3}{4}x + 1 = \frac{1}{2}x + 2$$

$$\frac{3}{4}x - \frac{1}{2}x + 1 = \frac{1}{2}x - \frac{1}{2}x + 2 \qquad \textit{Subtract } \tfrac{1}{2}x \textit{ from both sides.}$$

$$\frac{1}{4}x + 1 = 2$$

$$\frac{1}{4}x + 1 - 1 = 2 - 1 \qquad \textit{Subtract 1 from both sides.}$$

$$\frac{1}{4}x = 1$$

$$4 \cdot \frac{1}{4}x = 4 \cdot 1 \qquad \textit{Multiply both sides by 4.}$$

$$x = 4$$

The check is left to the reader.

8.

$$0.3x + 1.18 = 1.5x - 1.82$$

$$0.3x - 0.3x + 1.18 = 1.5x - 0.3x - 1.82 \qquad \textit{Subtract 0.3x from both sides.}$$

$$1.18 = 1.2x - 1.82$$

$$1.18 + 1.82 = 1.2x - 1.82 + 1.82 \qquad \textit{Add 1.82 to both sides.}$$

$$3.0 = 1.2x$$

$$\frac{3.0}{1.2} = \frac{1.2x}{1.2} \qquad \textit{Divide both sides by 1.2.}$$

$$2.5 = x$$

Again, the check is left to the reader.

REMARK Since division by a number is equivalent to multiplication by its multiplicative inverse, the multiplicative property of equality allows us to divide both sides by a nonzero number.

9.

$$\frac{2}{3}(2x - 9) = \frac{1}{2}(3x + 4) - 5$$

$$\frac{4}{3}x - 6 = \frac{3}{2}x + 2 - 5 \qquad \textit{Eliminate grouping symbols.}$$

$$\frac{4}{3}x - 6 = \frac{3}{2}x - 3 \qquad \textit{Combine like terms.}$$

$$6\left(\frac{4}{3}x - 6\right) = 6\left(\frac{3}{2}x - 3\right) \qquad \textit{Multiply both sides of the equation by the LCD of the fractions, 6, to eliminate the fractions.}$$

$$8x - 36 = 9x - 18$$

$$8x - 8x - 36 = 9x - 8x - 18 \qquad \textit{Subtract 8x from both sides.}$$
$$-36 = x - 18$$
$$-36 + 18 = x - 18 + 18 \qquad \textit{Add 18 to both sides.}$$
$$-18 = x$$

The check is left to the reader.

E xercises 2.1

Find the solutions of the following equations and check your solutions.

1. $5x = 100$

2. $8x = -144$

3. $-9x = 39$

4. $-2x = 0$

5. $0.3x = 51$

6. $-0.4x = 14$

7. $\frac{2}{3}x = 24$

8. $\frac{3}{5}x = -15$

9. $-\frac{1}{4}x = \frac{3}{8}$

10. $-\frac{7}{2}x = -49$

11. $3x + 7 = 15$

12. $5x - 9 = -23$

13. $-x + 3 = -8$

14. $-x - 5 = 3$

15. $-6x - 5 = 11$

16. $-7x - 2 = 12$

17. $8x + 7 = 7$

18. $\frac{5}{4}x - 3 = 7$

19. $\frac{3}{7}x + 4 = -\frac{1}{2}$

20. $-\frac{6}{5}x - \frac{3}{5} = -\frac{3}{2}$

21. $-\frac{2}{3}x + 1 = 5$

22. $0.01x + 3 = 7$

23. $0.2x - 1.31 = 2.39$

24. $3x + 1 = 5x - 13$

25. $12 - 4x = 5x + 3$

26. $-3x + 5 = 21 + x$

27. $-4x - 1 = -2x + 15$

28. $8x + 7 = 4x + 4$

29. $-5x + 7 = 10x + 11$

30. $16x - 1 = 8x + 2$

31. $5x - 7 = 5x + 2$

32. $3x + 1 = 4x + 1 - x$

33. $-6x - 4 = -9x - 3$

34. $\frac{3}{2}x + 5 = 5x - 2$

35. $\frac{2}{3}x - \frac{4}{9} = x - \frac{1}{6}$

36. $\frac{1}{5}x - 4 = 3 - \frac{1}{2}x$

37. $\frac{3}{4}x + 9 = \frac{1}{3}x + 4$

38. $5(2x - 1) = 5$

39. $-7(x + 2) = -14$

40. $3(4 - 3x) = 6 - 8x$

41. $-5(3x + 6) = x + 82$

42. $5(3x - 5) = 5x - 27$

43. $8(2x - 6) = 4x - 35$

44. $3(3 - 2x) = 2x + 65$

45. $4x - 6 = 3(x - 2) + x$

46. $5x + 1 = 2(2x - 3) + x$

47. $\frac{2(2x - 9)}{3} = x - 8$

48. $\frac{-3(3 - 2x)}{4} = 4 + \frac{2}{3}x$

49. $\frac{1}{2}(10 - 3x) = \frac{1}{3}x + 4$

50. $4\left(\frac{2}{5}x + 1\right) = \frac{1}{2}x + 15$

51. $-0.5x + 3.2 = 1.3x + 4.4$

52. $0.03x - 13 = 0.51x - 7$

53. $\frac{1}{2}x + \frac{3}{4} = \frac{1}{4}(2x - 3) + 1$

54. $\frac{1}{3}x - \frac{1}{2} = \frac{1}{3}(x - 1) - \frac{1}{6}$

55. $3(2x - 5) + 7 = 2(4x - 1) + 12$

56. $-4(8 - 7x) + 11 = 15(x - 2) - 4$

57. $2(4x - 8) + 9 = 5(2x + 1) - 12$

58. $-5(3 - x) + 3x = 6(2x - 3)$

59. $-3(x + 7) - 4x = 3(1 - 3x) - x$

60. $9(4x + 1) - 10x = 2(12x - 5) + 6$

61. $3(x - 7) - 4(2x + 1) = 2(x + 5)$

62. $-6(2x - 5) + 3(7 - 2x) = -7(3x - 8)$

63. $5(3x - 4) + 2(4x + 1) = 3(7x - 2)$

64. $4(2x + 1) + 2(2x + 1) = 5(x + 4)$

65. $10(3x - 5) - 7(3x - 5) = 4(2x + 6)$

66. $\frac{2}{3}(x - 7) - \frac{1}{2}(3x + 5) = 5x - 13$

67. $\frac{5(x + 1)}{6} + \frac{1}{3}(5x - 6) = \frac{1}{6}(7x - 2)$

68. $\frac{3x + 8}{3} + \frac{1}{2}(2x - 5) = \frac{1}{2}(2x - 1)$

69. $2.8(3x - 5.063) - 11.2(x + 3.75) = 0.24(6.5x - 486.04)$

70. $16.5(7.9x - 28.43) = 0.75(18x - 155.357) + 1.7(6x - 141.84)$

Write Algebra

71. Explain in your own words the steps used to solve a linear equation.

72. Explain what is meant by equivalent equations.

73. Explain how to solve the linear equation $ax + b = 0$ for x.

74. In solving an equation, if the variables drop out, what are the possible solutions?

2.2 *Linear Inequalities*

Recall from Chapter 1 that an inequality is a mathematical way of saying that one quantity is larger than another. Some examples of inequalities are:

$$x - 4 < 5 \qquad \text{\textit{read} x} - 4 \text{ \textit{is less than} 5}$$
$$7x \geq 23 \qquad \text{\textit{read} 7x \textit{ is greater than or equal to} 23}$$

NOTE ▶ Remember that the inequality symbol always points toward the smaller quantity.

In Section 2.1 we defined a linear equation in the variable x as any equation that could be written in the form $ax + b = 0$, $a \neq 0$. We have a similar definition for linear inequalities.

| Definition 2.2.1 | Any inequality that can be written in the form $ax + b > 0$ or $ax + b \geq 0$, where a and b are real numbers with $a \neq 0$, is called a **linear inequality** *in the variable x*. |

REMARK   It can be shown that the definitions and properties of this section apply to all inequalities.

An example of a linear inequality is $x \geq 3$. Some *solutions* of this inequality are $3\frac{1}{2}$, 4, 10, and 100. These numbers are called solutions because they yield a true statement when substituted for the variable in the inequality. In fact there are an infinite number of solutions to this inequality, and they are represented by the colored region on the following number line. $x \geq 3$ is represented by

which is called the **graph** of the solution set (or graph of the solutions). This graph says that x can be any number to the right of (and including) 3. The solid circle at 3 indicates that 3 is a solution of the inequality.

The inequality $x < 1$ is represented by

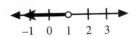

This graph says that x can be any number to the left of 1. The open circle at 1 indicates that 1 is not a solution of the inequality.

In finding the solutions of linear inequalities, we will employ tools similar to those used in solving linear equations and the following definition.

| **Definition 2.2.2** | Two inequalities are said to be **equivalent** if they have the same solution set. |

Consider the following. We know

$$-6 < 10$$
$$-6 + 8 \overset{?}{<} 10 + 8 \qquad \textit{Add 8 to both sides.}$$
$$2 < 18 \quad \checkmark \qquad \textit{A true statement}$$

NOTE ▶ The symbol $\overset{?}{<}$ indicates that we are checking whether the left side is less than the right side.

In similar fashion,

$$-6 < 10$$
$$-6 - 7 \overset{?}{<} 10 - 7 \qquad \textit{Now subtract 7.}$$
$$-13 < 3 \quad \checkmark \qquad \textit{A true statement}$$

These two facts illustrate the following property.

Additive Property of Inequality

> Let a, b, and c denote real numbers.
>
> If $a < b$, then $a + c < b + c$.

REMARK This property simply states that we can add or subtract the same number on both sides of an inequality and obtain a new inequality that is equivalent to the original one.

XAMPLE 2.2.1   Find and graph the solutions of the following inequalities.

1. $x - 3 > 1$

     $x - 3 + 3 > 1 + 3$ *Use the additive property of inequality to isolate* x.

         $x > 4$

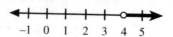

2. $x + 5 \leq 4$

     $x + 5 - 5 \leq 4 - 5$ *Use the additive property of inequality to isolate* x.

         $x \leq -1$

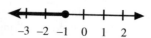

In solving linear equations, we often had to multiply both sides of the equation by some number. What would happen if we multiplied both sides of an inequality by some number? Again, consider the following. We know

$$-6 < 10$$
$$2(-6) \overset{?}{<} 2(10) \qquad \textit{Multiply both sides by 2.}$$
$$-12 < 20 \;\checkmark \qquad \textit{A true statement}$$

However,

$$-6 < 10$$
$$-3 \cdot (-6) \overset{?}{<} -3(10) \qquad \textit{Now multiply both sides by } -3.$$
$$18 > -30 \qquad \textit{The inequality sign must be reversed to make a true statement.}$$

These two facts illustrate the following property.

Multiplicative Property of Inequality

Let a, b, and c denote real numbers.

1. If $a < b$ and $c > 0$, then $ac < bc$.

2. If $a < b$ and $c < 0$, then $ac > bc$.

REMARK

This property says, in part 1, that we can multiply both sides of an inequality by any positive number and obtain a new inequality that is equivalent to the original one. *Part 2 says that if we multiply (or divide) both sides by a negative number, we must reverse the inequality sign to obtain an equivalent inequality.* Other than the exception in part 2, the steps used in solving linear inequalities are identical to those used in solving linear equations.

EXAMPLE 2.2.2   Find and graph the solutions of the following inequalities.

1. $2x < 12$

$$\frac{1}{2} \cdot 2x < \frac{1}{2} \cdot 12$$ *As with linear equations, we multiply both sides by the reciprocal of the coefficient.*

$$1 \cdot x < 6$$

$$x < 6$$

2. $12 \geq -4x$

$$-\frac{1}{4} \cdot 12 \leq -\frac{1}{4} \cdot (-4x)$$ *Here we must multiply both sides by $-\frac{1}{4}$. Remember to reverse the inequality symbol.*

$$-3 \leq x$$

which is the same as

$$x \geq -3$$

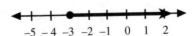

Suppose we have a more complicated inequality, such as $4x - 9 > 2x - 1$. What process is used to find the solution set?

3. $4x - 9 > 2x - 1$

$4x - 2x - 9 > 2x - 2x - 1$ *Notice that we followed exactly the same steps outlined in Section 2.1 to solve linear equations.*

$$2x - 9 > -1$$

$$2x - 9 + 9 > -1 + 9$$

$$2x > 8$$

$$\frac{1}{2} \cdot 2x > \frac{1}{2} \cdot 8$$

$$x > 4$$

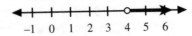

EXAMPLE 2.2.3   Find and graph the solutions of the following inequalities.

1. $4(2x - 1) + 7 < 5x - 6$

$$8x - 4 + 7 < 5x - 6$$

$$8x + 3 < 5x - 6$$

$$8x - 5x + 3 - 3 < 5x - 5x - 6 - 3$$

$$3x < -9$$

$$\frac{1}{3} \cdot 3x < \frac{1}{3} \cdot (-9)$$

$$x < -3$$

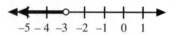

2. $5(2x + 1) < 3(4x - 1)$

$$10x + 5 < 12x - 3$$

$$10x - 12x + 5 - 5 < 12x - 12x - 3 - 5$$

$$-2x < -8$$

$$-\frac{1}{2} \cdot (-2x) > -\frac{1}{2} \cdot (-8) \qquad \textit{Don't forget to reverse the inequality symbol.}$$

$$x > 4$$

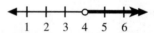

NOTE ▶   Remember when solving linear inequalities to follow the same steps used to solve linear equations. The only difference is in *reversing the inequality symbol when multiplying (or dividing) by a negative number.*

Sometimes two inequalities are used to define a solution set. An example of this is $1 < x < 6$. This says that x can be any number that is greater than 1 and at the same time less than 6. The solution set of the inequality $1 < x < 6$ is illustrated by the graph

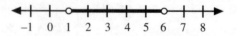

These types of inequalities always graph into a "bounded segment" of the number line. Solving an inequality such as $1 < 2x + 3 < 9$ is simply a matter of solving two inequalities at the same time. The goal is to isolate x in the middle.

E XAMPLE 2.2.4   Find and graph the solutions of the following inequalities.

1. $1 < 2x + 3 < 9$

$1 - 3 < 2x + 3 - 3 < 9 - 3$ *Subtract 3 from each member.*

$-2 < 2x < 6$

$\dfrac{1}{2} \cdot (-2) < \dfrac{1}{2} \cdot 2x < \dfrac{1}{2} \cdot 6$ *Multiply each member by $\frac{1}{2}$.*

$-1 < x < 3$

2. $-5 \le 1 - 3x \le 10$

$-5 - 1 \le 1 - 1 - 3x \le 10 - 1$

$-6 \le -3x \le 9$

$-\dfrac{1}{3} \cdot (-6) \ge -\dfrac{1}{3} \cdot (-3x) \ge -\dfrac{1}{3} \cdot 9$ *Reverse the inequality symbols.*

$2 \ge x \ge -3$

which is the same as

$-3 \le x \le 2$

REMARK   The inequalities of Example 2.2.4 can be written differently. For example $1 < 2x + 3 < 9$ can be written as $1 < 2x + 3$ and $2x + 3 < 9$. The word *and* indicates that the solutions we are looking for must satisfy both inequalities at the same time.

   In mathematics when two inequalities are connected with the words *and* or *or,* we call it a **compound inequality.** When we use the word *and,* we are looking for numbers that satisfy *both* inequalities. When we use the word *or,* we are looking for numbers that satisfy one inequality *or* the other but not necessarily both.

E XAMPLE 2.2.5   Find and graph the solutions of each compound inequality.

1. $2 + 3x < 4$ or $4 - x > 1$

$3x < 2$ $-x > -3$

$x < \dfrac{2}{3}$ $x < 3$ *Multiply by -1 and reverse the inequality sign.*

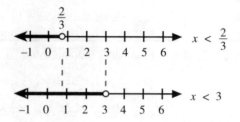

Every number to the left of 3 is in either one colored region or both. Thus the solution is $x < 3$, and its graph is

2. $5x + 1 \geq 3x \quad$ and $\quad x - 3 \leq -1$

$\quad 2x + 1 \geq 0 \qquad\qquad x \leq 2$

$\qquad 2x \geq -1$

$\qquad\quad x \geq -\dfrac{1}{2}$

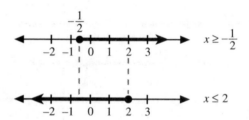

Every number between and including $-\frac{1}{2}$ and 2 is in both colored regions. Thus the solution is $-\frac{1}{2} \leq x \leq 2$ and its graph is

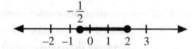

Solutions to inequalities can also be expressed using **interval notation.** For example, the solution set of the compound inequality $1 < x < 3$ is written as $(1, 3)$ using interval notation. The parentheses indicate that 1 and 3 are not members of the solution set. Because neither endpoint is included, this interval is said to be an **open interval.**

The solution set of the compound inequality $1 \leq x \leq 3$ is expressed using interval notation as $[1, 3]$. Square brackets are used to indicate that the endpoints 1 and 3 are included in this set. Because both endpoints are included, this interval is said to be a **closed interval.**

An interval may contain one endpoint and not the other. For example, the set of all real numbers x such that $1 < x \leq 3$ is represented by $(1, 3]$ and is sometimes referred to as a **half-open interval.**

Some intervals may have only one endpoint. The set of all real numbers x such that $x > 3$ has only one endpoint, 3 (which is not included in the set). This set is expressed using interval notation as $(3, \infty)$. The symbol ∞ (infinity) indicates that there is no finite upper bound for this set; that is, the ∞ symbol shows that the interval includes all real numbers greater than 3. The solution set of the inequality $x \leq 1$ is written as $(-\infty, 1]$. The symbol $-\infty$ (minus infinity) indicates that the interval includes all real numbers less than or equal to 1. The entire set of real numbers is represented by $(-\infty, \infty)$.

The symbols ∞ and $-\infty$ are not to be thought of as real numbers like 1 or 3. For this reason a parenthesis is always used with ∞ or $-\infty$ (never a square bracket). Also, when interval notation is used, the order in which the symbols appear matters. The notation $(1, 3)$ makes sense because there are real numbers greater than 1 and less than 3. The notation $(3, 1)$ would represent the set of real numbers greater than 3 and at the same time less than 1—there are no such numbers!

Various possibilities are summarized in the following chart:

INTERVAL NOTATION

Inequality	Graph	Notation	Type of Interval
$x < a$		$(-\infty, a)$	open
$x \leq a$		$(-\infty, a]$	closed
$x > a$		(a, ∞)	open
$x \geq a$		$[a, \infty)$	closed
$a < x < b$		(a, b)	open
$a \leq x < b$		$[a, b)$	half-open
$a < x \leq b$		$(a, b]$	half-open
$a \leq x \leq b$		$[a, b]$	closed

REMARKS

1. A square bracket indicates that an endpoint is included in the set.

2. A parenthesis immediately before or after a real number indicates that the number is not in the set.

3. A parenthesis is always used with ∞ or $-\infty$.

Let's now express the solution sets of some linear inequalities using interval notation.

XAMPLE 2.2.6   Express the solutions of the following inequalities using interval notation.

1. $1 < 3 - 2x \le 7$

$-2 < -2x \le 4$

$1 > x \ge -2$

which may be written

$$-2 \le x < 1$$

The solution set is $[-2, 1)$.

2. $4x + 3 < x$ or $1 - x < -3$

$\quad 3x + 3 < 0 \qquad\qquad -x < -4$

$\qquad 3x < -3 \qquad\qquad\quad x > 4$

$\qquad\quad x < -1$

The solution set may be written using interval notation as $(-\infty, -1)$ or $(4, \infty)$.

NOTE ▶   The symbol $\cup$ (used to represent the *union* of two sets) may be used to express the answer as $(-\infty, -1) \cup (4, \infty)$.

3. $3x \le 6$ and $5x > 0$

$\quad x \le 2$ and $\quad x > 0$

which is equivalent to

$$0 < x \le 2$$

In interval notation, the solution is $(0, 2]$.

NOTE ▶   The symbol $\cap$ (used to represent the *intersection* of two sets) may be used to express the answer as $(-\infty, 2] \cap (0, \infty)$, which is more simply written as $(0, 2]$.

Exercises 2.2

Find and graph the solutions of the following inequalities. Express your answers using interval notation.

1. $x + 5 \ge 10$

2. $x - 9 > -2$

3. $x - 8 \le 0$

4. $-x + 1 < -6$

5. $3x \le 27$

6. $-4x < 20$

7. $7x > 0$

8. $2x \ge 5$

9. $-x \le \dfrac{2}{3}$

10. $-x > 6$

11. $\dfrac{3}{4}x > -9$

12. $-\dfrac{1}{5}x \le \dfrac{3}{4}$

13. $2x + 1 \leq -7$

14. $3x - 4 < 8$

15. $4x - 2 > -10$

16. $3x - 7 > -7$

17. $\frac{1}{3}x - 2 < 1$

18. $-\frac{3}{4}x + 5 \geq 11$

19. $\frac{1}{5}x - \frac{1}{3} \geq -\frac{1}{2}$

20. $-\frac{1}{2}x + \frac{3}{4} \leq \frac{1}{4}$

21. $0.2x + 0.07 < 0.13$

22. $0.007x - 0.13 \geq 0.08$

23. $5x - 12 \leq 2x$

24. $8 - 4x < x$

25. $3x + 10 > 7x$

26. $2x + 15 \geq -3x$

27. $-6 - 2x > -x$

28. $\frac{1}{2}x - 3 < \frac{1}{4}x$

29. $\frac{2}{3}x - 1 \leq \frac{3}{4}x$

30. $\frac{1}{5}x + \frac{1}{2} \geq \frac{1}{10}x$

31. $2x + 5 < 4x - 9$

32. $7x + 11 \leq 3x + 19$

33. $8 - 2x \geq 2x + 7$

34. $-5x + 9 > 25 - x$

35. $-3x - 2 > -x + 14$

36. $\frac{3}{2}x + 3 > 5x - 4$

37. $\frac{1}{5}x - 3 \leq \frac{5}{4} - \frac{3}{2}x$

38. $\frac{1}{4}x - 3 < \frac{2}{3}x + 2$

39. $0.4x - 1.6 > 1.5x + 0.6$

40. $0.19 - 0.35x \leq 0.17x + 0.06$

41. $3(2x + 1) < 4(x - 3)$

42. $-2(x - 5) \geq 3(3x - 4)$

43. $4(2x - 3) - 7 > 3 + 6(1 - x)$

44. $3(2x - 4) + 5 < 4(x - 4) + 3$

45. $3(2x + 8) - 8x \geq 32$

46. $7(x + 2) - 6 > 4(2x + 3) - 5$

47. $4(2x - 3) - 5 < 2(3x - 1) - 15$

48. $\frac{2}{3}(x - 2) + 1 > \frac{1}{4}(3x - 4)$

49. $\frac{1}{5}(2x - 3) + 1 > \frac{1}{2}(2x - 1)$

50. $\frac{1}{3}(2x + 7) + x \leq \frac{3}{2}(x + 3)$

Find and graph the solutions of the following inequalities. Express your answers using interval notation.

51. $3 < x < 10$

52. $-8 \leq -x \leq 2$

53. $-7 < x - 4 < 1$

54. $3 \leq x + 2 < 7$

55. $2 \leq 6 - x \leq 8$

56. $-7 < 1 - x \leq -3$

57. $1 \leq 2x - 3 \leq 9$

58. $-2 \leq 3x + 4 < 16$

59. $7 < 5x + 2 \leq 27$

60. $-4 \leq 2x - 1 \leq 9$

61. $5 < 2 - 3x < 11$

62. $-5 < 3 - 2x < 6$

63. $1 < \frac{1}{3}x < 4$

64. $-2 \leq \frac{1}{2}x - 1 \leq 3$

65. $0 < \frac{3}{5}x + 2 < 8$

66. $-2 \leq 4 - \frac{3}{2}x < 1$

Find and graph the solutions of the following compound inequalities. Express your answers using interval notation.

67. $3x - 1 < 2$ and $5x + 4 > 9$

68. $2x + 5 \leq 1$ and $4x - 3 > 3$

69. $2x - 1 > -3$ or $x + 5 \leq 3x$

70. $3 - 3x < 1$ or $5x + 2 \geq 2$

71. $6x \leq 4$ or $4x - 3 \geq 7$

72. $2x - 1 > 6$ or $x + 5 < 2$

73. $5x - 2 < 4$ and $2x + 7 > 3$

74. $-4x < 1$ and $2x > 3x - 2$

75. $x - 1 \leq 4$ or $1 - x \leq 4$

76. $2 - x < 3$ or $x - 2 < 3$

Write Algebra

77. Explain when you must reverse the inequality sign in solving a linear inequality.

78. Explain why $x < 3$ is equivalent to $3 > x$.

79. If $1 < a < b$, what can we say about the relationship between $\frac{1}{a}$ and $\frac{1}{b}$?

2.3 Literal Equations and Applications

Consider the equation $D = RT$, where D represents the distance an object travels, R is its rate or speed, and T is the amount of time it was traveling. The equation $D = RT$ is generally referred to as the distance equation; it gives us a way of finding the distance an object travels when its rate and time are known.

However, suppose the distance and rate are known, and we want to determine the time. How can this formula aid us in finding the time? We know

$$D = RT$$

$$\frac{1}{R} \cdot D = \frac{1}{R} \cdot RT$$

$$\frac{D}{R} = T$$

We have now manipulated the distance equation so that we can find the time. We say that we have **solved the equation for T** because we have isolated T on one side of the equation. Now to apply this formula, suppose the distance was given as 270 mi and the rate as 45 mi/hr, and we want to find the time. We now know

$$T = \frac{D}{R}$$

$$T = \frac{270}{45}$$

$$T = 6 \text{ hr}$$

Another common formula is $P = 2L + 2W$. This formula states that the perimeter of a rectangle is equal to twice the length plus twice the width. Let's take this formula and solve for the width:

$$P = 2L + 2W$$

$$P - 2L = 2W \qquad \textit{Subtract 2L from both sides.}$$

$$\frac{1}{2} \cdot (P - 2L) = \frac{1}{2} \cdot 2W \qquad \textit{Multiply both sides by $\frac{1}{2}$.}$$

$$\frac{P - 2L}{2} = W$$

We now have a formula for the width in terms of the perimeter and length.

Equations such as $D = RT$ and $P = 2L + 2W$ are two examples of literal equations. A **literal equation** is any equation that uses letters to represent constants.

XAMPLE 2.3.1 Solve the following literal equations for the indicated variable.

1. $V = LWH$ for H

$$\frac{1}{LW} \cdot V = \frac{1}{LW} \cdot LWH \qquad \textit{Multiply both sides by } \tfrac{1}{LW}.$$

$$\frac{V}{LW} = H$$

2. The formula $C = \frac{5}{9} \cdot (F - 32)$ is used to convert degrees Fahrenheit (°F) to degrees Celsius (°C). Let's solve for F:

$$C = \frac{5}{9} \cdot (F - 32)$$

$$C = \frac{5}{9}F - \frac{160}{9} \qquad \textit{Eliminate grouping symbols.}$$

$$C + \frac{160}{9} = \frac{5}{9}F \qquad \textit{Add } \tfrac{160}{9} \textit{ to both sides.}$$

$$\frac{9}{5} \cdot \left(C + \frac{160}{9} \right) = \frac{9}{5} \cdot \frac{5}{9}F \qquad \textit{Multiply both sides by } \tfrac{9}{5}.$$

$$\frac{9}{5}C + 32 = F$$

We now have a formula for converting degrees Celsius to degrees Fahrenheit.

XAMPLE 2.3.2 The boiling point of water is 100°C. What is the boiling point of water in degrees Fahrenheit? From the previous example,

$$\frac{9}{5}C + 32 = F \qquad \textit{Here } C = 100.$$

$$\frac{9}{5} \cdot (100) + 32 = F$$

$$180 + 32 = F$$

$$212 = F$$

Thus the boiling point of water is 212°F.

EXAMPLE 2.3.3   The formula used to calculate simple interest is $I = PRT$. In this formula I represents the amount of interest earned, P represents the principal (i.e., the amount of money invested), R represents the rate of interest, and T stands for the time measured in years.

Suppose the town of Cut-And-Shoot, trying to hire a new sheriff, issued "law and order" bonds that pay a rate of 12% simple interest. What principal would you have to invest to earn $1000 in interest over the next 16 mo?

SOLUTION

$$I = PRT$$

$$\frac{1}{RT} \cdot I = \frac{1}{RT} \cdot PRT$$

$$\frac{I}{RT} = P$$

Now, $I = 1000$, $R = 0.12$, since $12\% = \frac{12}{100} = 0.12$, and $T = 16$ mo. *Do not forget to convert the time into years* (16 mo $= \frac{16}{12}$ yr $= \frac{4}{3}$ yr, so $T = \frac{4}{3}$). Substituting back into the formula yields:

$$\frac{1000}{(0.12)\left(\frac{4}{3}\right)} = P$$

$$\frac{1000}{\left(\frac{12}{100}\right)\left(\frac{4}{3}\right)} = P$$

$$\frac{1000}{\frac{4}{25}} = P$$

$$\frac{1000}{1} \cdot \frac{25}{4} = P$$

$$\$6250 = P$$

Exercises 2.3

Solve the following equations for the indicated variable.

1. $D = RT$ for R

2. $V = LWH$ for L

3. $A = LW$ for W

4. $I = PRT$ for T

5. $C = 2\pi r$ for r

6. $A = \frac{1}{2}bh$ for b

7. $V = \frac{1}{3}\pi r^2 h$ for h

8. $E = mc^2$ for m

9. $a = \frac{GM}{R^2}$ for G

10. $3y - 7x = 2$ for y

11. $P = 2L + 2W$ for L

12. $y + 5x = 4$ for y

13. $180 = x + y + z$ for z

14. $x + 2y - 3 = 0$ for y

15. $y = mx + b$ for m

16. $F = \frac{9}{5}C + 32$ for C

17. $m(x - x_1) = y - y_1$ for m

18. $m(x - x_1) = y - y_1$ for x_1

19. $S = 2x^2 + 4xy$ for y

20. $A = \frac{1}{2}h(b_1 + b_2)$ for b_1

21. $x = \frac{x_1 + x_2}{2}$ for x_2

22. $P = 2L + \pi r + 2r$ for L

23. $P = 2L + \pi r + 2r$ for r

24. $S = 2LW + 2LH + 2WH$ for L

25. $S = 2LW + 2LH + 2WH$ for W

26. $B = P + Prt$ for P

27. $T = \frac{1}{2}mv^2 + mad$ for m

Use a suitable formula and appropriate algebraic operations to solve the following problems.

28. The freezing point of water is 0° Celsius. What is the freezing point of water in degrees Fahrenheit?

29. Normal body temperature is 98.6° Fahrenheit. What is normal body temperature in degrees Celsius?

30. Harry Carey drives his car 333 mi in 4 hr 30 min. What is his average speed?

31. Pete Moss, a wealthy fertilizing merchant, drove his car 372 mi at a speed of 62 mi/hr. How many hours was Pete driving?

32. Jack Rabbit deposited $2400 in a savings account paying 8% simple interest. The amount of interest paid out was $1152. How long did Jack leave his money in the savings account?

33. Ima Saint deposits $7800 in a savings account that pays simple interest. At the end of 28 months she receives $1683.50 in interest. What is the interest rate?

34. The perimeter of a rectangular field is 142 yd. The length is 48 yd. Determine the width of the field.

35. The area of a rectangular field is 96 sq ft. If the width is 6 ft, determine the length of the field.

36. The volume of a box is 720 cu in. The length of the box is 12 in. and its width is 10 in. What is the height of the box? (*Hint:* Use the formulas on the inside front cover.)

37. A right triangle has an area of 68 sq m. If it has a base 8 m long, what is its height? (*Hint:* Use the formulas on the inside front cover.)

38. An equilateral triangle has a perimeter of 45 in. What is the length of one of the sides of the triangle?

39. Determine the radius of a circle with circumference 28π in.

40. Find the radius of a circle with area 36π sq in.

41. The surface area of a cube is 96 sq in. What is the length of one of the sides of the cube?

Write Algebra

42. Explain in your own words the steps used to solve $5y - 2x = 7$ for y.

2.4 *Applications*

One of the primary applications of algebra is to simplify and solve perplexing problems. In this section we will examine several key words and phrases together with their algebraic equivalents. We will then use the techniques discussed earlier in this chapter to solve problems.

Some commonly used words and phrases are:

- *sum*—indicates addition

- *difference*—indicates subtraction

- *product*—indicates multiplication

- *quotient*—indicates division

- *is*—indicates equality

- *is the same as*—indicates equality

Here are some examples of phrases you might encounter. To the right of the phrase is its algebraic equivalent.

Phrase	*Algebraic equivalent*
An unknown number	x
The sum of a number and 3	$x + 3$
A number increased by 4	$x + 4$
6 more than a number	$x + 6$
A number plus 5	$x + 5$
A number minus 1	$x - 1$
3 less than a number	$x - 3$
A number decreased by 7	$x - 7$
The difference of a number and 6	$x - 6$
2 minus a number	$2 - x$
The product of a number and 9	$9x$
A number multiplied by 2, or twice a number	$2x$
4 times a number	$4x$
A number divided by 6	$\dfrac{x}{6}$
7 divided by a number	$\dfrac{7}{x}$
The quotient of a number and 8	$\dfrac{x}{8}$
2 times a number plus 1	$2x + 1$
3 more than 5 times a number	$5x + 3$
4 less than twice a number	$2x - 4$
The sum of a number and 7, divided by 2	$\dfrac{x + 7}{2}$

Let's now use some algebraic expressions to find the solutions of the following problems.

EXAMPLE 2.4.1   Two more than three times a certain number is 14. What is the number?

SOLUTION   *Problems of this type are solved by letting a variable represent the number we want to find.* In this case let $x =$ a certain number. The algebraic expression $3x + 2$ represents 2 more than 3 times that certain number. So

$$3x + 2 = 14$$
$$3x = 12$$
$$x = 4$$

The number that satisfies the original problem is 4.

Often we are given a problem in which we are asked to find not just one number but perhaps two, three, or even more numbers. How can we use the tools of algebra in these cases? Consider the following examples.

EXAMPLE 2.4.2   The sum of two numbers is 19. The larger number is 1 more than twice the smaller number. What are the numbers?

SOLUTION   Since the larger number is described in terms of the smaller number, we let

$$x = \text{smaller number}$$
$$2x + 1 = \text{larger number}$$

We know that the sum

$$(\text{smaller number}) + (\text{larger number}) = 19$$
$$x \quad + \quad 2x + 1 \quad = 19$$
$$3x + 1 \quad = 19$$
$$3x = 18$$
$$x = 6$$

Do not stop here! All that we have determined is x (the smaller number). We still must find the larger number:

$$\text{larger number} = 2x + 1$$
$$= 2(6) + 1$$
$$= 12 + 1$$
$$= 13$$

Checking the solution, the sum of 6 and 13 is 19. So the two numbers are 6 and 13.

EXAMPLE 2.4.3   Find three consecutive integers whose sum is 36.

SOLUTION

Consecutive integers are integers that succeed one another in counting order. Examples of three consecutive integers are 26, 27, 28; 4, 5, 6; and -15, -14, -13. We let

$$x = \text{first integer}$$
$$x + 1 = \text{second integer}$$
$$x + 2 = \text{third integer}$$

Then

$$x + (x + 1) + (x + 2) = 36$$
$$x + x + 1 + x + 2 = 36$$
$$3x + 3 = 36$$
$$3x = 33$$
$$x = 11$$

The three consecutive integers are 11, 12, and 13. (The check is left to the reader.)

Number problems are merely one example of applications of algebraic expressions. Algebraic expressions may be used in many situations, as demonstrated in the remaining examples of this section.

EXAMPLE 2.4.4   The perimeter of a rectangle is 64 ft. The length of the rectangle is 3 ft less than 4 times the width. What are the dimensions of the rectangle?

SOLUTION

In geometric problems it is a good idea to draw a picture to represent the problem (see Figure 2.4.1). Since the length is described in terms of the width, we let

$$x = \text{width of the rectangle}$$
$$4x - 3 = \text{length of the rectangle}$$

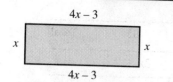

Figure 2.4.1

Now we can form an equation using the fact that the perimeter is the sum of the measures of the sides:

$$x + 4x - 3 + x + 4x - 3 = 64$$
$$10x - 6 = 64$$
$$10x = 70$$
$$x = 7$$

The width is *7 ft*. Then

$$\text{length} = 4x - 3$$
$$= 4(7) - 3$$
$$= 28 - 3$$
$$= 25\ ft$$

So we have a 7-ft by 25-ft rectangle. Checking the answer,

$$7 + 25 + 7 + 25 = 64$$

E XAMPLE 2.4.5   Ronnie has $92 in ones, fives, and tens. He has 4 more $1 bills than $10 bills. The number of $5 bills is 2 more than 3 times the number of $10 bills. How many of each type of bill does he have?

SOLUTION   Since the number of $1 and $5 bills is described in terms of the number of $10 bills, we let

$$x = \text{number of \$10 bills}$$
$$x + 4 = \text{number of \$1 bills}$$
$$3x + 2 = \text{number of \$5 bills}$$

If you had 7 $5 bills, you would have $35. The way we obtained the $35 was by multiplying 7 times 5 (the number of bills times the value of each bill). Repeating this process for each type of bill, we get the information in the following chart:

	$10	$1	$5	Total
Number of bills	x	$x + 4$	$3x + 2$	
Value	$10 \cdot x$	$1 \cdot (x + 4)$	$5 \cdot (3x + 2)$	92

The equation is generated by the bottom row:

$$\underset{\substack{\text{\$10 bills}}}{\text{value in}} + \underset{\substack{\text{\$1 bills}}}{\text{value in}} + \underset{\substack{\text{\$5 bills}}}{\text{value in}} = \underset{\substack{\text{value}}}{\text{total}}$$

$$10x + 1(x + 4) + 5(3x + 2) = 92$$
$$10x + x + 4 + 15x + 10 = 92$$
$$26x + 14 = 92$$
$$26x = 78$$
$$x = 3$$

From earlier, Ronnie has

$$x = \textit{3 tens}$$
$$x + 4 = 3 + 4 = \textit{7 ones}$$
$$3x + 2 = 3(3) + 2 = \textit{11 fives}$$

Check:

$$3 \text{ \$10 bills} = 3 \cdot \$10 = \$30$$
$$7 \text{ \$1 bills} = 7 \cdot \$1 = \$\ 7$$
$$11 \text{ \$5 bills} = 11 \cdot \$5 = \underline{\$55}$$
$$\text{total} = \$92$$

EXAMPLE 2.4.6

Heather has $2.40 in quarters and nickels. She has a total of 24 coins. How many quarters and how many nickels does she have?

SOLUTION

Since neither one of the coins is described in terms of the other, we can let x be the number of quarters or the number of nickels. Let

$$x = \text{number of quarters}$$
$$24 - x = \text{number of nickels}$$

because

$$\underset{\substack{\text{quarters}}}{\text{number of}} + \underset{\substack{\text{nickels}}}{\text{number of}} = \underset{\substack{\text{of coins}}}{\text{total number}}$$

$$x + \underset{\substack{\text{nickels}}}{\text{number of}} = 24$$

$$\underset{\substack{\text{nickels}}}{\text{number of}} = 24 - x$$

In this case our chart is:

	25	**5**	**Total**
Number of coins	x	$24 - x$	24
Value	$25 \cdot x$	$5 \cdot (24 - x)$	240

The bottom row gives the values in cents:

$$25 \cdot x + 5 \cdot (24 - x) = 240$$
$$25x + 120 - 5x = 240$$
$$20x + 120 = 240$$
$$20x = 120$$
$$x = 6$$

From earlier, Heather has

$$x = 6 \ quarters$$
$$24 - x = 24 - 6 = 18 \ nickels$$

The reader should check the solution.

EXAMPLE 2.4.7   What should the selling price of a house be in order to have $65,000 left after paying the real estate agent 6% of the selling price?

SOLUTION   Let x = selling price of the house. Then the real estate agent will be paid 6% of x, or $0.06x$.

Since,

$$\begin{pmatrix} \text{selling} \\ \text{price} \end{pmatrix} - \begin{pmatrix} \text{amount paid} \\ \text{to real estate} \\ \text{agent} \end{pmatrix} = \begin{pmatrix} \text{amount} \\ \text{left} \end{pmatrix}$$

we have

$$x - 0.06x = 65{,}000$$
$$0.94x = 65{,}000$$
$$x = \frac{65{,}000}{0.94}$$
$$x \doteq 69{,}148.936$$

So the selling price should be $69,150 in order to have $65,000 left.

EXAMPLE 2.4.8   Jake and Elwood leave Chicago traveling southwest at a rate of 50 mi/hr. Two hours later the sheriff of Cook County leaves Chicago traveling in the same direction at a rate of 75 mi/hr. How many hours will it take the sheriff to catch Jake and Elwood?

SOLUTION

Since we want to determine how many hours it will take the sheriff to catch Jake and Elwood, we let

$$t = \text{sheriff's traveling time}$$
$$t + 2 = \text{Jake and Elwood's traveling time}$$

When the sheriff catches Jake and Elwood, the distance the sheriff traveled must equal the distance Jake and Elwood traveled. *Recall: distance = rate · time.* See Figure 2.4.2.

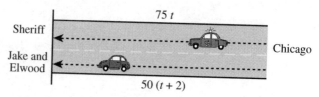

Figure 2.4.2

Now

$$\text{sheriff's distance} = 75 \cdot t$$
$$\text{Jake and Elwood's distance} = 50 \cdot (t + 2)$$

Since these distances must be the same, we obtain the equation

$$75t = 50(t + 2)$$
$$75t = 50t + 100$$
$$25t = 100$$
$$t = 4$$

So it takes the sheriff 4 hr to catch Jake and Elwood.

Perhaps an easier way to solve this problem is with the aid of the following chart:

	rate	·	time	=	distance
Jake and Elwood	50		$t + 2$		$50(t + 2)$
Sheriff	75		t		$75t$

Filling in the chart, we know Jake and Elwood's rate to be 50 mi/hr and the sheriff's rate to be 75 mi/hr. We decided to let t be the sheriff's time, so $t + 2$ is Jake and Elwood's time. The distance is found by multiplying the rate and time. Consequently, Jake and Elwood's distance is $50(t + 2)$, and the sheriff's distance is $75t$. Finally, we know that the distances must be the same when the sheriff catches Jake and Elwood. We then obtain precisely the same equation:

$$75t = 50(t + 2)$$

REMARK   Distance problems are usually easier to solve if we construct a chart of this form. The chart helps organize the given information in a clean and concise form.

The steps for solving word problems are now summarized.

Summary of Steps for Solving Word Problems

1. Determine what you are trying to find and let a variable represent that quantity. If the problem asks for more than one quantity, there must be something within the problem that tells how to represent all the different quantities in terms of one variable.

2. Draw a picture if appropriate.

3. Write an equation involving the variable, using the information given in the problem.

4. Solve the equation found in step 3.

5. Answer the question.

6. Check your solution. Go back to the statement of the problem and make sure your answer is reasonable.

Exercises 2.4

Use algebraic expressions to find the solutions of the following problems.

1. One more than twice a number is 43. What is the number?

2. Five more than 4 times a certain number is 29. What is the number?

3. One more than 3 times a number is -14. What is the number?

4. Five less than twice a number is 11. What is the number?

5. Three times a certain number decreased by one-half is 2. What is the number?

6. One minus 4 times a certain number is 13. What is the number?

7. The sum of two numbers is 13. The larger number is 1 more than 3 times the smaller number. What are the numbers?

8. The sum of two numbers is 18. The larger number is 4 more than the smaller number. What are the numbers?

9. The sum of two numbers is 5. The larger number is 17 more than twice the smaller number. What are the numbers?

10. The difference of two numbers is 8. The larger number is 4 more than 5 times the smaller number. What are the numbers?

11. The difference of two numbers is 3. The larger number is 33 minus twice the smaller number. What are the numbers?

12. The difference of two numbers is 7. The sum of the numbers is -1. What are the numbers?

13. Find three consecutive integers whose sum is 78.

14. Find three consecutive even integers whose sum is 48. (*Hint:* The three consecutive even integers can be represented by $x, x + 2,$ and $x + 4.$)

15. Find three consecutive odd integers whose sum is 111.

16. Find three consecutive integers with the property that 3 times the first integer plus the second integer minus 2 times the third integer is 9.

17. Find three consecutive even integers with the property that the second integer plus the third integer is 18 more than the first integer.

18. Find three consecutive odd integers with the property that the first integer plus twice the second integer is 3 more than twice the third integer.

19. The perimeter of a rectangle is 54 m. The length of the rectangle is 9 m more than twice the width. What are the dimensions of the rectangle?

20. The perimeter of a rectangle is 50 yd. The width is 1 yd more than one-third of the length. What are the dimensions of the rectangle?

21. The perimeter of a rectangle is 26 in. If 5 in. are added to the width of the rectangle, a square with perimeter 36 in. is obtained. What are the dimensions of the original rectangle?

22. A farmer wishes to fence a rectangular field by dividing it into three equal plots, as illustrated in Figure 2.4.3. We know that the farmer uses 170 yd of fencing and that the length of the field is 5 yd more than twice the width. What are the dimensions of this field?

Figure 2.4.3

23. The perimeter of a triangle is 17 cm. The second side is twice the length of the first side, and the third side is 1 cm less than 3 times the length of the first side. What are the lengths of the sides of the triangle?

24. A triangle is constructed so that the second angle is 8° larger than the first angle, and the third angle is 18° larger than 3 times the second. What are the angles of the triangle? (*Hint:* The sum of the three angles of any triangle is 180°.)

25. Roxanne has $4.65 in nickels and quarters. The number of quarters is 1 more than twice the number of nickels. How many quarters and how many nickels does Roxanne have?

26. Reginald has $2.45 in nickels and dimes. The number of nickels is 3 less than twice the number of dimes. How many nickels and how many dimes does Reginald have?

27. Ramon has $4.80 in quarters and dimes. He has a total of 30 coins. How many quarters and how many dimes does he have?

28. René has $2.63 in nickels and pennies. She has a total of 71 coins. How many nickels and how many pennies does she have?

29. Red has $545 in fives, tens, and twenties. He has 7 more $5 bills than $20 bills. The number of $10 bills is 3 less than twice the number of twenties. How many of each type of bill does he have?

30. Ray has $435 in fives, tens, and twenties. He has 2 more $10 bills than $5 bills. He has a total of 44 bills. How many of each type of bill does he have?

31. What should the selling price of a house be in order to have $80,000 left after paying the real estate agent 5% of the selling price?

32. What should the selling price of a house be in order to have $60,000 left after paying the real estate agent 3% of the selling price?

33. Moss Motors marks up each car 20% above its cost. They then advertise a sale of 15% off the listed price. If the sale price of a car is $11,515.80, what was the original cost of the car?

34. Fast Eddie's Used Cars marks up each car 60% above its cost. They then advertise a sale of 20% off the listed price. If the sale price of a car is $5376, what was the original cost of the car?

35. Henry leaves town at noon on a bicycle traveling west at a rate of 15 mi/hr. At 4:00 P.M. Sarah leaves town in a car traveling in the same direction at a rate of 45 mi/hr. How many hours will it take Sarah to catch Henry?

36. Dusty leaves the Circle K ranch at 1:00 P.M. on horseback traveling south at a rate of 12 mi/hr. At 3:30 P.M. Charley leaves the ranch in a jeep traveling in the same direction at a rate of 57 mi/hr. How many hours will it take Charley to catch Dusty?

37. Dottie leaves school in a car traveling north at a rate of 45 mi/hr. At the same time Gilbert leaves school in a car traveling south at a rate of 55 mi/hr. In how many hours will Gilbert and Dottie be 250 miles apart?

38. A train leaves Dallas at noon bound for Houston at a rate of 42 mi/hr. At the same time another train leaves Houston bound for Dallas (on the same route) at a rate of 38 mi/hr. The distance between Houston and Dallas is 260 mi. At what time will the crash occur?

Write Algebra

39. Write your own application problem that has 30 mi/hr and 40 mi/hr as its solution.

40. Write your own application problem that has a rectangle of length 12 ft and width of 5 ft as its solution.

2.5 *Absolute Value Equations*

The distance between any two distinct locations is always measured by some positive number. For example, the distance from Houston to Oklahoma City is 450 mi, and the distance from Oklahoma City to Houston is also 450 mi. The distance is not expressed as -450 mi. It does not matter where you start or end your trip; you will still travel the same 450 mi.

The notion that a quantity is always nonnegative is expressed mathematically by the concept of *absolute value*. Let's recall the definition from Chapter 1.

Definition 2.5.1	The **absolute value** of x, denoted $|x|$ and read "the absolute value of x," is defined as follows:

$$|x| = x \quad \text{if } x \geq 0$$

or

$$|x| = -x \quad \text{if } x < 0$$

REMARK   Remember that in Chapter 1 we saw that $|x|$ is the distance from zero to x on the real number line.

EXAMPLE 2.5.1   Evaluate the following.

1. $|3| = 3$ since $3 \geq 0$.
2. $|-3| = -(-3) = 3$ since $-3 < 0$.
3. $|0| = 0$ since $0 \geq 0$.

REMARK   This example illustrates that the absolute value of any number can never be negative (just as the distance between any two cities can never be negative).

Often, especially in this section, we encounter absolute value expressions in equations. How can we find solutions of this type of equation? Consider the following example.

EXAMPLE 2.5.2   Solve for x in the following equations.

1. $|x| = 3$ $x = 3$ or $x = -3$

NOTE ▶   There are *two* numbers that are 3 units from zero on the real number line: 3 and -3.

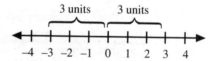

Thus the solution set is $\{3, -3\}$.

2. $|x + 1| = 7$

In this case the expression $x + 1$ represents a number that is 7 units from zero on the real number line. This number must be 7 or -7.

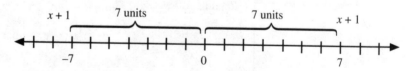

This fact yields two equations that we then solve for x:

$$
\begin{array}{ll}
x + 1 = 7 & \text{or} \quad x + 1 = -7 \\
x = 7 - 1 & \qquad x = -7 - 1 \\
x = 6 & \qquad x = -8
\end{array}
$$

$$\{6, -8\}$$

3. $|2x - 3| = 11$

$2x - 3$ represents a number that is 11 units from zero on the number line. This number must be 11 or -11.

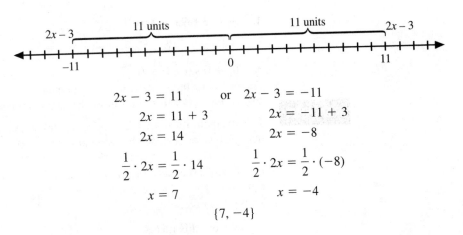

$$2x - 3 = 11 \quad \text{or} \quad 2x - 3 = -11$$
$$2x = 11 + 3 \qquad\qquad 2x = -11 + 3$$
$$2x = 14 \qquad\qquad\quad 2x = -8$$
$$\frac{1}{2} \cdot 2x = \frac{1}{2} \cdot 14 \qquad \frac{1}{2} \cdot 2x = \frac{1}{2} \cdot (-8)$$
$$x = 7 \qquad\qquad\quad x = -4$$

$$\{7, -4\}$$

It is a good idea to check your solution(s).

Checking the 7:   Checking the -4:

$$|2(7) - 3| \overset{?}{=} 11 \qquad |2(-4) - 3| \overset{?}{=} 11$$
$$|14 - 3| \overset{?}{=} 11 \qquad |-8 - 3| \overset{?}{=} 11$$
$$|11| \overset{?}{=} 11 \qquad |-11| \overset{?}{=} 11$$
$$11 = 11 \qquad\qquad 11 = 11$$

Try to avoid these mistakes:

Incorrect	Correct
1. $|2x - 3| = 11$ $2x - 3 = 11$ $2x = 14$ $x = 7$ Here we lost the solution $x = -4$. **2.** $|2x - 3| = 11$ $2x + 3 = 11$ Absolute value does not mean changing all minus signs to plus signs.	$\|2x - 3\| = 11$ $2x - 3 = 11 \quad \text{or} \quad 2x - 3 = -11$ $2x = 14 \qquad\qquad 2x = -8$ $x = 7 \qquad\qquad x = -4$

Sometimes an equation may have more than just the absolute value on one side. What is the procedure for finding solutions of such an equation?

XAMPLE 2.5.3   Find the solutions of the following equations.

1. $|x + 7| - 2 = 10$

$|x + 7| - 2 + 2 = 10 + 2$

$|x + 7| = 12$

As we isolate the variable in a linear equation, so must we isolate the absolute value expression. After this, we follow the steps outlined in the previous example.

$$x + 7 = 12 \qquad \text{or} \quad x + 7 = -12$$
$$x = 12 - 7 \qquad\qquad x = -12 - 7$$
$$x = 5 \qquad\qquad\qquad x = -19$$
$$\{5, -19\}$$

Try to avoid this mistake:

Incorrect	Correct
$|x + 7| - 2 = 10$	$|x + 7| - 2 = 10$
$x + 7 - 2 = 10$ or $x + 7 - 2 = -10$	$|x + 7| = 12$
$x + 5 = 10$ $\qquad\qquad$ $x + 5 = -10$	$x + 7 = 12$ or $x + 7 = -12$
$x = 5$ $\qquad\qquad\qquad$ $x = -15$	$x = 5$ $\qquad\qquad$ $x = -19$

2. $3|x + 4| = 15$

$\dfrac{1}{3} \cdot 3|x + 4| = 15 \cdot \dfrac{1}{3}$

$|x + 4| = 5$

We first multiply both sides by the reciprocal of the coefficient of the absolute value expression to isolate the absolute value. Then we follow the steps in the previous example.

$$x + 4 = 5 \qquad \text{or} \quad x + 4 = -5$$
$$x = 5 - 4 \qquad\qquad x = -5 - 4$$
$$x = 1 \qquad\qquad\qquad x = -9$$
$$\{1, -9\}$$

3. $|3x - 1| + 7 = 2$

$|3x - 1| + 7 - 7 = 2 - 7$

$|3x - 1| = -5$ $\qquad$ *Impossible!*

NOTE ▶   After isolating the absolute value, we obtain a negative number on the right-hand side of the equation. However, we know that absolute value can never be negative. Thus the solution set is the empty set; there are no solutions to this equation.

Still more complicated problems are those in which the absolute value occurs on both sides of the equation. These are very challenging problems, and we must be careful in applying the absolute value definition.

EXAMPLE 2.5.4

Find the solutions of the following equations.

1. $|2x + 1| = |x - 7|$

$2x + 1$ and $x - 7$ represent numbers whose distances from zero are the same. So these numbers must be equal or differ only in sign. Therefore, the problem becomes a matter of solving the following two equations:

$$2x + 1 = x - 7 \qquad \text{or} \qquad 2x + 1 = -(x - 7)$$
$$2x + 1 - x = x - 7 - x \qquad\qquad 2x + 1 = -x + 7$$
$$x + 1 = -7 \qquad\qquad 2x + x + 1 = -x + x + 7$$
$$x + 1 - 1 = -7 - 1 \qquad\qquad 3x + 1 = 7$$
$$x = -8 \qquad\qquad 3x + 1 - 1 = 7 - 1$$
$$3x = 6$$
$$\frac{1}{3} \cdot 3x = \frac{1}{3} \cdot 6$$
$$x = 2$$

The solution set is $\{-8, 2\}$. Checking the solutions:

$$x = -8 \qquad |2(-8) + 1| \overset{?}{=} |-8 - 7|$$
$$|-16 + 1| \overset{?}{=} |-15|$$
$$|-15| \overset{?}{=} 15$$
$$15 = 15$$

$$x = 2 \qquad |2(2) + 1| \overset{?}{=} |2 - 7|$$
$$|4 + 1| \overset{?}{=} |-5|$$
$$|5| \overset{?}{=} |-5|$$
$$5 = 5$$

Try to avoid this mistake:

Incorrect	Correct
$\|3x + 1\| = \|4 - 2x\|$	$\|3x + 1\| = \|4 - 2x\|$
$3x + 1 = 4 - 2x$	$3x + 1 = 4 - 2x$ or $3x + 1 = -(4 - 2x)$
$5x + 1 = 4$	$5x + 1 = 4$ $3x + 1 = -4 + 2x$
$5x = 3$	$5x = 3$ $x + 1 = -4$
$x = \dfrac{3}{5}$	$x = \dfrac{3}{5}$ $x = -5$
Here we lost the solution $x = -5$.	

2. $\|3x + 1\| = \|3x - 5\|$

$$3x + 1 = 3x - 5 \qquad \text{or} \qquad 3x + 1 = -(3x - 5)$$
$$3x - 3x + 1 = 3x - 3x - 5 \qquad\qquad 3x + 1 = -3x + 5$$
$$1 = -5 \qquad\qquad 3x + 3x + 1 = -3x + 3x + 5$$

This case generates no solutions.

$$6x + 1 = 5$$
$$6x + 1 - 1 = 5 - 1$$
$$6x = 4$$
$$\frac{1}{6} \cdot 6x = \frac{1}{6} \cdot 4$$
$$x = \frac{2}{3}$$

The solution set is $\{\frac{2}{3}\}$. The check is left to the reader.

Exercises 2.5

Find the solutions of the following equations.

1. $|x| = 2$ 　　　　2. $|y| = 7$ 　　　　3. $|z| = 23$ 　　　　4. $|-y| = 15$

5. $|s| = -13$ 　　　6. $|t| = -4$ 　　　7. $|-z| = -27$ 　　　8. $|x| = 0$

9. $|-s| = 0$ 　　　10. $|2y| = 14$ 　　　11. $|-5x| = 35$ 　　　12. $|3z| = -6$

13. $|-4y| = -48$ 　14. $|4r| = -20$ 　　15. $|5s| = 13$ 　　　16. $|2w| = 0$

17. $3\,|5p| = 0$ 　　18. $2|x| = 34$ 　　　19. $3|y| = 12$ 　　　20. $3|z| = 26$

21. $4|-s| = 24$ 　　22. $4|x| = -18$ 　　23. $-3|y| = -27$ 　　24. $-5|z| = 40$

25. $-4|-y| = -36$ 　26. $2|-z| = -15$ 　　27. $|x| + 5 = 9$ 　　28. $|r| - 8 = 3$

29. $|y| + 4 = 1$

30. $|x| + 6 = 6$

31. $|z| - 7 = -9$

32. $|2x + 1| = 5$

33. $|x - 7| = 9$

34. $|2 - 3x| = 12$

35. $|3x + 4| = -8$

36. $|2x - 6| = 0$

37. $2|x - 1| = 10$

38. $3|2x - 7| = 12$

39. $2|3 - 4x| = 15$

40. $4|2x - 5| = -16$

41. $-3|x + 4| = -21$

42. $-2|5x + 1| = -14$

43. $-6|1 - 3x| = 0$

44. $|x + 5| + 3 = 9$

45. $|2x - 4| - 2 = 11$

46. $|5 - 3x| + 1 = 10$

47. $|4x + 3| + 7 = 2$

48. $|2 - 3x| + 4 = 1$

49. $|6x - 3| - 8 = -8$

50. $|2x + 1| - 7 = -4$

51. $2|x + 3| - 5 = 7$

52. $3|2x - 1| + 1 = 19$

53. $5 - |3x - 4| = 1$

54. $4 - |x + 2| = 2$

55. $11 - 2|2x - 3| = 5$

56. $7 - 3|x - 2| = -4$

57. $10 - 5|6 - 4x| = 12$

58. $6 - 2|3x + 1| = 11$

59. $1.2|5.67x - 8.5035| + 6.451 = 18.7$

60. $1.8|23.9x + 13.1| - 6.005 = 394$

61. $|2x + 1| = |x - 4|$

62. $|3x - 2| = |2x + 5|$

63. $|6 - x| = |4x + 3|$

64. $|3x + 5| = |x + 7|$

65. $|5x - 1| = |4x - 9|$

66. $|3x - 1| = |5 - 2x|$

67. $|7 - x| = |9 - 3x|$

68. $|5x + 2| = |4 - 3x|$

69. $\left|\frac{1}{2}x + 5\right| = |x - 2|$

70. $\left|\frac{2}{3}x - \frac{1}{2}\right| = |3x + 1|$

71. $\left|\frac{3}{4}x + \frac{1}{3}\right| = \left|\frac{5}{2}x - 7\right|$

72. $|x + 2| = |x - 1|$

73. $|5 - 2x| = |2x + 7|$

74. $|3 - 4x| = |9 + 4x|$

75. $\left|\frac{1}{2}x - \frac{1}{3}\right| = \left|\frac{1}{2}x + \frac{1}{4}\right|$

Write Algebra

76. Explain why absolute value equations of the form $|ax + b| = c$ with $c > 0$ and $a \neq 0$ have two solutions.

77. Find the solutions of $|ax + b| = c$ with $c > 0$ and $a \neq 0$.

78. Explain why the solution set of $|x - 3| = -5$ is $\varnothing$.

79. Explain why equations of the form $|x - c| = 0$ have only one solution. What is that solution?

80. Explain why, if $|x - 2| = |3x + 5|$, then $x - 2 = -(3x + 5)$ as well as $x - 2 = 3x + 5$.

GROUP ACTIVITY

In exercises 81–86, find the solutions of the absolute value equations.

81. $|x - |3x + 1|| = 4$

82. $|x - |2x + 5|| = 6$

83. $|x - |5x + 7|| = 2$

84. $|x - |4x + 2|| = 3$

85. $|x - |2x + 3|| = 7$

86. $|x - |3x + 2|| = 9$

87. Describe your process for finding the solutions of these equations.

88. Predict the number of distinct solutions of the equation $|x - |ax + b|| = c$, where $a > 1$ and $b < ac$. What are these solutions?

2.6 *Absolute Value Inequalities*

In Section 2.2 simple linear inequalities were introduced. In this section we will investigate linear inequalities in which the variable expression is inside absolute value signs. What is the procedure for solving these types of problems? Consider the following example.

EXAMPLE 2.6.1   Find and graph the solutions of the following inequality.

$$|x| < 3$$

If we return to the notion that absolute value stands for distance, the inequality $|x| < 3$ indicates all numbers that are less than a distance of 3 units from zero, whether it is to the left or right of zero. The solutions are numbers satisfying $-3 < x < 3$, and the graph is the colored region on the following number line:

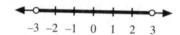

-3 -2 -1 0 1 2 3

Recall that our solution may be expressed using interval notation as $(-3, 3)$.

A generalization of the preceding argument leads to an important theorem that will aid us in finding the solutions of absolute value inequalities.

Theorem 2.6.1	If $|ax + b| < c$, where c is some positive number, then $-c < ax + b < c$.

EXAMPLE 2.6.2   Use Theorem 2.6.1 to find and graph the solutions of each of the following inequalities.

1. $|x| \le 5$
    $-5 \le x \le 5$ by Theorem 2.6.1, or, in interval notation, $[-5, 5]$.
    The graph is

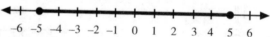

-6 -5 -4 -3 -2 -1 0 1 2 3 4 5 6

2. $\qquad |2x - 1| < 3$ *By Theorem 2.6.1 the variable expression, 2x − 1, must*
    $\qquad -3 < 2x - 1 < 3$ *be "trapped" between 3 and −3.*

    $-3 + 1 < 2x - 1 + 1 < 3 + 1$

    $\qquad -2 < 2x < 4$

    $\dfrac{1}{2} \cdot (-2) < \dfrac{1}{2} \cdot 2x < \dfrac{1}{2} \cdot 4$

    $\qquad -1 < x < 2$ or, using interval notation, $(-1, 2)$

The graph is

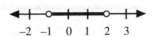

3.

$$5|4 - 3x| - 2 \leq 23$$

As with absolute value equations, we first isolate the absolute value expression.

$$5|4 - 3x| - 2 + 2 \leq 23 + 2$$

$$5|4 - 3x| \leq 25$$

$$\frac{1}{5} \cdot 5|4 - 3x| \leq \frac{1}{5} \cdot 25$$

$$|4 - 3x| \leq 5$$

$$-5 \leq 4 - 3x \leq 5 \qquad \textit{By Theorem 2.6.1.}$$

$$-5 - 4 \leq 4 - 4 - 3x \leq 5 - 4$$

$$-9 \leq -3x \leq 1$$

$$-\frac{1}{3} \cdot (-9) \geq -\frac{1}{3} \cdot (-3x) \geq -\frac{1}{3} \cdot (1) \qquad \textit{Remember to reverse the inequality signs.}$$

$$3 \geq x \geq -\frac{1}{3}$$

which is the same as $-\frac{1}{3} \leq x \leq 3$ or, using interval notation, $[-\frac{1}{3}, 3]$. The graph is

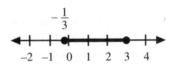

4. $|x + 3| < -4$

This is impossible! The absolute value can never be negative. Therefore, $|x + 3|$ cannot be less than -4. The solution set is $\varnothing$. To use Theorem 2.6.1, c must be positive.

Let's now consider the case $|x| > 3$. We can think of $|x| > 3$ as indicating all of the numbers whose distance from zero is greater than 3 units. These numbers can be more than 3 units to the right of zero, $x > 3$, or more than 3 units to the left of zero, $x < -3$. Putting this information together into one number line yields the following graph of $x < -3$ or $x > 3$:

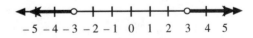

(Compare this number line with Example 2.6.1.)

A generalization of this example leads to a second theorem that will enable us to find solutions of inequalities similar to the previous example.

| **Theorem 2.6.2** | If $|ax + b| > c$, where c is some positive number, then $ax + b < -c$ or $ax + b > c$. |

XAMPLE 2.6.3   Use Theorem 2.6.2 to find and graph the solutions of each of the following inequalities.

1. $|x| \geq 2$

$x \leq -2$ or $x \geq 2$ by Theorem 2.6.2. The graph is

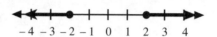

2. $|2x - 3| > 5$

$$
\begin{array}{lcl}
2x - 3 < -5 & \text{or} & 2x - 3 > 5 \\
2x - 3 + 3 < -5 + 3 & & 2x - 3 + 3 > 5 + 3 \\
2x < -2 & & 2x > 8 \\
\dfrac{1}{2} \cdot 2x < \dfrac{1}{2} \cdot (-2) & & \dfrac{1}{2} \cdot (2x) > \dfrac{1}{2} \cdot 8 \\
x < -1 & \text{or} & x > 4
\end{array}
$$

Theorem 2.6.2

or, using interval notation, $(-\infty, -1) \cup (4, \infty)$.
The graph is

Try to avoid these mistakes:

Incorrect	Correct
1. $\qquad |2x - 3| > 5$ $-5 > 2x - 3 > 5$ This says that $2x - 3$ is less than -5 and at the same time greater than 5, which is impossible.	$|2x - 3| > 5$ $2x - 3 < -5$ or $2x - 3 > 5$ We must solve two distinct inequalities. $\qquad 2x < -2 \qquad\qquad 2x > 8$ $\qquad\;\; x < -1 \quad$ or $\quad x > 4$
2. $|2x - 3| > 5$ $\qquad 2x - 3 > 5$ $\qquad\quad 2x > 8$ $\qquad\quad\; x > 4$ Here we lost the solution $x < -1$.	

3. $|-1 - 3x| > 8$

$$-1 - 3x < -8 \qquad \text{or} \qquad -1 - 3x > 8 \qquad \textit{Theorem 2.6.2}$$

$$-1 + 1 - 3x < -8 + 1 \qquad \qquad -1 + 1 - 3x > 8 + 1$$

$$-3x < -7 \qquad \qquad \qquad -3x > 9$$

$$-\frac{1}{3} \cdot (-3x) > -\frac{1}{3} \cdot (-7) \qquad -\frac{1}{3} \cdot (-3x) < -\frac{1}{3} \cdot (9) \qquad \begin{array}{l}\textit{Reverse the} \\ \textit{inequality signs.}\end{array}$$

$$x > \frac{7}{3} \qquad \qquad \text{or} \qquad \qquad x < -3$$

or, using interval notation, $(-\infty, -3) \cup (\frac{7}{3}, \infty)$.
The graph is

4. $|x + 2| > -3$

For any value of x, $|x + 2|$ is always zero or positive. Since zero or any positive number is always greater than -3, this inequality is always true. Thus the solution set is the set of all real numbers and the graph is

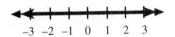

For convenience, the solutions and graphs of the two types of absolute value inequalities are summarized in the following chart. The letter c represents a positive number.

Type of Inequality	Solution	Graph
$|ax + b| < c$	$-c < ax + b < c$ One three-sided inequality	One region on the number line
$|ax + b| > c$	$ax + b < -c$ or $ax + b > c$ Two separate inequalities	Two distinct regions on the number line

Exercises 2.6

Find and graph the solutions of each of the following inequalities. Express your answers using interval notation.

1. $|x| < 1$

2. $|x| < 4$

3. $|x| > 2$

4. $|x| > 4$

5. $|x| \le \dfrac{3}{2}$

6. $|x| \ge \dfrac{1}{2}$

7. $|x| \ge -3$

8. $|x| \le 6$

9. $|x| + 5 < 10$

10. $|x| + 6 > 9$

11. $|x| + 1 \ge 2$

12. $|x| - \dfrac{1}{2} > 2$

13. $|x| - 2 < 0$

14. $|x| + 4 < 2$

15. $|x| - 6 < -3$

16. $|x| - 4 \ge 4$

17. $|x| + 7 \ge 7$

18. $|x| + 2 \le 2$

19. $|x - 1| \le 5$

20. $|x + 4| > 3$

21. $|x - 1| \ge 5$

22. $|3x + 2| \ge 7$

23. $|x - 3| < \dfrac{1}{2}$

24. $|2x + 1| < 7$

25. $|5x - 4| \le 9$

26. $|2x + 6| > 1$

27. $|1 - x| > \dfrac{3}{2}$

28. $|2 - x| < 3$

29. $|3 - 2x| \le 5$

30. $|5 - 3x| \ge 7$

31. $|1 - 3x| < 1$

32. $|-2 - 4x| \le 6$

33. $|-2 - 3x| > 5$

34. $|x + 2| + 1 > 4$

35. $|x + 2| - 1 < 3$

36. $|2x + 7| + 5 \le 10$

37. $|2x + 5| - 3 \ge 0$

38. $|7 - 3x| - 5 \ge -1$

39. $|4 - x| + 2 < 5$

40. $|5 - 3x| - 1 < 1$

41. $|4x - 3| + 6 > 1$

42. $3|x - 2| \ge 15$

43. $2|x + 7| < 8$

44. $3|2x - 4| \le 10$

45. $2|3x - 1| \ge 9$

46. $2|x + 1| - 5 \ge 7$

47. $2|x - 1| + 4 \le 8$

48. $3|2x + 9| - 1 < 11$

49. $3|2x + 7| + 1 > 10$

50. $2|3 - x| + 5 \ge 13$

51. $|x + 1| > 0$

52. $|x + 5| \le 0$

53. $|3x - 4| \le 0$

54. $|1 - 2x| < 0$

55. $|2x - 6| > 0$

56. $|3 - 4x| \ge 0$

57. $|x + 7| \ge -3$

58. $|x - 3| < -4$

59. $|5x - 7| \le -1$

60. $|2x - 10| \ge -7$

Write Algebra

61. Explain why the solution set of $|x| > -5$ is all real numbers.

62. Explain why the solution set of $|x| < -5$ is $\varnothing$.

63. Explain why the solution set of $|x - c| \le 0$ contains only one number. What is that number?

64. Explain why the solution set of $|x - c| > 0$ contains every real number except one number. What is that number?

G R O U P   A C T I V I T Y

In exercises 65–76, find and graph the solution sets of the following inequalities.

65. $3 \le |x| \le 7$

66. $1 \le |x| \le 6$

67. $2 < |x| < 5$

68. $3 < |x| < 4$

69. $4 \le |2x| \le 12$

70. $2 \le |2x| \le 10$

71. $3 \le |x - 1| \le 5$

72. $2 \le |x + 1| \le 6$

73. $4 < |x - 3| < 7$

74. $1 < |x - 2| < 4$

75. $1 \le |x + 2| \le 4$

76. $3 \le |x + 3| \le 6$

Applied Algebra

Bahar is the highway engineer for the Moss County Highway Department. A new 400-ft-long section of Main Street is to be constructed. This section of Main Street is a two-lane roadway. The roadway will be constructed in the spring when the temperature will average 70°F. Bahar anticipates that the temperature on the roadway will reach 110°F this summer. By how many inches will the roadway expand at 110°F?

SOLUTION

The amount of expansion of a roadway can be determined by the formula

$$E = K \cdot L \cdot (T - t)$$

where

E = the expansion of the roadway

K = the thermal coefficient of expansion

L = the length of the roadway being considered

T = the temperature on the roadway

t = the temperature at which the roadway was built

NOTES ▶ **1.** The units of measure for E and L must be the same.

2. For a two-lane roadway, $K = 0.000012$.

3. For the given value of K, the temperatures are measured in degrees Fahrenheit.

In Bahar's case

$$K = 0.000012; L = 400; T = 110; t = 70$$

Thus,

$$E = 0.000012(400)(110 - 70)$$
$$= 0.000012(400)(40)$$
$$= 0.0048(40)$$
$$= 0.192$$

The roadway will expand by 0.192 ft at 110°F.
However, Bahar wanted to know the expansion in inches, so

$$\text{expansion (in inches)} = (0.192)12$$
$$= 2.304 \text{ in.}$$

Your Turn

The following questions relate to the 400-ft-long section of Main Street.

1. By how many inches will the roadway expand at 95°F?

2. By how many inches will the roadway expand at 120°F?

3. By how many inches will the roadway contract at 50°F?

4. By how many inches will the roadway contract at 30°F?

   Bahar knows that if the roadway expansion is under 3.6 in. she will not have to install an expansion joint. What roadway temperature will produce an expansion of 3.6 in.?

> **SOLUTION**   Again we will use the formula

$$E = KL(T - t)$$

In our case

$$E = 3.6 \text{ in.}$$
$$K = 0.000012$$
$$L = 400 \cdot 12 = 4800 \text{ in.}$$
$$t = 70$$

Thus,

$$3.6 = 0.000012 \, (4800) \, (T - 70)$$
$$3.6 = 0.0576 \, (T - 70)$$
$$62.5 = T - 70 \qquad \textit{Divide both sides by 0.0576.}$$
$$132.5 = T$$

Hence a roadway temperature of 132.5°F will produce an expansion of 3.6 in.

Your Turn

The following questions relate to the 400-ft-long section of Main Street.

5. What roadway temperature will produce an expansion of 5.4 in.?

6. What roadway temperature will produce an expansion of 4.8 in.?

7. What roadway temperature will produce a contraction of 1.2 in.?

8. What roadway temperature will produce a contraction of 3.6 in.?

Chapter 2 Review

Terms to Remember

Coefficient	**p. 58**	Equivalent inequalities	**p. 67**
Like terms	**p. 58**	Compound inequality	**p. 71**
Solution	**p. 59**	Interval notation	**p. 72**
Solve an equation	**p. 59**	Open interval	**p. 72**
Solve for x	**p. 59**	Closed interval	**p. 72**
Solution set	**p. 59**	Half-open interval	**p. 72**
Equivalent equations	**p. 59**	Solve for a variable	**p. 76**
Linear equation	**p. 61**	Literal equation	**p. 76**
Linear inequality	**p. 66**	Absolute value	**p. 89**
Graph of the solution set	**p. 67**		

Properties

■ *Additive property of equality:* Let a, b, and c denote real numbers.

$$\text{If } a = b, \text{ then } a + c = b + c.$$

■ *Multiplicative property of equality:* Let a, b, and c denote real numbers, where $c \neq 0$.

$$\text{If } a = b, \text{ then } ac = bc.$$

■ *Additive property of inequality:* Let a, b, and c denote real numbers.

$$\text{If } a < b, \text{ then } a + c < b + c.$$

■ *Multiplicative property of inequality:* Let a, b, and c denote real numbers.

1. If $a < b$ and $c > 0$, then $ac < bc$.

2. If $a < b$ and $c < 0$, then $ac > bc$.

Equations and Inequalities

Let a, b, p, and q denote real numbers and c represent a positive real number.

Type	Solution	Graph
Linear equation $ax + b = 0$	$x = p$	
Absolute value equation $\|ax + b\| = c$	$x = p$ or $x = q$	
Linear inequality $ax + b < 0$	$x < p$ (or $x > p$)	
Absolute value inequality $\|ax + b\| < c$	$p < x < q$	
$\|ax + b\| > c$	$x < p$ or $x > q$	

Review Exercises

2.1   Find the solutions of the following equations and check your answers.

1. $7x = 56$

2. $\frac{4}{5}x = -64$

3. $3x - 1 = 15$

4. $\frac{1}{2}x - \frac{1}{3} = \frac{3}{4}$

5. $4x + 7 = 2x - 5$

6. $\frac{1}{2}x + 4 = \frac{2}{3}x - 1$

7. $2(3x - 7) = 4$

8. $2(x - 8) = 5(3x + 2)$

9. $3(5x - 2) + 3x = 12x + 4$

10. $\frac{3}{2}(4 - 5x) - 7 = 3 - 4x$

11. $5(2x - 1) - 7(x - 2) = 9$

12. $3(2x - 8) - 4(x + 2) = 2(3x + 4)$

2.2   Find and graph the solutions of the following inequalities. Express your answers using interval notation.

13. $x + 5 < 9$

14. $2 - x \geq 7$

15. $3x > -24$

16. $-\frac{2}{3}x \leq 0$

17. $2x + 1 > 10$

18. $-\frac{5}{4}x - 2 \leq 13$

19. $3x + 1 < 6x - 2$

20. $2(3x - 1) - 2 \geq x + 7$

21. $-5 \leq x \leq 3$

22. $1 \leq x + 2 \leq 6$

23. $-8 < 3x + 1 < 1$

24. $-1 < 7 - 2x < 5$

25. $1 - 6x > 4$ and $x - 6 > 4$

26. $7x + 5 < -2$ or $3x > 12$

2.3   **Solve the following equations for the indicated variables.**

27. $F = ma$ for a

28. $V = IR$ for R

29. $y = mx + b$ for x

30. $A = 4x^2 + 2xy$ for y

31. $y - 4x = 2$ for y

32. $m = \dfrac{y_2 - y_1}{x_2 - x_1}$ for y_1

33. $2x + 3y - 4z = 5$ for z

34. $2x + 5y - 6 = 0$ for x

Use suitable formulas and appropriate algebraic operations to solve the following.

35. The area of a rectangular field is 54 sq m. If the length is 9 m, determine the width of the field.

36. The perimeter of a rectangular field is 46 yd. If the width of the field is 7 yd, determine the length of the field.

37. To save energy, President Carter suggested that during the winter of 1977 patriotic Americans set their thermostats at 68° Fahrenheit. What is this temperature in degrees Celsius?

38. Sally deposited $2500 in a savings account paying 8% simple interest. The amount of interest paid out was $1275. How long did Sally leave her money in the savings account?

2.4   **Use algebraic expressions to find the solutions of the following problems.**

39. Three times a certain number decreased by 7 is 26. What is the number?

40. The sum of two numbers is 16. The larger number is 4 more than 5 times the smaller number. What are the numbers?

41. Find three consecutive integers with the property that 3 times the first integer minus 2 times the second integer plus the third integer is 22.

42. The perimeter of a rectangle is 74 m. The length is 4 m more than twice the width. What are the dimensions of the rectangle?

43. Takako has $4.30 in nickels, dimes, and quarters. The number of dimes is twice the number of quarters. The number of nickels is 2 more than 3 times the number of quarters. How many of each type of coin does Takako have?

44. What should the selling price of a house be in order to have $70,000 left after paying the real estate agent 6% of the selling price?

45. At 3:00 P.M. Felicia Foote leaves the gym jogging at a speed of 7 mi/hr. At 3:30 P.M. Harry Legs takes off running from the gym, following the same route, at a speed of 10 mi/hr. When will Harry overtake Felicia?

2.5   **Solve the following equations.**

46. $|5y| = 15$

47. $6|x| = 9$

48. $|m| - 9 = -4$

49. $|x - 4| = 9$

50. $|3x + 11| = 2$

51. $|1 - 5x| = 16$

52. $5|6x - 7| = 25$

53. $-3|x - 2| = 12$

54. $|4y + 3| + 10 = 10$

55. $13 - 2|x + 4| = 7$

56. $|3x + 8| = \left|\dfrac{1}{2}x + 2\right|$

57. $|3x + 5| = |3x - 7|$

2.6   **Find and graph the solutions of the following inequalities. Express your answers using interval notation.**

58. $|x + 2| < 6$

59. $|3x - 4| \leq 1$

60. $|7 - 4x| + 2 \leq 5$

61. $5|x - 2| < 10$

62. $|3x - 9| \leq 0$

63. $\left|2x - \dfrac{1}{2}\right| < -5$

64. $|x| - \dfrac{1}{2} > 2$

65. $|5x + 3| > 12$

66. $2|6 - x| \geq 4$

67. $\left|\dfrac{1}{2}x + 1\right| + 3 \geq 4$

68. $|8 - 2x| > 0$

69. $|10x - 3| \geq -6$

Chapter 2 Test

(You should be able to complete this test in 60 minutes.)

I. Find the solutions of the following equations and check your answers.

1. $4x - 3 = 61$

2. $4 - 6x = 3x + 67$

3. $4(2x + 6) - (7x - 5) = 2$

4. $\dfrac{3}{2}x - 4 = \dfrac{1}{5}x - \dfrac{1}{10}$

5. $9(2x - 3) - 4(5 + 3x) = 6(4 + x)$

6. $|2x - 3| = 7$

7. $|3x + 1| = |5x - 9|$

II. Find and graph the solutions of the following inequalities. Express your answers using interval notation.

8. $2(x + 2) \leq 3x - 4$

9. $-19 < 2x - 5 < -9$

10. $|5 - 3x| \leq 7$

11. $|4x + 3| > 11$

12. $x \geq 4 \text{ or } x > 2$

13. $3x - 6 < 2 \text{ and } 2x + 1 > -1$

III. Solve the following equations for the indicated variables.

14. $K = \dfrac{1}{2}mv^2$ for m

15. $A = \dfrac{h}{2}(a + b)$ for a

IV. Solve the following word problems.

16. The perimeter of a rectangle is 54 ft. The length is 1 ft less than 3 times the width. Find the dimensions of the rectangle.

17. A cashier has a total of 83 bills. He has twice as many fives as tens, with the rest of the bills twenties. The total value of the money is $660. How many of each type of bill does he have?

18. A car leaves Detroit at 12:00 P.M. traveling eastward at a rate of 36 mi/hr. At 1:00 P.M. another car leaves from the same point traveling eastward at 45 mi/hr. How long will it take for the second car to overtake the first car?

TEST YOUR MEMORY

These problems review Chapters 1 and 2.

I. Use the sets $A = \{0, 1, 2, 3, 4\}$, $B = \{1, 3, 5\}$, $E = \{2, 4, 6, \ldots\}$, N, I, Q, H, and R for problems 1–8. Are the following true or false?

1. $\sqrt{6} \in E$

2. $20 \in I$

3. $A \subseteq Q$

4. $E \subseteq H$

Find the following:

5. $A \cup B$

6. $A \cap B$

7. $B \cap E$

8. $A \cup E$

II. Name the property that justifies each step in the following.

9. $-2 + (x + 2) = (x + 2) + (-2)$
$$= x + (2 + (-2))$$
$$= x + 0$$
$$= x$$

III. Perform the indicated operations.

10. $-7 - 3 - (-8) - 9$

11. $32 \div 4 \cdot 2$

12. -6^2

13. $\dfrac{-\dfrac{6}{25}}{4}$

14. $(-1.2)(-3.14)(0.08)$

15. $\dfrac{1}{6} + \dfrac{3}{20} - 1$

16. $6 \cdot 4^2 \div 2 + 8$

17. $\sqrt{5 \cdot 8 + 7^2 - 8}$

18. $6|2^2 - 7|$

19. $-\dfrac{3}{8} + \left(-\dfrac{1}{6}\right) \div \dfrac{2}{3} + 7\left(-\dfrac{5}{8}\right)$

IV. Make the following conversions.

20. Fraction to a decimal: $\dfrac{9}{20}$

21. Decimal to a fraction in lowest terms: 0.48

V. Evaluate the following algebraic expressions for the given values of the variables.

22. $3x^2 - 7x + 9$ for $x = -3$

23. $\sqrt{100 - x^2}$ for $x = 8$

24. $7x - |4x - 6|$ for $x = \dfrac{1}{2}$

25. $3x^2 - 5xy - 2y^2$ for $x = -2, y = -3$

VI. Find the solutions of the following equations and check your answers.

26. $3x + 8 = -10$

27. $7x - 4 = 4x - 2$

28. $3(2x + 1) - 2(x - 4) = 11$

29. $4(2x - 3) - (x + 4) = 9x - 19$

30. $\dfrac{2}{3}x - 1 = \dfrac{1}{2}x + \dfrac{1}{6}$

31. $\dfrac{1}{4}(x + 1) - \dfrac{1}{3}(2x - 2) = 1$

32. $3\left(\dfrac{1}{2}x - 1\right) - 2\left(\dfrac{1}{3}x - 2\right) = \dfrac{3}{4}x$

33. $3(x - 5) - 4(x - 2) = 2(x + 7)$

34. $|2x - 5| = 9$

35. $|2x + 3| = |4x - 5|$

VII. Find and graph the solutions of the following inequalities. Express your answers using interval notation.

36. $3(x + 4) \leq 4x + 10$

37. $4(2x + 1) > 3(x - 2)$

38. $-5 < 3x - 2 < 8$

39. $|2x - 1| < 7$

40. $|x + 3| \geq 4$

41. $2x - 5 < 7$ and $3x - 4 \geq -7$

42. $x \geq 5$ or $x > -1$

VIII. Solve the following equations for the indicated variables.

43. $A = \dfrac{1}{2}bh$ for b

44. $h = vt - 16t^2$ for v

IX. Use algebraic expressions to find the solutions of the following problems.

45. Four times a certain number decreased by 1 is 15. What is the number?

46. The perimeter of a rectangle is 46 ft. The length is 5 ft more than twice the width. Find the dimensions of the rectangle.

47. What should the selling price of a house be in order to have $50,000 left after paying the real estate agent 4% of the selling price?

48. Kato has $3.05 in nickels, dimes, and quarters. The number of nickels is twice the number of dimes. The number of quarters is 4 less than the number of dimes. How many of each type coin does Kato have?

49. Afifa has $160 in fives and twenties. She has a total of 17 bills. How many of each type bill does she have?

50. Lech leaves the library in a car traveling north at a rate of 40 mi/hr. At the same time Mieczysław leaves the library in a car traveling south at a rate of 30 mi/hr. In how many hours will Lech and Mieczysław be 175 miles apart?

Polynomials and Factorable Quadratic Equations

Arlene, an architect, is building a new home along the coast of Louisiana. Arlene enjoys the view of the Gulf of Mexico and plans on including many large windows in her new home. However, Arlene is concerned about hurricanes and their devastating effects on Gulf Coast homes. She plans to design her home with windows that have a surface area of no more than 25 sq ft in each window. Arlene has already selected the thickness of glass that will be used in her windows. One of the problems facing Arlene is determining the maximum wind speed her windows can withstand before breaking.

At the end of this chapter we will be given a formula that will enable us to determine the maximum wind speed that Arlene's windows can withstand. In working with this formula we will use quadratic equations, which will be studied in this chapter.

3.1   Adding and Subtracting Polynomials

In Chapter 1 we worked with algebraic expressions. In this chapter we will study a special type of algebraic expression called a polynomial.

Definition 3.1.1

A **polynomial** is an algebraic expression that can be written as a finite sum of terms, each of which is a number or the product of a numerical factor and one or more variable factors raised to whole number powers.

REMARK

If an expression has variables under square root signs or in denominators, then the expression is not a polynomial. Thus the expressions $\sqrt{x} + 16$ and $4xy^2 - \frac{7}{y} + 3$ are not polynomials.

EXAMPLE 3.1.1

The following are polynomials.

1. $5xy + 3z^2$

2. $19mn$

3. $x^2 - 25$

This is a polynomial, since $x^2 - 25 = x^2 + (-25)$.

4. $\dfrac{ab}{9} + 2r^2h - \sqrt{17}k$

This is a polynomial, since $\frac{ab}{9} + 2r^2h - \sqrt{17}k = (\frac{1}{9})ab + 2r^2h + (-\sqrt{17}k)$. Only the 17 is under the square root sign, not the k.

Definition 3.1.2

If a polynomial contains only one term, it is called a **monomial;** if it contains two terms, it is called a **binomial;** and if it contains three terms, it is called a **trinomial.** If a polynomial has more than three terms, it has no special name.

REMARK

In Example 3.1.1, $19mn$ is a monomial, $5xy + 3z^2$ and $x^2 - 25$ are binomials, and $\frac{ab}{9} + 2r^2h - \sqrt{17}k$ is a trinomial.

It is standard practice to start each term of a polynomial with the numerical factor. Remember from Chapter 2 that this numerical factor is called the numerical coefficient, or just **coefficient,** of that term. The term $5xy$ has a coefficient of 5 and variable factors of x and y. The term $3z^2$ has a coefficient of 3 and two variable factors of z.

NOTE ▶ In the polynomial

$$x^4 - 5x^3 + \frac{2x^2}{3} - x + 22,$$

the coefficients are $1, -5, \frac{2}{3}, -1$, and 22, since

$$x^4 - 5x^3 + \frac{2x^2}{3} - x + 22 = 1x^4 + (-5)x^3 + \left(\frac{2}{3}\right) \cdot x^2 + (-1)x + 22$$

Another important property is the degree of a polynomial. We start by defining the degree of a term.

| Definition 3.1.3 | The **degree** of a term is the sum of the exponents of the variables. |

XAMPLE 3.1.2   Find the degree of each term.

1. $8xy^4z^5$

Since $x = x^1$, the degree is $1 + 4 + 5 = 10$.

2. -12

The degree of a constant term other than zero is said to be zero.

3. 3^5xy

The degree is 2, since $xy = x^1y^1$. Note that the exponent of the coefficient does not count toward the degree.

| Definition 3.1.4 | The **degree** of a polynomial is the same as that of its term of highest degree. |

XAMPLE 3.1.3   Find the degree of each of the following polynomials.

1. $3x^4 - xy + 2xy^5$

The degrees of the three terms are 4, 2, and 6, so the degree of the polynomial is 6.

2. $y^3 - 5y^2 + 3y + 8$

The degree is 3, the same as that of the first term.

When a polynomial contains only one variable, it is common practice to arrange the terms so that the degrees are in descending order. For example, the polynomial $y^3 - 5y^2 + 3y + 8$ is arranged in this order. We shall call this the

standard form for a polynomial in one variable. The coefficient of the highest degree term is called the **leading coefficient,** since it is written first when the polynomial is in standard form. The leading coefficient of $y^3 - 5y^2 + 3y + 8$ is 1.

EXAMPLE 3.1.4   Write the following polynomials in standard form and determine the leading coefficients.

1. $16 - x^2$

Standard form is $-x^2 + 16$ and the leading coefficient is -1.

2. $28x + 19x^3 + 2x^5 - x^2 - 3$

Standard form is $2x^5 + 19x^3 - x^2 + 28x - 3$ and the leading coefficient is 2.

Recall the statement of the distributive property from Chapter 1.

Distributive Property

Let a, b, and c denote real numbers; then

$$a(b + c) = ab + ac$$

or

$$ba + ca = (b + c)a$$

REMARK   If a rectangle has width a and length $b + c$, then its area is

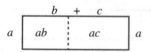

$$A = WL$$
$$= a(b + c)$$
$$= ab + ac$$

Let's now see how the distributive property applies to polynomials.

EXAMPLE 3.1.5   Use the distributive property to rewrite the following algebraic expressions.

1. $5(3x + 7z) = 5(3x) + 5(7z)$
                $= 15x + 35z$

NOTE ▶   It is sometimes said that the 5 is distributed across the polynomial $3x + 7z$.

2. $6(4a + b - 2c) = 6(4a) + 6(b) - 6(2c)$

$\qquad\qquad\qquad = 24a + 6b - 12c$

3. $-2(9r - 7s^2 + 3t) = (-2)(9r) - (-2)(7s^2) + (-2)(3t)$

$\qquad\qquad\qquad\qquad = -18r - (-14s^2) + (-6t)$

$\qquad\qquad\qquad\qquad = -18r + 14s^2 - 6t$

Recall from Chapter 2 that we used the distributive property to combine like terms. Now we will use the distributive property to simplify some polynomials by combining like terms.

EXAMPLE 3.1.6   Simplify each polynomial by combining like terms.

1. $14a - 5b + 2c - 12a + 4b + 15c$

To simplify, first write the polynomial with the like terms side by side.

$14a - 5b + 2c - 12a + 4b + 15c = 14a - 12a - 5b + 4b + 2c + 15c$

$\qquad\qquad\qquad\qquad\qquad\qquad\quad = 2a - b + 17c$

2. $5x^2 - 3x - 7x^2 + 3x + 4 + 10 = -2x^2 + 0x + 14$

$\qquad\qquad\qquad\qquad\qquad\qquad\quad = -2x^2 + 14 \qquad \textit{Since 0x = 0}$

3. $7rs^3 + 2x^2y - 3z + 10x^2y - 11rs^3 + 9 = -4rs^3 + 12x^2y - 3z + 9$

The $-3z$ and 9 are not like terms and cannot be combined.

We are now in a position to add and subtract polynomials or multiples of polynomials by using the distributive property.

EXAMPLE 3.1.7   Perform the indicated operations and simplify your answers by combining the like terms.

1. $(4x^2 + 2x - 11) + (x^2 - 7x - 4) = 4x^2 + x^2 + 2x - 7x - 11 - 4$

$\qquad\qquad\qquad\qquad\qquad\qquad\qquad = 5x^2 - 5x - 15$

NOTE ▶   In the polynomial $5x^2 - 5x - 15$, no like terms exist because no two terms contain the same number of factors of x.

2. $\left(\dfrac{1}{2}x + 2y\right) + 7\left(x - \dfrac{3}{4}y\right) = \dfrac{1}{2}x + 2y + 7x - \dfrac{21}{4}y$ $\qquad$ *Multiply first due to order of operations.*

$\qquad\qquad\qquad\qquad\qquad = \dfrac{15}{2}x - \dfrac{13}{4}y$

3. $(-3a - 4b + 2c) - (2a - 9b + c)$

This expression is the same as $(-3a - 4b + 2c) + (-1) \cdot (2a - 9b + c)$. *Now we simply distribute a coefficient of* -1 *across the second polynomial:*

$$(-3a - 4b + 2c) + (-1) \cdot (2a - 9b + c)$$
$$= -3a - 4b + 2c - 2a + 9b - c$$
$$= -3a - 2a - 4b + 9b + 2c - c$$
$$= -5a + 5b + c$$

NOTE ▶   When you subtract a polynomial, change all the signs of the terms within the parentheses in the second polynomial.

4. $5(3x^2 - 2x - 7) - 2(x^2 + 3x + 12)$
$$= 15x^2 - 10x - 35 - 2x^2 - 6x - 24$$
$$= 15x^2 - 2x^2 - 10x - 6x - 35 - 24$$
$$= 13x^2 - 16x - 59$$

NOTE ▶   We can think of the second expression either as distributing -2 across the polynomial or as distributing $+2$ and then changing the signs because of the subtraction. Either yields the same results.

Simple addition and subtraction problems can follow a vertical format similar to arithmetic. Consider the following example.

EXAMPLE 3.1.8   Perform the indicated operations.

1. $3x^2 - 7x + 4$
$+ \underline{(2x^2 + 8x - 27)}$ *Be sure the like terms are in the same column.*
     $5x^2 + x - 23$

2. $5y^2 - 9xy + 13x^2$
$- \underline{(8y^2 + xy - 2x^2)}$ *An easy way to do this subtraction is to change the signs of all the terms of the bottom polynomial and add.*

     $5y^2 - 9xy + 13x^2$
$+ \underline{(-8y^2 - xy + 2x^2)}$
     $-3y^2 - 10xy + 15x^2$

In Chapter 1 we worked with some expressions that are polynomials. For example, if r represents the radius of a circle, the polynomial πr^2 is used to calculate that circle's area. As we will see in the next example, polynomials can be used in many real-world settings.

EXAMPLE 3.1.9   Speedy Turner has determined that the following polynomial expresses his average monthly cost in dollars to operate his new car:

$$C = 0.08x + 50$$

where x is the number of miles he drives each month and C is the cost. Determine Speedy's costs if he drives 1000 mi in 1 mo; if he drives 1600 mi in 1 mo.

SOLUTION   If Speedy drives 1000 mi, then

$$C = 0.08(1000) + 50$$
$$= 80 + 50$$
$$= \$130$$

If Speedy drives 1600 mi, then

$$C = 0.08(1600) + 50$$
$$= 128 + 50$$
$$= \$178$$

Exercises 3.1

List the terms and coefficients of each of the following polynomials, and specify which are monomials, binomials, and trinomials. State the degree of each polynomial.

1. $9x^2 + 4x - 7$

2. $12x^3y - y^3 + 3x - \dfrac{1}{4}$

3. $23abc$

4. $100x^2 - y^2$

5. $27x^3 + 8y^3$

6. $-8 + m + 3k^2 - 15j^4$

7. 19

8. $\sqrt{5}x + 2r^4$

9. $\dfrac{1}{2}x^3yz^4$

10. $-r^2 + 2\pi r$

11. $\dfrac{2}{3}x + \dfrac{1}{8}y - 0.9z$

12. $0.2 - 7x + 1.7x^2$

13. $\dfrac{3x}{2} - \dfrac{8y^2}{5}$

14. $\dfrac{2a}{7} + \dfrac{3b}{5} - \dfrac{4c}{9} + \dfrac{5d}{8}$

15. $12x^5 - 9x^4 + 2x^2 + x + 61$

Write the following polynomials in standard form and determine the leading coefficient in each case.

16. $9x^2 + 4x - 7$

17. $81 - 100z^2$

18. $2 - 3x^2 + x^3 - 5x$

19. $x - 4x^4 + 17x^2 + 12$

20. $212b^{45} - 107b^3 + 19{,}017b^{215}$

21. 7

22. $\pi r^2 - 2\pi r$

23. $2y^3 + y^2 - 8y - 4$

24. $1.7x - 0.2x^2 + 0.1x^3$

25. $300 - 32t - 16t^2$

26. $\dfrac{1}{2} - \dfrac{2}{3}x + \dfrac{5}{4}x^2$

27. $\dfrac{2x^3}{7} - \dfrac{8x^5}{3}$

Use the distributive property to expand the following polynomials.

28. $3(12x - 7y)$

29. $-2(9a - 2b - 10)$

30. $\dfrac{1}{2}(-8x^2 - 4x + 16)$

31. $-0.4(7.1x - 9.2y - z)$

32. $0.8(-x^2 + 4.1y^2)$

33. $\dfrac{1}{5}(25x^3 + 47y^5 - z^2)$

34. $\left(-\dfrac{2}{3}\right)(4 + 3x - 2x^2)$

35. $-\left(19q + 15r - 101s - t^2\right)$

36. $1\left[\left(-\dfrac{4}{9}\right)m - 12hk + 5.7\right]$

37. $0\left(5r^2 + \dfrac{3}{8}s - 2t\right)$

Simplify each polynomial by combining like terms.

38. $5x - 23x + 9y + y$

39. $-2j + 11k - 14j$

40. $-8x + 3yz - 12zy + 19x$

41. $-12k + 13m + 42n$

42. $2.3r + 7s - r + 0.5s$

43. $\dfrac{1}{3}y + \dfrac{1}{4}z + \dfrac{2}{9}y - 3z$

44. $5x^2 + 7x + 9x^2 - 7x - 27$

45. $x^3 - 14x + 3 + 2x^2 - 15$

46. $-0.007x^2 + 2.1x - 4.09x^2 + 10.28x + 37.91$

47. $\dfrac{1}{5}x^2y - \dfrac{2}{3}yx^2 + \dfrac{1}{4}x$

Perform the indicated operations and simplify your answers by combining like terms.

48. $(3x^2 - x + 7) + (x^2 + 3x - 2)$

49. $(5x^3 - x + 2) + (2x^3 + 4x^2 - 2x + 9)$

50. $(2a^2 - 4ab - b^2) + (-3a^2 - ab + 4b^2)$

51. $(5x^2 - 2xy - 3y + y^2) + (2x - 5yx + y - 7y^2)$

52. $(3x + 7) - (x^2 - 5x + 8)$

53. $(x^3 + x^2 - 2) - (4x^3 - 2x^2 + 3x - 1)$

54. $(m^2 - 6mn - n - 12n^2) - (5n^2 + m - m^2)$

55. $(-4a^2 - 4a + b) - (2a + 3b - b^2 + 9a^2)$

56. $(2x^2 - 4x + 10) - (7x + 2) + (x^2 + x + 3)$

57. $(x^2 - 5x + 6) - (3x^2 + x - 8) + (4x^2 - 13)$

58. $\quad 2x^2 - 7x + \ 4$
$+ \ \underline{(3x^2 + 2x - 11)}$

59. $\quad\quad 5x^2 + 8x + 19$
$+ \ \underline{(-7x^2 - 9x + \ \ 1)}$

60. $\quad 8x^2 + 1$
$+ \ \underline{(2x^2 - 7)}$

61. $\quad\quad 8a - 3b + 4$
$+ \ \underline{(-9a + 5b - 4)}$

62. $\quad 5a^2 - 7a + 12$
$- \ \underline{(-a^2 + 2a - \ 9)}$

63. $\quad -3x^2 + 2x + 11$
$- \ \underline{(- x^2 + 2x + \ 9)}$

64. $\quad 5x^2y - 3y + \ 9$
$- \ \underline{(11x^2y + 2y - 16)}$

65. $\quad\quad 7ab + 8cd - 6$
$- \ \underline{(7ab - 8cd + 2)}$

66. $3(2x - 7) + 10(x + 2)$

67. $(7x + 3y - z) - (2x - 9y + 14z)$

68. $3(x^2 - x + 4) - 5(-2x^2 - 4x + 9)$

69. $-4(5x^2 + 3x - 1) + 2(5x^2 + 3x - 1)$

70. $\left(-\dfrac{1}{2}\right)\left(3p + \dfrac{9}{5}q\right) - \left(\dfrac{2}{3}\right)\left(\dfrac{1}{4}p - q\right)$

71. $3(x^2y - 4x + 3) - 2(y^2x + 3z - 7)$

72. $-5(2a + 3b - c) + 3(a + 5b - 12)$

73. $0.02(7x^2 - 0.01y^2) - 0.5(10x^2 + 2.1y^2)$

74. $-4(3a - 2b) + 6(-5c - 12d)$

75. $2(3m + 7n) - (9j - 10k)$

76. $\left(-\dfrac{3}{2}\right)(4x + y/6 - 3z) + 2\left(3x - \dfrac{5}{4}y + \dfrac{2x}{5}\right)$

77. $\dfrac{5}{6}\left(2a + \dfrac{3}{10}b\right) - \dfrac{1}{9}\left(5a - \dfrac{3}{2}b\right)$

Use polynomials to solve the following problems.

78. Dr. Susan Gillespie pays her monthly gas, water, and electric bills to the Moss County Public Utility Company. Her total bill in dollars is expressed by the polynomial

$$0.25x + 0.02y + 0.08z$$

where x is the number of cubic feet of gas she uses, y is the number of gallons of water, and z is the number of kilowatt-hours. Determine Dr. Gillespie's bill if she uses 120 cu ft of gas, 1000 gal of water, and 1500 kWh of power. What is her bill if she uses 100 cu feet of gas, 800 gal of water, and 1800 kWh of power?

79. A rectangular piece of cardboard measures y inches by z inches (Figure 3.1.1). Square pieces that measure x inches by x inches are cut out of each corner (Figure 3.1.2).
  a. Find a polynomial that represents the area of the resulting figure.
  b. If the sides are folded up (Figure 3.1.3), find an expression representing the volume of the resulting box.

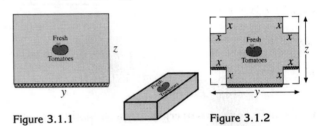

Figure 3.1.1   Figure 3.1.2

Figure 3.1.3

80. The Pumpkin Patch Doll Company has determined that the polynomial

$$x^2 + 400x - 10,000$$

yields the company's monthly income in dollars, where x is the number of dolls sold. The company's monthly cost in dollars is expressed by the polynomial

$$4x^2 - 320x + 7800$$

Find the income and cost when 100 dolls are produced. What is the profit or loss when 100 dolls are produced? Find a polynomial that represents the company's monthly profit. (*Hint:* Profit = income − cost.)

81. Bubba sells lemonade and chocolate chip cookies in front of his home. The polynomial

$$5x + 7y + 120$$

expresses his total cost, measured in cents, of selling x cups of lemonade and y cookies. His income measured in cents is represented by the polynomial

$$3x + 2y$$

Find Bubba's cost and income when he sells 20 cups of lemonade and 50 cookies. What is Bubba's profit or loss? Find a polynomial that represents Bubba's profit or loss.

3.2 *Multiplication of Polynomials*

Recall from Chapter 1 that the expression x^2 is simply a mathematical shorthand for $x \cdot x$; in a similar fashion z^3 stands for $z \cdot z \cdot z$. In general, the exponential expression x^n is defined to be

$$x^n = \underbrace{x \cdot x \cdot x \cdot \cdot \cdot x}_{n \text{ factors}}$$

Let's now see how exponential notation can be used to simplify some algebraic expressions. Suppose you were asked to simplify the expression $x^2 \cdot x^5$. We already know that $x^2 = x \cdot x$ and $x^5 = x \cdot x \cdot x \cdot x \cdot x$, so

$$x^2 \cdot x^5 = (x \cdot x) \cdot (x \cdot x \cdot x \cdot x \cdot x)$$
$$= x^7$$

Therefore, $x^2 \cdot x^5 = x^{2+5} = x^7$. This example leads to the following law.

*P*roduct Law of Exponents

Let m and n be positive integers; then

$$x^m \cdot x^n = x^{m+n}$$

NOTE ▶   By the product law,

1. $x^2 \cdot x^5 = x^{2+5} = x^7$;

2. $x^{71} \cdot x^{42} = x^{71+42} = x^{113}$;

3. $x^5 \cdot y^2$ cannot be simplified.

To use the product law of exponents *the bases must be exactly the same* and that is not the case in part 3.

Suppose we want to simplify the algebraic expression $(x^5)^3$. We know that

$$(x^5)^3 = x^5 \cdot x^5 \cdot x^5$$
$$= x^{5+5+5}$$
$$= x^{15}$$

Therefore, $(x^5)^3 = x^{5 \cdot 3} = x^{15}$. This example suggests the following law of exponents.

P ower to a Power Law of Exponents

Let m and n be positive integers; then

$$(x^m)^n = x^{m \cdot n}$$

NOTE ▶ By the power to a power law,

1. $(x^6)^4 = x^{6 \cdot 4} = x^{24}$

2. $(x^3)^{11} = x^{3 \cdot 11} = x^{33}$

Suppose we want to simplify the algebraic expression $(xy)^4$. We know that

$$(xy)^4 = (xy) \cdot (xy) \cdot (xy) \cdot (xy)$$
$$= x \cdot x \cdot x \cdot x \cdot y \cdot y \cdot y \cdot y \qquad \text{\textit{By the commutative and}}$$
$$= x^4 y^4 \qquad\qquad\qquad\qquad \text{\textit{associative properties of multiplication}}$$

Therefore, $(xy)^4 = x^4 y^4$. This example suggests the following law of exponents.

P roduct to a Power Law of Exponents

Let n be a positive integer; then

$$(xy)^n = x^n y^n$$

NOTE ▶ By the product to a power law,

1. $(xy)^8 = x^8 y^8$

2. $(x^2 y^3)^7 = (x^2)^7 (y^3)^7 \qquad \text{\textit{By the product to a power law}}$
   $$= x^{14} y^{21} \qquad\qquad \text{\textit{By the power to a power law}}$$

Let's now turn our attention back to polynomials. In Section 3.1 addition and subtraction of polynomials were covered. Now that we have the product laws of exponents, we can handle the multiplication of polynomials. Consider the following example.

EXAMPLE 3.2.1 Perform the indicated multiplications and simplify your answers by using the product laws of exponents.

1. $12x^5 \cdot 4x^9 = 12 \cdot 4 \cdot x^5 \cdot x^9$
   $$= 48x^{5+9}$$
   $$= 48x^{14}$$

2. $5x^3y^{10} \cdot 7xy^6 = 5 \cdot 7 \cdot x^3 \cdot x \cdot y^{10} \cdot y^6$

$\qquad\qquad\qquad = 35 \cdot x^{3+1} \cdot y^{10+6}$

$\qquad\qquad\qquad = 35x^4y^{16}$

3. $9a^3b^2 \cdot (-2ac^4) \cdot 6c^4d^5 = 9 \cdot (-2) \cdot 6 \cdot a^3 \cdot a \cdot b^2 \cdot c^4 \cdot c^4 \cdot d^5$

$\qquad\qquad\qquad\qquad = -108a^{3+1} \cdot b^2 \cdot c^{4+4} \cdot d^5$

$\qquad\qquad\qquad\qquad = -108a^4b^2c^8d^5$

4. $(5x^3y^4)^2 = 5^2(x^3)^2 \cdot (y^4)^2$

$\qquad\qquad\quad = 25x^6y^8$

5. $(2xy^3)^3(7x^2y^5) = 2^3x^3(y^3)^3 \cdot 7x^2y^5$

$\qquad\qquad\qquad = 8x^3y^9 \cdot 7x^2y^5$

$\qquad\qquad\qquad = 8 \cdot 7 \cdot x^3x^2y^9y^5$

$\qquad\qquad\qquad = 56x^5y^{14}$

Try to avoid these mistakes:

Incorrect	Correct
1. $3x^5 \cdot 8x^4 = 24x^{20}$	$3x^5 \cdot 8x^4 = 24x^9$
	Add the exponents.
2. $5x + 7x = 12x^2$	$5x + 7x = 12x$
	In the previous section we learned that when we *add* or *subtract* like terms, only the coefficient changes, not the variable factors.

In Example 3.2.1 only multiplication of monomials was considered. Suppose we want to multiply a polynomial by some monomial; what do we do?

EXAMPLE 3.2.2   Perform the indicated multiplications and simplify your answers.

1. $2x^3(5x + 7y)$

To do this multiplication we will use both the product law of exponents and the distributive property:

$$2x^3(5x + 7y) = 2x^3 \cdot 5x + 2x^3 \cdot 7y$$

$$= 2 \cdot 5 \cdot x^3 \cdot x + 2 \cdot 7 \cdot x^3 \cdot y$$

$$= 10x^4 + 14x^3y$$

2. $3x^2y^4(4x^5 - 7xy^2 + 9) = 3x^2y^4 \cdot 4x^5 - 3x^2y^4 \cdot 7xy^2 + 3x^2y^4 \cdot 9$

$= 3 \cdot 4 \cdot x^2 \cdot x^5 \cdot y^4 - 3 \cdot 7 \cdot x^2 \cdot x \cdot y^4 \cdot y^2$

$\qquad + 3 \cdot 9 \cdot x^2 \cdot y^4$

$= 12x^7y^4 - 21x^3y^6 + 27x^2y^4$

We are now in a position to consider multiplication problems in which none of the factors is a monomial.

E X A M P L E 3 . 2 . 3   Perform the indicated multiplications and simplify your answers.

1. $(3x + 2)(5x + 4) = 3x(5x + 4) + 2(5x + 4)$ *By the distributive property*

$= 3x \cdot 5x + 3x \cdot 4 + 2 \cdot 5x + 2 \cdot 4$

$= 15x^2 + 12x + 10x + 8$

$= 15x^2 + 22x + 8$

This is a very cumbersome and tedious way to work this problem. Fortunately there is a simpler method of multiplying, called the **FOIL Method:**

$$\qquad\qquad\qquad \text{F} \qquad \text{O} \qquad \text{I} \qquad \text{L}$$

$$(3x + 2)(5x + 4) = 3x \cdot 5x + 3x \cdot 4 + 2 \cdot 5x + 2 \cdot 4$$

$$= 15x^2 + 12x + 10x + 8$$

$$= 15x^2 + 22x + 8$$

This technique is called the FOIL Method because we multiply the *first* terms together, then the *outer* terms together, next the *inner* terms, and finally the *last* terms.

$$\qquad\qquad\qquad \text{F} \qquad \text{O} \qquad \text{I} \qquad \text{L}$$

2. $(2x + 7)(x - 10) = 2x^2 - 20x + 7x - 70$

$\qquad\qquad\qquad = 2x^2 - 13x - 70$

$$\qquad\qquad\qquad \text{F} \qquad \text{O} \qquad \text{I} \qquad \text{L}$$

3. $(5x - 4y)(6x - 5y) = 30x^2 - 25xy - 24xy + 20y^2$

$\qquad\qquad\qquad = 30x^2 - 49xy + 20y^2$

$$\qquad\qquad\qquad \text{F} \qquad \text{O} \qquad \text{I} \qquad \text{L}$$

4. $(4x + 7y)(4x - 7y) = 16x^2 - 28xy + 28xy - 49y^2$

$\qquad\qquad\qquad = 16x^2 - 49y^2$

This last example leads to a very important formula. Whenever the two binomials have the same first terms and the same last terms but opposite signs in the middle, the outer and inner terms combine to zero, leaving the difference of the squares of the first term and the last term. In general, we have the following formula.

Product Yielding the Difference of Two Squares

$$(a + b)(a - b) = a^2 - b^2$$

5. $(2x - 5)(2x + 5) = (2x)^2 - 5^2 = 4x^2 - 25$

6. $(9y + x^2)(9y - x^2) = (9y)^2 - (x^2)^2 = 81y^2 - x^4$

7. $(x + 3)^2 = (x + 3)(x + 3) = x^2 + 3x + 3x + 9$
$$= x^2 + 6x + 9$$

Part 7 leads us to another important formula, the formula for the **square of a binomial.** Whenever a binomial is squared, the product is a trinomial whose three terms are the square of the first term of the binomial, twice the product of the first and last terms, and the square of the last term. In general, we have the following formula.

Square of a Binomial

$$(a + b)^2 = a^2 + 2ab + b^2$$

8. $(3x + y)^2 = (3x)^2 + 2(3x)(y) + y^2$
$$= 9x^2 + 6xy + y^2$$

9. $(5x - 2)^2 = (5x)^2 + 2(5x)(-2) + (-2)^2$
$$= 25x^2 - 20x + 4$$

Try to avoid this mistake:

Incorrect	Correct
$(x + y)^2 = x^2 + y^2$	$(x + y)^2$
The square of a sum is not the sum of the squares. This error incorrectly assumes that an exponent distributes over a sum.	$= x^2 + 2xy + y^2$
To use the product to a power law of exponents, the operation within the parentheses must be multiplication.	$(xy)^2 = x^2y^2$

As the next example, consider the case when one of the factors is a trinomial.

E XAMPLE 3.2.4   Perform the indicated multiplications.

1. $(3x + 5)(2x^2 + x - 7)$

There are two possible ways to solve this problem. The first method is merely an extension of the FOIL Method. All that is done is to distribute the $3x$ across $2x^2 + x - 7$ and then distribute the 5. This is done in the following manner:

$$(3x + 5)(2x^2 + x - 7) = 6x^3 + 3x^2 - 21x + 10x^2 + 5x - 35$$
$$= 6x^3 + 3x^2 + 10x^2 - 21x + 5x - 35$$
$$= 6x^3 + 13x^2 - 16x - 35$$

Another technique is to use a vertical format similar to arithmetic. In this example we obtain

$$
\begin{array}{r}
2x^2 + x - 7 \\
3x + 5 \\
\hline
10x^2 + 5x - 35 \\
6x^3 + 3x^2 - 21x \\
\hline
6x^3 + 13x^2 - 16x - 35
\end{array}
$$

The idea is to first multiply through by 5 and then multiply through by $3x$, placing like terms underneath one another. When this is done, we add the like terms together and obtain precisely the same result.

2. $(x - y)(x^2 + xy + y^2)$

Let's use the vertical format:

$$
\begin{array}{r}
x^2 + xy + y^2 \\
x - y \\
\hline
-x^2y - xy^2 - y^3 \\
x^3 + x^2y + xy^2 \\
\hline
x^3 + 0x^2y + 0xy^2 - y^3 = x^3 - y^3
\end{array}
$$

XAMPLE 3.2.5   Perform the indicated multiplication.

$$(x + 3)(x - 2)(3x^2 - 5x + 7)$$

Multiply the first two factors using the FOIL Method.

$$= (x^2 - 2x + 3x - 6)(3x^2 - 5x + 7)$$
$$= (x^2 + x - 6)(3x^2 - 5x + 7)$$

Convert to the vertical format.

$$
\begin{array}{r}
3x^2 - 5x + 7 \\
x^2 + x - 6 \\
\hline
-18x^2 + 30x - 42 \\
3x^3 - 5x^2 + 7x \\
3x^4 - 5x^3 + 7x^2 \\
\hline
3x^4 - 2x^3 - 16x^2 + 37x - 42
\end{array}
$$

As an aid, all the laws of exponents and the multiplication formulas of this section are given here.

Let m and n be positive integers; then

$$x^m \cdot x^n = x^{m+n}$$ Product law

$$(x^m)^n = x^{m \cdot n}$$ Power to a power law

$$(xy)^n = x^n y^n$$ Product to a power law

$$(a + b)(a - b) = a^2 - b^2$$ Product yielding the difference of two squares

$$(a + b)^2 = a^2 + 2ab + b^2$$ Square of a binomial

E xercises 3.2

Perform the indicated multiplications.

1. $x^9 \cdot x^4$

2. $x^8 \cdot x^3$

3. $y^{17} \cdot y$

4. $y^{23} y^{24}$

5. $(x^2)^5$

6. $(x^{13})^2$

7. $(y^8)^7$

8. $(y^6)^6$

9. $(xy)^9$

10. $(ab)^{101}$

11. $(x^2 y^9)^4$

12. $(a^5 b^9)^5$

13. $5x^3 y \cdot 8x^4 y^9$

14. $3x^{10} y^{12} \cdot (-12x^2 yz^3)$

15. $-3x^5 y^2 \cdot 14xy^2$

16. $-5a^4 b^4 \cdot (-6a^5 b^5)$

17. $2x^2 y^5 \cdot 5xy^{17} \cdot 9x^6 y^8$

18. $3x^4 y^8 \cdot x^2 y^{23} \cdot 17xy$

19. $(3xy^3)^2$

20. $(4x^2y)^3$

21. $(-2x^3y^{12})^4$

22. $(-3x^6y^{15})^4$

23. $(2xy^2)^3(3x^2y^3)^3$

24. $(4xy^3)(8x^3y^4)^2$

25. $(5x^2y^3)^2(-2x^4y^2)^3$

26. $(-2x^3y)^2(-2x^{14}y^5)^3$

27. $6a^2b(2a - 10)$

28. $14x^2yz^4(-4xy^3 + 9y^4z^9)$

29. $10x^2(5x^2 - 3x + 4)$

30. $7x^2(2x^2 - 5x + 1)$

31. $-7m^3n^7(8m - 9n^2 + 15mn^3)$

32. $-3a^2b^3(9ab + 11b^2 - 4b^3)$

33. $(8x + 3)(9x + 4)$

34. $(2x + 9)(x - 7)$

35. $(5x - 11)(3x - 1)$

36. $(3x + 9)(2x - 7)$

37. $(5x + 2y)(x - 3y)$

38. $(11m - 4n)(2m - n)$

39. $(6m - 3n)(5m + 6n)$

40. $(y + 3)(y - 3)$

41. $(4a + 1)(4a - 1)$

42. $(x + 7)(x - 7)$

43. $(2x + 9)(2x - 9)$

44. $(3r + 2s)(3r - 2s)$

45. $(x^2 + y^2)(x^2 - y^2)$

46. $\left(\dfrac{x}{2} + \dfrac{2}{3}\right)\left(\dfrac{x}{2} - \dfrac{2}{3}\right)$

47. $\left(\dfrac{x}{2} - 2y\right)\left(\dfrac{x}{2} + 2y\right)$

48. $(a^2 + 5b^3)(a^2 - 5b^3)$

49. $(x + 4)^2$

50. $(y - 3)^2$

51. $(x - 7)^2$

52. $(x - 2y)^2$

53. $(3x + 4y)^2$

54. $(5a + 9b)^2$

55. $\left(\dfrac{1}{2}a - 3b\right)^2$

56. $\left(x + \dfrac{3}{4}\right)^2$

57. $(x^2 + 1)^2$

58. $(m^3 - 2n)^2$

59. $(x - 2)(x^2 + 2x + 4)$

60. $(x + 3y)(x^2 - 3xy + 9y^2)$

61. $(2x + 1)(x^2 - 3x + 2)$

62. $(5x - 3)(4x^2 - 7x + 9)$

63. $(9x - 7)(5x^2 + x - 6)$

64. $(x + 2y)(3x^2 - 4xy + 7y^2)$

65. $(3x - 5y)(x^2 + 2xy + y^2)$

66. $(x - 5)(3x^3 - 7x^2 + 6x + 2)$

67. $(x^2 + 2x - 1)(5x^2 - 4x + 3)$

68. $(3x^2 - 7x + 4)(11x^2 + 5x + 1)$

69. $(2x^2 - 4x - 9)(20x^2 + 7)$

70. $3x^2(2x - 1)(x^2 + 3x - 5)$

71. $2x^2(3x - 2)(x^2 - 5x - 4)$

72. $(x + 1)(x - 1)(2x + 3)$

73. $(2y + 5)(y - 3)(2y + 1)$

74. $(y^2 - 4)(2y - 5)(y - 2)$

75. $(x + 3)(x - 2)(2x^2 + x - 5)$

76. $(x - 3)(x + 4)(3x^2 + x + 2)$

77. $x^{2n+1} \cdot x^{n+4}$

78. $x^{3n-1} \cdot x^{2n+5}$

79. $(x^{2n})^{n+5}$

80. $(x^{3n})^{n-2}$

81. $(x^n + 2)(x^n - 4)$

82. $(x^n - 5)(x^n - 3)$

83. $(x^n + 3)^2$

84. $(x^n + 6)^2$

Write Algebra

85. Explain the formula for squaring $a + b$ in your own words.

86. Explain why, in general, $(a + b)^2 \neq a^2 + b^2$.

87. Explain the following laws of exponents in your own words: (a) Product law, (b) Power to a power law, and (c) Product to a power law.

3.3 Factoring Polynomials—Greatest Common Factor and Grouping

In Chapter 1 factors of integers were discussed. In this section we will learn how to find factors of polynomials. For the remainder of this chapter, we will consider only polynomials and factors of polynomials that have integer coefficients. Consider the following example.

XAMPLE 3.3.1   Find all the monomial factors of $5x^2 + 10x$.

$$5x^2 + 10x = 5 \cdot x \cdot x + 2 \cdot 5 \cdot x$$

We can observe that the two terms have common factors of 5 and x. Using the distributive property, we could rewrite this binomial as

$$5x^2 + 10x = 5(x^2 + 2x) \quad \text{or} \quad x(5x + 10) \quad \text{or} \quad 5x(x + 2)$$

Example 3.3.1 shows us an important way to determine the factors of polynomials. This method is to find the **greatest common factor.** In the example, $5x$ is called the greatest common factor because it is the "largest" monomial that is a factor of both $5x^2$ and $10x$. We find the greatest common factor as follows.

Factoring Out the Greatest Common Factor

1. Find the largest number that will divide into the coefficients of all the terms of the given polynomial. (Remember we are looking only for integral factors.)

2. Find the variable(s) that occur in *all* the terms and raise each to the *smallest* power to which it has been raised in any term.

3. Form the greatest common factor by multiplying the number found in part 1 and the variable factors determined in part 2.

XAMPLE 3.3.2   Write each of the following polynomials as a product with the greatest common factor as one of the factors.

1. $12x^3 - 18x^4$

The largest number that divides into the coefficients is 6. The variable x occurs in both terms, and its smallest power is 3. Thus the greatest common factor is $6x^3$:

$$12x^3 - 18x^4 = 6x^3 \cdot 2 - 6x^3 \cdot 3x$$
$$= 6x^3(2 - 3x)$$

2. $2b + 4bc - 10bd$

The greatest common factor is $2b$.

$$2b + 4bc - 10bd = 2b(1 + 2c - 5d)$$

Try to avoid this mistake:

Incorrect	Correct
$2b + 4bc - 10bd = 2b(2c - 5d)$	$2b + 4bc - 10bd$
	$\quad = 2b \cdot 1 + 2b \cdot 2c - 2b \cdot 5d$
	$\quad = 2b(1 + 2c - 5d)$
The understood 1 was omitted. You can always check your answer by multiplying:	Remember that $2b = 2b \cdot 1$.
	Check: $2b(1 + 2c - 5d)$
$2b(2c - 5d) = 4bc - 10bd$	$\quad = 2b + 4bc - 10bd$
$\qquad \neq 2b + 4bc - 10bd$	

3. $15m^4n^4 - 6m^2n^2 + 21m^5n$

The greatest common factor is $3m^2n$.

$$15m^4n^4 - 6m^2n^2 + 21m^5n = 3m^2n(5m^2n^3 - 2n + 7m^3)$$

NOTE ▶   Writing a polynomial as a product of factors is called **factoring** the polynomial.
The polynomials in Example 3.3.2 all had a common monomial factor. Suppose we have a common binomial factor. What is the procedure in that case?

EXAMPLE 3.3.3   Factor the following polynomials.

1. $x(x + 1) + 5(x + 1)$

Do not perform the indicated multiplication. Here we have two terms. The first term is $x(x + 1)$ and the second term is $5(x + 1)$. There is a common factor of $x + 1$ in each term. We can apply the distributive property to factor out $x + 1$:

$$x(x + 1) + 5(x + 1)$$
$$= (x + 1)(x \underline{\quad\quad} + 5 \underline{\quad\quad})$$
$$= (x + 1)(x + 5)$$

2. $3a(2a - 5b) - 7b(2a - 5b)$

Again we have a common binomial factor; this time it is $2a - 5b$. Applying the distributive property, we obtain

$$3a\underline{(2a - 5b)} - 7b\underline{(2a - 5b)} = (2a - 5b)(3a - 7b)$$

3. $2x(x + 3y) + (x + 3y)$

Since $(x + 3y) = 1(x + 3y)$,

$$2x\underline{(x + 3y)} + 1\underline{(x + 3y)} = (x + 3y)(2x + 1)$$

4. $5x^2(3x + 2) - 25x(3x + 2)$

There is a common monomial factor of $5x$ and a common binomial factor of $(3x + 2)$. Therefore the greatest common factor is $5x(3x + 2)$.

$$5x^2(3x + 2) - 25x(3x + 2) = x \cdot \underline{5x(3x + 2)} - 5 \cdot \underline{5x(3x + 2)}$$
$$= 5x(3x + 2)(x - 5)$$

Example 3.3.3 leads to another factoring technique called **factoring by grouping.** Here is how it works. Suppose you were asked to factor the following polynomial:

$$ax + bx + ay + by$$

Factoring by Grouping

1. Group the first two terms together and remove their common factor. In addition, group the last two terms together and remove their common factor.

2. Determine whether a common binomial factor has been generated. If it has, remove that common factor, as in Example 3.3.3.

3. If a common binomial factor has not been generated, interchange the second and third terms (reorder the polynomial) and repeat the process.

Now, returning to our problem:

$$ax + bx + ay + by = (ax + bx) + (ay + by) \qquad \textit{Step 1}$$
$$= x(a + b) + y(a + b) \qquad \textit{Step 1}$$
$$= (a + b)(x + y) \qquad \textit{Step 2}$$

NOTE ▶ Factoring by grouping is usually applied when the polynomial has four or more terms.

XAMPLE 3.3.4   Factor the following polynomials.

1. $2ac - 6ad + bc - 3bd = (2ac - 6ad) + (bc - 3bd)$
$$= 2a(c - 3d) + b(c - 3d)$$
$$= (c - 3d)(2a + b)$$

2. $m^2 - 4n - 4mn + m = (m^2 - 4n) + (-4mn + m)$
$$= 1(m^2 - 4n) + m(-4n + 1)$$

We have different binomial factors, so we must reorder the polynomial as indicated in step 3:

$$m^2 - 4n - 4mn + m = m^2 - 4mn - 4n + m$$
$$= (m^2 - 4mn) + (-4n + m)$$
$$= m(m - 4n) + 1(-4n + m)$$

(The binomials $m - 4n$ and $-4n + m$ are the same. Why?)

$$= (m - 4n)(m + 1)$$

3. $2x^3 + 3x^2 - 14x - 21 = (2x^3 + 3x^2) + (-14x - 21)$
$$= x^2(2x + 3) + 7(-2x - 3)$$

The binomials $2x + 3$ and $-2x - 3$ are opposites. Since we need them to be the same, we should factor out -7 instead of $+7$:

$$2x^3 + 3x^2 - 14x - 21 = (2x^3 + 3x^2) + (-14x - 21)$$
$$= x^2(2x + 3) - 7(2x + 3)$$
$$= (2x + 3)(x^2 - 7)$$

4. $18x^2y - 3xy^2 - 18xy + 3y^2$

First remove the command factor of $3y$:

$$18x^2y - 3xy^2 - 18xy + 3y^2 = 3y(6x^2 - xy - 6x + y)$$
$$= 3y[(6x^2 - xy) + (-6x + y)]$$
$$= 3y[x(6x - y) + 1(-6x + y)]$$

Again $6x - y$ and $-6x + y$ are opposites, so we should factor out -1 instead of $+1$:

$$= 3y[x(6x - y) - 1(6x - y)]$$
$$= 3y[(6x - y)(x - 1)]$$
$$= 3y(6x - y)(x - 1)$$

Exercises 3.3

Factor the following polynomials.

1. $10x^2 - 35$

2. $-16a - 24b + 8c$

3. $5x + 11y - 10z$

4. $8x^2 + 18xy + 2x$

5. $8x^3 + 32x^2y - 20x^2z$

6. $4x^3y^2 - 7x^2y^2 + 2xy^4$

7. $3x^3y^3 - 15x^5y^3 + 9x^4y^5$

8. $10ab - 3bc + 7cd$

9. $-6x^5 - 18x^3 - 10x^2$

10. $-7x^4y - 35x^2y^2 - 14x^2y^5$

11. $x(3x - 5) - 7(3x - 5)$

12. $5y(6y - 1) + 4(6y - 1)$

13. $2x(x - 3) + (x - 3)$

14. $z(8y + 5) + (8y + 5)$

15. $h(2i - 1) + 3k(2i - 1)$

16. $7y(3x + 9) - 14z(3x + 9)$

17. $4x(x - 2y) - 6y(x - 2y)$

18. $11x^2(5k - 7) + 33x(5k - 7)$

19. $15x^3(y + 8) + 5x^2(y + 8)$

20. $12a^3(b - 6) - 4a(b - 6)$

21. $8y^2z(1 - 5x) - 6yz^3(1 - 5x)$

22. $18n^3p^2(3t + 4) - 27np(3t + 4)$

23. $3x^2(x + y)^2 + 12x(x + y)$

24. $39x(7y + 5) + 13x^3(7y + 5)^2$

25. $2ax + 3bx + 2ay + 3by$

26. $10hn - 3km + 2hm - 15kn$

27. $12ac - 4ad - 3bc + bd$

28. $ax - 12by - 3ay + 4bx$

29. $6cf - 2cg + 6df - 2dg$

30. $3x^2 + xy + 9x + 3y$

31. $2x^2 - 4xy - x + 2y$

32. $4mn - 15n + 20m - 3n^2$

33. $4n + 15mn - 3n^2 - 20m$

34. $9x^2 - 3xy + 45x - 15y$

35. $6x^2 - 105y + 15xy - 42x$

36. $8x^2y - 16xy^2 + 12xy - 24y^2$

37. $3x^2y - 5x^2y^2 - 12xy^2 + 20xy^3$

38. $-42x^2z^3 + 6x^2z^2 + 14xz^4 - 2xz^3$

39. $12x^3 - 9x^2 + 4x - 3$

40. $5x^3 - 20x^2 + 2x - 8$

41. $3y^3 + 7y^2 + 18y + 42$

42. $10y^3 - 4y^2 + 25y - 10$

43. $6x^6 - 3x^5 - 12x^4 + 6x^3$

44. $72x^5 + 12x^4 - 36x^3 - 6x^2$

45. $8x^6 + 12x^4 + 40x^3 + 60x$

46. $36x^6 + 72x^4 + 27x^3 + 54x$

47. $12x^4 + 7xy - 21x^3 - 4x^2y$

48. $7xy^3 - 20xy + 2y^3 - 70x^2y$

49. $54b^3 + 6a^4 - 27a^3b - 12ab^2$

50. $8a^2 - 36a^2b - 48ab + 6a^3$

51. $x^{2n}y^{2n} - x^ny^n$

52. $x^{3n}y^{2n} + x^{2n}y^n$

53. $x^{n+4} + x^4$

54. $x^{x+2} + x^2$

Write Algebra

55. Describe how to determine the greatest common factor of two or more terms.

56. Describe the steps used in factoring by grouping.

3.4 Factoring Binomials

In this section we will consider certain binomials that can be factored using formulas. As we consider each polynomial in this section and the next, we will always begin by looking for a common factor.

XAMPLE 3.4.1   Factor the following binomials.

1. $x^2 - 25 = (x + 5)(x - 5)$

Notice that there is no common monomial factor, but the product on the right-hand side does yield $x^2 - 25$, which follows from the difference of two squares formula in Section 3.2. Thus, $(x + 5)(x - 5)$ is a factorization of $x^2 - 25$. The next two examples are similar.

2. $y^2 - 100 = y^2 - 10^2 = (y + 10)(y - 10)$

3. $9x^2 - 49y^2 = (3x)^2 - (7y)^2 = (3x + 7y)(3x - 7y)$

In each case we recognized that we had a *difference* of two terms, each of which was a *square*. Then we applied the difference of two squares formula.

Difference of Two Squares

$$a^2 - b^2 = (a + b)(a - b)$$

4. $x^2 - 5$

This binomial does not factor with integer coefficients because 5 is not a perfect square.

NOTE ▶ $x^2 - 5$ is called a *prime polynomial*.

Definition 3.4.1	A **prime polynomial** is a polynomial that can be factored with integer coefficients only as a product of one and the original polynomial.

5. $4x^2 + 25y^2$

NOTE ▶   This binomial is prime.

6. $75x^3 - 3x$

This binomial has a greatest common factor of $3x$.

$$75x^3 - 3x = 3x(25x^2 - 1)$$
$$= 3x(5x - 1)(5x + 1) \qquad \textit{Factoring the binomial}$$

7. $y^4 - 16 = (y^2 + 4)(y^2 - 4) \qquad \textit{Don't stop! We can factor } y^2 - 4.$
$$= (y^2 + 4)(y - 2)(y + 2)$$

NOTE ▶   None of the three factors can be factored any further. *Always carry your factorization to this point.* We say that $y^4 - 16$ has been **completely factored.** A polynomial is completely factored when all of its factors are prime polynomials.

8. $(x + 5)^2 - 81 = [(x + 5) + 9][(x + 5) - 9]$
$$= [x + 5 + 9][x + 5 - 9]$$
$$= (x + 14)(x - 4)$$

We noted in part 5 of Example 3.4.1 that some sums of two squares are not factorable. However, we do have formulas that enable us to factor both the sum and difference of two cubes.

▉um and Difference of Two Cubes

1. $x^3 + y^3 = (x + y)(x^2 - xy + y^2)$
2. $x^3 - y^3 = (x - y)(x^2 + xy + y^2)$

These formulas can be readily verified by performing the indicated multiplications.

XAMPLE 3.4.2

Completely factor the following polynomials.

1. $x^3 + 8$

This binomial is the same as $x^3 + 2^3$, so we can apply formula 1. In formula 1 the role of y is played by 2:

$$x^3 + 8 = x^3 + 2^3$$
$$= (x + 2)(x^2 - x \cdot 2 + 2^2)$$
$$= (x + 2)(x^2 - 2x + 4)$$

2. $27x^3 - 64$

This binomial is the same as $(3x)^3 - 4^3$, so that we can apply formula 2. In formula 2 the role of x is played by $3x$ and the role of y by 4:

$$27x^3 - 64 = (3x)^3 - 4^3$$
$$= (3x - 4)[(3x)^2 + 3x \cdot 4 + 4^2]$$
$$= (3x - 4)(9x^2 + 12x + 16)$$

3. $125r^3 + 8s^3$

This binomial is the same as $(5r)^3 + (2s)^3$, so again we are in a position to use formula 1. In formula 1 the role of x is played by $5r$ and the role of y by $2s$:

$$125r^3 + 8s^3 = (5r)^3 + (2s)^3$$
$$= (5r + 2s)[(5r)^2 - 5r \cdot 2s + (2s)^2]$$
$$= (5r + 2s)(25r^2 - 10rs + 4s^2)$$

4. $5a^3 - 135b^3$

First remove the common factor of 5:

$$5a^3 - 135b^3 = 5(a^3 - 27b^3)$$

Now the expression inside the parentheses is a difference of two cubes, and we can use formula 2:

$$
\begin{aligned}
5a^3 - 135b^3 &= 5(a^3 - 27b^3) \\
&= 5[a^3 - (3b)^3] \\
&= 5(a - 3b)[a^2 + a \cdot 3b + (3b)^2] \\
&= 5(a - 3b)(a^2 + 3ab + 9b^2)
\end{aligned}
$$

Try to avoid these mistakes:

Incorrect	Correct
$x^3 - y^3 = (x - y)^3$ $x^3 + y^3 = (x + y)^3$ An exponent cannot distribute over a sum or difference. $x^3 - y^3$ is "a difference of cubes," whereas $(x - y)^3$ is the "cube of a difference." Note the order of the words.	$x^3 - y^3 = (x - y)(x^2 + xy + y^2)$ $x^3 + y^3 = (x + y)(x^2 - xy + y^2)$ $\begin{aligned}(x - y)^3 &= (x - y)(x - y)(x - y) \\ &= (x - y)(x^2 - 2xy + y^2) \\ &= x^3 - 3x^2y + 3xy^2 - y^3 \\ &\neq x^3 - y^3\end{aligned}$ $\begin{aligned}(x + y)^3 &= (x + y)(x + y)(x + y) \\ &= (x + y)(x^2 + 2xy + y^2) \\ &= x^3 + 3x^2y + 3xy^2 + y^3 \\ &\neq x^3 + y^3\end{aligned}$

In the last example of this section, we examine some factoring by grouping problems that also involve a difference of two squares or a sum or difference of two cubes.

EXAMPLE 3.4.3   Factor the following polynomials.

1.
$$
\begin{aligned}
4x^3 + 8x^2 - 49x - 98 &= (4x^3 + 8x^2) + (-49x - 98) \\
&= 4x^2(x + 2) - 49(x + 2) \\
&= (x + 2)(4x^2 - 49)
\end{aligned}
$$

Don't stop here! The second factor is a difference of two squares.

$$= (x + 2)(2x + 7)(2x - 7)$$

Now the polynomial is completely factored.

$$\begin{aligned}
2. \quad 54y^4 + 81y^3 + 2y + 3 &= (54y^4 + 81y^3) + (2y + 3) \\
&= 27y^3(2y + 3) + 1(2y + 3) \\
&= (2y + 3)(27y^3 + 1)
\end{aligned}$$

This time the second factor is a sum of two cubes.

$$= (2y + 3)(3y + 1)(9y^2 - 3y + 1)$$

Exercises 3.4

Completely factor the following polynomials.

1. $x^2 - 9$

2. $y^2 - 121$

3. $z^2 - 64$

4. $4x^2 - 25y^2$

5. $36z^2 - 49b^2$

6. $100m^2 - 9n^2$

7. $2x^2 - 50$

8. $20m^2 - 45n^2$

9. $27x^3y - 12xy^3$

10. $x^2 - 3y^2$

11. $4x^2 + 9$

12. $25y^2 + 16x^2$

13. $3x^2 + 48$

14. $16x^4 - 1$

15. $81y^4 - 16z^4$

16. $(x + 2)^2 - 9$

17. $(x - 3)^2 - 49$

18. $(x - 1)^2 - y^2$

19. $(x + 4)^2 - 4y^2$

20. $(x + y)^2 - z^2$

21. $(2x - 3y)^2 - 9z^2$

22. $x^3 + 1$

23. $y^3 - 64$

24. $27y^3 - 125$

25. $8x^3 + 1$

26. $125a^3 - 27b^3$

27. $8m^3 + 27n^3$

28. $216h^3 + 125k^3$

29. $64x^3 - 27y^3$

30. $6x^3 + 48$

31. $375y^3 - 3$

32. $40y^3 - 135x^3$

33. $54r^3 + 128s^3$

34. $(x + 1)^3 + 8$

35. $(y - 2)^3 - 27$

36. $(3y + 2)^3 + (y - 1)^3$

37. $(2x - 1)^3 - (x + 5)^3$

38. $(2x + y)^3 - (x - 3y)^3$

39. $(x + y)^3 + (2x - y)^3$

40. $x^6 + y^6$

41. $x^6 - y^6$

42. $y^2(5y - 4) - 36(5y - 4)$

43. $x^2(2x - 3) - (2x - 3)$

44. $9a^2(7a + 2b) - 4b^2(7a + 2b)$

45. $r^3(r - 6) - 8(r - 6)$

46. $m^3(6m + 5n) - n^3(6m + 5n)$

47. $5x^3 + 2x^2 - 45x - 18$

48. $4y^3 + 12y^2 - y - 3$

49. $3a^3 + 4b^3 - 4a^2b - 3ab^2$

50. $6x^3 + 4x^2 - 24x - 16$

51. $3x^4 - 5x^3 - 3x + 5$

52. $2x^4 + 7x^3 + 2x + 7$

53. $4y^4 - 3y^3 + 32y - 24$

54. $y^4 - 2y^3 - 64y + 128$

55. $4x^5 - x^3 - 32x^2 + 8$

56. $x^5 - x^3y^2 + x^2y^3 - y^5$

57. $7x^5 - 7x^3 - 28x^3y^2 + 28xy^2$

58. $12x^5y - 75x^3y - 12x^3y^3 + 75xy^3$

59. $5x^3y^4 + 5x^3y - 40y^7 - 40y^4$

60. $54x^6 - 54x^3y^3 - 16x^3 + 16y^3$

61. $x^3 + 6x^2 + 12x + 8$

62. $27x^3 - 27x^2 + 9x - 1$

63. $x^{2n} - 1$

64. $x^{2n} - 25$

65. $x^{4n} - 81$

66. $x^{4n} - 16$

67. $x^{n+1} + 5x^n + 3x + 15$

68. $x^{n+1} + 2x^n + 7x + 14$

Completely factor the following polynomials. (These exercises review the factoring methods covered thus far.)

69. $3x^3 - 12x + x^2y - 4y$

70. $k^3 + 27$

71. $3x^3y^2 + 6xy^2 - 15xy^3$

72. $12x^2 - 24x - 6xy + 12y^2$

73. $3xy^4(2x - 5) + 12y^5(2x - 5)$

74. $49x^2 - y^2$

75. $5x^3 - 5x$

76. $x^4 - 2x^3 - 8x + 16$

77. $625 - m^4$

78. $20x^3 + 125x^2$

Write Algebra

79. In your own words, explain how to factor the sum of two cubes.

80. Explain how to recognize a difference of two squares.

81. Explain how to recognize a difference of two cubes.

82. Why can't the following sum of two squares be factored: $49x^2 + 9y^2$?

3.5 *Factoring Trinomials*

The trinomials that we consider in this section are for the most part second-degree trinomials of the form $ax^2 + bx + c$. Once we develop a process for factoring these trinomials, we will see that some trinomials of higher degree can be factored by the same process.

We start by considering *trinomials with leading coefficient 1,* such as $x^2 + 7x + 10$. They often have simple factorizations. We can readily verify that

$$(x + 5)(x + 2) = x^2 + 7x + 10$$

so $x^2 + 7x + 10$ has been factored. The question is: How did we know to use $x + 5$ and $x + 2$ as factors?

Let us consider what happens when we multiply two binomials of the type $(x + m)(x + n)$, where m and n are integers:

$$(x + m)(x + n) = x^2 + nx + mx + mn$$
$$= x^2 + (n + m)x + mn$$

The product is a trinomial with leading coefficient 1. To factor a trinomial of this type, we need to find the two numbers m and n whose product is the constant term and whose sum is the coefficient of x.

In the example of $x^2 + 7x + 10$, we are looking for two numbers whose product is 10 and whose sum is 7. We start the search by looking at the integer factors of 10.

Factors of 10	Sum of the Factors
$10 = \quad 1 \cdot 10$	$1 + 10 = 11$
$= (-1) \cdot (-10)$	$(-1) + (-10) = -11$
$= \quad 2 \cdot 5$	$2 + 5 = 7$
$= (-2) \cdot (-5)$	$(-2) + (-5) = -7$

By examining this list, we can determine that 2 and 5 satisfy our conditions. Thus the factorization is

$$x^2 + 7x + 10 = (x + 5)(x + 2)$$

EXAMPLE 3.5.1   Factor the following trinomials.

1. $x^2 + 7x + 6 = (x + m)(x + n)$

Here we are looking for two numbers whose product is 6 and whose sum is 7. After examining the factors of 6, we find the numbers to be 1 and 6. Thus

$$x^2 + 7x + 6 = (x + 1)(x + 6)$$

2. $x^2 + x - 20 = (x + \quad)(x + \quad)$

Here we want two numbers whose product is -20 and whose sum is 1. The two numbers, which must differ in sign, are 5 and -4. Thus

$$x^2 + x - 20 = (x + 5)(x - 4)$$

3. $x^2 + 10x + 25 = (x + \quad)(x + \quad)$
$$\qquad\qquad\quad = (x + 5)(x + 5)$$
$$\qquad\qquad\quad = (x + 5)^2$$

Here we want two numbers whose product is 25 and whose sum is 10. From Section 3.2 we know that $x^2 + 10x + 25$ is the square of a binomial.

NOTE ▶   The square of a binomial yields a trinomial that we call a **perfect square trinomial.** A perfect square trinomial is always of the form

$$a^2 + 2ab + b^2$$

and factors as

$$a^2 + 2ab + b^2 = (a + b)^2$$

4. $x^2 + 3x + 10 = (x +\ \ )(x +\ \ )$

In this case we must find two numbers whose product is 10 and whose sum is 3. By checking the factors of 10, we find that there are no such numbers. Thus $x^2 + 3x + 10$ is prime.

5. $2x^3 - 12x^2 + 16x$

NOTE ▶ *When factoring polynomials, always remove the greatest common factor first:*

$$2x^3 - 12x^2 + 16x = 2x(x^2 - 6x + 8)$$
$$= 2x(x +\ \ )(x +\ \ )$$

We need two negative numbers whose product is 8 and whose sum is -6. These two numbers are -4 and -2:

$$2x^3 - 12x^2 + 16x = 2x(x^2 - 6x + 8)$$
$$= 2x(x - 4)(x - 2)$$

The next group of trinomials to consider is that with *leading coefficient not equal to 1*. These trinomials do not factor as readily as those in the previous example.

EXAMPLE 3.5.2   Factor the following trinomials.

1. $2x^2 + 5x + 3$

Since the first term is $2x^2$ and all the terms are positive,

$$2x^2 + 5x + 3 = (2x + h)(x + k)$$

where h and k are positive numbers to be determined. We know $h \cdot k$ must be 3, so h and k are factors of 3. From this point it is just a matter of trial and error to show that

$$2x^2 + 5x + 3 \neq (2x + 1)(x + 3)$$
$$2x^2 + 5x + 3 = (2x + 3)(x + 1)$$

2. $8x^2 + 6x - 5$

Here we are not even sure how to start. We could use

$$8x^2 + 6x - 5 = (8x + h)(x + k)$$

or

$$8x^2 + 6x - 5 = (4x + h)(2x + k)$$

since by the FOIL Method either product would yield a leading term of $8x^2$. It is *usually* best to try the factorization that yields numbers of almost equal size. Thus, we first try

$$8x^2 + 6x - 5 = (4x + h)(2x + k)$$

Here $h \cdot k$ must be -5, so h and k must be 1 and -5, or -1 and 5. From this point, it is again a question of trial and error. Let's first try

$$8x^2 + 6x - 5 \stackrel{?}{=} (4x - 1)(2x + 5)$$
$$\stackrel{?}{=} 8x^2 + 20x - 2x - 5$$
$$\neq 8x^2 + 18x - 5 \qquad \textit{Incorrect}$$

Next,

$$8x^2 + 6x - 5 \stackrel{?}{=} (4x - 5)(2x + 1)$$
$$\stackrel{?}{=} 8x^2 + 4x - 10x - 5$$
$$\neq 8x^2 - 6x - 5 \qquad \textit{Incorrect}$$

The middle term $-6x$ is the opposite of what we want, $+6x$. When we get the opposite of what we are looking for, we must change the signs inside the factors:

$$8x^2 + 6x - 5 \stackrel{?}{=} (4x + 5)(2x - 1)$$
$$\stackrel{?}{=} 8x^2 - 4x + 10x - 5$$
$$= 8x^2 + 6x - 5$$

We have *finally* found the factorization.

3. $9x^2 - 12x + 4$

Since $9x^2 - 12x + 4 = (3x)^2 - 2(3x)(2) + (2)^2$, it is a perfect square trinomial. Thus

$$9x^2 - 12x + 4 = (3x - 2)^2$$

4. $9x^3 - 33x^2 - 12x$

First remove the common factor of $3x$ and obtain

$$9x^3 - 33x^2 - 12x = 3x(3x^2 - 11x - 4)$$
$$= 3x(3x + h)(x + k)$$

Again resorting to trial and error, we investigate h and k values that are factors of -4. After some searching, we are able to determine $h = 1$ and $k = -4$:

$$9x^3 - 33x^2 - 12x = 3x(3x^2 - 11x - 4)$$
$$= 3x(3x + 1)(x - 4)$$

5. $6x^2 + 10x - 5 \neq (3x + 5)(2x - 1)$
$\neq (3x - 1)(2x + 5)$
$\neq (3x - 5)(2x + 1)$
$\neq (3x + 1)(2x - 5)$
$\neq (6x + 5)(x - 1)$
$\neq (6x - 1)(x + 5)$
$\neq (6x - 5)(x + 1)$
$\neq (6x + 1)(x - 5)$

This is perhaps the most frustrating type of polynomial to try to factor. After trying every possibility, we determine that none of the potential factorizations works. This tells us that $6x^2 + 10x - 5$ is a prime polynomial.

EXAMPLE 3.5.3   Completely factor the following polynomials.

1. $x^4 - 4x^2 - 45$

Since $x^4 - 4x^2 - 45 = (x^2)^2 - 4(x^2) - 45$, the first term of each binomial factor is x^2.

$$x^4 - 4x^2 - 45 = (x^2 +\ \ )(x^2 -\ \ )$$
$$= (x^2 + 5)(x^2 - 9)$$
$$= (x^2 + 5)(x + 3)(x - 3)$$

2. $3(2x + 7)^2 - 2(2x + 7) - 8$

In this polynomial, make the substitution $y = 2x + 7$. Then

$$3(2x + 7)^2 - 2(2x + 7) - 8 = 3y^2 - 2y - 8$$
$$= (3y + 4)(y - 2)$$

Now substitute $2x + 7$ for y:

$$= [3(2x + 7) + 4][(2x + 7) - 2]$$
$$= [6x + 21 + 4][2x + 7 - 2]$$
$$= (6x + 25)(2x + 5)$$

3. $x^2 + 12x + 36 - 4y^2$

This polynomial has four terms instead of three. So we try factoring by grouping.

$$x^2 + 12x + 36 - 4y^2 = (x^2 + 12x) + (36 - 4y^2)$$
$$= x(x + 12) + 4(9 - y^2)$$

Since this did not generate a common binomial factor, we interchange the second and third terms.

$$x^2 + 12x + 36 - 4y^2 = x^2 + 36 + 12x - 4y^2$$
$$= (x^2 + 36) + (12x - 4y^2)$$
$$= 1(x^2 + 36) + 4(3x - y^2)$$

This can be frustrating! Since a common binomial factor was still not generated, we have to try something new. We group the first three terms of the original polynomial and obtain a perfect square trinomial. Factor this trinomial.

$$x^2 + 12x + 36 - 4y^2 = (x^2 + 12x + 36) - 4y^2$$
$$= (x + 6)^2 - 4y^2$$

We now have a difference of two squares, which can be factored using the technique illustrated in Example 3.4.1, part 8.

$$= [(x + 6) + 2y][(x + 6) - 2y]$$
$$= (x + 6 + 2y)(x + 6 - 2y)$$

NOTE ▶   If the first three terms in this example had not yielded a perfect square trinomial, we then would have tried grouping the last three terms.   ▬▬▬

The following facts are sometimes helpful in factoring trinomials *after the greatest common factor has been removed*.

$ax^2 + bx + c, \, a > 0$	$(dx + h)(ex + k)$
1. If b and c are positive,	then h and k are positive.
2. If c is positive but b is negative,	then h and k are negative.
3. If c is negative,	then h and k have opposite signs.
4. Don't even try a factorization that has a common monomial factor within a factor.	For example, $(2x + 6)(x + 1)$ has a factor of 2 in the first factor.
5. If b is odd,	then d and e can't both be even, and h and k can't both be even. For example, $(4x + h)(2x + k)$ yields a middle term of $4kx + 2hx$, which is even.

Alternative Method for Factoring Trinomials

In the last part of this section, we will consider a method for factoring trinomials with leading coefficient not equal to 1 that uses factoring by grouping.

The problem with trying to factor trinomials like $8x^2 + 6x - 5$ is that the trial-and-error method is so lengthy. We can shorten the guessing by using the following method.

1. Multiply the leading coefficient by the constant term. In this case the product is $8(-5) = -40$.

2. Find the two factors of this product whose sum is the coefficient of x. In this case $-4 \cdot 10 = -40$ and $-4 + 10 = 6$.

3. Split the x term into two terms whose coefficients are the two numbers found in step 2. In this example,

$$8x^2 + 6x - 5 = 8x^2 - 4x + 10x - 5$$

4. Factor the resulting polynomial by grouping:

$$8x^2 - 4x + 10x - 5 = 4x(2x - 1) + 5(2x - 1)$$
$$= (2x - 1)(4x + 5)$$

XAMPLE 3.5.4

Factor the following trinomials by grouping.

1. $12x^2 - 13x - 4$

Multiply $12(-4) = -48$. The factors of -48 that add up to -13 are -16 and 3. So

$$12x^2 - 13x - 4 = 12x^2 - 16x + 3x - 4$$
$$= 4x(3x - 4) + 1(3x - 4)$$
$$= (3x - 4)(4x + 1)$$

2. $15x^2 + 19x + 4$

Multiply $15 \cdot 4 = 60$. The factors of 60 that add up to 19 are 15 and 4. So

$$15x^2 + 19x + 4 = 15x^2 + 15x + 4x + 4$$
$$= 15x(x + 1) + 4(x + 1)$$
$$= (x + 1)(15x + 4)$$

3. $16x^2 + 34x + 15$

Multiply $16 \cdot 15 = 240$. The factors of 240 that add up to 34 are 24 and 10. So

$$16x^2 + 34x + 15 = 16x^2 + 24x + 10x + 15$$
$$= 8x(2x + 3) + 5(2x + 3)$$
$$= (2x + 3)(8x + 5)$$

Exercises 3.5

Completely factor the following polynomials.

1. $x^2 - 9x + 14$

2. $x^2 + 19x + 88$

3. $x^2 - 2x - 35$

4. $x^2 + 5x - 1$

5. $x^2 - 9x + 8$

6. $x^2 + 4x + 3$

7. $x^2 + 7x - 30$

8. $x^2 - 14x + 45$

9. $x^2 - x + 7$

10. $x^2 + 16x + 48$

11. $x^2 - 5x - 36$

12. $x^2 + 6x + 10$

13. $x^2 - 6x + 9$

14. $x^2 - 12x + 36$

15. $x^2 + 18x + 81$

16. $x^2 + 2x + 1$

17. $3x^2 + 24x + 21$

18. $5x^2 - 60x + 100$

19. $-4x^2 + 12x + 112$

20. $-8x^2 + 16x - 40$

21. $-2x^3 - 16x^2 - 32x$

22. $-4x^3 + 44x^2 - 120x$

23. $6x^3 + 18x^2 - 6x$

24. $5x^3y - 10x^2y - 75xy$

25. $3x^4y^2 + 9x^3y^2 + 3x^2y^2$

26. $x^2 + 6xy + 8y^2$

27. $x^2 - 5xy + 6y^2$

28. $x^2 + 3xy - 18y^2$

29. $x^2 + 6xy + 5y^2$

30. $x^2 - 15xy + 56y^2$

31. $x^2 + 8xy - 20y^2$

32. $2x^2 + 15x + 5$

33. $6x^2 - 13x - 5$

34. $6x^2 + 7x - 20$

35. $2x^2 + 5x - 14$

36. $4x^2 + 20x + 25$

37. $12x^2 + 25x - 7$

38. $20x^2 - 23x + 6$

39. $6x^2 - 9x - 4$

40. $7x^2 + 16x + 4$

41. $9x^2 - 15x + 4$

42. $5x^2 - 7x - 6$

43. $8x^2 + 30x - 27$

44. $8x^2 - 23x - 3$

45. $9x^2 + 30x + 25$

46. $2x^2 - 5x - 12$

47. $8x^2 - 7x + 20$

48. $10x^2 + 43x + 12$

49. $-2 + 17x + 9x^2$

50. $24 - 22x + 3x^2$

51. $10x + 25x^2 + 1$

52. $3x + 5x^2 + 9$

53. $6x^2 - 13x - 15$

54. $12x^2 - 40x - 32$

55. $16x^2 - 26x - 12$

56. $8x^2 + 20x + 28$

57. $20x^2 - 60x - 35$

58. $16x^4 - 44x^3 + 24x^2$

59. $10x^3 - 14x^2 + 8x$

60. $24x^3y + 72x^2y + 54xy$

61. $15x^3 + 42x^2 - 9x$

62. $2x^2 + 5xy + 3y^2$

63. $27x^2 + 72xy + 48y^2$

64. $8x^2 + 6xy - 5y^2$

65. $9x^2 - 16xy - 4y^2$

66. $4x^2 + 7xy + 2y^2$

67. $8x^2 - 14xy + 6y^2$

68. $7x^2 + 3xy - 4y^2$

69. $x^4 + 12x^2 + 27$

70. $x^4 + x^2 - 30$

71. $x^4 - x^2 - 12$

72. $x^4 - 11x^2 + 18$

73. $4x^4 + 4x^2 - 15$

74. $12x^4 + 17x^2 - 5$

75. $(3x + 1)^2 - 7(3x + 1) + 10$

76. $(5x + 2)^2 + 6(5x + 2) + 8$

77. $6(2x - 1)^2 + (2x - 1) - 12$

78. $3(x + 2)^2 - 8(x + 2) + 5$

79. $6(x + 2y)^2 - 19(x + 2y) + 8$

80. $8(2x - 3y)^2 + 23(2x - 3y) + 15$

81. $x^2 + 6x + 9 - y^2$

82. $4x^2 - 20x + 25 - 9y^2$

83. $4x^2 - y^2 - 14y - 49$

84. $16m^2 - 9n^2 + 6n - 1$

85. $4x^2 + 4xy + y^2 - z^2 + 2z - 1$

86. $x^2 + 6x + 9 - y^2 - 18yz - 81z^2$

87. $x^2 - 2x + 1 - y^2 - 2yz - z^2$

88. $y^2 - 8y + 16 - m^2 - 2mn - n^2$

89. $9x^2 + 12x + 4 - y^2 + 4yz - 4z^2$

90. $25x^2 + 10x + 1 - j^2 + 6jk - 9k^2$

91. $x^2 - x - 12 + 5xy - 20y$

92. $x^2 - 4x - 12 + 3xy + 6y$

93. $2x^2 - 11x + 5 - 6kx + 3k$

94. $3x^2 - 11x + 6 - 12kx + 8k$

95. $x^2 + 4xy + 3y^2 + xz + yz$

96. $x^2 + 7xy + 10y^2 + zx + 2zy$

97. $x^{2n} - 8x^n + 15$

98. $x^{2n} - 8x^n + 12$

99. $2x^{2n} - 5x^n - 3$

100. $2x^{2n} - 7x^n - 15$

Completely factor the following trinomials using factoring by grouping.

101. $3x^2 + 11x + 10$

102. $5x^2 + 18x + 9$

103. $4x^2 + x - 14$

104. $7x^2 - 20x - 3$

105. $6x^2 - 7x - 10$

106. $4x^2 - 4x - 15$

107. $9x^2 - 9x - 4$

108. $5x^2 - 2x - 24$

109. $4x^2 + 41x + 10$

110. $8x^2 + 26x + 15$

111. $12x^2 - 25x - 7$

112. $9x^2 - 10x - 16$

113. $8x^2 - 21x + 10$

114. $16x^2 + 32x + 15$

115. $12x^2 - 28x + 15$

116. $16x^2 - 22x + 7$

117. $12x^2 - 19x - 10$

Completely factor the following polynomials. (These exercises review all the factoring methods covered.)

118. $x^2 + 2x - 48$

119. $36 - 4y^2$

120. $2ax + xy + 10ay + 5y^2$

121. $12x^2 + 18x - 12$

122. $4x^3 - 24x^2 + 4x$

123. $y^3 + 64$

124. $2x^3(4x - y) - 2(4x - y)$

125. $y^3 + 8y^2 - y - 8$

126. $4x^2 + 28xy + 49y^2$

127. $9x^2 - 15x + 4$

Write Algebra

128. Describe the primary advantage of removing the greatest common factor at the start of a factorization problem.

129. Explain why the factorizations $-(x - 1)(x - 2)$ and $(-x + 1)(x - 2)$ are equivalent.

130. Explain how to recognize a perfect square trinomial.

The following example illustrates the technique of adding and subtracting the same expression, factoring by grouping, and then applying the difference of squares formula to factor some polynomials.

XAMPLE   Completely factor the following polynomial.

$$4x^4 + y^4 = 4x^4 + 4x^2y^2 + y^4 - 4x^2y^2$$
$$= (4x^4 + 4x^2y^2 + y^4) - 4x^2y^2$$
$$= (2x^2 + y^2)^2 - (2xy)^2$$
$$= (2x^2 + y^2 + 2xy)(2x^2 + y^2 - 2xy)$$

or
$$= (2x^2 + 2xy + y^2)(2x^2 - 2xy + y^2)$$

Completely factor the following polynomials.
Hint: First, add and subtract an appropriate term; next, use factoring by grouping; and finally, apply the difference of squares formula.

131. $x^4 + 4$ **132.** $y^4 + 64$

133. $64x^4 + 1$ **134.** $4y^4 + 1$

135. $x^4 + 13x^2 + 49$ **136.** $x^4 + 3x^2 + 4$

137. $x^4 - 14x^2 + 25$ **138.** $x^4 - 23x^2 + 49$

139. $4x^8 + y^4$ **140.** $64x^4 + y^8$

3.6 *Factoring Summary*

In this section we combine all the methods of factoring that we have covered, and we give a strategy for factoring an arbitrary polynomial.

Given a polynomial to factor:

1. Examine the terms to determine if there is a common factor. If so, factor out the greatest common factor.

2. Consider the number of terms.

Two terms	Difference of squares?
	Difference of cubes?
	Sum of cubes?
Three terms	Leading coefficient of 1?
	Leading coefficient not equal to 1?
Four or more terms	Factor by grouping

3. Examine all the factors to determine if any factor can be factored further.

Exercises 3.6

Completely factor the following polynomials.

1. $x^2 - 3x + 2$

2. $x^2 - 2x - 3$

3. $2x^2 + 7x + 3$

4. $16x^2 + 72x + 81$

5. $8x^3 + 27y^3$

6. $x^2 - 7xy - 8y^2$

7. $2x^3 + 5x^2 + 6x + 15$

8. $6(x + 3y)^2 + 13(x + 3y) + 2$

9. $3(2x - 1)^2 - 11(2x - 1) - 20$

10. $x^2 - 49$

11. $2x^3 - 20x^2 + 18x$

12. $125x^3 + 8y^3$

13. $5x^2 - 3x - 2$

14. $3x^3 + x^2 + 21x + 7$

15. $x^2 - xy - 12y^2$

16. $6x^3 + 27x^2 + 12x$

17. $5(x + 2y)^2 + 31(x + 2y) + 6$

18. $x^2 + 3xy - 10y^2$

19. $x^2 - 36$

20. $x^4 - 26x^2 + 25$

21. $18x^3y^3 - 6x^2y^4 - 4xy^5$

22. $64x^2 - 81y^2$

23. $4x^2 - (2y + 3)^2$

24. $x^2 + 10x + 25$

25. $x^2 + 6x + 9 - 16y^2$

26. $2x^3 + 5x^2 - 2x - 5$

27. $64x^3 - 1$

28. $2(3x - 4)^2 - (3x - 4) - 6$

29. $y^2 + 2y - 8$

30. $2x^2 - xy - 3y^2$

31. $4(x - y)^2 - 8(x - y) + 3$

32. $4x^2 + 4xy - 15y^2$

33. $x^2 + 6xy + 9y^2$

34. $x^2 + 14x + 49 - 4y^2$

35. $4x^4 - 17x^2 - 50$

36. $4x^2 + 4x - 3$

37. $x^2 - 4x + 4 - 9y^2$

38. $x^3 - 27$

39. $25x^2 - 16y^2$

40. $3x^4y - x^3y^2 + 5x^2y^3$

41. $3x^3 + x^2y - 3x - y$

42. $x^2 - 10x + 25 - 81y^2$

43. $(x + 3)^2 - 9y^2$

44. $9(2x - y)^2 - 27(2x - y) + 20$

45. $3x^2 - 8xy + 4y^2$

46. $6x^2 - 24x + 24$

47. $18a^3b^3 - 12ab^5 - 6ab^3$

48. $(x - 1)^2 - 4y^2$

49. $8x^2 - 6xy - 9y^2$

50. $(x + 2)^3 + 27y^3$

51. $4x^2 - 24x + 36$

52. $h^2 + 2hk + k^2 - x^2 - 8x - 16$

53. $x^4 - 10x^2 + 9$

54. $6m^5n^4 + 15m^3n^6 - 3m^2n^4$

55. $x^2 + 2xy + y^2 - m^2 - 6m - 9$

56. $9x^4 + 23x^2 - 12$

57. $(x + 4)^3 + 125y^3$

58. $25x^2 - (3y - 1)^2$

3.7 *Solving Quadratic Equations by Factoring*

In Chapter 2 we learned how to find the solutions of linear equations and inequalities. However, there are many circumstances that cannot be characterized by linear equations. An example is determining the height (above ground level) of a stuntman who jumps from a 400-ft-tall building. The equation that gives us the height is $h = -16t^2 + 400$, where h stands for the height measured in feet, and t for the time measured in seconds. Suppose we want to determine how long the stuntman has been falling when his height above ground level is 256 ft. We then want to solve the equation $256 = -16t^2 + 400$.

The equation $256 = -16t^2 + 400$ is an example of a **quadratic equation.** Other quadratic equations are:

$$x^2 + 7x + 10 = 0$$
$$x^2 - 25 = 0$$
$$2y^2 - 13y = 7$$

These examples lead us to the formal definition of a quadratic equation.

Definition 3.7.1	Any equation that can be written in the form

$$ax^2 + bx + c = 0 \qquad (1)$$

where a, b, and c are real numbers with $a \neq 0$, is called a **quadratic equation** in the variable x.

REMARK

a. If $a = 0$, then equation (1) becomes $bx + c = 0$ (a linear equation). Thus, we have the restriction $a \neq 0$.

b. The form $ax^2 + bx + c = 0$ is called **standard form.** We will see that it is usually easier to find solutions of a quadratic equation if we first place the quadratic equation in standard form.

c. When we have a quadratic equation in standard form, $ax^2 + bx + c = 0$, ax^2 is called the **quadratic term,** bx is called the **linear term,** and c is called the **constant term.**

EXAMPLE 3.7.1 Place the following quadratic equations in standard form. In addition, identify the quadratic term, the linear term, and the constant.

1.
$$2x^2 - 10 = x^2 - 3x$$
$$2x^2 - x^2 + 3x - 10 = x^2 - x^2 - 3x + 3x \qquad \text{\textit{Add 3x and subtract } x^2 \textit{ from both}}$$
$$\qquad\qquad\qquad\qquad\qquad\qquad\qquad\qquad \textit{sides. Then combine like terms.}$$
$$x^2 + 3x - 10 = 0$$

The quadratic term is x^2, the linear term is $3x$, and the constant is -10.

2.
$$(2x - 1)(x + 3) = 9$$
$$2x^2 + 6x - x - 3 = 9 \qquad \textit{Multiply.}$$
$$2x^2 + 5x - 3 = 9 \qquad \textit{Combine like terms.}$$
$$2x^2 + 5x - 3 - 9 = 9 - 9 \qquad \textit{Subtract 9 from both sides and combine like terms.}$$
$$2x^2 + 5x - 12 = 0$$

The quadratic term is $2x^2$, the linear term is $5x$, and the constant is -12.

To find solutions of quadratic equations, we use the following theorem.

Theorem 3.7.1 Zero Product theorem: Let a and b be real numbers. If $a \cdot b = 0$, then $a = 0$ or $b = 0$.

REMARK This theorem simply states that if the product of two numbers is zero, then one or both of the two numbers must be zero.

We may use the Zero Product theorem to solve factorable quadratic equations with the following steps.

1. Place the quadratic equation in standard form.

2. Factor the resulting polynomial.

3. Use the Zero Product theorem, and set each factor equal to zero.

4. Solve the resulting equations.

5. (Optional) Check your solutions.

XAMPLE 3.7.2   Find the solutions of the following quadratic equations; also check your solutions.

1. $x^2 + 3x - 10 = 0$

   $(x + 5)(x - 2) = 0$ *Factor.*

so

   $x + 5 = 0$ or $x - 2 = 0$ *Set each factor equal to zero.*

   $x = -5$ $x = 2$

To check:

When $x = -5$,

$$(-5)^2 + 3(-5) - 10 \overset{?}{=} 0$$
$$25 - 15 - 10 \overset{?}{=} 0$$
$$0 = 0 \qquad \text{x} = -5 \textit{ is a solution.}$$

When $x = 2$,

$$(2)^2 + 3(2) - 10 \overset{?}{=} 0$$
$$4 + 6 - 10 \overset{?}{=} 0$$
$$0 = 0 \qquad \text{x} = 2 \textit{ is a solution.}$$

Thus the solution set is $\{-5, 2\}$.

2. $256 = -16t^2 + 400$

   $16t^2 + 256 - 400 = -16t^2 + 16t^2 + 400 - 400$ *Place the equation in*
                                                            standard form.

   $16t^2 - 144 = 0$

   $16(t^2 - 9) = 0$

   $16(t + 3)(t - 3) = 0$ *Factor.*

Clearly $16 \neq 0$; thus,

   $t + 3 = 0$ or $t - 3 = 0$ *Set each factor equal to zero.*

   $t = -3$ $t = 3$

To check:

When $t = -3$,

$$256 \overset{?}{=} -16(-3)^2 + 400$$
$$256 \overset{?}{=} -16(9) + 400$$
$$256 \overset{?}{=} -144 + 400$$
$$256 = 256 \qquad \text{t} = -3 \textit{ is a solution.}$$

When $t = 3$,

$$256 \overset{?}{=} -16(3)^2 + 400$$
$$256 \overset{?}{=} -16(9) + 400$$
$$256 \overset{?}{=} -144 + 400$$
$$256 = 256 \qquad t = 3 \text{ is a solution.}$$

Therefore, the solution set is $\{-3, 3\}$.

3. $3x^2 - 2x - 8 = 0$

$(3x + 4)(x - 2) = 0$ *Factor.*

$3x + 4 = 0$ or $x - 2 = 0$ *Set each factor equal to zero.*

$\qquad 3x = -4 \qquad\qquad x = 2$

$\qquad x = -\dfrac{4}{3}$

By substituting these values of x, we can verify that the solution set is $\{-\frac{4}{3}, 2\}$.

4. $(2x - 1)(x + 3) = 9$

$2x^2 + 6x - x - 3 = 9$ *Place the equation in standard form.*

$2x^2 + 5x - 3 = 9$

$2x^2 + 5x - 12 = 0$

$(2x - 3)(x + 4) = 0$ *Factor.*

$2x - 3 = 0$ or $x + 4 = 0$ *Set each factor equal to zero.*

$\qquad 2x = 3 \qquad\qquad x = -4$

$\qquad x = \dfrac{3}{2}$

Thus, the solution set is $\{\frac{3}{2}, -4\}$. The check is left to the reader.

Incorrect	Correct
$(2x - 1)(x + 3) = 9$	$(2x - 1)(x + 3) = 9$
$2x - 1 = 9$ or $x + 3 = 9$	$2x^2 + 6x - x - 3 = 9$
$\quad 2x = 10 \qquad\qquad x = 6$	$\vdots$
$\quad x = 5$	$x = \frac{3}{2}$ or $x = -4$
We can apply the Zero Product theorem only when one side of the equation is equal to zero.	from part 4.

5. $x^2 - 2x + 2 = 0$

This quadratic equation is already in standard form. However, the polynomial $x^2 - 2x + 2$ cannot be factored (i.e., it is prime), so we cannot use the Zero Product theorem. This quadratic equation does have solutions, and we will learn how to find them in Chapter 7.

Exercises 3.7

Place the following quadratic equations in standard form. In addition, identify the quadratic term, the linear term, and the constant.

1. $5x^2 - 7x + 1 = 0$

2. $10 + 3x^2 - 5x = 0$

3. $2x - 6x^2 + 3x - 1 = 0$

4. $3x^2 + 2x - 4 = 5x - 4$

5. $9x^2 + 2 = 7x^2 - 12$

6. $x^2 + 3x - 4 = 5x^2 + x + 6$

7. $x - 7x^2 = 4 + 3x$

8. $9 - 6x + x^2 = -3x^2 - 6x + 7$

9. $4x^2 + 9x - 9 = 9x + 16$

10. $5(x^2 + 2x - 4) = 2 - 4x^2$

11. $x(x + 7) = 3x - 9$

12. $4x(2x + 1) = 2(1 - 4x)$

13. $(x + 3)(x - 4) = 0$

14. $(2x - 7)(4x + 5) = 0$

15. $2(3x + 1)(x - 7) = 0$

16. $(4x + 3)(4x - 3) = 7$

17. $(3x - 7)(x - 1) = 9x - 11$

18. $(4x - 5)(2x + 1) = 3x(x + 1)$

19. $(2x + 5)(2x + 5) = (x + 1)(7x - 4)$

20. $(x - 7)(5x - 2) = (3x + 4)(x - 2)$

Find the solutions of the following quadratic equations; also check your solutions.

21. $x^2 + 6x + 8 = 0$

22. $x^2 - 7x + 6 = 0$

23. $x^2 - 2x - 35 = 0$

24. $x^2 + x - 12 = 0$

25. $2x^2 + 12x - 32 = 0$

26. $-3x^2 + 9x + 12 = 0$

27. $x^2 + 2x = 0$

28. $5x^2 - 3x = 0$

29. $6x^2 - 2x = 0$

30. $-14x^2 + 21x = 0$

31. $x^2 - 25 = 0$

32. $x^2 - 49 = 0$

33. $4x^2 - 9 = 0$

34. $25x^2 - 16 = 0$

35. $32x^2 - 2 = 0$

36. $-3x^2 + 12 = 0$

37. $2x^2 - 7x - 4 = 0$

38. $3x^2 + x - 2 = 0$

39. $8x^2 + 19x + 6 = 0$

40. $6x^2 - 23x + 15 = 0$

41. $6x^2 - 41x - 7 = 0$

42. $6x^2 + x - 2 = 0$

43. $8x^2 + 2x - 3 = 0$

44. $10x^2 + 19x + 6 = 0$

45. $12x^2 - 31x + 9 = 0$

46. $12x^2 - 26x - 10 = 0$

47. $20x^2 - 28x + 8 = 0$

48. $12x^2 + 51x - 45 = 0$

49. $x^2 + 6x + 9 = 0$

50. $9x^2 - 12x + 4 = 0$

51. $2x^2 - 16x + 32 = 0$

52. $20x^2 + 20x + 5 = 0$

53. $x^2 + 5x - 4 = 10$

54. $x^2 - 13x + 19 = -21$

55. $2x^2 - x = 15$

56. $15x^2 + 47x + 53 = 17$

57. $8x^2 - 13x - 9 = 5x - 4$

58. $6x^2 - 32x + 30 = 9 - 7x$

59. $40 - 9x = 20x - 3x^2$

60. $2x + 5 = 40 + 25x - 6x^2$

61. $3x^2 + 5x + 20 = 2x^2 - 3x + 4$

62. $-2x^2 + 5x - 4 = -3x^2 + x + 8$

63. $10x^2 + 5x - 10 = x^2 + 5x - 6$

64. $3x^2 - x - 11 = -5x^2 - 3x + 4$

65. $9x^2 + 7x - 7 = 3x(x - 4)$

66. $6x^2 - 17x - 4 = x(x + 2)$

67. $3x^2 + 2x - 24 = x(2x - 3)$

68. $14x^2 + 16x - 6 = 3x(3 - 2x)$

69. $3x^2 - 8x - 31 = (2x + 3)(x - 7)$

70. $5x^2 + 8x + 14 = (x + 3)(x + 5)$

71. $21x^2 + 6x - 11 = (3x - 1)(2x - 1)$

72. $7x^2 + 23x + 4 = (x + 4)(x - 4)$

73. $(2x + 5)(x - 4) = (x + 7)(x - 4)$

74. $(x + 2)(3x - 1) = (x + 2)(x + 6)$

75. $(3x + 2)(3x - 2) = (3x + 2)(2x + 7)$

76. $(5x + 4)(3x - 5) = (3x - 5)(2x - 9)$

Write Algebra

77. Explain how to solve a quadratic equation by factoring.

G R O U P   A C T I V I T Y

In exercises 78–85, find the solutions of the quadratic equations, then find the sum of the squares of the solutions of each equation.

78. $x^2 - 10x + 21 = 0$

79. $x^2 - 9x + 8 = 0$

80. $x^2 - 8x + 12 = 0$

81. $x^2 - 11x + 30 = 0$

82. $x^2 - 13x + 12 = 0$

83. $x^2 - 7x + 12 = 0$

84. $x^2 - 13x + 42 = 0$

85. $x^2 - 8x + 15 = 0$

86. If x_1 and x_2 are the solutions of $x^2 - bx + c = 0$, predict the value of $x_1^2 + x_2^2$ (in terms of b and c). Verify your conjecture.

3.8 *Applications*

Numerous problems can be represented by quadratic equations. The following examples illustrate how we can use quadratic equations and the tools of this chapter to solve some of those problems.

EXAMPLE 3.8.1   Find two consecutive integers whose product is 56.

SOLUTION   Let

$$x = \text{first consecutive integer}$$
$$x + 1 = \text{second consecutive integer}$$

$$x(x + 1) = 56$$
$$x^2 + x = 56$$
$$x^2 + x - 56 = 0$$
$$(x + 8)(x - 7) = 0$$
$$x + 8 = 0 \quad \text{or} \quad x - 7 = 0$$
$$x = -8 \qquad\qquad x = 7$$

Thus, the two consecutive integers are -8 and -7, or 7 and 8. To verify the solutions, note that $(-8) \cdot (-7) = 56$ and $7 \cdot 8 = 56$.

EXAMPLE 3.8.2   The area of a rectangle is 36 sq ft. The length is 5 ft more than the width. What are the dimensions of the rectangle?

SOLUTION   Since the width describes the length, we let

$$x = \text{width of the rectangle}$$
$$x + 5 = \text{length of the rectangle}$$

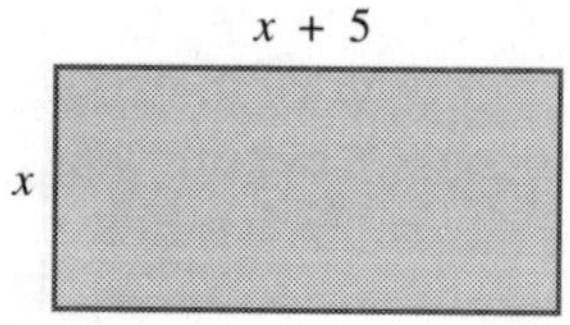

(See Figure 3.8.1.) Recall that the area of a rectangle is the width times the length; thus,

$x + 5$

x

Figure 3.8.1

$$x(x + 5) = 36$$
$$x^2 + 5x = 36$$
$$x^2 + 5x - 36 = 0$$
$$(x + 9)(x - 4) = 0$$
$$x + 9 = 0 \quad \text{or} \quad x - 4 = 0$$
$$x = -9 \qquad\qquad x = 4$$

We reject -9, since the width of a rectangle can never be negative. Consequently, the rectangle has

$$\text{width} = 4 \text{ ft}$$
$$\text{length} = 4 \text{ ft} + 5 \text{ ft} = 9 \text{ ft}$$

E X A M P L E   3 . 8 . 3

A cannonball is fired vertically upward, from ground level, with a speed of 160 ft/sec. The equation that gives the cannonball's height above ground level is $h = -16t^2 + 160t$, where h is the height (measured in feet) and t is the time (measured in seconds). How high is the cannonball after 3 sec? After 8 sec? When does it hit the ground?

SOLUTION

To determine the cannonball's height after 3 sec, we simply substitute $t = 3$ in the equation $h = -16t^2 + 160t$:

$$h = -16(3)^2 + 160(3)$$
$$h = -16(9) + 480$$
$$h = -144 + 480$$
$$h = 336$$

After 3 sec, the cannonball is 336 ft high.

In a similar fashion, to determine the height after 8 sec, we substitute $t = 8$:

$$h = -16(8)^2 + 160(8)$$
$$h = -16(64) + 1280$$
$$h = -1024 + 1280$$
$$h = 256$$

After 8 sec, the cannonball is 256 ft high.

The cannonball hits the ground when $h = 0$. In order to answer the last question, we must solve the equation

$$0 = -16t^2 + 160t$$
$$16t^2 - 160t = 0$$
$$16t(t - 10) = 0$$
$$16t = 0 \quad \text{or} \quad t - 10 = 0$$
$$t = 0 \qquad\qquad t = 10$$

Thus, the cannonball hits the ground after 10 sec. The solution $t = 0$ indicates that the cannonball was fired from ground level.

E X A M P L E   3 . 8 . 4

In the previous section we asked the question: How long has a stuntman been falling from a 400-ft-tall building when his height above ground level is 256 ft? Let's now answer that question.

Suppose Dangerous Dan claimed that he could jump from a 400-ft-tall building and land on a mattress. Furthermore, suppose we are given the equation for Dan's height as $h = -16t^2 + 400$. How long has Dan been falling when his height above ground level is 256 ft? When does Dan hit ground level, or maybe the mattress?

1. To answer the first question, we must substitute 256 for h in the equation $h = -16t^2 + 400$:

$$256 = -16t^2 + 400$$
$$16t^2 + 256 - 400 = 0$$
$$16t^2 - 144 = 0$$
$$16(t^2 - 9) = 0$$
$$16(t + 3)(t - 3) = 0$$
$$t + 3 = 0 \quad \text{or} \quad t - 3 = 0$$
$$t = -3 \qquad \qquad t = 3$$

Clearly, $t \neq -3$, since it does not make sense for Dan to be falling for -3 sec. Thus, the solution is $t = 3$; that is, Dan has been falling for 3 sec when he is 256 ft above ground level.

2. To answer the second question, we must substitute 0 for h. This substitution is made because, when Dan hits the ground (or mattress), his height above ground level will be 0:

$$0 = -16t^2 + 400$$
$$16t^2 - 400 = 0$$
$$16(t^2 - 25) = 0$$
$$16(t + 5)(t - 5) = 0$$
$$t + 5 = 0 \quad \text{or} \quad t - 5 = 0$$
$$t = -5 \qquad \qquad t = 5$$

Again, we reject $t = -5$. Hence, Dan is at ground level (one way or the other) after 5 sec.

XAMPLE 3.8.5   Mona and Maurice have a rectangular swimming pool that is 12 ft wide and 18 ft long. They are going to build a tile border of uniform width around the pool. They have 136 sq ft of tile. How wide is the border?

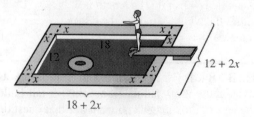

Figure 3.8.2

SOLUTION   Let x be the width of the border. By examining Figure 3.8.2, we can construct a quadratic equation that will enable us to find x, the width of the border. We can see that the swimming pool and its border determine a large rectangle whose width is $12 + 2x$ and length is $18 + 2x$. We can further observe that

$$\begin{pmatrix}\text{area of the large} \\ \text{rectangle}\end{pmatrix} - \begin{pmatrix}\text{area of the} \\ \text{swimming pool}\end{pmatrix} = \begin{pmatrix}\text{area of the} \\ \text{border}\end{pmatrix}$$

$$(12 + 2x)(18 + 2x) - \quad 12 \cdot 18 \quad = \quad 136$$

$$216 + 24x + 36x + 4x^2 - 216 = 136$$

$$4x^2 + 60x = 136$$

$$4x^2 + 60x - 136 = 0$$

$$4(x^2 + 15x - 34) = 0$$

$$4(x + 17)(x - 2) = 0$$

$$x + 17 = 0 \quad \text{or} \quad x - 2 = 0$$

$$x = -17 \qquad x = 2$$

We reject $x = -17$, since length is always positive, and conclude that the border must be 2 ft wide.

Exercises 3.8

Use quadratic equations to find the solutions of the following problems.

1. Find two consecutive positive integers whose product is 72.

2. Find two consecutive positive integers whose product is 156.

3. Find two consecutive even integers whose product is 48.

4. Find two consecutive odd integers whose product is 63.

5. The sum of the squares of two consecutive positive integers is 25. Find the integers.

6. A positive number is 5 larger than another positive number. The sum of the squares of the two numbers is 53. Find the numbers.

7. One number is 1 more than twice another number. The product of the two numbers is 21. What are the numbers?

8. Find two numbers whose sum is 16 and whose product is 48.

9. Find two numbers whose difference is 7 and whose product is -12.

10. The area of a rectangle is 72 sq ft. The length of the rectangle is twice the width. What are the dimensions of the rectangle?

11. The area of a rectangle is 42 sq ft. The length is 2 ft more than 4 times the width. What are the dimensions of the rectangle?

12. The sum of the areas of two squares is 73 sq yd. The sides of the second square are 5 yd longer than the sides of the first square. What are the dimensions of each square?

13. One leg of a right triangle is 2 in. longer than the other leg. The hypotenuse is 10 in. long. What are the lengths of the legs of the triangle? (*Hint:* Use the Pythagorean theorem.)

14. One leg of a right triangle is 1 cm less than twice the length of the other leg. The hypotenuse is 17 cm long. What are the lengths of the legs of the triangle?

15. The base of a triangle is 1 in. more than twice the height. Find the base and the height if the area of the triangle is 14 sq in.

16. A baseball is hit vertically upward with a speed of 40 ft/sec. The equation that gives the ball's height above ground level is $h = -16t^2 + 40t$. How high is the ball after 1 sec? After 2 sec? When does it hit the ground?

17. A bullet is fired from ground level vertically upward with a speed of 384 ft/sec. The equation that gives the bullet's height above ground level is $h = -16t^2 + 384t$. How high is the bullet after 5 sec? After 12 sec? When does it hit the ground?

18. An arrow is shot vertically upward from the roof of a 48-ft-tall building with a speed of 88 ft/sec. The equation that gives the arrow's height above ground level is $h = -16t^2 + 88t + 48$. How high is the arrow after 1 sec? After 3 sec? When does it hit the ground?

19. A piano is dropped from the top of a 256-ft-tall building. The equation that gives the piano's height above ground level is $h = -16t^2 + 256$. When does the piano hit the ground?

20. A ball is thrown downward from a 20-ft-tall house with a speed of 32 ft/sec. The equation that gives the ball's height above ground level is $h = -16t^2 - 32t + 20$. When does the ball hit the ground?

21. An astronaut on the surface of the moon jumps off a 41.6-ft-tall cliff. The equation that gives the astronaut's height above the lunar surface is $h = -2.6t^2 + 41.6$. When does the astronaut hit the lunar surface?

22. Tom and Mary have a rectangular swimming pool that is 9 ft wide and 12 ft long. They are going to build a tile border around the pool of uniform width. They have 162 sq ft of tile. How wide is the border?

23. Loretta and Charley have a rectangular swimming pool that is 8 ft wide and 14 ft long. They are going to build a tile border around the pool of uniform width. They have 75 sq ft of tile. How wide is the border?

24. A square piece of tin, 8 in. on a side, has four equal squares cut from its corners (see Figure 3.8.3). This new figure is to have an area of 55 sq in. What size squares should be cut to attain this area?

25. A painting and its frame cover 117 sq in. The frame is $\frac{3}{2}$ in. wide. The length of the painting is 4 in. greater than the width. What are the dimensions of the painting?

26. A painting and its frame cover 84 sq in. (see Figure 3.8.4). The frame is 2 in. wide at the top and bottom and 1 in. wide on the sides. The length of the painting is 3 in. greater than the width. What are the dimensions of the painting?

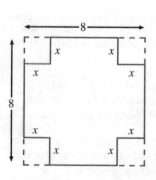

Figure 3.8.3

Figure 3.8.4

Applied Algebra

Arlene, an architect, is building a new home along the coast of Louisiana. Arlene enjoys the view of the Gulf of Mexico and plans on including many large windows in her new house. She plans on designing her home with windows that have a surface area of no more than 25 sq ft in each window. What is the maximum wind speed her windows can withstand before breaking?

SOLUTION   The maximum wind speed vertical windows can withstand before breaking can be determined by the formula

$$F_B = AS^2$$

In this formula

$$F_B = \text{breaking force constant}$$

NOTE ▶   This constant has different values for different thicknesses of glass.

$$A = \text{the area of the glass measured in square feet}$$
$$S = \text{the speed of the wind measured in miles per hour}$$

In Arlene's case we are given $F_B = 129{,}600$ and we know $A = 25$. Thus,

$$
\begin{aligned}
129{,}600 &= 25S^2 & &\text{\textit{A quadratic equation}} \\
0 &= 25S^2 - 129{,}600 & &\text{\textit{Write the equation in standard form.}} \\
0 &= 25(S^2 - 5184) \\
0 &= 25(S + 72)(S - 72) \\
S + 72 = 0 \quad &\text{or} \quad S - 72 = 0 & &\text{\textit{Use the Zero Product theorem.}} \\
S = -72 \qquad &\qquad\quad S = 72 & &\text{\textit{We reject the negative solution.}}
\end{aligned}
$$

Thus, Arlene's windows can withstand wind speeds of up to and including 72 mi/hr.

Your Turn

Josh is building a new home along the coast of Florida. He plans on building a home with windows that have a surface area of no more than 16 sq ft. Josh has already selected the thickness of glass that will be used for his windows. For the thickness of Josh's windows, $F_B = 102{,}400$. What is the maximum wind speed his windows can withstand before breaking?

Chapter 3 Review

Terms to Remember

Polynomial	**p. 110**	Factoring	**p. 127**
Monomial	**p. 110**	Factoring by grouping	**p. 128**
Binomial	**p. 110**	Difference of two squares	**p. 131**
Trinomial	**p. 110**	Prime polynomial	**p. 131**
Coefficient	**p. 110**	Completely factor	**p. 131**
Degree	**p. 111**	Sum of two cubes	**p. 132**
Standard form of a		Difference of two cubes	**p. 132**
polynomial	**p. 112**	Perfect square trinomial	**p. 136**
Leading coefficient	**p. 112**	Quadratic equation	**p. 146**
FOIL Method	**p. 121**	Standard form for a quadratic	
Product yielding the differ		equation	**p. 146**
ence of two squares	**p. 122**	Quadratic term	**p. 147**
Square of a binomial	**p. 122**	Linear term	**p. 147**
Greatest common factor	**p. 126**	Constant term	**p. 147**

Formulas

- *Product Law of Exponents:* $x^m \cdot x^n = x^{m+n}$
- *Power to a Power Law of Exponents:* $(x^m)^n = x^{mn}$
- *Product to a Power Law of Exponents:* $(xy)^n = x^n y^n$
- *Product yielding the difference of two squares:* $(a + b)(a - b) = a^2 - b^2$
- *Square of a binomial:* $(a + b)^2 = a^2 + 2ab + b^2$
- *Difference of two squares:* $a^2 - b^2 = (a + b)(a - b)$
- *Perfect square trinomial:* $a^2 + 2ab + b^2 = (a + b)^2$
- *Sum of two cubes:* $x^3 + y^3 = (x + y)(x^2 - xy + y^2)$
- *Difference of two cubes:* $x^3 - y^3 = (x - y)(x^2 + xy + y^2)$

Factoring Summary

1. Factor out the greatest common factor.

2. Consider the number of terms.

Two terms	Difference of squares?
	Difference of cubes?
	Sum of cubes?
Three terms	Leading coefficient of one?
	Leading coefficient not equal to one?
Four or more terms	Factor by grouping

Important Facts

■ *Zero Product theorem:* If a and b are real numbers and $a \cdot b = 0$, then either $a = 0$ or $b = 0$.

■ *Area of a rectangle:* area = (width)(length)

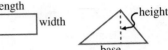

■ *Area of a triangle:* area = $\frac{1}{2}$(base)(height)

■ *Pythagorean theorem:* For a right triangle with legs a and b and hypotenuse c, $a^2 + b^2 = c^2$.

Review Exercises

3.1 **List the terms and coefficients of the following polynomials and specify which are monomials, binomials, and trinomials. State the degree of each polynomial.**

1. $5x^2 - x + 2$

2. $2\pi r^2 h$

3. $\frac{y}{3} + x$

4. $\frac{12x}{5} + y - 4$

5. $y^3 + \frac{3xy}{2} - \frac{x^2 y}{4} + \frac{2}{3}$

Use the distributive property to simplify the following polynomials.

6. $5x(2x - 3)$

7. $2.5(4x^2 - 0.6y)$

8. $\left(\frac{2}{9}\right)\left(6ax - \frac{1}{4}y + 15x^2\right)$

9. $-(y^3 + 5y^2 - 4y - 3)$

Simplify each polynomial by combining like terms.

10. $4x^2 + 7 - 3x - 5 + x^2 + 3x$

11. $-x^2 + 6xy + 4y + 2x^2 - 5y^2 + 3yx - 4y^2$

12. $\frac{2}{5}x + \frac{1}{3}y - \frac{4}{5} + \frac{1}{6}y - \frac{9}{10}x$

13. $1.8a - 3.5b + 11.6 - 7a + 0.25c$

Perform the indicated operations and simplify by combining like terms in the following.

14. $(5y^2 + 3y - 2) + (y^3 - 4y^2 - y + 7)$

15. $(x^2 - 3xy + 2y + 5y^2) - (4x^2 + 6xy - 2y^2 - 3x)$

16. $4(3x + 2) - 6\left(\frac{1}{2}x - 3\right)$

17. $-\left(5a + 3 - \frac{b}{2}\right) + \left(\frac{2}{3}\right)\left(9a - \frac{1}{4} - 3b\right)$

18. $0.05(4x^2 + 1.8x - 3) - 3(2.4x^2 - 1.04x + 10.2)$

19. $\quad 5x^2 - 8xy + 2y^2$
$\quad\; +(3x^2 + 8xy - 9y^2)$

20. $\quad 2a^2 - 3ab + c$
$\quad -(5a^2 + 6ab - 8c)$

3.2 Perform the indicated multiplications.

21. $x^{12} \cdot x^5$

22. $(x^8 y^3)^4$

23. $4a^2 b \cdot 7a^3 b^8$

24. $5a^3 b^4 \cdot (-3ab^2) \cdot 4a^5 b^3$

25. $(3xy^4)^2 \cdot (-2x^3 y^2)^3$

26. $3x^4 y^2 (2x^3 y - 6x^5 y^4)$

27. $-2x^2 (5x^2 - x + 3)$

28. $(3x - 4)(2x + 5)$

29. $(x + 5)(4x + 7y)$

30. $(y - 6)(y + 6)$

31. $(2a + 9)(2a - 9)$

32. $\left(\frac{1}{2}x + 3y\right)\left(\frac{1}{2}x - 3y\right)$

33. $(x - 5)(x + 5)$

34. $(x + 8)^2$

35. $(y - 1)^2$

36. $(2x + 3y)^2$

37. $(4x + 1)(16x^2 - 4x + 1)$

38. $(5x - 2y)(2x^2 + 3xy - y^2)$

39. $(y^2 + 2y + 4)(y^2 - 2y + 4)$

40. $(x + 5)(x - 2)(3x^2 + 4x - 1)$

3.3–3.6 Completely factor the following polynomials.

41. $10x^3 y^2 z^4 - 8x^2 y^5 z^3 - 14x^4 yz^7$

42. $2x^2 - 6xy + 2x$

43. $6a^3 x - 3a^2 x^2 + 3ax$

44. $4ax - 9by + 5cz$

45. $x^2 - 9x$

46. $y^2 + 4y$

47. $x^2 - 36$

48. $9x^2 - 25$

49. $y^2 + 16$

50. $3x^3 y - 3xy^3$

51. $2x^3 - 18x$

52. $2a^2 - 16$

53. $y^4 - 81$

54. $x^2 - 11x + 24$

55. $x^2 - 3x + 4$

56. $x^2 - x - 12$

57. $x^2 - 3xy - 10y^2$

58. $x^3 - 5x^2 - 6x$

59. $x^2 - 14x + 49$

60. $3x^4 - 18x^3 + 15x^2$

61. $3x^2 - 13x + 12$

62. $5x^2 - 3x - 8$

63. $18x^2 + 27x + 4$

64. $4x^2 + 4xy - 3y^2$

65. $18x^3 + 3x^2 - 45x$

66. $20x^3 - 32x^2 y + 12xy^2$

67. $x^4 + 4x^2 - 5$

68. $2x^4 + 7x^2 - 4$

69. $(3x + 1)^2 + 4(3x + 1) - 5$

70. $4(2x - 3y)^2 - 8(2x - 3y) - 21$

71. $x^3 + 27$

72. $64a^3 + 8$

73. $y^3 - 1$

74. $250x^3 - 2y^3$

75. $(x + 2y)^3 + (3x - 2y)^3$

76. $2x(y + 5) + 3(y + 5)$

77. $5xy^2 (3x + y) + (3x + y)$

78. $4x^2 (2x + 1) - (2x + 1)$

79. $8a^3 (a + 2b) + b^3 (a + 2b)$

80. $3ax - bx + 6ay - 2by$

81. $8xy + 15 + 6y + 20x$

82. $3x^3 - 7x^2 + 15x - 35$

83. $12x^2 y - 27y^3 - 8x^2 + 18y^2$

84. $12a^3 + 48a^2 b^2 + 2a^2 b + 8ab^3$

85. $x^2 - 8xy + 16y^2 - 25z^2$

86. $9x^5 + 8y^5 - 72x^2 y^3 - x^3 y^2$

87. $x^3 + 6x^2 + 12x + 8$

Completely factor the following trinomials using factoring by grouping.

88. $7x^2 - 3x - 4$

89. $12x^2 - 8x - 15$

90. $2x^2 + 13x + 11$

91. $16x^2 - 10x - 9$

92. $6x^2 - 3x - 9$

3.7 **Write the following quadratic equations in standard form. In addition, identify the quadratic term, the linear term, and the constant term.**

93. $7x^2 = 5x - 1$

94. $2x^2 - 3 + 2x = 0$

95. $(3x + 8)(x - 5) = 0$

96. $x^2 + 7x - 4 = 2x^2 + 7x - 11$

97. $3x = (5x - 1)(2x - 3)$

98. $x^2 + (x - 3)^2 = 27$

99. $(3x + 10)(x - 1) = 27$

100. $(2x - 5)^2 = 2x - 5$

Find the solutions of the following quadratic equations and check the solutions.

101. $x^2 - 11x - 26 = 0$

102. $x^2 + 12x + 27 = 0$

103. $2x^2 - 18x + 28 = 0$

104. $3x^2 + 45x + 108 = 0$

105. $4x^2 - 16x = 0$

106. $6x^2 + 9x = 0$

107. $x^2 - 36 = 0$

108. $121 - x^2 = 0$

109. $3x^2 - 12 = 0$

110. $18x^2 - 2 = 0$

111. $x^2 - 8x + 16 = 0$

112. $x^2 + 12x + 36 = 0$

113. $3x^2 - 17x + 24 = 0$

114. $5x^2 + 21x + 4 = 0$

115. $6x^2 + 17x = -7$

116. $8x^2 - 9 = 6x$

117. $(x + 4)^2 = 25$

118. $x^2 + (x - 2)^2 = 34$

119. $6x(x + 2) = x - 5$

120. $4x(x - 5) = x - 26$

121. $2x^2 - 9x - 22 = 6x + 5$

122. $2x^2 + 12x + 43 = x^2 - x + 3$

123. $8x^2 - 9x - 18 = (x - 3)(x + 2)$

124. $12x^2 - 19x - 7 = (x + 1)(x - 1)$

125. $(x + 6)(x - 5) = 5(x + 6)$

126. $(x + 2)(x - 6) = 10(x - 6)$

127. $(5x + 2)(x - 4) = (2x + 1)(x - 14)$

128. $(x + 9)(3x + 4) = (4x + 3)(x + 12)$

3.8 **Use quadratic equations to find the solutions of the following problems.**

129. Find two consecutive positive integers whose product is 132.

130. Find two consecutive even integers, the sum of whose squares is 340.

131. The area of a rectangle is 36 sq ft. The length is 1 ft less than twice the width. What are the dimensions of the rectangle?

132. The length of a rectangle is twice the width. If the width is increased by 4 m and the length is increased by 2 m, then this larger rectangle would have an area of 140 sq m. What are the dimensions of the original rectangle?

133. The area of a triangle is 28 sq cm. The height is 2 cm more than 3 times the base. Find the height and the base of the triangle.

134. One leg of a right triangle is 7 in. shorter than the other leg. The hypotenuse is 13 in. long. What are the lengths of the legs of the triangle?

135. A flea jumps vertically upward with a speed of 16 ft/sec. The equation that gives the flea's height above ground level is $h = -16t^2 + 16t$. How high is the flea after $\frac{1}{2}$ sec? When does it land back on the ground?

136. Little Adam is sitting in his high chair with a bowl of oatmeal in front of him, 4 ft above the floor. Instead of eating it, he drops it on the floor. The equation that gives the bowl's height above the floor is $h = -16t^2 + 4$. When does the bowl hit the ground?

137. Mr. and Mrs. Buck S. Bigg have a rectangular swimming pool that is 8 yd wide and 13 yd long. They are going to build a concrete border around the pool of uniform width. They can afford 100 sq yd of concrete. How wide will the border be?

138. A photograph and its frame cover 45 sq in. The frame is 1 in. wide. The length of the photograph is 1 in. longer than twice the width. What are the dimensions of the photograph?

Chapter 3 Test

(You should be able to complete this test in 60 minutes.)

I. List the terms and coefficients of the following polynomial. State the degree of the polynomial.

1. $x^3 + 5x^2 - x + 7$

II. Perform the indicated operations and simplify your answers by combining like terms.

2. $5x - 3y - xy - 2x - \dfrac{1}{2}y$

3. $(2a^2b + 5a - 7b - a^2) - (7a + 2ab^2 + 4a^2 - 5b + 3ab)$

4. $(5x^2y^3)^2(2x^4y)^3$

5. $-5x^2y^4(2x^3y^2 - 3x^6y)$

6. $(4x - 3y)(7x - 2y)$

7. $(3x + 9)(2x^2 - 6x + 11)$

III. Completely factor the following polynomials.

8. $x^2 + xy - 2y^2$

9. $3x^2 + 10x + 8$

10. $8p^3 + 1$

11. $x^2 - 4x + 5$

12. $2y - 128y^4$

13. $16x^3 - 48x^2 - x + 3$

14. $x^4 - 8x^2 + 7$

15. $9x^2 - 12xy + 4y^2 - 25$

16. $x^2 + 10x + 25$

17. $10a^4x^3 - 5a^2x + 5a^6$

18. $12xy^2 + 38xy - 50x$

19. $9x^2 - 4$

IV. Use factoring by grouping to factor the trinomial.

20. $3x^2 - 20x + 32$

V. Find the solutions of the following quadratic equations and check your solutions.

21. $x^2 - 8x - 20 = 0$

22. $(x - 5)^2 = 4$

23. $3x(3x + 5) = -4$

VI. Use quadratic equations to find the solutions of the following problems.

24. A rectangular door has a length that is 7 ft less than 4 times the width. The area is 30 sq ft. What are the dimensions of the door?

25. When King Kong was 800 ft up the side of the Empire State Building, a fighter plane got too close to him and was plucked out of the air by the irate primate. He smashed it and hurled it downward with a speed of 368 ft/sec. The equation that gives the plane's height above ground level is $h = -16t^2 - 368t + 800$. When will the plane hit the ground?

TEST YOUR MEMORY

These problems review Chapters 1–3.

I. Use the sets $D = \{0, 5, 10, 15, 20\}$, $G = \{0, 10, 20, 30\}$, $E = \{2, 4, 6, \ldots\}$, N, I, Q, H, and R for problems 1–4. Are the following true or false?

1. $90 \in Q$

2. $D \subseteq E$

Find the following:

3. $D \cup G$

4. $D \cap G$

II. Perform the indicated operations.

5. $-6 - 4 - (-2)$

6. $-3^2 \cdot 4 + 5 \cdot 12$

7. $\frac{2}{3} + \frac{3}{4} - 1$

8. $\frac{3}{8} \cdot \frac{2}{15} - \frac{3}{5} \div \frac{6}{7}$

III. Evaluate the following algebraic expressions for the given values of the variables.

9. $\sqrt{289 - 9x^2}$ for $x = 5$

10. $\sqrt{6x + 5} - \sqrt{3x + 14}$ for $x = \frac{2}{3}$

11. $4x^2 - |x - 5|$ for $x = -2$

12. $7x^2 - 29xy + 4y^2$ for $x = 1, y = -4$

IV. Perform the indicated operations and simplify your answers by combining like terms.

13. $(5x^3 - 3x^2 + 7) - (2x^2 - 8x - 4)$

14. $2x^3y^2(5x^2 - 8xy^4 - 2y)$

15. $(7x + 2y)(x - 3y)$

16. $(2x + 4)(3x^2 - 5x + 3)$

V. Completely factor the following polynomials.

17. $8x^2 - 14x - 15$

18. $9x^2 + 42x + 49$

19. $375a^4 + 24a$

20. $3x^3 + 5x^2 - 3x - 5$

21. $8x^3 + 12x^2y - 4x$

22. $9x^2 - 25y^2$

23. $4m^2 - 12mn + 9n^2 - 4$

24. $x^2 - 11xy + 10y^2$

VI. Find the solutions of the following equations and check your solutions.

25. $x^2 - 11x + 18 = 0$

26. $4 - 3x = -11$

27. $3x - 7 = 6x - 1$

28. $|3x + 1| = 13$

29. $\frac{1}{2}(2x + 3) - \frac{2}{3}(x - 5) = 5$

30. $(x - 4)^2 = 1$

31. $2(3x + 1) - 4(x - 3) = 7$

32. $5x^2 - 9x - 40 = 3x^2 - 5x + 8$

33. $|2x - 4| = |x + 7|$

34. $\frac{3}{4}\left(\frac{1}{2}x - 1\right) + \frac{1}{3}\left(\frac{3}{2}x + 2\right) = \frac{1}{2}$

35. $(2x + 1)(4x - 3) = -2$

36. $\frac{4}{3}x - \frac{3}{2} = \frac{1}{2}x - \frac{2}{3}$

37. $(5x + 1)(x - 2) = (2x + 2)(x - 1)$

38. $4x(x - 2) = 12x - 25$

VII. Find and graph the solutions of the following inequalities. Express your answers using interval notation.

39. $5 - 3x \geq 17$ **40.** $2 < 2x + 3 < 11$ **41.** $|x - 2| > 3$ **42.** $|3x - 1| \leq 5$

VIII. Use algebraic expressions to find the solutions of the following problems.

43. A rectangle has a length that is 2 ft less than 3 times the width. The area is 65 sq ft. What are the dimensions of the rectangle?

44. The perimeter of a rectangle is 34 ft. The length is 8 ft more than one-half the width. Find the dimensions of the rectangle.

45. One leg of a right triangle is 2 m longer than the other leg. The hypotenuse is 10 m long. What are the lengths of the legs of the triangle?

46. The area of a triangle is 16 sq m. The base is 4 m less than 3 times the height. Find the base and the height of the triangle.

47. Barbie and Ken have a rectangular swimming pool that is 10 ft wide and 14 ft long. They are going to build a tile border around the pool of uniform width. They have 81 sq ft of tile. How wide is the border?

48. Junior McBride uses his slingshot to shoot a stone vertically upward from the top of a 96-ft-tall cliff at a speed of 184 ft/sec. The equation that gives the stone's height above ground level is $h = -16t^2 + 184t + 96$. When does the stone hit the ground?

49. Charlotte has $2.20 in nickels and dimes. She has a total of 31 coins. How many nickels and how many dimes does she have?

50. Roberto leaves the "Good Eats" cafe at 1:00 P.M. traveling westward at a rate of 14 mi/hr. At 2:30 P.M. Raul leaves the "Good Eats" cafe and follows the same route as Roberto, but at a rate of 56 mi/hr. How long will it take Raul to catch Roberto?

CHAPTER 4

Rational Expressions

Conservationists in Moss County have become concerned about the wildlife population in the county. In particular, conservationists are worried that local fishing is too rapidly depleting the trout population in Lake Lurlene. Naturally, those who fish disagree, claiming that there are plenty of trout in the lake. Thus, a reasonable approximation of the trout population should be found.

In this chapter we will study rational expressions. We will see how rational equations (actually proportions) can be used to approximate the number of trout in Lake Lurlene.

167

4.1 Integer Exponents

In Chapter 3 the laws of exponents used in multiplying polynomials were introduced. These laws are listed again:

Let m and n be positive integers; then

$$x^m \cdot x^n = x^{m+n} \qquad \text{Product Law}$$
$$(x^m)^n = x^{m \cdot n} \qquad \text{Power to a Power Law}$$
$$(xy)^n = x^n y^n \qquad \text{Product to a Power Law}$$

In this section the laws of exponents used in dividing polynomials will be examined.

Consider the following problem from arithmetic.

$$\frac{2^5}{2^3} = \frac{\cancel{2} \cdot \cancel{2} \cdot \cancel{2} \cdot 2 \cdot 2}{\cancel{2} \cdot \cancel{2} \cdot \cancel{2}}$$
$$= 2^2, \quad \text{or} \quad 4$$

Suppose we want to simplify the algebraic expression

$$\frac{x^5}{x^3}$$

We know that

$$\frac{x^5}{x^3} = \frac{x \cdot x \cdot x \cdot x \cdot x}{x \cdot x \cdot x}$$
$$= \frac{\cancel{x} \cdot \cancel{x} \cdot \cancel{x} \cdot x \cdot x}{\cancel{x} \cdot \cancel{x} \cdot \cancel{x}} \qquad \textit{Generalize the preceding reasoning.}$$
$$= x^2$$

Therefore,

$$\frac{x^5}{x^3} = x^{5-3} = x^2$$

This example suggests the following law of exponents.

Quotient Law of Exponents

Let m and n be positive integers, with $m > n$ and $x \neq 0$; then

$$\frac{x^m}{x^n} = x^{m-n}$$

NOTE ▶   By the Quotient Law of Exponents,

1. $\dfrac{x^7}{x^2} = x^{7-2} = x^5$

2. $\dfrac{x^{12}}{x^3} = x^{12-3} = x^9$

REMARK   We will see how to handle the cases when $n \geq m$ later in this section.

Suppose we want to simplify the algebraic expression $\left(\dfrac{x}{y}\right)^2$. We know that

$$\left(\frac{x}{y}\right)^2 = \frac{x}{y} \cdot \frac{x}{y}$$

$$= \frac{x \cdot x}{y \cdot y}$$

$$= \frac{x^2}{y^2}$$

Therefore

$$\left(\frac{x}{y}\right)^2 = \frac{x^2}{y^2}$$

This example suggests the last law of exponents.

Quotient to a Power Law of Exponents

Let n be a positive integer and $y \neq 0$; then

$$\left(\frac{x}{y}\right)^n = \frac{x^n}{y^n}$$

NOTE ▶   By the Quotient to a Power Law,

1. $\left(\dfrac{x}{y}\right)^7 = \dfrac{x^7}{y^7}$

2. $\left(\dfrac{x^3}{y^2}\right)^5 = \dfrac{(x^3)^5}{(y^2)^5}$ *By the Quotient to a Power Law*

$$= \frac{x^{15}}{y^{10}} \qquad \textit{By the Power to a Power Law}$$

XAMPLE 4.1.1   Simplify each of the following.

1. $\dfrac{5x^3y^7}{4xy^2} = \dfrac{5}{4} \cdot \dfrac{x^3}{x} \cdot \dfrac{y^7}{y^2}$

$= \dfrac{5}{4} \cdot x^{3-1} \cdot y^{7-2}$

$= \dfrac{5}{4} x^2 y^5; \qquad x \neq 0, y \neq 0$

2. $\left(\dfrac{5x^2}{2y}\right)^3 = \dfrac{(5x^2)^3}{(2y)^3}$

$= \dfrac{5^3(x^2)^3}{2^3 y^3}$

$= \dfrac{125x^6}{8y^3}; \qquad y \neq 0$

3. $\left(\dfrac{-12a^8b^5c^6}{6a^2b^4c}\right)^2 = \left(\dfrac{-12}{6} \cdot \dfrac{a^8}{a^2} \cdot \dfrac{b^5}{b^4} \cdot \dfrac{c^6}{c}\right)^2$

$= (-2 \cdot a^{8-2}b^{5-4}c^{6-1})^2$

$= (-2a^6bc^5)^2$

$= (-2)^2(a^6)^2b^2(c^5)^2$

$= 4a^{12}b^2c^{10}; \qquad a \neq 0, b \neq 0, c \neq 0$

NOTE ▶   Although it is important to note the restrictions on the variable, such as $x \neq 0$, $y \neq 0$ in part 1, we usually do not write the restrictions. They are simply understood.

We have investigated cases when the exponents on algebraic expressions are positive integers. How do we handle algebraic expressions that have negative integers or zero as exponents? Let's return to the quotient law of exponents for the answer to this question.

We know that

$$\frac{x^n}{x^n} = 1 \qquad \text{when } x \neq 0$$

But

$$\frac{x^n}{x^n} = x^{n-n} \qquad \textit{By the Quotient Law of Exponents}$$

$$= x^0$$

Thus, for the Quotient Law of Exponents to be consistent, x^0 must be 1. This fact, plus the fact that division by zero is undefined, leads to the following definition.

Definition 4.1.1	If x is any real number other than zero, then
	$$x^0 = 1$$

NOTE ▶ 0^0 is undefined.

EXAMPLE 4.1.2   Evaluate each of the following.

1. $4^0 = 1$ **2.** $\left(-\dfrac{3}{2}\right)^0 = 1$

3. $-\dfrac{5^0}{8} = -\dfrac{(5)^0}{8}$ **4.** $\left(8x^5y^2\right)^0 = 1$ $x \neq 0, y \neq 0$

$\qquad\qquad = -\dfrac{1}{8}$

5. 0^0 is undefined.

The quotient law of exponents can also show us how to handle negative integer exponents. We know that

$$\frac{x^3}{x^7} = \frac{\not{x} \cdot \not{x} \cdot \not{x}}{\not{x} \cdot \not{x} \cdot \not{x} \cdot x \cdot x \cdot x \cdot x} = \frac{1}{x^4}$$

But

$$\frac{x^3}{x^7} = x^{3-7} \qquad \textit{By the Quotient Law of Exponents}$$

$$= x^{-4}$$

Here, for the quotient law of exponents to be consistent,

$$x^{-4} = \frac{1}{x^4}.$$

This observation leads to the final definition of this section.

Definition 4.1.2	If x is any real number other than zero, then $$x^{-n} = \frac{1}{x^n}$$

EXAMPLE 4.1.3   Evaluate each of the following.

1. $3^{-2} = \dfrac{1}{3^2}$

$\qquad = \dfrac{1}{9}$

Try to avoid this mistake:

Incorrect	Correct
$3^{-2} = -9$ A negative exponent does not make the answer negative.	$3^{-2} = \dfrac{1}{3^2} = \dfrac{1}{9}$

2. $(-2)^{-4} = \dfrac{1}{(-2)^4}$

$\qquad\quad = \dfrac{1}{16}$

3. $-5^{-2} = -(5)^{-2}$

$\qquad\quad = -\dfrac{1}{5^2}$

$\qquad\quad = -\dfrac{1}{25}$

4. $2^{-1} + 2^{-2} = \dfrac{1}{2^1} + \dfrac{1}{2^2}$

$\qquad\qquad\quad = \dfrac{1}{2} + \dfrac{1}{4}$

$\qquad\qquad\quad = \dfrac{3}{4}$

NOTE ▶ Using the definitions of this section, it can be shown that all the laws of exponents hold for all integer exponents.

E X A M P L E   4 . 1 . 4   Simplify each of the following. (Unless otherwise indicated, we will *always* write answers with only positive exponents. Also, we will assume that none of the variables is zero.)

1. $x^3 \cdot x^{-8} = x^{3+(-8)}$

$\qquad\quad\; = x^{-5}$

$\qquad\quad\; = \dfrac{1}{x^5}$

2. $(3^{-1} + 2^{-1})^{-1} = \left(\dfrac{1}{3} + \dfrac{1}{2}\right)^{-1}$

$\qquad\qquad\qquad\; = \left(\dfrac{5}{6}\right)^{-1}$

$\qquad\qquad\qquad\; = \dfrac{1}{\dfrac{5}{6}}$

$\qquad\qquad\qquad\; = \dfrac{6}{5}$

An exponent of negative one inverts a fraction.

Generalizing part 2 gives

$$\left(\frac{a}{b}\right)^{-1} = \frac{b}{a} \qquad \text{provided } a \neq 0 \text{ and } b \neq 0.$$

Try to avoid this mistake:

Incorrect	Correct
$(3^{-1} + 2^{-1})^{-1} = (3^{-1})^{-1} + (2^{-1})^{-1}$ $\qquad = 3^1 + 2^1$ $\qquad = 5$ The power of a sum is *not* the sum of the powers.	$(3^{-1} + 2^{-1})^{-1} = \left(\frac{1}{3} + \frac{1}{2}\right)^{-1}$ $\qquad = \frac{6}{5}$ (from part 2)

3. $\left(\dfrac{3}{5}\right)^{-2} = \dfrac{3^{-2}}{5^{-2}}$

$\qquad = \dfrac{\dfrac{1}{3^2}}{\dfrac{1}{5^2}}$

$\qquad = \dfrac{1}{3^2} \cdot \dfrac{5^2}{1}$

$\qquad = \dfrac{5^2}{3^2}$

$\qquad = \dfrac{25}{9}$

or $\left(\dfrac{3}{5}\right)^{-2} = \left(\dfrac{5}{3}\right)^2$ $\qquad$ *Invert and square.*

$\qquad = \dfrac{25}{9}$

Generalizing part 3 gives

$$\left(\frac{a}{b}\right)^{-n} = \frac{b^n}{a^n} \qquad \text{provided } a \neq 0 \text{ and } b \neq 0$$

NOTE ▶   This problem illustrates the following property. If a negative exponent is on a *factor* in the denominator of a fraction, the exponent can be made positive by taking that factor and making it a factor of the numerator. We have a similar property if the negative exponent is on a *factor* in the numerator.

4. $\dfrac{x^{-5}}{y^{-6}}$ =

$= \dfrac{y^6}{x^5}$

5. $\dfrac{2x^{-3}}{y^{-9}}$ =

$= \dfrac{2y^9}{x^3}$

Here we had negative exponents on factors and used the preceding rule.

6. $\dfrac{5x^4}{7x^{10}} = \dfrac{5}{7x^{10-4}}$

$= \dfrac{5}{7x^6}$ *Alternative Method*

7. $\left(\dfrac{t^{-4}}{t^{-10}}\right)^3 = \left(\dfrac{t^{10}}{t^4}\right)^3$ or $\left(\dfrac{t^{-4}}{t^{-10}}\right) = (t^{-4-(-10)})^3$

$\qquad\qquad = (t^{10-4})^3 \qquad\qquad\qquad = (t^6)^3$

$\qquad\qquad = (t^6)^3 \qquad\qquad\qquad\quad = t^{18}$

$\qquad\qquad = t^{18}$

8. $(2x^{-4}y^2)^{-2} \cdot (3xy^{-1})^3 = 2^{-2}(x^{-4})^{-2}(y^2)^{-2}3^3x^3(y^{-1})^3$

$\qquad\qquad\qquad\qquad = 2^{-2}x^8y^{-4}3^3x^3y^{-3}$

$\qquad\qquad\qquad\qquad = 2^{-2}3^3x^8x^3y^{-4}y^{-3}$

$\qquad\qquad\qquad\qquad = 2^{-2}3^3x^{11}y^{-7}$

$\qquad\qquad\qquad\qquad = \dfrac{1}{2^2} \cdot \dfrac{3^3}{1} \cdot \dfrac{x^{11}}{1} \cdot \dfrac{1}{y^7}$

$\qquad\qquad\qquad\qquad = \dfrac{27x^{11}}{4y^7}$

9. $\left(\dfrac{5x^2y^{-5}}{8x^6y^{-12}}\right)^2 = \left(\dfrac{5x^2y^{12}}{8x^6y^5}\right)^2$

$\qquad\qquad = \left(\dfrac{5y^7}{8x^4}\right)^2$

$\qquad\qquad = \dfrac{(5y^7)^2}{(8x^4)^2}$

$\qquad\qquad = \dfrac{5^2(y^7)^2}{8^2(x^4)^2}$

$\qquad\qquad = \dfrac{25y^{14}}{64x^8}$

10. $\left(\dfrac{2x^{-8}}{5y^{-6}}\right)^{-3}\left(\dfrac{3x^{-1}}{y^{-5}}\right)^{2} = \left(\dfrac{2y^{6}}{5x^{8}}\right)^{-3}\left(\dfrac{3y^{5}}{x}\right)^{2}$

$$= \frac{(2y^{6})^{-3}}{(5x^{8})^{-3}} \cdot \frac{(3y^{5})^{2}}{x^{2}}$$

$$= \frac{(5x^{8})^{3}}{(2y^{6})^{3}} \cdot \frac{(3y^{5})^{2}}{x^{2}}$$

$$= \frac{5^{3}(x^{8})^{3}}{2^{3}(y^{6})^{3}} \cdot \frac{3^{2}(y^{5})^{2}}{x^{2}}$$

$$= \frac{125x^{24}}{8y^{18}} \cdot \frac{9y^{10}}{x^{2}}$$

$$= \frac{1125x^{24}y^{10}}{8x^{2}y^{18}}$$

$$= \frac{1125x^{22}}{8y^{8}}$$

For convenience, the laws of exponents and the definitions of this section are given here.

Let m and n be integers and $x \neq 0$ and $y \neq 0$; then

$$\frac{x^{m}}{x^{n}} = x^{m-n} \qquad \textit{Quotient Law}$$

$$\left(\frac{x}{y}\right)^{n} = \frac{x^{n}}{y^{n}} \qquad \textit{Quotient to a Power Law}$$

$$x^{0} = 1$$

$$x^{-n} = \frac{1}{x^{n}}$$

E xercises 4.1

Evaluate each of the following.

1. $2^{2} \cdot 2^{3}$ **2.** $2^{5} \cdot 2^{-2}$ **3.** $\dfrac{5^{4}}{5^{2}}$ **4.** $\dfrac{8^{6}}{8^{3}}$

5. $\dfrac{(-7)^{4}}{(-7)^{5}}$ **6.** $\dfrac{(-2)^{2}}{(-2)^{6}}$ **7.** $(2^{2})^{4}$ **8.** $(3^{3})^{2}$

9. $(4^{-1})^{-2}$ **10.** $(5^{0})^{-4}$ **11.** 6^{-3} **12.** 4^{-5}

13. -8^{2} **14.** -5^{4} **15.** $(-11)^{2}$ **16.** $(-2)^{4}$

17. $(-3)^{-4}$

18. $(-5)^{-2}$

19. -4^{-3}

20. -1^{-6}

21. $2^{-3} + 4^{-1}$

22. $3^{-1} + 4^{-2}$

23. $4^{-1} + 2^{-2}$

24. $3^{-1} - 2^{-2}$

25. $(2^{-1} - 5^{-2})^{-1}$

26. $(3^{-2} - 3^{-1})^{-2}$

27. $8^{-1} + 8^{0} + 8$

28. $5^{-1} + 5^{0} + 5$

29. $\left(\dfrac{2}{3}\right)^3$

30. $\left(\dfrac{5}{7}\right)^2$

31. $\left(\dfrac{9}{8}\right)^0$

32. $\left(\dfrac{-3}{4}\right)^0$

33. $\left(\dfrac{5}{8}\right)^{-1}$

34. $\left(\dfrac{-9}{11}\right)^{-1}$

35. $\left(\dfrac{4}{3}\right)^{-2}$

36. $\left(\dfrac{2}{7}\right)^{-3}$

37. $\dfrac{5^{-2}}{3^{-3}}$

38. $\dfrac{4^{-2}}{5^{-1}}$

39. $\dfrac{5^2}{3^{-2}}$

40. $\dfrac{2^{-1}}{4^2}$

Simplify each of the following. (Remember to write your answers with only positive exponents.)

41. $\dfrac{7x^8y^2}{2x^6y}$

42. $\dfrac{15x^{11}y^7}{3x^2y^2}$

43. $\left(\dfrac{4x^3}{2y^5}\right)^3$

44. $\left(\dfrac{3y^9}{2x^{10}}\right)^4$

45. $\left(\dfrac{8x^9y^3}{4x^3y^2}\right)^2$

46. $\left(\dfrac{10x^7y^{15}z^4}{-5x^2y^5z}\right)^3$

47. $(3x^2y^{-4})(7x^{-5}y^{-1})$

48. $(2x^{-3}y)(5x^5y^{-4})$

49. $(3x^{-2}y^2)^{-1}(4xy^3)^{-2}$

50. $(3x^{-2}y^{-3})^2(5xy^{-1})^{-3}$

51. $(8x^0y^{-4})^0(2xy^{-2})^3$

52. $(5^0x^0y^{-2})^5(6x^{-2}y^{-4})^{-3}$

53. $\left(\dfrac{5x}{2y^3}\right)^2$

54. $\left(\dfrac{3x^4}{y}\right)^3$

55. $\left(\dfrac{2x^4}{5y^{-2}}\right)^3$

56. $\left(\dfrac{4xy^{-2}}{7x^3y^{-4}}\right)^2$

57. $\left(\dfrac{5x^{-3}y}{3x^2y^{-2}}\right)^3$

58. $\dfrac{5x^{-2}y^3}{3x^4y^{-1}}$

59. $\dfrac{8x^3y^{-2}}{10x^{-6}y^5}$

60. $\dfrac{9x^{-4}y^{-9}}{15x^{-7}y^{-2}}$

61. $\dfrac{2^{-2}x^{-3}y^{-4}}{8^{-1}x^{-1}y^{-2}}$

62. $\left(\dfrac{3y^2}{2x^3}\right)^{-3}$

63. $\dfrac{(2x^3y^{-4})^4}{(5x^0y^7)^2}$

64. $\dfrac{(3x^5y^{-4})^{-2}}{(2x^{-2}y^{-3})^3}$

65. $\dfrac{(-3x^4y^7)^3}{(2x^{-4}y^3)^{-4}}$

66. $\dfrac{(8xy^{-3})^{-1}}{(4x^{-2}y^6)^{-2}}$

67. $\left(\dfrac{5x^4}{3y}\right)^{-2}$

68. $\left(\dfrac{3x^2}{y^{-4}}\right)^2\left(\dfrac{6x}{y^3}\right)^{-2}$

69. $\left(\dfrac{3x^3}{y^{-2}}\right)^{-2}\left(\dfrac{4x^5}{y^2}\right)^2$

70. $\left(\dfrac{2x^{-2}}{3y^{-5}}\right)^{-3}\left(\dfrac{5x^{-1}}{y^{-2}}\right)^3$

71. $\left(\dfrac{-4x}{y^3}\right)^{-2}\left(\dfrac{6x^{-2}}{y^4}\right)^2$

72. $\left[\dfrac{4^0x^{-2}y^6}{8x^3y^4}\right]^{-2}$

73. $\left[\dfrac{6x^3y^{-2}}{9^0x^{-5}y^{-6}}\right]^{-3}$

74. $\left[\left(\dfrac{5x^3}{2y^4}\right)^{-2}\right]^{-1}$

75. $\left[\left(\dfrac{2x^2}{3y^3}\right)^{-1}\right]^{-3}$

76. $\dfrac{(x+7)^{-1}}{x+7}$

77. $\dfrac{(2x-y)^{-3}}{(2x-y)^{-4}}$

78. $(2x-5)^{-1}$

79. $(x+3)^{-2}$

80. $(x+y)^{-1}$

81. $(x+y)^{-2}$

82. $(x+y)(x^{-1}+y^{-1})$

83. $(x-y)(x^{-1}+y^{-1})$

84. $x^{2m}\cdot x^{4m}$

85. $(x^{5m})^2$

86. $\dfrac{x^{8m}}{x^{2m}}$

87. $\dfrac{x^{-3m}}{x^{5m}}$

88. $\dfrac{x^{m}}{x^{-m}}$

89. $\dfrac{x^{m} \cdot x^{5m}}{x^{2m}}$

90. $\left(\dfrac{x^{2m}x^{m}x^{-m}}{x^{3m}} \right)^{2}$

Simplify the following by factoring out the greatest common factor.

91. $5(2x - 3)^{4}(x - 7)^{4} + 8(2x - 3)^{3}(x - 7)^{5}$

92. $3(x + 2)^{2}(3x + 1)^{5} + 15(x + 2)^{3}(3x + 1)^{4}$

93. $\dfrac{4(x + 1)^{2} - 8x(x + 1)}{(x + 1)^{4}}$

94. $\dfrac{2x(x - 7)^{3} - 3x^{2}(x - 7)^{2}}{(x - 7)^{6}}$

95. $-(x + 1)(x - 2)^{-2} + (x - 2)^{-1}$

96. $-(2x - 3)(x + 5)^{-2} + 2(x + 5)^{-1}$

Write Algebra

97. Explain the following laws of exponents in your own words:
   a. the Quotient Law and   **b.** the Quotient to a Power Law.

98. Describe a process for raising a fraction to a power of -2.

4.2 *Reducing Rational Expressions to Lowest Terms*

In Chapter 1 we learned how to add, subtract, multiply, and divide numbers. Frequently, division problems are stated as fractions, such as $\frac{40}{5}$, $\frac{17}{3}$, $\frac{7}{10}$, and $-\frac{5}{2}$. You may remember that any number that can be written as the quotient of two integers (i.e., any number that can be written as a fraction with integral numerator and denominator) is called a *rational number.* A similar definition exists in algebra.

Definition 4.2.1	Any expression that can be written as the quotient of two polynomials, where the denominator is not equal to zero, is called a **rational expression.**

Examples of rational expressions are

$$\dfrac{x + 7}{x - 2}, \quad \dfrac{x^{2} - y^{2}}{x^{2} + 5xy + 4y^{2}}, \quad \dfrac{2x(x + 1)}{5x^{3}y^{2}(y + 4)}, \quad x^{2} - x - 12$$

The polynomial $x^{2} - x - 12$ is a rational expression because it is equivalent to

$$\dfrac{x^{2} - x - 12}{1}$$

Suppose you were asked to find the sum of $\frac{7}{4}$ and $\frac{3}{4}$. We know that $\frac{7}{4} + \frac{3}{4} = \frac{10}{4}$. However, $\frac{10}{4}$ is not the *simplified* answer to the problem. A **simplified fraction,** more commonly called a fraction reduced to its **lowest terms,** is a fraction in which the only common integer factor of both the numerator and denominator is 1.

To reduce $\frac{10}{4}$ to its lowest terms, we proceed as follows:

$$\frac{10}{4} = \frac{2 \cdot 5}{2 \cdot 2} \qquad \textit{Step 1: Factor numerator and denominator.}$$

$$= \frac{2}{2} \cdot \frac{5}{2} \qquad \textit{Step 2: Multiplication property of fractions.}$$

$$= 1 \cdot \frac{5}{2} \qquad \textit{Step 3: Since } \tfrac{2}{2} = 1.$$

$$= \frac{5}{2}$$

Thus, $\frac{10}{4}$ has been reduced to $\frac{5}{2}$.

We can achieve the same result by using the following reasoning:

$$\frac{10}{4} = \frac{2 \cdot 5}{2 \cdot 2} \qquad \textit{Factor numerator and denominator.}$$

$$= \frac{\cancel{2} \cdot 5}{\cancel{2} \cdot 2} \qquad \textit{Divide out the common factor of 2.}$$

$$= \frac{5}{2}$$

The expression "divide out the common factor of 2" replaces steps 2 and 3. Dividing out common factors is nothing more than a shortcut for reducing fractions. This idea is summarized in the following property.

Reduction Property of Fractions

$$\frac{ax}{ay} = \frac{\cancel{a}x}{\cancel{a}y} = \frac{x}{y} \qquad a \neq 0, y \neq 0$$

REMARKS

1. The reduction property of fractions says that we can divide out *factors* (not terms) that appear in both the numerator and the denominator.

2. For all values of x and y (except $y = 0$),

$$\frac{ax}{ay} = \frac{x}{y} \qquad \text{only if } a \neq 0$$

Let's now apply the reduction property of fractions to some rational expressions.

XAMPLE 4.2.1

Reduce the following rational expressions to lowest terms.

1. $\dfrac{21x^2y}{7xy^3} = \dfrac{\cancel{x} \cdot 3 \cdot x^{2-1}}{\cancel{x} \cdot y^{3-1}}$ ←*The numerator has more factors of* x.

← *The denominator has more factors of* y.

$= \dfrac{3x}{y^2}, \qquad x \neq 0, y \neq 0$ *(Although it is important that we note these restrictions, we usually do not write them. They are just understood.)*

2. $\dfrac{5x^8y^2z^6}{4x^5y^4z^9} = \dfrac{5x^{8-5}}{4y^{4-2}z^{9-6}}$

$= \dfrac{5x^3}{4y^2z^3}$

3. $\dfrac{6x^3y^2 + 3x^2y^2}{x^2y^4 - 4xy^4} = \dfrac{3x^2y^2(2x+1)}{xy^4(x-4)}$ *Before we can reduce this fraction, we must factor the numerator and denominator.*

$= \dfrac{3x^{2-1}(2x+1)}{y^{4-2}(x-4)}$

$= \dfrac{3x(2x+1)}{y^2(x-4)}$

4. $\dfrac{2x^2 + 3x - 35}{x^2 + 4x - 5} = \dfrac{(2x-7)\cancel{(x+5)}}{(x-1)\cancel{(x+5)}}$ *Factor the numerator and denominator. Divide out the common factor of* x + 5.

$= \dfrac{2x-7}{x-1}$

5. $\dfrac{x^2 + 7x + 12}{x^2 + 3x} = \dfrac{(x+4)\cancel{(x+3)}}{x\cancel{(x+3)}}$ *Factor the numerator and denominator. Divide out the common factor of* x + 3.

$= \dfrac{x+4}{x}$

6. $\dfrac{2x-5y}{5y-2x} = \dfrac{-1(-2x+5y)}{(5y-2x)}$

$= \dfrac{-1\cancel{(5y-2x)}}{\cancel{(5y-2x)}}$

$= \dfrac{-1}{1}$

$= -1$

There appears to be no common factor. However, 2x − 5y *and* 5y − 2x *are additive inverses. This tells us that if we factor* −1 *out of the numerator or denominator, we will generate a common factor.*

Try to avoid these mistakes:

Incorrect	Correct
1. $\dfrac{10}{15} = \dfrac{\cancel{5}+1}{\cancel{5}+6}$ $= \dfrac{1}{6}$	$\dfrac{10}{15} = \dfrac{\cancel{5}\cdot 2}{\cancel{5}\cdot 3}$ $= \dfrac{2}{3}$
Terms are being divided out.	
2. $\dfrac{x+8}{x+2} = \dfrac{\cancel{x}+8}{\cancel{x}+2}$ $= \dfrac{8}{2}$ $= 4$	$\dfrac{x+8}{x+2}$
Terms are being divided out. The reduction property of fractions says that we can divide out only common *factors*.	This fraction cannot be reduced; there is no common factor (other than 1) in the numerator and denominator.

7. $\dfrac{x^2 + 4x - 21}{9 - x^2} = \dfrac{(x+7)(x-3)}{(3+x)(3-x)}$ *Factor the numerator and denominator. As in part 6, there are factors that are additive inverses of each other.*

$\quad = \dfrac{(x+7)\cancel{(x-3)}}{(3+x)(-1)\cancel{(x-3)}}$ *Factor -1 out of $3 - x$ in the denominator. Divide out the common factor.*

$\quad = \dfrac{(x+7)}{-1(3+x)}$

$\quad = -\dfrac{x+7}{3+x}$

8. $\dfrac{x^3 - 1}{2x^2 + 3x - 5} = \dfrac{\cancel{(x-1)}(x^2 + x + 1)}{\cancel{(x-1)}(2x+5)}$ *Factor the numerator and denominator. Use the difference of two cubes formula to factor the numerator. Divide out the common factor.*

$\quad = \dfrac{x^2 + x + 1}{2x + 5}$

9. $\dfrac{ax + ay + bx + by}{ax - 2ay + bx - 2by} = \dfrac{a(x+y) + b(x+y)}{a(x-2y) + b(x-2y)}$ *Factor the numerator and denominator. Use factoring by grouping. Divide out the common factor.*

$\quad = \dfrac{\cancel{(a+b)}(x+y)}{\cancel{(a+b)}(x-2y)}$

$\quad = \dfrac{x+y}{x-2y}$

Exercises 4.2

Reduce the following rational expressions to lowest terms.

1. $\dfrac{8x^4y^2}{6xy^3}$

2. $\dfrac{-7x^2y}{4x^5y}$

3. $\dfrac{15x^5y^9}{3xy^3}$

4. $\dfrac{3x^4y}{5ab}$

5. $\dfrac{-9ab^2c^3}{4ab^3c}$

6. $\dfrac{8xy^5z^3}{-2x^3y^4z}$

7. $\dfrac{7x^3yz^4}{4xy^3z^8}$

8. $-\dfrac{8a^2b^3z^4}{12a^2b^3z^4}$

9. $\dfrac{-6xy^3z^2}{24x^3y^7z^8}$

10. $-\dfrac{10m^2n^5}{8mn^3}$

11. $\dfrac{3x^3 + x^2y}{x^2y^2 - 2xy^3}$

12. $\dfrac{2x^2y - 14xy}{x^2y + 3xy^2}$

13. $\dfrac{3x + 6y}{3x}$

14. $\dfrac{5x - 10y}{5x}$

15. $\dfrac{4x^3y - 4x^2y}{2xy^2 - 2x^2y^2}$

16. $\dfrac{8x^2y + 4x^2}{6x^2y + 3x^2}$

17. $\dfrac{x^2 + 5x + 6}{x^2 - 4x - 21}$

18. $\dfrac{2x^2 - 3x - 20}{x^2 - 2x - 8}$

19. $\dfrac{6x^2 + 7x - 3}{4x^2 - 9}$

20. $\dfrac{6x^2 + 7x - 5}{2x^2 - x}$

21. $\dfrac{x^2 - 2x - 3}{x^2 + 6x + 8}$

22. $\dfrac{6x^2 + 2x - 4}{3x^2 + 7x - 6}$

23. $\dfrac{x^2 + 3x - 10}{4 - x^2}$

24. $\dfrac{2x^2 + 5x - 25}{x^2 + 5x}$

25. $\dfrac{4x^2 - 12x + 9}{2x^2 - 11x + 12}$

26. $\dfrac{2x^2 + x - 3}{x^2 + x - 20}$

27. $\dfrac{3x^2 + 4x}{6x^2 + 5x - 4}$

28. $\dfrac{8x^2 + 28x + 12}{2x^2 - 4x - 30}$

29. $\dfrac{3x^2 + 3x - 6}{6x^2 + 30x}$

30. $\dfrac{3x^2 + 24x + 48}{6x^2 + 18x - 24}$

31. $\dfrac{2x^2 + 5x - 12}{9 - 4x^2}$

32. $\dfrac{2x^2 + 3x - 14}{6 - x - x^2}$

33. $\dfrac{x^3 + 8}{3x^2 + 2x - 8}$

34. $\dfrac{x^3 - 64}{x^2 - 8x + 16}$

35. $\dfrac{27x^3 - 1}{6x^2 + 4x - 2}$

36. $\dfrac{8x^3 + 27}{2x^2 - 3x - 9}$

37. $\dfrac{2ax - ay - 2bx + by}{ax + ay - bx - by}$

38. $\dfrac{2ax - 2a + 3x - 3}{6ax + 4a + 9x + 6}$

39. $\dfrac{6ax - 8x - 3ay + 4y}{ay + 2y - 2ax - 4x}$

40. $\dfrac{2x^2 - x - 3}{x^3 + x^2 + 4x + 4}$

41. $\dfrac{x^3 + xy + 3x^2y + 3y^2}{xy^2 + x^2 + 3y^3 + 3xy}$

42. $\dfrac{x^2 + x - 6}{x^3 + 3x^2 - 4x - 12}$

43. a. Reduce $\dfrac{x^2 - x - 6}{x^2 - 4}$.

   b. Evaluate your answer for $x = 1$, $x = 3$, and $x = -2$.

   c. Evaluate $\dfrac{x^2 - x - 6}{x^2 - 4}$ for $x = 1$, $x = 3$, and $x = -2$.

   d. Do you always get the same answer? Why or why not?

Write Algebra

44. When reducing a rational expression we divide out common factors, not terms. Explain the difference between a term and a factor in such a rational expression.

45. Explain how to reduce a rational expression in which the numerator and denominator contain factors that are additive inverses.

4.3 *Multiplying and Dividing Rational Expressions*

Suppose we are asked to calculate the product of $\frac{2}{3}$ and $\frac{5}{7}$; that is, we want to determine $\frac{2}{3} \cdot \frac{5}{7}$. In Chapter 1 we learned to find this product in the following fashion:

$$\frac{2}{3} \cdot \frac{5}{7} = \frac{2 \cdot 5}{3 \cdot 7} = \frac{10}{21}$$

This idea is summarized in the following property.

Multiplication Property of Fractions

$$\frac{c}{d} \cdot \frac{n}{m} = \frac{cn}{dm} \qquad d \neq 0, m \neq 0$$

Since algebra is merely a generalization of arithmetic, the multiplication property of fractions applies to rational expressions as well as rational numbers. The following example demonstrates how the multiplication property of fractions applies to rational expressions.

EXAMPLE 4.3.1   Perform the indicated multiplications and reduce the answers to lowest terms.

1. $\dfrac{5xy^2}{2ab} \cdot \dfrac{3x^2y^3}{4a^2b} = \dfrac{5xy^2 \cdot 3x^2y^3}{2ab \cdot 4a^2b}$

$\qquad\qquad = \dfrac{15x^3y^5}{8a^3b^2}$

Notice that this fraction cannot be reduced, since the greatest common factor of the numerator and denominator is one.

2. $\dfrac{4x^3}{9y^2} \cdot \dfrac{3y}{2x} = \dfrac{4x^3 \cdot 3y}{9y^2 \cdot 2x}$

$\qquad\quad = \dfrac{12x^3y}{18xy^2}$

$$= \frac{\cancel{6} \cdot 2 \cdot x^{3-1}}{\cancel{6} \cdot 3 \cdot y^{2-1}}$$

$$= \frac{2x^2}{3y}$$

REMARK As with the rational numbers, we will always reduce our answers to lowest terms.

3. $\dfrac{2x^2 - 7x - 4}{x^2 - x - 12} \cdot \dfrac{x^2 + x - 6}{2x^2 + x}$

$= \dfrac{(2x + 1)(x - 4)}{(x - 4)(x + 3)} \cdot \dfrac{(x + 3)(x - 2)}{x(2x + 1)}$ *Factor the numerator and denominator of both fractions.*

$= \dfrac{\cancel{(2x + 1)}\cancel{(x - 4)}\cancel{(x + 3)}(x - 2)}{\cancel{(x - 4)}\cancel{(x + 3)}x\cancel{(2x + 1)}}$ *Divide out the common factors.*

$= \dfrac{x - 2}{x}$

4. $\dfrac{x^2 + 5x + 6}{x^2 - x} \cdot \dfrac{x^2 + 3x - 4}{x^2 - 2x - 15}$

$= \dfrac{(x + 2)\cancel{(x + 3)}}{x\cancel{(x - 1)}} \cdot \dfrac{(x + 4)\cancel{(x - 1)}}{\cancel{(x + 3)}(x - 5)}$ *Observe that we can divide out a common factor from the first fraction's numerator with one in the second fraction's denominator.*

$= \dfrac{(x + 2)(x + 4)}{x(x - 5)}$ *The answer is in factored form.*
                     ←We may stop here

                    or

$= \dfrac{x^2 + 6x + 8}{x^2 - 5x}$ *←perform the multiplication.*

Later in this chapter we will see the usefulness of leaving an answer in factored form.

5. $(x + 5) \cdot \dfrac{x^2 + 6x - 7}{x^2 - 25} = \dfrac{x + 5}{1} \cdot \dfrac{(x + 7)(x - 1)}{(x + 5)(x - 5)}$ *Rewrite x + 5 as a fraction. Factor the numerator and denominator of the second fraction.*

$= \dfrac{\cancel{x + 5}}{1} \cdot \dfrac{(x + 7)(x - 1)}{\cancel{(x + 5)}(x - 5)}$ *Divide out common factors.*

$= \dfrac{(x + 7)(x - 1)}{x - 5}$

6. $\dfrac{x + 3}{2x^2 + x - 3} \cdot \dfrac{x - 1}{x^2 - 2x - 15} = \dfrac{\overset{1}{\cancel{x + 3}}}{(2x + 3)(\cancel{x - 1})} \cdot \dfrac{\overset{1}{\cancel{x - 1}}}{(x - 5)(\cancel{x + 3})}$

$= \dfrac{1}{(2x + 3)(x - 5)}$ *Don't forget the understood factor of 1 in the numerator.*

7. $\dfrac{x^2 - 2x - 15}{2x + 8} \cdot \dfrac{4x}{25 - x^2}$

$= \dfrac{(x - 5)(x + 3)}{2(x + 4)} \cdot \dfrac{4x}{(5 - x)(5 + x)}$

$= \dfrac{(\cancel{x - 5})(x + 3)}{\cancel{2}(x + 4)} \cdot \dfrac{\overset{2}{\cancel{4}x}}{(-1)(\cancel{x - 5})(5 + x)}$ *Factor −1 out of 5 − x in the denominator.*

$= \dfrac{2x(x + 3)}{(-1)(x + 4)(5 + x)} = -\dfrac{2x(x + 3)}{(x + 4)(x + 5)}$

8. $\dfrac{2ax + 2bx + a + b}{2ax - bx - 6a + 3b} \cdot \dfrac{ax - bx - 3a + 3b}{ax - 5a + bx - 5b}$

$= \dfrac{2x(a + b) + 1(a + b)}{x(2a - b) - 3(2a - b)} \cdot \dfrac{x(a - b) - 3(a - b)}{a(x - 5) + b(x - 5)}$ *Use factoring by grouping.*

$= \dfrac{(\cancel{a + b})(2x + 1)}{(2a - b)(\cancel{x - 3})} \cdot \dfrac{(a - b)(\cancel{x - 3})}{(x - 5)(\cancel{a + b})}$ *Divide out the common factors.*

$= \dfrac{(2x + 1)(a - b)}{(2a - b)(x - 5)}$

Let's now consider some division problems. Suppose you were asked to find $\frac{4}{3} \div \frac{5}{7}$. Again, from Chapter 1 we know that

$$\frac{4}{3} \div \frac{5}{7} = \frac{4}{3} \cdot \frac{7}{5} = \frac{28}{15}$$

The rule for the division of fractions is stated as follows.

Division Property of Fractions

$$\frac{a}{b} \div \frac{s}{t} = \frac{a}{b} \cdot \frac{t}{s} = \frac{at}{bs} \qquad b \neq 0, t \neq 0, s \neq 0$$

This property shows how to convert division problems into multiplication problems. To make this conversion, we simply invert the divisor and change the operation from division to mutiplication.

EXAMPLE 4.3.2

Perform the indicated divisions, and reduce the answers to lowest terms.

1. $\dfrac{5x^2}{12y} \div \dfrac{4y^3}{x} = \dfrac{5x^2}{12y} \cdot \dfrac{x}{4y^3}$ *Invert and multiply.*

$$= \dfrac{5x^2 \cdot x}{12y \cdot 4y^3}$$

$$= \dfrac{5x^3}{48y^4}$$

2. $\dfrac{8a^3}{3bc^2} \div \dfrac{2a}{5c^4} = \dfrac{8a^3}{3bc^2} \cdot \dfrac{5c^4}{2a}$ *Invert and multiply.*

$$= \dfrac{40a^3c^4}{6abc^2}$$

$$= \dfrac{\cancel{2} \cdot 20 \cdot a^{3-1}c^{4-2}}{\cancel{2} \cdot 3 \cdot b}$$

$$= \dfrac{20a^2c^2}{3b}$$

3. $\dfrac{x^2 - 5x - 14}{x^2 + 2x - 15} \div \dfrac{3x^2 + 6x}{x^2 - 9}$

$$= \dfrac{x^2 - 5x - 14}{x^2 + 2x - 15} \cdot \dfrac{x^2 - 9}{3x^2 + 6x}$$ *Invert and multiply.*

$$= \dfrac{(x - 7)\cancel{(x + 2)}}{(x + 5)\cancel{(x - 3)}} \cdot \dfrac{(x + 3)\cancel{(x - 3)}}{3x\cancel{(x + 2)}}$$ *Factor and divide out common factors.*

$$= \dfrac{(x - 7)(x + 3)}{3x(x + 5)}$$

 4. $\dfrac{2x^2 - 7x - 4}{x^2 - 8x} \cdot \dfrac{x + 5}{x - 4} \div \dfrac{2x^2 + 11x + 5}{3x - 2}$

$$= \dfrac{2x^2 - 7x - 4}{x^2 - 8x} \cdot \dfrac{x + 5}{x - 4} \cdot \dfrac{3x - 2}{2x^2 + 11x + 5}$$ *Invert and multiply.*

$$= \dfrac{(2x + 1)\cancel{(x - 4)}}{x(x - 8)} \cdot \dfrac{\cancel{x + 5}}{\cancel{x - 4}} \cdot \dfrac{3x - 2}{\cancel{(2x + 1)}\cancel{(x + 5)}}$$ *Factor and divide out common factors.*

$$= \dfrac{3x - 2}{x(x - 8)}$$

Exercises 4.3

Perform the indicated operations, and reduce the answers to lowest terms.

1. $\dfrac{5xy^2}{3a} \cdot \dfrac{2x^3y}{7a^4b}$

2. $\dfrac{4s^4t}{2m} \cdot \dfrac{9m^3n}{6t^8}$

3. $\dfrac{9xy^6}{5x^2b} \div \dfrac{6b^2y^8}{15b^4y}$

4. $\dfrac{xyz^2}{2xz^4} \div \dfrac{5x^3}{8z^9}$

5. $\dfrac{4xy^3}{2xy} \div \dfrac{12x^2y^3}{3xy}$

6. $\dfrac{6xy^2}{9x^2y^2} \div \dfrac{18y^3}{3x}$

7. $\dfrac{2ab^2}{7xb} \cdot \dfrac{2ab^2}{4a^3b^3}$

8. $\dfrac{25m^2n^5}{3m^5n^3} \cdot \dfrac{m^2n}{75n^6}$

9. $\dfrac{16xy^7}{12x^3y^2} \cdot \dfrac{3x^2y}{4y^6}$

10. $\dfrac{15abc}{6b^2c^3} \div \dfrac{10ac^3}{4bc^5}$

11. $\dfrac{10x^2y^2}{6xy^2} \div \dfrac{x^2y}{3x^2y^3}$

12. $\dfrac{28a^7b^3}{2a^3b^2} \cdot \dfrac{2ab}{4a^2b^2}$

13. $\dfrac{27ax^3}{5x^2} \div \dfrac{3ax^4}{5x^5y^4}$

14. $\dfrac{3a^2c^4}{12b^3c} \cdot \dfrac{4ab^4c}{a^2c^3}$

15. $2x^2y \cdot \dfrac{5ax^3}{8y^4}$

16. $4x^5y^2 \cdot \dfrac{3ab^2}{10xy^4}$

17. $a^2bc \div \dfrac{abc}{2xyz^2}$

18. $xy^3z \div \dfrac{3xy^4}{15ab}$

19. $\dfrac{10x^2y^3}{7a^2b} \div 5xy^5$

20. $\dfrac{12m^5n^2}{5ab^3} \div 4m^2n^2$

21. $\dfrac{5xy^2}{7a^3b^4} \div \dfrac{3x^4y^7}{6a^2b^6} \cdot \dfrac{14x^5y^2}{15a^4b}$

22. $\dfrac{8r^5s^8}{9m^5n} \div \dfrac{2r^5s^2}{12mn^7} \cdot \dfrac{3s^3}{16mn^5}$

23. $\dfrac{ab^2}{2x^4y^5} \cdot \dfrac{10xy^2}{9a^3b} \cdot 6xy^3$

24. $\dfrac{6a^3b^4}{7cd^5} \cdot \dfrac{21c^3d}{2ab^7} \cdot c^2d^2$

25. $\dfrac{5x^3y^5}{8a^3b^2} \cdot \dfrac{14ab^7}{9xy^8} \div \dfrac{a^2b^2}{12xy}$

26. $\dfrac{10m^6n^7}{9r^3s^2} \cdot 18r^5s^3 \div \dfrac{4r^2s}{3m^2n}$

27. $\dfrac{48a^2b^3}{35c^4d^5} \div 12b^2 \cdot 15c^2d$

28. $\dfrac{8x^3y^3}{27m^5n^7} \div \dfrac{4x^5y^6}{45mn^4} \cdot \dfrac{xy^2}{10mn}$

29. $\dfrac{2x + 2}{5x - 15} \cdot \dfrac{x - 3}{xy + y}$

30. $\dfrac{4 - x}{6x + 9} \cdot \dfrac{2xz + 3z}{5x - 20}$

31. $\dfrac{5m + n}{2y - 2xy} \cdot \dfrac{6x - 6}{20m + 4n}$

32. $\dfrac{p^3 + 4p}{3p - 15} \div \dfrac{8p^5}{2p - 10}$

33. $\dfrac{10x^2y^2 + 5xy^2}{8 - 4x} \div \dfrac{6x + 3}{12x - 24}$

34. $\dfrac{5xy - 2x}{7x + 28} \div \dfrac{10x^2y - 4x^2}{21x + 84}$

35. $\dfrac{2x^2 + 7x + 3}{2x^2 + 9x - 5} \cdot \dfrac{2x^2 - 7x + 3}{x^2 - 9}$

36. $\dfrac{x^2 + x - 20}{x^2 + x - 6} \cdot \dfrac{x^2 + 5x + 6}{x^2 + 7x + 10}$

37. $\dfrac{10x^2 + 17x + 3}{2x^2 + 7x + 6} \div \dfrac{5x^2 + 41x + 8}{2x^2 + x - 6}$

38. $\dfrac{9x^2 - 4}{7x^2 + 27x - 4} \div \dfrac{6x^2 - 13x + 6}{2x^2 + 5x - 12}$

39. $\dfrac{10x^2 - x - 2}{3x^2 + 19x - 14} \cdot \dfrac{x^2 + 11x + 28}{2x^2 + 7x - 4}$

40. $\dfrac{2x^2 + 9x + 4}{2x^2 + 13x + 6} \div \dfrac{2x^2 + 8x}{x^2 + 3x - 18}$

41. $\dfrac{x^2 - 1}{3x^2 - x - 4} \div \dfrac{5x - 5x^2}{3x^2 + 2x - 8}$

42. $\dfrac{2x^2 - x - 15}{2x^2 + 9x + 10} \cdot \dfrac{x^2 - 3x - 10}{21x - 7x^2}$

43. $\dfrac{2x^2 - 11x + 15}{2x^2 + 5x - 25} \div \dfrac{12x - 4x^2}{x + 5}$

44. $\dfrac{2x^2 - 8x}{2x - 1} \div \dfrac{x^2 - x - 12}{2x^2 + 5x - 3}$

45. $\dfrac{6 - 23x - 4x^2}{3x^2 + 10x - 48} \cdot \dfrac{3x^2 + x - 24}{8x^2 - 6x + 1}$

46. $\dfrac{6x^2 - 13x - 5}{5 + 3x - 2x^2} \cdot \dfrac{x^2 - 2x - 15}{3x^2 + 10x + 3}$

47. $\dfrac{x^2 - 4x + 4}{3x^2 - 5x - 2} \div \dfrac{16 - x^2}{3x^2 - 11x - 4}$

48. $\dfrac{5x^2 - 14x - 3}{2x^2 - 3x + 1} \div \dfrac{4x^2 + 7x + 3}{3 + x - 4x^2}$

49. $\dfrac{3x^2 - 10x - 8}{2x^2 - 3x - 20} \cdot (2x + 5)$

50. $\dfrac{8x^2 - 14x + 3}{3x^2 + 16x + 5} \cdot (3x + 1)$

51. $\dfrac{x^2 + 2x - 15}{2x^2 - 5x - 3} \div (x^2 + x - 20)$

52. $\dfrac{3x^2 + x - 4}{x + 5} \div (6x^2 + 11x + 4)$

53. $\dfrac{5x^2 - 3xy - 2y^2}{2x^2 + 3xy + y^2} \cdot \dfrac{2x^2 + 7xy + 3y^2}{5x^2 + 17xy + 6y^2}$

54. $\dfrac{4x^2 - 5xy - 6y^2}{4x^2 - 4xy + y^2} \cdot \dfrac{2x^2 + 3xy - 2y^2}{x^2 - 4y^2}$

55. $\dfrac{x^3 - 8}{x^2 + 2x - 3} \cdot \dfrac{3x^2 - 2x - 1}{3x^2 - 5x - 2}$

56. $\dfrac{x^3 - y^3}{2x^2 + xy - y^2} \cdot \dfrac{x^2 + 2xy + y^2}{x^2 - y^2}$

57. $\dfrac{x^3 + y^3}{4x^2 - 4xy + y^2} \div \dfrac{2x^2 - xy - 3y^2}{4x^2 - 8xy + 3y^2}$

58. $\dfrac{8x^3 + 27}{2x^2 - 5x + 3} \div \dfrac{2x^2 + 5x + 3}{2x^2 - x - 3}$

59. $\dfrac{ax + 2bx - 3a - 6b}{2ax - 2bx + a - b} \cdot \dfrac{2ax - 6bx + a - 3b}{ax + 2bx + a + 2b}$

60. $\dfrac{2mx - 2nx + 3m - 3n}{mx - nx - 4m + 4n} \cdot \dfrac{mx + 2nx - 4m - 8n}{2mx + 2nx + 3m + 3n}$

61. $\dfrac{2x^3 - 10x^2 - x + 5}{x^3 - 5x^2 + 3x - 15} \div \dfrac{2x^3 + 6x^2 - x - 3}{x^3 + 3x^2 + 3x + 9}$

62. $\dfrac{2x^3 + 3x^2 - 8x - 12}{2x^3 + x^2 - 2x - 1} \cdot \dfrac{2x^2 - x - 1}{x^2 - 3x - 10}$

63. $\dfrac{2x^2 + 3x + 1}{45x^4 + 30x^3} \div \dfrac{x^2 - 1}{2x^2 - 3x - 9} \div \dfrac{2x^2 - 5x - 3}{5x - 5x^2}$

64. $\dfrac{x^2 - 2xy - 3y^2}{2x^2 - xy - y^2} \div \dfrac{2x^2y - 6xy^2}{6x^2 + 11xy + 3y^2} \div \dfrac{2x^2 + 5xy + 3y^2}{16x^3 + 8x^2y}$

65. $\dfrac{x^3 - 1}{2x^2 - 11x + 12} \cdot \dfrac{2x^2 - 3x - 20}{8x^3 + 12x^2} \div \dfrac{2x^2 + 3x - 5}{4x^2 - 9}$

66. $\dfrac{x^3 + 2x^2 - 9x - 18}{2x^2 - 5x - 3} \cdot \dfrac{2x^2 + 3x + 1}{x^3 + 27} \div (x^2 + 3x + 2)$

4.4 Adding and Subtracting Rational Expressions

Suppose we are asked to calculate the sum of $\frac{2}{10}$ and $\frac{5}{10}$; that is, we want to determine $\frac{2}{10} + \frac{5}{10}$. In Chapter 1 we learned to find this sum in the following fashion:

$$\frac{2}{10} + \frac{5}{10} = \frac{2 + 5}{10} = \frac{7}{10}$$

This idea is summarized in the following property.

Addition Property of Fractions

$$\frac{a}{d} + \frac{b}{d} = \frac{a + b}{d} \qquad d \neq 0$$

We have a similar rule for finding the difference of two fractions with the same denominator.

Subtraction Property of Fractions

$$\frac{a}{d} - \frac{b}{d} = \frac{a - b}{d} \qquad d \neq 0$$

Let's now apply these properties to some rational expressions.

EXAMPLE 4.4.1   Perform the indicated additions and subtractions.

1. $\dfrac{5}{x} + \dfrac{7}{x} = \dfrac{5 + 7}{x}$ **2.** $\dfrac{3y}{2x} + \dfrac{5y}{2x} = \dfrac{3y + 5y}{2x}$

$\qquad\qquad = \dfrac{12}{x}$ $= \dfrac{8y}{2x}$

$\qquad\qquad\qquad\qquad\qquad\qquad\qquad = \dfrac{\cancel{2}(4y)}{\cancel{2}x}$

$\qquad\qquad\qquad\qquad\qquad\qquad\qquad = \dfrac{4y}{x}$

Always be sure to reduce your answer. After the addition, there is a common factor of 2 in the numerator and denominator that should be eliminated.

3. $\dfrac{x}{x^2 - x - 12} + \dfrac{3}{x^2 - x - 12} = \dfrac{x + 3}{x^2 - x - 12}$

$\qquad\qquad\qquad\qquad\qquad\qquad = \dfrac{\overset{1}{\cancel{x + 3}}}{\underset{1}{\cancel{(x + 3)}}(x - 4)}$ *Factor the denominator and reduce.*

$\qquad\qquad\qquad\qquad\qquad\qquad = \dfrac{1}{x - 4}$

4. $\dfrac{11}{3xy^2} - \dfrac{4}{3xy^2} = \dfrac{11 - 4}{3xy^2} = \dfrac{7}{3xy^2}$

5. $\dfrac{4x - 2}{3x^2 - 4x + 5} - \dfrac{3x + 8}{3x^2 - 4x + 5} = \dfrac{4x - 2 - (3x + 8)}{3x^2 - 4x + 5}$

$\qquad\qquad\qquad\qquad\qquad\qquad\qquad = \dfrac{4x - 2 - 3x - 8}{3x^2 - 4x + 5}$

$\qquad\qquad\qquad\qquad\qquad\qquad\qquad = \dfrac{x - 10}{3x^2 - 4x + 5}$

6. $\dfrac{5y - 3}{y + 2} + \dfrac{8}{y + 2} - \dfrac{2y - 1}{y + 2}$

$= \dfrac{5y - 3 + 8 - (2y - 1)}{y + 2}$

$= \dfrac{5y - 3 + 8 - 2y + 1}{y + 2}$ *Don't forget to change the sign.*

$= \dfrac{3y + 6}{y + 2}$

$= \dfrac{3(y + 2)}{y + 2}$

$= 3$

Try to avoid this mistake:

Incorrect	Correct
$\dfrac{2x + 3}{x + 7} - \dfrac{x + 2}{x + 7}$	$\dfrac{2x + 3}{x + 7} - \dfrac{x + 2}{x + 7}$
$= \dfrac{2x + 3 - x + 2}{x + 7}$	$= \dfrac{2x + 3 - (x + 2)}{x + 7}$
$= \dfrac{x + 5}{x + 7}$	$= \dfrac{2x + 3 - x - 2}{x + 7}$
	$= \dfrac{x + 1}{x + 7}$
Here we have forgotten to change the sign of the 2. Remember to subtract the entire numerator of the second fraction.	In subtraction problems, you must change the sign of every term in the numerator of the fraction that you are subtracting because the fraction line is a grouping symbol.

7. $\dfrac{7}{x - 5} + \dfrac{4}{5 - x} = \dfrac{7}{x - 5} + \dfrac{4}{-1(x - 5)}$

$= \dfrac{7}{x - 5} - \dfrac{4}{x - 5}$

$= \dfrac{7 - 4}{x - 5} = \dfrac{3}{x - 5}$

Since the denominators are not exactly the same, it might seem as if you cannot add these fractions. However, the denominators are additive inverses. Factoring -1 out of either denominator generates a common denominator.

The problems we just investigated, except for part 7, all had the same denominator. Suppose we want to add two fractions with different denominators. How do we solve that sort of problem? Let's return to arithmetic to find the answer.

The addition property of fractions states that to add two fractions, the denominators must be exactly the same. We can make the denominators the same by first finding the **least common denominator** (abbreviated LCD) and then converting the fractions to equivalent fractions with the LCD as their denominators.

How to Find the LCD

1. Factor the denominators into products of prime numbers.

2. *Determine the LCD by forming a product of all the distinct factors and raising each factor to its highest power found in any denominator.*

The following chart illustrates how to add fractions with unlike denominators.

Arithmetic	Algebra
$\dfrac{7}{12} + \dfrac{3}{10}$	$\dfrac{5}{2x^2y^3} + \dfrac{1}{6x^5y^2}$
$12 = 2^2 \cdot 3$ *Factor the*	$2x^2y^3 = 2 \cdot x^2y^3$
$10 = 2 \cdot 5$ *denominators.*	$6x^5y^2 = 2 \cdot 3 \cdot x^5y^2$
$\text{LCD} = 2^2 \cdot 3 \cdot 5$	$\text{LCD} = 2 \cdot 3 \cdot x^5 \cdot y^3$
$= 60$	$= 6x^5y^3$
$\dfrac{7}{12} + \dfrac{3}{10} = \dfrac{7}{12} \cdot \dfrac{5}{5} + \dfrac{3}{10} \cdot \dfrac{6}{6}$	$\dfrac{5}{2x^2y^3} + \dfrac{1}{6x^5y^2} = \dfrac{5}{2x^2y^3} \cdot \dfrac{3x^3}{3x^3}$
$= \dfrac{35}{60} + \dfrac{18}{60}$	$+ \dfrac{1}{6x^5y^2} \cdot \dfrac{y}{y}$
$= \dfrac{53}{60}$	$= \dfrac{15x^3}{6x^5y^3} + \dfrac{y}{6x^5y^3}$
	$= \dfrac{15x^3 + y}{6x^5y^3}$

EXAMPLE 4.4.2   Perform the indicated additions and subtractions.

1. $\dfrac{2}{3x} + \dfrac{7}{5x^3} = \dfrac{2}{3x} \cdot \dfrac{5x^2}{5x^2} + \dfrac{7}{5x^3} \cdot \dfrac{3}{3}$ $\text{LCD} = 3 \cdot 5 \cdot x^3$
$= 15x^3$

$= \dfrac{10x^2}{15x^3} + \dfrac{21}{15x^3} = \dfrac{10x^2 + 21}{15x^3}$.

2. $\dfrac{4}{x-1} + \dfrac{5}{x+3}$ $LCD = (x-1)(x+3)$

$= \dfrac{4}{x-1} \cdot \dfrac{(x+3)}{(x+3)} + \dfrac{5}{x+3} \cdot \dfrac{(x-1)}{(x-1)}$

$= \dfrac{4x+12}{(x-1)(x+3)} + \dfrac{5x-5}{(x-1)(x+3)} = \dfrac{9x+7}{(x-1)(x+3)}$

3. $\dfrac{4x+1}{x^2-3x} - \dfrac{3x}{x^2-x-6}$

$= \dfrac{4x+1}{x(x-3)} - \dfrac{3x}{(x-3)(x+2)}$ *We must first factor both denominators in order to find* $LCD = x(x-3)(x+2)$.

$= \dfrac{(4x+1)}{x(x-3)} \cdot \dfrac{(x+2)}{(x+2)} - \dfrac{3x}{(x-3)(x+2)} \cdot \dfrac{x}{x}$

$= \dfrac{4x^2+9x+2}{x(x-3)(x+2)} - \dfrac{3x^2}{x(x-3)(x+2)} = \dfrac{x^2+9x+2}{x(x-3)(x+2)}$

4. $\dfrac{3x+2}{x^2-8x+16} - \dfrac{x}{x^2-6x+8}$

$= \dfrac{3x+2}{(x-4)^2} - \dfrac{x}{(x-4)(x-2)}$ *Factor each denominator.* $LCD = (x-4)^2(x-2)$

$= \dfrac{(3x+2)}{(x-4)^2} \cdot \dfrac{(x-2)}{(x-2)} - \dfrac{x}{(x-4)(x-2)} \cdot \dfrac{(x-4)}{(x-4)}$

$= \dfrac{3x^2-4x-4}{(x-4)^2(x-2)} - \dfrac{x^2-4x}{(x-4)^2(x-2)}$

$= \dfrac{3x^2-4x-4-x^2+4x}{(x-4)^2(x-2)}$

$= \dfrac{2x^2-4}{(x-4)^2(x-2)}$ or $\dfrac{2(x^2-2)}{(x-4)^2(x-2)}$

5. $\dfrac{x+4}{2x^2-5x-3} + \dfrac{x-7}{x^2-2x-3}$

$= \dfrac{x+4}{(2x+1)(x-3)} + \dfrac{x-7}{(x-3)(x+1)}$ *Factor each denominator.* $LCD = (2x+1)(x-3)(x+1)$

$= \dfrac{(x+4)}{(2x+1)(x-3)} \cdot \dfrac{(x+1)}{(x+1)} + \dfrac{(x-7)}{(x-3)(x+1)} \cdot \dfrac{(2x+1)}{(2x+1)}$

$= \dfrac{x^2+5x+4}{(2x+1)(x-3)(x+1)} + \dfrac{2x^2-13x-7}{(2x+1)(x-3)(x+1)}$

$= \dfrac{x^2+5x+4+2x^2-13x-7}{(2x+1)(x-3)(x+1)}$

$= \dfrac{3x^2-8x-3}{(2x+1)(x-3)(x+1)}$

$$= \frac{(3x + 1)(x - 3)}{(2x + 1)(x - 3)(x + 1)}$$

Factor the numerator and reduce your answer.

$$= \frac{3x + 1}{(2x + 1)(x + 1)}$$

6. $\dfrac{x + y}{2x^2 + 7xy + 3y^2} + \dfrac{2x - 5y}{2x^2 - xy - y^2}$

$$= \frac{x + y}{(2x + y)(x + 3y)} + \frac{2x - 5y}{(2x + y)(x - y)}$$

Factor each denominator.
LCD = $(2x + y)(x + 3y)(x - y)$

$$= \frac{(x + y)}{(2x + y)(x + 3y)} \cdot \frac{(x - y)}{(x - y)} + \frac{(2x - 5y)}{(2x + y)(x - y)} \cdot \frac{(x + 3y)}{(x + 3y)}$$

$$= \frac{x^2 - y^2}{(2x + y)(x + 3y)(x - y)} + \frac{2x^2 + xy - 15y^2}{(2x + y)(x + 3y)(x - y)}$$

$$= \frac{x^2 - y^2 + 2x^2 + xy - 15y^2}{(2x + y)(x + 3y)(x - y)}$$

$$= \frac{3x^2 + xy - 16y^2}{(2x + y)(x + 3y)(x - y)}$$

Exercises 4.4

Perform the indicated operations.

1. $\dfrac{3}{2x} + \dfrac{7}{2x}$

2. $\dfrac{8}{3xy^2} + \dfrac{10}{3xy^2}$

3. $\dfrac{4x}{3a} - \dfrac{2x}{3a}$

4. $\dfrac{5m}{2d} - \dfrac{3m}{2d}$

5. $\dfrac{2k}{st} - \dfrac{9k}{st}$

6. $\dfrac{xy}{4x^2y^3} + \dfrac{3xy}{4x^2y^3}$

7. $\dfrac{5x}{2x + 3} + \dfrac{7x}{2x + 3}$

8. $\dfrac{3x + 1}{5x - 2} + \dfrac{4x - 6}{5x - 2}$

9. $\dfrac{x}{4x - 7} - \dfrac{3x + 5}{4x - 7}$

10. $\dfrac{5}{x - 1} + \dfrac{3}{1 - x}$

11. $\dfrac{2x}{x + 1} + \dfrac{2}{x + 1}$

12. $\dfrac{2x + 3}{x^2 + 5x + 1} + \dfrac{x - 7}{x^2 + 5x + 1}$

13. $\dfrac{5x + 1}{2x^2 + x - 11} - \dfrac{2x + 9}{2x^2 + x - 11}$

14. $\dfrac{x + 8}{x^2 + 7x + 10} - \dfrac{3}{x^2 + 7x + 10}$

15. $\dfrac{x + 3}{4x^2 - 1} + \dfrac{x - 2}{4x^2 - 1}$

16. $\dfrac{2x^2}{2x + 3} - \dfrac{x + 6}{2x + 3}$

17. $\dfrac{x^2 + 2x}{x + 1} + \dfrac{1}{x + 1}$

18. $\dfrac{7x}{y - 2} - \dfrac{3x}{2 - y}$

19. $\dfrac{2x^2 - 8}{x - 5} + \dfrac{x^2 + x + 12}{5 - x}$

20. $\dfrac{5}{2x} - \dfrac{1}{3x}$

21. $\dfrac{8}{5x} + \dfrac{3}{2y}$

22. $\dfrac{a}{x} + \dfrac{b}{3y}$

23. $\dfrac{11}{5x} - \dfrac{3c}{10y}$

24. $\dfrac{5}{4x^3} - \dfrac{1}{8x}$

25. $\dfrac{3}{4xy^2} + \dfrac{7}{3x^2y}$

26. $\dfrac{5}{6x^3y} + \dfrac{9}{4xy^5}$

27. $\dfrac{7x}{y} + 2$

28. $\dfrac{2}{5mn} - 1$

29. $\dfrac{5}{x + 2} + \dfrac{2}{x + 3}$

30. $\dfrac{3}{x - 1} + \dfrac{6}{x + 4}$

31. $\dfrac{3}{2x + 5} - \dfrac{5}{x - 3}$

32. $\dfrac{2}{2x - 1} - \dfrac{1}{2x + 1}$

33. $\dfrac{2x}{x + 4} + \dfrac{3x}{3x - 4}$

34. $\dfrac{5x}{3x + 2} - \dfrac{x}{2x - 1}$

35. $\dfrac{2x}{5x + 3} - \dfrac{4x}{2x - 3}$

36. $\dfrac{3x}{2x + 1} + \dfrac{3x}{3x - 1}$

37. $\dfrac{x + 2}{2x - 3} - \dfrac{x - 3}{x + 6}$

38. $\dfrac{2x + 1}{3x + 2} + \dfrac{x - 4}{3x - 2}$

39. $\dfrac{x - 1}{2x + 5} + \dfrac{3x - 2}{2x - 3}$

40. $\dfrac{3x + 1}{x + 3} - \dfrac{x - 4}{3x - 4}$

41. $\dfrac{2}{x + 1} + 3$

42. $\dfrac{5x}{x - 1} - 1$

43. $\dfrac{2x + 3}{x - 4} + 2$

44. $\dfrac{2}{x^2 + 4x} + \dfrac{x - 3}{x^2 + 3x - 4}$

45. $\dfrac{3}{x^2 - 5x} + \dfrac{7}{25 - x^2}$

46. $\dfrac{2x + 5}{x^2 - 3x} - \dfrac{x - 2}{2x^2 - 7x + 3}$

47. $\dfrac{x + 1}{2x^2 - x} + \dfrac{x - 11}{2x^2 + 5x - 3}$

48. $\dfrac{5x + 2}{6x^2 + 2x} - \dfrac{x + 2}{3x^2 - 8x - 3}$

49. $\dfrac{2}{x^2 - 4x - 5} + \dfrac{5}{x^2 - 2x - 15}$

50. $\dfrac{2x}{2x^2 + 5x - 3} + \dfrac{1}{2x^2 - 9x + 4}$

51. $\dfrac{4x}{3x^2 - 5x - 2} - \dfrac{1}{3x^2 + 13x + 4}$

52. $\dfrac{2x}{x^2 - 1} - \dfrac{3}{x^2 + 5x + 4}$

53. $\dfrac{5x}{x^2 - x - 6} + \dfrac{3}{x^2 - 7x + 12}$

54. $\dfrac{5}{x^2 - 4x + 4} - \dfrac{7}{x^2 + x - 6}$

55. $\dfrac{x + 4}{x^2 + 4x + 3} + \dfrac{x - 3}{2x^2 - x - 3}$

56. $\dfrac{3x - 5}{x^2 - x - 12} - \dfrac{x + 1}{x^2 + 5x + 6}$

57. $\dfrac{2x + 3}{x^2 - 1} + \dfrac{x - 2}{x^2 - 6x + 5}$

58. $\dfrac{5x + 1}{2x^2 - 7x - 15} - \dfrac{3x + 2}{2x^2 + x - 3}$

59. $\dfrac{4x - 5}{x^2 - 2x - 3} - \dfrac{3x + 4}{x^2 - 6x + 9}$

60. $\dfrac{5x - 8}{3x^2 - 8x + 4} - \dfrac{2x - 9}{3x^2 + 13x - 10}$

61. $\dfrac{2x + 9}{3x^2 - 7x - 20} + \dfrac{x - 12}{x^2 - 16}$

62. $\dfrac{3x + 11}{x^2 + 8x + 15} - \dfrac{5x + 7}{2x^2 + 11x + 5}$

63. $\dfrac{3x - 2}{2x^2 - 9x + 10} - \dfrac{x + 6}{x^2 - 6x + 8}$

64. $\dfrac{x - 1}{5x^2 + 18x + 9} + \dfrac{4x + 7}{3x^2 + 10x + 3}$

65. $\dfrac{x - 1}{6x^2 - 7x + 2} + \dfrac{x + 2}{2x^2 - 7x + 3}$

66. $\dfrac{8x + 3}{4x^2 - 1} - \dfrac{5x - 1}{2x^2 - 15x + 7}$

67. $\dfrac{5x + 1}{x^2 + x - 2} - \dfrac{3x + 5}{x^2 + 2x - 3}$

68. $\dfrac{2x - 7}{x^2 + 3x - 4} + \dfrac{x + 31}{x^2 - x - 20}$

69. $\dfrac{x + 1}{4x^2 + 4x - 15} - \dfrac{4x + 5}{8x^2 - 10x - 3}$

70. $\dfrac{4x - 1}{3x^2 + x - 2} + \dfrac{5x - 6}{3x^2 + 4x - 4}$

71. $\dfrac{3x - 4}{2x^2 - 3x - 5} - \dfrac{4x + 3}{3x^2 + 5x + 2}$

72. $\dfrac{3x - 1}{2x^2 - 3x - 2} + \dfrac{x + 2}{4x^2 + 4x + 1}$

73. $\dfrac{2x + y}{x^2 - y^2} - \dfrac{x - y}{2x^2 + 3xy + y^2}$

74. $\dfrac{7x + 6y}{2x^2 + 7xy + 3y^2} - \dfrac{11x + 9y}{3x^2 + 10xy + 3y^2}$

75. $\dfrac{2x + y}{x^2 + 5xy + 6y^2} - \dfrac{x - 4y}{x^2 + 6xy + 8y^2}$

76. $\dfrac{x - 10y}{3x^2 - 8xy + 4y^2} + \dfrac{2x + y}{3x^2 - 5xy + 2y^2}$

77. $\dfrac{5x - y}{2x^2 + xy - 3y^2} - \dfrac{3x + 2y}{x^2 - 2xy + y^2}$

78. $\dfrac{7x - 3y}{4x^2 + 4xy + y^2} + \dfrac{x - 2y}{2x^2 - xy - y^2}$

79. $\dfrac{5}{2x + 2} + \dfrac{7}{3x + 3} - \dfrac{1}{12}$

80. $\dfrac{8}{5x - 10} - \dfrac{3}{2x - 4} + \dfrac{5}{6}$

81. $\dfrac{x}{2x^2 + 5x - 3} + \dfrac{2x - 1}{x^2 + 2x - 3} - \dfrac{1}{2x^2 - 3x + 1}$

82. $\dfrac{x - 5}{x^2 - x - 2} + \dfrac{2}{x^2 - 5x + 6} + \dfrac{x + 3}{x^2 - 2x - 3}$

83. $\dfrac{3x + 1}{x^2 + x - 20} - \dfrac{7}{x^2 + 9x + 20} - \dfrac{x - 3}{x^2 - 16}$

84. $\dfrac{2x - 3}{2x^2 + 5x - 3} + \dfrac{6x - 5}{2x^2 - 9x + 4} + \dfrac{2x - 1}{x^2 - x - 12}$

85. $\dfrac{x - 2}{x^2 + 3x + 2} - \dfrac{x + 1}{x^2 - 4} + \dfrac{2x - 7}{x^2 - x - 2}$

86. $\dfrac{5x + 3}{3x^2 + 5x - 2} + \dfrac{x + 7}{3x^2 - 13x + 4} - \dfrac{2x - 1}{x^2 - 2x - 8}$

Write Algebra

87. Describe the process of finding the LCD for rational expressions with unlike denominators.

4.5 *Complex Fractions*

In Section 4.3 we examined the problem $\frac{4}{3} \div \frac{5}{7}$. The division problem could also be stated as

$$\dfrac{\dfrac{4}{3}}{\dfrac{5}{7}}$$

This fraction is an example of what is called a *complex fraction.*

Definition 4.5.1	Any fraction that contains a fraction in its numerator and/or denominator is called a **complex fraction.**

The following fractions are complex fractions:

$$\dfrac{\dfrac{2}{3}}{7}, \quad \dfrac{4}{\dfrac{5}{9}}, \quad \dfrac{\dfrac{12}{5}}{\dfrac{3}{10}}, \quad \dfrac{x + \dfrac{1}{3}}{y - \dfrac{1}{5}}, \quad \dfrac{\dfrac{2}{x} + \dfrac{7}{y}}{12}$$

$$\dfrac{\dfrac{5}{2x^2}}{\dfrac{9}{4a^3x}}, \quad \dfrac{\dfrac{1}{x} + \dfrac{1}{y}}{\dfrac{1}{x^2} - \dfrac{1}{y^2}}$$

In this section we will learn how to convert complex fractions into **simple fractions.** (A simple fraction does not have a fraction in its numerator or denominator.) There are two methods for converting complex fractions into simple fractions. One method uses the following steps:

1. Combine all the expressions in the numerator of the complex fraction.

2. Combine all the expressions in the denominator of the complex fraction.

3. Convert the complex fraction into a division problem and simplify using the techniques developed in Section 4.3.

The following example illustrates this method for simplifying complex fractions.

E X A M P L E   4 . 5 . 1   Simplify the following complex fractions:

1. $\dfrac{\dfrac{4}{3}}{\dfrac{5}{7}} = \dfrac{4}{3} \div \dfrac{5}{7}$ *Convert to a division problem.*

$= \dfrac{4}{3} \cdot \dfrac{7}{5}$ *Invert and multiply by the divisor.*

$= \dfrac{28}{15}$

2. $\dfrac{\dfrac{1}{2} - \dfrac{1}{3}}{\dfrac{5}{8}} = \dfrac{\dfrac{3}{6} - \dfrac{2}{6}}{\dfrac{5}{8}}$

$= \dfrac{\dfrac{1}{6}}{\dfrac{5}{8}}$ *Combine the numbers in the numerator.*

$= \dfrac{1}{6} \div \dfrac{5}{8}$

$= \dfrac{1}{6} \cdot \dfrac{8}{5}$

$= \dfrac{8}{30}$

$= \dfrac{4}{15}$

3. $\dfrac{\dfrac{4}{x} + \dfrac{3}{y}}{\dfrac{8}{xy^2}} = \dfrac{\dfrac{4y}{xy} + \dfrac{3x}{xy}}{\dfrac{8}{xy^2}}$

$= \dfrac{\dfrac{4y + 3x}{xy}}{\dfrac{8}{xy^2}}$ *Combine the expressions in the numerator.*

$= \dfrac{4y + 3x}{xy} \div \dfrac{8}{xy^2}$

$= \dfrac{4y + 3x}{xy} \cdot \dfrac{xy^2}{8}$

$= \dfrac{xy^2(4y + 3x)}{8xy}$

$= \dfrac{y(4y + 3x)}{8}$ *Reduce the answer.*

4. $\dfrac{\dfrac{x + 2}{x^2 - 9}}{\dfrac{x^2 - 4}{2x^2 + x - 15}} = \dfrac{x + 2}{x^2 - 9} \div \dfrac{x^2 - 4}{2x^2 + x - 15}$

$= \dfrac{x + 2}{x^2 - 9} \cdot \dfrac{2x^2 + x - 15}{x^2 - 4}$

$= \dfrac{\cancel{x + 2}}{\cancel{(x + 3)}(x - 3)} \cdot \dfrac{(2x - 5)\cancel{(x + 3)}}{\cancel{(x + 2)}(x - 2)}$ *Factor numerators and denominators and divide out common factors.*

$= \dfrac{2x - 5}{(x - 3)(x - 2)}$

5. $\dfrac{2}{\dfrac{3}{x + 4} - \dfrac{1}{x - 2}} = \dfrac{2}{\dfrac{3(x - 2)}{(x + 4)(x - 2)} - \dfrac{1(x + 4)}{(x - 2)(x + 4)}}$ *Combine the expressions in the denominator.*

$= \dfrac{2}{\dfrac{3x - 6}{(x + 4)(x - 2)} - \dfrac{x + 4}{(x - 2)(x + 4)}}$

$= \dfrac{2}{\dfrac{2x - 10}{(x + 4)(x - 2)}}$

$= 2 \div \dfrac{2x - 10}{(x + 4)(x - 2)}$

$= \dfrac{2}{1} \cdot \dfrac{(x + 4)(x - 2)}{2x - 10}$

$$= \frac{\cancel{2}}{1} \cdot \frac{(x+4)(x-2)}{\cancel{2}(x-5)}$$

$$= \frac{(x+4)(x-2)}{x-5}$$

6. $\dfrac{\dfrac{1}{x} + \dfrac{1}{y}}{\dfrac{1}{x^2} - \dfrac{1}{y^2}} = \dfrac{\dfrac{y}{xy} + \dfrac{x}{xy}}{\dfrac{y^2}{x^2y^2} - \dfrac{x^2}{x^2y^2}}$ *Combine the expressions in both the numerator and the denominator.*

$$= \frac{\dfrac{y+x}{xy}}{\dfrac{y^2-x^2}{x^2y^2}}$$

$$= \frac{y+x}{xy} \div \frac{y^2-x^2}{y^2x^2}$$

$$= \frac{y+x}{xy} \cdot \frac{x^2y^2}{y^2-x^2}$$

$$= \frac{\cancel{x+y}}{xy} \cdot \frac{x^2y^2}{\cancel{(y+x)}(y-x)}$$

$$= \frac{x^2y^2}{xy(y-x)}$$

$$= \frac{xy}{y-x}$$

The second method for simplifying complex fractions uses the following steps:

> 1. Find the LCD of all the fractions within the complex fraction.
>
> 2. Multiply the numerator and denominator of the complex fraction by this LCD.

The following example illustrates this method for simplifying fractions.

EXAMPLE 4.5.2   Simplify the following complex fractions.

1. $\dfrac{\dfrac{4}{3}}{\dfrac{5}{7}} = \dfrac{\dfrac{4}{3} \cdot 21}{\dfrac{5}{7} \cdot 21}$ *Multiply numerator and denominator by the LCD, 21.*

$$= \frac{28}{15}$$

2. $\dfrac{\dfrac{1}{y} + \dfrac{2}{x}}{\dfrac{3}{y} - \dfrac{1}{x}} = \dfrac{\left(\dfrac{1}{y} + \dfrac{2}{x}\right) \cdot xy}{\left(\dfrac{3}{y} - \dfrac{1}{x}\right) \cdot xy}$ *Multiply numerator and denominator by the LCD,* xy.

$$= \dfrac{\dfrac{1}{y} \cdot (xy) + \dfrac{2}{x} \cdot (xy)}{\dfrac{3}{y} \cdot (xy) - \dfrac{1}{x} \cdot (xy)}$$

$$= \dfrac{x + 2y}{3x - y}$$

3. $\dfrac{2 + \dfrac{3}{x + 1}}{1 - \dfrac{5}{x + 1}} = \dfrac{\left(2 + \dfrac{3}{x + 1}\right) \cdot (x + 1)}{\left(1 - \dfrac{5}{x + 1}\right) \cdot (x + 1)}$ *Multiply numerator and denominator by the LCD,* x + 1.

$$= \dfrac{2(x + 1) + \dfrac{3}{x + 1} \cdot (x + 1)}{1(x + 1) - \dfrac{5}{x + 1} \cdot (x + 1)}$$

$$= \dfrac{2x + 2 + 3}{x + 1 - 5}$$

$$= \dfrac{2x + 5}{x - 4}$$

4. $\dfrac{\dfrac{1}{x} + \dfrac{1}{y}}{\dfrac{1}{x^2} - \dfrac{1}{y^2}} = \dfrac{\left(\dfrac{1}{x} + \dfrac{1}{y}\right) \cdot (x^2 y^2)}{\left(\dfrac{1}{x^2} - \dfrac{1}{y^2}\right) \cdot (x^2 y^2)}$

$$= \dfrac{\dfrac{1}{x} \cdot (x^2 y^2) + \dfrac{1}{y} \cdot (x^2 y^2)}{\dfrac{1}{x^2} \cdot (x^2 y^2) - \dfrac{1}{y^2} \cdot (x^2 y^2)}$$

$$= \dfrac{xy^2 + x^2 y}{y^2 - x^2}$$

$$= \dfrac{xy(y + x)}{(y + x)(y - x)}$$

$$= \dfrac{xy}{y - x}$$

5. $\dfrac{\dfrac{8}{x-5} - \dfrac{4}{x+2}}{\dfrac{4}{x+2} - \dfrac{2}{x-5}} = \dfrac{\left(\dfrac{8}{x-5} - \dfrac{4}{x+2}\right) \cdot (x-5)(x+2)}{\left(\dfrac{4}{x+2} - \dfrac{2}{x-5}\right) \cdot (x-5)(x+2)}$

$= \dfrac{\dfrac{8}{x-5} \cdot (x-5)(x+2) - \dfrac{4}{x+2} \cdot (x-5)(x+2)}{\dfrac{4}{x+2} \cdot (x-5)(x+2) - \dfrac{2}{x-5} \cdot (x-5)(x+2)}$

$= \dfrac{8(x+2) - 4(x-5)}{4(x-5) - 2(x+2)}$

$= \dfrac{8x + 16 - 4x + 20}{4x - 20 - 2x - 4}$

$= \dfrac{4x + 36}{2x - 24}$

$= \dfrac{4(x+9)}{2(x-12)}$

$= \dfrac{2(x+9)}{x-12}$

Some fractions with negative exponents can be written as complex fractions.

EXAMPLE 4.5.3

Simplify the following.

$\dfrac{x^{-1} - y^{-1}}{x^{-3} - y^{-3}} = \dfrac{\dfrac{1}{x} - \dfrac{1}{y}}{\dfrac{1}{x^3} - \dfrac{1}{y^3}}$ *Use the definition of negative exponents.*

$= \dfrac{\left(\dfrac{1}{x} - \dfrac{1}{y}\right) \cdot \dfrac{x^3 y^3}{1}}{\left(\dfrac{1}{x^3} - \dfrac{1}{y^3}\right) \cdot \dfrac{x^3 y^3}{1}}$ *Multiply by the LCD.*

$= \dfrac{x^2 y^3 - x^3 y^2}{y^3 - x^3}$

$= \dfrac{x^2 y^2 (y - x)}{(y - x)(y^2 + yx + x^2)}$

$= \dfrac{x^2 y^2}{y^2 + yx + x^2}$

Try to avoid this mistake:

Incorrect	Correct
$$\frac{x^{-1} - y^{-1}}{x^{-3} - y^{-3}} = \frac{x^3 - y^3}{x^1 - y^1}$$ The negative exponents are on *terms*, not *factors*. We cannot make the exponents positive by moving the terms from the numerator to the denominator or vice versa.	$$\frac{x^{-1} - y^{-1}}{x^{-3} - y^{-3}} = \frac{\dfrac{1}{x} - \dfrac{1}{y}}{\dfrac{1}{x^3} - \dfrac{1}{y^3}}$$ $$\vdots \text{ (see Example 4.5.3)}$$ $$= \frac{x^2 y^2}{y^2 + yx + x^2}$$

Exercises 4.5

Simplify the following fractions.

1. $\dfrac{\dfrac{5}{7}}{\dfrac{45}{16}}$

2. $\dfrac{\dfrac{9}{8}}{\dfrac{27}{48}}$

3. $\dfrac{\dfrac{5x}{2y}}{2}$

4. $\dfrac{\dfrac{7x^2}{5y^3}}{3}$

5. $\dfrac{\dfrac{3x}{2y^2}}{\dfrac{5x^4}{4y^3}}$

6. $\dfrac{\dfrac{21x^4}{8y}}{\dfrac{7x^3}{16y^2}}$

7. $\dfrac{\dfrac{2ab}{5xy}}{\dfrac{11x^3y}{4a^2b}}$

8. $\dfrac{\dfrac{9mn}{4s^2t}}{\dfrac{7st^3}{10m^4}}$

9. $\dfrac{\dfrac{5x+3}{2x-1}}{\dfrac{x+4}{x-2}}$

10. $\dfrac{\dfrac{3x-4}{x-2}}{\dfrac{x-1}{x+6}}$

11. $\dfrac{\dfrac{2x+6}{5x-7}}{\dfrac{x+3}{5x-7}}$

12. $\dfrac{\dfrac{x-1}{4x+24}}{\dfrac{1-x}{3x+18}}$

13. $\dfrac{\dfrac{2x^2+3x}{y^2-y}}{\dfrac{8x^4+12x^3}{3y^3-3y^2}}$

14. $\dfrac{\dfrac{5x^3-5x^2}{xy^2+7y}}{\dfrac{x-1}{3xy^3+21y^2}}$

15. $\dfrac{\dfrac{x^2+x-12}{x^2+9x+20}}{\dfrac{9-x^2}{x^2+8x+15}}$

16. $\dfrac{\dfrac{x^2-5x+6}{10x+5}}{\dfrac{4-x^2}{6x+3}}$

17. $\dfrac{\dfrac{x+7}{4x^2}}{\dfrac{x-1}{8x}}$

18. $\dfrac{\dfrac{2x+5}{9x^3}}{\dfrac{x-2}{6x^5}}$

19. $\dfrac{\dfrac{2x-6}{5xy^3}}{\dfrac{12-4x}{15x^2y^5}}$

20. $\dfrac{\dfrac{5x^2-5xy}{3y}}{\dfrac{7y^2-7xy}{2x^2}}$

21. $\dfrac{\dfrac{8xy-4y}{6xy^2}}{\dfrac{12x^2y-6xy}{9x^2y^2}}$

22. $\dfrac{\dfrac{9x^3y^2}{xy^2+2xy}}{\dfrac{3x^3-9x^2}{xy+2x}}$

23. $\dfrac{\dfrac{x^2+5x}{x^2+x-6}}{\dfrac{2x^3}{x^2+2x-8}}$

24. $\dfrac{\dfrac{6x^2}{2x^2+5x+3}}{\dfrac{2x-2x^2}{x^2-1}}$

25. $\dfrac{2 + \dfrac{3}{x}}{1 - \dfrac{7}{x}}$

26. $\dfrac{4 - \dfrac{7}{x}}{3 - \dfrac{2}{x}}$

27. $\dfrac{\dfrac{2}{x} + \dfrac{7}{y}}{12}$

28. $\dfrac{\dfrac{4}{x} - \dfrac{8}{y}}{2}$

29. $\dfrac{\dfrac{2}{x} + \dfrac{2}{y}}{\dfrac{4}{x^2 y^2}}$

30. $\dfrac{\dfrac{3}{x} - \dfrac{1}{y}}{\dfrac{6}{5x^3 y}}$

31. $\dfrac{\dfrac{3}{x} - \dfrac{1}{2}}{\dfrac{2}{y} - \dfrac{1}{5}}$

32. $\dfrac{\dfrac{1}{x} - \dfrac{1}{2}}{\dfrac{5}{7x} - \dfrac{5}{14xy}}$

33. $\dfrac{\dfrac{1}{x} - \dfrac{1}{y}}{\dfrac{1}{x^2} - \dfrac{1}{y^2}}$

34. $\dfrac{\dfrac{1}{y^2} - \dfrac{1}{x^2}}{\dfrac{1}{y^3} - \dfrac{1}{x^3}}$

35. $\dfrac{\dfrac{1}{x^4} - \dfrac{1}{y^4}}{\dfrac{1}{x^2} + \dfrac{1}{y^2}}$

36. $\dfrac{3 + \dfrac{11}{x} - \dfrac{4}{x^2}}{6 + \dfrac{7}{x} - \dfrac{3}{x^2}}$

37. $\dfrac{1 - \dfrac{5}{2x} - \dfrac{3}{2x^2}}{2 - \dfrac{15}{2x} + \dfrac{9}{2x^2}}$

38. $\dfrac{2 - \dfrac{7}{3x} - \dfrac{10}{3x^2}}{4 - \dfrac{8}{3x} - \dfrac{5}{x^2}}$

39. $\dfrac{1}{\dfrac{1}{x} + \dfrac{1}{y}}$

40. $\dfrac{x - 3y}{\dfrac{2}{x} - \dfrac{3}{y}}$

41. $\dfrac{2x - y}{\dfrac{2}{y} - \dfrac{1}{x}}$

42. $\dfrac{3x - 5}{\dfrac{5}{x} - 3}$

43. $\dfrac{\dfrac{5}{x} - \dfrac{2}{y}}{x - 1}$

44. $\dfrac{\dfrac{2}{x} + \dfrac{1}{y}}{x + 3}$

45. $\dfrac{1 - \dfrac{1}{9x^2}}{1 - 3x}$

46. $\dfrac{\dfrac{1}{4} - \dfrac{1}{m^2}}{m + 2}$

47. $\dfrac{\dfrac{1}{x} - \dfrac{1}{4}}{x - 4}$

48. $\dfrac{\dfrac{2}{x - 3} - 1}{x - 5}$

49. $\dfrac{\dfrac{3}{x^2} - \dfrac{1}{3}}{x - 3}$

50. $\dfrac{\dfrac{x + 1}{x} - \dfrac{1}{2}}{x + 2}$

51. $\dfrac{\dfrac{4}{2 + h} - 2}{h}$

52. $\dfrac{\dfrac{5 + h}{3 + h} - \dfrac{5}{3}}{h}$

53. $\dfrac{\dfrac{1}{x} - \dfrac{1}{y}}{\dfrac{x - y}{5}}$

54. $\dfrac{\dfrac{8}{y^3} + \dfrac{1}{x^3}}{\dfrac{2x + y}{y}}$

55. $\dfrac{\dfrac{1}{x - 2} + 1}{\dfrac{1}{x - 2} - 1}$

56. $\dfrac{\dfrac{4}{2x + 1} + 3}{\dfrac{5}{2x + 1} - 2}$

57. $\dfrac{4}{\dfrac{1}{x + 3} + \dfrac{5}{x - 2}}$

58. $\dfrac{3}{\dfrac{2}{x - 1} - \dfrac{1}{x + 3}}$

59. $\dfrac{\dfrac{1}{x - 5} + \dfrac{2}{5 - x}}{\dfrac{4}{x^2 - 25}}$

60. $\dfrac{\dfrac{1}{x - 3} - 2}{5 + \dfrac{1}{3 - x}}$

61. $\dfrac{3 - \dfrac{6}{x - 2}}{\dfrac{2}{2 - x} + 1}$

62. $\dfrac{\dfrac{2a^2}{a - b} + b}{a + \dfrac{2b^2}{b - a}}$

63. $\dfrac{\dfrac{2}{x} + \dfrac{4}{x - 1}}{\dfrac{1}{x} - \dfrac{3}{x - 1}}$

64. $\dfrac{\dfrac{5}{2x} - \dfrac{1}{2x - 1}}{\dfrac{2}{3x} + \dfrac{3}{2x - 1}}$

65. $\dfrac{\dfrac{9}{x^2} + \dfrac{2}{x - 2}}{\dfrac{1}{x^2} + \dfrac{1}{x - 2}}$

66. $\dfrac{\dfrac{1}{x^2} - \dfrac{2}{2x + 1}}{\dfrac{5}{x^2} - \dfrac{3}{2x + 1}}$

67. $\dfrac{\dfrac{2}{x + 1} + \dfrac{1}{x - 2}}{\dfrac{3}{x + 1} + \dfrac{1}{x - 2}}$

68. $\dfrac{\dfrac{3}{x + 2} + \dfrac{1}{2x + 3}}{\dfrac{5}{x + 2} - \dfrac{2}{2x + 3}}$

69. $\dfrac{\dfrac{x}{x - 1} - \dfrac{8}{x + 2}}{\dfrac{-3x}{x - 1} + \dfrac{5x + 4}{x + 2}}$

70. $\dfrac{\dfrac{2x}{x + 3} - \dfrac{1}{x - 2}}{\dfrac{-2x}{x + 3} + \dfrac{3x - 8}{x - 2}}$

71. $\dfrac{x^{-1} - y^{-1}}{x^{-2} - y^{-2}}$

72. $\dfrac{y^{-2} - x^{-2}}{y^{-3} - x^{-3}}$

73. $\dfrac{x^{-1} - (x + 5)^{-1}}{5}$

74. $\dfrac{x^{-1} - (x + 2)^{-1}}{2}$

75. $\dfrac{5x^{-1} + 15x^{-2}}{2 + 6x^{-1}}$

76. $\dfrac{3 - 27x^{-2}}{7x - 63x^{-1}}$

77. $\dfrac{1 - 2x^{-1} - 15x^{-2}}{1 - 3x^{-1} - 10x^{-2}}$

78. $\dfrac{1 - 36x^{-2}}{2 + 17x^{-1} + 30x^{-2}}$

Write Algebra

79. Define a complex fraction.

80. Describe two methods for simplifying a complex fraction.

4.6 *Division of Polynomials*

The first part of this chapter illustrated several similarities between arithmetic and algebra. In this section we will see how the division of polynomials is similar to the division of integers. Let's first examine the case in which a polynomial is divided by a monomial.

 EXAMPLE 4.6.1   Perform the indicated division.

1. $\dfrac{21x^2y^3 - 12x^3 + 6x^2}{3x^2}$

There are two methods for solving this problem. The first is to factor the numerator and then reduce the fraction:

$$\frac{21x^2y^3 - 12x^3 + 6x^2}{3x^2} = \frac{3x^2(7y^3 - 4x + 2)}{3x^2}$$

$$= \frac{7y^3 - 4x + 2}{1} = 7y^3 - 4x + 2$$

The second method makes use of the following property:

$$\frac{a + b + c}{d} = \frac{a}{d} + \frac{b}{d} + \frac{c}{d}$$

Applying this property to our problem, we get

$$\frac{21x^2y^3 - 12x^3 + 6x^2}{3x^2} = \frac{21x^2y^3}{3x^2} - \frac{12x^3}{3x^2} + \frac{6x^2}{3x^2}$$

$$= 7y^3 - 4x + 2 \qquad \textit{Now reduce each fraction.}$$

When the divisor is a monomial, this technique is usually the best way to evaluate division problems.

2. $\dfrac{10xy - 4y + 5}{5xy} = \dfrac{10xy}{5xy} - \dfrac{4y}{5xy} + \dfrac{5}{5xy}$

$$= \dfrac{2}{1} - \dfrac{4}{5x} + \dfrac{1}{xy} = 2 - \dfrac{4}{5x} + \dfrac{1}{xy}$$

Suppose the divisor is a binomial or a trinomial. What process do we follow to evaluate the division problem? Before answering this question, let's refresh our memory on long division problems from arithmetic by considering the problem $6708 \div 31$.

$31\overline{)6708}$ *First determine how many integral times 31 will divide into 67; it will go 2 times.*

$\begin{array}{r} 2 \\ 31\overline{)6708} \\ \underline{62} \\ 50 \end{array}$ *Place the 2 in the quotient and multiply 2 by 31, obtaining 62. Subtract 62 from 67, obtaining 5. Bring down the next digit, 0, and divide 31 into 50, obtaining 1.*

$\begin{array}{r} 21 \\ 31\overline{)6708} \\ \underline{62} \\ 50 \\ \underline{31} \\ 198 \end{array}$ *Place the 1 in the quotient to the right of 2, and multiply 1 by 31, obtaining 31. Subtract 31 from 50, yielding 19. Bring down the last digit, 8, and divide 31 into 198, obtaining 6.*

$\begin{array}{r} 216 \\ 31\overline{)6708} \\ \underline{62} \\ 50 \\ \underline{31} \\ 198 \\ \underline{186} \\ 12 \end{array}$ *Place the 6 in the quotient to the right of 1, and multiply 6 by 31, obtaining 186. Subtract 186 from 198 to determine the remainder, 12. We can now state the answer as $216\frac{12}{31}$.*

Recall that $216\frac{12}{31} = 216 + \frac{12}{31}$. This is the form that will be used in algebraic problems. We will use this process from arithmetic to divide polynomials by binomials and trinomials.

EXAMPLE 4.6.2

Perform the indicated divisions.

1. $\dfrac{5x^2 + 13x + 8}{x + 2}$ *Rewrite in long division format.*

$x + 2\overline{)5x^2 + 13x + 8}$ *Determine the quotient of $5x^2$ and x, which is 5x.*

$$\begin{array}{r} 5x \\ x + 2\overline{)5x^2 + 13x + 8} \\ \underline{5x^2 + 10x} \end{array}$$

Place 5x in the quotient, and multiply 5x by x + 2, obtaining 5x² + 10x.

$$\begin{array}{r} 5x \\ x + 2\overline{)5x^2 + 13x + 8} \\ \underline{5x^2 + 10x} \\ 3x + 8 \end{array}$$

Subtract 5x² + 10x from 5x² + 13x, obtaining 3x. Bring down the next entry, 8, and divide 3x by x, obtaining +3.

$$\begin{array}{r} 5x + 3 \\ x + 2\overline{)5x^2 + 13x + 8} \\ \underline{5x^2 + 10x} \\ 3x + 8 \\ \underline{3x + 6} \\ 2 \end{array}$$

Place +3 in the quotient to the right of 5x, and multiply +3 by x + 2, obtaining 3x + 6. Subtract 3x + 6 from 3x + 8, yielding the remainder of 2. Thus, the answer to this division problem is

$$5x + 3 + \frac{2}{x + 2}$$

REMARK

To use this process on long division problems, both the divisor polynomial and the dividend polynomial must be arranged in descending powers of the variable. If a power of the variable is missing in the dividend, insert that power into the polynomial with a coefficient of zero. This zero term acts as a placeholder, just as zero does in arithmetic problems.

2. $\dfrac{8x^2 + 3}{2x - 1}$

$$2x - 1\overline{)8x^2 + 0x + 3}$$

Insert 0x as a placeholder.

$$\begin{array}{r} 4x \\ 2x - 1\overline{)8x^2 + 0x + 3} \\ \underline{8x^2 - 4x} \\ 4x + 3 \end{array}$$

Be careful here; remember that you are subtracting 8x² − 4x from 8x² + 0x. This subtraction yields 4x.

$$\begin{array}{r} 4x + 2 \\ 2x - 1\overline{)8x^2 + 0x + 3} \\ \underline{8x^2 - 4x} \\ 4x + 3 \\ \underline{4x - 2} \\ 5 \end{array}$$

Again, you are subtracting 4x − 2 from 4x + 3, yielding 5.

Thus the solution is

$$4x + 2 + \frac{5}{2x - 1}$$

Try to avoid this mistake:

Incorrect	Correct
$$\begin{array}{r} 4x \\ 2x - 1\overline{)8x^2 + 0x + 3} \\ \underline{8x^2 - 4x} \\ -4x + 3 \end{array}$$	$$\begin{array}{r} 4x \\ 2x - 1\overline{)8x^2 + 0x + 3} \\ \underline{8x^2 - 4x} \\ 4x + 3 \end{array}$$
Remember that we are *subtracting* the bottom polynomial from the top polynomial.	*An easy way to do the subtraction correctly is to change the signs of all the terms of the bottom polynomial and add. In this case,* $$\begin{array}{r} 4x \\ 2x - 1\overline{)8x^2 + 0x + 3} \\ \underline{{}^-8x^2 - {}^+4x} \\ 4x + 3 \end{array}$$

3. $\dfrac{3x^3 - 10x^2 + 2x + 7}{x - 2}$

$$
\begin{array}{r}
3x^2 - 4x - 6 \\
x - 2\overline{)3x^3 - 10x^2 + 2x + 7} \\
\underline{{}^-3x^3 - {}^+6x^2} \\
-4x^2 + 2x \\
\underline{- {}^+4x^2 + {}^-8x} \\
-6x + 7 \\
\underline{- {}^+6x + {}^-12} \\
-5
\end{array}
$$

Change the signs and add.

Change the signs and add.

Change the signs and add.

Hence, the solution is $3x^2 - 4x - 6 + \frac{-5}{x-2}$, or $3x^2 - 4x - 6 - \frac{5}{x-2}$.

4. $\dfrac{2x^4 - 15x^2 + 2x + 6}{x^2 + 3x - 2}$

$$
\begin{array}{r}
2x^2 - 6x + 7 \\
x^2 + 3x - 2\overline{)2x^4 + 0x^3 - 15x^2 + 2x + 6} \\
\underline{{}^-2x^4 + {}^-6x^3 - {}^+4x^2} \\
-6x^3 - 11x^2 + 2x \\
\underline{- {}^+6x^3 - {}^+18x^2 + {}^-12x} \\
7x^2 - 10x + 6 \\
\underline{{}^-7x^2 + {}^-21x - {}^+14} \\
-31x + 20
\end{array}
$$

Insert $0x^3$ as a placeholder.

Hence, the solution is

$$2x^2 - 6x + 7 + \frac{-31x + 20}{x^2 + 3x - 2}$$

5. $\dfrac{10x^3 + 8x^2 - 15x - 9}{2x^2 - 3}$

$$
\begin{array}{r}
5x + 4 \\
2x^2 - 3\overline{\smash{\big)}\ 10x^3 + 8x^2 -\ \ 15x - 9} \\
\underline{^-10x^3\ \ \ \ \ \ \ \ -\ ^+15x} \\
8x^2 +\ \ \ 0x - 9 \\
\underline{^-8x^2\ \ \ \ \ \ -\ ^+12} \\
3
\end{array}
$$

Hence, the solution is $5x + 4 + \dfrac{3}{2x^2 - 3}$

E xercises 4.6

Perform the indicated divisions.

1. $\dfrac{35x^3 - 49x^2 + 28}{7}$

2. $\dfrac{-9x^2 + 12x + 30}{3}$

3. $\dfrac{5x^3 + 10x^2 - 40x}{5x}$

4. $\dfrac{6x^4 - 15x^3 + 9x^2}{3x^2}$

5. $\dfrac{20x^5 - 28x^4 - 4x^3}{4x^3}$

6. $\dfrac{9x^2y^2 - 12xy^2 + 6xy}{3xy}$

7. $\dfrac{10x^4y^2 - 16x^3y^3 - 2x^2y^4}{2x^2y^2}$

8. $\dfrac{15x^3y - 2x^2 + 12y}{3x^2y}$

9. $\dfrac{-7x^2y^3 + 2x^2 + y}{xy^2}$

10. $\dfrac{60xy - 15x^2 + 4y^2}{5xy}$

11. $\dfrac{6y^4 + 45x^2 - 7y}{10xy^2}$

12. $\dfrac{48y^2z^2 - 27xz^3 - 10x^2y}{6xyz^2}$

13. $\dfrac{3x - 10x^4 - 72z^4}{8x^2yz}$

14. $\dfrac{15y^3z + 18xz^2 - 4y^2}{9xy^2z}$

15. $\dfrac{15y^2z^5 + 7x^4z^3 + 2xy^6}{6x^2y^3z^4}$

16. $\dfrac{2x^2 + 9x + 1}{x + 5}$

17. $\dfrac{5x^2 - 11x - 14}{x - 3}$

18. $\dfrac{6x^2 + 9x - 6}{2x - 1}$

19. $\dfrac{24x^2 - 62x + 36}{4x - 7}$

20. $\dfrac{14x^2 + 29x - 18}{2x + 5}$

21. $\dfrac{40x^2 - 44x + 17}{5x - 3}$

22. $\dfrac{12x^2 + 34x + 13}{3x + 7}$

23. $\dfrac{18x^2 - 43x + 24}{2x - 3}$

24. $\dfrac{2x^3 - x^2 - 10x + 11}{x - 2}$

25. $\dfrac{2x^3 - 13x^2 - 10x + 19}{2x + 3}$

26. $\dfrac{15x^3 - 5x^2 + 12x - 2}{3x - 1}$

27. $\dfrac{-19 - 8x + 21x^2 + 12x^3}{4x + 7}$

28. $\dfrac{5 + 33x - 27x^2 + 4x^3}{x - 5}$

29. $\dfrac{16x + 31x^2 + 8 + 40x^3}{5x + 2}$

30. $\dfrac{-7x + 12x^2 - 21 + 4x^3}{x + 3}$

31. $\dfrac{5x^2 - 13}{x - 2}$

32. $\dfrac{2x^2 - 15}{x + 3}$

33. $\dfrac{4x^2 - 1}{2x - 1}$

34. $\dfrac{4x^2 - 7}{2x - 3}$

35. $\dfrac{9x^2 - 5}{3x - 2}$

36. $\dfrac{8x^2 - 39}{2x - 5}$

37. $\dfrac{25x^2 - 4}{5x + 2}$

38. $\dfrac{16x^2 - 9}{4x - 3}$

39. $\dfrac{4x^3 - x - 7}{2x - 3}$

40. $\dfrac{4x^3 - 17x^2 + 19}{x - 4}$

41. $\dfrac{6x^3 + 20x^2 - 19}{2x + 6}$

42. $\dfrac{4x^4 + 10x^3 + 20x + 1}{2x + 6}$

43. $\dfrac{2x^4 + x^3 - 3x^2 + 9x - 13}{x^2 + x - 3}$

44. $\dfrac{6x^4 - 2x^3 + x^2 + 6}{2x^2 - 2x + 1}$

45. $\dfrac{2x^4 + 3x^3 + 3x^2 - 4}{2x^2 - x - 1}$

46. $\dfrac{9x^4 + 6x^3 + 12x^2 + 3x + 7}{3x^2 + 2x + 3}$

47. $\dfrac{x^3 - 8}{x - 2}$

48. $\dfrac{8x^3 - 125}{2x - 5}$

49. $\dfrac{27x^3 + 64}{3x + 4}$

50. $\dfrac{x^3 + 27}{x + 3}$

51. $\dfrac{x^4 - 16}{x + 2}$

52. $\dfrac{16x^4 - 1}{2x - 1}$

53. $\dfrac{3x^3 + x^2 - 7x - 3}{x^2 + 2x + 1}$

54. $\dfrac{4x^3 + 5x - 1}{2x^2 - x + 3}$

55. $\dfrac{12x^3 - 29x^2 + 6x + 14}{3x^2 - 2x - 2}$

56. $\dfrac{5x^3 - 27x^2 + 14x - 26}{5x^2 - 2x + 4}$

57. $\dfrac{2x^4 - 12x^3 + 27x - 3}{2x^2 - 4}$

58. $\dfrac{14x^4 + 21x^3 - 57x^2 - 8x + 25}{7x^2 - 4}$

59. $\dfrac{2x^6 + 3x^4 - x^2 + 5x - 11}{2x^2 - 3}$

60. $\dfrac{6x^6 - 3x^5 + 8x^4 - 4x^3 + 11x^2 - x + 1}{3x^2 + 1}$

4.7 *Synthetic Division (Optional)*

In this section we will learn a condensed division method called **synthetic division.** This technique will enable us to perform some division problems quicker and, we hope, with fewer errors. One note of caution: The synthetic method can be used only when the divisor polynomial is written in the form $x - r$, where r is any real number.

Before we derive the synthetic division process, let's review Example 4.6.2, part 3:

$$
\begin{array}{r}
3x^2 - 4x - 6 \\
x - 2 \overline{\smash{\big)}\, 3x^3 - 10x^2 + 2x + 7} \\
\underline{3x^3 - 6x^2} \\
-4x^2 + 2x \\
\underline{-4x^2 + 8x} \\
-6x + 7 \\
\underline{-6x + 12} \\
-5
\end{array}
$$

Thus the answer to this division problem is

$$3x^2 - 4x - 6 - \frac{5}{x - 2}$$

In division the coefficients of the variables and the constants determine the quotient polynomial. Let's now drop all the variables in the preceding division to obtain a simpler arrangement:

$$
\begin{array}{r}
3 - 4 - 6 \\
1 - 2\,\overline{\big)\,3 - 10 + 2 + 7} \\
③- 6 \\
- 4 +② \\
-④+ 8 \\
- 6 + ⑦ \\
-⑥+ 12 \\
\hline
- 5
\end{array}
$$

The numbers in the circles are simply duplicates of the numbers above them.

Let's now drop all the circled numbers plus the 1 in the divisor. We drop the 1 in the divisor because we are considering only divisors of the form $x - r$, and we will always have an understood 1 in this position.

$$
\begin{array}{r}
3 - 4 - \ 6 \\
-2\,\overline{\big)\,3 - 10 \ \ \ 2 \ \ \ \ 7} \\
- 6 \\
\hline
- 4 \\
8 \\
\hline
- 6 \\
12 \\
\hline
-5
\end{array}
$$

We will shorten this form even further to:

$$
\begin{array}{r}
3 - 4 - \ 6 \\
-2\,\overline{\big)\,3 - 10 \ \ \ 2 \ \ \ \ \ 7} \\
- 6 \ \ \ 8 \ \ \ 12 \\
\hline
- 4 - 6 - \ 5
\end{array}
$$

$$
\begin{array}{r}
3 - 4 - \ 6 \\
-2\,\overline{\big)\,3 - 10 \ \ \ 2 \ \ \ \ \ 7} \\
- 6 \ \ \ 8 \ \ \ 12 \\
\hline
3 - \ 4 - 6 - \ 5
\end{array}
$$

Now bring down the leading coefficient of the divided polynomial.

Notice that the first three numbers of the bottom row are the same as the quotient and that the last number, -5, is the remainder. Since the quotient polynomial (represented by the top row) is the same as the first three entries of the bottom row, we will now drop the top row:

$$
\begin{array}{r}
-2\,\overline{\big)\,3 - 10 \ \ \ 2 \ \ \ \ \ 7} \\
- 6 \ \ \ 8 \ \ \ 12 \\
\hline
3 - \ 4 - 6 - \ 5
\end{array}
$$

Observe that the middle row is obtained by multiplying -2 by the first three entries in the last row. Recall that the bottom row was obtained by subtracting the middle row from the top row. However, if we could add the top two rows rather than subtract, our chances of making a careless error would be greatly decreased. We can convert the problem to addition by changing the -2 to $+2$ and changing the signs of the entries in the middle row.

$$\begin{array}{r|rrrr} 2 & 3 & -10 & 2 & 7 \\ & & 6 & -8 & -12 \\ \hline & 3 & -4 & -6 & -5 \end{array}$$

Notice that the middle row is now obtained by multiplying 2 by the first three entries in the last row. The last row gives us the answer to the problem. *The answer is always a polynomial of one degree less than the dividend polynomial. The last row gives the coefficients of the variables and the remainder of the answer.* Since the dividend polynomial is of degree 3—that is, it has an x^3 term—the quotient polynomial is of degree 2. Thus the answer is

$$3x^2 - 4x - 6 - \frac{5}{x-2}$$

The explanation and development of synthetic division are more difficult than the process itself. The following examples show how easy it really is.

XAMPLE 4.7.1   Perform the indicated divisions using synthetic division.

1. $\dfrac{4x^3 - 3x^2 - 11x + 14}{x - 2} \rightarrow x - 2\overline{\smash{\big)}4x^3 - 3x^2 - 11x + 14}$

   In the divisor position, place the number that is being subtracted from the variable.

   $$\begin{array}{r|rrrr} 2 & 4 & -3 & -11 & 14 \\ & \downarrow & 8 & 10 & -2 \\ \hline & 4 & 5 & -1 & 12 \end{array}$$

   Bring down the 4 and multiply 4 by the divisor 2, obtaining 8. Now place the 8 under the -3 and add, yielding 5. Multiply 5 by the divisor 2, obtaining 10. Place the 10 under the -11 and add to yield -1. Finally, multiply -1 by 2 and obtain -2. Place the -2 under the 14 and add to determine the remainder, 12. Since the dividend polynomial is of degree 3, the quotient polynomial must be of degree 2. Thus the answer is

   $$4x^2 + 5x - 1 + \frac{12}{x-2}$$

2. $\dfrac{2x^4 - 9x^3 + 18x - 3}{x - 4} \rightarrow x - 4\overline{\smash{\big)}2x^4 - 9x^3 + 0x^2 + 18x - 3}$

REMARK As in long division, we must insert $0x^2$ to act as a placeholder.

$$
\begin{array}{r|rrrrr}
4 & 2 & -9 & 0 & 18 & -3 \\
  & \downarrow & 8 & -4 & -16 & 8 \\
\hline
  & 2 & -1 & -4 & 2 & 5 \\
\end{array}
$$

Since the dividend polynomial is of degree 4, the quotient polynomial must be of degree 3. Thus the answer is

$$2x^3 - x^2 - 4x + 2 + \frac{5}{x - 4}$$

3. $\dfrac{5x^3 + 12x^2 - 3x - 7}{x + 3} \rightarrow x + 3 \overline{\smash{\big)}\, 5x^3 + 12x^2 - 3x - 7}$

$$
\begin{array}{r|rrrr}
-3 & 5 & 12 & -3 & -7 \\
   & \downarrow & -15 & 9 & -18 \\
\hline
   & 5 & -3 & 6 & -25 \\
\end{array}
$$

We have -3 in the divisor position. This is because $x + 3 = x - (-3)$. Remember that we place, in the divisor position, the number that we are subtracting from the variable. Using the synthetic division chart, we find the answer to be

$$5x^2 - 3x + 6 - \frac{25}{x + 3}$$

REMARK We can check the answer using the fact that

$$
\begin{aligned}
\text{(dividend)} \quad &= \quad \text{(quotient)} \cdot \text{(divisor)} \quad + \text{(remainder)} \\
5x^3 + 12x^2 - 3x - 7 &= (5x^2 - 3x + 6) \cdot (x + 3) \quad + \quad (-25) \\
&= 5x^3 - 3x^2 + 6x + 15x^2 - 9x + 18 + (-25) \\
&= 5x^3 + 12x^2 - 3x + 18 + (-25) \\
&= 5x^3 + 12x^2 - 3x - 7
\end{aligned}
$$

4. $\dfrac{6x^3 + x^2 + 19x + 14}{x + \dfrac{2}{3}} \rightarrow x + \dfrac{2}{3} \overline{\smash{\big)}\, 6x^3 + x^2 + 19x + 14}$

$$
\begin{array}{r|rrrr}
-\dfrac{2}{3} & 6 & 1 & 19 & 14 \\
   & \downarrow & -4 & 2 & -14 \\
\hline
   & 6 & -3 & 21 & 0 \\
\end{array}
$$
The remainder is 0.

Thus, the answer is $6x^2 - 3x + 21$.

Exercises 4.7

Perform the indicated divisions using synthetic division.

1. $\dfrac{x^2 - 3x + 5}{x - 2}$

2. $\dfrac{3x^2 + 5x + 2}{x + 1}$

3. $\dfrac{2x^2 - 9}{x + 2}$

4. $\dfrac{2x^2 - 37}{x - 5}$

5. $\dfrac{9x^2 - 3x + 4}{x - \dfrac{1}{3}}$

6. $\dfrac{4x^2 + 4x - 5}{x + \dfrac{3}{2}}$

7. $\dfrac{2x^3 - 7x^2 - x + 6}{x - 3}$

8. $\dfrac{x^3 + 4x^2 - 5x - 9}{x + 5}$

9. $\dfrac{x^3 - 7x - 4}{x + 2}$

10. $\dfrac{2x^3 - 5x^2 - 1}{x - 1}$

11. $\dfrac{3x^3 + 11x^2 - 5x}{x + 4}$

12. $\dfrac{x^3 - 4x^2 - 15x + 18}{x - 6}$

13. $\dfrac{4x^3 - 5x + 1}{x + \dfrac{1}{2}}$

14. $\dfrac{3x^3 - 5x^2 + x + 1}{x + \dfrac{1}{3}}$

15. $\dfrac{6x^3 - 4x^2 + 9x - 8}{x - \dfrac{2}{3}}$

16. $\dfrac{4x^3 - x^2 - 8x + 5}{x - \dfrac{1}{4}}$

17. $\dfrac{x^4 - 2x^3 - 5x^2 + x + 5}{x + 2}$

18. $\dfrac{x^4 - 5x^2 + 10x + 1}{x + 3}$

19. $\dfrac{2x^4 - 3x^3 + x^2 + 7x - 5}{x - 1}$

20. $\dfrac{2x^4 - 9x^3 + 10x + 24}{x - 4}$

21. $\dfrac{16x^4 - 9x^2 + 4x - 2}{x + \dfrac{3}{4}}$

22. $\dfrac{6x^4 - 7x^3 - 11x^2 + 2x + 3}{x - \dfrac{1}{6}}$

23. $\dfrac{x^3 + 125}{x + 5}$

24. $\dfrac{x^4 - 81}{x + 3}$

25. $\dfrac{x^6 - 64}{x - 2}$

26. $\dfrac{x^3 - a^3}{x - a}$

27. $\dfrac{x^2 - 5}{x - \sqrt{5}}$

28. $\dfrac{x^2 - 6}{x + \sqrt{6}}$

29. $\dfrac{x^4 - 9}{x + \sqrt{3}}$

30. $\dfrac{x^4 - 4}{x - \sqrt{2}}$

Write Algebra

31. Explain why zeros are inserted as placeholders in the synthetic division process.

32. Describe how you could use synthetic division to divide by $x^2 - 4$.

G R O U P   A C T I V I T Y

In exercises 33–42, use synthetic division to determine the value of k so that the numerator polynomial is divisible by the denominator polynomial, that is, the remainder is zero. Write the quotient polynomial. The numerator is also divisible by the quotient. Why? Factor the quotient polynomial completely. Write the complete factorization of the dividend polynomial.

33. $\dfrac{2x^4 - 13x^3 + 18x^2 + 13x + k}{x - 4}$

34. $\dfrac{4x^4 - 23x^3 + 11x^2 + 23x + k}{x - 5}$

35. $\dfrac{4x^4 + 12x^3 - 36x^2 + kx}{x + 3}$

36. $\dfrac{6x^4 - 15x^3 - 24x^2 + kx}{x - 2}$

37. $\dfrac{5x^4 + 10x^3 + kx^2 - 10x}{x + 2}$

38. $\dfrac{6x^4 + 14x^3 + kx^2 + 16x}{x - 1}$

39. $\dfrac{9x^4 + kx^3 + 35x^2 - 5x - 4}{x + 4}$

40. $\dfrac{4x^4 + kx^3 - 25x^2 + x + 6}{x + 2}$

41. $\dfrac{kx^4 - 5x^3 - 16x^2 + 45x - 18}{x - 2}$

42. $\dfrac{kx^4 - 8x^3 - 15x^2 + 32x + 12}{x - 3}$

4.8 *Equations Involving Rational Expressions*

Sarah's car averages 20 mi per gallon of gasoline. She is preparing to make a 100-mi trip. How many gallons of gasoline will Sarah's car burn on the trip? We know that

$$\frac{\text{number of miles}}{\text{number of gallons}} = \text{miles per gallon}$$

If we let x be the number of gallons that Sarah's car will burn, we obtain the equation

$$\frac{100}{x} = 20$$

In this section we will learn how to find solutions to equations like $\frac{100}{x} = 20$; that is, we will find solutions to equations that contain fractions. To solve **fractional equations,** multiply the equation by the LCD of all the fractions on either side of the equation. When this multiplication is done, all the fractions are eliminated. When the LCD contains variables, however, multiplying by the LCD can cause problems. Consider the following reasoning: Given

$$x = 7$$

then

$$x^2 = 7x \qquad \textit{Multiply both sides by x.}$$
$$x^2 - 7x = 0$$
$$x(x - 7) = 0$$
$$x = 0 \quad \text{or} \quad x - 7 = 0 \qquad \textit{By the Zero Product theorem}$$
$$x = 0 \qquad\qquad x = 7$$

However, 0 is not a solution of the original equation and is called an **extraneous solution.** When both sides of an equation are multiplied by a variable expression, extraneous solutions *may* be introduced.

NOTE ▶   *When solving fractional equations, always check each solution to verify that it is not an extraneous solution.*

XAMPLE 4.8.1   Solve for x in each of the following equations.

1. $$\frac{100}{x} = 20$$

$$x \cdot \left(\frac{100}{x}\right) = 20 \cdot x \qquad \textit{The LCD is x. Multiply both sides of the equation by x.}$$
$$100 = 20x$$
$$5 = x$$

Let's check the answer:

$$\frac{100}{5} \overset{?}{=} 20$$
$$20 = 20$$

Therefore, {5} is the solution set. (Sarah will need 5 gal of gas for the trip.)

2. $$\frac{5}{2x} + \frac{1}{3x} = \frac{17}{18}$$

$$\frac{18x}{1}\left(\frac{5}{2x} + \frac{1}{3x}\right) = \frac{17}{18} \cdot \frac{18x}{1} \qquad \textit{The LCD is 18x.}$$

(It sometimes makes the multiplication easier if the LCD is written as a fraction with denominator one.)

$$\frac{18x}{1} \cdot \frac{5}{2x} + \frac{18x}{1} \cdot \frac{1}{3x} = \frac{17}{18} \cdot \frac{18x}{1}$$
$$45 + 6 = 17x$$
$$51 = 17x$$
$$3 = x$$

Checking the solution is left to the reader.

3.
$$\frac{2}{3} - \frac{5}{3x} = \frac{1}{x^2}$$

$$\frac{3x^2}{1}\left(\frac{2}{3} - \frac{5}{3x}\right) = \frac{1}{x^2} \cdot \frac{3x^2}{1} \qquad \textit{The LCD is } 3x^2.$$

$$\frac{3x^2}{1} \cdot \frac{2}{3} - \frac{3x^2}{1} \cdot \frac{5}{3x} = \frac{1}{x^2} \cdot \frac{3x^2}{1}$$

$$2x^2 - 5x = 3 \qquad \leftarrow \textit{A quadratic equation}$$

$$2x^2 - 5x - 3 = 0$$

$$(2x + 1)(x - 3) = 0$$

$$2x + 1 = 0 \quad \text{or} \quad x - 3 = 0$$

$$2x = -1 \qquad\qquad x = 3$$

$$x = -\frac{1}{2}$$

Again, the check is left to the reader.

4.
$$\frac{2}{x - 1} - \frac{7}{x + 2} = \frac{1}{4} \qquad \textit{The LCD is } 4(x - 1)(x + 2).$$

$$\frac{4(x - 1)(x + 2)}{1} \cdot \left(\frac{2}{x - 1} - \frac{7}{x + 2}\right) = \frac{1}{4} \cdot \frac{4(x - 1)(x + 2)}{1}$$

$$\frac{4\cancel{(x - 1)}(x + 2)}{1} \cdot \frac{2}{\cancel{x - 1}}$$

$$- \frac{4(x - 1)\cancel{(x + 2)}}{1} \cdot \frac{7}{\cancel{x + 2}} = \frac{1}{\cancel{4}} \cdot \frac{\cancel{4}(x - 1)(x + 2)}{1}$$

$$8(x + 2) - 28(x - 1) = (x - 1)(x + 2)$$

$$8x + 16 - 28x + 28 = x^2 + x - 2$$

$$44 - 20x = x^2 + x - 2$$

$$0 = x^2 + 21x - 46$$

$$0 = (x - 2)(x + 23)$$

$$x - 2 = 0 \quad \text{or} \quad x + 23 = 0$$

$$x = 2 \qquad\qquad x = -23$$

Again, the check is left to the reader.

5.
$$\frac{2x + 7}{x + 5} - \frac{x - 8}{x - 4} = \frac{x + 18}{x^2 + x - 20} \qquad \textit{The LCD is } (x + 5)(x - 4).$$

$$\frac{(x + 5)(x - 4)}{1} \cdot \left(\frac{2x + 7}{x + 5} - \frac{x - 8}{x - 4}\right)$$

$$= \frac{x + 18}{(x + 5)(x - 4)} \cdot \frac{(x + 5)(x - 4)}{1}$$

$$\frac{\cancel{(x+5)}(x-4)}{1} \cdot \frac{2x+7}{\cancel{x+5}} - \frac{(x+5)\cancel{(x-4)}}{1} \cdot \frac{x-8}{\cancel{x-4}}$$

$$= \frac{x+18}{\cancel{(x+5)}\cancel{(x-4)}} \cdot \frac{\cancel{(x+5)}\cancel{(x-4)}}{1}$$

$$(x-4)(2x+7) - (x+5)(x-8) = x+18$$

$$2x^2 - x - 28 - x^2 + 3x + 40 = x+18$$

$$x^2 + 2x + 12 = x + 18$$

$$x^2 + x - 6 = 0$$

$$(x+3)(x-2) = 0$$

$$x + 3 = 0 \quad \text{or} \quad x - 2 = 0$$

$$x = -3 \qquad x = 2$$

The check is left to the reader.

6.
$$1 + \frac{5}{x-5} = \frac{2x-5}{x-5} \qquad \textit{The LCD is } x - 5.$$

$$\frac{(x-5)}{1} \cdot \left(1 + \frac{5}{x-5}\right) = \frac{2x-5}{x-5} \cdot \frac{x-5}{1}$$

$$\frac{(x-5)}{1} \cdot 1 + \frac{\cancel{x-5}}{1} \cdot \frac{5}{\cancel{x-5}} = \frac{2x-5}{\cancel{x-5}} \cdot \frac{\cancel{x-5}}{1}$$

$$x - 5 + 5 = 2x - 5$$

$$x = 2x - 5$$

$$5 = x$$

When we check the solution, we obtain

$$1 + \frac{5}{5-5} \overset{?}{=} \frac{2(5)-5}{5-5}$$

$$1 + \frac{5}{0} \overset{?}{=} \frac{5}{0} \qquad \leftarrow\textit{Division by zero}$$

Since division by zero is undefined, the solution set is the empty set, $\varnothing$.

Exercises 4.8

Solve for x in each of the following equations.

1. $\dfrac{-3}{2x} = \dfrac{9}{8}$

2. $\dfrac{5}{x-3} = \dfrac{2}{x}$

3. $\dfrac{3}{2x} + \dfrac{1}{3x} = \dfrac{11}{12}$

4. $\dfrac{7}{6x} - \dfrac{5}{8x} = \dfrac{13}{12}$

5. $\dfrac{12x+3}{3x+2} = 3 - \dfrac{5}{3x+2}$

6. $\dfrac{5x+2}{x-1} = 2 + \dfrac{7}{x-1}$

7. $\dfrac{x + 1}{2} = \dfrac{x - 5}{4}$

8. $\dfrac{3}{2x - 4} = \dfrac{2}{4x + 1}$

9. $\dfrac{2}{2x - 1} + \dfrac{3}{x - 4} = \dfrac{5}{2x - 1}$

10. $\dfrac{x + 2}{x + 3} = 1 - \dfrac{1}{x + 3}$

11. $\dfrac{2}{x + 1} - \dfrac{1}{3x - 2} = \dfrac{3}{3x - 2}$

12. $\dfrac{3}{x - 1} + \dfrac{4}{x + 2} = \dfrac{16}{x^2 + x - 2}$

13. $\dfrac{5}{2x + 1} - \dfrac{1}{2x - 1} = \dfrac{6}{4x^2 - 1}$

14. $\dfrac{3}{x - 2} - \dfrac{2}{x} = \dfrac{6}{x^2 - 2x}$

15. $1 = \dfrac{x + 5}{x - 3} - \dfrac{8}{x - 3}$

16. $\dfrac{x + 1}{x + 2} + \dfrac{2}{x^2 + 3x + 2} = 1$

17. $\dfrac{1}{x} - \dfrac{2}{4x + 1} = \dfrac{1}{8x^2 + 2x}$

18. $\dfrac{x + 5}{x + 2} = 1 + \dfrac{4}{x^2 + 4x + 4}$

19. $\dfrac{2x + 1}{x - 2} = \dfrac{2x + 1}{x + 2} + \dfrac{10}{x^2 - 4}$

20. $\dfrac{3x - 2}{x + 1} = \dfrac{3x - 2}{x + 2} - \dfrac{7}{x^2 + 3x + 2}$

21. $\dfrac{1}{x^2} - \dfrac{1}{6x} = \dfrac{1}{6}$

22. $\dfrac{1}{3} - \dfrac{5}{6x} = \dfrac{2}{x^2}$

23. $\dfrac{5}{x^2} - \dfrac{1}{3x} = 2$

24. $\dfrac{5}{2x^2} + \dfrac{2}{x} = \dfrac{-2}{5}$

25. $\dfrac{2}{x} - \dfrac{3}{x^2} = \dfrac{1}{3}$

26. $\dfrac{x}{x + 1} - \dfrac{2}{x - 2} = \dfrac{3}{x^2 - x - 2}$

27. $\dfrac{2}{2x - 1} + \dfrac{x}{x + 4} = \dfrac{36}{2x^2 + 7x - 4}$

28. $\dfrac{6x}{3x + 2} + \dfrac{1}{x - 4} = \dfrac{-28x}{3x^2 - 10x - 8}$

29. $\dfrac{4x}{x - 1} - \dfrac{x}{x + 3} = \dfrac{-12}{x^2 + 2x - 3}$

30. $\dfrac{x + 5}{x + 2} = \dfrac{6}{x}$

31. $\dfrac{3x + 6}{2x + 5} = \dfrac{-1}{x}$

32. $\dfrac{2}{2x + 5} = \dfrac{2x - 3}{2x - 7}$

33. $\dfrac{3x}{x - 2} = \dfrac{x + 4}{2 - x}$

34. $\dfrac{-1}{2x - 1} = \dfrac{4x - 5}{1 - 2x}$

35. $\dfrac{x + 3}{9x + 1} = \dfrac{x + 1}{7}$

36. $\dfrac{5}{5x + 3} = \dfrac{5x - 2}{x - 1}$

37. $\dfrac{x}{x - 4} = \dfrac{x + 2}{3x - 10}$

38. $\dfrac{2x}{x - 3} = \dfrac{x + 5}{x + 9}$

39. $\dfrac{9x + 3}{10x - 30} = \dfrac{x}{x - 3}$

40. $\dfrac{3}{x + 1} = \dfrac{2x + 5}{x^2 + x}$

41. $\dfrac{2}{x + 3} + \dfrac{4}{x + 1} = \dfrac{4}{3}$

42. $\dfrac{3}{2x - 1} + \dfrac{2}{x + 3} = \dfrac{7}{5}$

43. $\dfrac{1}{x - 5} - \dfrac{5}{x - 2} = \dfrac{3}{2}$

44. $\dfrac{5}{3x + 1} - \dfrac{4}{x + 1} = \dfrac{-3}{4}$

45. $\dfrac{x - 2}{x - 3} + \dfrac{1}{x + 2} = \dfrac{13}{6}$

46. $\dfrac{x + 1}{x} + \dfrac{x - 1}{3x - 2} = \dfrac{7}{4}$

47. $\dfrac{x + 2}{2x - 1} + \dfrac{x + 5}{x + 3} = \dfrac{5}{3}$

48. $\dfrac{2x + 4}{x + 2} - \dfrac{x + 3}{x - 2} = \dfrac{1}{2}$

49. $\dfrac{x + 3}{x + 4} - \dfrac{5}{x - 1} = \dfrac{5x - 3}{x^2 + 3x - 4}$

50. $\dfrac{2x - 3}{x + 2} + \dfrac{1}{x - 2} = \dfrac{3x - 1}{x^2 - 4}$

51. $\dfrac{x + 2}{2x + 3} + \dfrac{3x + 1}{x - 3} = \dfrac{9x + 5}{2x^2 - 3x - 9}$

52. $\dfrac{x - 1}{x - 5} - \dfrac{2x - 1}{x - 3} = \dfrac{7x - 3}{x^2 - 8x + 15}$

53. $\dfrac{x}{3x-2} - \dfrac{8}{2x+3} = \dfrac{2x^2 - 24x + 18}{6x^2 + 5x - 6}$

54. $\dfrac{2}{x^2-1} + \dfrac{x+2}{x^2 - x - 2} = \dfrac{x}{x^2 - 3x + 2}$

55. $\dfrac{x-7}{2x^2+9x-5} - \dfrac{2x-6}{x^2-25} = \dfrac{x+1}{2x^2-11x+5}$

56. $\dfrac{x+6}{x^2+5x+6} + \dfrac{6}{x^2+x-2} = \dfrac{2x+4}{x^2+2x-3}$

57. $\dfrac{3x-1}{x^2+4x+3} - \dfrac{3}{x^2-x-12} = \dfrac{2x-5}{x^2-3x-4}$

58. $\dfrac{x-2}{4x^2-29x+30} - \dfrac{x+1}{20x^2-13x-15} = \dfrac{x+2}{5x^2-27x-18}$

Sometimes students confuse the question of adding and subtracting rational expressions with that of finding the solutions of equations involving rational expressions. This error is illustrated next.

Try to avoid this mistake:

Incorrect

$$\dfrac{5x+15}{x^2+3x+2} - \dfrac{4x}{x^2-1}$$

$$\dfrac{5x+15}{(x+1)(x+2)} - \dfrac{4x}{(x+1)(x-1)} \qquad LCD = (x+1)(x+2)(x-1)$$

$$\dfrac{(x+1)(x+2)(x-1)}{1}\left(\dfrac{5x+15}{(x+1)(x+2)} - \dfrac{4x}{(x+1)(x-1)}\right) \qquad \textit{Multiply by the LCD.}$$

$$\dfrac{\cancel{(x+1)}\cancel{(x+2)}(x-1)}{1}\cdot\dfrac{5x+15}{\cancel{(x+1)}\cancel{(x+2)}} - \dfrac{\cancel{(x+1)}(x+2)\cancel{(x-1)}}{1}\cdot\dfrac{4x}{\cancel{(x+1)}\cancel{(x-1)}}$$

$$(x-1)(5x+15) - 4x(x+2)$$

$$5x^2 + 10x - 15 - 4x^2 - 8x$$

$$x^2 + 2x - 15$$

$$(x+5)(x-3)$$

$$x+5=0 \quad \text{or} \quad x-3=0$$

$$x=-5 \qquad\qquad x=3$$

NOTE ▶ There was *no equation given!* Hence it does not make any sense to "solve for *x*."

Correct

$$\frac{5x + 15}{x^2 + 3x + 2} - \frac{4x}{x^2 - 1}$$

$$= \frac{5x + 15}{(x + 1)(x + 2)} - \frac{4x}{(x + 1)(x - 1)} \qquad LCD = (x + 1)(x + 2)(x - 1).$$

$$= \frac{5x + 15}{(x + 1)(x + 2)} \cdot \frac{(x - 1)}{(x - 1)} - \frac{4x}{(x + 1)(x - 1)} \cdot \frac{(x + 2)}{(x + 2)}$$

$$= \frac{5x^2 + 10x - 15}{(x + 1)(x + 2)(x - 1)} - \frac{4x^2 + 8x}{(x + 1)(x - 1)(x + 2)}$$

$$= \frac{5x^2 + 10x - 15 - 4x^2 - 8x}{(x + 1)(x + 2)(x - 1)}$$

$$= \frac{x^2 + 2x - 15}{(x + 1)(x + 2)(x - 1)}$$

$$= \frac{(x + 5)(x - 3)}{(x + 1)(x + 2)(x - 1)}$$

NOTE ▶   In general, when we add or subtract two rational expressions, we obtain another rational expression.

In the following exercises, either perform the indicated operation or solve the equation, whichever is appropriate.

59. $\dfrac{1}{2x} + \dfrac{3}{5x^2 - 10x}$

60. $\dfrac{5}{2x + 1} = \dfrac{2}{x - 3}$

61. $\dfrac{2x - 1}{x - 5} = \dfrac{x + 2}{x + 3}$

62. $\dfrac{4}{x} - \dfrac{1}{2} + \dfrac{7}{3x}$

63. $\dfrac{x^2 + 3x + 2}{2x^2 + 7x - 15} \cdot \dfrac{3x^2 + 11x - 20}{x^2 + 4x + 4}$

64. $\dfrac{x}{x + 4} - 2 = \dfrac{-3}{x - 2}$

65. $\dfrac{x + 1}{2x^2 - 3x - 2} + \dfrac{x - 5}{x^2 + x - 6} = 0$

66. $\dfrac{4}{x - 6} - \dfrac{2}{x + 3}$

67. $\dfrac{x}{2} + \dfrac{x}{3} = \dfrac{25}{6}$

68. $\dfrac{3x + 4}{x + 8} = \dfrac{x - 1}{x - 2}$

69. $\dfrac{8}{x + 1} = \dfrac{9}{x + 2}$

70. $\dfrac{5}{x - 1} - \dfrac{3}{1 - x}$

71. $\dfrac{5}{2x + 1} - \dfrac{3}{x - 4}$

72. $\dfrac{3}{x + 5} + \dfrac{1}{x + 2} = \dfrac{7}{4}$

73. $\dfrac{1}{x} - \dfrac{4}{5} = \dfrac{3}{2x}$

74. $\dfrac{x}{4} + \dfrac{x}{5} = \dfrac{9}{10}$

75. $\dfrac{3}{x - 2} - \dfrac{4}{2 - x}$

76. $\dfrac{x^2 + 2x - 24}{3x^2 + 5x - 2} \cdot \dfrac{2x^2 - x - 10}{x^2 + 12x + 36}$

77. $\dfrac{2}{x+3} + \dfrac{7}{x+5} = \dfrac{5}{3}$

78. $\dfrac{x+1}{x^2-4} - \dfrac{x-3}{2x^2-3x-2}$

79. $\dfrac{25x^2-9}{3x^2+12x} \div \dfrac{10x^3-6x^2}{10x+40}$

80. $\dfrac{3}{x} - \dfrac{2}{3} = \dfrac{4}{5x}$

81. $\dfrac{2}{x} - \dfrac{1}{3} + \dfrac{5}{2x}$

82. $\dfrac{1}{3x} + \dfrac{5}{6x^2+6x}$

83. $\dfrac{x+3}{x^2-1} - \dfrac{x-5}{2x^2+3x+1}$

84. $\dfrac{49x^2-16}{2x^2+10x} \div \dfrac{21x^4-12x^3}{4x+20}$

85. $\dfrac{x^2-25}{x^2+10x+25} \cdot \dfrac{8x^2}{10-2x}$

86. $\dfrac{x-10}{x^2-2x-8} + \dfrac{x+12}{x^2-x-6} = 0$

87. $\dfrac{x}{x+3} - \dfrac{3}{2} = \dfrac{-1}{x-2}$

88. $\dfrac{x^2-9}{x^2+6x+9} \cdot \dfrac{10x^2}{6-2x}$

Write Algebra

89. Explain the difference between adding and subtracting rational expressions and finding the solutions of equations involving rational expressions.

90. Explain why we check for extraneous solutions when we solve equations involving rational expressions.

91. Why can't 3 or -1 be a solution of $\dfrac{a}{x-3} = \dfrac{b}{x+1}$?

4.9 *Applications*

In the previous section we found solutions to equations that contained fractions. We are now prepared to examine application problems where fractional equations must be used. Consider the following examples.

XAMPLE 4.9.1

What number must be added to the numerator and subtracted from the denominator of $\frac{5}{7}$ to obtain $\frac{1}{2}$?

SOLUTION   Let $x =$ the number; then

$$\frac{5+x}{7-x} = \frac{1}{2}$$

$$2(7-x) \cdot \frac{5+x}{7-x} = \frac{1}{2} \cdot 2(7-x) \qquad \textit{Multiply by the LCD, } 2(7-x).$$

$$2(5+x) = 1(7-x)$$

$$10+2x = 7-x$$

$$3x = -3$$

$$x = -1$$

Thus, the required number is -1. The check is left to the reader.

EXAMPLE 4.9.2   The sum of a number and its reciprocal is $\frac{13}{6}$. What is the number?

SOLUTION   Let

$$x = \text{the number}$$

$$\frac{1}{x} = \text{reciprocal of the number}$$

$$x + \frac{1}{x} = \frac{13}{6}$$

$$\frac{6x}{1} \cdot \left(x + \frac{1}{x} \right) = \frac{13}{6} \cdot \frac{6x}{1} \qquad \textit{Multiply by the LCD, 6x.}$$

$$\frac{6x}{1} \cdot x + \frac{6x}{1} \cdot \frac{1}{x} = \frac{13}{6} \cdot \frac{6x}{1}$$

$$6x^2 + 6 = 13x$$

$$6x^2 - 13x + 6 = 0$$

$$(3x - 2)(2x - 3) = 0$$

$$3x - 2 = 0 \quad \text{or} \quad 2x - 3 = 0$$

$$3x = 2 \qquad\qquad 2x = 3$$

$$x = \frac{2}{3} \qquad\qquad x = \frac{3}{2}$$

Here we have two solutions, $\frac{2}{3}$ and $\frac{3}{2}$. Why?

EXAMPLE 4.9.3   Elmer, using a spray painter, can paint his house in 4 days. His son, Elmer, Jr., using a brush, can paint the house in 12 days. Working together, how long will it take them to paint the house?

SOLUTION   To solve work problems we will use this fundamental rule.

If it takes x days (hours) to complete a job, then in 1 day (hour) we complete $\frac{1}{x}$ part of the job. For example, if it takes 3 hr to mow a yard, then in 1 hr $\frac{1}{3}$ of the yard has been mowed.

In our case, let $x =$ number of days required to paint the house working together. The following chart describes the part of the house painted by Elmer, Elmer, Jr., and both:

	Elmer	**Elmer, Jr.**	**Both**
$\dfrac{\text{part}}{\text{day}}$	$\dfrac{1}{4}$	$\dfrac{1}{12}$	$\dfrac{1}{x}$

Then,

$$\left(\begin{array}{c}\text{part painted by}\\ \text{Elmer in 1 day}\end{array}\right) + \left(\begin{array}{c}\text{part painted by}\\ \text{Elmer, Jr., in 1 day}\end{array}\right) = \left(\begin{array}{c}\text{part painted by both Elmer}\\ \text{and Elmer, Jr., in 1 day}\end{array}\right)$$

$$\frac{1}{4} \qquad + \qquad \frac{1}{12} \qquad = \qquad \frac{1}{x}$$

$$\frac{12x}{1} \cdot \left(\frac{1}{4} + \frac{1}{12}\right) = \frac{1}{x} \cdot \frac{12x}{1} \qquad \textit{Multiply by the LCD, 12x.}$$

$$\frac{12x}{1} \cdot \frac{1}{4} + \frac{12x}{1} \cdot \frac{1}{12} = \frac{1}{x} \cdot \frac{12x}{1}$$

$$3x + x = 12$$

$$4x = 12$$

$$x = 3$$

Working together, they can paint the house in 3 days.

NOTE ▶   3 days is a reasonable answer, since it should take less time for both to paint the house than it would take Elmer working alone.

XAMPLE 4.9.4   A swimming pool can be filled by an inlet pipe in 12 hr. The drain can empty the pool in 20 hr. How long will it take to fill the pool if the drain is left open?

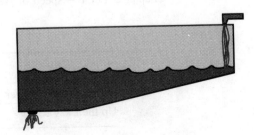

SOLUTION

Let x = number of hours to fill the pool with the drain open. Then

	Inlet Pipe	Drain	Both
$\dfrac{\text{part}}{\text{hour}}$	$\dfrac{1}{12}$	$\dfrac{1}{20}$	$\dfrac{1}{x}$

$$\begin{pmatrix} \text{part of the pool} \\ \text{filled in 1 hr with} \\ \text{the drain closed} \end{pmatrix} - \begin{pmatrix} \text{part of the} \\ \text{pool emptied} \\ \text{in 1 hr} \end{pmatrix} = \begin{pmatrix} \text{part of the pool} \\ \text{filled in 1 hr with} \\ \text{the drain open} \end{pmatrix}$$

$$\frac{1}{12} \quad - \quad \frac{1}{20} \quad = \quad \frac{1}{x}$$

$$\frac{60x}{1} \cdot \left(\frac{1}{12} - \frac{1}{20} \right) = \frac{1}{x} \cdot \frac{60x}{1} \qquad \textit{Multiply by the LCD, 60x.}$$

$$\frac{60x}{1} \cdot \frac{1}{12} - \frac{60x}{1} \cdot \frac{1}{20} = \frac{1}{x} \cdot \frac{60x}{1}$$

$$5x - 3x = 60$$

$$2x = 60$$

$$x = 30$$

It takes 30 hr to fill the pool with the drain open.

EXAMPLE 4.9.5

Henry's motorboat can travel 14 mi down the Lazy River in the same time that it can travel 12 mi up the Lazy River. If the speed of the current in the Lazy River is 2mi/hr, what is the speed of Henry's boat in still water?

SOLUTION

To find solutions to distance-rate-time problems, we use the tables developed in Chapter 2. Let x = speed of Henry's boat in still water. The chart is:

	Distance	÷ Rate	= Time
Down the Lazy River	14	$x + 2$	
Up the Lazy River	12	$x - 2$	

$x + 2$ is Henry's downstream rate, since the current will increase his speed by 2 mi/hr. Likewise, $x - 2$ is Henry's rate traveling upstream. How do we fill in the time column? Since $d = r \cdot t$, $\frac{d}{r} = t$ (solving for t). Thus, the completed chart is:

	Distance	÷ Rate	= Time
Down the Lazy River	14	$x + 2$	$\dfrac{14}{x + 2}$
Up the Lazy River	12	$x - 2$	$\dfrac{12}{x - 2}$

We are given that the two times are the same:

$$\frac{14}{x + 2} = \frac{12}{x - 2}$$

$$(x + 2)(x - 2) \cdot \frac{14}{x + 2} = \frac{12}{x - 2} \cdot (x + 2)(x - 2) \qquad \textit{Multiply by the LCD,}$$
$$\textit{(x + 2)(x - 2).}$$

$$14(x - 2) = 12(x + 2)$$
$$14x - 28 = 12x + 24$$
$$2x = 52$$
$$x = 26$$

Henry's speed in still water is 26 mi/hr.

EXAMPLE 4.9.6

Sue hikes up a 6-mi-long mountain trail. At the top of the mountain she rents a pair of skis and skis back down the trail. Her skiing speed is 4 mi/hr faster than her hiking speed. Sue spends a total of 4 hr on the trail. What is her hiking speed? What is her skiing speed?

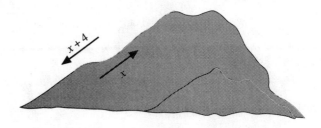

SOLUTION   Let

$$x = \text{Sue's hiking speed}$$
$$x + 4 = \text{Sue's skiing speed}$$

Distance	÷ Rate	= Time
Hiking	6	x	$\dfrac{6}{x}$
Skiing	6	$x + 4$	$\dfrac{6}{x + 4}$

Now

$$(\text{time hiking}) + (\text{time skiing}) = (\text{total time})$$

$$\frac{6}{x} + \frac{6}{x + 4} = 4$$

$$\frac{x(x + 4)}{1} \cdot \left(\frac{6}{x} + \frac{6}{x + 4}\right) = 4 \cdot \frac{x(x + 4)}{1}$$

$$\frac{x(x + 4)}{1} \cdot \frac{6}{x} + \frac{x(x + 4)}{1} \cdot \frac{6}{x + 4} = 4x(x + 4)$$

$$6(x + 4) + 6x = 4x(x + 4)$$

$$6x + 24 + 6x = 4x^2 + 16x$$

$$12x + 24 = 4x^2 + 16x$$

$$0 = 4x^2 + 4x - 24$$

$$0 = x^2 + x - 6$$

$$0 = (x + 3)(x - 2)$$

$$x + 3 = 0 \quad \text{or} \quad x - 2 = 0$$

$$x = -3 \qquad x = 2$$

Thus, Sue's hiking speed is 2 mi/hr, and her skiing speed is 6 mi/hr.

NOTE ▶ We reject the solution of -3, since Sue could not hike at a rate of -3 mi/hr.

Exercises 4.9

Find the solutions to the following problems.

1. What number must be subtracted from the numerator and added to the denominator of $\frac{9}{11}$ to obtain $\frac{3}{7}$?

2. What number must be added to both the numerator and the denominator of $\frac{1}{3}$ to obtain $\frac{5}{7}$?

3. What number must be subtracted from both the numerator and the denominator of $\frac{3}{5}$ to obtain $\frac{3}{4}$?

4. The numerator of a fraction is 4 more than the denominator. If 2 is added to the denominator and 5 is subtracted from the numerator, we obtain $\frac{2}{5}$. What is the value of the original fraction?

5. The denominator of a fraction is 4 more than the numerator. If 3 is added to both the numerator and the denominator, we obtain $\frac{1}{2}$. What is the value of the original fraction?

6. The denominator of a fraction is 1 more than twice the numerator. If 6 is subtracted from both the numerator and the denominator, we obtain $\frac{5}{3}$. What is the value of the original fraction?

7. The sum of a number and its reciprocal is $\frac{25}{12}$. What is the number?

8. The sum of a number and its reciprocal is $\frac{29}{10}$. What is the number?

9. The sum of the reciprocals of two consecutive integers is $\frac{7}{12}$. What are the integers?

10. The difference of the reciprocals of two consecutive positive integers is $\frac{1}{2}$. What are the integers?

11. The sum of the reciprocals of two consecutive even integers is $\frac{7}{24}$. What are the integers?

12. The sum of the reciprocals of two consecutive even integers is $\frac{9}{40}$. What are the integers?

13. Jolene can mow her yard in 2 hr. Her son Danny can mow the yard in 3 hr. Working together, how long will it take them to mow the yard?

14. The Moss County *Tribune*'s old printing press prints the Sunday newspaper in 6 hr. The new printing press prints the Sunday newspaper in 2 hr. Working together, how long will it take the two presses to print the Sunday edition?

15. Irene can prepare the Moss County Electric Company's annual report in 45 min. It takes Jerry, her boss, 6 hr to prepare the report. Working together, how long will it take them to prepare the report?

16. Jimmy can wash his car in 30 min. His 4-yr-old son Jimmy, Jr. takes 2 hr to wash the car. Working together, how long will it take them to wash the car?

17. Working together Debbie and Jim can dust and vacuum their home in 4 hr. Working alone Debbie can dust and vacuum the house in 5 hr. How long does it take Jim to dust and vacuum the house?

18. Working together, Larry, Joe, and Shirley can wash the windows at the Moss County Courthouse in 1 hr. Working alone, Larry can wash the windows in 2 hr and Joe can wash the windows in 3 hr. How long does it take Shirley to wash the windows?

19. Gary takes twice as long as Barbara to sort the Moss County mail. Together, they can sort the mail in 8 hr. How long does it take each of them to sort the mail?

20. A swimming pool can be filled by an inlet pipe in 12 hr. The drain can empty the pool in 36 hr. How long will it take to fill the pool if the drain is left open?

21. At Four Mile Island, there are two pipes that carry water to the nuclear reactor chamber. The first pipe can fill the reactor chamber in 4 min; the second pipe can fill the reactor chamber in 6 min. The drain can empty the reactor chamber in 12 min. How long will it take both pipes to fill the chamber if the drain is left open?

22. One pipe can fill a reservoir 3 hr faster than another pipe. Together, they fill the reservoir in 2 hr. How long does it take each pipe to fill the reservoir?

23. One pipe can fill an oil tank 1 hr faster than another pipe. The tanker trucks can empty the tank in 3 hr. With both pipes open and the tanker trucks draining the tank, it takes 2 hr to fill the tank. How long does it take each pipe to fill the tank?

24. Sally's motorboat can travel 15 mi downstream in the same time that it can travel 12 mi upstream. If the speed of the current is 3 mi/hr, what is the speed of Sally's boat in still water?

25. Tim's rowboat can travel 7 mi downstream in the same time that it can travel 3 mi upstream. If the speed of the current is 1 mi/hr, what is Tim's rowing speed in still water?

26. Colonel Scanlon can fly his plane 300 mi with the wind in the same time that he can fly it 240 mi against the wind. If the speed of his plane is 180 mi/hr, what is the wind speed?

27. Maureen can fly her plane 250 mi with the wind in the same time that she can fly it 200 mi against the wind. If the speed of her plane is 90 mi/hr, what is the wind speed?

28. Ragu takes his old motorboat to his favorite fishing spot 5 mi downstream. The trip downstream and back takes 4 hr. If the speed of the current is 3 mi/hr, what is the speed of Ragu's boat in still water?

29. Patty flies her airplane 280 mi to the Moss County International Airport and picks up a load of bananas. She then flies back home. The trip to the airport was with the wind, and the trip home was against the wind. The speed of the wind was 30 mi/hr. If Patty's total flying time was 7 hr, what is the speed of Patty's plane in still air?

30. Sam hikes up a 6-mi-long mountain trail. His speed going down the trail is 1 mi/hr faster than his speed going up the trail. Sam spends a total of 5 hr on the trail. What is Sam's hiking speed up the trail? Down the trail?

31. Roachman climbs to the top of a $\frac{3}{4}$-mi-high cliff. He then parachutes down. His parachuting speed is 4 mi/hr faster than his climbing speed. His total climbing and parachuting time is $1\frac{2}{3}$ hr. What is Roachman's climbing time?

32. Matthew walked from his home to Dodge City at the rate of 4 mi/hr. When he got to Dodge City, Miss Kitty offered to take him home in her buggy. The speed of Miss Kitty's buggy was 12 mi/hr. Matthew spent a total of $5\frac{1}{3}$ hr walking and riding. What is the distance from Matthew's home to Dodge City?

33. Cathy Brewerton runs from her home to the local ice cream parlor at a rate of 8 mi/hr. She walks back home at a speed of 2 mi/hr. Cathy spends a total time of $1\frac{1}{4}$ hr walking and running. What is the distance from Cathy's home to the ice cream parlor?

34. Daniel can make the trip from his home to his mother-in-law's home by car or bus. If he goes by car, his speed will be 42 mi/hr; the speed of the bus will be 28 mi/hr. Daniel chooses to go by bus. The trip by bus is $1\frac{1}{2}$ hr longer than by car. How far is it from Daniel's home to his mother-in-law's home?

35. Rayburn's boat has a speed of 14 mi/hr in still water. The total time required for Rayburn to make a round trip from his home on the river to his favorite fishing hole is 7 hr. If the speed of the current is 2 mi/hr, what is the distance from Rayburn's home to the fishing hole?

36. Mrs. Ray's private plane can travel at 120 mi/hr in still air. She flies from her ranch to Dallas against a 20-mi/hr wind. She then flies back to her ranch with the wind. If the trip home takes her 36 fewer minutes, how far is it from her ranch to Dallas? (*Hint:* Convert 36 min to hours.)

Applied Algebra

Conservationists in Moss County have become concerned about the wildlife population in the county. In particular, conservationists are worried that local fishing is too rapidly depleting the trout population in Lake Lurlene. Using proportions, estimate the trout population of Lake Lurlene.

SOLUTION   Suppose 400 tagged trout are released into Lake Lurlene and that all fishing is suspended for 1 month. At the end of the month fish are caught at a number of sites on Lake Lurlene. The number of tagged and untagged trout are counted and then the fish are released back into the lake. Of the 248 trout caught, 66 were tagged.

We can now estimate the trout population in Lake Lurlene using the following proportion:

$$\frac{\text{number of tagged trout caught}}{\text{number of tagged trout released}} = \frac{\text{number of trout caught}}{\text{number of trout in lake}}$$

To simplify let

$$x = \text{the number of trout in Lake Lurlene}$$

$$\frac{66}{400} = \frac{248}{x}$$

$$66x = 400 \cdot 248$$

$$66x = 99{,}200$$

$$x = \frac{99{,}200}{66}$$

$$x = 1503.\overline{03}$$

So a good approximation would be 1500 trout.

Your Turn

Given the following data, use proportions to estimate the following populations in Lake McMurry.

1. Estimate the bass population given that 250 bass are tagged and released into Lake McMurry. One month later, 32 of 104 bass caught are tagged. (*Note:* After counting, the bass are released back into the lake.)

2. Estimate the turtle population given that 70 turtles are tagged and released into Lake McMurry. One month later, 16 of 43 turtles caught are tagged. (Note: After counting, the turtles are released back into the lake.)

Chapter 4 Review

Terms to Remember

Rational expression	p. 177	Complex fraction	p. 194
Simplified fraction or lowest terms	p. 177	Simple fraction	p. 194
		Synthetic division	p. 207
Least common denominator	p. 190	Fractional equation	p. 212
		Extraneous solution	p. 213

Definitions

1. $x^0 = 1$, $x \neq 0$; 0^0 is undefined.

2. Let n be a positive integer.

$$x^{-n} = \frac{1}{x^n} \qquad x \neq 0$$

Laws of Exponents

Let m and n be integers and x and y be real numbers with $x \neq 0$ and $y \neq 0$.

$x^m \cdot x^n = x^{m+n}$	Product Law	$\dfrac{x^m}{x^n} = x^{m-n}$	Quotient Law
$(x^m)^n = x^{m \cdot n}$	Power to a Power Law		
$(xy)^n = x^n y^n$	Product to a Power Law	$\left(\dfrac{x}{y}\right)^n = \dfrac{x^n}{y^n}$	Quotient to a Power Law

Properties of Fractions

■ *Reduction Property of Fractions:*

$$\frac{ax}{ay} = \frac{\cancel{a}x}{\cancel{a}y} = \frac{x}{y} \qquad a \neq 0$$

■ *Multiplication Property of Fractions:*

$$\frac{c}{d} \cdot \frac{n}{m} = \frac{cn}{dm}$$

■ *Division Property of Fractions:*

$$\frac{a}{b} \div \frac{s}{t} = \frac{a}{b} \cdot \frac{t}{s} = \frac{at}{bs}$$

■ *Addition Property of Fractions:*

$$\frac{a}{d} + \frac{b}{d} = \frac{a+b}{d}$$

■ *Subtraction Property of Fractions:*

$$\frac{a}{d} - \frac{b}{d} = \frac{a-b}{d}$$

Review Exercises

4.1   Evaluate each of the following:

1. $5^6 \cdot 5^{-2}$

2. $4 \cdot 4^3$

3. $\dfrac{6^5}{6^2}$

4. $\dfrac{7^3}{7^5}$

5. $(2^3)^2$

6. $(3^2)^{-2}$

7. $(8^0)^{-3}$

8. $(-6^3)^0$

9. -9^2

10. -3^6

11. $(-4)^4$

12. $(-6)^{-2}$

13. -5^{-3} **14.** $(-3^{-2})^{-2}$ **15.** $[(2^{-2})^3]^2$ **16.** $2^{-3} + 3^{-2}$

17. $2^{-1} + 4^{-1}$ **18.** $(6^{-1} - 3^{-2})^{-1}$ **19.** $4^{-2} - 4^{-1} - 4^0$ **20.** $3^{-2} - 3^{-1} + 3^0$

21. $\left(\dfrac{4}{5}\right)^3$ **22.** $\left(\dfrac{9}{13}\right)^2$ **23.** $\left(\dfrac{5}{6}\right)^{-3}$ **24.** $\left(\dfrac{2}{3}\right)^{-4}$

25. $\left(\dfrac{5}{8}\right)^0$ **26.** $\left(\dfrac{-7}{6}\right)^0$ **27.** $\left(\dfrac{-3}{5}\right)^{-2}$ **28.** $\left(\dfrac{-4}{5}\right)^{-4}$

29. $\dfrac{4^{-2}}{4^{-6}}$ **30.** $\dfrac{7^{-5}}{7^{-3}}$

Simplify each of the following. (Remember to write your answers with only positive exponents.)

31. $(5x^{-4}y^2)^{-2}(2x^3y^{-4})^3$ **32.** $(4x^{-3}y)^3(8x^{-1}y^{-3})^{-2}$ **33.** $(9x^5y^{-4})^0(6x^{-5}y^0)^{-2}$

34. $(-7^0x^{-3}y^2)^4(3x^0y^{-2})^3$ **35.** $\left(\dfrac{7x^2}{4y}\right)^2$ **36.** $\left(\dfrac{-5a}{2b^2}\right)^3$

37. $\left(\dfrac{2x^{-5}}{5y^4}\right)^2$ **38.** $\left(\dfrac{3x^4}{4y^{-2}}\right)^3$ **39.** $\left(\dfrac{-6x^2}{5y^{-3}}\right)^{-3}$

40. $\left(\dfrac{11x^{-2}}{4y}\right)^{-2}$ **41.** $\left(\dfrac{x^7}{2y^5}\right)^{-2}\left(\dfrac{8x^2}{3y^{-1}}\right)^{-1}$ **42.** $\left(\dfrac{-5x^{-3}}{8y}\right)^2\left(\dfrac{x^5}{6y^{-3}}\right)^{-2}$

43. $\dfrac{6x^{-3}y^7}{-3x^2y^2}$ **44.** $\dfrac{8x^5y^{-4}}{4x^{-4}y^{-6}}$ **45.** $\left(\dfrac{6^0x^5y^{-7}}{2x^2y^2}\right)^{-2}$

46. $\left(\dfrac{5x^{-2}y^{-1}}{2^0x^3y^{-5}}\right)^{-3}$ **47.** $\left[\left(\dfrac{3x^4}{5y^2}\right)^{-2}\right]^{-1}$ **48.** $\dfrac{(3x-1)^{-2}}{(3x-1)^{-5}}$

49. $(y + 4)^{-2}$ **50.** $(x + y)(x^{-1} - y^{-1})$ **51.** $(x^{3m})^2$

52. $\dfrac{x^{2m}}{x^{-4m}}$ **53.** $\dfrac{x^{-m} \cdot x^{3m}}{x^{-5m}}$ **54.** $\dfrac{9x^2(x-2)^2 - 6x^3(x-2)}{(x-2)^4}$

55. $-(2x + 1)(x + 3)^{-2} + 2(x + 3)^{-1}$ **56.** $6(x + 5)^3(3x - 1) + 3(x + 5)^2(3x - 1)^2$

4.2   Reduce the following rational expressions to lowest terms.

57. $\dfrac{12x^5y}{15xy^6}$ **58.** $\dfrac{7a^3b^2}{14a^4b^5}$ **59.** $-\dfrac{xy^4z^7}{5xy^2z^6}$

60. $\dfrac{40a^2b^2c^3}{-16a^8b^2c^7}$ **61.** $\dfrac{6x^2y - 4x^3y^5}{8xy^3 + 12x^4y^2}$ **62.** $\dfrac{9x^5y^3 + 3x^4y^4}{15x^4y^2 + 5x^3y^3}$

63. $\dfrac{x^2 - 8x + 15}{5x - x^2}$ **64.** $\dfrac{7x^2 + 21x}{x^2 + 6x + 9}$ **65.** $\dfrac{2x^2 + 6xy - 20y^2}{x^2 - 25y^2}$

66. $\dfrac{2y^2 - 8x^2}{12x^2 + 30xy - 18y^2}$ **67.** $\dfrac{2x^2 - x - 15}{4x^2 + 4x - 15}$ **68.** $\dfrac{3x^2 - 13x + 4}{9x^2 - 15x + 4}$

69. $\dfrac{8a^3 + b^3}{8a^2 + 6ab + b^2}$ **70.** $\dfrac{a^3 - 27b^3}{a^2 - 6ab + 9b^2}$ **71.** $\dfrac{x^3 - 2x^2y - 4xy^2 + 8y^3}{2y^2 + yx - x^2}$

72. $\dfrac{3a^3 + a^2b - 3ab^2 - b^3}{3b^2 - 2ba - a^2}$

4.3 Perform the indicated operations and reduce the answers to lowest terms.

73. $\dfrac{6x^4y^3}{5xy^2} \cdot \dfrac{10xy^5}{4x^7y}$

74. $6ax^2 \cdot \dfrac{3a^3y^2}{4x^5y}$

75. $10a^3bc^4 \div \dfrac{2ab^3}{5b^2c^5}$

76. $\dfrac{8x^4y^3}{3xyz^2} \div 9x^3y^5z$

77. $\dfrac{4a^2x^4}{5ab^3} \cdot \dfrac{15x^2y^3}{2xz^5} \div \dfrac{3ay^6}{7b^2z}$

78. $\dfrac{6x^5y^2}{7a^3b^4} \cdot \dfrac{14ay^5}{9a^2z} \div \dfrac{3x^3y^7}{4a^2b^4}$

79. $\dfrac{x^2 + 2x - 15}{3x^2 + 15x} \cdot \dfrac{2x^3 + 6x^2}{x^2 - 6x + 9}$

80. $\dfrac{4x^2 - 16x}{x^2 - 16} \cdot \dfrac{x^2 + 2x - 8}{3x^3 + 6x^2}$

81. $\dfrac{4x^2 - y^2}{3x^2 + 10xy + 3y^2} \div \dfrac{y^2 - yx - 2x^2}{x^2 + 4xy + 3y^2}$

82. $\dfrac{2y^2 - 7yx - 4x^2}{4x^2 + 11xy + 6y^2} \div \dfrac{4x^2 + 11xy - 3y^2}{3x^2 + 13xy + 12y^2}$

83. $\dfrac{6x^2 + 7x - 3}{8x^3 + 10x^2 - 3x} \cdot \dfrac{8x^2 - 2x}{6x^2 + 10x - 4}$

84. $\dfrac{12x^2y - 6xy^2}{18x^3 + 21x^2y + 6xy^2} \cdot \dfrac{3x^2 - 4xy - 4y^2}{4x^2y - 10xy^2 + 4y^3}$

85. $\dfrac{a^3 - 8b^3}{6b^2 - ba - a^2} \div \dfrac{3a^3b + 6a^2b^2 + 12ab^3}{a^2 - 9b^2}$

86. $\dfrac{5b^2 - 3ba - 2a^2}{18a^3 - 12a^2b + 8ab^2} \div \dfrac{3a^2 - ab - 2b^2}{27a^3 + 8b^3}$

87. $\dfrac{2ax + xy - 6a - 3y}{4a^2 + 2ay + 10a + 5y} \cdot \dfrac{6ax - 4ay + 15x - 10y}{3x^3 - 2x^2y - 27x + 18y}$

88. $\dfrac{2y^3 + 3y^2 - 8y - 12}{ay + 2a + 2y^2 + 4y} \cdot \dfrac{4ax + ay + 8xy + 2y^2}{8xy + 2y^2 + 12x + 3y}$

89. $\dfrac{x^2 + 2x - 35}{2x^2 - 3x - 2} \cdot \dfrac{x^2 - 9x + 14}{2x^3 + 10x^2} \div \dfrac{x^2 - 49}{8x^2 + 4x}$

90. $\dfrac{x^2 - 10x + 24}{4x^2 + 5x + 1} \cdot \dfrac{4x^3 + x^2 - 4x - 1}{6x^2 - 37x + 6} \div (x^2 - 5x + 4)$

4.4 Perform the indicated additions and subtractions.

91. $\dfrac{5x}{4x^2y} + \dfrac{7x}{4x^2y}$

92. $\dfrac{11y}{6x^3} - \dfrac{2y}{6x^3}$

93. $\dfrac{2x^2}{5x - 1} + \dfrac{8x^2}{5x - 1}$

94. $\dfrac{4x}{x - 2} - \dfrac{8}{x - 2}$

95. $\dfrac{5x}{x^2 - 6x - 7} + \dfrac{5}{x^2 - 6x - 7}$

96. $\dfrac{3x - 1}{x^2 - 9} + \dfrac{10}{x^2 - 9}$

97. $\dfrac{2y}{y^2 - y - 12} - \dfrac{y - 3}{y^2 - y - 12}$

98. $\dfrac{3y + 2}{2y^2 - y - 1} - \dfrac{2y + 3}{2y^2 - y - 1}$

99. $\dfrac{4}{x^2} + \dfrac{5}{3x}$

100. $\dfrac{6x}{5y} - \dfrac{11}{2x}$

101. $\dfrac{2x}{x - 5} + \dfrac{3}{5 - x}$

102. $\dfrac{7}{2x - y} + \dfrac{3 - y}{y - 2x}$

103. $\dfrac{5x}{5x - 1} + \dfrac{2}{x + 4}$

104. $\dfrac{7}{x - 3} - \dfrac{2x}{3x + 4}$

105. $\dfrac{9}{2x - 5} + 2$

106. $5 - \dfrac{x + 1}{4x - 1}$

107. $\dfrac{6}{2x - 4} + \dfrac{5}{3x - 6}$

108. $\dfrac{2}{x^2 + 3x} + \dfrac{8}{5x + 15}$

109. $\dfrac{x + 2}{x^2 - 5x - 14} - \dfrac{x}{x^2 - 49}$

110. $\dfrac{x - 3}{4x^2 - 9} + \dfrac{3x}{2x^2 + 9x + 9}$

111. $\dfrac{11x - 11}{x^2 - x - 12} - \dfrac{8x - 4}{x^2 - 5x - 24}$

112. $\dfrac{x + 1}{x^2 + 5x + 6} - \dfrac{3}{x^2 + x - 2}$

113. $\dfrac{x + 2y}{2x^2 - 7xy + 6y^2} + \dfrac{3y}{4x^2 - 4xy - 3y^2}$

114. $\dfrac{x + 9y}{12x^2 - xy - y^2} - \dfrac{8y}{9x^2 - y^2}$

115. $\dfrac{10x + 23}{4x^2 + 4x - 3} - \dfrac{4x + 11}{2x^2 + x - 3}$

116. $\dfrac{x}{6x^2 - 5x - 4} + \dfrac{2}{4x^2 + 4x + 1}$

117. $\dfrac{x + 2}{x^2 - 9} + \dfrac{x - 4}{2x^2 + 7x + 3} - \dfrac{3x}{2x^2 - 5x - 3}$

118. $\dfrac{x + 1}{3x^2 + 5x - 2} - \dfrac{x - 2}{3x^2 + 7x + 2} + \dfrac{2x}{9x^2 - 1}$

4.5 Simplify the following fractions.

119. $\dfrac{\dfrac{3xy^3}{7y}}{\dfrac{9x^5}{14xy}}$

120. $\dfrac{\dfrac{5a^2b}{4x^3}}{\dfrac{3ab^4}{2ax}}$

121. $\dfrac{\dfrac{3x + 7}{x - 2}}{\dfrac{x + 7}{2 - x}}$

122. $\dfrac{\dfrac{3x + 6}{5 - x}}{\dfrac{9x}{x - 5}}$

123. $\dfrac{\dfrac{2x^2 - 7x - 4}{4x^2 - 1}}{\dfrac{x^2 - 7x + 12}{x^3 - 27}}$

124. $\dfrac{\dfrac{x^2 + 3x - 10}{2x^2 + 13x + 15}}{\dfrac{2x^2 - 9x + 9}{4x^2 - 9}}$

125. $\dfrac{\dfrac{6x - 15}{6x - 24}}{\dfrac{10x - 25}{2}}$

126. $\dfrac{\dfrac{4x - 8}{2x}}{\dfrac{x^2 - 4}{x + 2}}$

127. $\dfrac{4 + \dfrac{3}{x}}{\dfrac{2}{x} - 3}$

128. $\dfrac{\dfrac{2}{y} - 1}{\dfrac{5}{y} - 2}$

129. $\dfrac{\dfrac{3}{x^2} + \dfrac{2}{y}}{\dfrac{5}{2x^3y^2}}$

130. $\dfrac{\dfrac{4}{a} - \dfrac{7}{b}}{\dfrac{2}{ab^2}}$

131. $\dfrac{2x}{\dfrac{1}{x} - \dfrac{3}{y}}$

132. $\dfrac{5x^2y}{\dfrac{2}{x^2} + \dfrac{y}{x}}$

133. $\dfrac{\dfrac{1}{x} - \dfrac{1}{y}}{\dfrac{1}{x^4} - \dfrac{1}{y^4}}$

134. $\dfrac{\dfrac{1}{9} - \dfrac{1}{y^2}}{y - 3}$

135. $\dfrac{\dfrac{4x}{x - 1} - \dfrac{16}{3}}{x - 4}$

136. $\dfrac{1 + \dfrac{5x}{x + 2}}{\dfrac{1}{x} + 3}$

137. $\dfrac{\dfrac{3}{5x} + \dfrac{2}{x - 3}}{\dfrac{1}{5x} - \dfrac{4}{x - 3}}$

138. $\dfrac{\dfrac{2}{x} - \dfrac{2}{2x + 1}}{\dfrac{3}{x^2} + \dfrac{3}{2x + 1}}$

139. $\dfrac{\dfrac{4 + h}{6 + 3h} - \dfrac{2}{3}}{h}$

140. $\dfrac{x^2 - \dfrac{1}{x}}{\dfrac{1}{x + 1} + x}$

141. $\dfrac{\dfrac{8}{2 - x} + x}{2 + \dfrac{2x^2}{x - 2}}$

142. $\dfrac{x^{-4} - y^{-4}}{x^{-2} + y^{-2}}$

143. $\dfrac{x^{-1} - (x + 3)^{-1}}{3}$

4.6 Perform the indicated divisions.

144. $\dfrac{7x^5 - 21x^4 + 14x^2}{7x}$

145. $\dfrac{12x^2y^4 - 18x^3y^3 - 15x^4y^2}{3xy^2}$

146. $\dfrac{8a^3b - 6a^2b^2 + 12ab^4}{10a^2b^2}$

147. $\dfrac{16a^4b^5 + 20a^3b^3 + 18ab^2}{4a^2b^3}$

148. $\dfrac{4x^2 + 11x - 7}{x + 4}$

149. $\dfrac{2x^2 - 9x + 5}{2x + 1}$

150. $\dfrac{2x^3 + 3x^2 - 11x - 3}{2x - 3}$

151. $\dfrac{5x^3 - 8x^2 + 2x + 4}{x - 2}$

152. $\dfrac{6x^2 + 5}{3x + 9}$

153. $\dfrac{12x^2 - 11}{2x + 5}$

154. $\dfrac{12x^3 - 35x^2 + 2}{4x - 1}$

155. $\dfrac{18x^3 - 14x + 6}{3x - 2}$

156. $\dfrac{5x^4 + 17x^3 - 22x^2 + 41x + 20}{5x - 3}$

157. $\dfrac{6x^4 - 8x^2 + 22x - 20}{2x + 4}$

158. $\dfrac{2x^4 - 5x^3 - 5x^2 + 27x - 15}{2x^2 + 3x - 3}$

159. $\dfrac{6x^4 - x^3 - 8x^2 - 10}{3x^2 + x + 4}$

160. $\dfrac{64x^3 + 27}{4x + 3}$

161. $\dfrac{x^4 - 81}{x - 3}$

162. $\dfrac{2x^3 - x^2 - 8x + 4}{2x^2 - 5x + 2}$

163. $\dfrac{12x^4 + 9x^3 - 26x^2 - 15x + 10}{3x^2 - 5}$

164. $\dfrac{4x^4 - 10x^3 - 25x - 30}{2x^2 + 6}$

165. $\dfrac{4x^5 + 5x^3 - 16x^2 - 6x + 12}{4x^2 - 3}$

4.7 Perform the indicated divisions using synthetic division.

166. $\dfrac{2x^2 - 7x + 8}{x - 2}$

167. $\dfrac{3x^2 + 5}{x + 3}$

168. $\dfrac{x^3 + 3x^2 - 6x + 5}{x + 4}$

169. $\dfrac{2x^3 - x^2 - 15x - 3}{x - 4}$

170. $\dfrac{x^4 - 8x^2 + x - 12}{x - 3}$

171. $\dfrac{x^4 - 2x^3 + 5x^2 - 4x}{x - 2}$

172. $\dfrac{3x^3 - 7x^2 - x + 2}{x - \dfrac{1}{3}}$

173. $\dfrac{6x^4 - 3x^3 - 7x - 3}{x + \dfrac{3}{2}}$

174. $\dfrac{64x^3 - 1}{x - \dfrac{1}{4}}$

175. $\dfrac{x^5 - 32}{x - 2}$

4.8 Solve for x in each of the following equations.

176. $\dfrac{4}{2x - 1} = \dfrac{12}{5x}$

177. $\dfrac{2}{5x} - \dfrac{3}{4x} = \dfrac{7}{20}$

178. $\dfrac{3}{x - 2} - 1 = \dfrac{x + 1}{x - 2}$

179. $\dfrac{6x - 5}{3x - 2} = 1 - \dfrac{1}{3x - 2}$

180. $\dfrac{4}{x + 1} + \dfrac{2}{3x - 1} = \dfrac{5}{x + 1}$

181. $\dfrac{7}{2x + 3} = \dfrac{3}{x - 4} - \dfrac{x + 5}{2x^2 - 5x - 12}$

182. $\dfrac{x - 2}{x + 1} = 1 - \dfrac{9}{x^2 + 7x + 6}$

183. $\dfrac{x + 1}{x - 2} = 2 - \dfrac{x - 2}{x - 4}$

184. $\dfrac{1}{2} = \dfrac{7}{4x} + \dfrac{1}{x^2}$

185. $\dfrac{3}{25} + \dfrac{2}{5x} = \dfrac{1}{x^2}$

186. $\dfrac{2x}{x + 2} - \dfrac{1}{x + 3} = \dfrac{1}{x^2 + 5x + 6}$

187. $\dfrac{x}{x - 2} - \dfrac{3}{3x + 2} = \dfrac{8}{3x^2 - 4x - 4}$

188. $\dfrac{x + 1}{3x - 11} = \dfrac{3}{2x - 8}$

189. $\dfrac{3x}{x - 8} = \dfrac{-1}{x + 3}$

190. $\dfrac{5}{x + 6} + \dfrac{1}{x - 2} = 1$

191. $\dfrac{2x + 3}{2x + 2} = \dfrac{14x - 41}{2x^2 - 4x - 6} + \dfrac{3x - 9}{x + 1}$

4.9   Find the solutions to the following problems.

192. What number must be added to both the numerator and denominator of $\frac{7}{3}$ to obtain $\frac{5}{3}$?

193. The numerator of a fraction is 1 more than twice the denominator. If 21 is subtracted from the numerator, we obtain -2. What is the value of the original fraction?

194. The sum of a number and its reciprocal is $\frac{53}{14}$. What is the number?

195. The sum of the reciprocals of two consecutive even integers is $\frac{13}{84}$. What are the integers?

196. A farmer can plow a field in 4 days using a tractor. Her hired hand can plow the same field in 400 days using a hoe. How many days will be required for the plowing if they work together?

197. A recent transportation study indicates that in the event of a threat of nuclear attack, Houston could be evacuated in 1 day via IH 10, IH 45, and Highway 59. If only IH 10 was used, it would take 3 days, and if only IH 45 was used, it would take 2 days. How long would it take to evacuate Houston if only Highway 59 was used?

198. A certain toilet tank can be emptied in 6 sec through its drain when it is flushed. This happens at the same time that water is flowing in through the inlet pipe. With the drain closed, the inlet pipe can fill the tank in 30 sec. How long would it take to drain the tank if water was not flowing in through the inlet pipe?

199. Doris can shell a quart of pecans twice as fast as her daughter Dottie. Working together they can shell a quart of pecans in 20 min. How long would it take Dottie to shell a quart of pecans by herself?

200. Tarzan can swim 7 mi up the Umgawa River in the same time that he can swim 13 mi down the Umgawa River. If he can swim 5 mi/hr in still water, how fast is the current moving in the Umgawa River?

201. One day Buzzardman flew 90 mi against the winds of Hurricane Wadsworth to rescue Anne Lane. He then flew 75 mi with the wind to capture his archenemy Crazy Ragu. The hurricane's winds were traveling at 90 mi/hr. If both trips took a total of $\frac{1}{2}$ hr, how fast can Buzzardman fly in still air?

202. Robin Hood and his band of merry men can travel twice as fast through Sherwood Forest as the Sheriff of Nottingham. One evening Robin Hood traveled 4 mi through the forest. The next morning it took the sheriff 20 min longer to travel the same 4 mi. How fast can Robin Hood travel through the forest?

203. Hermes can travel 2 mi up Mt. Olympus at a rate that is 18 mi/hr slower than when he rides Pegasus down the same distance. The entire trip takes him 14 min. How fast can he go up? How fast can he ride down? (*Hint:* Change 14 min to hours.)

Chapter 4 Test

(You should be able to complete this test in 60 min.)

I. Simplify each of the following. (Remember to write your answer with only positive exponents.)

1. $(-3)^{-2}$

2. $\left(\dfrac{2y^{-3}}{x^{10}}\right)^2$

3. $4a^2b^0(3a^2b^3)^2$

4. $\left(\dfrac{3x^3y^{-4}}{x^2y^4}\right)^{-2}$

II. Reduce the following rational expression to lowest terms.

5. $\dfrac{27a^3 - 8b^3}{8b^3 - 12ab^2 + 18a^2b - 27a^3}$

III. Perform the indicated operations and reduce the answers to lowest terms.

6. $\dfrac{2x - 21}{x^2 + 9x + 18} + \dfrac{5x - 21}{x^2 + 2x - 3}$

7. $\dfrac{4x^2 - 13x - 12}{x^2 + 3x - 28} \cdot \dfrac{x^2 + 9x + 14}{x^2 - 4x - 12}$

8. $\dfrac{a^2b - 3a^3 - 4b^3 + 12ab^2}{a^2 - 2ab - 8b^2} \div \dfrac{6a^2 + ab - b^2}{2a^2 - 7ab - 4b^2}$

9. $\dfrac{3x - 5}{x^2 - 9} - \dfrac{x + 1}{2x^2 - 6x} + \dfrac{5}{2x + 6}$

IV. Simplify the following fractions.

10. $\dfrac{\dfrac{2x + 2}{3x}}{\dfrac{x^2 - 1}{9x}}$

11. $\dfrac{\dfrac{2x + 1}{x} - \dfrac{5}{2}}{x - 2}$

12. $\dfrac{x^{-1} - y^{-1}}{x^{-4} - y^{-4}}$

V. Perform the indicated divisions.

13. $\dfrac{5x^3 - 17x^2 + 26x - 8}{5x - 2}$

14. $\dfrac{3x^4 + 7x^3 - 15x^2 + 9x + 10}{x^2 + 2x - 5}$

15. $\dfrac{2x^5 - 7x^4 - x^3 + 12x^2 - 5}{x - 3}$

16. $\dfrac{x^4 - 14x^2 + 9x + 3}{x + 4}$

VI. Solve for x in each of the following equations.

17. $\dfrac{5}{3x + 2} = \dfrac{5}{x - 2}$

18. $\dfrac{x - 1}{x - 4} - \dfrac{4}{x + 2} = \dfrac{24}{x^2 - 2x - 8}$

VII. Find the solutions to the following problems.

19. It takes Nathaniel 18 min to stand up all his toy soldiers. It takes his sister Elizabeth 24 min to knock them all down. If Elizabeth starts knocking them over as soon as Nathaniel starts standing them up, how long will it take for Nathaniel to stand up all his soldiers?

20. Fat Chance can row his boat 9 mi up Moss Creek in the same time that it takes him to row 15 mi down the creek. If Fat can row 2 mi/hr in still water, how fast is the current in Moss Creek?

TEST YOUR MEMORY

These problems review Chapters 1–4.

I. Perform the indicated operations.

1. $\dfrac{5}{8} \cdot \dfrac{4}{15} + \dfrac{1}{2} - \dfrac{2}{3} \div \dfrac{1}{9}$

2. $10 - 2[12 - 4^2]$

II. Simplify each of the following. (Remember to write your answer with only positive exponents.)

3. $3^{-1} + (-3)^{-2}$

4. $5x^0 y^3 (2x^4 y^2)^2$

5. $\left(\dfrac{6x^{-3} y^4}{x^2 y} \right)^2$

III. Completely factor the following polynomials.

6. $256m^3 - 4n^3$

7. $12x^2 + 8xy - 15y^2$

8. $2ax + 4bx + ay + 2by$

9. $4x^2 + 100y^2$

10. $x^4 - 10x^2 + 9$

11. $5x^3 - 4x^2 - 20x + 16$

IV. Perform the indicated operations.

12. $2(4x^3 - 7x + 1) - 3(2x^2 - 4x - 6)$

13. $(2x^2 y^4)^3 (3xy^5)^2$

14. $5xy^4(2x^2 - 7xy + 3y^4 - 1)$

15. $(x - 4)(7x^2 - x - 2)$

16. $\dfrac{x^2 - x - 30}{10x^4 - 20x^3} \cdot \dfrac{2x^2 - 8}{x^2 + 7x + 10}$

17. $\dfrac{x^3 - 27}{3x^3 - x^2 - 12x + 4} \div \dfrac{x^2 + 3x - 18}{3x^2 + 5x - 2}$

18. $\dfrac{3x^2 - 2}{x - 2} + \dfrac{x^2 + 3x}{2 - x}$

19. $\dfrac{5}{x + 3} - \dfrac{2}{x - 1}$

20. $\dfrac{x - 1}{x^2 - 4} + \dfrac{x + 3}{x^2 - 3x + 2}$

21. $\dfrac{5x + 2}{x + 2} - \dfrac{4x}{x^2 + 5x + 6} - \dfrac{2x + 4}{x + 3}$

V. Simplify the following fractions.

22. $\dfrac{\dfrac{x^2 - x - 12}{4x^3 + 12x^2}}{\dfrac{x^2 - 5x + 4}{12x^2 - 12x}}$

23. $\dfrac{\dfrac{2x + 7}{x} - \dfrac{1}{4}}{x + 4}$

24. $\dfrac{x^{-2} - y^{-2}}{x^{-1} - y^{-1}}$

VI. Perform the indicated divisions.

25. $\dfrac{2x^4 - 5x^3 - x^2 + 25x - 26}{x^2 - 3x + 4}$

26. $\dfrac{3x^3 + 2x^2 - 14x - 17}{x + 2}$

VII. Find the solutions of the following equations and check your solutions.

27. $6x^2 + 11x + 4 = 0$

28. $5 - 3(2x + 1) = 4x + 8$

29. $(x + 2)(2x - 5) = -7$

30. $\dfrac{1}{x + 3} + 1 = \dfrac{7x + 10}{x^2 + 3x}$

31. $|3x - 2| = |x + 7|$

32. $(x + 3)(x - 5) = (2x - 9)(x - 3)$

33. $\dfrac{6}{x-2} = \dfrac{3}{x+1}$

34. $\dfrac{2}{3}(x+6) - \dfrac{3}{4}(x+7) = -1$

35. $2(x-4) + 5(2x-4) = 0$

36. $|2x+7| = 8$

37. $\dfrac{2x+3}{x+4} - \dfrac{4}{2x+5} = \dfrac{-1}{2x^2+13x+20}$

38. $\dfrac{x-1}{x+3} + \dfrac{x-3}{x-4} = \dfrac{7}{x^2-x-12}$

39. $\dfrac{1}{2}x + \dfrac{2}{3} = \dfrac{2}{3}x - \dfrac{5}{6}$

40. $(3x-2)(x-1) = 2(2x+1)$

VIII. **Find and graph the solutions of the following inequalities. Express your answers using interval notation.**

41. $5x + 7 \le 8x + 13$

42. $-5 < 3x + 4 < 19$

IX. **Use algebraic expressions to find the solutions of the following problems.**

43. A rectangle has a length that is 2 ft more than four times the width. The area is 72 sq ft. What are the dimensions of the rectangle?

44. The area of a triangle is 25 sq cm. The base is 5 cm less than three times the height. Find the base and the height of the triangle.

45. Aretha has $350 in fives and tens. The number of $10 bills is 2 more than five times the number of fives. How many of each type bill does she have?

46. Red and Esther have a rectangular swimming pool that is 10 ft wide and 12 ft long. They are going to build a tile border of uniform width around the pool. They have 168 sq ft of tile. How wide is the border?

47. Lucinda can type a 10-page report in 2 hr. Reza can type the report in 6 hr. Working together, how long will it take them to type the report?

48. Lucy and Ricky have a big argument. Lucy leaves home traveling north at a rate of 58 mi/hr. At the same time Ricky leaves home traveling south at a rate of 62 mi/hr. In how many hours will Lucy and Ricky be 180 mi apart?

49. Big Oak McKenzie's rowboat can travel 12 mi downstream in the same time that it can travel 4 mi upstream. If the speed of the current is 2 mi/hr, what is Big Oak's rowing speed in still water?

50. Bartolo walks from his home to school at a rate of 3 mi/hr. He rides his little brother's bicycle back home at a rate of 15 mi/hr. Bartolo spends a total time of $\frac{4}{5}$ hr walking and riding. What is the distance from Bartolo's home to school?

Exponential and Radical Expressions

When trying to put out a fire, fire-fighters sometimes work with three mathematical quantities. The first of these quantities we will investigate is the flow rate of the water available at the site of the fire. The second quantity measures the maximum horizontal distance a stream of water will extend. The third quantity is called the friction loss. The friction loss measures the amount of water pressure that is lost as the water passes through a hose. At the end of this chapter we will see how firefighters determine flow rate. At the end of Chapter 6 we will see how firefighters determine the maximum horizontal distance of a stream of water. In Chapter 7 we will examine the friction loss as water passes through a fire hose.

5.1 *Rational Exponents*

In previous chapters we learned how to work with integer exponents. In this chapter we will extend our work to include rational exponents. We will also investigate the connection between rational exponents and radicals.

The laws of exponents introduced in Chapters 3 and 4 are summarized here.

Let m and n be integers and $x \neq 0$ and $y \neq 0$. Then

$$x^m \cdot x^n = x^{m+n} \qquad \textit{Product Law}$$

$$\frac{x^m}{x^n} = x^{m-n} \qquad \textit{Quotient Law}$$

$$(x^m)^n = x^{mn} \qquad \textit{Power to a Power Law}$$

$$(xy)^n = x^n \cdot y^n \qquad \textit{Product to a Power Law}$$

$$\left(\frac{x}{y}\right)^n = \frac{x^n}{y^n} \qquad \textit{Quotient to a Power Law}$$

Consider the following line of reasoning. We know that

$$9 = 9^1$$
$$= 9^{2/2}$$
$$= (9^{1/2})^2 \qquad \textit{By the power to a power law of exponents}$$

Now, $9^{1/2}$ should be some real number with the property that $(9^{1/2})^2 = 9$. Thus, for the power to a power law of exponents to be consistent, $9^{1/2}$ could be 3, or $9^{1/2}$ could be -3. The value of $9^{1/2}$ is determined by the following definition.

Definition 5.1.1	Let $a > 0$ and n be a positive even integer. The **principal nth root of a,** denoted $a^{1/n}$, is the *positive* number such that $(a^{1/n})^n = a$.

XAMPLE 5.1.1   Evaluate each of the following.

1. $9^{1/2} = 3$, since $3 > 0$ and $3^2 = 9$.

2. $16^{1/4} = 2$, since $2 > 0$ and $2^4 = 16$.

3. $\left(\dfrac{4}{9}\right)^{1/2} = \dfrac{2}{3}$, since $\dfrac{2}{3} > 0$ and $\left(\dfrac{2}{3}\right)^2 = \dfrac{4}{9}$.

4. $(-36)^{1/2}$ This expression does not represent a real number, since no real number squared is -36. It violates the previously stated condition that $a > 0$.

NOTE ▶   When $a < 0$ and n is a positive even integer, $a^{1/n}$ is not a real number.

5. $-36^{1/2} = -(36^{1/2}) = -6$

In this problem the base is positive 36, unlike in part 4.

We now know how to handle expressions like $a^{1/n}$ when n is even. What do we do when n is an odd integer? Again consider the following line of reasoning. We know that

$$8 = 8^1$$
$$= 8^{3/3}$$
$$= (8^{1/3})^3$$

Now, $8^{1/3}$ should be some real number with the property that $(8^{1/3})^3 = 8$. Since $2^3 = 8$, it would be reasonable to expect that $8^{1/3} = 2$.

Definition 5.1.2 | Let a be a real number and n be a positive odd integer. The **nth root of a**, denoted $a^{1/n}$, is the real number such that $(a^{1/n})^n = a$.

EXAMPLE 5.1.2   Evaluate each of the following.

1. $27^{1/3} = 3$, since $3^3 = 27$.

2. $(-64)^{1/3} = -4$, since $(-4)^3 = -64$.

3. $32^{1/5} = 2$, since $2^5 = 32$.

Note that $0^{1/n} = 0$, where n is any positive integer. This fact can be seen from a couple of examples:

$$0^{1/2} = 0 \qquad since\ 0^2 = 0$$
$$0^{1/5} = 0 \qquad since\ 0^5 = 0$$

The rational exponents we have been investigating have all had a numerator of 1. Suppose we had a rational exponent with a numerator not equal to 1—for example, $4^{3/2}$. We know that

$$4^{3/2} = (4^{1/2})^3$$
$$= 2^3$$
$$= 8$$

smaller #'s
this way

Also,

$$4^{3/2} = (4^3)^{1/2}$$
$$= 64^{1/2}$$
$$= 8$$

Notice that either path leads to the same answer. This observation suggests the following definition.

Definition 5.1.3	Let m and n be positive integers with $\frac{m}{n}$ being reduced to lowest terms; then

$$a^{m/n} = (a^{1/n})^m$$

or

$$a^{m/n} = (a^m)^{1/n} \quad \text{provided } a^{1/n} \text{ is a real number.}$$

REMARK   It is usually easiest to use the first form of this definition, since $a^{1/n}$ will, in general, be a smaller number than a^m.

EXAMPLE 5.1.3   Evaluate each of the following.

1. $16^{3/2} = (16^{1/2})^3$
   $\quad\quad = 4^3$
   $\quad\quad = 64$

2. $(-27)^{4/3} = [(-27)^{1/3}]^4$
   $\quad\quad\quad\quad = [-3]^4$
   $\quad\quad\quad\quad = 81$

3. $\left(\dfrac{1}{4}\right)^{3/2} = \left[\left(\dfrac{1}{4}\right)^{1/2}\right]^3$
   $\quad\quad\quad = \left(\dfrac{1}{2}\right)^3$
   $\quad\quad\quad = \dfrac{1}{8}$

Finally, suppose that the rational exponent is negative. To be consistent with the quotient law of exponents, we define negative fractional exponents as follows.

Definition 5.1.4	Let m and n be positive integers with $\frac{m}{n}$ being reduced to lowest terms and $a \neq 0$; then

$$a^{-m/n} = \frac{1}{a^{m/n}}$$

NOTE ▶   Using the definitions of this section, it can be shown that the laws of exponents hold for all rational exponents.

EXAMPLE 5.1.4

Evaluate each of the following.

1. $27^{-2/3} = \dfrac{1}{(27)^{2/3}}$

$\quad\quad = \dfrac{1}{(27^{1/3})^2}$

$\quad\quad = \dfrac{1}{3^2}$

$\quad\quad = \dfrac{1}{9}$

2. $(-8)^{-5/3} = \dfrac{1}{(-8)^{5/3}}$

$\quad\quad = \dfrac{1}{[(-8)^{1/3}]^5}$

$\quad\quad = \dfrac{1}{[-2]^5}$

$\quad\quad = \dfrac{1}{-32}$

3. $\left(\dfrac{4}{9}\right)^{-3/2} = \dfrac{1}{\left(\dfrac{4}{9}\right)^{3/2}}$

$\quad\quad = \dfrac{1}{\left[\left(\dfrac{4}{9}\right)^{1/2}\right]^3}$

$\quad\quad = \dfrac{1}{\left(\dfrac{2}{3}\right)^3}$

$\quad\quad = \dfrac{1}{\dfrac{8}{27}}$

$\quad\quad = \dfrac{27}{8}$

As a final example we will consider the cases in which rational exponents are on variables as well as numbers. We will write our answers with only positive exponents, and we will assume that all the variables represent positive real numbers.

EXAMPLE 5.1.5

Simplify each of the following. Assume that all the variables represent positive real numbers.

1. $(3^{1/2}x)^3 \cdot (x^{-1/4})^2 = (3^{1/2})^3 x^3 x^{-2/4}$

$\quad\quad\quad\quad\quad\quad\quad = 3^{3/2} x^3 x^{-1/2}$

$\quad\quad\quad\quad\quad\quad\quad = 3^{3/2} x^{3-(1/2)}$

$\quad\quad\quad\quad\quad\quad\quad = 3^{3/2} x^{5/2}$

2. $\dfrac{(x \cdot x^{1/2})^3}{x^{5/3}} = \dfrac{(x^{1+(1/2)})^3}{x^{5/3}}$

$\quad\quad\quad = \dfrac{(x^{3/2})^3}{x^{5/3}}$

$\quad\quad\quad = \dfrac{x^{9/2}}{x^{5/3}}$

$\quad\quad\quad = x^{(9/2)-(5/3)}$

$\quad\quad\quad = x^{(27/6)-(10/6)}$

$\quad\quad\quad = x^{17/6}$

3. $\dfrac{8^{1/2}x^{-1/3}y^{5/8}}{8^{1/6}x^{5/3}y^{1/4}} = 8^{(1/2)-(1/6)}x^{(-1/3)-(5/3)}y^{(5/8)-(1/4)}$

$\qquad = 8^{1/3}x^{-2}y^{3/8}$

$\qquad = 2 \cdot \dfrac{1}{x^2} \cdot y^{3/8}$

$\qquad = \dfrac{2y^{3/8}}{x^2}$

4. $x^{1/2}(x^2 + 3x^{-1/2}) = x^{1/2} \cdot x^2 + 3x^{-1/2} \cdot x^{1/2}$

$\qquad = x^{(1/2)+2} + 3x^{(-1/2)+(1/2)}$

$\qquad = x^{5/2} + 3x^0$

$\qquad = x^{5/2} + 3$

5. $(x^{1/2} + y^{1/2})(x^{1/2} - y^{1/2}) = x^{1/2}x^{1/2} - x^{1/2}y^{1/2} + x^{1/2}y^{1/2} - y^{1/2}y^{1/2}$

$\qquad = x^{(1/2)+(1/2)} - y^{(1/2)+(1/2)}$

$\qquad = x - y$

6. $(x^{1/3} + 1)(x^{2/3} - x^{1/3} + 1) = x^{1/3}x^{2/3} - x^{1/3}x^{1/3} + x^{1/3} + x^{2/3} - x^{1/3} + 1$

$\qquad = x^{(1/3)+(2/3)} - x^{(1/3)+(1/3)} + x^{1/3} + x^{2/3} - x^{1/3} + 1$

$\qquad = x - x^{2/3} + x^{1/3} + x^{2/3} - x^{1/3} + 1$

$\qquad = x + 1$

E xercises 5.1

Evaluate each of the following.

1. $36^{1/2}$

2. $121^{1/2}$

3. $196^{1/2}$

4. $400^{1/2}$

5. $\left(\dfrac{25}{36}\right)^{1/2}$

6. $\left(\dfrac{9}{49}\right)^{1/2}$

7. $\left(\dfrac{144}{25}\right)^{1/2}$

8. $\left(\dfrac{225}{64}\right)^{1/2}$

9. $(-9)^{1/2}$

10. $(-4)^{1/2}$

11. $(-1)^{1/2}$

12. $-9^{1/2}$

13. $-4^{1/2}$

14. $-1^{1/2}$

15. $0^{1/2}$

16. $1^{1/4}$

17. $-1^{1/4}$

18. $(-1)^{1/4}$

19. $256^{1/4}$

20. $81^{1/4}$

21. $(-16)^{1/4}$

22. $-16^{1/4}$

23. $\left(\dfrac{16}{81}\right)^{1/4}$

24. $\left(\dfrac{625}{256}\right)^{1/4}$

25. $64^{1/3}$

26. $-8^{1/3}$

27. $(-125)^{1/3}$

28. $343^{1/3}$

29. $\left(\dfrac{-8}{125}\right)^{1/3}$

30. $-\left(\dfrac{27}{8}\right)^{1/3}$

31. $32^{1/5}$

32. $1^{1/5}$

33. $243^{1/5}$

34. $\left(\dfrac{243}{1024}\right)^{1/5}$

35. $\left(\dfrac{-1}{32}\right)^{1/5}$

36. $0^{1/5}$

37. $4^{5/2}$ **38.** $9^{3/2}$ **39.** $(-81)^{3/4}$ **40.** $-16^{5/4}$

41. $-8^{2/3}$ **42.** $(-8)^{2/3}$ **43.** $32^{3/5}$ **44.** $\left(\dfrac{4}{25}\right)^{3/2}$

45. $-\left(\dfrac{1}{16}\right)^{3/2}$ **46.** $\left(\dfrac{-8}{27}\right)^{4/3}$ **47.** $\left(\dfrac{16}{81}\right)^{3/4}$ **48.** $-\left(\dfrac{1}{32}\right)^{2/5}$

49. $49^{-1/2}$ **50.** $-100^{-1/2}$ **51.** $(-64)^{-2/3}$ **52.** $(-32)^{-3/5}$

53. $25^{-3/2}$ **54.** $(-49)^{-3/2}$ **55.** $\left(\dfrac{16}{25}\right)^{-1/2}$ **56.** $\left(\dfrac{81}{49}\right)^{-3/2}$

57. $\left(\dfrac{1}{27}\right)^{-2/3}$ **58.** $\left(\dfrac{-8}{125}\right)^{-4/3}$ **59.** $\left(\dfrac{16}{81}\right)^{-3/4}$ **60.** $-\left(\dfrac{27}{125}\right)^{-2/3}$ $\dfrac{25}{9}$

Simplify each of the following. (Remember to write your answers with only positive exponents.) Assume that all of the variables represent positive real numbers.

61. $25^{1/3} \cdot 25^{1/6}$ **62.** $8^{1/2} \cdot 8^{1/6}$ **63.** $(7^{-1/2})^4$

64. $(3^{1/4})^{12}$ **65.** $(9^{2/5} \cdot 9^{11/10})^2$ **66.** $(32^{-3/5} \cdot 32^{1/2})^4$

67. $\dfrac{27^{1/2}}{27^{1/6}}$ **68.** $\dfrac{8^{5/12}}{8^{3/4}}$ **69.** $5^{1/3} \cdot 5^{1/5}$

70. $\dfrac{8^{-5/8}}{8^{-3/4}}$ **71.** $\dfrac{9^{1/6} \cdot 9^{-2}}{9^{2/3}}$ **72.** $\dfrac{16^{-1/2} \cdot 16^{5/12}}{16^{2/3}}$

73. $(6^{-3/4})^{-2/3}$ **74.** $(7^{3/4} \cdot 7^{-1/5})^2$ **75.** $(9^{1/20} \cdot 32^{1/50})^{10}$

76. $(8^{1/18} \cdot 9^{1/12})^6$ **77.** $x^{3/4} \cdot x^{1/12}$ **78.** $2x^{1/3} \cdot 5x^{1/6}$

79. $(27x^4)^{2/3}$ **80.** $(-32x^{-5/3})^{3/5}$ **81.** $(4x^3)^{1/2}(8x^{1/3})$

82. $(16x^3)^{1/2}(16x^3)^{1/4}$ **83.** $\dfrac{x^{5/3}}{x^{1/2}}$ **84.** $\dfrac{x^{3/4}}{x^{4/5}}$

85. $\dfrac{2x^{-2/3}}{3x^{1/4}}$ **86.** $\dfrac{15x^{1/8}}{20x^{-1/6}}$ **87.** $\dfrac{(x^2 \cdot x^{1/2})^3}{x^5}$

88. $\dfrac{(x \cdot x^{-2/3})^{-4}}{x^{3/4}}$ **89.** $\dfrac{-6x^3y^{-1/2}}{4x^{1/3}y^2}$ **90.** $\dfrac{-5x^{-2}y}{10x^{1/4}y^{2/3}}$

91. $\dfrac{16^{3/8}x^{-1/2}y^{5/3}}{16^{1/8}x^{1/4}y^{1/2}}$ **92.** $\dfrac{32^{1/10}x^2y^{-2/3}}{32^{3/10}x^{1/2}y^{-3/2}}$ **93.** $\left(\dfrac{4x^4}{9y^{1/3}}\right)^{1/2}$

94. $\left(\dfrac{-27x^{-1}}{8y^{-3/2}}\right)^{2/3}$ **95.** $\left(\dfrac{8x^{-1/2}}{125y^{1/4}}\right)^{-1/3}$ **96.** $\left(\dfrac{25x^3}{49y^{-1/3}}\right)^{-1/2}$

97. $\dfrac{(-8x^3y^7)^{2/3}}{6x^4y^{-1/3}}$ **98.** $\dfrac{(-64x^2y^6)^{-1/3}}{2x^{1/3}y^{-2}}$ **99.** $\left(\dfrac{4x}{y^3}\right)^{-3/2}\left(\dfrac{8x^4}{y^{-2}}\right)^{1/3}$

100. $\left(\dfrac{9x^{-3}}{y^{4/3}}\right)^{1/2}\left(\dfrac{27x}{y^6}\right)^{-2/3}$ **101.** $x^{1/6}(3x^{1/3} - 7)$ **102.** $x^{1/2}(2x^{3/2} - x^{1/2})$

103. $5x^{-3/2}(2x + 5x^{1/2})$ **104.** $3x^{-1/3}(7x^{-5/3} - 2x^2)$ **105.** $7x^{-2/3}(3x^{-1/6} + 4xy^{3/4})$

106. $5x^{3/4}(2x^{1/8}y - 3xy^2)$ **107.** $x^{1/4}(x^4 - 3x^{3/4} + x^{-1/4})$ **108.** $2x^{-3/2}(5x^2 - 7x + 4)$

109. $(x^{1/4} + 2)(x^{1/4} - 2)$

110. $(x^{-1/2} + y^{-1/2})(x^{1/2} - y^{1/2})$

111. $(x^{1/2} - 5)^2$

112. $(x^{3/2} + 9)^2$

113. $(x^{1/3} - 2)(x^{2/3} + 2x^{1/3} + 4)$

114. $(x^{2/3} + 3)(x^{4/3} - 3x^{2/3} + 9)$

115. $x^{m/2} \cdot x^{m/2}$

116. $(x^{3m})^{-1/3}$

117. $\dfrac{x^{m/3}}{x^{m/4}}$

118. $\dfrac{x^{-m/5}}{x^{-m/2}}$

119. $\left(\dfrac{x^{m/3} \cdot x^{-m/3}}{x^{3m}}\right)^3$

120. $(49x^{3m/2})^{-1/2}$

121. $(2^{1/2}x^{m/4})^{1/2}(2^{1/4}x^{m/6})^3$

122. $(x^{-m}y^{3m})^{1/2}(x^{m/4}y^{-3m/5})^{1/2}$

123. $\dfrac{1}{2} \cdot \left(\dfrac{x+1}{x}\right)^{-1/2} \cdot \dfrac{-1}{x^2}$

124. $\dfrac{1}{2}\left(\dfrac{x+1}{x-3}\right)^{-1/2} \cdot \dfrac{-4}{(x-3)^2}$

Simplify the following by factoring out the greatest common factor.

125. $4(2x + 1)^{3/2}(3x - 1)^{1/3} + 3(2x + 1)^{1/2}(3x - 1)^{4/3}$

126. $9(6x + 7)^{1/2}(4x - 5)^{5/2} + 10(6x + 7)^{3/2}(4x - 5)^{3/2}$

127. $(2x + 1)^{1/2}(3x - 2)^{-2/3} + (2x + 1)^{-1/2}(3x - 2)^{1/3}$

128. $2(4x - 1)^{1/2}(3x + 5)^{-1/3} + 2(4x - 1)^{-1/2}(3x + 5)^{2/3}$

Write Algebra

129. What three steps will you perform to evaluate $8^{-2/3}$?

130. Explain why $(-4)^{1/2}$ is not a real number.

I N T E R N E T   C O N N E C T I O N

The table on page 247 contains statistics obtained from the Internet on annual U.S. energy supply and demand. The web address is http://www.eia.doe.gov/emeu/steo/pub/a1tab.html. In the following example we will use some of the information in this table to construct a model to predict future energy supply and demand levels.

NOTE ▶   In this Internet Connection we will use a calculator to perform most of our calculations.

EXAMPLE   Use the formula $P = \dfrac{a}{(y - 1980)^{0.28}}$ to create a model that predicts U.S. crude oil production (measured in million barrels per day).

Continued

| SOLUTION | Given the formula

$$P = \frac{a}{(y - 1980)^{0.28}}$$

we will let

$y =$ the year

$P =$ U.S. crude oil production measured in million barrels per day during year y.

Since our model contains one unknown, a, we need to use one set of data to determine a. We can choose any year to generate this data. Let's choose 1991. Thus the data for our model are $y = 1991$, $P = 7.42$. Substituting these data into our model, we obtain

$$7.42 = \frac{a}{(1991 - 1980)^{0.28}}$$

$$7.42 = \frac{a}{11^{0.28}}$$

$$7.42(11^{0.28}) = a \qquad \qquad \textit{Multiply both sides by } 11^{0.28}.$$

$$7.42(1.957) \doteq a$$

$$14.52 \doteq a$$

NOTE ▶   0.28 is a rational exponent on 11. Using a scientific calculator we approximate it by pressing 11 $\boxed{x^y}$ 0.28 $\boxed{=}$.

Hence, our model is now given by

$$P = \frac{14.52}{(y - 1980)^{0.28}}$$

Using this model let's predict the U.S. crude oil production in millions of barrels per day for the year 2001.

$$P = \frac{14.52}{(2001 - 1980)^{0.28}}$$

$$P = \frac{14.52}{21^{0.28}}$$

$$P \doteq \frac{14.52}{2.345}$$

$$P \doteq 6.19$$

Thus, using our model we predict that U.S. crude oil production will be 6.19 millions of barrels per day during 2001.

Continued

Exercises

1. Construct a model using the equation

$$P = \frac{a}{(y - 1980)^{0.28}}$$

   that predicts U.S. crude oil production (measured in millions of barrels per day).

2. Predict the U.S. crude oil production for the year 2004 using your results from exercise 1.

3. Construct a model using the equation

$$D = a(y - 1980)^{0.12}$$

   that predicts U.S. petroleum demand (measured in millions of barrels per day).

4. Predict the U.S. petroleum demand for the year 2004 using your results from exercise 3.

Group Activity

5. Working in groups use different sets of data to generate models of the form

$$P = \frac{a}{(y - 1980)^{0.28}}$$

   to predict U.S. crude oil production.

6. Working in groups use different sets of data to generate models of the form

$$D = a(y - 1980)^{0.12}$$

   to predict U.S. petroleum demand.

7. Discuss the reasons why a particular model might yield more accurate results for a given year than another model.

Continued

Table A1. Annual U.S. Energy Supply and Demand
Energy Information Administration/Short-Term Energy Outlook—January 1998

	1985	1986	1987	1988	1989	1990	1991	1992	1993	1994	1995	1996	1997	1998	1999
Real Gross Domestic Product (GDP) (billion chained 1992 dollars)	5324	5488	5649	5865	6062	6136	6079	6244	6390	6611	6742	6928	7187	*7345*	*7465*
Imported Crude Oil Price (nominal dollars per barrel)	26.99	14.00	18.13	14.57	18.08	21.75	18.70	18.20	16.14	15.52	17.14	20.61	18.62	*16.73*	*17.51*
Petroleum Supply Crude Oil Production (million barrels per day)	8.97	8.68	8.35	8.14	7.61	7.36	7.42	7.17	6.85	6.66	6.56	6.46	6.40	*6.42*	*6.39*
Total Petroleum Net Imports (including SPR) (million barrels per day)	4.29	5.44	5.91	6.59	7.20	7.16	6.63	6.94	7.62	8.05	7.89	8.50	8.94	*9.25*	*9.51*
Energy Demand World Petroleum (million barrels per day)	59.9	60.2	61.8	63.1	64.9	65.9	66.0	66.6	66.8	67.0	68.3	70.1	71.9	*73.7*	*75.5*
U.S. Petroleum (million barrels per day)	15.78	16.33	16.72	17.34	17.37	17.04	16.77	17.10	17.24	17.72	17.72	18.31	18.61	*18.97*	*19.22*
Natural Gas (trillion cubic feet)	17.28	16.22	17.21	18.03	18.80	18.72	19.03	19.54	20.28	20.71	21.58	21.96	22.10	*22.87*	*23.26*
Coal (million short tons)	818	804	837	884	891	897	894	907	944	951	962	1007	1034	*1053*	*1072*
Electricity (billion kilowatthours): Utility Sales	2324	2369	2457	2578	2647	2713	2762	2763	2861	2935	3013	3085	3115	*3205*	*3252*
Nonutility Own Use	NA	NA	NA	NA	108	113	122	132	138	150	158	164	169	*173*	*178*
Total	2324	2369	2457	2578	2755	2826	2884	2895	3000	3085	3171	3249	3284	*3378*	*3430*
Total Energy Demand (quadrillion Btu)	74.0	74.3	76.9	80.2	81.3	81.2	81.1	82.4	84.2	85.9	87.5	89.7	90.9	*92.8*	*94.1*
Total Energy Demand per Dollar of GDP (thousand Btu per 1992 Dollar)	13.90	13.54	13.61	13.68	13.42	13.23	13.33	13.20	13.17	12.99	12.98	12.95	12.65	*12.64*	*12.60*
Adjusted Total Energy Demand (quadrillion Btu)	NA	NA	NA	NA	NA	84.1	84.0	85.5	87.3	89.2	90.9	93.9	94.8	*96.5*	*97.8*
Adjusted Total Energy Demand per Dollar of GDP (thousand Btu per 1992 Dollar)	NA	NA	NA	NA	NA	13.70	13.82	13.70	13.67	13.49	13.49	13.56	13.19	*13.14*	*13.11*

5.2 Radicals

In Chapter 1 we evaluated expressions like $\sqrt{9}$. The symbol $\sqrt{}$ is called a **radical sign.** In the preceding section we learned how to handle expressions with rational exponents. We are now ready to learn how to convert rational exponents to radicals. This change in notation is illustrated as follows.

Let n be a positive integer greater than 1 and a be a real number; then

$$a^{1/n} = \sqrt[n]{a}$$

REMARKS

1. The expression $\sqrt[n]{a}$ is read "the nth root of a."

2. a is called the **radicand.**

3. n is called the **index** (or **order**) of the radical.

4. If no index is stated, the index is 2; for example,

$$\sqrt{9} = \sqrt[2]{9}$$

XAMPLE 5.2.1

Evaluate each of the following radical expressions.

1. $\sqrt{9} = 9^{1/2}$ 2. $\sqrt[3]{-125} = (-125)^{1/3}$
   $\phantom{\sqrt{9}} = 3$ $\phantom{\sqrt[3]{-125}} = -5$

3. $\sqrt[4]{16} = 16^{1/4}$ 4. $\sqrt{-36} = (-36)^{1/2}$
   $\phantom{\sqrt[4]{16}} = 2$

As we mentioned, this expression is not a real number.

Suppose we have exponents on factors in the radicand. What steps do we follow to evaluate those radical expressions? Consider the next example.

XAMPLE 5.2.2

Evaluate each of the following radical expressions.

1. $\sqrt{3^2} = \sqrt{9} = 3$ 2. $\sqrt{(-3)^2} = \sqrt{9} = 3$

3. $\sqrt[3]{2^3} = \sqrt[3]{8} = 2$ 4. $\sqrt[3]{(-2)^3} = \sqrt[3]{-8} = -2$

These examples illustrate that $\sqrt{x^2} = |x|$ and $\sqrt[3]{x^3} = x$; they are generalized as follows:

If n is a positive even integer and a is a real number, then

$$\sqrt[n]{a^n} = |a|$$

If n is a positive odd integer and a is a real number, then

$$\sqrt[n]{a^n} = a$$

EXAMPLE 5.2.3   Simplify each of the following.

1. $\sqrt{x^2} = |x|$

2. $\sqrt[3]{y^3} = y$

3. $\sqrt{(5x)^2} = |5x|$

NOTE ▶   For ease in expression, we will assume that the variables represent positive numbers in the remainder of this example. Consequently,

$$\sqrt{(5x)^2} = 5x$$

4. $\sqrt{49x^2y^8} = \sqrt{(7xy^4)^2}$ 5. $\sqrt[4]{16a^4b^{12}} = \sqrt[4]{(2ab^3)^4}$

$\qquad\quad = 7xy^4$ $\qquad\qquad\quad = 2ab^3$

6. $\sqrt[3]{27a^6b^{12}} = \sqrt[3]{(3a^2b^4)^3}$ 7. $\sqrt{x^2 + 2x + 1} = \sqrt{(x + 1)^2}$

$\qquad\quad = 3a^2b^4$ $\qquad\qquad\quad = x + 1$

We know from Section 5.1 that

$$a^{m/n} = (a^m)^{1/n}$$

or

$$a^{m/n} = (a^{1/n})^m$$

This fact leads to the following notation.

Let m and n be positive integers and a be a real number; then

$$a^{m/n} = \sqrt[n]{a^m} \quad \text{or} \quad a^{m/n} = (\sqrt[n]{a})^m$$

provided that $\sqrt[n]{a}$ is a real number.

> **REMARK**   A radical is changed to a rational exponent by the following steps:
>
> 1. Make the index of the radical the denominator of the rational exponent.
> 2. Make the power of the radicand (or the power of the radical) the numerator of the rational exponent.

E XAMPLE 5.2.4   Convert the following expressions to radicals.

1. $5^{3/2} = \sqrt{5^3} = \sqrt{125}$

2. $x^{5/7} = \sqrt[7]{x^5}$

3. $(3xy^2)^{2/9} = \sqrt[9]{(3xy^2)^2} = \sqrt[9]{9x^2y^4}$

E XAMPLE 5.2.5   Convert the following radicals to expressions with rational exponents and simplify where possible.

1. $\sqrt[3]{6^2} = 6^{2/3}$

2. $\sqrt{x} = x^{1/2}$

3. $(\sqrt[3]{y})^5 = y^{5/3}$

4. $\sqrt{(x + y)^7} = (x + y)^{7/2}$

5. $\sqrt{x} \cdot \sqrt[3]{x} = x^{1/2} \cdot x^{1/3}$
   $= x^{1/2+1/3}$
   $= x^{5/6}$

6. $\dfrac{\sqrt{x^5}}{\sqrt{x}} = \dfrac{x^{5/2}}{x^{1/2}}$
   $= x^{5/2-1/2}$
   $= x^{4/2}$
   $= x^2$

7. $\sqrt[3]{\sqrt[4]{x}} = \sqrt[3]{(x)^{1/4}}$
   $= (x^{1/4})^{1/3}$
   $= x^{1/12}$

E xercises 5.2

Evaluate each of the following radical expressions, if possible.

1. $\sqrt{100}$

2. $\sqrt{225}$

3. $\sqrt{\dfrac{36}{25}}$

4. $\sqrt{\dfrac{121}{4}}$

5. $\sqrt{0}$

6. $\sqrt{1}$

7. $\sqrt{-1}$

8. $\sqrt{-4}$

9. $\sqrt[3]{27}$

10. $\sqrt[3]{125}$

11. $\sqrt[3]{-1}$

12. $\sqrt[3]{-64}$

13. $\sqrt[3]{\dfrac{8}{27}}$ **14.** $\sqrt[3]{-\dfrac{1}{216}}$ **15.** $\sqrt{4^2}$ **16.** $\sqrt{5^2}$

17. $\sqrt{(-4)^2}$ **18.** $\sqrt{(-5)^2}$ **19.** $\sqrt[3]{4^3}$ **20.** $\sqrt[3]{5^3}$

21. $\sqrt[3]{(-4)^3}$ **22.** $\sqrt[3]{(-5)^3}$ **23.** $\sqrt[4]{0}$ **24.** $\sqrt[4]{-81}$

25. $\sqrt[4]{(-8)^4}$ **26.** $\sqrt[5]{-1024}$ **27.** $\sqrt[5]{\dfrac{32}{243}}$ **28.** $\sqrt[4]{-\dfrac{1}{16}}$

29. $\sqrt{9}\cdot\sqrt{4}$ **30.** $\sqrt[3]{8}\cdot\sqrt[3]{125}$ **31.** $\sqrt[3]{-27}\cdot\sqrt{81}$ **32.** $\sqrt[5]{32}\cdot\sqrt[4]{625}$

33. $\dfrac{\sqrt{100}}{\sqrt{49}}$ **34.** $\dfrac{\sqrt[3]{27}}{\sqrt[3]{64}}$ **35.** $\dfrac{\sqrt[3]{125}}{\sqrt{9}}$ **36.** $\dfrac{\sqrt{4}}{\sqrt[3]{8}}$

37. $\sqrt{\sqrt{81}}$ **38.** $\sqrt{\sqrt[3]{64}}$

Convert the following expressions to radicals.

39. $7^{2/3}$ **40.** $8^{4/9}$ **41.** $(-5)^{1/5}$ **42.** $(-9)^{1/3}$

43. $\left(\dfrac{3}{4}\right)^{1/2}$ **44.** $\left(\dfrac{5}{8}\right)^{3/7}$ **45.** $x^{3/2}$ **46.** $(2x)^{1/2}$

47. $(4xy)^{3/4}$ **48.** $(5x^2y)^{2/5}$ **49.** $(x+3y)^{1/2}$ **50.** $(2x-5y)^{2/3}$

51. $(16x^2-25y^2)^{1/2}$ **52.** $(x^3+y^3)^{1/3}$

Convert the following radicals to expressions with rational exponents and simplify where possible. Assume that the variables represent positive numbers.

53. $\sqrt[2]{7}$ **54.** $\sqrt[3]{10}$ **55.** $\sqrt[3]{4^2}$ **56.** $\sqrt{5^3}$

57. $\sqrt{x}$ **58.** $\sqrt{x^5}$ **59.** $\sqrt{x^{12}}$ **60.** $\sqrt[3]{t^4}$

61. $(\sqrt[3]{t})^8$ **62.** $\sqrt[3]{t^{12}}$ **63.** $\sqrt[3]{(-x)^6}$ **64.** $\sqrt[5]{p^{20}}$

65. $\sqrt[4]{k^4}$ **66.** $\sqrt[6]{n^{12}}$ **67.** $\sqrt[3]{a^{15}}$ **68.** $\sqrt[4]{(3xy)^3}$

69. $\sqrt[5]{(5xy^2)^2}$ **70.** $\sqrt{x^2+y^2}$ **71.** $\sqrt[3]{x^3+1}$ **72.** $\sqrt{(5xy^3)^4}$

73. $\sqrt{(7x^3y)^6}$ **74.** $\sqrt{(x+y)^2}$ **75.** $\sqrt[3]{(-2xy^3)^6}$ **76.** $\sqrt[3]{(5x^2y^2)^{12}}$

77. $(\sqrt{x+3y})^2$ **78.** $(\sqrt[3]{2x+1})^3$ **79.** $(\sqrt[3]{x+y})^6$ **80.** $\sqrt[4]{x}\cdot\sqrt[3]{x}$

81. $\sqrt{x^3}\cdot\sqrt[6]{x}$ **82.** $\sqrt{x}\cdot\sqrt{x^3}$ **83.** $\sqrt{5y^3}\cdot\sqrt{5y^7}$ **84.** $\sqrt[3]{2x}\cdot\sqrt[3]{4x^2}$

85. $\sqrt[3]{-3x^4}\cdot\sqrt[3]{9x^8}$ **86.** $\dfrac{\sqrt{x}}{\sqrt[3]{x}}$ **87.** $\dfrac{\sqrt[4]{x}}{\sqrt[3]{x}}$ **88.** $\dfrac{\sqrt[6]{x}}{\sqrt{x}}$

89. $\dfrac{\sqrt{x^5}}{\sqrt{x}}$ **90.** $\dfrac{\sqrt{x}}{\sqrt{x^4}}$ **91.** $\dfrac{\sqrt[3]{x^7}}{\sqrt[3]{x^4}}$ **92.** $\sqrt{\sqrt[3]{x}}$

93. $\sqrt[4]{\sqrt[5]{x}}$ **94.** $\sqrt[6]{\sqrt{x}}$ **95.** $\sqrt[5]{\sqrt[3]{x^2}}$ **96.** $\sqrt[4]{\sqrt{x^3}}$

97. $\sqrt{\sqrt[3]{\sqrt[4]{x}}}$

Simplify each of the following. Assume that the variables represent positive real numbers.

98. $\sqrt{25x^2y^2}$

99. $\sqrt{81x^4y^8}$

100. $\sqrt{9x^2y^4z^6}$

101. $\sqrt{100x^2y^{10}z^{12}}$

102. $\sqrt[3]{x^3y^3}$

103. $\sqrt[3]{8x^6y^9}$

104. $\sqrt[3]{-27x^3y^6z^{12}}$

105. $\sqrt[3]{-64x^9y^{30}z^3}$

106. $\sqrt[4]{x^8y^{12}}$

107. $\sqrt[4]{16x^4y^{16}z^{24}}$

108. $\sqrt{x^2 + 8x + 16}$

109. $\sqrt{4x^2 + 4x + 1}$

110. $\sqrt{x^2 + 2xy + y^2}$

111. $\sqrt{x^2 + 6xy + 9y^2}$

112. $\sqrt{5} = 5^{1/2}$ cannot be simplified to a rational number because 5 is not a perfect square. Using a calculator with a $\boxed{\sqrt{}}$ key, we can approximate $\sqrt{5}$ as 2.236067977. A scientific calculator can be used to approximate other roots. If your calculator has an $\boxed{x^{1/y}}$ key, then the following sequence will approximate $\sqrt[3]{7} = 7^{1/3}$, 7 $\boxed{x^{1/y}}$ 3 $\boxed{=}$ (Answer = 1.912931183). If your calculator does not have $\boxed{x^{1/y}}$, it should have $\boxed{x^y}$, which together with an $\boxed{\text{INV}}$ or $\boxed{\text{2nd}}$ key can approximate any root. Follow the sequence 7 $\boxed{\text{INV}}$ $\boxed{x^y}$ 3 $\boxed{=}$. Use a calculator to approximate the following roots.

a. $\sqrt[3]{11}$ **b.** $\sqrt[4]{6}$ **c.** $2^{1/5}$ **d.** $3^{1/6}$
e. $13^{2/3}$ (*Hint:* Recall that $13^{2/3} = (13^{1/3})^2$)

113. By selecting values for x and y, show that $\sqrt{x^2 + y^2} \ne x + y$.

114. By selecting positive and negative values for x, show why it is necessary to write $\sqrt{x^2} = |x|$.

5.3 *Simplifying Radical Expressions*

Radical expressions can be changed by applying the laws of exponents and the facts that we learned in Section 5.2. We know

$$\sqrt[n]{ab} = (ab)^{1/n}$$
$$= a^{1/n}b^{1/n} \qquad \text{\textit{By the product to a power}}$$
$$\qquad\qquad\qquad \text{\textit{law of exponents}}$$
$$= \sqrt[n]{a}\sqrt[n]{b}$$

The observation that $\sqrt[n]{ab} = \sqrt[n]{a}\sqrt[n]{b}$ is nothing more than the product to a power law of exponents in radical form. We will call this form the multiplication property of radicals.

Multiplication Property of Radicals

Let n be a natural number; then

$$\sqrt[n]{ab} = \sqrt[n]{a}\sqrt[n]{b}$$

provided that a and b are not both negative.

REMARK We have observed that $\sqrt[n]{a}$ is not a real number when $a < 0$ and n is even. We will consider products like $\sqrt[n]{a}\sqrt[n]{b}$, where a and/or b are negative, later in this chapter.

EXAMPLE 5.3.1 Multiply the following radicals. Assume that all variables represent positive numbers.

1. $\sqrt{3} \cdot \sqrt{7} = \sqrt{3 \cdot 7}$
$= \sqrt{21}$

2. $\sqrt{5x} \cdot \sqrt{6y} = \sqrt{5x \cdot 6y}$
$= \sqrt{30xy}$

3. $\sqrt[3]{-4} \cdot \sqrt[3]{16} = \sqrt[3]{-4 \cdot 16}$
$= \sqrt[3]{-64}$
$= -4$

4. $\sqrt{x} \cdot \sqrt{x^5} = \sqrt{x \cdot x^5}$
$= \sqrt{x^6}$
$= x^{6/2}$
$= x^3$

From the previous section we know that

$$\sqrt{25} = 5$$
$$\sqrt[3]{64} = 4$$
$$\sqrt[4]{16} = 2$$

In each case the radical expression simplifies to a rational number because:

■ The number under $\sqrt{}$, $25 = 5^2$, is a perfect square.
■ The number under $\sqrt[3]{}$, $64 = 4^3$, is a perfect cube.
■ The number under $\sqrt[4]{}$, $16 = 2^4$, is a perfect fourth power.

This is not always the case. Using the multiplication property of radicals in the reverse direction, we know that

$$\sqrt{75} = \sqrt{25 \cdot 3}$$
$$= \sqrt{25} \cdot \sqrt{3}$$
$$= 5\sqrt{3}$$

$5\sqrt{3}$ is called the **simplified** form of $\sqrt{75}$.

To simplify a radical (which is free of fractions) we use the following steps:

Simplifying a Radical

1. If the radical is a square root, factor the radicand so that the largest perfect square factor within the radicand is one of the factors. If the radical is a cube root, factor the radicand so that the largest perfect cube factor within the radicand is one of the factors, and so on.
2. Rewrite the radical as a product of radicals of the factors found in step 1.
3. Simplify the radical of the perfect square factor, perfect cube factor, and so on.

E XAMPLE 5.3.2   Simplify the following radicals. Assume that all variables represent positive numbers.

1. $\sqrt{700} = \sqrt{100 \cdot 7}$
$= \sqrt{100} \cdot \sqrt{7}$
$= 10\sqrt{7}$

2. $\sqrt{48x^2y^3} = \sqrt{16x^2y^2 \cdot 3y}$
$= \sqrt{16x^2y^2} \cdot \sqrt{3y}$
$= 4xy\sqrt{3y}$

3. $\sqrt{18xy^8} = \sqrt{9y^8 \cdot 2x}$ *Why is $9y^8$ a perfect square?*
$= \sqrt{9y^8} \cdot \sqrt{2x}$
$= 3y^4 \cdot \sqrt{2x}$

NOTE ▶ $\sqrt{y^8} = y^{8/2} = y^4$. If a factor of the radicand has an even exponent, then it is a perfect square. To find the square root, divide the exponent by 2. Thus $\sqrt{x^{10}} = x^5$, $\sqrt{y^{16}} = y^8$, and so on.

4. $\sqrt[3]{40x^3y^5} = \sqrt[3]{8x^3y^3 \cdot 5y^2}$
$= \sqrt[3]{8x^3y^3} \cdot \sqrt[3]{5y^2}$
$= 2xy\sqrt[3]{5y^2}$

5. $\sqrt[3]{81x^4y^5z^6} = \sqrt[3]{27x^3y^3z^6 \cdot 3xy^2}$ *Why is $27x^3y^3z^6$ a perfect cube?*
$= \sqrt[3]{27x^3y^3z^6} \cdot \sqrt[3]{3xy^2}$
$= 3xyz^2\sqrt[3]{3xy^2}$

NOTE ▶ $\sqrt[3]{z^6} = z^{6/3} = z^2$. If a factor of the radicand has an exponent that is a multiple of 3, then it is a perfect cube. To find the cube root, divide the exponent by 3. Thus $\sqrt[3]{x^{12}} = x^4$, $\sqrt[3]{y^{27}} = y^9$, and so on. Similarly, if the radicand of a fourth root contains a factor with an exponent that is a multiple of 4, then the fourth root can be found by dividing by 4. Thus $\sqrt[4]{x^8} = x^2$, $\sqrt[4]{y^{16}} = y^4$, and so forth.

6. $\sqrt[4]{48x^7y^{12}} = \sqrt[4]{16x^4y^{12} \cdot 3x^3}$
$= \sqrt[4]{16x^4y^{12}} \cdot \sqrt[4]{3x^3}$
$= 2xy^3\sqrt[4]{3x^3}$

REMARK   When the multiplication property of radicals is used to simplify a radical, the radical with the perfect squares, cubes, and so on will drop out. The exponents on the factors in the remaining radicand will always be less than the order of the radical.

In addition to a rule for multiplying radical expressions, there is a rule for dividing radical expressions. We know

$$\sqrt[n]{\frac{a}{b}} = \left(\frac{a}{b}\right)^{1/n}$$

$$= \frac{a^{1/n}}{b^{1/n}} \qquad \textit{By the quotient to a power law of exponents}$$

$$= \frac{\sqrt[n]{a}}{\sqrt[n]{b}}$$

This reasoning leads to the next property.

□ivision Property of Radicals

Let n be a natural number; then

$$\sqrt[n]{\frac{a}{b}} = \frac{\sqrt[n]{a}}{\sqrt[n]{b}}$$

provided that a and b are not both negative and $b \neq 0$.

 XAMPLE 5.3.3   Divide the following radicals. Assume that all variables represent positive numbers.

1. $\sqrt{\frac{25}{36}} = \frac{\sqrt{25}}{\sqrt{36}}$

$= \frac{5}{6}$

2. $\frac{\sqrt{8}}{\sqrt{18}} = \sqrt{\frac{8}{18}}$

$= \sqrt{\frac{4}{9}}$

$= \frac{2}{3}$

3. $\frac{\sqrt{5x^3}}{\sqrt{20x}} = \sqrt{\frac{5x^3}{20x}}$

$= \sqrt{\frac{x^2}{4}}$

$= \frac{\sqrt{x^2}}{\sqrt{4}}$

$= \frac{x}{2}$

4. $\sqrt{\frac{2x}{25}} = \frac{\sqrt{2x}}{\sqrt{25}}$

$= \frac{\sqrt{2x}}{5}$

5. $\sqrt[3]{\frac{5x}{8y^6}} = \frac{\sqrt[3]{5x}}{\sqrt[3]{8y^6}}$

$= \frac{\sqrt[3]{5x}}{2y^2}$

In some mathematics courses we come across numbers like $\dfrac{1}{\sqrt{2}}$, real numbers with radicals in the denominator. Expressions with radicals in the denominator or with a fraction in the radicand are *not* considered simplified. How can we eliminate the radical from the denominator of $\dfrac{1}{\sqrt{2}}$?

$$\frac{1}{\sqrt{2}} = \frac{1}{\sqrt{2}} \cdot \frac{\sqrt{2}}{\sqrt{2}} \qquad \textit{Multiply the numerator and the}$$
$$\textit{denominator by the same quantity.}$$
$$= \frac{\sqrt{2}}{\sqrt{4}}$$
$$= \frac{\sqrt{2}}{2}$$

Thus, $\dfrac{1}{\sqrt{2}}$ has been simplified to $\dfrac{\sqrt{2}}{2}$.

REMARK

This process of eliminating radicals from the denominator is called **rationalizing the denominator.**

Rationalizing the Denominator

1. Simplify the radical in the denominator, if possible.
2. If the radical in the denominator is a square root, multiply the numerator and denominator by that square root. If the radical in the denominator is a cube root, multiply the numerator and denominator by a cube root expression that will make the radicand in the denominator a perfect cube. Follow a similar process for higher order radicals.
3. Simplify the radical expressions obtained in step 2.

EXAMPLE 5.3.4   Simplify the following radicals. Assume that all variables represent positive numbers.

1.
$$\sqrt{\frac{2}{3}} = \frac{\sqrt{2}}{\sqrt{3}}$$
$$= \frac{\sqrt{2}}{\sqrt{3}} \cdot \frac{\sqrt{3}}{\sqrt{3}}$$
$$= \frac{\sqrt{6}}{\sqrt{9}}$$
$$= \frac{\sqrt{6}}{3}$$

2.
$$\frac{5}{\sqrt{10}} = \frac{5}{\sqrt{10}} \cdot \frac{\sqrt{10}}{\sqrt{10}}$$
$$= \frac{5 \cdot \sqrt{10}}{\sqrt{100}}$$
$$= \frac{5 \cdot \sqrt{10}}{10}$$
$$= \frac{\sqrt{10}}{2}$$

3. $\dfrac{7}{\sqrt{12}} = \dfrac{7}{\sqrt{4} \cdot \sqrt{3}}$

$= \dfrac{7}{2\sqrt{3}}$

$= \dfrac{7}{2\sqrt{3}} \cdot \dfrac{\sqrt{3}}{\sqrt{3}}$

$= \dfrac{7\sqrt{3}}{2\sqrt{9}}$

$= \dfrac{7\sqrt{3}}{2 \cdot 3}$

$= \dfrac{7\sqrt{3}}{6}$

4. $\sqrt{\dfrac{5x}{3y}} = \dfrac{\sqrt{5x}}{\sqrt{3y}}$

$= \dfrac{\sqrt{5x}}{\sqrt{3y}} \cdot \dfrac{\sqrt{3y}}{\sqrt{3y}}$

$= \dfrac{\sqrt{15xy}}{\sqrt{9y^2}}$

$= \dfrac{\sqrt{15xy}}{3y}$

5. $\sqrt{\dfrac{4x^2}{5y^5}} = \dfrac{\sqrt{4x^2}}{\sqrt{5y^5}}$

$= \dfrac{2x}{\sqrt{y^4} \cdot \sqrt{5y}}$

$= \dfrac{2x}{y^2\sqrt{5y}} \cdot \dfrac{\sqrt{5y}}{\sqrt{5y}}$

$= \dfrac{2x\sqrt{5y}}{y^2\sqrt{25y^2}}$

$= \dfrac{2x\sqrt{5y}}{y^2 \cdot 5y}$

$= \dfrac{2x\sqrt{5y}}{5y^3}$

6. $\sqrt[3]{\dfrac{5y}{4x}} = \dfrac{\sqrt[3]{5y}}{\sqrt[3]{4x}}$

$= \dfrac{\sqrt[3]{5y}}{\sqrt[3]{4x}} \cdot \dfrac{\sqrt[3]{2x^2}}{\sqrt[3]{2x^2}}$

$= \dfrac{\sqrt[3]{10x^2y}}{\sqrt[3]{8x^3}}$

$= \dfrac{\sqrt[3]{10x^2y}}{2x}$

To simplify a cube root, as in number 6, we need a perfect cube in the radicand. Since $\sqrt[3]{4x} = \sqrt[3]{2^2 x^1}$, we need only multiply by $\sqrt[3]{2x^2}$ to get $\sqrt[3]{2^2 x^1} \cdot \sqrt[3]{2^1 x^2} = \sqrt[3]{2^3 x^3}$, which is $2x$.

7. $\sqrt[4]{\dfrac{16x}{3y^3}} = \dfrac{\sqrt[4]{16x}}{\sqrt[4]{3y^3}}$

$= \dfrac{2\sqrt[4]{x}}{\sqrt[4]{3y^3}} \cdot \dfrac{\sqrt[4]{27y}}{\sqrt[4]{27y}}$ or $\dfrac{2\sqrt[4]{x}}{\sqrt[4]{3y^3}} \cdot \dfrac{\sqrt[4]{3^3 y}}{\sqrt[4]{3^3 y}}$

$= \dfrac{2\sqrt[4]{27xy}}{\sqrt[4]{81y^4}}$ $= \dfrac{2\sqrt[4]{3^3 xy}}{\sqrt[4]{3^4 y^4}}$

$= \dfrac{2\sqrt[4]{27xy}}{3y}$ $= \dfrac{2\sqrt[4]{27xy}}{3y}$

We have discussed properties for multiplying and dividing radicals with different radicands. However, no such properties exist for adding and subtracting radicals with different radicands. This is illustrated below.

Try to avoid these mistakes:

Incorrect	Correct
1. $\sqrt{9 + 16} = \sqrt{9} + \sqrt{16}$ $= 3 + 4$ $= 7$	$\sqrt{9 + 16} = \sqrt{25}$ $= 5$
2. $\sqrt{169 - 25} = \sqrt{169} - \sqrt{25}$ $= 13 - 5$ $= 8$	$\sqrt{169 - 25} = \sqrt{144}$ $= 12$

Remember, in general:

$$\sqrt[n]{a + b} \neq \sqrt[n]{a} + \sqrt[n]{b}$$
$$\sqrt[n]{a - b} \neq \sqrt[n]{a} - \sqrt[n]{b}$$

One final comment: Occasionally the index of the radical can be reduced. For example,

$$\sqrt[4]{x^2} = x^{2/4}$$
$$= x^{1/2}$$
$$= \sqrt{x}$$

In addition to our other rules, a radical is considered simplified if the index of the radical and the exponents of the factors in the radicand have no common factors.

EXAMPLE 5.3.5   Simplify the following radicals. Assume that all variables represent positive numbers.

1. $\sqrt[6]{x^4} = x^{4/6}$
    $= x^{2/3}$
    $= \sqrt[3]{x^2}$

2. $\sqrt[4]{x^{10}} = \sqrt[4]{x^8} \cdot \sqrt[4]{x^2}$
    $= x^2 \cdot x^{2/4}$
    $= x^2 \cdot x^{1/2}$
    $= x^2 \cdot \sqrt{x}$

Let us summarize the rules for simplifying radicals:

Simplifying a Radical

1. Remove all perfect square factors from square roots, perfect cube factors from cube roots, and so on.
2. Remove all fractions from the radicand.
3. Rationalize all denominators.
4. If the index and an exponent on a factor in the radicand have a common factor, eliminate the common factor.

Exercises 5.3

Simplify the following radicals. Assume that all variables represent positive numbers.

1. $\sqrt{5}\sqrt{2}$

2. $\sqrt{7}\sqrt{11}$

3. $\sqrt{15}\sqrt{10}$

4. $\sqrt{35}\sqrt{14}$

5. $\sqrt{13}\sqrt{13}$

6. $\sqrt{17}\sqrt{17}$

7. $\sqrt[3]{4}\sqrt[3]{6}$

8. $\sqrt[3]{10}\sqrt[3]{12}$

9. $\sqrt[3]{15}\sqrt[3]{50}$

10. $\sqrt[4]{18}\sqrt[4]{4}$

11. $\sqrt[4]{7}\sqrt[4]{2}$

12. $\sqrt{x^3}\sqrt{x^7}$

13. $\sqrt{48}$

14. $\sqrt{98}$

15. $\sqrt{150}$

16. $\sqrt{63}$

17. $\sqrt{80}$

18. $\sqrt{112}$

19. $\sqrt{108}$

20. $\sqrt{128}$

21. $\sqrt[3]{16}$

22. $\sqrt[3]{54}$

23. $\sqrt[3]{-40}$

24. $\sqrt[3]{-128}$

25. $\sqrt[4]{80}$

26. $\sqrt[4]{243}$

27. $\sqrt{32x^3}$

28. $\sqrt{25x^4}$

29. $\sqrt[3]{128x^7}$

30. $\sqrt[3]{125x^2}$

31. $\sqrt[4]{64x^7}$

32. $\sqrt[4]{81x^{10}}$

33. $\sqrt{x^2y^2}$

34. $\sqrt{25x^6y^8}$

35. $\sqrt{45x^3y^9}$

36. $\sqrt{200xy^5}$

37. $\sqrt{32x^7y^4}$

38. $\sqrt{98x^{11}y^6}$

39. $\sqrt{75x^4y^5z^6}$

40. $\sqrt{72x^8y^9z^{12}}$

41. $\sqrt{144x^5y^7z^9}$

42. $\sqrt{169x^3y^4z^4}$

43. $\sqrt[3]{64x^3y^{10}}$

44. $\sqrt[3]{54x^8y^7}$

45. $\sqrt[3]{81x^9y^2z^4}$

46. $\sqrt[3]{32x^3y^6z^9}$

47. $\sqrt[4]{32x^5y^7}$

48. $\sqrt[4]{64x^9y^5}$

49. $\dfrac{1}{\sqrt{3}}$

50. $\dfrac{2}{\sqrt{3}}$

51. $\sqrt{\dfrac{1}{2}}$

52. $\sqrt{\dfrac{3}{5}}$

53. $\dfrac{\sqrt{3}}{\sqrt{12}}$

54. $\dfrac{\sqrt{50}}{\sqrt{32}}$

55. $\dfrac{3}{\sqrt{6}}$

56. $\dfrac{2}{\sqrt{8}}$

57. $\dfrac{5}{\sqrt{x}}$

58. $\dfrac{3}{\sqrt{2y}}$

59. $\dfrac{4}{\sqrt{8x}}$

60. $\dfrac{5}{\sqrt{10y}}$

61. $\sqrt[3]{\dfrac{1}{4}}$

62. $\sqrt[3]{\dfrac{4}{9}}$

63. $\sqrt{\dfrac{9x^2}{16y^4}}$

64. $\sqrt{\dfrac{81x^{10}}{25y^2}}$

65. $\dfrac{\sqrt{6x^3}}{\sqrt{24x}}$

66. $\dfrac{\sqrt{5x}}{\sqrt{20x^2}}$

67. $\sqrt{\dfrac{2}{9x}}$

68. $\sqrt{\dfrac{5}{4x}}$

69. $\sqrt{\dfrac{7}{2x}}$

70. $\sqrt{\dfrac{1}{12x}}$

71. $\sqrt{\dfrac{7x}{2y}}$

72. $\sqrt{\dfrac{11x}{15y}}$

73. $\sqrt{\dfrac{2x}{3y^3}}$

74. $\sqrt{\dfrac{10x}{7y^5}}$

75. $\sqrt{\dfrac{25x^4}{8y^7}}$

76. $\sqrt{\dfrac{49x^3}{9y^3}}$

77. $\sqrt{\dfrac{8x^4}{5y^2}}$

78. $\sqrt{\dfrac{12x^5}{7y^7}}$

79. $\sqrt[3]{\dfrac{8}{27x^3}}$

80. $\sqrt[3]{\dfrac{125}{8y^9}}$

81. $\sqrt[3]{\dfrac{7}{8x}}$

82. $\sqrt[3]{\dfrac{x}{2y^3}}$

83. $\sqrt[3]{\dfrac{5}{4x^2}}$

84. $\sqrt[3]{\dfrac{6}{3y}}$

85. $\sqrt[3]{\dfrac{16}{3x^2}}$

86. $\sqrt[3]{\dfrac{54}{4z}}$

87. $\sqrt[3]{\dfrac{3y^2}{5x^4}}$

88. $\sqrt[3]{\dfrac{2y}{9x^5}}$

89. $\sqrt[4]{\dfrac{16}{x^8}}$

90. $\sqrt[4]{\dfrac{625x^4}{81y^{12}}}$

91. $\sqrt[4]{\dfrac{3}{5}}$

92. $\sqrt[4]{\dfrac{2}{9}}$

93. $\sqrt[4]{\dfrac{7x}{2y}}$

94. $\sqrt[4]{\dfrac{11z}{5y^4}}$

95. $\sqrt[4]{\dfrac{5}{4y^3}}$

96. $\sqrt[4]{\dfrac{2}{3x^2}}$

97. $\sqrt[6]{x^3}$

98. $\sqrt[6]{x^2}$

99. $\sqrt[4]{16x^2}$

100. $\sqrt[4]{81x^4y^3}$

101. $\sqrt[n]{x^{4n}y^{2n}}$

Write Algebra

102. Explain the calculator steps needed to evaluate $\dfrac{5}{\sqrt{2}}$.

103. Explain two ways to simplify $\dfrac{5}{\sqrt{20}}$.

G R O U P   A C T I V I T Y

In exercises 104–123, simplify each of the radicals. Assume that the variables represent *any* real number. Recall, that for *all* real numbers $\sqrt{x^2} = |x|$. Explain why some of your answers do not need absolute value signs.

104. $\sqrt{x^6}$

105. $\sqrt{x^8}$

106. $\sqrt{x^7}$

107. $\sqrt{x^{11}}$

108. $\sqrt{36x^2}$

109. $\sqrt{16x^2}$

110. $\sqrt{40x^3}$

111. $\sqrt{50x^5}$

112. $\sqrt{x^4y^3}$

113. $\sqrt{xy^7}$

114. $\sqrt{45x^5y^9}$

115. $\sqrt{200xy^3}$

116. $\sqrt{x^4 + x^2}$

117. $\sqrt{x^2 - x^8}$

118. $\sqrt{4x^2 - 20x^6}$

119. $\sqrt{9x^2 + 18x^6}$

120. $\sqrt{x^2 - 10x + 25}$

121. $\sqrt{x^2 - 14x + 49}$

122. $\sqrt{4x^2 + 12xy + 9y^2}$

123. $\sqrt{9x^2 + 30xy + 25y^2}$

5.4 *Operations with Radical Expressions*

We learned how to combine like terms in Chapter 2 by using the distributive property. In this section we will learn how to combine like radical terms by again using the distributive property. First we must define the phrase *like radical.*

Definition 5.4.1	**Like radicals** are radicals that have the same index and radicand.

EXAMPLE 5.4.1

Perform the indicated operations. Simplify your answer by combining like radical terms.

1. $2\sqrt{5} + 8\sqrt{5} - 3\sqrt{5} = (2 + 8 - 3)\sqrt{5}$ *By the distributive property*

$= 7\sqrt{5}$

2. $\sqrt[3]{6} - 9\sqrt[3]{6} - 2\sqrt[3]{6} = (1 - 9 - 2)\sqrt[3]{6}$

$= -10\sqrt[3]{6}$

3. $2\sqrt{x} + 7\sqrt{x} - \sqrt{2x} = (2 + 7)\sqrt{x} - \sqrt{2x}$

$= 9\sqrt{x} - \sqrt{2x}$

4. $3\sqrt{5x} + 4\sqrt{5x} + 9\sqrt{y} - 11\sqrt{y} = (3 + 4)\sqrt{5x} + (9 - 11)\sqrt{y}$

$= 7\sqrt{5x} + (-2)\sqrt{y}$

$= 7\sqrt{5x} - 2\sqrt{y}$

This expression cannot be simplified further because the remaining radicals are not like radicals.

Sometimes we come across radical expressions that apparently have no like radical terms. After simplification, however, some like radical terms may be found. These types of problems are illustrated in the next example.

EXAMPLE 5.4.2

Perform the indicated operations. Assume that all variables represent positive numbers.

1. $7\sqrt{12} + \sqrt{75} = 7\sqrt{4}\sqrt{3} + \sqrt{25}\sqrt{3}$

$= 7 \cdot 2\sqrt{3} + 5\sqrt{3}$

$= 14\sqrt{3} + 5\sqrt{3}$

$= 19\sqrt{3}$

2. $\sqrt[3]{54} - 5\sqrt[3]{16} = \sqrt[3]{27}\sqrt[3]{2} - 5\sqrt[3]{8}\sqrt[3]{2}$

$= 3\sqrt[3]{2} - 5 \cdot 2\sqrt[3]{2}$

$= 3\sqrt[3]{2} - 10\sqrt[3]{2}$

$= -7\sqrt[3]{2}$

3. $2\sqrt{27} - \sqrt{50} = 2\sqrt{9}\sqrt{3} - \sqrt{25}\sqrt{2}$
$$= 2 \cdot 3\sqrt{3} - 5\sqrt{2}$$
$$= 6\sqrt{3} - 5\sqrt{2}$$

These radicals cannot be combined because they are not like radicals.

4. $\sqrt{\dfrac{1}{3}} + \sqrt{12} = \dfrac{1}{\sqrt{3}} + \sqrt{12}$

$$= \dfrac{1}{\sqrt{3}} \cdot \dfrac{\sqrt{3}}{\sqrt{3}} + \sqrt{4}\sqrt{3} \qquad \textit{Rationalize the denominator.}$$

$$= \dfrac{\sqrt{3}}{3} + 2\sqrt{3}$$

$$= \dfrac{\sqrt{3}}{3} + \dfrac{6\sqrt{3}}{3}$$

$$= \dfrac{7\sqrt{3}}{3}$$

5. $3\sqrt{2x^3} - x\sqrt{200x} + \sqrt{32x^3} = 3\sqrt{x^2}\sqrt{2x} - x\sqrt{100}\sqrt{2x} + \sqrt{16x^2}\sqrt{2x}$
$$= 3x\sqrt{2x} - 10x\sqrt{2x} + 4x\sqrt{2x}$$
$$= -3x\sqrt{2x}$$

6. $\sqrt{20x} - \sqrt{48y} + \sqrt{125x} + \sqrt{3y} = \sqrt{4}\sqrt{5x} - \sqrt{16}\sqrt{3y} + \sqrt{25}\sqrt{5x} + \sqrt{3y}$
$$= 2\sqrt{5x} - 4\sqrt{3y} + 5\sqrt{5x} + \sqrt{3y}$$
$$= 7\sqrt{5x} - 3\sqrt{3y}$$

Exercises 5.4

Perform the indicated operations. Assume that all variables represent positive numbers.

1. $3\sqrt{7} - 5\sqrt{7} - 8\sqrt{7}$

2. $4\sqrt{2} + 3\sqrt{2} - \sqrt{2}$

3. $6\sqrt[3]{10} - 11\sqrt[3]{10}$

4. $-9\sqrt[3]{4} + \sqrt[3]{4} + 3\sqrt[3]{4}$

5. $2\sqrt{15x} - 7\sqrt{15x} + 5\sqrt{15x}$

6. $3x\sqrt{2x} + x\sqrt{2x}$

7. $x\sqrt[3]{10y} - 8x\sqrt[3]{10y}$

8. $-2x\sqrt[3]{6x^2} - 3x\sqrt[3]{6x^2} - 4x\sqrt[3]{6x^2}$

9. $5\sqrt{12} - 2\sqrt{27} + 3\sqrt{3}$

10. $4\sqrt{20} + \sqrt{45} - \sqrt{80}$

11. $5\sqrt{8} - \sqrt{18} + 2\sqrt{32}$

12. $-\sqrt{40} - \sqrt{90} - \sqrt{160}$

13. $\sqrt[3]{32} + \sqrt[3]{108}$

14. $4\sqrt[3]{250} - 3\sqrt[3]{128}$

15. $\sqrt[4]{32} - \sqrt[4]{2}$

16. $\sqrt[4]{32} - \sqrt[4]{162}$

17. $\sqrt{27x} - 2\sqrt{12x} + 2\sqrt{48x}$

18. $\sqrt{28x} - 3\sqrt{112x} + 4\sqrt{63x}$

19. $x\sqrt{18x} + 5\sqrt{2x^3} + 2x\sqrt{8x}$

20. $y\sqrt{108y} - 2\sqrt{3y^3} - 3y\sqrt{27y}$

21. $\sqrt[3]{108x} - \sqrt[3]{32x}$

22. $x\sqrt[3]{250x} - \sqrt[3]{16x^4}$

23. $\sqrt[4]{162x} + \sqrt[4]{32x}$

24. $\sqrt[4]{48x^5} - \sqrt[4]{3x^5}$

25. $\sqrt{27x} + 2\sqrt{12x} - \sqrt{150y} - 4\sqrt{24y}$

26. $5\sqrt{5x} - \sqrt{98y} - 2\sqrt{32y} + 3\sqrt{20x}$

27. $5x\sqrt{3x^2} - x^2\sqrt{300} + 2y\sqrt{8y^2} + \sqrt{8y^4}$

28. $x^2\sqrt{18y} - 2x\sqrt{2x^2y} + 4y\sqrt{12x} - \sqrt{3xy^2}$

29. $\sqrt{112} + 2\sqrt{180}$

30. $4\sqrt{50} - 2\sqrt{147}$

31. $\sqrt[3]{54} - \sqrt[3]{40}$

32. $3\sqrt[3]{40} + \sqrt[3]{162}$

33. $\sqrt{90x} - 2\sqrt{40x} + 5\sqrt{10x^3}$

34. $2x\sqrt{3x} + \sqrt{108x^3} - 3\sqrt{75x}$

35. $y\sqrt{32y} + 5\sqrt{2y^3} - \sqrt{98y}$

36. $\sqrt{2x^4} - 2x\sqrt{8} + \sqrt{128}$

37. $x\sqrt{20x^2} - x\sqrt{45} + 2\sqrt{500}$

38. $\sqrt{3x^5} + \sqrt{12xy^4} + \sqrt{27xz^4}$

39. $\sqrt[3]{54x^2y} - \sqrt[3]{2x^2y}$

40. $\sqrt[3]{250xy^2} - \sqrt[3]{54xy^2}$

41. $4\sqrt[3]{3x^4y} - x\sqrt[3]{24xy}$

42. $\sqrt[3]{2x^5y^2} - 3\sqrt[3]{2x^2y^5}$

43. $\sqrt[4]{80xy} - 3\sqrt[4]{5xy}$

44. $\sqrt[4]{162xy^3} - \sqrt[4]{2xy^3}$

45. $y^2\sqrt[4]{16xy} - y\sqrt[4]{xy^5}$

46. $2\sqrt[4]{x^6y^2} + x\sqrt[4]{81x^2y^2}$

47. $\dfrac{1}{\sqrt{2}} + 5\sqrt{2}$

48. $\dfrac{4}{\sqrt{3}} - 7\sqrt{3}$

49. $\dfrac{3}{\sqrt{5}} + \sqrt{80}$

50. $\dfrac{5}{\sqrt{6}} + \sqrt{24}$

51. $\dfrac{1}{\sqrt[3]{2}} - \sqrt[3]{108}$

52. $\dfrac{2}{\sqrt[3]{3}} - \sqrt[3]{72}$

53. $\sqrt{3} + \dfrac{5}{\sqrt{3}} + \dfrac{2\sqrt{27}}{3}$

54. $\sqrt{2} - \dfrac{4}{\sqrt{2}} - \dfrac{\sqrt{50}}{2}$

55. $\sqrt{\dfrac{2}{3}} + 5\sqrt{6} - \sqrt{\dfrac{3}{2}}$

56. $\sqrt{\dfrac{5}{3}} - 2\sqrt{60} + \sqrt{\dfrac{3}{5}}$

57. $\sqrt{\dfrac{x}{2}} + 5\sqrt{2x}$

58. $\sqrt{\dfrac{3x}{5}} - 7\sqrt{15x}$

59. $\sqrt[3]{\dfrac{x}{4}} - \sqrt[3]{16x}$

60. $\sqrt[3]{\dfrac{x}{3}} + \sqrt[3]{72x}$

61. $\sqrt{\dfrac{2}{x}} + \sqrt{\dfrac{x}{2}} - \sqrt{2x}$

62. $\sqrt{\dfrac{3}{y}} - \sqrt{\dfrac{y}{3}} - \sqrt{3y}$

63. $\sqrt{\dfrac{2}{x}} + \sqrt{32x} - \sqrt{\dfrac{1}{2x}}$

64. $\sqrt{\dfrac{5}{x}} - \sqrt{125x} + \sqrt{\dfrac{4}{5x}}$

5.5 More Operations with Radical Expressions

Let's now examine multiplication of radical expressions. We will see that multiplication of radical expressions is very similar to multiplication of polynomials.

XAMPLE 5.5.1   Perform the indicated multiplications and simplify your answers. Assume that all variables represent positive numbers.

1. $\sqrt{3}(2 + \sqrt{15}) = 2\sqrt{3} + \sqrt{15}\sqrt{3}$

$\qquad\qquad\qquad = 2\sqrt{3} + \sqrt{45}$

$\qquad\qquad\qquad = 2\sqrt{3} + \sqrt{9}\sqrt{5}$

$\qquad\qquad\qquad = 2\sqrt{3} + 3\sqrt{5}$

how do u know use those mult.

2. $2\sqrt{5}(3 - \sqrt{7} + 4\sqrt{5}) = 2 \cdot 3\sqrt{5} - 2\sqrt{5}\sqrt{7} + 2 \cdot 4\sqrt{5}\sqrt{5}$

$$= 6\sqrt{5} - 2\sqrt{35} + 8 \cdot 5$$

$$= 6\sqrt{5} - 2\sqrt{35} + 40$$

3. $(2\sqrt{5} + \sqrt{3})(\sqrt{5} + 4\sqrt{3}) = 2\sqrt{5}\sqrt{5} + 2 \cdot 4\sqrt{5}\sqrt{3} + \sqrt{3}\sqrt{5} + 4\sqrt{3}\sqrt{3}$

$$= 2 \cdot 5 + 8\sqrt{15} + \sqrt{15} + 4 \cdot 3$$

$$= 10 + 9\sqrt{15} + 12$$

$$= 22 + 9\sqrt{15}$$

In this problem we used the FOIL Method to perform the multiplication.

4. $(\sqrt{2x} - 7)^2 = (\sqrt{2x} - 7)(\sqrt{2x} - 7)$

$$= (\sqrt{2x})^2 - 7\sqrt{2x} - 7\sqrt{2x} + 7^2$$

$$= 2x - 14\sqrt{2x} + 49$$

5. $(2 + \sqrt{x - 1})^2 = (2 + \sqrt{x - 1})(2 + \sqrt{x - 1})$

$$= 4 + 2\sqrt{x - 1} + 2\sqrt{x - 1} + (\sqrt{x - 1})^2$$

$$= 4 + 4\sqrt{x - 1} + x - 1$$

$$= x + 3 + 4\sqrt{x - 1}$$

Try to avoid this mistake:

Incorrect	Correct
$(\sqrt{x} + \sqrt{y})^2 = x + y$	$(\sqrt{x} + \sqrt{y})^2 = (\sqrt{x} + \sqrt{y})(\sqrt{x} + \sqrt{y})$
	$= \sqrt{x}\sqrt{x} + \sqrt{x}\sqrt{y}$
	$+ \sqrt{x}\sqrt{y} + \sqrt{y}\sqrt{y}$
	$= x + 2\sqrt{xy} + y$

6. $(3 + \sqrt{5})(3 - \sqrt{5}) = 9 - 3\sqrt{5} + 3\sqrt{5} - (\sqrt{5})^2$

$$= 9 - 5$$

$$= 4$$

NOTE ▶ The product $(3 + \sqrt{5})(3 - \sqrt{5})$ yields a number free of radicals. The numbers $3 + \sqrt{5}$ and $3 - \sqrt{5}$ are called **conjugates** of each other. To find the conjugate of a two-termed expression involving only square roots, simply change the sign between the terms. In general, the product of conjugates will be free of radicals (provided that the radicals are *square roots*), because of the formula for the difference of two squares.

7. $(\sqrt{x} - \sqrt{y})(\sqrt{x} + \sqrt{y}) = (\sqrt{x})^2 + \sqrt{x}\sqrt{y} - \sqrt{x}\sqrt{y} - (\sqrt{y})^2$

$$= x - y$$

We are now in a position to simplify radical expressions like $\dfrac{1}{3 + \sqrt{5}}$. Remember that a radical expression is *not* considered simplified if radicals are present in the denominator. To rationalize the denominator of $\dfrac{1}{3 + \sqrt{5}}$, apply part 6 of Example 5.5.1:

$$\frac{1}{3 + \sqrt{5}} = \frac{1}{(3 + \sqrt{5})} \cdot \frac{(3 - \sqrt{5})}{(3 - \sqrt{5})}$$

$$= \frac{1(3 - \sqrt{5})}{(3 + \sqrt{5})(3 - \sqrt{5})}$$

$$= \frac{3 - \sqrt{5}}{9 - 5} = \frac{3 - \sqrt{5}}{4}$$

The denominator has been rationalized.

EXAMPLE 5.5.2   Rationalize the denominator of the following radical expressions. Assume that all variables represent positive numbers.

1. $\dfrac{5}{2 - \sqrt{3}} = \dfrac{5}{(2 - \sqrt{3})} \cdot \dfrac{(2 + \sqrt{3})}{(2 + \sqrt{3})}$

$$= \frac{5(2 + \sqrt{3})}{(2 - \sqrt{3})(2 + \sqrt{3})}$$

$$= \frac{10 + 5\sqrt{3}}{4 - 3}$$

$$= 10 + 5\sqrt{3}$$

2. $\dfrac{8}{\sqrt{5} + \sqrt{3}} = \dfrac{8}{(\sqrt{5} + \sqrt{3})} \cdot \dfrac{(\sqrt{5} - \sqrt{3})}{(\sqrt{5} - \sqrt{3})}$

$$= \frac{8(\sqrt{5} - \sqrt{3})}{(\sqrt{5} + \sqrt{3})(\sqrt{5} - \sqrt{3})}$$

$$= \frac{8(\sqrt{5} - \sqrt{3})}{5 - 3}$$

$$= \frac{8(\sqrt{5} - \sqrt{3})}{2}$$

$$= 4(\sqrt{5} - \sqrt{3})$$

$$= 4\sqrt{5} - 4\sqrt{3}$$

3. $\dfrac{2 + \sqrt{3}}{\sqrt{6} - 1} = \dfrac{(2 + \sqrt{3})}{(\sqrt{6} - 1)} \cdot \dfrac{(\sqrt{6} + 1)}{(\sqrt{6} + 1)}$

$$= \frac{(2 + \sqrt{3})(\sqrt{6} + 1)}{(\sqrt{6} - 1)(\sqrt{6} + 1)}$$

$$= \frac{2\sqrt{6} + 2 + \sqrt{18} + \sqrt{3}}{6 - 1}$$

$$= \frac{2\sqrt{6} + 2 + 3\sqrt{2} + \sqrt{3}}{5}$$

4. $\dfrac{5\sqrt{5} - \sqrt{2}}{2\sqrt{5} - \sqrt{2}} = \dfrac{(5\sqrt{5} - \sqrt{2})}{(2\sqrt{5} - \sqrt{2})} \cdot \dfrac{(2\sqrt{5} + \sqrt{2})}{(2\sqrt{5} + \sqrt{2})}$

$$= \frac{50 + 5\sqrt{10} - 2\sqrt{10} - 2}{20 - 2}$$

$$= \frac{48 + 3\sqrt{10}}{18}$$

$$= \frac{3(16 + \sqrt{10})}{18}$$

$$= \frac{16 + \sqrt{10}}{6}$$

5. $\dfrac{\sqrt{x}}{\sqrt{x} + \sqrt{y}} = \dfrac{\sqrt{x}}{(\sqrt{x} + \sqrt{y})} \cdot \dfrac{(\sqrt{x} - \sqrt{y})}{(\sqrt{x} - \sqrt{y})}$

$$= \frac{\sqrt{x}(\sqrt{x} - \sqrt{y})}{(\sqrt{x} + \sqrt{y})(\sqrt{x} - \sqrt{y})}$$

$$= \frac{x - \sqrt{xy}}{x - y}$$

Exercises 5.5

trouble.

Perform the indicated multiplications and simplify your answers. Assume that all variables represent positive numbers.

1. $\sqrt{3}(2 + \sqrt{5})$

2. $\sqrt{5}(1 - \sqrt{3} - \sqrt{10})$

3. $2\sqrt{6}(7 - \sqrt{5} + \sqrt{2})$

4. $3\sqrt{3}(2\sqrt{3} + \sqrt{6} - 4)$

5. $\sqrt{x}(3 + \sqrt{x})$

6. $\sqrt{x}(2 - \sqrt{3} - \sqrt{y})$

7. $2\sqrt{3y}(4 - y + \sqrt{6})$

8. $3\sqrt{5y}(\sqrt{5y} + \sqrt{x} - 2\sqrt{3})$

9. $(2 + \sqrt{7})(\sqrt{8} - 3)$

10. $(\sqrt{3} - 4)(\sqrt{5} + 6)$

11. $(\sqrt{5} + \sqrt{7})(\sqrt{10} + \sqrt{2})$

12. $(\sqrt{3} - \sqrt{2})(\sqrt{6} - \sqrt{2})$

13. $(2\sqrt{3} + 4\sqrt{5})(3\sqrt{2} - 6\sqrt{7})$

14. $(4\sqrt{2} - 2\sqrt{3})(5\sqrt{3} - 8\sqrt{6})$

15. $(2\sqrt{x} - 1)(3\sqrt{x} + 4)$

16. $(\sqrt{x} + \sqrt{y})(\sqrt{2x} - \sqrt{3y})$

17. $(2\sqrt{x} + 3\sqrt{y})(4\sqrt{x} + 5\sqrt{y})$

18. $(\sqrt{2x} - \sqrt{5y})(\sqrt{2x} - 1)$

19. $(\sqrt{5} - \sqrt{3})^2$

20. $(\sqrt{7} - \sqrt{2})^2$

21. $(2\sqrt{3} + 1)^2$

22. $(5\sqrt{2} - 7)^2$

23. $(3\sqrt{2} - 6\sqrt{3})^2$

24. $(4\sqrt{3} + 7\sqrt{5})^2$

25. $(\sqrt{x} - \sqrt{2y})^2$

26. $(\sqrt{3x} + \sqrt{5y})^2$

27. $(2\sqrt{x} + 7)^2$

28. $(4\sqrt{3x} - 1)^2$

29. $(1 + \sqrt{x - 1})^2$

30. $(5 + \sqrt{x + 3})^2$

31. $(3 - \sqrt{x + 2})^2$

32. $(3 + 4\sqrt{2x - 6})^2$

33. $(\sqrt{x} - \sqrt{3x - 2})^2$

34. $(\sqrt{2x - 3} + \sqrt{x + 5})^2$

35. $(\sqrt{5} - \sqrt{3})(\sqrt{5} + \sqrt{3})$

36. $(\sqrt{2} - \sqrt{7})(\sqrt{2} + \sqrt{7})$

37. $(5\sqrt{3} + \sqrt{6})(5\sqrt{3} - \sqrt{6})$

38. $(3\sqrt{2} + 2)(3\sqrt{2} - 2)$

39. $(\sqrt{x} + 7)(\sqrt{x} - 7)$

40. $(\sqrt{x} - \sqrt{3})(\sqrt{x} + \sqrt{3})$

41. $(\sqrt{2x} + \sqrt{3y})(\sqrt{2x} - \sqrt{3y})$

42. $(2\sqrt{x} - 5\sqrt{y})(2\sqrt{x} + 5\sqrt{y})$

43. $(\sqrt[3]{3} + 2)(\sqrt[3]{4} - 5)$

44. $(\sqrt[3]{2} - 3)(\sqrt[3]{20} + 1)$

45. $(\sqrt[3]{2} + \sqrt[3]{3})^2$

46. $(\sqrt[3]{4} - \sqrt[3]{5})(\sqrt[3]{2} + \sqrt[3]{25})$

47. $(\sqrt[3]{2x} - 1)(\sqrt[3]{2x} + 1)$

48. $(\sqrt[3]{4x} + 5)(\sqrt[3]{x} - 2)$

49. $(\sqrt[3]{3} - 1)(\sqrt[3]{9} + \sqrt[3]{3} + 1)$

50. $(\sqrt[3]{2} - 3)(\sqrt[3]{4} + 3\sqrt[3]{2} + 9)$

51. $(\sqrt[3]{x} - \sqrt[3]{y})(\sqrt[3]{x^2} + \sqrt[3]{xy} + \sqrt[3]{y^2})$

52. $(\sqrt[3]{x} + \sqrt[3]{y})(\sqrt[3]{x^2} - \sqrt[3]{xy} + \sqrt[3]{y^2})$

Rationalize the denominator of the following radicals. Assume that all variables represent positive numbers.

53. $\dfrac{1}{\sqrt{3} + 2}$

54. $\dfrac{1}{\sqrt{5} - 1}$

55. $\dfrac{2}{\sqrt{3} + \sqrt{5}}$

56. $\dfrac{\sqrt{3}}{\sqrt{6} - \sqrt{2}}$

57. $\dfrac{x}{\sqrt{x} + \sqrt{y}}$

58. $\dfrac{\sqrt{x}}{\sqrt{x} - \sqrt{y}}$

59. $\dfrac{2 + \sqrt{3}}{2 - \sqrt{3}}$

60. $\dfrac{\sqrt{5} + 2}{\sqrt{5} - 2}$

61. $\dfrac{\sqrt{x} + 5}{\sqrt{x} - 4}$

62. $\dfrac{x - \sqrt{3}}{2x - \sqrt{6}}$

63. $\dfrac{\sqrt{3} + \sqrt{2}}{\sqrt{6} - \sqrt{3}}$

64. $\dfrac{\sqrt{8} - \sqrt{2}}{\sqrt{6} - \sqrt{8}}$

65. $\dfrac{\sqrt{x} + \sqrt{y}}{\sqrt{x} - \sqrt{y}}$

66. $\dfrac{\sqrt{2x} - \sqrt{y}}{\sqrt{3x} + \sqrt{5y}}$

67. $\dfrac{5\sqrt{3} + 7}{2\sqrt{3} - 4}$

68. $\dfrac{3\sqrt{6} - 4}{2\sqrt{2} - 3}$

69. $\dfrac{2\sqrt{3} + 3\sqrt{2}}{4\sqrt{2} + 2\sqrt{3}}$

70. $\dfrac{5\sqrt{2} - 3\sqrt{3}}{2\sqrt{2} + 3\sqrt{3}}$

71. $\dfrac{2\sqrt{x} - 3\sqrt{y}}{3\sqrt{x} + 2\sqrt{y}}$

72. $\dfrac{2\sqrt{3x} + 4\sqrt{2y}}{\sqrt{3x} - \sqrt{6y}}$

73. $\dfrac{2\sqrt{2x} - 5\sqrt{3y}}{\sqrt{3x} + \sqrt{6y}}$

74. $\dfrac{3\sqrt{5x} - \sqrt{8y}}{\sqrt{2x} - 2\sqrt{10y}}$

33.) $(\sqrt{x} - \sqrt{3x-2})^2$

G R O U P   A C T I V I T Y

In exercises 75–80, perform the indicated operations and simplify your answers.

75. $\sqrt{4 + \sqrt{7}} - \sqrt{4 - \sqrt{7}}$

76. $\sqrt{2 + \sqrt{3}} + \sqrt{2 - \sqrt{3}}$

77. $\sqrt{6 + \sqrt{11}} + \sqrt{6 - \sqrt{11}}$

78. $\sqrt{7 + \sqrt{13}} - \sqrt{7 - \sqrt{13}}$

79. $\sqrt{5 + 2\sqrt{6}} + \sqrt{5 - 2\sqrt{6}}$

80. $\sqrt{3 + 2\sqrt{2}} - \sqrt{3 - 2\sqrt{2}}$

5.6 *Radical Equations*

Equations in which the variable appears in at least one radicand are called radical equations. Some examples are:

$$\sqrt{x} = 7, \qquad \sqrt[3]{x^2 + 2} = 3, \qquad \sqrt{x - 3} + \sqrt{x} = 3$$

By applying the following law, we will learn how to find the solutions of radical equations.

> If $a = b$, then $a^n = b^n$, when n is any natural number.

REMARK

This law says that we can raise both sides of an equation to any natural number power. For example,

$$\sqrt{x} = 7$$
$$(\sqrt{x})^2 = 7^2$$
$$x = 49$$

So 49 is the solution. Let's check it.

$$\sqrt{49} \overset{?}{=} 7$$
$$7 = 7$$

Suppose we square both sides of the equation $x = 10$. We obtain

$$x^2 = 100$$
$$x^2 - 100 = 0$$
$$(x + 10)(x - 10) = 0$$
$$x + 10 = 0 \quad \text{or} \quad x - 10 = 0$$
$$x = -10 \qquad\quad x = 10$$

Clearly, -10 is not a solution of our original equation. -10 is called an **extraneous solution.** When both sides of an equation are raised to a natural number power, we run the risk of generating extraneous solutions. *Therefore, whenever you raise both sides of an equation to a natural number power, you must always check your solutions.*

The following steps enable us to find the solution(s) of any radical equation:

> 1. Isolate one of the radical terms; that is, place one of the radical terms on one side of the equation and all of the remaining terms on the other side of the equation.
> 2. Raise both sides of the equation to the power equal to the index of the radical.
> 3. If no radicals remain, find the solution(s) of the resulting equation. If radicals are still present, repeat steps 1 and 2.
> 4. Check your potential solutions.

E X A M P L E   5 . 6 . 1   Find the solution(s) of the following radical equations.

1.
$$\sqrt[3]{x^2 + 2} = 3$$
$$(\sqrt[3]{x^2 + 2})^3 = 3^3$$
$$x^2 + 2 = 27$$
$$x^2 - 25 = 0$$
$$(x + 5)(x - 5) = 0$$
$$x + 5 = 0 \quad \text{or} \quad x - 5 = 0$$
$$x = -5 \qquad\qquad x = 5$$

When $x = -5$,

$$\sqrt[3]{(-5)^2 + 2} \stackrel{?}{=} 3$$
$$\sqrt[3]{25 + 2} \stackrel{?}{=} 3$$
$$\sqrt[3]{27} \stackrel{?}{=} 3$$
$$3 = 3$$

When $x = 5$,

$$\sqrt[3]{5^2 + 2} \stackrel{?}{=} 3$$
$$\sqrt[3]{25 + 2} \stackrel{?}{=} 3$$
$$\sqrt[3]{27} \stackrel{?}{=} 3$$
$$3 = 3$$

Hence -5 and 5 are the solutions.

2.
$$\sqrt{2x^2 - x - 1} = \sqrt{x^2 + x + 2}$$
$$(\sqrt{2x^2 - x - 1})^2 = (\sqrt{x^2 + x + 2})^2$$
$$2x^2 - x - 1 = x^2 + x + 2$$
$$x^2 - 2x - 3 = 0$$
$$(x - 3)(x + 1) = 0$$
$$x = 3 \quad \text{or} \quad x = -1$$

When $x = 3$,

$$\sqrt{2 \cdot (3)^2 - 3 - 1} \stackrel{?}{=} \sqrt{3^2 + 3 + 2}$$
$$\sqrt{18 - 3 - 1} \stackrel{?}{=} \sqrt{9 + 3 + 2}$$
$$\sqrt{14} = \sqrt{14}$$

When $x = -1$,

$$\sqrt{2(-1)^2 - (-1) - 1} \stackrel{?}{=} \sqrt{(-1)^2 + (-1) + 2}$$
$$\sqrt{2 + 1 - 1} \stackrel{?}{=} \sqrt{1 - 1 + 2}$$
$$\sqrt{2} = \sqrt{2}$$

Thus, 3 and -1 are the solutions.

3. $\sqrt{x^2 + 3x - 17} + 2 = x$
   $\sqrt{x^2 + 3x - 17} = x - 2$ *Isolate the radical term.*
   $(\sqrt{x^2 + 3x - 17})^2 = (x - 2)^2$
   $x^2 + 3x - 17 = x^2 - 4x + 4$
   $3x - 17 = -4x + 4$
   $7x = 21$
   $x = 3$

When $x = 3$,

$$\sqrt{3^2 + 3 \cdot 3 - 17} + 2 \stackrel{?}{=} 3$$
$$\sqrt{9 + 9 - 17} + 2 \stackrel{?}{=} 3$$
$$\sqrt{1} + 2 = 3$$

Thus, 3 is the solution.

4. $\sqrt{x + 7} - 1 = x$
   $\sqrt{x + 7} = x + 1$ *Isolate the radical term.*
   $(\sqrt{x + 7})^2 = (x + 1)^2$
   $x + 7 = x^2 + 2x + 1$
   $0 = x^2 + x - 6$
   $0 = (x + 3)(x - 2)$
   $x = -3$ or $x = 2$

When $x = 2$,

$$\sqrt{2 + 7} - 1 \stackrel{?}{=} 2$$
$$\sqrt{9} - 1 \stackrel{?}{=} 2$$
$$3 - 1 = 2$$

When $x = -3$,

$$\sqrt{-3 + 7} - 1 \overset{?}{=} -3$$
$$\sqrt{4} - 1 \overset{?}{=} -3$$
$$2 - 1 \neq -3$$

-3 is an extraneous solution. The *only* solution is 2.

5. $\sqrt{x - 3} + \sqrt{x} = 3$

$\qquad \sqrt{x - 3} = 3 - \sqrt{x} \qquad$ *Isolate a radical term.*

$\qquad (\sqrt{x - 3})^2 = (3 - \sqrt{x})^2$

$\qquad x - 3 = 9 - 6\sqrt{x} + x$

$\qquad -12 = -6\sqrt{x} \qquad$ *Isolate the remaining radical.*

$\qquad 2 = \sqrt{x} \qquad$ *Divide both sides by -6.*

$\qquad 2^2 = (\sqrt{x})^2$

$\qquad 4 = x$

When $x = 4$.

$$\sqrt{4 - 3} + \sqrt{4} \overset{?}{=} 3$$
$$\sqrt{1} + \sqrt{4} \overset{?}{=} 3$$
$$1 + 2 = 3$$

Thus, 4 is the solution.

6. $\sqrt{4x + 1} = \sqrt{x - 2} + \sqrt{x + 3}$

$\qquad (\sqrt{4x + 1})^2 = (\sqrt{x - 2} + \sqrt{x + 3})^2$

$\qquad 4x + 1 = x - 2 + 2\sqrt{x - 2}\sqrt{x + 3} + x + 3$

$\qquad 4x + 1 = x - 2 + 2\sqrt{(x - 2)(x + 3)} + x + 3$

$\qquad 4x + 1 = 2x + 1 + 2\sqrt{x^2 + x - 6}$

$\qquad 2x = 2\sqrt{x^2 + x - 6} \qquad$ *Isolate the remaining radical.*

$\qquad x = \sqrt{x^2 + x - 6} \qquad$ *Divide both sides by 2.*

$\qquad x^2 = (\sqrt{x^2 + x - 6})^2$

$\qquad x^2 = x^2 + x - 6$

$\qquad 0 = x - 6$

$\qquad 6 = x$

When $x = 6$,

$$\sqrt{4(6) + 1} \overset{?}{=} \sqrt{6 - 2} + \sqrt{6 + 3}$$
$$\sqrt{25} \overset{?}{=} \sqrt{4} + \sqrt{9}$$
$$5 = 2 + 3$$

Thus, 6 is the solution.

Try to avoid this mistake:

Incorrect	Correct
$\sqrt{4x + 1} = \sqrt{x - 2}$ $+ \sqrt{x + 3}$ $(\sqrt{4x + 1})^2 = (\sqrt{x - 2}$ $+ \sqrt{x + 3})^2$ $4x + 1 = x - 2 + x + 3$	$\sqrt{4x + 1} = \sqrt{x - 2}$ $+ \sqrt{x + 3}$ $(\sqrt{4x + 1})^2 = (\sqrt{x - 2}$ $+ \sqrt{x + 3})^2$ $4x + 1 = x - 2$ $+ 2\sqrt{x - 2}\sqrt{x + 3}$ $+ x + 3$ *Remember:* $(a + b)^2 = a^2 + 2ab + b^2$

7. $\sqrt[4]{3x + 1} = -2$

$(\sqrt[4]{3x + 1})^4 = (-2)^4$

$3x + 1 = 16$

$3x = 15$

$x = 5$

When $x = 5$,

$$\sqrt[4]{3(5) + 1} \stackrel{?}{=} -2$$

$$\sqrt[4]{15 + 1} \stackrel{?}{=} -2$$

$$\sqrt[4]{16} \neq -2 \qquad \textit{Since } \sqrt[4]{16} = 2$$

Since 5 is not a solution, the solution set of the original equation is $\varnothing$, the empty set.

8. $\sqrt{x + \sqrt{x + 2}} = 2$

$(\sqrt{x + \sqrt{x + 2}})^2 = 2^2$ *Square both sides to eliminate the outside radical.*

$x + \sqrt{x + 2} = 4$

$\sqrt{x + 2} = 4 - x$ *Isolate the radical.*

$(\sqrt{x + 2})^2 = (4 - x)^2$ *Square both sides.*

$x + 2 = 16 - 8x + x^2$

$0 = x^2 - 9x + 14$

$0 = (x - 2)(x - 7)$

$x = 2 \quad \text{or} \quad x = 7$

When $x = 2$,

$$\sqrt{2 + \sqrt{2 + 2}} \stackrel{?}{=} 2$$
$$\sqrt{2 + \sqrt{4}} \stackrel{?}{=} 2$$
$$\sqrt{2 + 2} \stackrel{?}{=} 2$$
$$\sqrt{4} = 2$$

When $x = 7$,

$$\sqrt{7 + \sqrt{7 + 2}} \stackrel{?}{=} 2$$
$$\sqrt{7 + \sqrt{9}} \stackrel{?}{=} 2$$
$$\sqrt{7 + 3} \stackrel{?}{=} 2$$
$$\sqrt{10} \neq 2$$

Thus, 2 is the only solution.

9.
$$\frac{2}{\sqrt{x + 5}} = \sqrt{x + 5} + 1$$

$$\sqrt{x + 5} \cdot \frac{2}{\sqrt{x + 5}} = \sqrt{x + 5} \cdot \sqrt{x + 5}$$
$$+ \sqrt{x + 5} \cdot 1 \qquad \textit{Eliminate the fraction.}$$
$$2 = x + 5 + \sqrt{x + 5}$$
$$-x - 3 = \sqrt{x + 5} \qquad \textit{Isolate the radical.}$$
$$(-x - 3)^2 = (\sqrt{x + 5})^2 \qquad \textit{Square both sides.}$$
$$x^2 + 6x + 9 = x + 5$$
$$x^2 + 5x + 4 = 0$$
$$(x + 4)(x + 1) = 0$$
$$x + 4 = 0 \quad \text{or} \quad x + 1 = 0$$
$$x = -4 \qquad \qquad x = -1$$

When $x = -4$,

$$\frac{2}{\sqrt{-4 + 5}} \stackrel{?}{=} \sqrt{-4 + 5} + 1$$

$$\frac{2}{\sqrt{1}} \stackrel{?}{=} \sqrt{1} + 1$$

$$\frac{2}{1} \stackrel{?}{=} 1 + 1$$

$$2 = 2$$

When $x = -1$,

$$\frac{2}{\sqrt{-1 + 5}} \overset{?}{=} \sqrt{-1 + 5} + 1$$

$$\frac{2}{\sqrt{4}} \overset{?}{=} \sqrt{4} + 1$$

$$\frac{2}{2} \overset{?}{=} 2 + 1$$

$$1 \neq 3$$

Thus, -4 is the only solution.

Exercises 5.6

Find the solution(s) of the following radical equations.

1. $\sqrt{2x + 1} = 3$

2. $\sqrt{5x - 4} = 6$

3. $\sqrt{x^2 + 7} - 4 = 0$

4. $\sqrt{x^2 + x - 5} - 1 = 0$

5. $\sqrt{3x - 7} + 2 = 0$

6. $\sqrt{x^2 + 16} + 6 = 1$

7. $\sqrt[3]{3x + 1} = 4$

8. $\sqrt[3]{5 - 6x} = -3$

9. $\sqrt[3]{3x^2 - 10x} + 2 = 0$

10. $\sqrt[3]{x^2 + 4} - 1 = 4$

11. $\sqrt[4]{3x + 5} = -3$

12. $\sqrt[4]{x^2 + 16} + 3 = 1$

13. $\sqrt[4]{2x - 3} - 1 = 0$

14. $\sqrt[4]{x^2 + x - 4} = 2$

15. $\sqrt{2x + 3} = \sqrt{7 - x}$

16. $\sqrt{4x - 12} = \sqrt{3x + 2}$

17. $\sqrt{3x^2 - x - 6} = \sqrt{2x^2 + x + 9}$

18. $\sqrt{3x^2 - 11x - 12} = \sqrt{x^2 - 4x + 3}$

19. $\sqrt[3]{3x - 2} = \sqrt[3]{5x + 7}$

20. $\sqrt[3]{5x^2 + 3x - 9} = \sqrt[3]{x^2 + 3x - 8}$

21. $\sqrt[4]{4x - 3} = \sqrt[4]{x + 8}$

22. $\sqrt[4]{3x^2 - 7x + 2} = \sqrt[4]{x^2 - 2x + 5}$

23. $x = \sqrt{3x + 40}$

24. $2x = \sqrt{4x + 15}$

25. $\sqrt{2x^2 + 2x - 3} = x$

26. $\sqrt{2x^2 - 3x - 28} = x$

27. $\sqrt{4x^2 - x - 1} + 1 = 2x$

28. $\sqrt{x^2 - 2x - 3} - x = 3$

29. $\sqrt{x^2 - 8x + 26} + 5 = x$

30. $\sqrt{x^2 - x + 5} + 1 = x$

31. $\sqrt{2x + 5} - x = 3$

32. $\sqrt{x + 1} + 1 = x$

33. $\sqrt{x + 8} - x = -4$

34. $\sqrt{3x + 4} - 2 = x$

35. $\sqrt{3x + 5} + 1 = 3x$

36. $\sqrt{2x + 3} + 6x = 5$

37. $\sqrt{3 - x} - x = 3$

38. $\sqrt{2x + 7} + 1 = -2x$

39. $\sqrt{x - 4} + x = 6$

40. $\sqrt{x + 2} + \sqrt{x} = 2$

41. $\sqrt{x - 16} - \sqrt{x} = -2$

42. $\sqrt{3x + 1} - \sqrt{x} = 1$

43. $\sqrt{2x + 7} - \sqrt{x} = 2$

44. $\sqrt{x + 7} + \sqrt{x + 4} = 3$

45. $\sqrt{x - 1} - \sqrt{x + 4} = -1$

46. $\sqrt{2x - 3} - \sqrt{x - 5} = 2$

47. $\sqrt{2x + 3} + \sqrt{x + 2} = 2$

48. $\sqrt{2 - x} - \sqrt{3x + 1} = 1$

49. $\sqrt{5 - 2x} - \sqrt{x + 6} = 1$

50. $\sqrt{x + 1} + \sqrt{2x + 2} = \sqrt{3x + 3}$

51. $\sqrt{x + 2} + \sqrt{x - 1} = \sqrt{4x + 1}$

52. $\sqrt{x + 6} - \sqrt{x + 2} = \sqrt{4x}$

53. $\sqrt{2x + 2} - \sqrt{5x + 1} = \sqrt{-4x}$

54. $\sqrt{x-2} + \sqrt{x+6} = \sqrt{4x+4}$

56. $\sqrt{\sqrt{x+2}-2x} = 1$

58. $\sqrt{2x + \sqrt{x+6}} = 3$

60. $\sqrt[3]{3x + \sqrt{x+4}} = -2$

62. $\dfrac{2x}{\sqrt{x-1}} = \sqrt{4x-4} + 1$

64. $\dfrac{1}{\sqrt{x-2}} + \sqrt{x+1} = 3\sqrt{x-2}$

55. $\sqrt{x + \sqrt{x+11}} = 1$

57. $\sqrt{\sqrt{2x+3} - 3x} = 2$ $2x+3$

59. $\sqrt[3]{2x + \sqrt{x+1}} = 2$

61. $\dfrac{x}{\sqrt{2x-5}} = 3$

63. $\dfrac{1}{\sqrt{x+2}} - \sqrt{x+1} = \sqrt{x+2}$

mult.
× den.

Write Algebra

65. Describe the steps used to solve a radical equation.

66. Explain why we check for extraneous solutions when we solve radical equations.

67. Explain why, in general, $(\sqrt{x+2} + 3)^2 \neq x + 2 + 9$.

5.7 *Complex Numbers*

Finding the solutions of the equation $x^2 = 1$, should, by now, be an easy task:

$$x^2 = 1$$
$$x^2 - 1 = 0$$
$$(x - 1)(x + 1) = 0$$
$$x = 1 \quad \text{or} \quad x = -1$$

Another method for finding the solutions of the equation $x^2 = 1$ is as follows:

$$x^2 = 1$$
$$x = \pm\sqrt{1}$$
$$x = \pm 1$$

(The expression $x = \pm 1$ means $x = 1$ or $x = -1$.) Let's apply this second method to the equation $x^2 = -1$.

$$x^2 = -1$$
$$x = \pm\sqrt{-1}$$

The solution $\sqrt{-1}$ is not a real number, since no real number squared is -1. The number $\sqrt{-1}$ will be denoted by the symbol i. As will be shown later in this section, i is the base unit for a new set of numbers:

Definition 5.7.1	$i = \sqrt{-1}$ and $i^2 = -1$

By applying the preceding definition and the multiplication property of radicals, the square root of any negative number can be written in terms of i as follows.

If $r > 0$,

$$\sqrt{-r} = \sqrt{r} \cdot \sqrt{-1}$$
$$= \sqrt{r}\,i \quad \text{or} \quad i\sqrt{r}$$

Let us use the above reasoning to express the square root of the following negative numbers in terms of i:

$$\sqrt{-9} = \sqrt{9}\sqrt{-1}, \qquad -\sqrt{-100} = -\sqrt{100}\sqrt{-1}, \qquad \sqrt{-21} = \sqrt{21}\sqrt{-1}$$
$$= 3i \qquad\qquad\qquad = -10i \qquad\qquad\qquad = i\sqrt{21}$$

EXAMPLE 5.7.1

Express the following in terms of i and simplify.

1. $\sqrt{-16} = \sqrt{16}\sqrt{-1}$
   $\qquad = 4i$

2. $\sqrt{-4} + \sqrt{-49} = \sqrt{4}\sqrt{-1} + \sqrt{49}\sqrt{-1}$
   $\qquad\qquad\qquad = 2i + 7i$
   $\qquad\qquad\qquad = 9i \qquad$ *Combine like terms.*

3. $\sqrt{64} + \sqrt{-1} = 8 + i$

NOTE ▶ The multiplication and division properties of radicals are not defined when both radicands are negative. Therefore, always convert expressions of the form $\sqrt{-r}$, where $r > 0$, to $i\sqrt{r}$ before applying either of these properties.

4. $\sqrt{-4} \cdot \sqrt{-25} = \sqrt{4}\sqrt{-1} \cdot \sqrt{25}\sqrt{-1}$
   $\qquad\qquad\qquad = 2 \cdot i \cdot 5 \cdot i$
   $\qquad\qquad\qquad = 10i^2$
   $\qquad\qquad\qquad = 10 \cdot (-1)$
   $\qquad\qquad\qquad = -10$

Try to avoid this mistake:

Incorrect	Correct
$\sqrt{-4}\sqrt{-25} = \sqrt{-4\cdot-25}$ $= \sqrt{100}$ $= 10$ because $\sqrt{a}\sqrt{b} \neq \sqrt{ab}$ if both a and b are negative.	$\sqrt{-4}\sqrt{-25} = \sqrt{4}\sqrt{-1}$ $\cdot \sqrt{25}\sqrt{-1}$ $= 2i \cdot 5i$ $= 10i^2$ $= -10$

5. $\dfrac{\sqrt{-36}}{\sqrt{-49}} = \dfrac{\sqrt{36}\sqrt{-1}}{\sqrt{49}\sqrt{-1}}$

$= \dfrac{6i}{7i}$

$= \dfrac{6}{7}$

The following definition shows how i is used to generate a new set of numbers.

Definition 5.7.2   Any number that can be expressed in the form $a + bi$, where a and b are real numbers and $i = \sqrt{-1}$, is called a **complex number.**

REMARKS

1. a is called the **real part** of the complex number $a + bi$.

2. b is called the **imaginary part** of the complex number $a + bi$.

3. If $a = 0$ and $b \neq 0$, $a + bi$ becomes bi, which is called a **pure imaginary number.**

4. If $b = 0$, $a + bi$ becomes a, a real number. This fact demonstrates that the real numbers are a subset of the complex numbers.

5. The set of complex numbers is labeled with a C. We have shown that $R \subseteq C$. The set of complex numbers is the last set of numbers we will examine.

Examples of complex numbers are $2 + i, \frac{3}{8} - \frac{7}{8}i$, and $\sqrt{2} + \frac{1}{2}i$. The following rules show how to add and subtract complex numbers.

Addition and Subtraction of Complex Numbers

Let a, b, c, and d be real numbers and $i = \sqrt{-1}$.

$$(a + bi) + (c + di) = (a + c) + (b + d)i$$
$$(a + bi) - (c + di) = (a - c) + (b - d)i$$

EXAMPLE 5.7.2   Perform the indicated operations.

1. $(2 + 3i) + (7 - 4i) = (2 + 7) + (3 - 4)i$ *Combine like terms.*
$$= 9 + (-1)i$$
$$= 9 - i$$

2. $(3 - 4i) - (-8 + 9i) = (3 - (-8)) + (-4 - 9)i$
$$= 11 + (-13)i$$
$$= 11 - 13i$$

Don't be intimidated by the following multiplication formula. Multiplication of complex numbers is not as complicated as it looks.

Multiplication of Complex Numbers

Let a, b, c, and d be real numbers and $i = \sqrt{-1}$.

$$(a + bi)(c + di) = (ac - bd) + (ad + bc)i$$

This multiplication formula can be derived by using the FOIL Method for multiplying binomials:

$$(a + bi)(c + di) = ac + adi + bci + bdi^2$$
$$= ac + adi + bci + bd(-1)$$
$$= ac + adi + bci - bd$$
$$= (ac - bd) + (ad + bc)i$$

NOTE ▶   The process of multiplying complex numbers is the same as the process of multiplying polynomials.

EXAMPLE 5.7.3   Perform the indicated multiplications.

1. $(5 + 3i)(2 - 4i) = 5 \cdot 2 - 5 \cdot 4i + 2 \cdot 3i - 3i \cdot 4i$
$$= 10 - 20i + 6i - 12i^2$$
$$= 10 - 14i - 12(-1)$$
$$= 10 - 14i + 12$$
$$= 22 - 14i$$

2. $3i(6 - 7i) = 3i \cdot 6 - 3i \cdot 7i$

$\qquad\qquad = 18i - 21i^2$

$\qquad\qquad = 18i - 21(-1)$

$\qquad\qquad = 21 + 18i$

3. $(2 + 3i)(2 - 3i) = 2 \cdot 2 - 2 \cdot 3i + 2 \cdot 3i - 3i \cdot 3i$

$\qquad\qquad\qquad = 4 - 6i + 6i - 9i^2$

$\qquad\qquad\qquad = 4 - 9(-1)$

$\qquad\qquad\qquad = 4 + 9$

$\qquad\qquad\qquad = 13$

Here the product of two complex numbers is a real number.

NOTE ▶ The product $(2 - 3i)(2 + 3i)$ yields a real number. The numbers $2 - 3i$ and $2 + 3i$ are called **complex conjugates** of each other. To find the conjugate of a complex number, simply change the sign of the imaginary part of the complex number. Division of complex numbers by using conjugates is illustrated in the following example.

A complex number is considered simplified if it can be written in the form $a + bi$. Clearly, the complex number $\frac{1}{2 - 3i}$ is not simplified. To determine how to simplify $\frac{1}{2 - 3i}$, refer to part 3 of Example 5.7.3:

$$\frac{1}{2 - 3i} = \frac{1}{(2 - 3i)} \cdot \frac{(2 + 3i)}{(2 + 3i)}$$

$$= \frac{1(2 + 3i)}{(2 - 3i) \cdot (2 + 3i)}$$

$$= \frac{2 + 3i}{4 - 6i + 6i - 9i^2}$$

$$= \frac{2 + 3i}{4 + 9}$$

$$= \frac{2 + 3i}{13}$$

$$= \frac{2}{13} + \frac{3}{13}i$$

The complex number has been simplified. By multiplying the top and bottom by the conjugate of the denominator, we obtained an equivalent fraction with a real number in the denominator.

E XAMPLE 5.7.4   Perform the indicated divisions.

1. $\dfrac{2}{5 + i} = \dfrac{2}{(5 + i)} \cdot \dfrac{(5 - i)}{(5 - i)}$

$= \dfrac{2(5 - i)}{(5 + i)(5 - i)}$

$= \dfrac{10 - 2i}{25 - 5i + 5i - i^2}$

$= \dfrac{10 - 2i}{25 + 1}$

$= \dfrac{10 - 2i}{26}$

$= \dfrac{10}{26} - \dfrac{2}{26}i$

$= \dfrac{5}{13} - \dfrac{1}{13}i$

2. $\dfrac{5 - 5i}{3 + i} = \dfrac{(5 - 5i)}{(3 + i)} \cdot \dfrac{(3 - i)}{(3 - i)}$

$= \dfrac{(5 - 5i)(3 - i)}{(3 + i)(3 - i)}$

$= \dfrac{15 - 5i - 15i + 5i^2}{9 - i^2}$

$= \dfrac{15 - 20i - 5}{9 + 1}$

$= \dfrac{10 - 20i}{10}$

$= \dfrac{10}{10} - \dfrac{20i}{10}$

$= 1 - 2i$

3. $\dfrac{2 + 4i}{3 - 5i} = \dfrac{(2 + 4i)}{(3 - 5i)} \cdot \dfrac{(3 + 5i)}{(3 + 5i)}$

$= \dfrac{(2 + 4i)(3 + 5i)}{(3 - 5i)(3 + 5i)}$

$= \dfrac{6 + 10i + 12i + 20i^2}{9 - 25i^2}$

$= \dfrac{6 + 22i - 20}{9 + 25}$

$= \dfrac{-14 + 22i}{34}$

$= \dfrac{-14}{34} + \dfrac{22}{34}i$

$= \dfrac{-7}{17} + \dfrac{11}{17}i$

We now state the formula for division of complex numbers. However, as was the case with multiplication, the formula is more complicated than the actual process.

D*ivision of Complex Numbers*

Let a, b, c, and d be real numbers and $i = \sqrt{-1}$.

$$\frac{a + bi}{c + di} = \frac{(a + bi)(c - di)}{(c + di)(c - di)} = \frac{(a + bi)(c - di)}{c^2 + d^2}$$

Since $i^2 = -1$, the natural number powers of i form a pattern. Consider the following list.

$$i = i$$
$$i^2 = -1$$
$$i^3 = i^2 \cdot i = (-1) \cdot i = -i$$
$$i^4 = i^2 \cdot i^2 = (-1) \cdot (-1) = 1$$

$$i^5 = i^4 \cdot i = 1 \cdot i = i$$
$$i^6 = i^4 \cdot i^2 = 1 \cdot (-1) = -1$$
$$i^7 = i^4 \cdot i^3 = 1 \cdot (-i) = -i$$
$$i^8 = i^4 \cdot i^4 = 1 \cdot 1 = 1$$

From this pattern, i to any natural number power can be expressed as $1, -1, i,$ or $-i$.

EXAMPLE 5.7.5   Simplify the following.

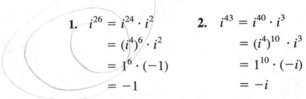

1. $i^{26} = i^{24} \cdot i^2$
$$= (i^4)^6 \cdot i^2$$
$$= 1^6 \cdot (-1)$$
$$= -1$$

2. $i^{43} = i^{40} \cdot i^3$
$$= (i^4)^{10} \cdot i^3$$
$$= 1^{10} \cdot (-i)$$
$$= -i$$

NOTE ▶ i raised to a power that is a multiple of 4 is always 1. Why? To simplify i to any natural number power, first determine the largest multiple of 4 within the power. Then follow the steps demonstrated in the previous example.

Definition 5.7.3

1. $i^0 = 1$

2. If n is any natural number, $i^{-n} = \dfrac{1}{i^n}$.

By applying Definition 5.7.3, we can simplify i to any integer power.

EXAMPLE 5.7.6   Simplify the following.

1. $i^{-6} = \dfrac{1}{i^6}$

$$= \dfrac{1}{i^4 \cdot i^2}$$

$$= \dfrac{1}{1 \cdot (-1)}$$

$$= -1$$

2. $i^{-5} = \dfrac{1}{i^5}$

$$= \dfrac{1}{i^4 \cdot i}$$

$$= \dfrac{1}{1 \cdot i}$$

$$= \dfrac{1}{i} \cdot \dfrac{-i}{-i} \qquad \text{\textit{Multiply top and bottom by the conjugate of the denominator.}}$$

$$= \dfrac{-i}{-i^2}$$

$$= \dfrac{-i}{-(-1)}$$

$$= -i$$

E xercises 5.7

Express the following in the form $a + bi$ and simplify.

1. $\sqrt{-81}$

2. $-\sqrt{-121}$

3. $-\sqrt{-20}$

4. $\sqrt{-75}$

5. $3\sqrt{-48}$

6. $-5\sqrt{-98}$

7. $\sqrt{-9} + \sqrt{-16}$

8. $\sqrt{-4} - \sqrt{-100}$

9. $2\sqrt{-25} - 3\sqrt{-49}$

10. $5\sqrt{-1} + 6\sqrt{-16}$

11. $\sqrt{-12} + \sqrt{-27}$

12. $\sqrt{-20} - \sqrt{-125}$

13. $\sqrt{81} + \sqrt{-9}$

14. $\sqrt{25} - \sqrt{-144}$

15. $\sqrt{48} - \sqrt{-98}$

16. $-\sqrt{50} + \sqrt{-12}$

17. $3\sqrt{18} - 2\sqrt{-24}$

18. $5\sqrt{72} + 3\sqrt{-75}$

19. $\sqrt{-16} \cdot \sqrt{-9}$

20. $\sqrt{-25} \cdot \sqrt{-100}$

21. $\sqrt{-13} \cdot \sqrt{-13}$

22. $\sqrt{-21} \cdot \sqrt{-21}$

23. $-\sqrt{-3} \cdot \sqrt{-3}$

24. $-\sqrt{-8} \cdot (-\sqrt{-8})$

25. $-\sqrt{5} \cdot \sqrt{-5}$

26. $-\sqrt{17} \cdot \sqrt{-17}$

27. $3\sqrt{-4}\sqrt{8}$

28. $5\sqrt{-12}\sqrt{6}$

29. $\dfrac{\sqrt{-81}}{\sqrt{-4}}$

30. $\dfrac{\sqrt{-9}}{\sqrt{-121}}$

31. $\dfrac{-\sqrt{-25}}{\sqrt{-1}}$

32. $\dfrac{-\sqrt{-36}}{\sqrt{-100}}$

33. $\dfrac{\sqrt{-4}\sqrt{-9}}{\sqrt{-16}}$

34. $\dfrac{\sqrt{-25}\sqrt{-49}}{\sqrt{-36}}$

Perform the indicated operations. Express all answers in the form $a + bi$.

35. $(2 + 3i) + (5 - 6i)$

36. $(-4 + 7i) + (3 + 2i)$

37. $\left(\dfrac{1}{4} + \dfrac{1}{8}i\right) + \left(-\dfrac{1}{3} + \dfrac{1}{6}i\right)$

38. $\left(-\dfrac{3}{8} - \dfrac{5}{8}i\right) + \left(\dfrac{3}{2} - \dfrac{1}{2}i\right)$

39. $(-5 - 2i) - (3 + i)$

40. $(6 + 11i) - (-4 - 2i)$

41. $\left(-\dfrac{2}{3} + \dfrac{1}{9}i\right) - \left(\dfrac{3}{4} - \dfrac{1}{4}i\right)$

42. $\left(\dfrac{3}{5} - \dfrac{1}{5}i\right) - \left(-\dfrac{3}{10} - \dfrac{7}{10}i\right)$

43. $7(3 + 9i)$

44. $-11(2 - i)$

45. $2i(4 - 7i)$

46. $-5i(-6 - 8i)$

47. $(6 + i)(3 - 4i)$

48. $(2 + 4i)(5 + 6i)$

49. $(5 - 3i)(2 - 4i)$

50. $(-8 + i)(9 - 3i)$

51. $(2 - 5i)(2 + 5i)$

52. $(5 - 6i)(5 + 6i)$

53. $(\sqrt{3} + i)(\sqrt{3} - i)$

54. $(\sqrt{5} + 2i)(\sqrt{5} - 2i)$

55. $(2 + 3i)^2$

56. $(-7 - 4i)^2$

57. $(\sqrt{3} + i)^2$

58. $(2\sqrt{2} - i\sqrt{5})^2$

Perform the indicated divisions. Express all answers in the form $a + bi$.

59. $\dfrac{5}{3i}$

60. $\dfrac{7}{-4i}$

61. $\dfrac{5 + 3i}{i}$

62. $\dfrac{-6 + 2i}{i}$

63. $\dfrac{4 + i}{2i}$

64. $\dfrac{3 - 2i}{-3i}$

65. $\dfrac{1}{4 + 2i}$

66. $\dfrac{1}{3 - 5i}$

67. $\dfrac{3}{2 - 3i}$

68. $\dfrac{i}{5 + 2i}$

69. $\dfrac{2i}{1 + 3i}$

70. $\dfrac{25i}{3 + 4i}$

71. $\dfrac{-4i}{-3 - 5i}$

72. $\dfrac{2 + 3i}{5 - 2i}$

73. $\dfrac{3 - 4i}{3 + i}$

74. $\dfrac{2 + 5i}{2 - 5i}$

75. $\dfrac{9 - 17i}{2 - i}$

76. $\dfrac{3 - 7i}{3 + 7i}$

77. $\dfrac{-2 - 5i}{1 - 4i}$

78. $\dfrac{16 - 15i}{6 - i}$

79. $\dfrac{-3 + 2i}{-5 - 4i}$

help

Simplify the following.

80. i^{30}

81. i^{91}

82. i^{12}

83. $-i^{27}$

84. $-i^{41}$

85. $-6i^{65}$

86. $4i^{44}$

87. $2i^{22}$

88. $3i^{2525}$

89. i^{-15}

90. i^{-22}

91. i^{-37}

92. $-i^{-29}$

93. $-i^{-56}$

94. $-5i^{-1}$

95. $2i^{-3}$

96. $-7i^{0}$

97. $3i^{-10}$

98. $-10i^{-15}$

Write Algebra

99. In Section 5.3 we stated $\sqrt[n]{ab} = \sqrt[n]{a} \cdot \sqrt[n]{b}$, provided that a and b are not both negative. Explain why we need this restriction.

100. Describe the relationship between complex numbers, pure imaginary numbers, and real numbers.

Applied Algebra

When trying to put out a fire, firefighters are interested in the flow rate of the water available at the site of the fire. The flow rate of water is determined by the equation

$$Q = 29.7D^2\sqrt{P}$$

In this equation Q represents the flow rate of water measured in gallons per minute (gal/min), D is the nozzle diameter measured in inches, and P is the static pressure of the water measured in pounds per square inch (lb/sq in.)

1. Find the flow rate of water from a $1\frac{1}{2}$-in.-diameter nozzle at 60 lb/sq in.

SOLUTION

$$Q = 29.7D^2\sqrt{P}$$

In our case $D = 1.5$ and $P = 60$:

$$Q = 29.7(1.5)^2\sqrt{60}$$
$$= (29.7)(2.25)\sqrt{60}$$
$$= 66.825 \cdot \sqrt{60}$$
$$\doteq 518 \text{ gal/min}$$

2. Find the static pressure if the flow rate is 751 gal/min from a nozzle with a diameter of 2 in.

SOLUTION

$$Q = 29.7D^2\sqrt{P}$$

In our case $Q = 751$ and $D = 2$:

$$751 = 29.7(2)^2\sqrt{P}$$
$$751 = 29.7(4)\ \sqrt{P}$$
$$751 = 118.8\ \sqrt{P}$$
$$\frac{751}{118.8} = \sqrt{P}$$
$$6.32 \doteq \sqrt{P}$$
$$(6.32)^2 \doteq (\sqrt{P})^2 \qquad \textit{Now square both sides.}$$
$$40 \doteq P$$

Thus, the static pressure is approximately 40 lb/sq in.

Your Turn

1. Find the flow rate of water from a $1\frac{3}{4}$-in.-diameter nozzle at 25 lb/sq in.

2. Find the flow rate of water from a $2\frac{1}{4}$-in.-diameter nozzle at 50 lb/sq in.

3. Find the static pressure if the flow rate is 248 gal/min from a nozzle with a diameter of $1\frac{1}{2}$ in.

4. Find the static pressure if the flow rate is 824 gal/min from a nozzle with a diameter of $2\frac{1}{4}$ in.

Chapter 5 Review

Terms to Remember

Principal nth root of a	p. 238	(Radical) conjugates	p. 264
Radical sign	p. 248	Extraneous solution	p. 268
Radicand	p. 248	Complex numbers	p. 277
Index or order	p. 248	Real part	p. 277
Simplified radical	p. 253	Imaginary part	p. 277
Rationalize the denominator	p. 256	Pure imaginary number	p. 277
Like radicals	p. 261	Complex conjugates	p. 279

Notation

Let n and m be natural numbers:

n is even $\quad x^{1/n} =$ principal nth root $\qquad (x^{1/n})^n = x$

n is odd $\quad x^{1/n} = n$th root

$$x^{m/n} = (x^{1/n})^m$$

$$x^{-m/n} = \frac{1}{x^{m/n}} \qquad x \neq 0$$

$$\sqrt[n]{x} = x^{1/n}$$

$$\sqrt[n]{x^m} = (\sqrt[n]{x})^m = x^{m/n}, \text{ provided } \sqrt[n]{x} \text{ is a real number}$$

Laws and Properties

■ *Laws of exponents:* Let m and n be rational numbers and $x \neq 0$ and $y \neq 0$. Then

$$x^m \cdot x^n = x^{m+n}$$

$$\frac{x^m}{x^n} = x^{m-n}$$

$$(x^m)^n = x^{m \cdot n}$$

$$(xy)^n = x^n \cdot y^n$$

$$\left(\frac{x}{y}\right)^n = \frac{x^n}{y^n}$$

■ *Properties of radicals:* Let n be a positive integer and a and b be real numbers.

If n is even,

$$\sqrt[n]{a^n} = |a|$$

If n is odd,

$$\sqrt[n]{a^n} = a$$

If a and b are not both negative,

$$\sqrt[n]{ab} = \sqrt[n]{a}\sqrt[n]{b}$$

$$\sqrt[n]{\frac{a}{b}} = \frac{\sqrt[n]{a}}{\sqrt[n]{b}} \qquad b \neq 0$$

■ *Radical equations:* If $a = b$, then $a^n = b^n$, where n is any natural number.

■ *Arithmetic of complex numbers:* Let a, b, c, and d be real numbers and $i = \sqrt{-1}$.

$$(a + bi) + (c + di) = (a + c) + (b + d)i$$

$$(a + bi) - (c + di) = (a - c) + (b - d)i$$

$$(a + bi)(c + di) = (ac - bd) + (ad + bc)i$$

$$\frac{a + bi}{c + di} = \frac{(a + bi)}{(c + di)} \cdot \frac{(c - di)}{(c - di)} \cdot = \frac{(a + bi)(c - di)}{c^2 + d^2}$$

Review Exercises

5.1 **Evaluate each of the following, if possible.**

1. $169^{1/2}$ **2.** $100^{1/2}$ **3.** $\left(\dfrac{9}{25}\right)^{1/2}$ **4.** $\left(\dfrac{4}{81}\right)^{1/2}$ **5.** $-16^{1/2}$

6. $(-16)^{1/2}$ **7.** $1^{1/3}$ **8.** $1^{1/4}$ **9.** $(-1)^{1/3}$ **10.** $(-1)^{1/4}$

11. $81^{1/4}$ **12.** $\left(\dfrac{8}{125}\right)^{1/3}$ **13.** $\left(-\dfrac{27}{64}\right)^{1/3}$ **14.** $-32^{1/5}$ **15.** $-243^{1/5}$

16. $8^{2/3}$ **17.** $-27^{2/3}$ **18.** $16^{3/2}$ **19.** $27^{4/3}$ **20.** $1^{5/2}$

21. $32^{2/5}$ **22.** $\left(\dfrac{125}{27}\right)^{4/3}$ **23.** $-4^{3/2}$ **24.** $(-4)^{3/2}$ **25.** $8^{-4/3}$

26. $-81^{-3/2}$ **27.** $-64^{-2/3}$ **28.** $36^{-1/2}$ **29.** $\left(\dfrac{1}{4}\right)^{-1/2}$ **30.** $\left(\dfrac{49}{25}\right)^{-3/2}$

31. $\left(-\dfrac{8}{27}\right)^{-2/3}$ **32.** $\left(\dfrac{1}{16}\right)^{-3/4}$ **33.** $\left(-\dfrac{1}{32}\right)^{-3/5}$

Simplify each of the following. (Remember to write your answers with only positive exponents.) Assume that all of the variables represent positive real numbers.

34. $9^{2/5} \cdot 9^{1/10}$ **35.** $8^{7/12} \cdot 8^{-1/4}$ **36.** $(16^{1/8})^2$ **37.** $(25^{-1/4} \cdot 25^{2/3})^6$

38. $\dfrac{4^{1/3}}{4^{-1/6}}$ **39.** $\dfrac{64^{-1} \cdot 64^{7/15}}{64^{-1/5}}$ **40.** $(6^{-2/5})^{-10}$ **41.** $(8^{1/12} \cdot 25^{-1/8})^4$

42. $x^{2/5} \cdot x^{1/4}$ **43.** $(9x^{2/7})^{1/2}$ **44.** $(27x^6)^{-2/3}$ **45.** $\dfrac{x^{3/8}}{x^{2/3}}$

46. $\dfrac{4x^{-1/4}}{6x^{-4/5}}$ **47.** $\dfrac{(x^{-1} \cdot x^{3/4})^{-2/3}}{x^{-1/12}}$ **48.** $\dfrac{-8x^2y^{-1/4}}{2x^{-1/3}y^{-3/5}}$ **49.** $\dfrac{27^{4/9}x^{-1/3}y^{5/6}}{27^{1/9}x^{-2/7}y^{-3/4}}$

50. $\left(\dfrac{8x^{-2}}{125y^{1/5}}\right)^{-5/3}$ **51.** $\left(\dfrac{2x^{-3/7}}{8y^{-1/2}}\right)^{3/2}$ **52.** $\dfrac{(36x^4y^{-2/3})^{1/2}}{4x^{-1/4}y^{-1/6}}$ **53.** $\left(\dfrac{8x^2}{y^{-3}}\right)^{1/3}\left(\dfrac{9x^4}{y^{-1}}\right)^{-3/2}$

54. $x^{1/4}(5 - x^{1/2})$ **55.** $3x^{-1/6}(2x^{1/2} + x^{1/6})$ **56.** $5x^{2/3}(3x^{4/3} + 7x^{1/3} - x^{-2/3})$

57. $(x^{1/3} + 3)(x^{1/3} - 3)$ **58.** $(x^{-1/2} + 3)^2$ **59.** $(x^{1/3} + 3)(x^{2/3} - 3x^{1/3} + 9)$

60. $x^{m/3} \cdot x^{2m}$ **61.** $(x^{2m})^{3/2}$ **62.** $\dfrac{x^{m/2}}{x^{-2m/3}}$

63. $\dfrac{(x^{m/4} \cdot x^{-m})^3}{x^{m/2}}$ **64.** $(16x^{m/3}y^{-m})^{1/2}$ **65.** $(5^{1/3}x^{m/2})^{3/2}(5^{1/2}x^{m/4})^3$

66. $\dfrac{1}{2}\left(\dfrac{x}{x-1}\right)^{-1/2}\dfrac{-1}{(x-1)^2}$ **67.** $15(6x - 1)^{3/2}(x + 3)^{5/3} + \dfrac{5}{3}(6x - 1)^{5/2}(x + 3)^{2/3}$

68. $2(3x + 1)^{-1/3}(3 - 4x)^{1/2} - 2(3x + 1)^{2/3}(3 - 4x)^{-1/2}$

5.2 **Evaluate each of the following radical expressions, if possible.**

69. $\sqrt{441}$ **70.** $\sqrt{7^2}$ **71.** $\sqrt[4]{-9}$ **72.** $\sqrt[3]{125}$ **73.** $\sqrt[3]{-8}$

74. $\sqrt[3]{8^3}$

75. $\sqrt[4]{16}$

76. $\sqrt[4]{-1}$

77. $\sqrt[4]{9^4}$

78. $\sqrt[5]{-32}$

79. $\sqrt[5]{3^5}$

80. $\sqrt[5]{1}$

81. $\sqrt[3]{\dfrac{8}{27}}$

82. $\sqrt{\dfrac{1}{16}}$

83. $\sqrt{9} \cdot \sqrt{16}$

84. $\sqrt{4} \cdot \sqrt[3]{8}$

85. $\dfrac{\sqrt[3]{27}}{\sqrt{36}}$

86. $\sqrt[4]{\sqrt{256}}$

Convert the following expressions to radicals.

87. $6^{2/5}$

88. $a^{2/3}$

89. $(7x)^{3/4}$

90. $(-5ab^2)^{4/3}$

91. $(3x + 5)^{3/2}$

92. $(x^2 + y^2)^{1/2}$

Convert the following radicals to expressions with rational exponents and simplify where possible. Assume that the variables represent positive numbers.

93. $\sqrt{11}$

94. $\sqrt[3]{x}$

95. $\sqrt[4]{a^3}$

96. $\sqrt[3]{x^6}$

97. $\sqrt{m^7}$

98. $(\sqrt[4]{y})^5$

99. $\sqrt[4]{p^{12}}$

100. $\sqrt[5]{7x^3}$

101. $\sqrt{(3x^2y)^3}$

102. $\sqrt[3]{a^3 + b^3}$

103. $\sqrt[4]{(a + b)^4}$

104. $(\sqrt{a + b})^6$

105. $\sqrt[3]{y} \cdot \sqrt[4]{y^3}$

106. $\sqrt{3x} \cdot \sqrt{3x^5}$

107. $\dfrac{\sqrt[3]{x}}{\sqrt[6]{x}}$

108. $\dfrac{\sqrt[4]{x}}{\sqrt[4]{x^3}}$

109. $\sqrt[3]{\sqrt{x}}$

110. $\sqrt[4]{\sqrt[3]{x^2}}$

Simplify each of the following. Assume that the variables represent positive real numbers.

111. $\sqrt{16x^2y^4}$

112. $\sqrt{36x^6y^8}$

113. $\sqrt[3]{27x^3y^6}$

114. $\sqrt[4]{16x^8y^4}$

115. $\sqrt{x^2 + 2x + 1}$

116. $\sqrt{y^2 + 8y + 16}$

5.3   Simplify the following radicals. Assume that all variables represent positive numbers.

117. $\sqrt{6} \cdot \sqrt{5}$

118. $\sqrt{15} \cdot \sqrt{20}$

119. $\sqrt{7} \cdot \sqrt{7}$

120. $\sqrt[3]{2} \cdot \sqrt[3]{11}$

121. $\sqrt[3]{6} \cdot \sqrt[3]{18}$

122. $\sqrt[4]{8} \cdot \sqrt[4]{10}$

123. $\sqrt{75}$

124. $\sqrt{216}$

125. $\sqrt{99}$

126. $\sqrt[3]{56}$

127. $\sqrt[3]{81}$

128. $\sqrt[3]{-250}$

129. $\sqrt[4]{48}$

130. $\sqrt[4]{4x^6}$

131. $\sqrt[4]{2x^4y^9}$

132. $\sqrt{18x^9}$

133. $\sqrt{24x^4y^{12}}$

134. $\sqrt{25x^3y^6z^9}$

135. $\sqrt[3]{24x^4y^{12}}$

136. $\sqrt[3]{81x^2y^3z^8}$

137. $\dfrac{2}{\sqrt{x}}$

138. $\sqrt{\dfrac{3}{7}}$

139. $\dfrac{5}{\sqrt{3y}}$

140. $\dfrac{x}{\sqrt{4z}}$

141. $\sqrt{\dfrac{6x^5}{y^3}}$

142. $\sqrt{\dfrac{9x}{2y^4}}$

143. $\sqrt{\dfrac{5x^2}{28y^3}}$

144. $\dfrac{5}{\sqrt[3]{y}}$

145. $\dfrac{x}{\sqrt[3]{2}}$

146. $\dfrac{3x}{\sqrt[3]{y^2}}$

147. $\sqrt[3]{\dfrac{2}{5}}$

148. $\sqrt[3]{\dfrac{8}{3x}}$

149. $\sqrt[3]{\dfrac{7x^6}{40}}$

150. $\sqrt[3]{\dfrac{2x^8}{9y}}$

151. $\dfrac{3ax}{\sqrt[4]{y}}$

152. $\sqrt[4]{\dfrac{3x}{2y^2}}$

153. $\sqrt[4]{x^6}$

154. $\sqrt[4]{x^2y^8z^{14}}$

155. $\sqrt[6]{x^4y^9}$

5.4   Perform the indicated operations. Assume that all variables represent positive numbers.

156. $5\sqrt{3} + 8\sqrt{3} - 9\sqrt{3}$

157. $15\sqrt[3]{6} + \sqrt[3]{6} - 20\sqrt[3]{6}$

158. $8\sqrt{7x} - 11\sqrt{7x}$

159. $9x^2\sqrt{2y} + x^2\sqrt{2y}$

160. $3x\sqrt[3]{5x^2} - 7x\sqrt[3]{5x^2}$

161. $\sqrt{18} - \sqrt{98} + 5\sqrt{8}$

162. $3\sqrt{24} + \sqrt{36} - 8\sqrt{54}$

163. $\sqrt[3]{40} + 3\sqrt[3]{72}$

164. $\sqrt[4]{243} + 5\sqrt[4]{48}$

165. $\sqrt{16x^3} - x\sqrt{8x} - 2x\sqrt{9x}$

166. $\sqrt{27y^5} + y\sqrt{12y^3} - 6y\sqrt{75y}$

167. $4\sqrt[3]{2p^4} + p\sqrt[3]{54p} - \sqrt[3]{16p}$

168. $\sqrt{6x^2y} + 3\sqrt{xy^3} + x\sqrt{96y} - 5y\sqrt{49xy}$

169. $\sqrt{20x^5y^6} - 7xy\sqrt{45x^3y^4} + y^2\sqrt{40x^4y}$

170. $4y\sqrt[3]{81x^5y^9} + 3x\sqrt[3]{24x^2y^{12}}$

171. $7x\sqrt[4]{16x^9y^4} - 3y\sqrt[4]{x^{13}}$

172. $\dfrac{3}{\sqrt{2}} - 3\sqrt{2}$

173. $\dfrac{2}{\sqrt[3]{5}} - \sqrt[3]{200}$

174. $\sqrt{5} + \dfrac{8}{\sqrt{5}} - \sqrt{80}$

175. $\sqrt{\dfrac{2}{5}} + \sqrt{40} - \sqrt{\dfrac{5}{2}}$

176. $\sqrt{\dfrac{x}{3}} + 7\sqrt{3x} - \sqrt{\dfrac{3}{x}}$

177. $\sqrt[3]{\dfrac{x}{2}} - \sqrt[3]{108x}$

5.5   Perform the indicated multiplications and simplify your answers. Assume that all variables represent positive numbers.

178. $\sqrt{6}(\sqrt{3} + 2)$

179. $3\sqrt{2}(5 + \sqrt{10} - 3\sqrt{2})$

180. $\sqrt{3x}(\sqrt{x} - \sqrt{3} + x)$

181. $(\sqrt{3} + 5)(\sqrt{7} - 1)$

182. $(\sqrt{8} + 3)(5 - \sqrt{2})$

183. $(3\sqrt{5} - 2)(5\sqrt{5} + 6)$

184. $(4\sqrt{x} - 1)(\sqrt{x} + 3)$

185. $(3\sqrt{x} - \sqrt{y})(3\sqrt{x} + 4\sqrt{y})$

186. $(\sqrt{6} - \sqrt{2})^2$

187. $(2\sqrt{y} + 1)^2$

188. $(\sqrt{7} + \sqrt{3})(\sqrt{7} - \sqrt{3})$

189. $(\sqrt{x} - 2)(\sqrt{x} + 2)$

190. $(\sqrt[3]{5x} + 2)(\sqrt[3]{5x} + 3)$

191. $(\sqrt[3]{5} + 1)(\sqrt[3]{25} - \sqrt[3]{5} + 1)$

Rationalize the denominator of the following radical expressions. Assume that all variables represent positive numbers.

192. $\dfrac{3}{\sqrt{5} + 2}$

193. $\dfrac{\sqrt{3}}{\sqrt{7} - 1}$

194. $\dfrac{x}{\sqrt{x} + 3}$

195. $\dfrac{\sqrt{6}}{\sqrt{2} - \sqrt{3}}$

196. $\dfrac{\sqrt{10} + 2}{2 + \sqrt{2}}$

197. $\dfrac{\sqrt{3x} + \sqrt{y}}{\sqrt{5x} + \sqrt{2y}}$

198. $\dfrac{3\sqrt{7} + 1}{\sqrt{3} - 2\sqrt{5}}$

199. $\dfrac{4\sqrt{3x} - 5\sqrt{2y}}{3\sqrt{6x} - 6\sqrt{8y}}$

200. $\dfrac{5\sqrt{x} + 3\sqrt{2y}}{\sqrt{3} - 1}$

201. $\dfrac{4\sqrt{3x} - 7\sqrt{2y}}{\sqrt{5} - 2}$

5.6   Find the solution(s) of the following radical equations.

202. $\sqrt{3x + 1} = 4$

203. $\sqrt{4x + 5} - 3 = 0$

204. $\sqrt[3]{2x - 7} + 6 = 5$

205. $\sqrt[4]{x^2 - 6x} = 2$

206. $\sqrt{6x - 1} = \sqrt{x^2 + 4x}$

207. $\sqrt[3]{2 - x^2} = \sqrt[3]{3x - 2}$

208. $\sqrt{3x^2 - 5x - 3} = x$

209. $\sqrt{x^2 - 3x + 1} + 2 = x$

210. $\sqrt{6x + 19} = x + 2$

211. $\sqrt{13 - 3x} - 3 = 2x$

212. $\sqrt{x + 6} + \sqrt{x} = 3$

213. $\sqrt{x + 11} - \sqrt{2x - 1} = 1$

214. $\sqrt{3x + 7} + \sqrt{2 - x} = 3$

215. $\sqrt{5x - 4} - \sqrt{x + 5} = \sqrt{x - 3}$

216. $\sqrt{x - \sqrt{3x + 7}} = 1$

217. $\dfrac{x}{\sqrt{2x + 5}} = 2$

5.7 **Express the following in the form $a + bi$ and simplify.**

218. $-\sqrt{-16}$

219. $\sqrt{-25} + \sqrt{-81}$

220. $-\sqrt{-4} - \sqrt{-100}$

221. $\sqrt{-49} \cdot \sqrt{-36}$

222. $-\sqrt{-4} \cdot \sqrt{-64}$

223. $\dfrac{\sqrt{-36}}{\sqrt{-9}}$

224. $\dfrac{\sqrt{-144}}{\sqrt{-16}}$

225. $\sqrt{-18} + \sqrt{-8}$

226. $\sqrt{-12} \cdot \sqrt{-6}$

227. $\sqrt{-7} \cdot \sqrt{7}$

228. $\dfrac{\sqrt{-64} \cdot \sqrt{-9}}{\sqrt{-36}}$

Perform the indicated operations. Express all answers in the form $a + bi$.

229. $(5 + 7i) + (8 - 2i)$

230. $\left(\dfrac{1}{2} + 3i\right) + (2 + i)$

231. $(4 - 5i) - (1 + 7i)$

232. $\left(\dfrac{2}{3} - i\right) - \left(\dfrac{1}{6} - i\right)$

233. $6(5 + 8i)$

234. $3i(1 - 4i)$

235. $(5 + 2i)(7 + i)$

236. $(-2 + 6i)(4 + 9i)$

237. $(6 - 5i)(6 + 5i)$

238. $(2 + 7i)^2$

Perform the indicated divisions. Express all answers in the form $a + bi$.

239. $\dfrac{5 + 3i}{2i}$

240. $\dfrac{1 + 4i}{-3i}$

241. $\dfrac{4}{1 + 3i}$

242. $\dfrac{5i}{1 - 2i}$

243. $\dfrac{3 - i}{4 - i}$

244. $\dfrac{32 - 4i}{1 - 5i}$

Simplify the following.

245. i^{20}

246. i^{53}

247. i^{35}

248. $-5i^0$

249. i^{-2}

250. $3i^{-25}$

Chapter 5 Test

(You should be able to complete this test in 60 minutes.)

I. Simplify each of the following. (Remember to write your answer with only positive exponents.)

1. $9^{-1/2}$

2. $-8^{2/3}$

3. $(4x^6y^3)^{3/2} \cdot 3xy^2$

4. $\dfrac{(p^{1/3})^{-4} \cdot p^{5/3}}{p^{-2/3}}$

II. Simplify each of the following. Assume that all variables represent positive numbers.

5. $\sqrt{25^{-3}}$

6. $\sqrt{50x^3}$

7. $\sqrt{72x^2y} + 5x\sqrt{8y}$

8. $\sqrt[3]{24x^8y^5}$

9. $\dfrac{2}{\sqrt[3]{4x}}$

10. $\sqrt{\dfrac{a}{x^4y}}$

11. $\dfrac{\sqrt{2}}{\sqrt{6}+2}$

12. $\dfrac{\sqrt{60}}{\sqrt{15}}$

13. $(2\sqrt{x}-3)(\sqrt{x}+4)$

14. $(\sqrt{7}-2)^2$

III. Find the solution(s) of the following radical equations.

15. $\sqrt{y^2-3y-9}-3=y$

16. $\sqrt{2x-5}=\sqrt{3x}-2$

IV. Perform the indicated operations. Express all answers in the form $a+bi$.

17. $(3+7i)+(2-i)$

18. $\sqrt{-16}+\sqrt{-144}$

19. $\left(\dfrac{1}{2}-3i\right)-(2+5i)$

20. $\sqrt{-25}\cdot\sqrt{-4}$

21. i^{-5}

22. $\dfrac{\sqrt{-36}}{\sqrt{-9}}$

23. $(2+9i)(1-2i)$

24. $(4+3i)^2$

25. $\dfrac{4+12i}{2i}$

26. $\dfrac{34i}{1-4i}$

TEST YOUR MEMORY

These problems review Chapters 1–5.

I. Simplify each of the following.

1. $-\left(\dfrac{2}{3}\right)^2 + \dfrac{5}{8} \div \dfrac{1}{4} - \dfrac{5}{6}$

2. $2^{-1} + \left(\dfrac{2}{3}\right)^{-2}$

3. $25^{1/2} - 27^{-2/3}$

II. Simplify each of the following. (Remember to write your answer with only positive exponents.)

4. $(3xy^3)^{-2}(x^{-4}y^5)^3$

5. $\left(\dfrac{2x^{-3}y^5}{x^{-6}y^{-1}}\right)^2$

6. $(9x^4y^5)^{3/2} \cdot 2x^2y$

7. $\dfrac{(x^{-2/3})^2 x^{1/2} \cdot}{x^{-1}}$

III. Completely factor the following polynomials.

8. $p^4 - 16q^4$

9. $1 - 125t^3$

10. $x^2 + 2xy + y^2 - 25$

11. $(2x + y)^2 + 7(2x + y) + 12$

IV. Perform the indicated operations.

12. $(3x^2 - 5)(4x^2 + x - 2)$

13. $\dfrac{x^2 - 1}{8x^4 - 8x^3} \div \dfrac{x^2 + 2x + 1}{2x^3 - 8x^2 - 10x}$

14. $\dfrac{2x}{x - 4} + \dfrac{5}{x + 6}$

15. $\dfrac{2x + 3}{x^2 - 4x + 4} - \dfrac{x + 1}{x^2 + x - 6}$

V. Simplify the following fractions.

16. $\dfrac{\dfrac{2x}{x - 3} + 1}{\dfrac{4}{x - 3} + 2}$

17. $\dfrac{x^{-1} - y^{-1}}{x^{-4} - y^{-4}}$

VI. Perform the indicated division.

18. $\dfrac{3x^4 - 14x^3 - 3x^2 + 24x + 12}{x^2 - 4x - 2}$

VII. Simplify each of the following. Assume that all variables represent positive numbers.

19. $\sqrt{36^{-2}}$

20. $\sqrt{18x^3y^{16}}$

21. $x\sqrt{72y} + 2x\sqrt{32y} - 10\sqrt{2x^2y}$

22. $\sqrt[3]{24x} - 3\sqrt[3]{81x}$

23. $\sqrt{\dfrac{3x}{5y}}$

24. $\sqrt[3]{\dfrac{5x}{2y^2}}$

25. $(2\sqrt{x} - 7)^2$

26. $\dfrac{\sqrt{2}}{\sqrt{10} - 2}$

VIII. Perform the indicated operations. Express all answers in the form $a + bi$.

27. $\sqrt{-20} + 2\sqrt{-45}$

28. $\sqrt{-9} \cdot \sqrt{-25}$

29. $(5 + i)(2 - 4i)$

30. $\dfrac{8 - i}{2 + 3i}$

IX. Find the solutions of the following equations and check your solutions.

31. $\dfrac{4}{x-1} = \dfrac{5}{x+2}$

32. $\dfrac{1}{2x} + \dfrac{5}{3} = \dfrac{8}{x}$

33. $2x^2 - 5x - 12 = 0$

34. $4x(x+4) = 4x - 9$

35. $\sqrt{3x+7} = 4$

36. $\sqrt[3]{2x-1} = -2$

37. $|3x-4| = |2x-1|$

38. $\sqrt{4x+11} = 2x + 4$

39. $\dfrac{2x}{x+3} - \dfrac{1}{x-2} = \dfrac{-5}{x^2+x-6}$

40. $(x-2)(x+3) = -6$

41. $\sqrt{2x+3} - 1 = \sqrt{x+5}$

42. $\dfrac{x+3}{x+1} + \dfrac{2}{x-1} = \dfrac{2x-1}{x^2-1}$

43. $(x+8)(x+5) = (2x+13)(x+4)$

44. $|4-3x| = 7$

X. Use algebraic expressions to find the solutions of the following problems.

45. A rectangle has a length that is 1 ft less than 3 times the width. The perimeter of the rectangle is 94 ft. What are the dimensions of the rectangle?

46. One leg of a right triangle is 5 m less than twice the other leg. The hypotenuse is 5 m long. What are the lengths of the legs of the triangle?

47. Wilma has $4.30 in dimes and quarters. She has a total of 22 coins. How many dimes and how many quarters does she have?

48. A cannonball is fired from ground level vertically upward with a speed of 288 ft/sec. The equation that gives the cannonball's height above ground level is $h = -16t^2 + 288t$. When does the cannonball hit the ground?

49. Working together Fred and Barney can sweep out the Moss County Zoo in 3 hr. Working alone Fred can sweep out the zoo in 4 hr. How long does it take Barney to sweep out the zoo?

50. Colonel Newhouse flies his airplane 100 mi to the Moss County International Airport and picks up a load of peanuts. He then flies back home. The trip to the airport was with the wind, and the trip home was against the wind. The speed of the wind was 10 mi/hr. If Colonel Newhouse's total flying time was $1\frac{5}{6}$ hr, what is the speed of Colonel Newhouse's plane in still air?

Relations and Functions

A three-story building is on fire. When the firefighters arrive at the scene of the fire, it is determined that (for safety purposes) all firefighters must be at least 70 ft from the fire. Suppose that the fire engines can generate a nozzle pressure of only 48 lb/sq in. What is the minimum nozzle diameter that will enable a stream of water to reach the edge of the fire? At the end of this chapter we will apply the general formula firefighters use to determine the maximum horizontal distance of a stream of water.

6.1 *The Cartesian Coordinate System*

Every year in August, residents of the Gulf Coast area become increasingly concerned about the possibility of a hurricane striking there. As they listen to their radios and televisions, they usually hear the location of a tropical storm or hurricane given in terms of its *latitude* and *longitude*. The latitude of the hurricane indicates its distance north or south of the equator, and the longitude of the hurricane indicates its distance east or west of the prime meridian through Greenwich, England (see Figure 6.1.1). Notice that it takes two numbers, usually called coordinates, to determine the location of a hurricane. Also, the order in which the numbers are given is important. If a hurricane is located at latitude 28° N and 87° W, then it threatens the city of New Orleans, Louisiana, whose coordinates are latitude 30° N and longitude 90° W. But if the coordinates of the hurricane are switched to 87° N and 28° W, then it would be closer to Santa Claus at the North Pole.

In algebra we can locate points on a type of "map" in a similar fashion. We start with the basic building block, an **ordered pair.** An ordered pair is two elements enclosed in parentheses separated by a comma, in which the order of the elements is important—for example, $(-3, 2)$, $(2, -3)$, $(-1, \frac{1}{2})$, and (x, y). In an ordered pair the element on the left side is called the **first coordinate** or **abscissa.** The element on the right side is called the **second coordinate** or **ordinate.** In the ordered pair (x, y), x is the first coordinate and y is the second coordinate. The ordered pairs $(-3, 2)$ and $(2, -3)$ are different. They contain the same elements, but the order in which they are listed is different.

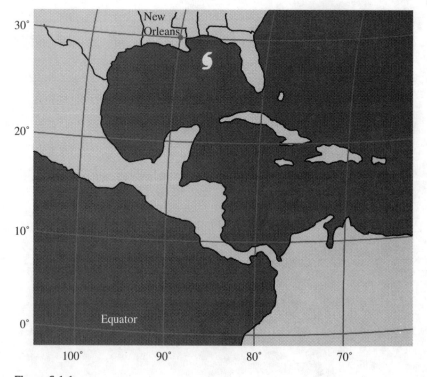

Figure 6.1.1

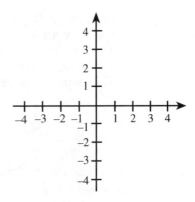

Figure 6.1.2

The next building block we need is a "map," which we call a **rectangular co-ordinate system** or the **Cartesian coordinate system,** named in honor of René Descartes (1596–1650). Legend has it that one afternoon Descartes was lying on his bed watching a fly crawl on the ceiling and he noticed that he could describe the fly's location in terms of its distance from two adjacent walls. This provided the motivation for Descartes to construct a coordinate system in the following manner (see Figure 6.1.2). He took a horizontal real number line and laid across it another real number line so that the two lines crossed at a right angle. In addition he positioned this now vertical number line so that its zero point was over the zero point on the horizontal line. The positive numbers went to the right of zero on the horizontal line; the negative numbers went to the left. On the vertical number line, the positive numbers went up from zero, while the negative numbers went down.

The point of intersection of these two lines is called the **origin.** To make things simpler, the horizontal number line is usually called the **x-axis** and the vertical number line is called the **y-axis.** With this system any point in the plane can be expressed in terms of its distances from the axes, which is given by an ordered pair of *real* numbers. In Figure 6.1.3 we can locate point *A* by starting at the origin and going 2 units to the right along the *x*-axis and 4 units up parallel to the *y*-axis. Mathematicians associate with point *A* the ordered pair (2, 4). We can locate point *B* by going 4 units along the *x*-axis and 2 units parallel to the *y*-axis. Notice that (2, 4)

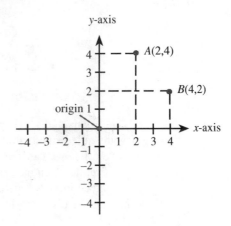

Figure 6.1.3

and (4, 2) are different ordered pairs and that A and B are different points. The horizontal movement along the x-axis is the first coordinate, and the movement parallel to the y-axis is the second coordinate. Therefore, we interchangeably use the terms first or **x-coordinate** and second or **y-coordinate.** When we draw a point associated with an ordered pair (x, y), we say we are **graphing,** or **plotting,** the ordered pair.

XAMPLE 6.1.1   Graph the following ordered pairs:

$$A = (1, 4) \qquad E = (-2, -4)$$
$$B = (0, 2) \qquad F = (0, -3)$$
$$C = (-3, 2) \qquad G = (3, -1)$$
$$D = (-4, 0) \qquad H = (5, 0)$$

See Figure 6.1.4.

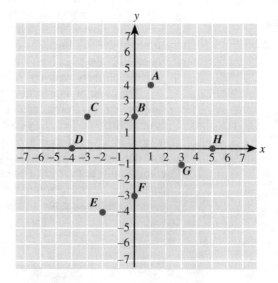

Figure 6.1.4

The x-axis and y-axis partition the plane into four distinct regions. These regions are called **quadrants** and are identified in Figure 6.1.5. From our work in Example 6.1.1, we can see that if a point is in quadrant I both of its coordinates are positive, if it is in quadrant II the first coordinate (x-coordinate) is negative and the second coordinate (y-coordinate) is positive, if it is in quadrant III both coordinates are negative, and if it is in quadrant IV the first is positive and the second negative.

To graph a set of ordered pairs, we graph all the ordered pairs in the set.

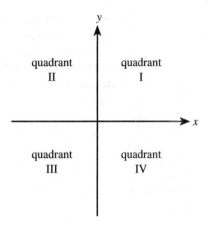

Figure 6.1.5

E X A M P L E   6 . 1 . 2   Graph the following set of ordered pairs:

$$\left\{(-1, 3), \left(\frac{2}{3}, 4\right), (0, 0), \left(-\frac{3}{2}, -4\right)(2.5, 0)\right\}$$

See Figure 6.1.6.

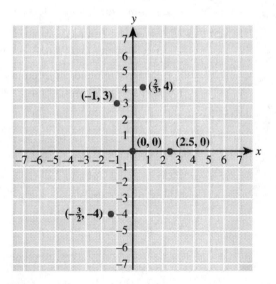

Figure 6.1.6

A **solution to an equation in two variables,** such as $y = 2x + 1$, is an ordered pair of real numbers with the property that when the x-coordinate is substituted for x and the y-coordinate is substituted for y in the equation, we obtain a true statement. For example, $(3, 7)$ satisfies the equation, since if we substitute the

x-coordinate 3 for x and the y-coordinate 7 for y, we get a true statement, $7 = 2(3) + 1$. We can generate as many ordered pair solutions as we wish by substituting numbers for one of the variables and solving the resulting equation for the other variable. The table here lists some ordered pairs that you should verify are solutions to $y = 2x + 1$.

x	y
0	1
1	3
2	5
-1	-1
-2	-3
-3	-5

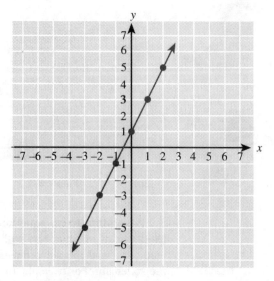

Figure 6.1.7(a)

If we graph these ordered pairs, we can get a "picture" of the solutions to this equation [see Figure 6.1.7(a)].

NOTE ▶ The graph of an ordered pair is a point. It is common practice to use the expressions "ordered pair" and "point" interchangeably, depending on the context.

If we plotted more points, we would soon have enough evidence to indicate that we could join the points with a straight line [see Figure 6.1.7(b)] and have a

Figure 6.1.7(b)

picture of all the ordered pairs that satisfy the equation. This picture is called the **graph of the equation.** The arrows at the end of the line indicate that it continues indefinitely in both directions.

XAMPLE 6.1.3

Graph the following equations.

1. $y = x^2$

Since this equation is solved for y, we can generate ordered pairs by substituting values for x and calculating the corresponding values of y. Verify that the ordered pairs in the table are indeed solutions to the equation $y = x^2$. By graphing these points and then connecting them with a smooth curve, we generate the curve shown in Figure 6.1.8.

x	y
-2	4
-1	1
$-\frac{1}{2}$	$\frac{1}{4}$
0	0
$\frac{1}{2}$	$\frac{1}{4}$
1	1
2	4

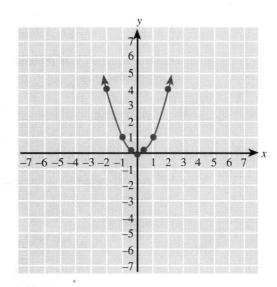

Figure 6.1.8

NOTE ▶ The graph of an equation of the form $y = x^2 + c$ is called a *parabola* and is ∪-shaped. The graph of an equation of the form $y = -x^2 + c$ is ∩-shaped.

2. $y = \sqrt{x - 1}$

Notice that if we substituted $x = 0$, then $y = \sqrt{0 - 1} = \sqrt{-1} = i$. Since y must be a real number, we must have $x - 1 \geq 0$ or $x \geq 1$. So the table does not contain values of x that are less than 1. The curve is shown in Figure 6.1.9.

NOTE ▶ The graph of an equation of the form $y = \sqrt{x + c}$ is ⌐-shaped.

x	y
1	0
$\frac{5}{4}$	$\frac{1}{2}$
2	1
5	2

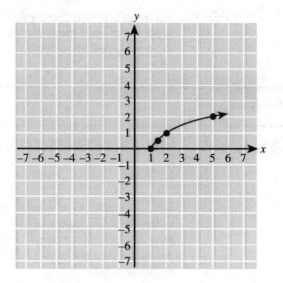

Figure 6.1.9

3. $y^2 = 3 - x$

For this equation it is probably easier to substitute values for y and calculate the corresponding values for x, since the equation can be solved for x:

$$y^2 = 3 - x$$
$$x + y^2 = 3$$
$$x = 3 - y^2$$

Fill in those values for x that are missing in the table. The curve is shown in Figure 6.1.10.

x	y
-1	2
2	1
$2\frac{3}{4}$	$\frac{1}{2}$
3	0
$2\frac{3}{4}$	$-\frac{1}{2}$
2	-1
	-2
	-3

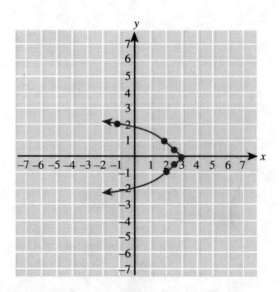

Figure 6.1.10

NOTE ▶   The graph of an equation of the form $x = y^2 + c$ is called a *parabola* and is C-shaped. The graph of an equation of the form $x = -y^2 + c$ is ⊃-shaped.

Exercises 6.1

1. Find the ordered pairs associated with the given points in Figure 6.1.11.

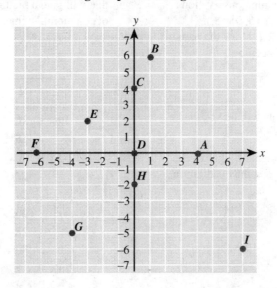

Figure 6.1.11

2. Graph the following set of ordered pairs.

$\{A(-3, 4), B(2, -1.5), C\left(\sqrt{3}, \frac{2}{3}\right), D(0, -3), E\left(8, \frac{1}{4}\right), F(5, 0), G(-2, -0.1), H(\pi, -1), I(-2.4, -6), \text{ and } J(0, 0)\}.$

Graph the following equations.

3. $y = 2x - 3$ **4.** $y = -x + 4$ **5.** $x + 2y = 4$ **6.** $3x + y = 6$

7. $\frac{1}{2}x + 2 = y$ **8.** $x = \frac{2}{3}y - \frac{1}{3}$ **9.** $2x + 3y = 0$ **10.** $5x - 2y = 0$

11. $y = x^2 - 5$ **12.** $y = x^2 + 1$ **13.** $y = 9 - x^2$ **14.** $y = 4 - x^2$

15. $x = 1 - y^2$ **16.** $x = 5 - y^2$ **17.** $y = \sqrt{x - 4}$ **18.** $y = \sqrt{x + 2}$

19. $x = y^2 + 1$ **20.** $x = y^2 - 3$

21. Mr. Descartes (a man with extremely accurate vision) made the following observations about the fly on his ceiling. At 1:00 P.M. the fly landed on a spot 2 ft from the western wall and 1 ft from the southern wall. Five seconds later the fly moved to (2, 3), where the first coordinate indicates the distance from the western wall measured in feet and the second coordinate indicates the distance from the southern wall measured in feet. The following set indicates Mr. Descartes's remaining observations taken every 5 sec: {(3, 2), (4, 1), (6, 0), (8, 1), (9, 2), (8, 3), (6, 4), (4, 3), (3, 2)}. Draw the path the fly followed on Descartes's ceiling.

22. Draw the first quadrant of a rectangular coordinate system, labeling the horizontal number line as the t-axis (for time) and the vertical number line as the p-axis (for the price of a share of stock). The table below contains ordered pairs, each of which gives a date and the closing price on that date of a share of stock in the Stop-er-up Company (which manufactures an antidiarrhea drug). Graph the ordered pairs and connect the points with a curve that will give a picture of the behavior of this stock over a month's time.

Write Algebra

23. Explain how to tell if a parabola will be ∪-, ∩-, ⊂-, or ⊃-shaped.

24. Explain the relationships between the signs of the coordinates in an ordered pair and the quadrant in which the point is located.

	t	p	
June	16	20	
	18	19	
	20	22	
	22	24	(A stomach virus reaches
	24	35	epidemic proportions
	26	39	across the country.)
	28	40	
	30	34	
July	2	35	
	4	35	
	6	33	(A Canadian study reports
	8	8	that after being fed
	10	7	Stop-er-up for 2 wk,
	12	8	laboratory mice exploded.)

Exercises 25–29 require the use of a graphing calculator.

25. Graph $y = x^3$. Next graph $y = x^3 + 2$ and $y = x^3 + 4$. Predict the graph of $y = x^3 + 6$. Graph $y = x^3 + 6$. Describe the pattern you see in the graphs.

26. Graph $y = x^3$. Next graph $y = x^3 - 3$ and $y = x^3 - 5$. Predict the graph of $y = x^3 - 7$. Graph $y = x^3 - 7$. Describe the pattern you see in the graphs.

27. Graph $y = x^3$. Next graph $y = (x - 2)^3$ and $y = (x - 4)^3$. Predict the graph of $y = (x - 6)^3$. Graph $y = (x - 6)^3$. Describe the pattern you see in the graphs.

28. Graph $y = x^3$. Next graph $y = (x + 2)^3$ and $y = (x + 4)^3$. Predict the graph of $y = (x + 6)^3$. Graph $y = (x + 6)^3$. Describe the pattern you see in the graphs.

29. Graph $y = x^3$. Next graph $y = (\frac{x}{2})^3$ and $y = (\frac{x}{3})^3$. Predict the graph of $y = (\frac{x}{4})^3$. Graph $y = (\frac{x}{4})^3$. Describe the pattern you see in the graphs.

GROUP   ACTIVITY

In exercises 30–34, make a table of ordered pairs for each equation using $x = -4, -3, -2, -1, 0, 1,$ $2, 3, 4$. Use these ordered pairs to graph the equations. Explain why each pair of graphs coincide for some points but not for others.

30. $y = x$
$\quad y = \sqrt{x^2}$

31. $y = 2x$
$\quad y = x + \sqrt{x^2}$

32. $y = -2x$
$\quad y = -x - \sqrt{x^2}$

33. $y = x + 1$
$\quad y = \sqrt{x^2 + 2x + 1}$

34. $y = x - 1$
$\quad y = \sqrt{x^2 - 2x + 1}$

6.2 *Relations and Functions*

Suppose we had two groups of people and we wanted to describe a relationship between a person in the first group (group X) and a person in the second group (group Y). We could do this in two ways. First, we could give a rule or formula that tells us how to link two people together. For example, we could say the mother of a person in group X is a certain person in group Y. Second, we could write a set listing pairs of people that are linked together by this relationship. For example, {(Jimmy Carter, Lillian Carter), (Prince Charles, Queen Elizabeth), (Liza Minelli, Judy Garland), . . .} is a set of pairs of people who satisfy the condition that the mother of the first person is the second. Another example might be the son of, with ordered pairs like {(Queen Elizabeth, Prince Charles), (Bing Crosby, Gary Crosby), (Bing Crosby, Nathaniel Crosby), . . .}.

In mathematics a set of ordered pairs is called a **relation.** The **domain** of the relation is the set of all first coordinates, and the **range** is the set of all second coordinates. In the first example the domain is the set of all people and the range is the set of all mothers. In the second example the domain is the set of all parents with sons, and the range is the set of all sons.

In algebra we also have two ways of describing a correspondence between two sets of *real numbers*. Consider a set of natural numbers, $X = \{1, 2, 3, \ldots, 10\}$ and a set of perfect squares, $Y = \{1, 4, 9, \ldots, 100\}$. To describe a correspondence that takes a number from X and links it with its square from Y, either we could use a rule or equation such as $y = x^2$, where x is an element of X, or we could make a roster of the ordered pairs that satisfy this equation: $\{(1, 1), (2, 4), (3, 9), (4, 16), \ldots,$ $(10, 100)\}$. This relation, call it S, can also be written as a set using the equation $y = x^2$, which determines the ordered pairs that belong to it:

$$S = \{(x, y): y = x^2, x \in X\}$$

This equation is read "S equals the set of all ordered pairs (x, y) such that $y = x^2$, where x is an element of X." Set X is the domain of S and set Y is the range.

EXAMPLE 6.2.1   Given $T = \{(x, y): y = 3x + 1, x = -2, -1, 0, 6\}$, write T as a set of ordered pairs.

SOLUTION

■ If $x = -2$, then $y = 3(-2) + 1 = -5$, so we get the ordered pair $(-2, -5)$.

■ If $x = -1$, then $y = 3(-1) + 1 = -2$, so we get the ordered pair $(-1, -2)$.

■ If $x = 0$, then $y = 3(0) + 1 = 1$, so we get the ordered pair $(0, 1)$.

■ If $x = 6$, then $y = 3(6) + 1 = 19$, so we get the ordered pair $(6, 19)$.

Now we can rewrite T in roster form:

$$T = \{(-2, -5), (-1, -2), (0, 1), (6, 19)\}$$

Note that the domain of T is $\{-2, -1, 0, 6\}$, and the range of T is $\{-5, -2, 1, 19\}$.

We are now in a position to define a function.

Definition of Function

> A **function** consists of two sets of real numbers called the domain and range, and a rule or formula that assigns to each element in the domain one and only one element in the range.

Alternative Definition

> A **function** is a relation such that no two distinct ordered pairs have the same first coordinate.

This is a most important definition, so let's clarify it by returning to our two groups of people and the two relations we gave them. In the first example, when we picked a person from the first group, there was *only one* person from the second group who was the first person's mother. So the mother of is an example of a functional relationship. But in the second example, when we picked a person in the first group, such as Bing Crosby, there was more than one person from the second group that could be named as his son—namely, Gary Crosby and Nathaniel Crosby, to list just two. So the son of is not an example of a functional relationship.

The next relation, S, that we considered is a function, since given any number from X, there is only one number from Y that is its square. Is T a function? Why or why not?

XAMPLE 6.2.2   Find the domain and range of $F = \{(x, y): y = x^2\}$ and determine whether F is a function.

SOLUTION F resembles S except that nothing is stated about its domain or range. Whenever a relation is stated in this way, *the domain will always be the largest subset of the real numbers that can be substituted for x into the equation and yield a real solution for y.* So the domain of F is {all real numbers}, since any value for x, when it is squared, will yield a real value for y. Similarly *the range will always be the largest subset of the real numbers that can be substituted for y in the equation and yield a real solution for x.* So the range of F is $\{y: y \geq 0\}$, since if y were negative, such as $y = -4$, then $x^2 = -4$ has the imaginary solutions $\pm 2i$. F is a function for the same reason that S was a function, since given any real number for x, there is only one number that is its square.

The equation for F is the same as that of Example 6.1.3, part 1. There is an infinite number of ordered pairs that satisfy this equation, so we cannot write a set listing all the ordered pairs that belong to this function. But we can draw a "picture" of this set, as in Figure 6.2.1. This picture is now called the **graph of the function.** It gives us some information about the function F. The domain of F is measured by the extent to which the graph extends to the left and right along the x-axis. The range of F is measured by the extent to which the graph extends up and down along the y-axis.

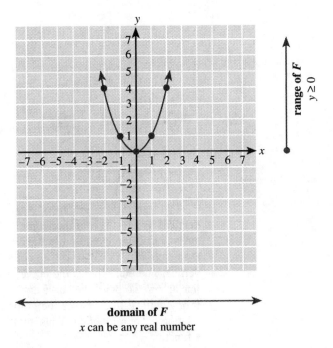

domain of F
x can be any real number

Figure 6.2.1

EXAMPLE 6.2.3   Find the domain and range of $G = \{(x, y): y^2 = 3 - x\}$ and determine whether or not G is a function.

SOLUTION

We saw the equation of G in Example 6.1.3, part 3, and we have reproduced its graph in Figure 6.2.2. From the graph we can see that the curve extends only to the left of $x = 3$. Indeed if x were bigger than 3—say, $x = 4$—then $y^2 = 3 - 4 = -1$ and so $y = \pm i$. Thus the domain of G is $\{x: x \leq 3\}$. The range of G is {all real numbers}, since any number can be substituted for y. G is *not* a function, since for the one value $x = 2$, there are two values for y, 1 and -1. *If any element of the domain is assigned to more than one element of the range, the relation is not a function.*

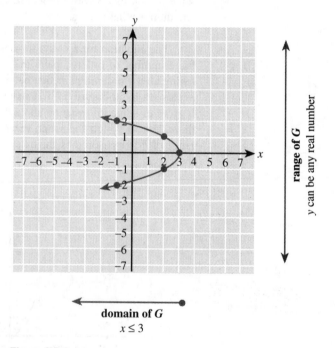

domain of *G*
$x \leq 3$

Figure 6.2.2

We can use the graphs of F and G in Figures 6.2.3 and 6.2.4 to motivate **the vertical line test.** From the graph of F (recall that F is a function) we can see that any vertical line would cross the curve at only one point. On the other hand, from the graph of G (recall that G is not a function) we can see that there is at least one vertical line that crosses the curve at more than one point. For example, a vertical line at $x = 2$ crosses the curve at $(2, 1)$ and $(2, -1)$. The *vertical line test* states that if *any* vertical line crosses the graph of a given relation more than once, then that relation is not a function. Thus, by the vertical line test F is a function and G is not a function.

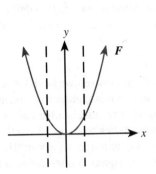

Figure 6.2.3

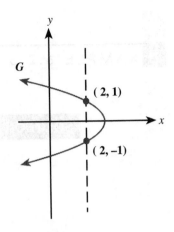

Figure 6.2.4

XAMPLE 6.2.4   Using the vertical line test, determine which of the following is the graph of a function.

a.

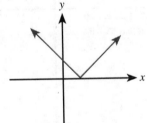

b.

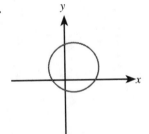

SOLUTIONS

a.

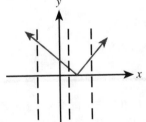

b.

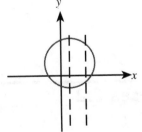

A *function* because no vertical line crosses the curve more than once

Not a function, since there is at least one vertical line that crosses the curve more than once

EXAMPLE 6.2.5

Find the domain and range of the relation defined by $y = \sqrt{x - 1}$ and use the vertical line test to determine whether or not it is a function.

SOLUTION

Here, as we will do from now on, we have given only the equation and omitted the set notation. The graph of this equation is given in Figure 6.2.5. For y to be a real number the radicand must be nonnegative, so $x - 1 \geq 0$ or $x \geq 1$. Thus the domain is $\{x: x \geq 1\}$. Since the radical sign stands for the nonnegative square root of $x - 1$, y cannot be negative. So the range is $\{y: y \geq 0\}$. Since any vertical line would cross the curve only once, this equation does define a function.

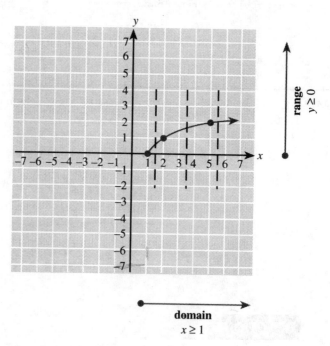

Figure 6.2.5

Exercises 6.2

Write the following sets as sets of ordered pairs and identify the domain and range.

1. $V = \{(x, y): y = -2x + 7, x = -5, 0, \frac{3}{2}, 2.1\}$

2. $W = \{(x, y): 3x + 2y = 6, x = -\frac{5}{2}, 0, 2, \frac{1}{4}\}$

3. $X = \{(x, y): y = \sqrt{2x - 1}, x = \frac{1}{2}, 1, 3, 5\}$

4. $Y = \{(x, y): y = \sqrt[3]{x + 1}, x = -9, 0, 26\}$

5. $Z = \{(x, y): y = |x| + 3, x = -2, 0, 2\}$

6. $B = \{(x, y): y = |3x - 4|, x = -4, 1, \frac{5}{2}\}$

For the relations defined by the following equations, find the domain and range and determine whether each relation is a function.

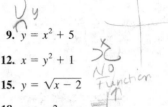

7. $y = 3x - 4$ $D: -\infty, \infty$

8. $2x - y = 5$ $D: -\infty, \infty$

9. $y = x^2 + 5$

10. $y = 3 - x^2$

11. $x = 2 - y^2$

12. $x = y^2 + 1$ NO function

13. $y = \sqrt{x + 3}$

14. $y = \sqrt{x - 4}$

15. $y = \sqrt{x - 2}$

16. $y = \sqrt{x + 5}$

17. $y = 4$

18. $y = -3$

Determine the domain and range of each relation whose graph is given.

19.

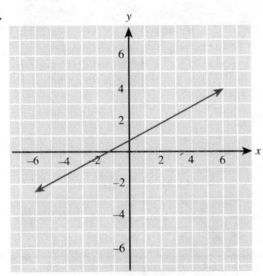

20.

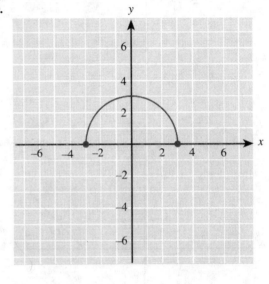

21.

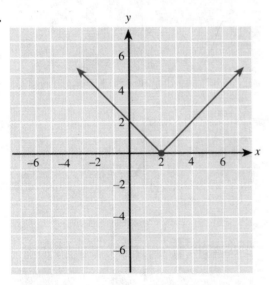

22.

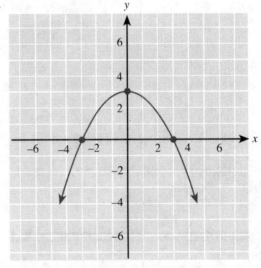

HW go over

23.

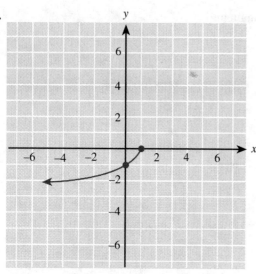

24.

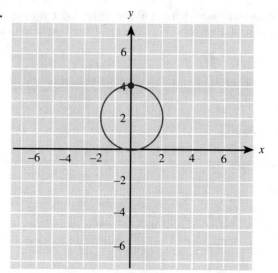

25.

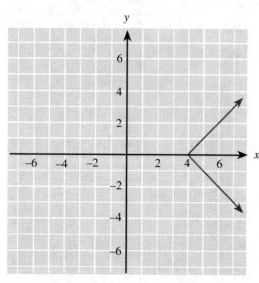

26.

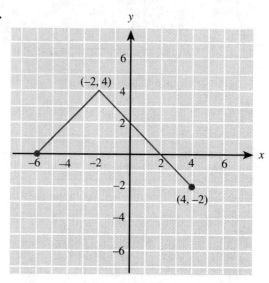

27.

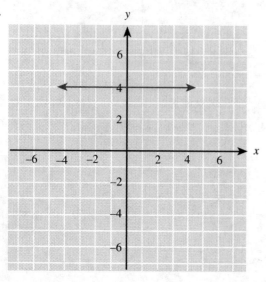

28.

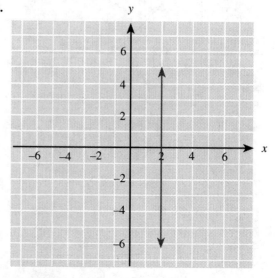

Using the vertical line test, determine which of the following are graphs of functions.

29.

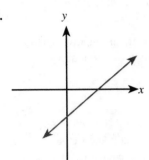

30.

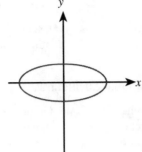

31.

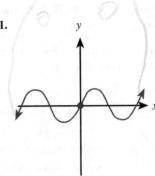

32.

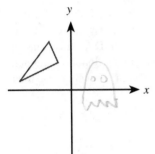

33.

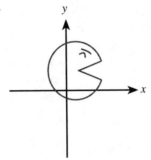

34.

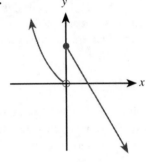

35.

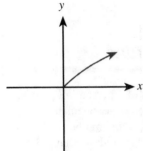

36.

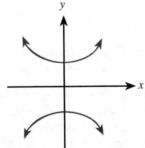

37.

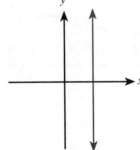

38. Consider the set of all ordered pairs where the first coordinate is a president of the United States and the second coordinate is a vice president who served with the president. Is this relation a function?

39. Listed in the table are the earnings from 1978 to 1983 for Jim Pluto, a mathematics instructor at Moss County Community College.

Year	1978	1979	1980	1981	1982	1983
Income	15,620	16,550	18,105	18,782	18,105	20,750

40. In 1982 the Moss County Swamp Commission found 47 toads in Bogg's Creek. Every year thereafter the toad population increased by 7.
  a. Find a function that will determine the number of toads present in any subsequent year.
  b. How many toads will be present in Bogg's Creek in 1992, 1995, 2001, and 2525?

a. Construct a set of ordered pairs where the first coordinate is Pluto's income for that year. Is this set a function?
b. Construct a set of ordered pairs where the first coordinate is the year in which he earned it. Is this set a function?

Write Algebra

41. In your own words, state the two definitions of a function.

42. Define the domain and range of a relation.

43. Explain how we can find the domain and range of a relation from its graph.

44. Explain how to use the vertical line test to determine if a graph is the graph of a function.

⊞ **The following exercises require the use of a graphing calculator.**

45. Graph $y = |x|$. Does this equation define a function? What are the domain and range of this relation? Express your answers using interval notation.

46. Graph $y = |x - 3|$. Does this equation define a function? What are the domain and range of this relation? Express your answers using interval notation.

47. Graph $y = |x| + 3$. Does this equation define a function? What are the domain and range of this relation? Express your answers using interval notation.

48. Compare the graphs of exercises 46 and 47 with exercise 45. Predict the graph of $y = |x - 1| + 2$. Now graph $y = |x - 1| + 2$. Does this equation define a function? What are the domain and range of this relation? Express your answers using interval notation.

49. Graph $y = |x^2 - 4|$. Does this equation define a function? What are the domain and range of this relation? Express your answers using interval notation.

6.3 *Function Notation and Combinations of Functions*

Let's now return to the functional relation, the mother of a person in group X, given in Section 6.2. We could use special notation to shorten a statement such as the mother of Liza Minnelli is Judy Garland. Let's replace "the mother of" by $M(\ \ )$. Then the statement would look like M(Liza Minnelli) = Judy Garland. Each time we choose a person x from the first group, $M(x)$ represents that person's mother from the second group.

In algebra, whenever we are dealing with a function, it is common to use this notation, which is called **functional notation.** In the last example of Section 6.2 we had a function defined by the equation $y = \sqrt{x - 1}$. We can use functional notation by choosing a letter—say, f (f, g, h, and k are common choices)—and *replacing* y *by* $f(x)$. This symbol $f(x)$ is read "f of x" and is used to stand for the second coordinate in an ordered pair when x is the first coordinate. The equation that now defines our function is $f(x) = \sqrt{x - 1}$. We still get the same function, the same set of ordered pairs, regardless of which equation we use. The $f(x)$ notation simply stresses the idea that we are dealing with a function and that the second coordinate depends upon the first coordinate. In our example $f(2)$ is the second coordinate when the first coordinate is 2; thus $f(2) = \sqrt{2 - 1} = \sqrt{1} = 1$. $f(5) = \sqrt{5 - 1} = \sqrt{4} = 2$ is the second coordinate when the first coordinate is 5.

Sometimes we think of the equation that defines f as a function machine (see Figure 6.3.1). We choose a value for x, such as $x = 5$, as the input, then turn the

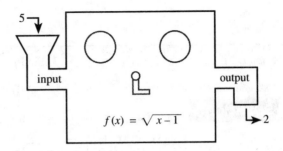

Figure 6.3.1

crank, and the machine gives us the output $y = 2$. Because f is a function, the machine gives only one output for each input. Find the output when the input is

$$10, \quad 3, \quad a, \quad a + 4, \quad a + h, \quad -2$$

Answers: $\quad 3, \quad \sqrt{2}, \quad \sqrt{a - 1}, \quad \sqrt{a + 3}, \quad \sqrt{a + h - 1}, \quad$ Tilt

The idea of a function machine may seem silly, but you may actually own one. If you have a handheld calculator with keys like $\boxed{\sqrt{x}}$, $\boxed{x^2}$, or $\boxed{1/x}$, you have a function machine. When you punch in a number, like 6, that is the input. Then if you press the name of a function, like $\boxed{x^2}$, that is "turning the crank." The display will then read a number, 36, and that is the output.

XAMPLE 6.3.1   Let $f(x) = 2x - 7$. Find $f(-2)$, $f(0)$, $f(\frac{3}{2})$, $f(a)$, $f(a + h)$, and $\dfrac{f(a + h) - f(a)}{h}$.

SOLUTIONS

$$f(-2) = 2(-2) - 7 = -4 - 7 = -11$$

$$f(0) = 2(0) - 7 = 0 - 7 = -7$$

$$f\left(\frac{3}{2}\right) = 2\left(\frac{3}{2}\right) - 7 = 3 - 7 = -4$$

$$f(a) = 2(a) - 7 = 2a - 7$$

$$f(a + h) = 2(a + h) - 7 = 2a + 2h - 7$$

$\dfrac{f(a + h) - f(a)}{h}$ *This expression is called the difference quotient and has special significance in calculus.*

$$= \frac{(2a + 2h - 7) - (2a - 7)}{h} \quad \text{[From f(a) and f(a + h) above]}$$

$$= \frac{2a + 2h - 7 - 2a + 7}{h}$$

$$= \frac{2h}{h}$$

$$= 2$$

XAMPLE 6.3.2   Let $g(x) = x^2 + x - 3$. Find $g(-1)$, $g(5)$, $g(2x)$, $g(a)$, $g(a + h)$, and $\frac{g(a + h) - g(a)}{h}$.

SOLUTIONS

$g(-1) = (-1)^2 + (-1) - 3 = 1 - 1 - 3 = -3$

$g(5) = (5)^2 + (5) - 3 = 25 + 5 - 3 = 27$

$g(2x) = (2x)^2 + (2x) - 3 = 4x^2 + 2x - 3$

$g(a) = (a)^2 + (a) - 3 = a^2 + a - 3$

$g(a + h) = (a + h)^2 + (a + h) - 3$

$\qquad = a^2 + 2ah + h^2 + a + h - 3$

$\dfrac{g(a + h) - g(a)}{h} = \dfrac{(a^2 + 2ah + h^2 + a + h - 3) - (a^2 + a - 3)}{h}$ *[From* g(a + h) *and* g(a) *above]*

$\qquad = \dfrac{a^2 + 2ah + h^2 + a + h - 3 - a^2 - a + 3}{h}$

$\qquad = \dfrac{2ah + h^2 + h}{h}$

$\qquad = 2a + h + 1$

Try to avoid these mistakes:

Incorrect	Correct
If $g(x) = x^2 + x - 3$, then $g(2x) = (x^2 + x - 3)(2x)$ $\qquad = 2x^3 + 2x^2 - 6x$ $g(a + h) = g(a) + h$ $\qquad = a^2 + a - 3 + h$	If $g(x) = x^2 + x - 3$, then $g(2x) = 4x^2 + 2x - 3$ (from Example 6.3.2) $g(a + h) = a^2 + 2ah + h^2$ $\qquad\qquad + a + h - 3$ (from Example 6.3.2)

The following table shows us how any two functions (let's call them f and g) can be combined using the elementary operations of addition, subtraction, multiplication, and division to generate new functions.

NOTE ▶   The following combinations apply only for values of x in both the domain of f and the domain of g.

Combinations of Functions

Sum: $(f + g)(x) = f(x) + g(x)$

Difference: $(f - g)(x) = f(x) - g(x)$

Product: $(f \cdot g)(x) = f(x) \cdot g(x)$

Quotient: $\left(\dfrac{f}{g}\right)(x) = \dfrac{f(x)}{g(x)}, \qquad g(x) \neq 0$

REMARK   In the quotient function $\left(\frac{f}{g}\right)(x)$, we have the restriction $g(x) \neq 0$. Why?

The next example illustrates how we can find combinations of functions.

XAMPLE 6.3.3   Let $f(x) = x^2 + 3x - 1$ and $g(x) = x - 4$. Find $(f + g)(x)$, $(f - g)(x)$, $(f \cdot g)(x)$, $\left(\frac{f}{g}\right)(x)$, $(f + g)(3)$ and $f(g(3))$.

SOLUTIONS

$$
\begin{aligned}
(f + g)(x) &= f(x) + g(x) \\
&= (x^2 + 3x - 1) + (x - 4) \\
&= x^2 + 3x - 1 + x - 4 \\
&= x^2 + 4x - 5
\end{aligned}
$$

$$
\begin{aligned}
(f - g)(x) &= f(x) - g(x) \\
&= (x^2 + 3x - 1) - (x - 4) \\
&= x^2 + 3x - 1 - x + 4 \\
&= x^2 + 2x + 3
\end{aligned}
$$

$$
\begin{aligned}
(f \cdot g)(x) &= f(x) \cdot g(x) \\
&= (x^2 + 3x - 1) \cdot (x - 4) \\
&= x^3 - 4x^2 + 3x^2 - 12x - x + 4 \\
&= x^3 - x^2 - 13x + 4
\end{aligned}
$$

$$
\begin{aligned}
\left(\frac{f}{g}\right)(x) &= \frac{f(x)}{g(x)} \\
&= \frac{x^2 + 3x - 1}{x - 4}, \qquad x \neq 4
\end{aligned}
$$

$$
(f + g)(3) = \,?
$$

We can work this problem two ways:

Method I	Method II
$(f + g)(3) = f(3) + g(3)$	$(f + g)(x) = x^2 + 4x - 5$
$\qquad = 17 + (-1)$ *Why?*	from above, so
$\qquad = 16$	$(f + g)(3) = 3^2 + 4 \cdot 3 - 5$
	$\qquad = 9 + 12 - 5$
	$\qquad = 16$

NOTE ▶   Using either method, we find $(f + g)(3) = 16$.

$$f(g(3)) = f(-1) \qquad \textit{From above, note that g(3) = -1.}$$
$$= (-1)^2 + 3(-1) - 1$$
$$= 1 - 3 - 1$$
$$= -3$$

Finally, let's consider an example in which we have some unusual combinations of functions.

EXAMPLE 6.3.4   Let $h(x) = 2x^2 - 1$ and $k(x) = 3x + 5$. Find $h(3)$, $5h(3)$, $[h(3)]^2$, $k(7)$, and $6k(7) + 5h(3)$.

SOLUTION

$$h(3) = 2(3)^2 - 1 = 2 \cdot 9 - 1 = 18 - 1 = 17$$
$$5h(3) = 5 \cdot 17 = 85$$
$$[h(3)]^2 = [17]^2 = 289$$
$$k(7) = 3(7) + 5 = 21 + 5 = 26$$
$$6k(7) + 5h(3) = 6 \cdot 26 + 5 \cdot 17$$
$$= 156 + 85$$
$$= 241$$

Exercises 6.3

For the given functions, find the following values: $f(-2), f(-\frac{1}{2}), f(0), f(\frac{1}{3}), f(1),$
$f(a),$ and $f(a + h).$

1. $f(x) = 2x - 3$

2. $f(x) = 3x + 1$

3. $f(x) = x^2 - 2$

4. $f(x) = 4 - x^2$

5. $f(x) = 2x^3$

6. $f(x) = x^3 - 1$

7. $f(x) = -2$

8. $f(x) = 4$

In exercises 9–16, find $(f + g)(x)$, $(f - g)(x)$, $(f \cdot g)(x)$, and $(\frac{f}{g})(x)$.

9. $f(x) = 2x^2 + 5x - 1$, $g(x) = x - 2$

10. $f(x) = 3x^2 + x + 4$, $g(x) = x - 1$

11. $f(x) = x^2 + 5$, $g(x) = x^2 - 9$

12. $f(x) = x^2 - 7$, $g(x) = x^2 - 16$

13. $f(x) = 2x + 3$, $g(x) = x - 11$

14. $f(x) = 5x + 1$, $g(x) = x - 6$

15. $f(x) = x^3 + 3x - 5$, $g(x) = 2x + 1$

16. $f(x) = 2x^3 - x + 2$, $g(x) = 3x - 4$

In exercises 17–54, let $f(x) = 3x - 1$, $g(x) = 3x^2 + 5x - 1$, $m(x) = x^2 - 4$, and $n(x) = 2x + 1$. Find the following:

17. $f(-2)$

18. $g(3)$

19. $m(-3)$

20. $3m(-3) - 7f(-2)$

21. $5f(-2) - 4g(3)$

22. $g(3) \cdot f(-2)$

23. $[f(-2)]^3$

24. $[m(-3)]^4$

25. $\left(\dfrac{m(-3)}{g(3)}\right)^2$

26. $(f + g)(2)$

27. $(m \cdot n)(-1)$

28. $\left(\dfrac{m}{n}\right)(6)$

29. $(g - f)(0)$

30. $f(x) + g(x)$

31. $g(x) - m(x)$

32. $n(x) \cdot m(x)$

33. $\dfrac{m(x)}{n(x)}$

34. $f(2x)$

35. $g(2x)$

36. $m(3x)$

37. $n(3x)$

38. $f(a)$

39. $f(a) + f(h)$

40. $f(a + h)$

41. $\dfrac{f(a + h) - f(a)}{h}$

42. $n(a)$

43. $n(a) + n(h)$

44. $n(a + h)$

45. $\dfrac{n(a + h) - n(a)}{h}$

46. $g(a)$

47. $g(a) + g(h)$

48. $g(a + h)$

49. $\dfrac{g(a + h) - g(a)}{h}$

50. $g(f(-2))$

51. $f(g(3))$

52. $g(f(3))$

53. $f(m(-3))$

54. $m(f(-3))$

Write Algebra

55. Explain why $f(x + h) \neq f(x) + h$.

56. Explain the difference between $f(2x)$ and $2f(x)$.

G R O U P   A C T I V I T Y

In exercises 57–64, find the indicated function values. Assume that *a, b, c,* and *d* are constants.

57. If $f(x) = ax^6 + bx^4 + cx^2 + 9$
and $f(3) = 14$, find $f(-3)$.

58. If $f(x) = ax^6 + bx^4 + cx^2 + 4$
and $f(-7) = 8$, find $f(7)$.

59. If $f(x) = ax^6 + bx^4 + cx^2 + d$
and $f(-3) = 6$, find $f(3)$.

60. If $f(x) = ax^6 + bx^4 + cx^2 + d$
and $f(4) = 10$, find $f(-4)$.

61. If $f(x) = ax^5 + bx^3 + cx$
and $f(-2) = 7$, find $f(2)$.

62. If $f(x) = ax^5 + bx^3 + cx$
and $f(-8) = 13$, find $f(8)$.

63. If $f(x) = ax^5 + bx^3 + cx + 8$
and $f(-3) = 6$, find $f(3)$.

64. If $f(x) = ax^5 + bx^3 + cx - 2$
and $f(-4) = 13$, find $f(4)$.

I N T E R N E T   C O N N E C T I O N

The table on page 320 contains statistics from the Internet on annual U.S. city average costs for a gallon of gasoline. The Web address is http://146.142.4.24/cgi-bin/surveymost. The table lists the average cost for a gallon of gasoline for each month from January 1976 until December 1997. In the following exercises we will use this table to compute additional averages and rates of change in the average price of a gallon of gasoline.

Exercises

I. For the years 1976, 1979, 1989, and 1994 compute the annual average cost of a gallon of gasoline.
II. The table on page 320 can be used to define a function. Every month has only one average price. For example, the average price of a gallon of gasoline during April 1980 was $1.264. We can use function notation to represent this fact as follows:

$$F(\text{April}/1980) = \$1.264$$

Here the expression $F(\text{month}/\text{year})$ represents the U.S. city average price of a gallon of gasoline for a particular month and year.

1. Evaluate each of the following:

$$F(\text{July}/1979), F(\text{May}/1982), F(\text{December}/1991) \quad \text{and} \quad F(\text{August}/1995)$$

Continued

2. Find the domain of F.
3. Describe the range of F.
4. Describe why F defines a function.

III. In mathematics we are sometimes interested in finding what is called an *average rate of change*. For example, suppose we were asked to find the average price change per month of a gallon of gasoline between January 1979 and January 1980.

$$
\begin{aligned}
\text{The total price change} \\
\text{between January 1980 and} &= F(\text{January}/1980) - F(\text{January}/1979) \\
\text{January 1979} &= 1.131 - 0.716 \\
&= 0.415
\end{aligned}
$$

Since we want to find an average price change per month between January 1980 and January 1979, we must divide 0.415 by 12.

$$
\begin{aligned}
\text{Average price change per month} &= \frac{F(\text{January}/1980) - F(\text{January}/1979)}{12} \\
&= \frac{0.415}{12} \\
&\doteq 0.0346
\end{aligned}
$$

This result tells us that, on the average, gasoline prices increased approximately $3\frac{1}{2}¢$ per month between January 1979 and January 1980.

Generalizing the preceding argument we can say

$$
\text{Average rate of change} = \frac{\text{change in price}}{\text{change in time}}
$$

$$
\text{(in the price of a gallon of gasoline)}
$$

Find the following average rates of change.

1. Average price change per month between October 1979 and December 1980

2. Average price change per month between February 1989 and October 1993

3. Average price change per month between May 1995 and August 1997

4. Average price change per year between July 1988 and July 1989

5. Average price change per year between May 1992 and May 1994

6. Average price change per year between March 1990 and September 1991

IV.

1. What year starting in January and ending in December had the largest increase in the average price of a gallon of gasoline?

Continued

2. What year starting in January and ending in December had the largest decrease in the average price of a gallon of gasoline?

3. Describe any of the historic or economic factors that generated the price changes in a gallon of gasoline.

Bureau of Labor Statistics Data
Average Price Data: U.S. City Average for Gasoline, unleaded regular (cost per gallon/3.8 liters)

Year	Jan	Feb	Mar	Apr	May	Jun	Jul	Aug	Sep	Oct	Nov	Dec	Ann
1976	0.605	0.600	0.594	0.592	0.600	0.616	0.623	0.628	0.630	0.629	0.629	0.626	n/a
1977	0.627	0.637	0.643	0.651	0.659	0.665	0.667	0.667	0.666	0.665	0.664	0.665	n/a
1978	0.648	0.647	0.647	0.649	0.655	0.663	0.674	0.682	0.688	0.690	0.695	0.705	n/a
1979	0.716	0.730	0.755	0.802	0.844	0.901	0.949	0.988	1.020	1.028	1.041	1.065	n/a
1980	1.131	1.207	1.252	1.264	1.266	1.269	1.271	1.267	1.257	1.250	1.250	1.258	n/a
1981	1.298	1.382	1.417	1.412	1.400	1.391	1.382	1.376	1.376	1.371	1.369	1.365	n/a
1982	1.358	1.334	1.284	1.225	1.237	1.309	1.331	1.323	1.307	1.295	1.283	1.260	n/a
1983	1.230	1.187	1.152	1.215	1.259	1.277	1.288	1.285	1.274	1.255	1.241	1.231	n/a
1984	1.216	1.209	1.210	1.227	1.236	1.229	1.212	1.196	1.203	1.209	1.207	1.193	n/a
1985	1.148	1.131	1.159	1.205	1.231	1.241	1.242	1.229	1.216	1.204	1.207	1.208	n/a
1986	1.194	1.120	0.981	0.888	0.923	0.955	0.890	0.843	0.860	0.831	0.821	0.823	n/a
1987	0.862	0.905	0.912	0.934	0.941	0.958	0.971	0.995	0.990	0.976	0.976	0.961	n/a
1988	0.933	0.913	0.904	0.930	0.955	0.955	0.967	0.987	0.974	0.957	0.949	0.930	n/a
1989	0.918	0.926	0.940	1.065	1.119	1.114	1.092	1.057	1.029	1.027	0.999	0.980	n/a
1990	1.042	1.037	1.023	1.044	1.061	1.088	1.084	1.190	1.294	1.378	1.377	1.354	n/a
1991	1.247	1.143	1.082	1.104	1.156	1.160	1.127	1.140	1.143	1.122	1.134	1.123	n/a
1992	1.073	1.054	1.058	1.079	1.136	1.179	1.174	1.158	1.158	1.154	1.159	1.136	n/a
1993	1.117	1.108	1.098	1.112	1.129	1.130	1.109	1.097	1.085	1.127	1.113	1.070	n/a
1994	1.043	1.051	1.045	1.064	1.080	1.106	1.136	1.182	1.177	1.152	1.163	1.143	n/a
1995	1.129	1.120	1.115	1.140	1.200	1.226	1.195	1.164	1.148	1.127	1.101	1.101	n/a
1996	1.129	1.124	1.162	1.251	1.323	1.299	1.272	1.240	1.234	1.227	1.250	1.260	n/a
1997	1.261	1.255	1.235	1.231	1.226	1.229	1.205	1.253	1.277	1.242	1.213	1.177	

6.4 *Linear Functions*

In this section we will be concerned with graphing equations of the form $Ax + By + C = 0$, where A, B, and C are real numbers. We will examine some of the characteristics of these graphs and see what conclusions we are able to draw.

EXAMPLE 6.4.1   Sketch the graph of the equation $2x - 3y = 12$.

SOLUTION

The only graphing technique we have, at this point, is to find several points on the curve and then smooth the graph. In general we have been picking values for *x* and then solving for y, which generates ordered pairs (points on the curve). When ordered pairs are generated in this fashion, x is called the **independent variable** and y is called the **dependent variable.**

■ If $x = 0$, then $2(0) - 3y = 12$, so that $y = -4$. This says $\underline{(0, -4)}$ is on the curve.

■ If $x = 3$, then $2(3) - 3y = 12$, so that $y = -2$. This says $\underline{(3, -2)}$ is on the curve.

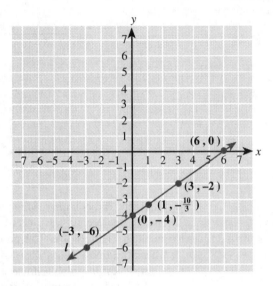

Figure 6.4.1

In a similar fashion we can obtain the ordered pairs $(6, 0)$, $(-3, -6)$, and $(1, -\frac{10}{3})$. Plotting these points and smoothing the curve yield the graph in Figure 6.4.1. Notice that these points all lie on the same straight line. In fact all the points that satisfy $2x - 3y = 12$ are on line l. This leads to an important definition.

Definition 6.4.1	Any equation that can be written in the form $Ax + By + C = 0$, where A, B, and C are real numbers and with the property that not both A and B are zero, is called a **linear equation.**

REMARK   Equations of this type are called linear equations because they generate straight line graphs in the Cartesian coordinate system.

EXAMPLE 6.4.2   Sketch the graph of $3x - y = 6$.

SOLUTION   Since we have a linear equation, only two points are necessary to define the graph. However, it is usually a good idea to get a third point as a check to make sure no

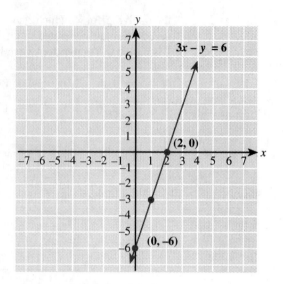

Figure 6.4.2

arithmetic errors were made in the calculations of the other two points. Three points on our line are $(0, -6)$, $(2, 0)$, and $(1, -3)$. Plotting these points, we obtain the graph in Figure 6.4.2. The point $(0, -6)$ is called the **y-intercept** because this is where the line crosses the y-axis. Likewise $(2, 0)$ is called the **x-intercept.** It is often easiest to graph a linear equation by finding its intercepts. The intercepts are easily found by substituting zero for one of the variables and then solving the equation for the remaining variable.

XAMPLE 6.4.3   Find the intercepts and graph the equation $3y - 2x = 4$.

SOLUTION   If $x = 0$, then

$$3y - 2(0) = 4$$
$$3y = 4$$
$$y = \frac{4}{3}$$

If $y = 0$, then

$$3(0) - 2x = 4$$
$$-2x = 4$$
$$x = -2$$

Again it is a good idea to find a third point; for example, $(1, 2)$ satisfies the equation. [If the points are close together, a fourth point wouldn't hurt—say, $(4, 4)$.] The line is shown in Figure 6.4.3.

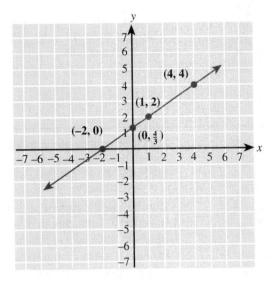

Figure 6.4.3

EXAMPLE 6.4.4   Graph the equation $y = 2x$.

SOLUTION   This time when we find the x-intercept and y-intercept, we get only one point, since if $x = 0$, then $y = 0$ also. We should still find two other points. Letting $x = -1$, we get $y = 2(-1) = -2$, and letting $x = 2$, we get $y = 2(2) = 4$. So $(0, 0)$, $(-1, -2)$, and $(2, 4)$ are points on the line graphed in Figure 6.4.4.

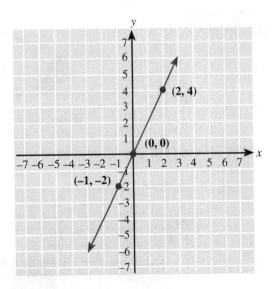

Figure 6.4.4

There are two special types of linear equations, $x = a$ and $y = b$, where a and b are real numbers. Let's investigate what these lines look like by graphing some particular equations.

EXAMPLE 6.4.5   Sketch the graphs of the following equations.

1. $y = -4$

If we follow the pattern of Example 6.4.2, we will pick some values for x and then solve for y. Let's choose $x = -1$, 0, and 4. How can we substitute these into the equation, since no x is present? An easy way to interpret the equation $y = -4$ is by saying y is always -4 while x can be any real number. This is because the equation $y = -4$ is equivalent to $y = 0x - 4$. Thus, we get the ordered pairs $(-1, -4)$, $(0, -4)$, and $(4, -4)$. Plotting these points yields the horizontal line shown in Figure 6.4.5. In fact any equation of the form $y = b$ generates a horizontal line.

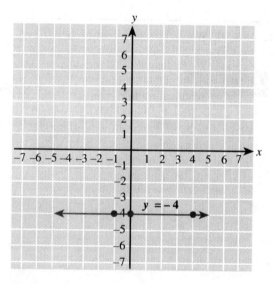

Figure 6.4.5

2. $x = 3$

The equation $x = 3$ says x must always be 3 while y can be any real number. This is because the equation $x = 3$ is equivalent to $x = 0y + 3$. Some ordered pairs that satisfy this equation are $(3, -2)$, $(3, 0)$, and $(3, \frac{1}{2})$. Plotting these points yields the vertical line shown in Figure 6.4.6. Any equation of the form $x = a$ generates a vertical line.

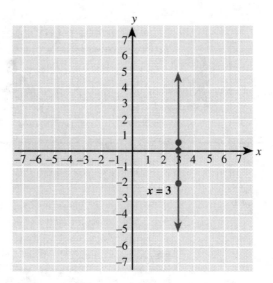

Figure 6.4.6

NOTE ▶ Note that vertical lines are the only type of straight lines that are not graphs of functions. Why? This fact leads to the following definition.

| **Definition 6.4.2** | Any equation that can be written in the form $Ax + By + C = 0$, where A, B, and C are real numbers and with the property that $B \neq 0$, is said to define a **linear function.** |

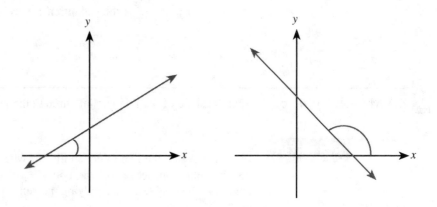

Figure 6.4.7

A very significant characteristic of a line is the angle that the line makes with the x-axis (see Figure 6.4.7). Some lines are very steep and form an angle close to 90 degrees, whereas others rise quite slowly and make a small angle (see Figure 6.4.8). Fortunately, mathematicians have devised a useful and practical way of measuring this characteristic. This "steepness" is generally referred to as the slope of the line and is defined on the next page.

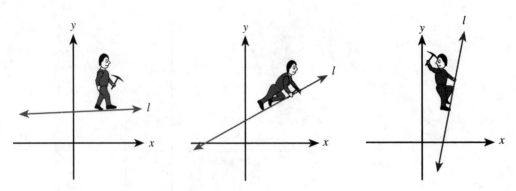

Figure 6.4.8

$\mathbf{S}$lope of a Line

Let (x_1, y_1) and (x_2, y_2) be *any* two distinct points on a line; then the **slope** of that line is defined to be

$$\frac{y_2 - y_1}{x_2 - x_1}$$

REMARK Is is common practice to let the letter m stand for the slope of a line.

Let's examine what this definition means by looking at a particular problem.

XAMPLE 6.4.6   Find the slope of the line passing through the points $(1, 2)$ and $(4, 4)$.

SOLUTION Let's first sketch the line to get an idea of what it looks like (see Figure 6.4.9). To use the definition, let one of the points equal (x_1, y_1)—say, $(1, 2)$—and the other point, $(4, 4)$, equal (x_2, y_2). Applying the definition of slope, we determine the slope of this line to be

$$m = \frac{y_2 - y_1}{x_2 - x_1} = \frac{4 - 2}{4 - 1} = \frac{2}{3}$$

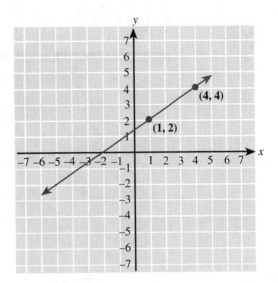

Figure 6.4.9

REMARKS

1. What if we had chosen (x_1, y_1) to be $(4, 4)$ and (x_2, y_2) to be $(1, 2)$? Then

$$m = \frac{y_2 - y_1}{x_2 - x_1} = \frac{2 - 4}{1 - 4} = \frac{-2}{-3} = \frac{2}{3}$$

We get the same answer. Why?

2. This is the same line we graphed in Example 6.4.3. There we saw that the points $(-2, 0)$ and $(0, \frac{4}{3})$ are also on the line. If we used these two points to find the slope of this same line, shouldn't we get the same slope? Let (x_1, y_1) be $(-2, 0)$ and let (x_2, y_2) be $(0, \frac{4}{3})$. Then

$$m = \frac{y_2 - y_1}{x_2 - x_1} = \frac{\frac{4}{3} - 0}{0 - (-2)} = \frac{\frac{4}{3}}{2} = \frac{2}{3}$$

So we do get the same answer for m. In fact, regardless of which two points we use to find m, we always get the same answer.

Try to avoid the mistake:

Incorrect	Correct
$(x_1, y_1) = (1, 2), (x_2, y_2) = (4, 4)$	$m = \dfrac{y_2 - y_1}{x_2 - x_1} = \dfrac{4 - 2}{4 - 1} = \dfrac{2}{3}$
$m = \dfrac{y_2 - y_1}{x_1 - x_2} = \dfrac{4 - 2}{1 - 4} = \dfrac{2}{-3}$	or if
	$(x_1, y_1) = (4, 4)$
	$(x_2, y_2) = (1, 2)$
	$m = \dfrac{y_2 - y_1}{x_2 - x_1} = \dfrac{2 - 4}{1 - 4}$
	$= \dfrac{-2}{-3} = \dfrac{2}{3}$
The x's were subtracted in an order opposite that of the y's.	In either case the x's and y's were subtracted in the same order.

Geometrically here is what happens when you are determining the slope. Return to the line of Example 6.4.6 and construct the right triangle shown in Figure 6.4.10. We can see that the distance between the points $(4, 4)$ and $(4, 2)$ is 2. In a similar fashion we can see that the distance between $(1, 2)$ and $(4, 2)$ is 3. Notice

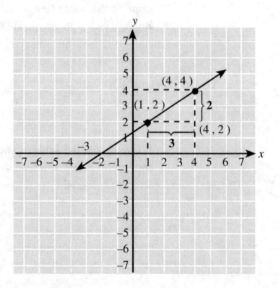

Figure 6.4.10

that the slope $\frac{2}{3}$ is the length of the vertical leg of the right triangle divided by the length of the horizontal leg. Often the slope is referred to as the $\frac{\text{rise}}{\text{run}}$.

In our example this means if we start at the point $(1, 2)$ and run over 3 units (to the right) and rise (up) 2 units, we will arrive at another point on the line $(4, 4)$.

EXAMPLE 6.4.7

Graph the following lines.

1. A line with $m = \frac{3}{4}$ through the point $(-1, 1)$

Using the $\frac{\text{rise}}{\text{run}}$ interpretation of slope, start at the point $(-1, 1)$, run (to the right) 4 units, and rise (up) 3 units, and we have another point on the line. Now with two points we can draw the line as in Figure 6.4.11.

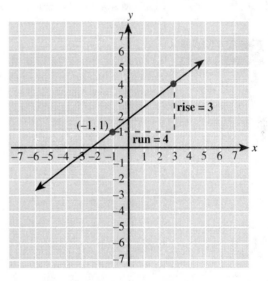

Figure 6.4.11

2. A line with $m = -\frac{2}{5}$ through $(-3, 1)$

Here we must decide whether to place the negative sign in the numerator or the denominator, since $-\frac{2}{5} = \frac{-2}{5} = \frac{2}{-5}$. It makes no difference where we choose to place the negative sign, the results will yield the same line. If we place the negative sign with the 2, we will start at $(-3, 1)$ and run 5 (to the right), but the rise will be *down* 2, since down is the negative direction on the y-axis. On the other hand, if we place the negative sign with the 5, we will start at $(-3, 1)$, run 5 to the *left* (since to the left is the negative direction on the x-axis), and rise (up) 2. Observe from Figure 6.4.12 that either way the same line is determined.

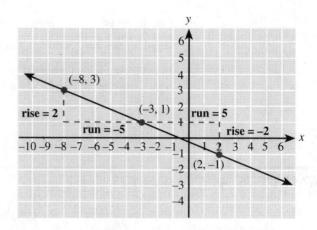

Figure 6.4.12

EXAMPLE 6.4.8

Find the slope of the line $3x - 4y = 12$.

SOLUTION

To find the slope of a line, we may use *any* two points on the line. As mentioned earlier, the easiest points to find on a line are often the x- and y-intercepts. The two intercepts here are $(0, -3)$ and $(4, 0)$. Now we can apply the definition of slope. Let $(x_1, y_1) = (0, -3)$ and $(x_2, y_2) = (4, 0)$, so that

$$m = \frac{0 - (-3)}{4 - 0} = \frac{3}{4}$$

EXAMPLE 6.4.9

Find the slopes of the following lines.

1. $y = -4$

From Example 6.4.5 we know that $(-1, -4)$ and $(4, -4)$ are two points on this line. Let $(x_1, y_1) = (-1, -4)$ and $(x_2, y_2) = (4, -4)$, so that

$$m = \frac{-4 - (-4)}{4 - (-1)} = \frac{0}{5} = 0$$

This says that as we run across a horizontal line there is no rise. *The slope of any horizontal line is zero.*

2. $x = 3$

Again from Example 6.4.5 we know that $(3, -2)$ and $(3, \frac{1}{2})$ are two points on this line. Letting $(x_1, y_1) = (3, -2)$ and $(x_2, y_2) = (3, \frac{1}{2})$, we get

$$m = \frac{\frac{1}{2} - (-2)}{3 - 3} = \frac{\frac{5}{2}}{0}$$

which is undefined. *The slope of any vertical line is undefined.* We will avoid the use of the ambiguous expression "no slope."

EXAMPLE 6.4.10

Use a graphing calculator to graph the equation $3x + 2y = 4$.

SOLUTION

When using a graphing calculator to graph an equation, we must first solve for y. Thus,

$$3x + 2y = 4$$
$$2y = -3x + 4$$
$$y = \frac{-3}{2}x + 2$$

Now graph the equation $y = -\frac{3}{2}x + 2$. Your graph should look something like Figure 6.4.13.

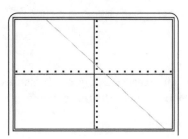

Figure 6.4.13

Exercises 6.4

Sketch the graphs of the following linear equations, and find all intercepts.

1. $2x - y = 6$

2. $4y + 3x = -12$

3. $y = -2x$

4. $y - x = 0$

5. $x + 3y = 0$

6. $-2x = 3y + 8$

7. $5x - 10y + 1 = 0$

8. $\frac{1}{3}y + \frac{3}{4}x - 2 = 0$

9. $\frac{2}{3}x - \frac{1}{5}y = -3$

10. $y = 5x + 2$

11. $y = 2x - \frac{8}{5}$

12. $y = \frac{1}{4}x - \frac{2}{3}$

13. $y = -5$

14. $y = 0$

15. $x = \frac{7}{4}$

16. $x = 0$

Find the slopes of the lines passing through the following points.

17. $(6, 0)$ and $(0, -3)$ **18.** $(0, 4)$ and $(7, 0)$ **19.** $(-4, 1)$ and $(3, -5)$

20. $(-2, -5)$ and $(1, 9)$ **21.** $(3, -7)$ and $(6, 2)$ **22.** $(-9, 11)$ and $(-3, 5)$

23. $(0, 0)$ and $\left(\frac{1}{3}, -6\right)$ **24.** $\left(\frac{1}{4}, -\frac{2}{3}\right)$ and $\left(\frac{1}{2}, \frac{5}{9}\right)$ **25.** $\left(\frac{1}{5}, \frac{3}{8}\right)$ and $\left(-\frac{7}{15}, -\frac{1}{4}\right)$

26. $(-2.1, 0.5)$ and $(1.4, -3)$ **27.** $(5, 2)$ and $(9, 2)$ **28.** $(-3, 1)$ and $(-3, 10)$

Graph the following lines with the given slopes and passing through the indicated points.

29. $m = \frac{5}{2}$, through $(1, 4)$ **30.** $m = \frac{1}{3}$, through $(-2, 5)$ **31.** $m = 2$, through $(3, -1)$

32. $m = -\frac{3}{4}$, through $(-5, 7)$ **33.** $m = -\frac{2}{5}$, through $(-2, -6)$ **34.** $m = -4$, through $(1, -3)$

35. $m = 0$, through $(5, 1)$ **36.** m undefined, through $(-2, 3)$

Find the slopes of the following lines.

37. $2x - y = 6$ **38.** $4y + 3x = -12$ **39.** $y = -2x$ **40.** $y - x = 0$

41. $x + 3y = 0$ **42.** $-2x = 3y + 8$ **43.** $5x - 10y + 1 = 0$ **44.** $\frac{1}{3}y + \frac{3}{4}x - 2 = 0$

45. $\frac{2}{3}x - \frac{1}{5}y = -3$ **46.** $y = -5x + 2$ **47.** $y = 2x - \frac{8}{5}$ **48.** $y = \frac{1}{4}x - \frac{2}{3}$

49. $y = -5$ **50.** $y = 0$ **51.** $x = \frac{7}{4}$ **52.** $x = 0$

For each exercise graph all three equations on the same coordinate axes. What conclusions appear to be true about the three lines?

53. $y = \frac{1}{2}x - 2$

$y = \frac{1}{2}x + 0$

$y = \frac{1}{2}x + 3$

54. $y = -2x + 4$

$y = -2x$

$y = -2x - 1$

Graph each pair of lines on the same coordinate axes. What conclusions appear to be true about each pair of lines?

55. $y = \frac{2}{3}x - 2$

$y = -\frac{3}{2}x + 1$

56. $y = -3x + 4$

$y = \frac{1}{3}x + 1$

Graph the following lines on the same coordinate axis.

57. $y = 2$ **58.** $y = \frac{1}{3}x + 2$

59. $y = 5x + 2$

60. $y = -\dfrac{1}{3}x + 2$

61. $y = -5x + 2$

62. Graph the equation that you found in the toad problem in Exercises 6.2 (exercise 40). What is the slope of the line?

63. The Gold Fool Mining Company has a gold mine in Moss County with a proven reserve as of 1990 of 748 lb of gold. The company is able to extract 22 lb of gold per year.
   a. What will the mine's reserves be in 1996, 2000, and 2004?
   b. Find a function that will determine the mine's reserves in any subsequent year.
   c. Graph this function.
   d. What is the slope of the line?
   e. In what year will the company mine the last pound of gold?

Write Algebra

64. Describe what it means geometrically when the slope of a line is:
   a. positive,   **b.** negative,
   c. zero,   **d.** undefined.

65. Describe the procedure for graphing a line given a point and the slope of the line.

66. Describe how to find the intercepts of a line.

67. Does every vertical line have a y-intercept? Why or why not?

68. Does every horizontal line have an x-intercept? Why or why not?

◫ **Use a graphing calculator for the following exercises.**

69. Graph $y = x$. Next graph $y = x + 2$ and $y = x - 3$. Predict the graph of $y = x + 5$. Now graph $y = x + 5$. Describe the pattern you see in the graphs.

70. Graph $x + 2y = 4$. Next graph $2x + 2y = 4$ and $3x + 2y = 4$. Predict the graph of $4x + 2y = 4$. Now graph $4x + 2y = 4$. Describe the pattern you see in the graphs.

71. Graph $y = 3x + 1$. Next graph $y = 3x - 4$ and $y = 3x + 5$. Predict the graph of $y = 3x - 1$. Now graph $y = 3x - 1$. Describe the pattern you see in the graphs.

72. Graph $y = 2x + 1$ and $y = -\frac{1}{2}x - 3$. (In this problem, square up the grid; that is, on a T1-81 press Zoom 5.) Predict the graphs of $y = 2x - 1$ and $y = -\frac{1}{2}x + 4$. Describe the patterns of the graphs.

In exercises 73–76, determine the slope of line l in Figure 6.4.14 so that the shaded region (a trapezoid) will have the given area. Recall that the formula for the area of a trapezoid is

$$A = \tfrac{1}{2}(b_1 + b_2)h$$

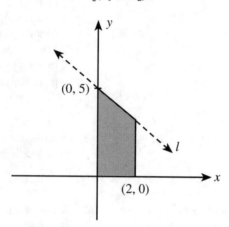

Figure 6.4.14

73. Area $= 6$

74. Area $= 9$

75. Area $= 13$

76. Area $= 17$

77. Suppose that the area of the shaded region is A. What is the slope of l in terms of A?

78. Suppose that the slope of l is m. What is the area of the shaded region in terms of m?

6.5 *Equations of Lines*

In Section 6.4 we were given equations of lines and we then graphed those lines. Additionally we determined the slopes of those lines. In this section we will go in the reverse direction; that is, given points that determine a line, we want to find the equation of the line. There are two important forms that make finding equations of lines relatively simple. The first of these is called the **point-slope form.** This is used when a point on the line and the slope of the line are known. It can also be used when two points on the line are known.

Suppose we want to find the equation of the line with $m = 2$ and passing through the point $(4, 5)$. Recall

$$m = \frac{y_2 - y_1}{x_2 - x_1}$$

Let $(x, y) = (x_2, y_2)$, where (x, y) is any point on the desired line. Let $(x_1, y_1) = (4, 5)$, the given point. Substituting in the slope formula, we obtain

$$\frac{y - 5}{x - 4} = 2$$

$$(x - 4) \cdot \frac{y - 5}{x - 4} = 2 \cdot (x - 4) \qquad \textit{Multiply by the LCD.}$$

$$y - 5 = 2(x - 4)$$

$$y - 5 = 2x - 8$$

$$y = 2x - 3 \qquad \textit{Solve for y.}$$

A generalization of the preceding argument justifies the following theorem.

Theorem 6.5.1

Point-slope form: Given a line with slope m and containing the point (x_1, y_1), the equation of that line is

$$y - y_1 = m(x - x_1)$$

Proof

Let (x_1, y_1) be a particular point on line l and (x, y) be any other point (see Figure 6.5.1). The slope of l is $m = \frac{y_2 - y_1}{x_2 - x_1}$. In the slope formula let $(x_2, y_2) = (x, y)$; then $m = \frac{y - y_1}{x - x_1}$. Clearing the equation $m = \frac{y - y_1}{x - x_1}$ of fractions yields $m(x - x_1) = y - y_1$, or $y - y_1 - m(x - x_1)$.

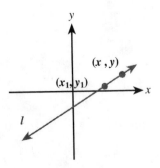

Figure 6.5.1

EXAMPLE 6.5.1   Find the equation of the line with $m = -\frac{3}{5}$ and passing through the point $(2, -6)$.

SOLUTION   Use the point-slope form: $y - y_1 = m(x - x_1)$. In our case $m = -\frac{3}{5}$ and $(x_1, y_1) = (2, -6)$. Substituting these into the point-slope form we get

$$y - (-6) = -\frac{3}{5}(x - 2)$$

Do not stop at this point. Simplify the equation and combine like terms.

$$y + 6 = -\frac{3}{5}x + \frac{6}{5}$$

$$y = -\frac{3}{5}x - \frac{24}{5} \tag{1}$$

$$5y = -3x - 24$$

$$3x + 5y + 24 = 0 \tag{2}$$

In this example the final answer in equation (2) is in the form $Ax + By + C = 0$, which was given in Definition 6.4.1. This form is called **standard form.** But there are two reasons why we might have preferred to stop at equation (1). One reason will be illustrated in Theorem 6.5.2. The other reason goes back to Definition 6.4.2, where we said that an equation of the form $Ax + By + C = 0$ defines a linear *function*. If we want to use functional notation, we will need the equation solved for y, as in equation (1), and then we can substitute $f(x)$ for y and write $f(x) = -\frac{3}{5}x - \frac{24}{5}$. Let us use this equation to graph the line, as in Figure 6.5.2.

$$f(-3) = -\frac{3}{5}(-3) - \frac{24}{5}$$

$$= \frac{9}{5} - \frac{24}{5}$$

$$= -\frac{15}{5} = -3$$

$$f(0) = -\frac{3}{5}(0) - \frac{24}{5}$$

$$= 0 - \frac{24}{5} = -\frac{24}{5}$$

$$f(2) = -\frac{3}{5}(2) - \frac{24}{5}$$

$$= -\frac{6}{5} - \frac{24}{5} = -\frac{30}{5} = -6$$

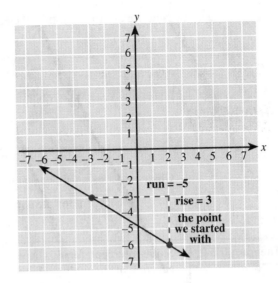

Figure 6.5.2

EXAMPLE 6.5.2   Find the equation of the line passing through the points $(-7, 6)$ and $(5, -3)$.

SOLUTION   To find the equation of any line we must know the slope. In this example we let $(x_1, y_1) = (-7, 6)$ and $(x_2, y_2) = (5, -3)$; now we can obtain the slope

$$m = \frac{-3 - 6}{5 - (-7)} = \frac{-9}{12} = \frac{-3}{4}$$

After calculating the slope, we can use the point-slope form. A natural question might be which of the two points should be used in the point-slope form. It turns out that either point will generate the same equation.

First we use $(-7, 6)$ in the point-slope form; then the equation is

$$y - 6 = -\frac{3}{4}(x - (-7))$$

$$y - 6 = -\frac{3}{4}(x + 7)$$

$$y - 6 = -\frac{3}{4}x - \frac{21}{4}$$

$$y = -\frac{3}{4}x + \frac{3}{4}$$

Now we use $(5, -3)$; then the equation is

$$y - (-3) = -\frac{3}{4}(x - 5)$$

$$y + 3 = -\frac{3}{4}x + \frac{15}{4}$$

$$y = -\frac{3}{4}x + \frac{3}{4}$$

Observe that using either point yields the same equation.

Another form for the equation of a line is called the *slope-intercept form*.

Theorem 6.5.2

Slope-intercept form: Given a line l with slope m and y-intercept $(0, b)$, the equation of l is

$$y = mx + b$$

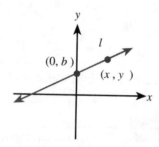

Figure 6.5.3

Proof

Assume that line l in Figure 6.5.3 has y-intercept $(0, b)$ and slope m. Then letting $(x_1, y_1) = (0, b)$ and using the point-slope form, we get

$$y - b = m(x - 0)$$
$$y - b = mx$$
$$y = mx + b$$

REMARK

Since the x-coordinate of the y-intercept is always zero, we call the y-intercept b instead of $(0, b)$.

EXAMPLE 6.5.3   Find the equation of the line with slope $\frac{5}{3}$ and y-intercept -2, and graph the line.

SOLUTION

Using the slope-intercept form we are given $m = \frac{5}{3}$ and $b = -2$. Hence the desired equation is $y = \frac{5}{3}x + (-2)$ or $y = \frac{5}{3}x - 2$. To graph the line we will use as our first point the y-intercept $(0, -2)$. Then we will use the slope to run to the right 3 units and rise up 5 units. This yields another point on the line, $(3, 3)$ (see Figure 6.5.4).

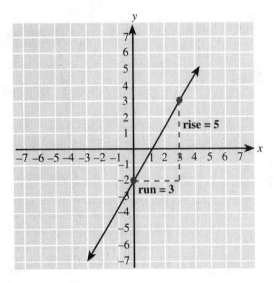

Figure 6.5.4

XAMPLE 6.5.4   Find the slope and y-intercept of the line $7x - 2y = 6$; then graph the line.

SOLUTION

To find the slope and y-intercept, all that we need to do is solve the given equation for y. Solving for y, we obtain

$$7x - 2y = 6 \rightarrow -2y = -7x + 6$$

$$y = \left(\frac{-7}{-2}\right)x + \left(\frac{6}{-2}\right)$$

$$y = \frac{7}{2}x - 3$$

Using the slope-intercept form, we know that the slope is $\frac{7}{2}$ and the y-intercept is -3. With this information we can graph the line, as in Figure 6.5.5. Start with the y-intercept and use the slope to find another point.

NOTE ▶   When you solve a linear equation for y, the coefficient of x is the slope and the constant term is the y-intercept.

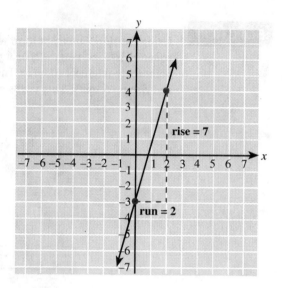

Figure 6.5.5

An important characteristic of pairs of lines is the angle that they form. If two lines never cross, they are called **parallel.** If two lines cross at a right angle, they are called **perpendicular.** We can use the slopes of lines to tell us whether a given pair of nonvertical lines are parallel or perpendicular.

Properties of Parallel and Perpendicular Lines

1. Two distinct nonvertical lines are *parallel* if and only if the lines have the same slope.
2. Two distinct nonvertical lines are *perpendicular* if and only if the slopes of the lines are negative reciprocals of each other. (An easy way to remember this fact is that the product of the slopes of perpendicular lines is equal to -1.)

See exercises 53–56 in Exercises 6.4.

EXAMPLE 6.5.5

1. Show that the following pair of lines are parallel: $6x + 2y = 10$ and $y = -3x + 7$. Graph the lines on the same axes.

Solving for y and using the slope-intercept form, we get

$$
\begin{array}{c|c}
6x + 2y = 10 & y = -3x + 7 \\
2y = -6x + 10 & m = -3 \\
y = -3x + 5 & \\
m = -3 &
\end{array}
$$

Since these lines have the same slope, they are parallel. Again using the y-intercepts and the slopes, we graph the lines in Figure 6.5.6.

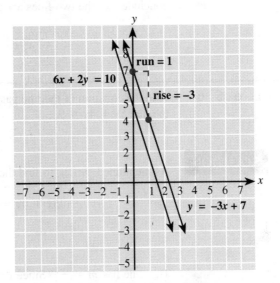

Figure 6.5.6

2. Show that the following pair of lines are perpendicular: $y = 2x + 3$ and $x + 2y = 14$. Graph the lines on the same axes.

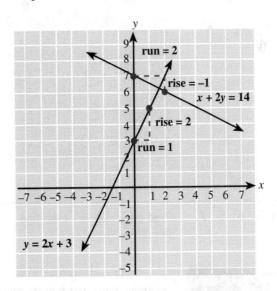

Figure 6.5.7

$$
\begin{array}{l|l}
y = 2x + 3 & x + 2y = 14 \\
m = 2 & 2y = -x + 14 \\
 & y = -\dfrac{1}{2}x + 7 \\
 & m = -\dfrac{1}{2}
\end{array}
$$

Since the slopes are negative reciprocals, or since they have a product of -1, $(2 \cdot (-\frac{1}{2}) = -1)$, then by the Properties of Parallel and Perpendicular Lines we can conclude that the two lines are perpendicular (see Figure 6.5.7).

EXAMPLE 6.5.6   Find the equation of the line passing through the point $(-1, 5)$ and perpendicular to the line $5x + 4y = 9$.

SOLUTION

As mentioned earlier, to find the equation of a line we must know the slope:

$$5x + 4y = 9 \quad \text{implies} \quad 4y = -5x + 9$$

$$y = -\frac{5}{4}x + \frac{9}{4}$$

This line has $m = -\frac{5}{4}$. Since the lines are to be perpendicular, the slope we want must be $\frac{4}{5}$. Why?

Now with $m = \frac{4}{5}$ and $(x_1, y_1) = (-1, 5)$, we can use the point-slope form. Substituting, we obtain

$$y - 5 = \frac{4}{5}(x - (-1))$$

$$y - 5 = \frac{4}{5}(x + 1)$$

$$y - 5 = \frac{4}{5}x + \frac{4}{5}$$

$$y = \frac{4}{5}x + \frac{29}{5}$$

EXAMPLE 6.5.7   Find the equation of the line passing through the point $(6, -4)$ and parallel to the line $2x - 3y = 11$.

SOLUTION

Again we must first find the slope of our given line, $2x - 3y = 11$:

$$2x - 3y = 11 \quad \text{implies} \quad -3y = -2x + 11$$

$$y = \frac{2}{3}x - \frac{11}{3}$$

This line has $m = \frac{2}{3}$. Since the lines are to be parallel, the slope we want must also be $\frac{2}{3}$.

Now with $m = \frac{2}{3}$ and $(x_1, y_1) = (6, -4)$ we can use the point-slope form. Substituting, we obtain

$$y - (-4) = \frac{2}{3}(x - 6)$$

$$y + 4 = \frac{2}{3}x - 4$$

$$y = \frac{2}{3}x - 8$$

XAMPLE 6.5.8   Plot the following three points and show that they are vertices of a right triangle: $A = (-2, 0)$, $B = (2, -3)$, and $C = (8, 5)$.

SOLUTION   Since a right triangle is merely a triangle that has a right angle, all we must show is that two sides of the triangle are perpendicular:

$$m \text{ of } \overline{AB} = \frac{-3 - 0}{2 - (-2)} = \frac{-3}{4}$$

$$m \text{ of } \overline{AC} = \frac{5 - 0}{8 - (-2)} = \frac{5}{10} = \frac{1}{2}$$

$$m \text{ of } \overline{BC} = \frac{5 - (-3)}{8 - 2} = \frac{8}{6} = \frac{4}{3}$$

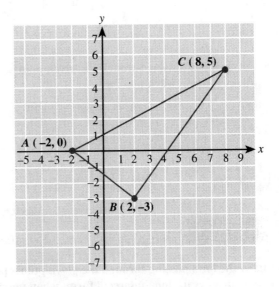

Figure 6.5.8

Since the slopes of $\overline{AB}$ and $\overline{BC}$ are negative reciprocals, we know that those lines are perpendicular. Thus a right triangle is formed (see Figure 6.5.8).

Exercises 6.5

Find the equations of the following lines. Write the answers in slope-intercept form.

1. $m = 5$; y-intercept $= \frac{1}{2}$

2. $m = -\frac{3}{4}$; y-intercept $= 7$

3. $m = 1$; y-intercept $= -9$

4. $m = -8$; y-intercept $= \frac{5}{4}$

5. $m = -\frac{1}{2}$; y-intercept $= -3$

6. $m = \frac{9}{7}$; y-intercept $= \frac{1}{3}$

7. $m = 0$; y-intercept $= 11$

8. $m = 0$; y-intercept $= -5$

Find the slopes and y-intercepts of the following lines.

9. $y = 3x - 4$

10. $y = \frac{1}{5}x + 2$

11. $3x - 5y = 7$

12. $7x + 3y = 12$

13. $5y - 6x = 8$

14. $x = 9y + 4$

15. $y = \frac{2x - 4}{7}$

16. $\frac{2x + y}{5} = 1$

17. $y = 3$

18. $x = -7$

Find the equations of the following lines. Write the answers in slope-intercept form, when possible.

19. $m = 3$, through $(-1, -2)$

20. $m = -\frac{1}{4}$, through $(5, -7)$

21. $m = \frac{9}{2}$, through $(3, 8)$

22. $m = -1$, through $\left(\frac{1}{2}, -\frac{3}{5}\right)$

23. $m = \frac{4}{3}$, through $\left(\frac{5}{8}, -\frac{1}{3}\right)$

24. $m = -\frac{2}{5}$, through $\left(5, -\frac{1}{4}\right)$

25. $m = 0$, through $(-6, 5)$

26. m undefined, through $(1, -3)$

Find the equations of the lines passing through the given points. Write the answers in slope-intercept form, when possible.

27. $(-2, 4)$ and $(-5, 7)$

28. $(3, 5)$ and $(-3, -10)$

29. $(-8, 6)$ and $(4, -3)$

30. $(3, -8)$ and $(-4, 6)$

31. $(0, 0)$ and $(-2, 3)$

32. $(0, 0)$ and $\left(-\frac{1}{2}, \frac{3}{8}\right)$

33. $\left(2, -\frac{1}{8}\right)$ and $\left(\frac{1}{2}, \frac{5}{4}\right)$

34. $\left(\frac{3}{4}, \frac{1}{3}\right)$ and $\left(\frac{1}{2}, -\frac{5}{9}\right)$

35. $(8, -4)$ and $(-3, -4)$

36. $(5, 3)$ and $(5, -6)$

37. x-intercept $= 3$, y-intercept $= -2$

38. x-intercept $= -4$, y-intercept $= 2$

Find the equations of the lines passing through the given points parallel to the given lines. Write the answers in slope-intercept form, when possible.

39. Through $(1, -2)$ parallel to $y = 3x + 4$ $m = 3$ $(1, -2)$

40. Through $(-5, 7)$ parallel to $y = \frac{1}{2}x + 8$

41. Through $(3, 11)$ parallel to $5x + 4y = -8$

42. Through $(-8, -1)$ parallel to $2y - 8x + 7 = 0$

[Handwritten margin notes near top right: "Plus w/slope formula into Slope/point form", "y = mx+b", "2x+3y=7", "3y = -2x+7", over 3, $-\frac{2}{3}x + \frac{1}{3}$]

43. Through $(\frac{1}{4}, -\frac{3}{5})$ parallel to $y = \frac{6x - 13}{8}$

44. Through $(\frac{1}{2}, -\frac{1}{3})$ parallel to $2x + 3y = 7$

45. Through $(-8, -9)$ parallel to $x = 5$

46. Through $(1, 5)$ parallel to the x-axis

47. Through $(3, -7)$ parallel to $y = 4$

48. Through $(-6, 4)$ parallel to the y-axis

Find the equations of the lines passing through the given points perpendicular to the given lines. Write the answers in slope-intercept form, when possible.

49. Through $(1, -2)$ perpendicular to $y = 3x + 4$ *[handwritten: recip., m = $\frac{-1}{3}$]*

50. Through $(-5, 7)$ perpendicular to $y = -\frac{1}{2}x + 8$ *[handwritten: slope]*

51. Through $(3, 11)$ perpendicular to $5x + 4y = -8$

52. Through $(-8, -1)$ perpendicular to $2y - 8x + 7 = 0$

53. Through $(\frac{1}{4}, -\frac{3}{5})$ perpendicular to $y = \frac{6x - 13}{8}$

54. Through $(\frac{1}{2}, -\frac{1}{3})$ perpendicular to $2x + 3y = 7$

55. Through $(-8, -9)$ perpendicular to $x = 5$

56. Through $(1, 5)$ perpendicular to the x-axis

57. Through $(3, -7)$ perpendicular to $y = 4$

58. Through $(-6, 4)$ perpendicular to the y-axis

59. Plot the following three points and show that they are vertices of a right triangle: $A = (5, 3)$, $B = (-1, -4)$, $C = (-8, 2)$.

60. Plot the following three points and show that they are vertices of a right triangle: $A = (6, -1)$, $B = (3, 4)$, $C = (-7, -2)$.

61. Plot the following four points and show that they are vertices of a rectangle: $A = (-2, -5)$, $B = (4, -2)$, $C = (2, 2)$, $D = (-4, -1)$.

62. Plot the following four points and show that they are vertices of a rectangle: $A = (1, -1)$, $B = (6, 2)$, $C = (3, 7)$, $D = (-2, 4)$.

63. Plot the following four points and show that they are vertices of a parallelogram: $A = (1, 6)$, $B = (3, 3)$, $C = (-2, -7)$, $D = (-4, -4)$.

64. It costs the Pete Moss Fertilizer Company $135 to process its first 100-lb bag of fertilizer. The cost of processing two bags is $150, the cost of processing three bags is $165, and so on. Find an equation that will describe the cost of processing x bags of fertilizer. (Assume the relationship is linear.) What is the overhead of this company (i.e., the cost of operations even when no bags of fertilizer are processed)? Sketch the graph of the cost equation. Note that this graph cannot leave quadrant I. Why?

65. Assuming the relationship between degrees Fahrenheit and degrees Celsius is linear, derive the formula converting degrees Fahrenheit into degrees Celsius. Use the fact that the freezing point of water is $32°F$, or $0°C$, and the boiling point of water is $212°F$, or $100°C$.

66. Hawkeye Yamamoto, a pilot on an aircraft carrier, takes off and rises in a straight path for the first 42 sec. When his horizontal distance from the carrier is 100 ft, he is 100 ft above sea level. When his horizontal distance from the carrier is 1000 ft, he is 640 ft above sea level. Find the equation that describes Hawkeye's altitude for the first 42 sec. How high above sea level is the flight deck?

67. What will be the form of equations of lines parallel to
   a. the x-axis,   **b.** the y-axis,
   c. $x = 7$, and   **d.** $y = 8$?

68. What will be the form of equations of lines perpendicular to
   a. the x-axis,   **b.** the y-axis,
   c. $x = 7$, and   **d.** $y = 8$?

69. What is the equation of the x-axis? The y-axis?

⊞ **Use a graphing calculator for the following exercises.**

70. Graph $y = x + 2$. Next graph $y = 2x + 2$ and $y = 3x + 2$. Predict the graph of $y = 4x + 2$. Graph $y = 4x + 2$. Describe the pattern you see in the graphs.

71. Graph $y = x + 2$. Next graph $y = \frac{x}{2} + 2$ and $y = \frac{x}{3} + 2$. Predict the graph of $y = \frac{x}{4} + 2$. Graph $y = \frac{x}{4} + 2$. Describe the pattern you see in the graphs.

72. Graph $y = -x - 1$. Next graph $y = -2x - 1$ and $y = -3x - 1$. Predict the graph of $y = -4x - 1$. Graph $y = -4x - 1$. Describe the pattern you see in the graphs.

73. Graph $y = -x$. Next graph $y = -\frac{x}{2}$ and $y = -\frac{x}{3}$. Predict the graph of $y = -\frac{x}{4}$. Graph $y = -\frac{x}{4}$. Describe the pattern you see in the graphs.

74. Graph $y = 3x + 2$ and $y = -\frac{1}{3}x - 4$. (In this problem, square up the grid; i.e., on a TI-81, press zoom 5.) Predict the graphs of $y = 3x - 1$ and $y = -\frac{1}{3}x + 5$. Graph $y = 3x - 1$ and $y = -\frac{1}{3}x + 5$. Describe the patterns you see in the graphs.

G R O U P   A C T I V I T Y

In exercises 75–78, determine the equation of line *l* in slope-intercept form. Line *l* determines one side of the shaded triangle whose area is given.

75.

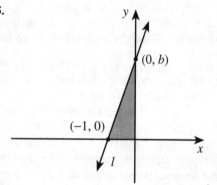

(a) Area = 2

(b) Area = 7

(c) Area = $\frac{1}{4}$

(d) Area = A

76.

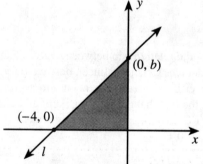

(a) Area = 4

(b) Area = 10

(c) Area = 3

(d) Area = A

Continued

77.

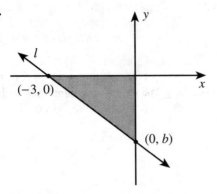

(a) Area = 6

(b) Area = 15

(c) Area = 2

(d) Area = A

78.

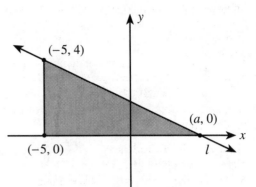

(a) Area = 16

(b) Area = 24

(c) Area = 12

(d) Area = A

I N T E R N E T   C O N N E C T I O N

The table on page 349 contains statistics obtained from the Internet on the number of resident and nonresident housing units in the state of Hawaii. The Web address is http://www.hawaii.gov/dbedt/srs/sr230.html [select Table 3]. In the following example we will use some of the information in this table to construct a model to predict the future number of resident housing units in the state of Hawaii.

E X A M P L E   Construct a linear model to predict the number of resident housing units in the state of Hawaii.

SOLUTION   Since we are *assuming* a linear model, we will use the equation

$$y = mx + b$$

Continued

In this equation we will let

$$x = \text{the year}$$
$$y = \text{the number of resident housing units in the state of Hawaii}$$

To build our model we pick any two years. Let's choose 1988 and 1993. Thus, the two data points for our model are (1988, 354314) and (1993, 392749).

Hence, $m = \dfrac{y_2 - y_1}{x_2 - x_1} = \dfrac{392749 - 354314}{1993 - 1988} = \dfrac{38435}{5} = 7687$

Now to find our linear model we use the point-slope form and obtain

$$y - 354314 = 7687\,(x - 1988)$$
$$y - 354314 = 7687x - 15,281,756$$
$$y = 7687x - 14,927,442$$

Using this model let's predict the number of resident housing units in the state of Hawaii for the year 2001.

$$y = 7687x - 14,927,442$$

For the year 2001 we obtain

$$y = 7687(2001) - 14,927,442$$
$$y = 15,381,687 - 14,927,442$$
$$y = 454,245$$

Exercises

I. In the following problems construct a *linear* model to predict the number of housing units for the indicated category. Use the equation $y = mx + b$ and let $x = $ the year and $y = $ the number of housing units.

1. The number of resident housing units in the state of Hawaii.

2. The number of resident housing units in the city and county of Honolulu.

3. The number of resident housing units in the county of Kauai.

4. The number of resident housing units in the county of Maui.

5. The number of nonresident housing units in the state of Hawaii.

6. The number of nonresident housing units in the city and county of Honolulu.

Continued

7. The number of nonresident housing units in the county of Kauai.

8. The number of nonresident housing units in the county of Maui.

II. Predict the number of housing units for all the categories in exercise I for the year 2004.

Table 3. Resident and Nonresident Housing Units, by County: Annually, 1984 to 1994

Category and Year	State Total	City and County of Honolulu	Other Counties			
			Total	Hawaii	Kauai	Maui[a]
RESIDENT[b]						
1984	336,882	255,450	81,432	37,262	14,720	29,450
1985	336,777	256,396	80,381	37,775	14,352	28,254
1986	340,417	258,713	81,704	38,954	13,982	28,768
1987	347,529	262,898	84,631	39,920	14,059	30,652
1988	354,314	267,885	86,429	41,043	14,016	31,370
1989	362,912	272,272	90,640	43,502	14,139	32,999
1990	370,682	276,618	94,064	46,138	15,047	32,879
1991	378,438	280,167	98,271	48,178	16,114	33,979
1992	384,961	280,672	104,289	51,087	16,752	36,450
1993	392,749	285,200	107,549	53,176	16,526	37,847
1994	399,501	289,864	109,637	54,721	15,878	39,038
NONRESIDENT[c]						
1984	17,602	6,887	10,715	1,902	1,598	7,215
1985	21,804	9,016	12,788	2,279	2,076	8,433
1986	22,999	9,838	13,161	2,052	2,574	8,535
1987	21,896	9,236	12,660	2,102	2,631	7,927
1988	21,120	7,635	13,485	2,354	2,855	8,276
1989	19,140	5,919	13,221	2,018	3,034	8,169
1990	19,128	5,065	14,063	2,115	2,566	9,382
1991	19,887	4,668	15,219	2,401	2,535	10,283
1992	22,645	8,133	14,512	2,334	2,687	9,491
1993	21,081	7,820	13,261	2,220	1,669	9,372
1994	21,247	6,835	14,412	2,483	2,880	9,049

[a]Including Kalawao County (Kalaupapa Settlement).
[b]Estimated as of April 1. Includes all housing units other than condominium units in rental pools and intended for transient occupancy.
[c]Condominium units in rental pools and intended for transient occupancy, based on survey data from the Hawaii Visitors and Convention Bureau. Includes condo/hotel units.
SOURCE: Hawaii State Department of Business, Economic Development, and Tourism.

GROUP ACTIVITY

1. Working in groups, use different data points to generate linear models for the exercises in part I. Try to determine which models yield more accurate predictions for particular years.

2. Discuss the reasons why a particular model might yield more accurate results for a given year than another model.

FURTHER EXPLORATION

1. When you use the model $y = mx + b$ to predict the number of housing units, what does m measure?

2. Surf the Internet to see if you find housing unit data for the state of Hawaii for the years after 1994. If you are successful, test the accuracy of your models from part I.

3. Use a graphing calculator to find the linear regression equations for the exercises from part I. Compare your equations with the linear regression equations.

6.6 *The Distance and Midpoint Formulas*

Knowing the distance between any two points in the Cartesian plane can often yield useful and practical results. In this section we will discover a formula to calculate that distance. Before examining the formula, let's look at some particular problems.

EXAMPLE 6.6.1   Find the distance between the points (5, 5) and (5, 2).

SOLUTION

We can see from Figure 6.6.1 that the distance between these two points is 3. One might observe from this example that the distance between any two points on the same vertical line is simply the y-coordinate of the top point minus the y-coordinate of the bottom point.

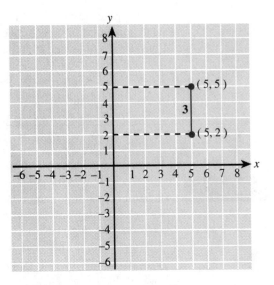

Figure 6.6.1

XAMPLE 6.6.2   Find the distance between the points (5, 2) and (1, 2).

SOLUTION   Here we can see from Figure 6.6.2 that the distance between these two points is 4. Again one might observe from this example that the distance between any two points on the same horizontal line is simply the *x*-coordinate of the right-hand point minus the *x*-coordinate of the left-hand point.

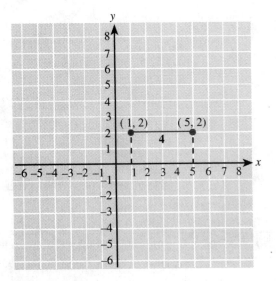

Figure 6.6.2

Suppose we want to find the distance between two points that are not on the same horizontal or vertical line. What steps do we take to calculate that distance? Consider the following example.

EXAMPLE 6.6.3   Find the distance between the points (1, 2) and (5, 5).

SOLUTION

Construct the right triangle indicated in Figure 6.6.3. Let's represent the distance between (1, 2) and (5, 5) with the letter d. Now, from Examples 6.6.1 and 6.6.2 we know that the distance from (5, 5) to (5, 2) is 3, and the distance from (5, 2) to (1, 2) is 4. Since we have a right triangle, by the Pythagorean theorem $4^2 + 3^2 = d^2$. (Remember that the sum of the squares of the legs of a right triangle equals the length of the hypotenuse squared.) This implies that $16 + 9 = d^2$, so that $25 = d^2$ and $d = 5$.

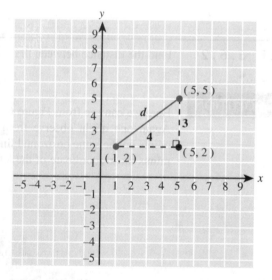

Figure 6.6.3

This is a very long and tedious way of finding the distance between any two points. The Distance Formula, which has the Pythagorean theorem built in, will enable us to calculate the distance between any two points.

Theorem 6.6.1

Distance Formula: Given any two points $A = (x_1, y_1)$ and $B = (x_2, y_2)$, the distance between A and B, denoted $d(A, B)$, is

$$\sqrt{(x_2 - x_1)^2 + (y_2 - y_1)^2}$$

> **Proof**   Construct the right triangle in Figure 6.6.4. The right angle is at $C = (x_2, y_1)$. We know that $d(B, C) = y_2 - y_1$ and $d(A, C) = x_2 - x_1$. Apply the Pythagorean theorem: $[d(A, B)]^2 = (x_2 - x_1)^2 + (y_2 - y_1)^2$, so that $d(A, B) = \sqrt{(x_2 - x_1)^2 + (y_2 - y_1)^2}$.

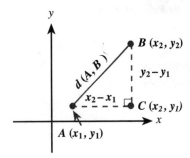

Figure 6.6.4

EXAMPLE 6.6.4

Using the Distance Formula, find the distance between $(1, 2)$ and $(5, 5)$.

SOLUTION

From Example 6.6.3 we know that the distance between these points is 5. We are simply going to use the Distance Formula to verify that result. In the Distance Formula we let $(x_1, y_1) = (1, 2)$ and $(x_2, y_2) = (5, 5)$. Then

$$\begin{aligned} d(A, B) &= \sqrt{(x_2 - x_1)^2 + (y_2 - y_1)^2} \\ &= \sqrt{(5 - 1)^2 + (5 - 2)^2} \\ &= \sqrt{4^2 + 3^2} \\ &= \sqrt{16 + 9} = \sqrt{25} = 5 \end{aligned}$$

NOTE ▶   We would have arrived at the same result if we had let $(x_1, y_1) = (5, 5)$ and $(x_2, y_2) = (1, 2)$.

Try to avoid this mistake:

Incorrect	Correct
$\sqrt{16 + 9} = \sqrt{16} + \sqrt{9}$ $= 4 + 3 = 7$	$\sqrt{16 + 9} = \sqrt{25} = 5$
The square root of a sum is *not* the sum of the square roots.	Always add the numbers and then take the square root.

XAMPLE 6.6.5 Find the distance between $(-2, 6)$ and $(5, -2)$.

SOLUTION Let $(x_1, y_1) = (-2, 6)$ and $(x_2, y_2) = (5, -2)$. Then

$$
\begin{aligned}
d &= \sqrt{(5 - (-2))^2 + (-2 - 6)^2} \\
&= \sqrt{(5 + 2)^2 + (-8)^2} \\
&= \sqrt{7^2 + (-8)^2} \\
&= \sqrt{49 + 64} = \sqrt{113}
\end{aligned}
$$

 XAMPLE 6.6.6 Plot the following three points, and using the Distance Formula, show that they are vertices of a right triangle: $A = (-5, -2)$, $B = (5, 3)$, $C = (3, 7)$.

SOLUTION

$$
\begin{aligned}
d(A, B) &= \sqrt{(5 - (-5))^2 + (3 - (-2))^2} \\
&= \sqrt{10^2 + 5^2} \\
&= \sqrt{100 + 25} = \sqrt{125} \\
d(B, C) &= \sqrt{(3 - 5)^2 + (7 - 3)^2} \\
&= \sqrt{(-2)^2 + 4^2} \\
&= \sqrt{4 + 16} = \sqrt{20} \\
d(A, C) &= \sqrt{(3 - (-5))^2 + (7 - (-2))^2} \\
&= \sqrt{8^2 + 9^2} \\
&= \sqrt{64 + 81} = \sqrt{145}
\end{aligned}
$$

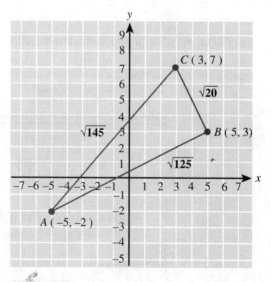

Figure 6.6.5

Thus, $[d(A, C)]^2 = [d(B, C)]^2 + [d(A, B)]^2$ because

$$(\sqrt{145})^2 = (\sqrt{20})^2 + (\sqrt{125})^2.$$

We can conclude that points A, B, and C are the vertices of a right triangle (see Figure 6.6.5).

The midpoint of a line segment has several interesting applications. How can we find that midpoint? Consider the following example.

 EXAMPLE 6.6.7   Find the midpoint M of the line segment from $A = (1, 1)$ to $B = (5, 9)$.

SOLUTION   The midpoint appears to be $(3, 5)$, since 3 is the midpoint of the x-coordinates and 5 is the midpoint of the y-coordinates (Figure 6.6.6). If $(3, 5)$ really is the midpoint, $d(A, M) = d(B, M) = \frac{1}{2}d(A, B)$. Let's verify this fact:

$$d(A, M) = \sqrt{(3 - 1)^2 + (5 - 1)^2}$$
$$= \sqrt{2^2 + 4^2} = \sqrt{4 + 16}$$
$$= \sqrt{20} = \sqrt{4 \cdot 5} = 2\sqrt{5}$$
$$d(B, M) = \sqrt{(3 - 5)^2 + (5 - 9)^2}$$
$$= \sqrt{(-2)^2 + (-4)^2} = \sqrt{4 + 16}$$
$$= \sqrt{20} = \sqrt{4 \cdot 5} = 2\sqrt{5}$$
$$d(A, B) = \sqrt{(5 - 1)^2 + (9 - 1)^2}$$
$$= \sqrt{4^2 + 8^2} = \sqrt{16 + 64}$$
$$= \sqrt{80} = \sqrt{16 \cdot 5} = 4\sqrt{5}$$

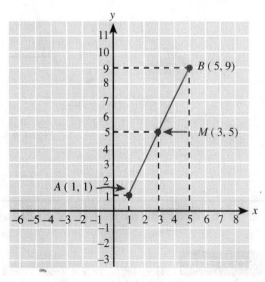

Figure 6.6.6

Since $d(A, M) = d(B, M) = \frac{1}{2}d(A, B)$, $(3, 5)$ is the midpoint of $\overline{AB}$.

As in the case of the Distance Formula, we can derive a formula that will always yield the midpoint of a given line segment.

Theorem 6.6.2

Midpoint Formula: Given any line segment from $A = (x_1, y_1)$ to $B = (x_2, y_2)$, the coordinates of the midpoint M of $\overline{AB}$ are

$$M = \left(\frac{x_1 + x_2}{2}, \frac{y_1 + y_2}{2} \right).$$

REMARK

To find a number that is exactly midway between two given numbers, we take the average of the two numbers. So $\frac{x_1 + x_2}{2}$ is halfway between x_1 and x_2, and $\frac{y_1 + y_2}{2}$ is halfway between y_1 and y_2 (Figure 6.6.7). Thus it seems reasonable that the point $M\left(\frac{x_1 + x_2}{2}, \frac{y_1 + y_2}{2} \right)$ would be halfway between A and B. This does not constitute a proof, which would be a bit laborious and not very enlightening at this point.

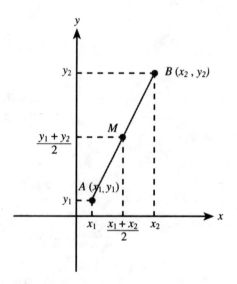

Figure 6.6.7

XAMPLE 6.6.8   Find the midpoint of the line segment from $(-1, 3)$ to $(7, -10)$.

SOLUTION   Here let $(x_1, y_1) = (-1, 3)$ and $(x_2, y_2) = (7, -10)$; then the midpoint is

$$M = \left(\frac{-1 + 7}{2}, \frac{3 + (-10)}{2} \right)$$

$$= \left(\frac{6}{2}, \frac{-7}{2} \right)$$

$$= \left(3, \frac{-7}{2} \right)$$

NOTE ▶   We would have arrived at the same result if we had let $(x_1, y_1) = (7, -10)$ and $(x_2, y_2) = (-1, 3)$.

XAMPLE 6.6.9   We have already shown in Example 6.6.6 that $A = (-5, -2)$, $B = (5, 3)$, and $C = (3, 7)$ form a right triangle. Show that the midpoint of the hypotenuse is equidistant from each of the points A, B, and C.

SOLUTION   From Example 6.6.6 we know that $\overline{AC}$ is the hypotenuse. Thus, all we must show is that the midpoint of $\overline{AC}$ is equidistant from each of the points A, B, and C. See Figure 6.6.8. The midpoint M of $\overline{AC}$ is

$$M = \left(\frac{3 + (-5)}{2}, \frac{7 + (-2)}{2} \right)$$

$$= \left(\frac{-2}{2}, \frac{5}{2} \right) = \left(-1, \frac{5}{2} \right)$$

Now

$$d(A, M) = \sqrt{(-1 - (-5))^2 + \left(\frac{5}{2} - (-2) \right)^2}$$

$$= \sqrt{4^2 + \left(\frac{9}{2} \right)^2} = \sqrt{16 + \frac{81}{4}} = \sqrt{\frac{145}{4}} = \frac{\sqrt{145}}{2}$$

$$d(M, C) = \sqrt{(3 - (-1))^2 + \left(7 - \frac{5}{2} \right)^2}$$

$$= \sqrt{4^2 + \left(\frac{9}{2} \right)^2} = \sqrt{16 + \frac{81}{4}} = \sqrt{\frac{145}{4}} = \frac{\sqrt{145}}{2}$$

$$d(M, B) = \sqrt{(5 - (-1))^2 + \left(3 - \frac{5}{2} \right)^2}$$

$$= \sqrt{6^2 + \left(\frac{1}{2} \right)^2} = \sqrt{36 + \frac{1}{4}} = \sqrt{\frac{145}{4}} = \frac{\sqrt{145}}{2}$$

Since all these distances are equal, we can conclude that the midpoint of the hypotenuse is equidistant from the points A, B, and C.

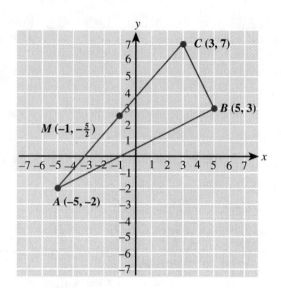

Figure 6.6.8

Exercises 6.6

In exercises 1–18, determine the length and midpoint of the line segments with the given endpoints.

1. $(6, 3)$ and $(6, -5)$

2. $(-2, -4)$ and $(8, -4)$

3. $(-1, 3)$ and $(5, 11)$

4. $(-9, -8)$ and $(-1, 7)$

5. $(1, -5)$ and $(-2, 5)$

6. $(5, 3)$ and $(3, -2)$

7. $(-5, 9)$ and $(3, -1)$

8. $(-3, 4)$ and $(1, -4)$

9. $(0, 3)$ and $(-2, -6)$

10. $(2, -11)$ and $(9, 13)$

11. $(-2, 1)$ and $(4, 7)$

12. $(0, -5)$ and $(-8, 3)$

13. $\left(2, \dfrac{1}{3}\right)$ and $\left(-\dfrac{1}{2}, -3\right)$

14. $\left(-\dfrac{4}{3}, 1\right)$ and $\left(-\dfrac{1}{2}, -\dfrac{1}{4}\right)$

15. $(-2.1, 4.3)$ and $(-8.5, -6)$

16. $(4.89, -3.17)$ and $(12.4, 6.92)$

17. $(\sqrt{12}, \sqrt{20})$ and $(\sqrt{3}, \sqrt{5})$

18. $(\sqrt{24}, \sqrt{8})$ and $(\sqrt{54}, \sqrt{50})$

19. If the midpoint of $\overline{AB}$ is $(4, -1)$ and $A = (9, 4)$, find B.

20. If the midpoint of $\overline{AB}$ is $(1, 2)$ and $A = (8, -5)$, find B.

21. Find the point that is three-fourths of the way from A to B, where $A = (-2, -6)$ and $B = (8, 10)$.

22. Find the point that is three-fourths of the way from B to A, where $A = (-1, 7)$ and $B = (7, -9)$.

23. Plot the following three points, and using the Distance Formula, show that they are vertices of a right triangle: $A = (-1, 6)$, $B = (1, 2)$, and $C = (-5, -1)$.

24. Show that the midpoint of the hypotenuse in problem 23 is equidistant from the vertices.

25. Plot the following three points, and using the Distance Formula, show that they are vertices of an isosceles triangle: $A = (2, 5)$, $B = (-4, -1)$, and $C = (-6, 7)$.

26. Plot the following three points, and using the Distance Formula, show that they are vertices of a right isosceles triangle: $A = (-3, 7)$, $B = (3, 2)$, and $C = (-8, 1)$.

27. Plot the following four points and show that they are vertices of a rectangle: $A(2, 3)$, $B(4, 2)$, $C(8, 10)$, and $D(6, 11)$.

28. Plot the following four points and show that they are vertices of a square: $A(-3, 2)$, $B(-1, 0)$, $C(-3, -2)$, and $D(-5, 0)$.

29. Find all points of the form $(x, -3)$ that are 10 units from $(2, 5)$.

30. Find all points of the form $(1, y)$ that are 5 units from $(-2, 3)$.

31. Show that the points $P(-1, 3)$, $Q(7, -1)$, and $R(6, 2)$ all lie on a circle whose center is the point $C(2, -1)$. Find an equation that would determine whether an arbitrary point (x, y) lies on this circle.

33. Find the coordinates of M, the midpoint of $\overline{AB}$, and N, the midpoint of $\overline{BC}$. Show $\overline{MN}$ is half as long as $\overline{AC}$.

32. Show that the diagonals have the same length.

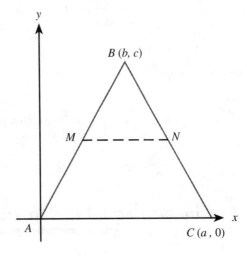

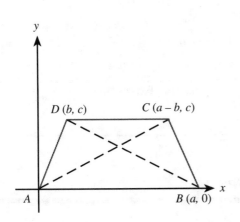

34. Suppose a circle is drawn in the Cartesian coordinate system with a diameter that has end points $A(-2, 3)$ and $B(4, 0)$. What are the coordinates of the center of the circle and what is the length of its radius?

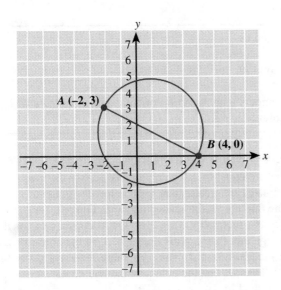

Write Algebra

35. Explain the midpoint formula in your own words.

36. Show that $(\frac{a+b}{2}, \frac{c+d}{2})$ lies on the line joining (a, c) to (b, d). (*Hint:* Find the equation of the line joining (a, c) and (b, d) and show that $(\frac{a+b}{2}, \frac{c+d}{2})$ is a solution of that equation.)

G R O U P   A C T I V I T Y

In exercises 37–42, find the equation of the line passing through the two given points. Write your answer in slope-intercept form. Next, find the distance between the points.

37. $(1, 5)$ and $(4, 11)$

38. $(-1, -7)$ and $(3, 5)$

39. $(-2, -8)$ and $(1, 1)$

40. $(4, 3)$ and $(6, 7)$

41. $(-3, -11)$ and $(0, 1)$

42. $(2, 13)$ and $(-3, -7)$

43. Let (x_1, y_1) and (x_2, y_2) be any two distinct points on the line $y = mx + b$. Express the distance between these two points in terms of x_1, x_2, and m.

6.7 *Linear Inequalities in Two Variables*

As we saw in Section 6.4, graphing linear equations is a relatively simple task. In this section we will extend our work from graphing linear equations to graphing **linear inequalities in two variables.** In addition, graphing linear inequalities will lead us to solving systems of linear inequalities.

Definition 6.7.1	An inequality that can be written as

$$Ax + By < C \quad \text{or} \quad Ax + By \leq C$$

where A, B, and C are real numbers with A and B not both zero, is called a **linear inequality in two variables.**

Consider the inequality $y \geq 3$. Now $y = 3$ is a horizontal line that consists of all the points whose y-coordinates are equal to 3. The points that satisfy the inequality $y > 3$ are those whose y-coordinates are greater than 3. Therefore the graph of $y \geq 3$ is all the points on or above the line $y = 3$, the shaded region in Figure 6.7.1.

Similarly, if we consider the inequality $x > 2$, its graph consists of all the points whose x-coordinates are greater than 2. This time $x \neq 2$, so the points on the line $x = 2$ do not satisfy the inequality $x > 2$. Thus the graph is the region to the right of the vertical line $x = 2$. The fact that the points on the line do not satisfy the inequality is indicated by the dashed line in Figure 6.7.2.

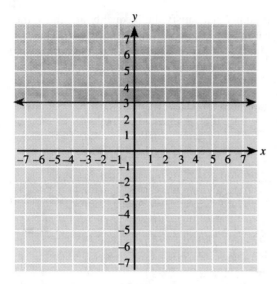

Figure 6.7.1

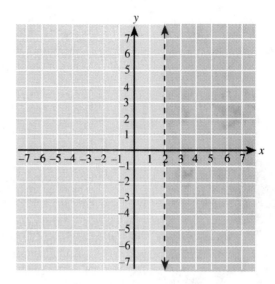

Figure 6.7.2

NOTE ▶   The lines $y = 3$ and $x = 2$ in the preceding two examples are called **boundary lines.** *The equation of the boundary line is determined by replacing the inequality symbol by an equals sign.*

Now let us consider the case where the boundary line is neither horizontal nor vertical.

Consider an inequality such as $2x - 3y \leq 6$. As in the previous two examples, the boundary line, defined by the equation $2x - 3y = 6$, divides the Cartesian plane into two distinct regions called **half-planes.** And as in the previous examples, the points in one of the half-planes satisfy the inequality.

Since $2x - 3y \leq 6$, all the points on the line $2x - 3y = 6$ satisfy the inequality. Let's pick some points in each half-plane determined by the boundary line and decide whether they satisfy our given inequality. Let's try $(0, 0)$, $(1, 4)$, $(5, 0)$, and $(4, -2)$.

Point	Substituted into Inequality	True or False	Point Satisfies Inequality
$(0, 0)$	$2(0) - 3(0) \leq 6$		
	$0 \leq 6$	True	Yes
$(1, 4)$	$2(1) - 3(4) \leq 6$		
	$-10 \leq 6$	True	Yes
$(5, 0)$	$2(5) - 3(0) \leq 6$		
	$10 \leq 6$	False	No
$(4, -2)$	$2(4) - 3(-2) \leq 6$		
	$14 \leq 6$	False	No

Notice in Figure 6.7.3 that the points that satisfy the inequality are above the line $2x - 3y = 6$, whereas those that fail are below the line $2x - 3y = 6$. Thus all the points above and including the line $2x - 3y = 6$ determine the graph of the solution set of $2x - 3y \leq 6$. This is the shaded region in Figure 6.7.4.

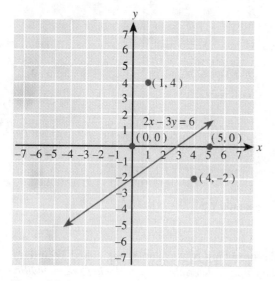

Figure 6.7.3

Figure 6.7.4

The steps used in graphing linear inequalities are as follows:

1. Graph the boundary line. If the inequality is $\geq$ or $\leq$, the line is solid. If the inequality is $>$ or $<$, the boundary line is dashed. (A dashed boundary line indicates that points on the line do not satisfy the inequality.)

2. The boundary line divides the Cartesian plane into two half-planes. Pick a point in one half-plane (be sure that the point is not on the boundary line) and substitute its coordinates into the given inequality. If that point yields a true statement, then the half-plane that includes this point is the graph of the solution set. If that point does not yield a true statement, then the other half-plane is the graph of the solution set.

XAMPLE 6.7.1

Sketch the graph of the solution set of $x + 2y < -4$.

SOLUTION

First sketch the boundary line $x + 2y = -4$. Since the inequality is $<$, the boundary line is dashed. Now we pick some point that is not on the boundary line. The easiest point to use is the origin. Substituting $(0, 0)$ into the given inequality, we obtain $0 + 2(0) < -4$; but $0 \nless -4$. Thus the half-plane that contains $(0, 0)$ is *not* the graph of the solution set. We indicate this by shading in the region below the boundary line in Figure 6.7.5.

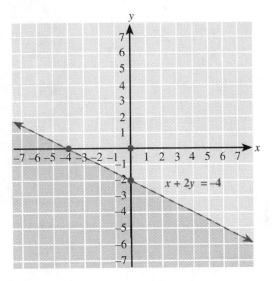

Figure 6.7.5

When we refer to a **system of inequalities,** we are simply considering two or more inequalities together. The solution set of a system of inequalities is the set of points that satisfy all of the given inequalities. Consider the following examples.

XAMPLE 6.7.2

Sketch the graph of the solution set of the following system of inequalities:

$$\begin{cases} x + y < 2 & \text{(1)} \\ 2x - y \geq 0 & \text{(2)} \end{cases}$$

SOLUTION

First determine the solution set of $x + y < 2$. The boundary line is the dashed line in Figure 6.7.6. Picking the point $(0, 0)$ and substituting into inequality (1), we get $0 + 0 < 2$, a true statement, so that the graph of the solution set of inequality (1) is the region below $x + y = 2$.

Next we graph the boundary line $2x - y = 0$. The boundary line in this case is solid. Next we pick some point that is not on the boundary line; let's try $(4, 0)$. Substituting $(4, 0)$ into inequality (2), we obtain $2(4) - 0 \geq 0$; this is equivalent to $8 \geq 0$, a true statement. Thus the points in the half-plane to the right of the line $2x - y = 0$ satisfy inequality (2). Finally, the solution set of the system of inequalities consists of the points that satisfy both inequalities (the intersection of the solution sets), and these points are indicated by the shaded region in Figure 6.7.7.

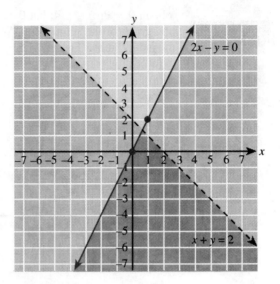

Figure 6.7.6

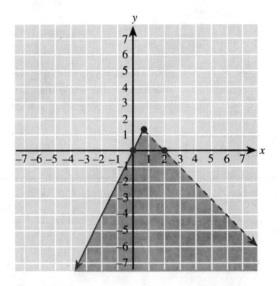

Figure 6.7.7

As a final example let's investigate a system containing three inequalities.

XAMPLE 6.7.3   Sketch the graph of the solution set of the following system of inequalities:

$$\begin{cases} x - y < 2 & (1) \\ 2x + y < 4 & (2) \\ x \ge -1 & (3) \end{cases}$$

SOLUTION   Graph $x - y = 2$ as a dashed line. Now trying $(0, 0)$ in inequality (1), we get $0 - 0 < 2$, so the region above $x - y = 2$ satisfies inequality (1). Now graph $2x + y = 4$, again with a dashed line. Try $(0, 0)$ in inequality (2): $2(0) + 0 < 4$. This is equivalent to $0 < 4$, so the region below $2x + y = 4$ satisfies inequality (2). Finally graph the line $x = -1$. This boundary line is solid. To solve inequality (3) we want points with x values greater than or equal to -1. These points are to the right of the boundary line. See Figure 6.7.8.

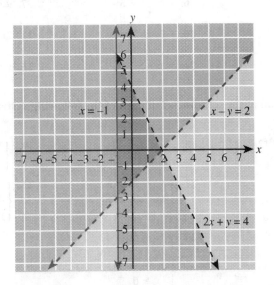

Figure 6.7.8

We can now determine the solution set of the system of inequalities. This solution set consists of all points that satisfy all three inequalities, and its graph is indicated by the triangular region in Figure 6.7.9.

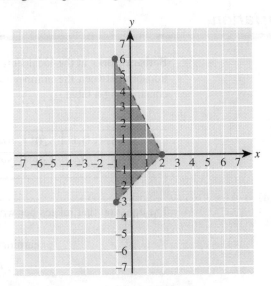

Figure 6.7.9

Exercises 6.7

Sketch the graph of the solution set of each of the following inequalities.

1. $y - 2x < 6$

2. $3x + 2y \geq 12$

3. $y \leq \frac{1}{3}x - 1$

4. $y > 4x - 5$

5. $x \leq 2y - 1$

6. $x > \frac{3}{4}y + 3$

7. $9 < 3x - y$

8. $8 > 4x + 3y$

9. $5x > -3y + 7$

10. $-2y \leq 7x - 8$

11. $2x > 3y$

12. $x + 4y \geq 0$

13. $y < 3$

14. $y \geq -5$

15. $x \leq 0$

16. $x > \frac{7}{8}$

Sketch the graph of the solution set of each of the following systems of inequalities.

17. $\begin{cases} 3x + 4y \leq 12 \\ y \geq -5 \end{cases}$

18. $\begin{cases} y < 2x - 3 \\ x \leq 4 \end{cases}$

19. $\begin{cases} 2x + y < 5 \\ 3x - y < 1 \end{cases}$

20. $\begin{cases} 3x \geq y \\ 2x - 5y < 10 \end{cases}$

21. $\begin{cases} y + 3x \leq -5 \\ 21 \leq 2x - 3y \end{cases}$

22. $\begin{cases} 3y - 2x > 3 \\ 4x - 6y > 5 \end{cases}$

23. $\begin{cases} y - x < 1 \\ 4y + 5x < 20 \\ y > -1 \end{cases}$

24. $\begin{cases} 7y - 2x < 27 \\ 3x - 2y < 2 \\ x + 5y > 12 \end{cases}$

25. $\begin{cases} 2x + 3y \leq 9 \\ 4x + y \leq 8 \\ x \geq 0 \\ y \geq 0 \end{cases}$

26. $\begin{cases} x + 6y \leq 42 \\ 2x + y \leq 18 \\ 4x + y \leq 32 \\ x \geq 0 \\ y \geq 0 \end{cases}$

27. Describe all the points in the first quadrant by using a system of linear inequalities.

28. Describe all the points in the third quadrant by using a system of linear inequalities.

6.8 *Variation*

Barbara works at the local library earning $7.00 an hour. Let us consider the relationship between the number of hours that she works, H, and her earnings, E, by constructing a table.

H	5	10	20	40
E	35	70	140	280

The equation that relates her working hours and her earnings is

$$\text{earnings} = 4 \cdot (\text{number of hours})$$
$$E = 4H$$

We can see that the larger H is, the larger E will be.

In algebra this type of relationship between two variables is called **direct variation,** or *direct proportion.* This leads us to the following definition.

| Definition 6.8.1 | *y* **varies directly** as *x*, or *y* is *directly proportional* to *x*, if $y = kx$, where *k* is a constant, called the **constant of variation** or *constant of proportionality.* |

In the previous example the constant of variation is 7.

Let us consider another example of direct variation. The area of a circle can be found by using the formula $A = \pi r^2$, where *r* is the radius of the circle. This time we have one quantity, the area, varying directly as the *square* of another quantity, the radius. In this example the constant of variation is π. We use Definition 6.8.1 to work the following example.

EXAMPLE 6.8.1

1. *y* varies directly as *x*. If $y = 24$ when $x = 3$, find *k*, the constant of variation.

SOLUTION

Since

$$\underbrace{y \text{ varies directly as } x}$$
$$y \quad = k \cdot \quad x$$

Now we substitute the given values for *x* and *y* and solve for *k*.

$$24 = k \cdot 3$$
$$8 = k$$

2. *y* varies directly as the cube of *x*. If $y = 2$ when $x = 2$, find *k*, the constant of variation.

SOLUTION

Since *y* varies directly as the cube of *x*,

$$y = k \cdot x^3$$

Now we substitute the given values for *x* and *y* and solve for *k*.

$$2 = k \cdot 2^3$$
$$2 = k \cdot 8$$
$$\frac{1}{4} = k$$

The next example takes variation problems a little further.

EXAMPLE 6.8.2 y varies directly as the square of x. $y = 25$ when $x = 10$. Find y when $x = 8$.

SOLUTION   Since y varies directly as the square of x,

$$y = k \cdot x^2$$

Now we substitute 25 for y and 10 for x and solve for k.

$$25 = k \cdot 10^2$$
$$25 = k \cdot 100$$
$$\frac{1}{4} = k$$

We now know

$$y = \frac{1}{4}x^2$$

Thus, when $x = 8$,

$$y = \frac{1}{4} \cdot 8^2$$
$$= \frac{1}{4} \cdot 64$$
$$= 16$$

Solving variation problems is basically a five-step process.

1. Find an equation that relates the variables.
2. Substitute the first set of given values for the variables and solve the resulting equation for the constant of variation.
3. Substitute the value of the constant of variation into the equation from step 1.
4. Substitute the new values for the variables into the equation from step 3.
5. Solve the resulting equation.

Let's now investigate an application problem involving direct variation.

EXAMPLE 6.8.3   Hooke's law states that the force required to stretch a spring beyond its natural length varies directly as the increase in the spring's length. If a force of 18 lb is nec-

essary to stretch a spring from its natural length of 20 in. to a length of 23 in., how much force will be required to stretch the spring from its natural length to a length of 25 in.?

SOLUTION

Analyzing the first sentence, we can extract an equation as follows:
Force varies directly as the increase in the spring's length.

$$F = kx \qquad \text{\textit{Set up the equation.}}$$

See Figure 6.8.1. In this equation $F = kx$, F and x are variables, and k is the constant of variation that needs to be determined.

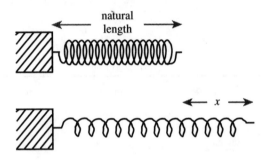

Figure 6.8.1

We are told that when the force is 18 lb, the spring is stretched from 20 to 23 in., an increase of 3 in. So

$$F = kx$$
$$18 = k \cdot 3 \qquad \text{\textit{Substitute the given values for the variables.}}$$
$$6 = k$$

Now we can rewrite the equation, substituting the value for k:

$$F = 6x \qquad \text{\textit{Substitute for the constant of variation.}}$$

The question is, What force is required to stretch the spring from 20 to 25 in., an increase of 5 in.

$$F = 6x$$
$$F = 6 \cdot 5 \qquad \text{\textit{Substitute the new value for x.}}$$
$$F = 30 \text{ lb}$$

NOTE ▶ Stretching the spring a greater distance requires a larger force; that is, as x increases, F increases.

There is an opposite type of relationship that can exist between two quantities. Suppose Randy is building a patio in the shape of a rectangle, and he can afford to buy only enough concrete to make a patio with an area of 144 sq ft. He could make the width 9 ft, and then the length would be 16 ft. If he increased the width to 12 ft, the length would decrease from 16 to 12 ft. On the other hand, if he decreased the width from 9 to 8 ft, the length would increase from 16 to 18 ft. In either case, changes in the width and length behave the opposite or inversely of each other. The larger one dimension is, the smaller the other dimension is. An equation that relates the length, L, and the width, W, is $L = \frac{144}{W}$. This leads to the next definition.

Definition 6.8.2	y **varies inversely** as x, or y is *inversely proportional* to x, if $y = \frac{k}{x}$, where k is the constant of variation.

In the previous example, the constant of variation is 144. Let's investigate an example involving inverse variation.

EXAMPLE 6.8.4 y varies inversely as the cube of x. $y = 3$ when $x = 2$. Find y when $x = 3$.

SOLUTION   Since y varies inversely as the cube of x,

$$y = \frac{k}{x^3} \qquad \text{Set up the equation.}$$

Now substitute 3 for y and 2 for x and solve for k.

$$3 = \frac{k}{2^3} \qquad \text{Substitute the given values for the variables.}$$

$$3 = \frac{k}{8}$$

$$24 = k$$

We now know

$$y = \frac{24}{x^3} \qquad \text{Substitute for the constant of variation.}$$

Thus, when $x = 3$,

$$y = \frac{24}{3^3} \qquad \text{Substitute the new value for x.}$$

$$= \frac{24}{27}$$

$$= \frac{8}{9}$$

The following example illustrates an application of inverse variation.

EXAMPLE 6.8.5

The like poles of two magnets repel each other with a force that is inversely proportional to the square of the distance between them. If two magnets are 3 cm apart and repel each other with a force of 0.6 newton (N), with how much force do they repel each other when they are 2 cm apart?

REMARK

Don't let the formulas and strange units in these problems scare you. You don't have to know any physics or chemistry to solve these problems. They can all be worked in five steps, involving only some simple algebra.

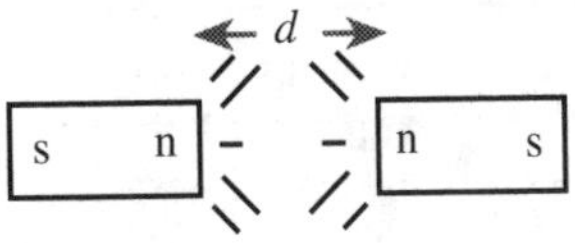

Figure 6.8.2

SOLUTION

See Figure 6.8.2. From the first sentence we get the equation

$$F = \frac{k}{d^2}$$ *Set up the equation.*

Then we are given 3 cm for d and 0.6 N for F:

$$0.6 = \frac{k}{3^2}$$ *Substitute the given values for the variables.*

$$9(0.6) = k$$

$$5.4 = k \quad \text{so} \quad F = \frac{5.4}{d^2}$$ *Substitute for the constant of variation.*

Now we find F when d is 2 cm:

$$F = \frac{5.4}{2^2} = \frac{5.4}{4} = 1.35 \ N$$ *Substitute new value for d and find F.*

When the distance between the magnets is smaller, the force is larger; that is, as d decreases, F increases.

Sometimes we have situations in which more than two variables are varying with respect to one another. For example, consider the can shown in Figure 6.8.3, which is in the shape of a right circular cylinder. The volume, V, depends on the height, h, of the can and the radius, r, of the (circular) base. The relation between

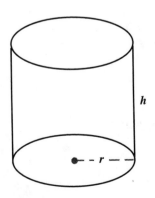

Figure 6.8.3

the three variables is given by the formula $V = \pi r^2 h$. We can see that as either the radius or the height increases, the volume also increases. So the volume varies directly with respect to both the radius and the height. This is an example of what is called **joint variation.** It leads to the next definition.

Definition 6.8.3	y **varies jointly** as x and w if $y = kxw$, where k is the constant of variation.

In the previous example we would say that V varies jointly as h and the square of r. The constant of variation is π.

Joint variation is a natural extension of our previous work in this section. Consider the next example.

EXAMPLE 6.8.6 y varies jointly as x and w^2. $y = 108$ when $x = 2$ and $w = 9$. Find y when $x = 4$ and $w = 3$.

SOLUTION   Since y varies jointly as x and w^2,

$$y = k \cdot x \cdot w^2 \qquad \textit{Set up the equation.}$$

Now we substitute 108 for y, 2 for x, and 9 for w.

$$108 = k \cdot 2 \cdot 9^2 \qquad \textit{Substitute the given values for the variables.}$$
$$108 = k \cdot 162$$
$$\frac{108}{162} = k$$
$$\frac{2}{3} = k$$

We now know

$$y = \frac{2}{3} xw^2 \qquad \textit{Substitute for the constant of variation.}$$

Thus, when $x = 4$ and $w = 3$,

$$y = \frac{2}{3} \cdot 4 \cdot 3^2 \qquad \textit{Substitute new values for } \text{x} \textit{ and } \text{w.}$$

$$= \frac{2}{3} \cdot 4 \cdot 9$$

$$= 24$$

Let's investigate an application involving both direct and inverse variation. In this application, we are working with Boyle's law and Charles's law.

XAMPLE 6.8.7   The volume of a certain gas varies directly as the temperature and inversely as the pressure. When the temperature is 480 kelvins and the pressure is 12 lb/sq in., the gas occupies a volume of 200 cu in. Find the volume when the temperature is 360 kelvins and the pressure is 15 lb/sq in.

SOLUTION   From the first sentence we get the equation

$$V = \frac{k \cdot T}{P} \qquad \text{Set up the equation.}$$

To determine k, we take the values for T, P, and V from the second sentence:

$$200 = \frac{k \cdot 480}{12} \qquad \text{Substitute the given values for the variables.}$$

$$\frac{200 \cdot 12}{480} = k$$

$$5 = k$$

So

$$V = \frac{5T}{P} \qquad \text{Substitute for the constant of variation.}$$

Now we find V using the new values for T and P from the last sentence:

$$V = \frac{5 \cdot 360}{15} \qquad \text{Substitute new values for T and P.}$$

$$= 120 \text{ cu in.}$$

E xercises 6.8

Find the constant of variation in each of the following.

1. y varies directly as x, and $y = 32$ when $x = 4$.

2. y varies directly as x, and $y = 35$ when $x = 7$.

3. y varies directly as the cube root of x, and $y = 2$ when $x = 64$.

4. y varies directly as the square root of x, and $y = \frac{5}{2}$ when $x = 100$.

5. y varies inversely as x^2, and $y = \frac{1}{4}$ when $x = 6$.

6. y varies inversely as x^4, and $y = 1$ when $x = 2$.

7. y varies jointly as $\sqrt{x}$ and $\sqrt[3]{w}$; $y = -30$ when $x = 25$ and $w = 27$.

8. y varies jointly as x^2 and w^3; $y = -216$ when $x = 3$ and $w = 2$.

9. y varies directly as x^3 and inversely as $\sqrt{w}$; $y = 3$ when $x = 2$ and $w = 16$.

10. y varies directly as $\sqrt[3]{x}$ and inversely as w^2; $y = \frac{1}{2}$ when $x = 27$ and $w = 2$.

Write an equation for each of the following sentences, and determine the constant of variation.

11. The circumference of a circle varies directly as the radius of the circle. The circumference is 10π in. when the radius is 5 in.

12. The hypotenuse of an isosceles right triangle is directly proportional to one of its legs. The hypotenuse is $3\sqrt{2}$ ft when the leg is 3 ft.

13. The surface area of a cube varies directly as the square of an edge. The surface area is 96 sq m when the edge is 4 m.

14. The volume of a rectangular box varies jointly as its length, width, and height. The volume is 75 cu in. when the length is 6 in., the width is 2.5 in., and the height is 5 in.

15. The area of a trapezoid varies jointly as its height and the sum of its bases. The area is 18 sq cm when the height is 3 cm and the bases are 5 cm and 7 cm.

16. The base of a triangle is directly proportional to the area and inversely proportional to the height. The base is 12 in. when the area is 30 sq in. and the height is 5 in.

17. The height of a right circular cone varies directly as the volume and inversely as the square of the radius of the base. The height is 7 ft when the volume is 84π cu ft and the radius is 6 ft.

18. The radius of a sphere varies directly as the square root of the surface area. The radius is 5 cm when the surface area is 100π sq cm.

Solve the following problems.

19. y varies directly as x^3. $y = 32$ when $x = 2$. Find y when $x = 3$.

20. y varies directly as x^4. $y = 48$ when $x = 2$. Find y when $x = 3$.

21. y varies inversely as x. $y = 2$ when $x = 5$. Find y when $x = 4$.

22. y varies inversely as x. $y = 3$ when $x = 4$. Find y when $x = 6$.

23. y varies jointly as x and w^3. $y = 162$ when $x = 1$ and $w = 3$. Find y when $x = 3$ and $w = 2$.

24. y varies jointly as x^2 and w^2. $y = 252$ when $x = 2$ and $w = 3$. Find y when $x = 3$ and $w = 3$.

25. y varies directly as x and inversely as $\sqrt{w}$. $y = 6$ when $x = 3$ and $w = 16$. Find y when $x = 6$ and $w = 25$.

26. y varies directly as x and inversely as $\sqrt{w}$. $y = 6$ when $x = 2$ and $w = 9$. Find y when $x = 5$ and $w = 16$.

27. y varies jointly as x and w^2 and inversely as z. $y = 4$ when $x = 3$, $w = 2$, and $z = 6$. Find y when $x = 4$, $w = 3$, and $z = 12$.

28. y varies jointly as x^2 and w^2 and inversely as z. $y = 6$ when $x = 2$, $w = 3$, and $z = 18$. Find y when $x = 5$, $w = 2$, and $z = 15$.

Solve the following application problems.

29. When an object is dropped, the distance that it falls in t seconds varies directly as the square of t. If an object falls 4 ft in $\frac{1}{2}$ sec, how far will it fall in 3 sec?

30. Hooke's law states that the force required to stretch a spring beyond its natural length varies directly as the increase in the spring's length. If a force of 30 lb is necessary to stretch a spring from its natural length of 17 in. to a length of 19.5 in., to what length will a force of 64 lb stretch the spring?

31. Hooke's law also applies when a spring is being compressed; that is, the force required to compress the spring varies directly as the amount the spring has been compressed. If a force of 12 lb will compress a spring from a natural length of 32 in. to a length of 30 in., how much force will be required to compress the spring to a length of 25 in.?

32. In a certain city the sales tax is directly proportional to the amount purchased. If the tax on a tape recorder that costs $49.83 is $2.99, what will be the total cost for a camera priced at $189.95?

33. When a principal is invested in a bank, the amount of simple interest it earns varies jointly as the rate of interest and the time, in years, that it is left in the bank. If a certain principal invested at a rate of 5% earns $180 of interest in 3 yr, how much interest will the same principal earn in 15 mo in another bank at a rate of 6%? (*Hint:* Remember that time is measured in years.)

35. In Miss Grundy's algebra class everybody watches TV for at least 2 hr every day. It has been determined that the grade a student earns on an algebra test varies inversely as the number of hours that the student spent watching TV the day before the test. If Red Eyes watched TV for 8 hr and got 25% on his algebra test, what should Ima Saint get on her test after watching 2 hr of TV?

37. The volume of a certain gas varies directly as the temperature and inversely as the pressure. When the temperature is 400 kelvins and the pressure is 12 lb/sq in., the volume is 80 cu in. Find the pressure on the same gas when the volume is 54 cu in. and the temperature is 450 kelvins.

39. The force of gravitational attraction between two bodies varies jointly as the mass m_1 and the mass m_2 of each of the bodies, and inversely as the square of the distance between them. The force of attraction between Bea Flea, whose mass is 1.50×10^{-6} kg, and Lee Flea, whose mass is 1.80×10^{-6} kg, is 5.00×10^{-18} N when they are 6.00×10^{-6} m apart. Find the force of attraction between Wanda Lean, whose mass is 50.00 kg, and Bill Bulk, whose mass is 90.5 kg, when they are 5.79×10^{-2} m apart.

34. The period of a pendulum (that is, the time it takes to swing back and forth once) varies directly as the square root of the length of the pendulum. If a pendulum that is 9 ft long takes $3\frac{1}{3}$ sec to swing back and forth, what will be the period of a pendulum that is 4 ft long?

36. According to Ohm's law, the resistance in a simple electric circuit is directly proportional to the voltage and inversely proportional to the current. If the resistance is 8 ohms (Ω) when the voltage is 48 V and the current is 6 amperes (A), find the resistance when the voltage is 136 V and the current is 8 A.

38. The kinetic energy of a moving object varies jointly as its mass and the square of its velocity. If the kinetic energy is 10 J when an object of mass 5 kg is traveling at a velocity of 2 m/sec, find the kinetic energy of an object of mass 4000 kg that is traveling at a velocity of 5 m/sec.

40. The maximum load that can be supported by a beam with a rectangular cross section varies jointly as the width and the square of the depth of the cross section and inversely as the length of the beam. A beam that is $1\frac{1}{2}$ in. wide, $3\frac{1}{2}$ in. deep, and 6 ft long can support a load of 1000 lb. What load can be supported by an 8-ft-long beam that is $3\frac{1}{2}$ in. wide and $5\frac{1}{2}$ in. deep?

6.9 *Branch Functions (Optional)*

Oftentimes, especially in higher mathematics, a function will not be defined by just one equation. It might turn out that two or more equations are needed to represent some situation. Consider overnight package delivery rates charged by Otis's Overnight Express of Moss County. The charge is $20 for next-day delivery (within Moss County) of a package weighing 1 lb or less, $25 if the package weighs more than 1 lb while not exceeding 3 lb, $35 if the package weighs more than 3 lb while not exceeding 5 lb, and $50 if the package weighs more than 5 lb up to 10 lb. (*Note:* Otis does not deliver packages weighing more than 10 lb.) Figure 6.9.1 represents pictorially the graph of Otis's prices.

This graph shows that if our package weighs 4.25 lb, the charge will be $35. If the package weighs 7.85 lb, the charge will be $50. In Figure 6.9.1 we let the *x*-coordinate represent the weight and the *y*-coordinate the charge. This is due to the

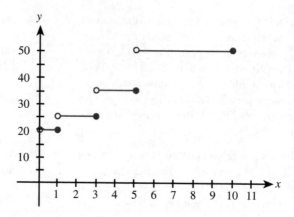

Figure 6.9.1

fact that the charge is a function of how much the package weighs. Notice that this figure has open circles, called **boundary points,** at (0, 20), (1, 25), (3, 35), and (5, 50). Near these points the graph changes. The closed circles at (1, 20), (3, 25), (5, 35) and (10, 50) are also called boundary points. From the graph, we can see that as soon as the package weighs a little more than 3 lb the price jumps to $35. The graph in Figure 6.9.1 can now be described by four equations:

1. $y = 20$ if $0 < x \leq 1$

2. $y = 25$ if $1 < x \leq 3$

3. $y = 35$ if $3 < x \leq 5$

4. $y = 50$ if $5 < x \leq 10$

Remember y is standing for the charge and x the weight of the package. We can combine these four equations into one expression in the following manner:

$$y = \begin{cases} 20 & \text{if } 0 < x \leq 1 \\ 25 & \text{if } 1 < x \leq 3 \\ 35 & \text{if } 3 < x \leq 5 \\ 50 & \text{if } 5 < x \leq 10 \end{cases}$$

Functions defined in the preceding manner are called **piecewise functions,** or **branch functions.**

EXAMPLE 6.9.1   Sketch the graph of the following branch function:

$$f(x) = \begin{cases} 2x + 1 & \text{if } x \le 3 \\ 5 & \text{if } x > 3 \end{cases}$$

To sketch the graph of this branch function, we first consider it as two equations.

1. $y = 2x + 1$ if $x \le 3$

2. $y = 5$ if $x > 3$

From our work in Section 6.4 we know that the graph of each of these equations is a straight line. In equation 1 we can pick only x values less than or equal to 3. By the definition of the function, if we choose x values larger than 3 we must go to equation 2. Thus, some of the ordered pairs that satisfy equation 1 are (3, 7), (2, 5), (1, 3), and (0, 1). In a similar fashion some of the ordered pairs that satisfy equation 2 are (4, 5), (5, 5), and (6, 5). The point (3, 5) is a *boundary point* and is marked with an open circle due to the fact that when $x = 3$ we must use equation 1. In addition, (3, 7) is a boundary point; it is marked with a closed circle because our function does pass through this point. Now the graph of the function can be sketched by drawing in the lines and *stopping* at the boundary points (see Figure 6.9.2).

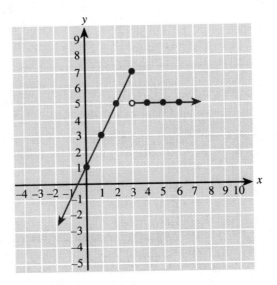

Figure 6.9.2

Try to avoid this mistake:

Incorrect	Correct
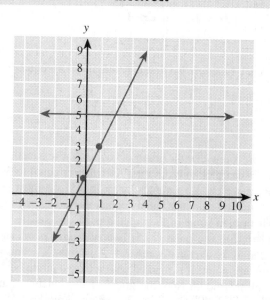	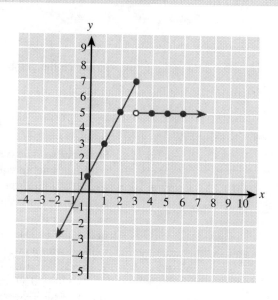
This error assumes that the domain of each equation is all real numbers.	The domain of the top equation is $\{x: x \leq 3\}$. Thus $(3, 7)$ is the rightmost point on the graph of the top equation.
	The domain of the bottom equation is $\{x: x > 3\}$. The graph of the bottom equation will not contain a *leftmost* point. The graph of the bottom equation extends up to but does not include the point $(3, 5)$.

EXAMPLE 6.9.2

Sketch the graph of the following branch function:

$$f(x) = \begin{cases} -3x + 5 & \text{if } x \leq 1 \\ x + 1 & \text{if } x > 1 \end{cases}$$

Again we consider this branch function as two equations:

1. $y = -3x + 5$ if $x \leq 1$

2. $y = x + 1$ if $x > 1$

Points satisfying equation 1 are $(1, 2)$, $(0, 5)$, and $(-1, 8)$ with $(1, 2)$ a boundary point with a closed circle. Some points satisfying equation 2 are $(2, 3)$, $(3, 4)$, and $(4, 5)$ with an open circle at $(1, 2)$ (see Figure 6.9.3). Notice that in this case our two branches lined up and the graph of our function did not have any jumps or open circles as in Example 6.9.1. In addition, note that the open circle at (1, 2) was "filled in" by the graph of the top branch of our function.

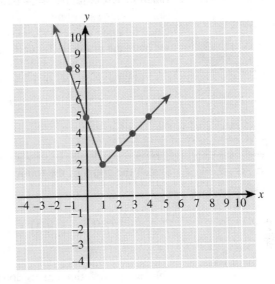

Figure 6.9.3

Branch functions may have their domains broken in many different fashions. Consider the next example.

EXAMPLE 6.9.3   Sketch the graph of the following branch function:

$$f(x) = \begin{cases} 2x + 3 & \text{if } x \neq 1 \\ -2 & \text{if } x = 1 \end{cases}$$

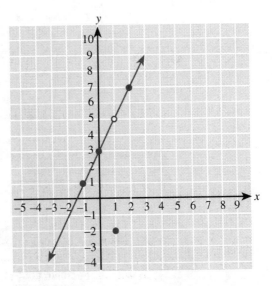

Figure 6.9.4

An easy way to interpret this function is as follows: $y = 2x + 3$ for every value of x except 1, and when $x = 1$ the function defines y to be -2. Hence the graph of our function is the line $y = 2x + 3$ without the point $(1, 5)$, plus the point $(1, -2)$. Some points satisfying $y = 2x + 3$ are $(-1, 1)$, $(0, 3)$, and $(2, 7)$. Remember to include an open circle at $(1, 5)$. In addition we must include the point $(1, -2)$. Thus, we have the graph in Figure 6.9.4.

Finally, let's consider a branch function with three pieces.

E X A M P L E   6 . 9 . 4   Sketch the graph of the following branch function:

$$f(x) = \begin{cases} 4 & \text{if } x < -3 \\ -x + 1 & \text{if } -3 \le x \le 0 \\ 2x - 1 & \text{if } x > 0 \end{cases}$$

This branch function breaks down into three equations:

1. $y = 4$ if $x < -3$

2. $y = -x + 1$ if $-3 \le x \le 0$

3. $y = 2x - 1$ if $x > 0$

Points satisfying equation 1 are $(-4, 4)$, $(-5, 4)$, and $(-6, 4)$ with an open circle at $(-3, 4)$. Points satisfying equation 2 are $(-2, 3)$, and $(-1, 2)$ with closed circles at $(-3, 4)$ and $(0, 1)$. Points satisfying equation 3 are $(1, 1)$, $(2, 3)$, and $(3, 5)$ with an open circle at $(0, -1)$. Connecting the points and stopping at the boundary points yields the graph in Figure 6.9.5.

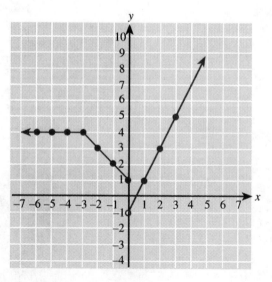

Figure 6.9.5

Exercises 6.9

Sketch the graphs of the following functions.

1. $f(x) = \begin{cases} 5 & \text{if } x \le -4 \\ -3 & \text{if } x > -4 \end{cases}$

2. $f(x) = \begin{cases} -1 & \text{if } x \le 0 \\ 4 & \text{if } x > 0 \end{cases}$

3. $f(x) = \begin{cases} -3 & \text{if } x < 2 \\ -x + 5 & \text{if } x \ge 2 \end{cases}$

4. $f(x) = \begin{cases} 4 & \text{if } x < -1 \\ -x - 3 & \text{if } x \ge -1 \end{cases}$

5. $f(x) = \begin{cases} x - 1 & \text{if } x < 3 \\ -2x + 8 & \text{if } x \ge 3 \end{cases}$

6. $f(x) = \begin{cases} 2x - 1 & \text{if } x < -1 \\ x + 6 & \text{if } x \ge -1 \end{cases}$

7. $f(x) = \begin{cases} -x + 4 & \text{if } x < 0 \\ 2x & \text{if } x \ge 0 \end{cases}$

8. $f(x) = \begin{cases} -2x + 9 & \text{if } x \le 4 \\ 2x - 7 & \text{if } x > 4 \end{cases}$

9. $f(x) = \begin{cases} -5 & \text{if } x \ne 3 \\ 2 & \text{if } x = 3 \end{cases}$

10. $f(x) = \begin{cases} -1 & \text{if } x \ne 2 \\ 4 & \text{if } x = 2 \end{cases}$

11. $f(x) = \begin{cases} -3x + 1 & \text{if } x \ne 2 \\ 4 & \text{if } x = 2 \end{cases}$

12. $f(x) = \begin{cases} x - 2 & \text{if } x \ne -3 \\ 0 & \text{if } x = -3 \end{cases}$

13. $f(x) = \begin{cases} 2x & \text{if } x \ne -1 \\ 3 & \text{if } x = -1 \end{cases}$

14. $f(x) = \begin{cases} 2x + 3 & \text{if } x \ne 1 \\ -1 & \text{if } x = 1 \end{cases}$

15. $f(x) = \begin{cases} x - 4 & \text{if } x \ne 1 \\ -3 & \text{if } x = 1 \end{cases}$

16. $f(x) = \begin{cases} x + 1 & \text{if } x \ne 2 \\ 3 & \text{if } x = 2 \end{cases}$

17. $f(x) = \begin{cases} \frac{1}{2}x + 5 & \text{if } x < -2 \\ -3x + 7 & \text{if } -2 \le x \le 3 \\ x + 1 & \text{if } x > 3 \end{cases}$

18. $f(x) = \begin{cases} x + 6 & \text{if } x < -4 \\ \frac{1}{2}x + 7 & \text{if } -4 \le x \le 2 \\ -x + 5 & \text{if } x > 2 \end{cases}$

19. $f(x) = \begin{cases} -2x + 5 & \text{if } x < 0 \\ 5 & \text{if } 0 \le x < 4 \\ x - 2 & \text{if } x \ge 4 \end{cases}$

20. $f(x) = \begin{cases} -1 & \text{if } x < -1 \\ x & \text{if } -1 \le x < 1 \\ 1 & \text{if } x \ge 1 \end{cases}$

21. $f(x) = \begin{cases} -x + 1 & \text{if } x \le -2 \\ 3 & \text{if } -2 < x < 3 \\ -2x + 9 & \text{if } x \ge 3 \end{cases}$

22. $f(x) = \begin{cases} 2x + 4 & \text{if } x \le -1 \\ 2 & \text{if } -1 < x < 2 \\ -x + 4 & \text{if } x \ge 2 \end{cases}$

23. $f(x) = \dfrac{x}{|x|}$

24. $f(x) = \dfrac{|2x|}{x}$

25. The rates charged by Leroy's Laundromat are determined by weight. If your load is less than 6 lb, Leroy will charge you $8. If your load is at least 6 lb but less than 10 lb, the charge is $20. If the load is at least 10 lb but no more than 14 lb, the charge is $30. (Leroy doesn't handle loads greater than 14 lb.) Convert Leroy's charges into a branch function. Sketch the function.

27. Write a branch function whose graph is a line with a hole in it.

28. Write a branch function for the graph shown.

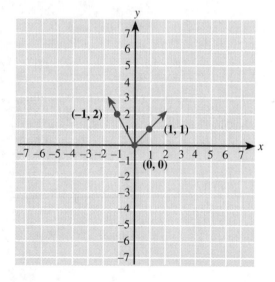

26. Malcolm flies in his hot-air balloon every weekend. The following branch function gives Malcolm's height above ground level. In this function S stands for distance (measured in feet) and t is the amount of flying time (measured in seconds).

$$S = \begin{cases} 2t & \text{if } 0 \le t \le 250 \\ 500 & \text{if } 250 < t \le 1000 \\ 1000 - \frac{1}{2}t & \text{if } t > 1000 \end{cases}$$

a. Graph this function.
b. How high is Malcolm after 40 sec? After 200 sec? When does Malcolm reach his maximum height? What is this maximum height? When does Malcolm land the balloon?

Applied Algebra

A three-story building is on fire. When the firefighters arrive at the scene of the fire, it is determined (for safety purposes) that all firefighters must be at least 70 ft from the fire. Suppose that the fire engines can generate a nozzle pressure of only 48 lb/sq in. What is the minimum nozzle diameter that will enable a stream of water to reach the edge of the fire?

SOLUTION   The maximum horizontal distance of a stream of water is determined by the equation

$$S = 0.5N + 26$$

In this equation S represents the maximum horizontal distance measured in feet, and N is the nozzle pressure in pounds per square inch for a $\frac{3}{4}$-in. diameter nozzle. The horizontal distance of the stream of water will increase 10 ft for every $\frac{1}{4}$-in. increase in nozzle diameter (if the pressure remains constant).

In our case $N = 48$.

$$S = 0.5(48) + 26$$
$$= 24 + 26$$
$$= 50$$

Now when $N = 48$ lb/sq in., $S = 50$ ft for a $\frac{3}{4}$-in.-diameter nozzle. However, our distance must be 70 ft. Since a $\frac{3}{4}$-in. nozzle at 48 lb/sq in. will extend a stream of water only 50 ft, we must increase the distance of the stream by 20 ft. Recall that (if the pressure remains constant) we increase the distance of the stream 10 ft by increasing the diameter of the nozzle by $\frac{1}{4}$ in. Since $\frac{20}{10} = 2$, if we increase the diameter of the nozzle by $2 \cdot (\frac{1}{4})$ in.—i.e., by $\frac{1}{2}$ in.—the stream of water will reach the edge of the fire. Thus we need a nozzle of diameter $\frac{3}{4}$ in. $+ \frac{1}{2}$ in. $= 1\frac{1}{4}$ in. to reach the edge of the fire.

Your Turn

1. Find the minimum nozzle diameter that will enable a stream of water to extend 100 ft if the nozzle pressure is 68 lb/sq in.

2. Find the maximum horizontal distance of a stream of water if the nozzle pressure is 42 lb/sq in. from a $1\frac{1}{2}$-in.-diameter nozzle.

Chapter 6 Review

Terms to Remember

Ordered pair	p. 294	Graph of an equation	p. 299
First coordinate	p. 294	Relation	p. 303
Abscissa	p. 294	Domain	p. 303
Second coordinate	p. 294	Range	p. 303
Ordinate	p. 294	Function	p. 304
Rectangular coordinate system	p. 295	Graph of a function	p. 305
Cartesian coordinate system	p. 295	Vertical line test	p. 306
Origin	p. 295	Functional notation	p. 312
x-axis	p. 295	Independent variable	p. 321
y-axis	p. 295	Dependent variable	p. 321
x-coordinate	p. 296	Linear equation	p. 321
y-coordinate	p. 296	y-intercept	p. 322
Graphing an ordered pair	p. 296	x-intercept	p. 322
Plotting an ordered pair	p. 296	Linear function	p. 325
Quadrant	p. 296	Slope	p. 326
Solution to an equation in two variables	p. 297	Rise	p. 329
		Run	p. 329

Point-slope form	p. 334, 335	Half-plane	p. 361
Standard form	p. 336	System of linear inequalities	p. 363
Slope-intercept form	p. 338	Direct variation	p. 367
Parallel lines	p. 340	Constant of variation	p. 367
Perpendicular lines	p. 340	Inverse variation	p. 370
Distance Formula	p. 352	Joint variation	p. 372
Midpoint Formula	p. 356	Boundary points	p. 376
Linear inequality	p. 360	Piecewise, or branch, function	p. 376
Boundary line	p. 361		

Formulas and Notation

■ *Functional notation:* $f(x)$, read as "f of x," replaces y in the equation defining the function.

■ *Slope:* The line passing through the points (x_1, y_1) and (x_2, y_2) has a slope

$$m = \frac{y_2 - y_1}{x_2 - x_1}$$

■ *Point-slope form:* The equation of the line passing through the point (x_1, y_1) and having slope m is

$$y - y_1 = m(x - x_1)$$

■ *Slope-intercept form:* The equation of the line with slope m and y-intercept $(0, b)$ is

$$y = mx + b$$

■ *Horizontal line:* Every horizontal line has slope $m = 0$. The equation of the horizontal line with y-intercept $(0, b)$ is

$$y = b$$

■ *Vertical line:* The slope of every vertical line is undefined. The equation of the vertical line with x-intercept $(a, 0)$ is

$$x = a$$

■ *Distance Formula:* The distance between two points $A(x_1, y_1)$ and $B(x_2, y_2)$, denoted $d(A, B)$, is

$$d(A, B) = \sqrt{(x_2 - x_1)^2 + (y_2 - y_1)^2}$$

■ *Midpoint Formula:* Given any line segment from $A(x_1, y_1)$ to $B(x_2, y_2)$ the coordinates of the midpoint M of $\overline{AB}$ are

$$M = \left(\frac{x_1 + x_2}{2}, \frac{y_1 + y_2}{2}\right)$$

■ *Direct variation:* y varies directly as x if $y = kx$, where k is the constant of variation.

■ *Inverse variation:* y varies inversely as x if $y = \frac{k}{x}$, where k is the constant of variation.

■ *Joint variation:* y varies jointly as x and w if $y = kxw$, where k is the constant of variation.

Review Exercises

6.1 Graph the following equations.

1. $y = 3x + 2$

2. $y - 2x = 8$

3. $2x - 3y = 6$

4. $y = x^2 + 4$

5. $y = 1 - x^2$

6. $x = y^2 + 1$

7. $y = \sqrt{x + 1}$

8. $y = \sqrt{x - 5}$

6.2 For the relations defined by the following equations, find the domains and ranges, and determine whether each relation is a function.

9. $y = 2x - 7$

10. $3x + 2y = 6$

11. $y = x^2 + 3$

12. $x = y^2 - 2$

13. $y = \sqrt{x - 3}$

14. $y = \sqrt{x - 5}$

6.3 In the following problems let $f(x) = 4x^2 - 3x + 7$ and $g(x) = 5x - 2$. Find the following.

15. $g(2)$

16. $f(-3)$

17. $[f(-3)]^2$

18. $4f(-3) - 5g(2)$

19. $f(x) + g(x)$

20. $f(x) \cdot g(x)$

21. $g(a) + g(h)$

22. $g(a + h)$

23. $\dfrac{g(a + h) - g(a)}{h}$

24. $f(g(2))$

6.4 Sketch the graphs of the following linear equations, and find all intercepts.

25. $y = 2x - 1$

26. $3y - 4x = 12$

27. $x + 2y = 5$

28. $y - 3x = 0$

29. $x = -2$

30. $y = 4$

Find the slope of each line passing through the following points.

31. $(0, 2)$ and $(-3, 0)$

32. $(1, -3)$ and $(7, 6)$

33. $(0, 0)$ and $(3, 7)$

34. $\left(\frac{1}{2}, \frac{2}{3}\right)$ and $\left(-\frac{3}{4}, \frac{1}{9}\right)$

35. $(2, -4)$ and $(2, 7)$

36. $(-1, -3)$ and $(5, -3)$

Find the slope of each of the following lines.

37. $y = 2x - 1$

38. $3y - 4x = 12$

39. $x + 2y = 5$

40. $y - 3x = 0$

41. $x = -2$

42. $y = 4$

6.5 Find the equations of the following lines. Write the answers in slope-intercept form, if possible.

43. $m = 4$, through $(5, -3)$

44. $m = -\frac{3}{2}$, through $(0, 7)$

45. $m = 0$, through $(-2, 1)$

46. m undefined, through $(-3, -5)$

47. Through $(0, 2)$ and $(-3, 0)$

48. Through $(1, -3)$ and $(7, 6)$

49. Through $(0, 0)$ and $(3, 7)$

50. Through $(\frac{1}{2}, \frac{2}{3})$ and $(-\frac{3}{4}, \frac{1}{9})$

51. Through $(2, -4)$ and $(2, 7)$

52. Through $(-1, -3)$ and $(5, -3)$

53. Through $(3, -4)$ parallel to $y = 3x + 7$

54. Through $(0, 3)$ parallel to $2x + 5y = 11$

55. Through $(7, -1)$ parallel to the x-axis

56. Through $(\frac{1}{3}, -\frac{3}{4})$ parallel to $9x - 3y = 4$

57. Through $(-5, 2)$ perpendicular to $y = \frac{5}{4}x - 8$

58. Through $(-1, -4)$ perpendicular to $3y - 4x = 10$

59. Through $(4, -6)$ perpendicular to the x-axis

60. Through $(-\frac{3}{2}, -\frac{1}{3})$ perpendicular to $x - 4y = 6$

6.6   **Determine the length and the midpoint of the line segments with the given endpoints.**

61. $(-1, 2)$ and $(5, 10)$

62. $(3, -7)$ and $(8, 5)$

63. $(-2, -5)$ and $(4, 3)$

64. $(-3, 5)$ and $(6, 4)$

65. $(1, -3)$ and $(9, -3)$

66. $\left(\frac{2}{3}, 0\right)$ and $\left(-\frac{2}{9}, 8\right)$

6.7   **Sketch the graph of the solution set of each of the following inequalities.**

67. $y < \dfrac{5}{2}$

68. $x \le -3$

69. $3x + y > 5$

70. $\dfrac{1}{2}x \le y + 4$

Sketch the graph of the solution set of each of the following systems of inequalities.

71. $\begin{cases} 3y - 4x \le 12 \\ 6x - y \le 4 \end{cases}$

72. $\begin{cases} y < x + 3 \\ x - 3y < 3 \\ 3y - 1 < 0 \end{cases}$

6.8   **Solve the following problems involving variation.**

73. a. The area of an equilateral triangle varies directly as the square of the side. Write an equation for this sentence.

    b. The area is $9\sqrt{3}$ sq in. when the side is 6 in. long. Determine the constant of variation.

74. When a ball rolls down an inclined plane, the distance it travels is directly proportional to the time it traveled. If a ball rolls 3 ft in $1\frac{1}{2}$ sec, how far will it travel in 3 sec?

75. The yearly increase in population of a certain city varies directly as that city's population at the beginning of the year. At the beginning of 1985 the population was 100,000, and that year the city experienced an increase of 2000 people. What will be the increase in population in 2010 if the population at the beginning of the year is 165,000?

76. The weight of a body near the surface of the earth varies inversely as the square of the distance of that body from the center of the earth. A man standing on the surface of the earth is approximately 4000 mi from the center of the earth. If he weighs 200 lb there, how much would he weigh if he were 400 mi above the surface of the earth?

6.9   **Sketch the graphs of the following functions.**

77. $f(x) = \begin{cases} -3 & \text{if } x \le 1 \\ 4 & \text{if } x > 1 \end{cases}$

78. $f(x) = \begin{cases} 1 - 2x & \text{if } x < -2 \\ x + 3 & \text{if } x \ge -2 \end{cases}$

79. $f(x) = \begin{cases} 3x - 2 & \text{if } x \le 3 \\ 13 - 2x & \text{if } x > 3 \end{cases}$

80. $f(x) = \begin{cases} 5 - 2x & \text{if } x \ne 0 \\ -3 & \text{if } x = 0 \end{cases}$

81. $f(x) = \begin{cases} x - 3 & \text{if } x \neq 1 \\ 4 & \text{if } x = 1 \end{cases}$

82. $f(x) = \begin{cases} x + 5 & \text{if } x < -2 \\ 3 & \text{if } -2 \leq x \leq 1 \\ x + 2 & \text{if } x > 1 \end{cases}$

83. Willie Makit, a runner in the seventh annual Moss County Marathon, paced himself as follows: for the first 5 mi he averaged 12 min 30 sec per mile, for the next 15 mi he averaged 11 min 30 sec per mile, for the next 5 mi he averaged 14 min/mi. On the last mile Willie began his kick and ran in world-class time of 4 min.
   a. Convert this information into a branch function. (*Hint:* His pace is a function of the distance he has traveled.)
   b. Graph this function.

84. The Moss County Agriculture Agent conducted the following study concerning the growth of watermelons. On May 1, she began her study of one 10-ft vine in T. C. McDonald's field. Listed below is a branch function describing the number of pounds of watermelons on the vine.

$$W(t) = \begin{cases} 15 + \dfrac{3}{2}t, & 0 \leq t \leq 60 \\ \\ 245 - \dfrac{7}{3}t, & 60 < t \end{cases}$$

Here t stands for the number of days after May 1.

   a. Graph this branch function.
   b. How many pounds of watermelon are on the vine on May 1?
   c. When does the number of pounds of watermelon on the vine reach its maximum, and what is this maximum?
   d. When do all the watermelons rot?

Chapter 6 Test

(You should be able to complete this test in 60 minutes.)

I. For the relations defined by the following equations, find the domains and ranges and graph them.

   1. $y = x^2 - 3$

   2. $y = \sqrt{x + 3}$

II. Let $f(x) = \dfrac{3}{x - 2}$ and $g(x) = x^2 + 1$. Find the following.

   3. $f(-3)$

   4. $f(0)$

   5. $[f(3)]^2$

   6. $g(3x)$

   7. Domain of f

   8. $f(x) \cdot g(x)$

   9. $\dfrac{g(a + h) - g(a)}{h}$

III. Graph the following lines.

   10. $3x - 4y = 8$; also find both intercepts and the slope.

   11. Passing through $(-1, 4)$ with slope $m = -\frac{3}{2}$. Also find the equation.

IV. Find the equations of the following lines. Write the equations in slope-intercept form, if possible.

   12. Through $(3, -1)$ and $(-5, 5)$

   13. Through $(-4, 1)$ and parallel to $x - 4y = 12$

14. Through $(4, 6)$ and perpendicular to $2x + 5y = 10$

15. Find the length and midpoint of the line segment with endpoints $(1, -2)$ and $(\frac{7}{2}, 4)$.

V. Graph the solution sets of the following inequalities.

16. $2x < 5 - y$

17. $\begin{cases} x + 3y \geq 6 \\ x \leq 4 \end{cases}$

18. The intensity of illumination from a light source varies inversely as the square of the distance from the source. If the intensity is 6 foot-candles from a light source that is 5 ft away, what is the intensity when the same light source is 10 ft away?

TEST YOUR MEMORY

These problems review Chapters 1–6.

I. Simplify each of the following.

1. $3^0 + 2^{-1} - \left(\dfrac{2}{3}\right)^{-2}$

2. $36^{-1/2} - 8^{2/3} + 4^{-1}$

II. Simplify each of the following. Assume that all variables represent positive numbers.

3. $\dfrac{(3x^{-2}y^3)^2}{(2x^4y^{-3})^{-3}}$

4. $\left(\dfrac{4x^3y^2}{25x^{-4}y^{1/3}}\right)^{1/2}$

III. Perform the indicated operations.

5. $\dfrac{3x^3 - 9x^2 + 27x}{27x^4 - 90x^3 + 63x^2} \cdot \dfrac{3x^3 - 7x^2 - 27x + 63}{x^3 + 27}$

6. $\dfrac{10x}{x^2 + 3x - 4} - \dfrac{9x - 20}{x^2 + x - 12}$

IV. Simplify the following fractions.

7. $\dfrac{\dfrac{3x}{x+1} - 4}{\dfrac{x}{x+1} - 1}$

8. $\dfrac{x^{-4} - y^{-4}}{x^{-2} - y^{-2}}$

V. Simplify each of the following. Assume that all variables represent positive numbers.

9. $2\sqrt{45x} - \sqrt{500x} + \sqrt{20x}$

10. $2\sqrt[3]{32y} + \sqrt[3]{108y}$

11. $\sqrt{\dfrac{7x}{4y}}$

12. $\sqrt[3]{\dfrac{2}{9y}}$

13. $(4\sqrt{x} + 3)^2$

14. $\dfrac{\sqrt{10}}{\sqrt{6} - 2}$

VI. Perform the indicated operations. Express all answers in the form $a + bi$.

15. $\sqrt{-4} \cdot \sqrt{-36}$

16. $\dfrac{13 + 11i}{1 - 3i}$

VII. For the relations defined by the following equations, find the domains and ranges and sketch the graphs. Express your answers using interval notation.

17. $x = y^2 - 4$

18. $y = \sqrt{x - 1}$

VIII. Let $f(x) = 5x - 4$ and $g(x) = x^2 + 7$. Find the following.

19. $f(-2)$

20. $g(5)$

21. $[f(3)]^2$

22. $4f(4) - 3g(-1)$

23. $g(2a)$

24. $f(x) \cdot g(x)$

IX. Graph the following lines.

25. $2x + 5y = 5$. Also find both intercepts and the slope.

26. Passing through $(3, -1)$ with slope $m = \frac{3}{2}$. Also find the equation.

X. Find the equations of the following lines. Write the equations in slope-intercept form, if possible.

27. Through $(3, 9)$ and $(-5, -1)$

28. Through $(-1, 2)$ and parallel to $y = 3x + 7$

29. Through $(-4, -7)$ and perpendicular to $3x - 4y = 12$

30. Through $(6, -2)$ and parallel to the x-axis

XI. Find the solutions of the following inequalities and graph the solution sets on number lines. Express your answers using interval notation.

31. $4x - 1 < 9$

32. $-4 \leq 3x + 5 \leq 17$

XII. Graph the solution sets of the following inequalities.

33. $x + 3y \leq 6$

34. $\begin{cases} x - 2y > 2 \\ y > -2x + 4 \end{cases}$

XIII. Find the solutions of the following equations and check your solutions.

35. $\sqrt{2x - 3} = 7$

36. $\sqrt{3x + 1} + 4 = 0$

37. $|5x - 2| = 6$

38. $\sqrt{x + 6} = x$

39. $\dfrac{3x - 1}{x + 2} + \dfrac{3}{x - 1} = \dfrac{x + 15}{x^2 + x - 2}$

40. $\dfrac{3}{x} - \dfrac{1}{4} = \dfrac{7}{3x}$

41. $\sqrt{2x} + 1 = 4x$

42. $|2x - 5| = |3x - 8|$

43. $1 + \dfrac{6}{x + 2} = \dfrac{2x + 10}{x + 2}$

44. $\dfrac{1}{2}\left(\dfrac{2}{3}x + 4\right) - \dfrac{1}{3}\left(\dfrac{1}{2}x - 6\right) = 3$

XIV. Use algebraic expressions to find the solutions of the following problems.

45. The area of a triangle is 63 sq in. The base is 4 in. more than twice the height. Find the base and the height of the triangle.

46. Wanda and Sam have a rectangular swimming pool that is 8 ft wide and 11 ft long. They are going to build a tile border of uniform width around the pool. They have 92 sq ft of tile. How wide is the border?

47. Wanda's and Sam's swimming pool can be filled by an inlet pipe in 8 hr. The drain can empty the pool in 10 hr. How long will it take to fill the pool if the drain is left open?

48. Kwan hikes up a 2-mi mountain trail. He then skis down the mountain trail. His skiing speed is 3 mi/hr faster than his hiking speed. His total time hiking and skiing is $2\frac{1}{2}$ hr. What is Kwan's skiing time?

49. When an object is dropped, the distance that it falls in t seconds varies directly as the square of t. If an object falls 144 ft in 3 sec, how far will it fall in 5 sec?

50. The number of hours Bonnie sleeps every night varies inversely as the square of the number of cups of coffee that she drinks during the day. If Bonnie drinks 2 cups of coffee, she gets 8 hr. of sleep. If Bonnie drinks 4 cups of coffee, how many hours should she sleep? (*Note:* Bonnie drinks at least 2 cups of coffee every day.)

Quadratic and Higher-Degree Equations and Inequalities

The amount of water pressure that is lost as water passes through a fire hose is called *friction loss*. For a hose of given length and diameter, the friction loss is a function of the flow rate of the water through the hose. If a hose is 100 ft long and $2\frac{1}{2}$ in. in diameter, then the equation determining the friction loss is

$$F_L = 2Q^2 + Q$$

In this equation F_L measures the friction loss in pounds per square inch and Q represents the flow rate in 100 gal/min. This equation determines a quadratic function. In this chapter we will study quadratic functions and quadratic equations in one variable. We will then be able to find either the friction loss or the flow rate, given the other quantity.

7.1 *Graphing Quadratic Functions*

If x represents the length of a side of a square, then the equation that determines the area, y, of the square is

$$y = x^2$$

In this case, x must be positive, since it represents a length. The graph of the equation $y = x^2$, where x can be any real number, was examined in Chapter 6. Some of the ordered pairs that satisfy the equation $y = x^2$ are given in the chart accompanying Figure 7.1.1. By plotting these points and then connecting them with a smooth curve, we generate the curve shown in Figure 7.1.1.

The curve in Figure 7.1.1 is called a **parabola.** The point $(0, 0)$ is called the **vertex.** In this type of parabola, which opens upward, the vertex is the lowest point on the parabola. The vertex will be the highest point if the parabola opens downward. The vertical line passing through the vertex is called the **axis** of the parabola. In this case the axis of the parabola is the y-axis. Note that the graph of the parabola is symmetric about its axis; that is, if a mirror is placed along the axis, one side is the mirror image of the other side. The equation $y = x^2$ defines a function and is an example of a special group of functions.

x	y
-2	4
-1	1
$-\dfrac{1}{2}$	$\dfrac{1}{4}$
0	0
$\dfrac{1}{2}$	$\dfrac{1}{4}$
1	1
2	4

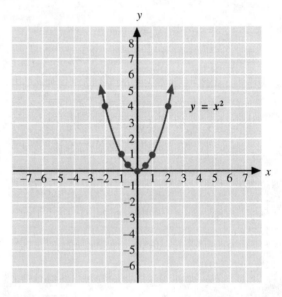

Figure 7.1.1

Definition 7.1.1	Any equation that can be written in the form $y = ax^2 + bx + c$, where a, b, and c are real numbers with $a \neq 0$, is said to define a **quadratic function.**

REMARK   The graph of a quadratic function is a parabola that opens upward or downward.

EXAMPLE 7.1.1   Sketch the graphs of the following quadratic functions.

1. $y = 2x^2$

Some of the ordered pairs that satisfy the equation $y = 2x^2$ are given in the chart. Plot these points and draw a smooth curve, as in Figure 7.1.2.

x	y
-2	8
-1	2
$-\dfrac{1}{2}$	$\dfrac{1}{2}$
0	0
$\dfrac{1}{2}$	$\dfrac{1}{2}$
1	2
2	8

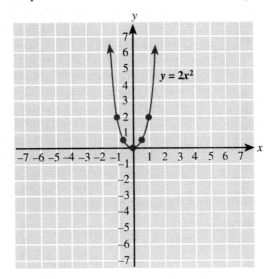

Figure 7.1.2

2. $y = \dfrac{1}{2}x^2$

Some of the ordered pairs that satisfy the equation $y = \frac{1}{2}x^2$ are given in the chart below. Plot these points and draw a smooth curve, as in Figure 7.1.3.

x	y
-2	2
-1	$\dfrac{1}{2}$
$-\dfrac{1}{2}$	$\dfrac{1}{8}$
0	0
$\dfrac{1}{2}$	$\dfrac{1}{8}$
1	$\dfrac{1}{2}$
2	2

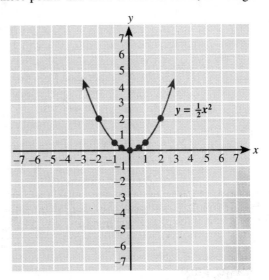

Figure 7.1.3

3. $y = -x^2$

Some of the ordered pairs that satisfy the equation $y = -x^2$ are given in the chart below. Plot these points and draw a smooth curve, as in Figure 7.1.4.

x	y
-2	-4
-1	-1
$-\dfrac{1}{2}$	$-\dfrac{1}{4}$
0	0
$\dfrac{1}{2}$	$-\dfrac{1}{4}$
1	-1
2	-4

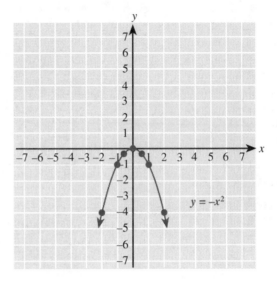

Figure 7.1.4

REMARKS   Given an equation in the form

$$y = ax^2 + bx + c, \qquad a \neq 0$$

1. When $|a| > 1$, the parabola is stretched as compared to $y = x^2$, which makes it look thinner than $y = x^2$.

2. When $|a| < 1$, the parabola is wider than $y = x^2$.

3. When $a > 0$, the parabola is $\cup$-shaped (opens upward) and the vertex is the lowest point on the parabola.

4. When $a < 0$, the parabola is $\cap$-shaped (opens downward) and the vertex is the highest point on the parabola.

The axis of each of the parabolas in Example 7.1.1 was the y-axis. In addition, each of the parabolas had its vertex as the origin. In the next example we will investigate parabolas that have their vertices elsewhere on the y-axis.

EXAMPLE 7.1.2

Sketch the graph and determine the vertex of the following quadratic functions.

1. $y = 2x^2 - 4$

Some points on the parabola are given in the chart. This parabola is graphed in Figure 7.1.5. It has the same shape as the parabola in Example 7.1.1, part 1, except that it has been lowered 4 units.

x	y
-2	4
-1	-2
$-\dfrac{1}{2}$	$-3\dfrac{1}{2}$
0	-4
$\dfrac{1}{2}$	$-3\dfrac{1}{2}$
1	-2
2	4

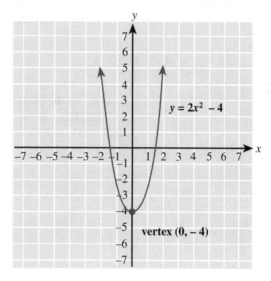

Figure 7.1.5

x	y
-2	1
-1	4
$-\dfrac{1}{2}$	$4\dfrac{3}{4}$
0	5
$\dfrac{1}{2}$	$4\dfrac{3}{4}$
1	4
2	1

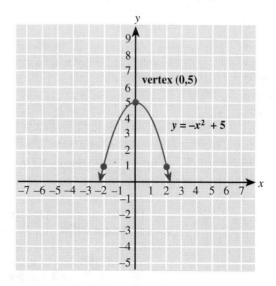

Figure 7.1.6

2. $y = -x^2 + 5$

Some points on the parabola are given in the chart. This parabola is graphed in Figure 7.1.6. It has the same shape as the parabola in Example 7.1.1, part 3, except that it has been raised 5 units.

In the next example parabolas with vertices on the x-axis are examined.

EXAMPLE 7.1.3

Sketch the graph and determine the vertex of the following quadratic functions.

1. $y = \dfrac{1}{2}(x + 1)^2$

Some points on the parabola are given in the chart. This parabola is graphed in Figure 7.1.7. It has the same shape as the parabola in Example 7.1.1, part 2, except that it has been shifted 1 unit to the left, so its axis is the vertical line $x = -1$.

x	y
-3	2
-2	$\dfrac{1}{2}$
-1	0
0	$\dfrac{1}{2}$
1	2
2	$4\dfrac{1}{2}$
3	8

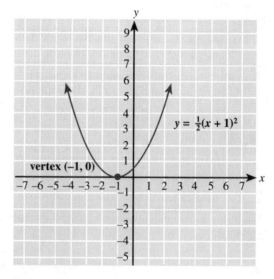

Figure 7.1.7

2. $y = -(x - 2)^2$

Some of the points on the parabola are given in the chart. This parabola is graphed in Figure 7.1.8. It has the same shape as the parabola in Example 7.1.1, part 3, except that it has been shifted 2 units to the right, so its axis is the vertical line $x = 2$.

x	y
-2	-16
-1	-9
0	-4
1	-1
2	0
3	-1
4	-4

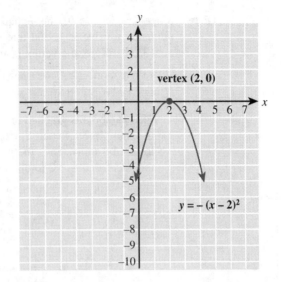

Figure 7.1.8

As a final example we will examine parabolas with vertices not on the x- or y-axis.

XAMPLE 7.1.4   Sketch the graph and determine the vertex of the following quadratic functions.

1. $y = 2(x - 3)^2 + 1$

This parabola is graphed in Figure 7.1.9. It has the same shape as the parabola in Example 7.1.1, part 1, except that it has been shifted 3 units to the right and raised 1 unit. Thus the vertex is at $(3, 1)$. We choose values of x on either side of $x = 3$ to obtain the ordered pairs in the chart.

x	y
1	9
2	3
3	1
4	3
5	9

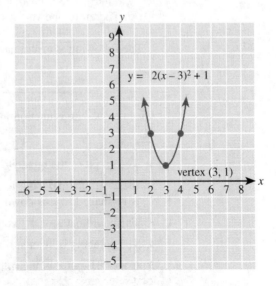

Figure 7.1.9

NOTE ▶ Since $(x - 3)^2 \geq 0$, the equation $y = 2(x - 3)^2 + 1$ implies that $y \geq 1$. The *minimum* y value is 1, and that minimum occurs when $x - 3 = 0$—that is, when $x = 3$. Therefore, the vertex is at $(3, 1)$.

2. $y = \dfrac{1}{2}(x + 2)^2 - 3$

This parabola is graphed in Figure 7.1.10. It has the same shape as the parabola in Example 7.1.1, part 2, except that it has been shifted 2 units to the left and lowered 3 units. Thus the vertex is at $(-2, -3)$. We choose values of x on either side of $x = -2$ to obtain the ordered pairs in the chart.

x	y
-4	-1
-3	$-2\frac{1}{2}$
-2	-3
-1	$-2\frac{1}{2}$
0	-1

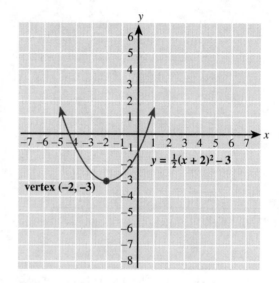

Figure 7.1.10

NOTE ▶ Since $(x + 2)^2 \geq 0$, the equation $y = \frac{1}{2}(x + 2)^2 - 3$ implies that $y \geq -3$. The *minimum* y value is -3, and that minimum occurs when $x + 2 = 0$—that is, when $x = -2$. Therefore, the vertex is at $(-2, -3)$.

The following is a summary of the information developed in this section.

Given a quadratic function in the form
$$y = a(x - h)^2 + k \qquad a \neq 0$$

1. The graph of the quadratic function is a parabola.

2. If $a > 0$, the parabola is ∪-shaped (opens upward). If $a < 0$, the parabola is ∩-shaped (opens downward).

3. The coordinates of the vertex are (h, k).

4. The parabola is symmetric with respect to its axis, the vertical line $x = h$.

5. If $a > 0$, $y = k$ is the minimum y-value. If $a < 0$, $y = k$ is the maximum y-value.

Exercises 7.1

Sketch the graph and determine the vertex of the following quadratic functions.

1. $y = 3x^2$

2. $y = 4x^2$

3. $y = -5x^2$

4. $y = \frac{1}{4}x^2$

5. $y = \frac{2}{3}x^2$

6. $y = -\frac{1}{3}x^2$

7. $y = x^2 + 1$

8. $y = -x^2 + 3$

9. $y = 3x^2 - 2$

10. $y = 4x^2 - 1$

11. $y = -5x^2 - 4$

12. $y = \frac{1}{4}x^2 - \frac{1}{2}$

13. $y = \frac{2}{3}x^2 + \frac{1}{3}$

14. $y = \frac{1}{3}x^2 + 2$

15. $y = (x - 4)^2$

16. $y = -(x - 1)^2$

17. $y = 3(x - 3)^2$

18. $y = 4(x + 1)^2$

19. $y = -5\left(x + \frac{1}{2}\right)^2$

20. $y = \frac{1}{4}(x + 2)^2$

21. $y = \frac{2}{3}(x - 5)^2$

22. $y = -\frac{1}{3}(x + 2)^2$

23. $y = (x - 4)^2 + 1$

24. $y = -(x - 1)^2 + 3$

25. $y = 3(x - 3)^2 - 2$

26. $y = 4(x + 1)^2 - 1$

27. $y = -5\left(x + \frac{1}{2}\right)^2 - 4$

28. $y = \frac{1}{4}(x + 2)^2 - \frac{1}{2}$

29. $y = \frac{2}{3}(x - 5)^2 + \frac{1}{3}$

30. $y = -\frac{1}{3}(x + 2)^2 + 2$

31. $y = (x + 3)^2 + 1$

32. $y = (x - 2)^2 + 3$

33. $y = (x - 5)^2 - 6$

34. $y = (x + 1)^2 - 4$

35. $y = 2(x - 3)^2 - 4$

36. $y = 3(x + 1)^2 + 6$

37. $y = -2(x + 2)^2 + 1$

38. $y = -\frac{1}{2}(x - 1)^2 - 3$

39. $y = 4\left(x - \frac{1}{2}\right)^2 - 2$

40. $y = -4\left(x + \frac{5}{2}\right)^2 + 3$

41. $y = \frac{1}{4}(x + 3)^2 + \frac{1}{4}$

42. $y = -\frac{1}{3}(x - 2)^2 - \frac{2}{3}$

Write Algebra

43. Given a quadratic function in the form

$$y = a(x - h)^2 + k, \qquad a \neq 0$$

a. Explain how the value of a affects the shape of the graph.

b. If $a > 0$ and $k > 0$, explain why the graph has no x-intercept.

44. Explain why the graph of any quadratic function has only one y-intercept.

7.2 *Completing the Square to Find the Vertex*

All the quadratic functions in Section 7.1 were in the form $y = a(x - h)^2 + k$. This is called **standard form.** When a quadratic function is in standard form, its graph is easily sketched using the techniques of the previous section. In this section we will learn how to transform the equation of a quadratic function from

$$y = ax^2 + bx + c$$

into the standard form

$$y = a(x - h)^2 + k$$

First let us review some terminology.

Perfect Square Trinomial	Examples
Any trinomial that can be expressed as the square of a binomial is called a **perfect square trinomial.**	$4x^2 + 12x + 9$, since $4x^2 + 12x + 9 = (2x + 3)^2$ $x^2 + 4x + 4$, since $x^2 + 4x + 4 = (x + 2)^2$ $x^2 - 2hx + h^2$, since $x^2 - 2hx + h^2 = (x - h)^2$

Suppose we were required to determine the number that must be added to $x^2 + 8x$ to form a perfect square trinomial. We want

$$x^2 + 8x + \underline{\hspace{1cm}} = (x + t)^2$$
$$x^2 + 8x + \underline{\hspace{1cm}} = x^2 + 2tx + t^2$$

This implies

$$8 = 2t$$
$$4 = t$$

Thus,

$$16 = t^2$$

It appears that the required number is 16. Let's check:

$$x^2 + 8x + 16 = (x + 4)^2$$

The process of finding the number required to make a polynomial a perfect square trinomial is called **completing the square.**

NOTE ▶   To complete the square on any polynomial of the form $x^2 + bx$, add $(\frac{1}{2}b)^2 = (\frac{1}{2}$ the coefficient of $x)^2$ to that polynomial.

XAMPLE 7.2.1   Determine the number that must be added to the following polynomials to form a perfect square trinomial.

1. $x^2 - 20x$

$$b = -20$$
$$\left(\frac{1}{2}b\right)^2 = \left[\frac{1}{2} \cdot (-20)\right]^2$$
$$= (-10)^2$$
$$= 100$$

So $x^2 - 20x + 100$ should be a perfect square trinomial:

$$x^2 - 20x + 100 = (x - 10)^2 \;\checkmark$$

2. $x^2 + 5x$

$$b = 5$$
$$\left(\frac{1}{2}b\right)^2 = \left(\frac{1}{2} \cdot 5\right)^2$$
$$= \left(\frac{5}{2}\right)^2$$
$$= \frac{25}{4}$$

So $x^2 + 5x + \dfrac{25}{4}$ should be a perfect square trinomial:

$$x^2 + 5x + \frac{25}{4} = \left(x + \frac{5}{2}\right)^2 \;\checkmark$$

Consider the quadratic function

$$y = x^2 - 6x + 10$$

To place this quadratic function in standard form we first rewrite it as

$$y = (x^2 - 6x + \underline{\hphantom{000}}) + 10$$

We must choose the _____ so that $x^2 - 6x +$ _____ is a perfect square trinomial. From our work in Example 7.2.1,

$$\underline{\hspace{1cm}} = \left[\frac{1}{2} \cdot (-6)\right]^2$$
$$= (-3)^2$$
$$= 9$$

Now

$$y = x^2 - 6x + 10$$
$$y = (x^2 - 6x + \underline{\hspace{0.5cm}}) + 10$$
$$y = (x^2 - 6x + 9) + 10 - 9$$

NOTE ▶ We added 9 inside the parentheses to form a perfect square trinomial. Notice that since 9 is added inside the parentheses, 9 must be subtracted outside the parentheses to preserve the value of the right-hand side of the equation.

$$y = (x - 3)^2 + 1$$

The quadratic function is in standard form. Recall from Section 7.1 that the vertex is (3, 1). We choose values of x on either side of $x = 3$ to obtain the ordered pairs in the chart. Plot these points and smooth in the curve as shown in Figure 7.2.1.

x	y
1	5
2	2
3	1
4	2
5	5

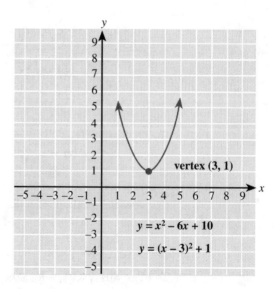

Figure 7.2.1

EXAMPLE 7.2.2   Place the following quadratic functions in standard form. Also find the vertex and sketch the graph.

1. $y = x^2 + 4x + 1$

$y = (x^2 + 4x + \underline{\quad}) + 1$ $\underline{\quad} = (\frac{1}{2} \cdot 4)^2 = 2^2 = 4$

$y = (x^2 + 4x + 4) + 1 - 4$ *We must subtract 4, since 4 is added inside the parentheses.*

$y = (x + 2)^2 - 3$

The vertex is $(-2, -3)$. Some points on the parabola, which is graphed in Figure 7.2.2, are shown in the chart.

x	y
-4	1
-3	-2
-2	-3
-1	-2
0	1

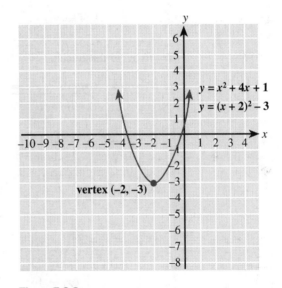

$y = x^2 + 4x + 1$
$y = (x + 2)^2 - 3$
vertex $(-2, -3)$

Figure 7.2.2

2. $y = 2x^2 - 4x - 3$

$y = 2(x^2 - 2x) - 3$ *Make the leading coefficient 1 by factoring a 2 out of the x terms.*

$y = 2(x^2 - 2x + \underline{\quad}) - 3$ $\underline{\quad} = [\frac{1}{2} \cdot (-2)]^2 = (-1)^2 = 1$

$y = 2(x^2 - 2x + 1) - 3 - 2$ *We must subtract 2, since $2 \cdot 1$ is added to the right-hand side.*

$y = 2(x - 1)^2 - 5$

The vertex is $(1, -5)$. Some points on the parabola, which is graphed in Figure 7.2.3, are shown in the following chart.

x	y
-1	3
0	-3
1	-5
2	-3
3	3

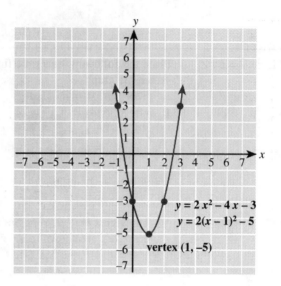

$$y = 2x^2 - 4x - 3$$
$$y = 2(x - 1)^2 - 5$$
vertex $(1, -5)$

Figure 7.2.3

Try to avoid these mistakes:

Incorrect	Correct
1. $y = 2x^2 - 4x - 3$ $y = 2(x^2 - 2x) - 3$ $y = 2(x^2 - 2x + 1) - 3 - 1$	$y = 2x^2 - 4x - 3$ $y = 2(x^2 - 2x) - 3$ $y = 2(x^2 - 2x + 1) - 3 - 2$ Remember that we are adding $2 \cdot 1$, not just 1.
2. $y = 2x^2 + 12x + 32$ $y = 2(x^2 + 6x + 16)$ $y = 2(x^2 + 6x) + 16$	$y = 2x^2 + 12x + 32$ $y = 2(x^2 + 6x) + 32$ Remove the factor of 2 only from the x terms.

3. $y = -x^2 + 6x - 5$

     $y = -(x^2 - 6x) - 5$ *Make the leading coefficient 1 by factoring -1 out of the x terms.*

     $y = -(x^2 - 6x +$ _____$) - 5$ _____ $= [\frac{1}{2} \cdot (-6)]^2 = (-3)^2 = 9$

     $y = -(x^2 - 6x + 9) - 5 + 9$ *We must add 9, since $-1 \cdot 9$ is added to the right-hand side.*

     $y = -(x - 3)^2 + 4$

The vertex is $(3, 4)$. Some points on the parabola, which is graphed in Figure 7.2.4, are shown in the following chart.

x	y
1	0
2	3
3	4
4	3
5	0

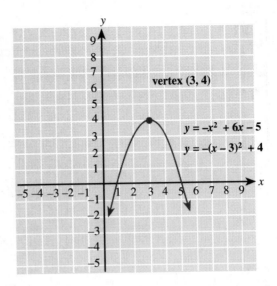

vertex (3, 4)

$y = -x^2 + 6x - 5$

$y = -(x - 3)^2 + 4$

Figure 7.2.4

Try to avoid this mistake:

Incorrect	Correct
$y = -x^2 + 6x - 5$ $y = -(x^2 - 6x) - 5$ $y = -(x^2 - 6x + 9) - 5 - 9$	$y = -x^2 + 6x - 5$ $y = -(x^2 - 6x) - 5$ $y = -(x^2 - 6x + 9) - 5 + 9$ We must *add* 9, since $-1 \cdot 9$ has already been added to the right-hand side.

4. $y = 3x^2 + 4x + 3$

$y = 3\left(x^2 + \dfrac{4}{3}x\right) + 3$ *Make the leading coefficient 1 by factoring 3 out of the x terms.*

$y = 3\left(x^2 + \dfrac{4}{3}x + \underline{}\right) + 3$ $\underline{} = (\tfrac{1}{2} \cdot \tfrac{4}{3})^2$
 $= (\tfrac{2}{3})^2 = \tfrac{4}{9}$

$y = 3\left(x^2 + \dfrac{4}{3}x + \dfrac{4}{9}\right) + 3 - \dfrac{4}{3}$ *We must subtract $\tfrac{4}{3}$, since $3(\tfrac{4}{9})$ is added to the right-hand side.*

$y = 3\left(x + \dfrac{2}{3}\right)^2 + \dfrac{5}{3}$

The vertex is $(-\tfrac{2}{3}, \tfrac{5}{3})$. Some points on the parabola, which is graphed in Figure 7.2.5, are shown in the following chart.

x	y
-2	7
-1	2
$-\dfrac{2}{3}$	$\dfrac{5}{3}$
0	3
1	10

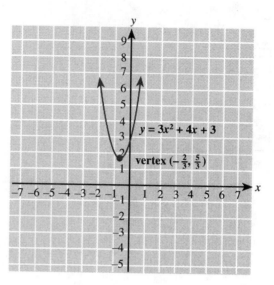

$y = 3x^2 + 4x + 3$

vertex $\left(-\frac{2}{3}, \frac{5}{3}\right)$

Figure 7.2.5

Use the following steps to place any quadratic function in standard form.

1. Place the quadratic function in the form

$$y = ax^2 + bx + c$$

2. Make the leading coefficient 1 by factoring a out of the x terms:

$$y = a\left(x^2 + \frac{b}{a}x\right) + c$$

3. Determine the number needed to complete the square:

$$\left(\frac{1}{2} \text{ coefficient of } x\right)^2 = \left(\frac{1}{2} \cdot \frac{b}{a}\right)^2 = \left(\frac{b}{2a}\right)^2 = \frac{b^2}{4a^2}$$

4. Inside the parentheses, add the number from step 3,

$$\frac{b^2}{4a^2}$$

Outside the parentheses, subtract

$$a \cdot (\text{the number from step 3}) = a\left(\frac{b^2}{4a^2}\right) = \frac{b^2}{4a}$$

5. Rewrite the perfect square trinomial inside the parentheses as a binomial squared, and combine the numbers outside the parentheses.

As we have seen, the vertex is the key point in sketching the graph of a parabola. In the previous examples we located the vertex by placing the quadratic function in standard form. We will now derive a formula that will enable us to find quickly the vertex of any quadratic function.

Consider the general quadratic function

$$y = ax^2 + bx + c$$

Let's now place this function in standard form.

$$y = ax^2 + bx + c$$

$$y = a\left(x^2 + \frac{b}{a}x\right) + c \qquad \textit{Make the leading coefficient 1 by factoring a out of the x terms:}$$

$$\underline{} = \left(\frac{1}{2} \cdot \frac{b}{a}\right)^2$$

$$y = a\left(x^2 + \frac{b}{a}x + \underline{}\right) + c \qquad = \left(\frac{b}{2a}\right)^2 = \frac{b^2}{4a^2}$$

$$y = a\left(x^2 + \frac{b}{a}x + \frac{b^2}{4a^2}\right) + c - \frac{b^2}{4a} \qquad \textit{We must subtract } \frac{b^2}{4a}, \textit{ since a } \left(\frac{b^2}{4a^2}\right)$$

$$\qquad\qquad\qquad\qquad\qquad\qquad\qquad \textit{is added to the right-hand side.}$$

$$y = a\left(x + \frac{b}{2a}\right)^2 + \frac{4ac - b^2}{4a}$$

Thus, the vertex is

$$\left(-\frac{b}{2a}, \frac{4ac - b^2}{4a}\right)$$

Given this formula, we can use the following steps to sketch quickly the graph of any quadratic function.

1. Place the quadratic function in the form

$$y = ax^2 + bx + c$$

2. Find $-\dfrac{b}{2a}$, the x-coordinate of the vertex.

3. Find the y-coordinate of the vertex by substituting the x value obtained in step 2 into the equation of step 1. (We could also use $\dfrac{4ac - b^2}{4a}$ to determine the y-coordinate of the vertex.)

4. Determine a few points on each side of the vertex, and sketch the graph of the parabola.

We illustrate this procedure in the following example.

EXAMPLE 7.2.3   Sketch the graph of the following quadratic function. Find the vertex by using the expression $-\dfrac{b}{2a}$ to calculate the x-coordinate of the vertex.

$$y = 2x^2 - 12x + 13$$

SOLUTION   In this case $a = 2$, $b = -12$, and $c = 13$, so

$$\frac{-b}{2a} = \frac{-(-12)}{2 \cdot 2} = \frac{12}{4} = 3$$

Thus the x-coordinate of the vertex is 3.

To find the y-coordinate of the vertex, substitute 3 for x in the given equation:

$$y = 2 \cdot 3^2 - 12 \cdot 3 + 13$$
$$y = 18 - 36 + 13$$
$$y = -5$$

The vertex is $(3, -5)$. Finding two points on either side of the vertex, we can graph the parabola. See Figure 7.2.6.

x	y
1	3
2	-3
3	-5
4	-3
5	3

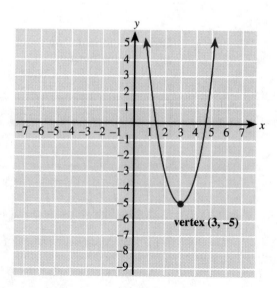

Figure 7.2.6

The remaining examples of this section will illustrate some applications of quadratic functions.

EXAMPLE 7.2.4　The Moss County Public Works Department has determined that the annual cost of keeping the county's parks clean is

$$y = 6x^2 - 840x + 42{,}600$$

where y is the cost in dollars and x is the number of "Litter, and we will fine you!" signs. Find the number of signs that will minimize the county's cleaning cost. What is the minimum cost?

SOLUTION　Graphically $y = 6x^2 - 840x + 42{,}600$ is a parabola that opens upward (see Figure 7.2.7). Therefore, this equation has its minimum y value—its minimum cost—at the vertex:

$$\begin{aligned}
y &= 6x^2 - 840x + 42{,}600 \\
&= 6(x^2 - 140x) + 42{,}600 \\
&= 6(x^2 - 140x + 4900) + 42{,}600 - 29{,}400 \\
&= 6(x^2 - 140x + 4900) + 13{,}200 \\
&= 6(x - 70)^2 + 13{,}200
\end{aligned}$$

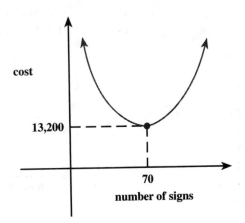

Figure 7.2.7

The vertex is (70, 13,200). Consequently, 70 signs will produce a minimum cost of $13,200.

EXAMPLE 7.2.5　Amanda wants to plant a rectangular garden next to her house. She has 40 ft of fencing and will not fence the side of the garden by her house. What dimensions will produce the largest garden?

SOLUTION   Let

$$x = \text{width of the garden}$$
$$40 - 2x = \text{length of the garden}$$

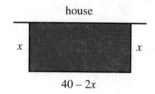

Figure 7.2.8

These dimensions are illustrated in Figure 7.2.8.

$$\text{area} = x(40 - 2x)$$
$$= 40x - 2x^2$$
$$= -2x^2 + 40x$$

The graph of this equation is a parabola opening downward. See Figure 7.2.9.

$$\text{area} = -2(x^2 - 20x)$$
$$= -2(x^2 - 20x + 100) + 200$$
$$= -2(x - 10)^2 + 200$$

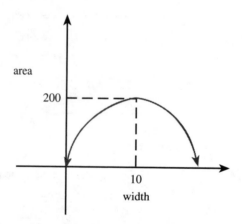

Figure 7.2.9

From the vertex, $(10, 200)$, we find that the maximum area is 200 sq ft when $x = 10$.

$$\text{width} = x$$
$$= 10 \text{ ft}$$
$$\text{length} = 40 - 2x$$
$$= 40 - 2(10)$$
$$= 20 \text{ ft}$$

From the graph we can see that if x is smaller or larger than 10 ft, the area is less than 200 sq ft.

Exercises 7.2

Determine the number that must be added to each of the following polynomials to form a perfect square trinomial.

1. $x^2 + 6x$

2. $x^2 + 14x$

3. $x^2 - 22x$

4. $x^2 - 30x$

5. $x^2 + 3x$

6. $x^2 + x$

7. $x^2 - 7x$

8. $x^2 - x$

Place the quadratic functions in exercises 9–54 in standard form. Also find the vertex and sketch the graph.

9. $y = x^2 - 2x + 5$

10. $y = x^2 - 6x + 11$

11. $y = x^2 + 2x + 1$

12. $y = x^2 - 8x + 16$

13. $y = x^2 + 10x + 22$

14. $y = x^2 + 4x + 11$

15. $y = x^2 - 4x + 2$

16. $y = x^2 + 6x + 12$

17. $y = -x^2 + 6x - 7$

18. $y = -x^2 - 2x - 8$

19. $y = -x^2 - 6x + 1$

20. $y = -x^2 + 8x - 18$

21. $y = -x^2 + 6x - 9$

22. $y = -x^2 - 10x - 25$

23. $y = -x^2 - 4x - 3$

24. $y = -x^2 - 12x - 40$

25. $y = 2x^2 - 12x + 17$

26. $y = 4x^2 + 8x + 7$

27. $y = 2x^2 - 16x + 32$

28. $y = 5x^2 + 30x + 45$

29. $y = 3x^2 - 12x + 17$

30. $y = 2x^2 - 16x + 27$

31. $y = 3x^2 + 12x + 5$

32. $y = -2x^2 - 4x - 6$

33. $y = -2x^2 + 12x - 23$

34. $y = -3x^2 + 6x - 2$

35. $y = -4x^2 - 40x - 92$

36. $y = -2x^2 + 8x - 8$

37. $y = -3x^2 - 6x - 3$

38. $y = -5x^2 - 10x - 4$

39. $y = x^2 - x + 1$

40. $y = x^2 - 3x + 2$

41. $y = x^2 + x + 2$

42. $y = x^2 + 5x + 5$

43. $y = x^2 - \dfrac{4}{3}x + 1$

44. $y = x^2 + \dfrac{2}{3}x - 1$

45. $y = x^2 + \dfrac{5}{2}x + 2$

46. $y = x^2 - \dfrac{3}{2}x + \dfrac{1}{2}$

47. $y = 2x^2 + 2x - 1$

48. $y = 2x^2 - 6x + 5$

49. $y = 3x^2 - 3x - 1$

50. $y = 3x^2 + 8x + 6$

51. $y = -2x^2 - x - 1$

52. $y = -3x^2 - 10x - 8$

53. $y = -2x^2 + \dfrac{8}{3}x + \dfrac{1}{9}$

54. $y = -3x^2 + 3x - \dfrac{11}{4}$

Sketch the graphs of the following quadratic functions. Find each vertex by using the expression $-\dfrac{b}{2a}$ to calculate the x-coordinate of the vertex.

55. $y = x^2 - 6x + 7$

56. $y = x^2 - 10x + 24$

57. $y = 2x^2 + 16x + 29$

58. $y = 2x^2 + 8x + 4$

59. $y = -x^2 - 12x - 35$

60. $y = -x^2 - 2x + 3$

61. $y = -3x^2 + 12x - 12$

62. $y = -3x^2 + 24x - 48$

63. $y = \dfrac{1}{2}x^2 - 2x + 3$

64. $y = \dfrac{1}{2}x^2 - 4x + 10$

65. The Pumpkin Patch Doll Company has determined that the profit the company makes on Pumpkin Patch dolls is given by the equation $y = -3x^2 + 720x - 17,800$, where y is the profit measured in dollars and x is the number of Pumpkin Patch dolls. Find the number of dolls that will maximize the company's profit. What is the maximum profit?

66. The Big Time Oil Company has determined that the cost of operating its corporate headquarters is given by the equation $y = 5x^2 - 70x + 2505$, where y is the cost measured in dollars and x is the number of hours per day that the headquarters is open. Find the number of hours that will minimize the cost of operating the headquarters. What is the minimum cost?

67. Ann wants to plant a rectangular garden next to a straight river. She has 120 ft of fencing and will not fence the side of the garden by the river. What dimensions will produce the largest garden?

68. Crazy Jay wants to plant a rectangular garden next to an abandoned mine field. He has 220 ft of fencing and will not fence the side of the garden by the mine field. What dimensions will produce the largest garden?

69. Farmer Jones wants to build a rectangular pen with three equal parts (see Figure 7.2.10). He has 200 ft of fencing. What dimensions will produce the largest pen? [*Hint:* Let x = the width of the pen; then $\frac{1}{2}(200 - 4x) = 100 - 2x$ = the length of the pen.]

70. Farmer Odell wants to build a rectangular pen with two equal parts (see Figure 7.2.11). She has 180 ft of fencing. What dimensions will produce the largest pen? [*Hint:* Let x = the width of the pen; then $\frac{1}{2}(180 - 3x) = 90 - \frac{3}{2}x$ = the length of the pen.]

Figure 7.2.10

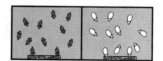

Figure 7.2.11

71. A bullet is fired from ground level upward with a speed of 384 ft/sec. The equation that gives the bullet's height above ground level is $h = -16t^2 + 384t$. When does the bullet reach its maximum height? What is the maximum height? When does it hit the ground?

72. An arrow is shot upward from the roof of a 112-ft-tall building with a speed of 96 ft/sec. The equation that gives the arrow's height above ground level is $h = -16t^2 + 96t + 112$. When does the arrow reach its maximum height? What is the maximum height? When does it hit the ground?

Write Algebra

73. Explain, in your own words why, in the equation $y = a(x - h)^2 + k$, (h, k) is the highest or lowest point on the graph.

🖩 **The following exercises require the use of a graphing calculator.**

74. Graph $y = x^2$. Next graph $y = x^2 + x$ and $y = x^2 + 2x$. Predict the graph of $y = x^2 + 3x$. Graph $y = x^2 + 3x$. Describe the pattern you see in the graphs.

75. Graph $y = x^2$. Next graph $y = x^2 - 2x$ and $y = x^2 - 4x$. Predict the graph of $y = x^2 - 6x$. Graph $y = x^2 - 6x$. Describe the pattern you see in the graphs.

76. Graph $y = (x - 1)(x + 3)$. Next graph $y = 2(x - 1)(x + 3)$ and $y = -(x - 1)(x + 3)$. Predict the graph of $y = -2(x - 1)(x + 3)$. Graph $y = -2(x - 1)(x + 3)$. Describe the pattern you see in the graphs.

77. Use the trace and zoom keys of a graphing calculator to approximate (to two decimal places) the vertex of the parabolas determined by the following equations.
    a. $y = x^2 + 3x - 5$
    b. $y = 2x^2 - 6x + 3$
    c. $y = -1.5x^2 - 11x - 13$
    d. $y = -2.3x^2 + 6.2x - 1.09$

In exercises 78 and 79, (a) complete each table for the given values of x, (b) determine the value of x that produces a minimum value of z, and (c) verify your conjecture by rewriting the radicand as a function of x.

78. $z = \sqrt{x^2 + y}$

$y = -4x + 13$

x	z
-2	
-1	
0	
1	
2	
3	
4	
5	
6	

79. $z = \sqrt{x^2 - 2y}$

$y = 6x - 18$

x	z
2	
3	
4	
5	
6	
7	
8	
9	
10	

7.3 *Solving Quadratic Equations by Completing the Square*

The process of completing the square was used in the previous section to find the vertex of a parabola. In this section we will learn how to find the solutions of any quadratic equation by completing the square. However, we also need another process, called the **extraction of roots** (it really isn't as painful as it sounds), which works as follows:

$$x^2 = 4$$
$$x = \pm\sqrt{4}$$
$$x = \pm 2$$

The extraction of roots technique is generalized in the following theorem.

Theorem 7.3.1	Extraction of Roots theorem: If $x^2 = a$, where a is any real number, then $x = \pm\sqrt{a}$.

Proof   Given

$$x^2 = a$$

then

$$x^2 - a = 0$$
$$(x + \sqrt{a})(x - \sqrt{a}) = 0$$
$$x + \sqrt{a} = 0 \quad \text{or} \quad x - \sqrt{a} = 0$$
$$x = -\sqrt{a} \qquad x = \sqrt{a}$$

REMARK   In general, this theorem states that when we have

$$(\text{polynomial})^2 = \text{real number}$$

then

$$\text{polynomial} = \pm\sqrt{\text{real number}}$$

EXAMPLE 7.3.1   Use the Extraction of Roots theorem to find the solutions of the following quadratic equations.

1. $x^2 = 36$

$$x = \pm\sqrt{36} \qquad \textit{By Theorem 7.3.1}$$
$$x = \pm 6$$

Recall that $x = \pm 6$ means $x = 6$ or $x = -6$.

NOTE ▶   In this section, checking solutions is left to the reader.

2. $9x^2 = 4$

$$x^2 = \frac{4}{9}$$

$$x = \pm\sqrt{\frac{4}{9}} \qquad \textit{By Theorem 7.3.1}$$

$$x = \pm\frac{2}{3}$$

3. $x^2 - 7 = 0$

$$x^2 = 7$$

$$x = \pm\sqrt{7} \qquad \textit{By Theorem 7.3.1}$$

4. $x^2 + 16 = 0$

$\qquad x^2 = -16$

$\qquad\quad x = \pm\sqrt{-16}$ *By Theorem 7.3.1*

$\qquad\quad x = \pm 4i$

5. $2(x - 2)^2 = 50$

$\qquad (x - 2)^2 = 25$

$\qquad\quad x - 2 = \pm\sqrt{25}$ *By Theorem 7.3.1*

$\qquad\quad x - 2 = \pm 5$

So

$$x - 2 = 5 \quad \text{or} \quad x - 2 = -5$$
$$x = 7 \qquad\qquad x = -3$$

6. $(2x + 3)^2 - 1 = 12$

$\qquad (2x + 3)^2 = 13$

$\qquad\quad 2x + 3 = \pm\sqrt{13}$ *By Theorem 7.3.1*

$\qquad\qquad 2x = -3 \pm\sqrt{13}$

$\qquad\qquad\quad x = \dfrac{-3 \pm \sqrt{13}}{2}$

Suppose we were required to find the solutions of the quadratic equation

$$x^2 - 6x + 4 = 0$$

The only tools we have for solving quadratic equations are the Zero Product theorem from Chapter 3 and the Extraction of Roots theorem from this section. We cannot use the Zero Product theorem, since the left-hand side is a prime polynomial. To use the Extraction of Roots theorem, we must have (polynomial)^2 = real number. To obtain a polynomial squared, we will use the *completing the square* process:

$$x^2 - 6x + 4 = 0$$

$\qquad\quad x^2 - 6x = -4$ *Isolate the x terms.*

$\quad x^2 - 6x + \underline{\quad\quad} = -4$ $\underline{\quad\quad} = [\frac{1}{2}\cdot(-6)]^2 = (-3)^2 = 9$

$\qquad x^2 - 6x + 9 = -4 + 9$ *Add 9 to both sides of the equation.*

$\qquad\quad (x - 3)^2 = 5$

$\qquad\qquad x - 3 = \pm\sqrt{5}$ *By Theorem 7.3.1*

$\qquad\qquad\quad x = 3 \pm \sqrt{5}$

E X A M P L E   7 . 3 . 2 Find the solutions of the following quadratic equations by completing the square.

1. $x^2 + 2x + 5 = 0$

$$x^2 + 2x = -5 \qquad \text{Isolate the x terms.}$$

$$x^2 + 2x + \underline{\quad} = -5 \qquad \underline{\quad} = (\tfrac{1}{2} \cdot 2)^2 = 1^2 = 1$$

$$x^2 + 2x + 1 = -5 + 1 \qquad \text{Add 1 to both sides of the equation.}$$

$$(x + 1)^2 = -4$$

$$x + 1 = \pm\sqrt{-4} \qquad \text{By Theorem 7.3.1}$$

$$x + 1 = \pm 2i$$

$$x = -1 \pm 2i$$

2. $2x^2 - 6x + 5 = 0$

$$2x^2 - 6x = -5 \qquad \text{Isolate the x terms.}$$

$$x^2 - 3x = -\frac{5}{2} \qquad \begin{array}{l}\text{Make the leading coefficient 1 by dividing both sides}\\ \text{by 2.}\end{array}$$

$$x^2 - 3x + \underline{\quad} = -\frac{5}{2} \qquad \underline{\quad} = [\tfrac{1}{2} \cdot (-3)]^2 = (-\tfrac{3}{2})^2 = \tfrac{9}{4}$$

$$x^2 - 3x + \frac{9}{4} = -\frac{5}{2} + \frac{9}{4}$$

$$\left(x - \frac{3}{2}\right)^2 = -\frac{1}{4}$$

$$x - \frac{3}{2} = \pm\sqrt{-\frac{1}{4}} \qquad \text{By Theorem 7.3.1}$$

$$x - \frac{3}{2} = \pm\frac{1}{2}i$$

$$x = \frac{3}{2} \pm \frac{1}{2}i$$

or

$$x = \frac{3 \pm i}{2}$$

3. $3x^2 - 12x + 11 = 0$

$$3x^2 - 12x = -11 \qquad \text{Isolate the x terms.}$$

$$x^2 - 4x = -\frac{11}{3} \qquad \begin{array}{l}\text{Make the leading coefficient 1 by dividing both}\\ \text{sides by 3.}\end{array}$$

$$x^2 - 4x + \underline{\quad} = -\frac{11}{3} \qquad \underline{\quad} = [\tfrac{1}{2} \cdot (-4)]^2 = (-2)^2 = 4$$

$$x^2 - 4x + 4 = -\frac{11}{3} + 4$$

$$(x - 2)^2 = \frac{1}{3}$$

$$x - 2 = \pm\sqrt{\frac{1}{3}} \qquad \textit{By Theorem 7.3.1}$$

$$x - 2 = \pm\frac{\sqrt{3}}{3} \qquad \textit{Rationalize the denominator.}$$

$$x = 2 \pm \frac{\sqrt{3}}{3}$$

or

$$x = \frac{6}{3} \pm \frac{\sqrt{3}}{3}$$

$$x = \frac{6 \pm \sqrt{3}}{3}$$

The following steps will enable us to find the solutions of *any* quadratic equation (i.e., any equation of the form $ax^2 + bx + c = 0$, where $a \neq 0$) by completing the square.

1. Isolate the x terms; that is, rewrite the equation as

$$ax^2 + bx = -c$$

2. Make the leading coefficient 1 by dividing both sides of the equation by a.
3. Determine the number needed to complete the square on the x terms. [Remember that this number is $(\frac{1}{2} \cdot \text{coefficient of } x)^2$.]
4. Add the number found in step 3 to both sides of the equation.
5. Rewrite the perfect square trinomial on the left-hand side as a binomial squared, and combine the numbers on the right-hand side.
6. Apply the Extraction of Roots theorem and solve for x.

Exercises 7.3

Use the Extraction of Roots theorem to find the solutions of the following quadratic equations.

1. $x^2 = 25$ 2. $x^2 = 144$ 3. $x^2 = 169$ 4. $x^2 = 81$

5. $x^2 = 80$ 6. $x^2 = 48$ 7. $x^2 = 98$ 8. $x^2 = 150$

9. $x^2 = -196$ 10. $x^2 = -49$ 11. $x^2 = -121$ 12. $x^2 = -225$

13. $x^2 - 64 = 0$ 14. $x^2 - 100 = 0$ 15. $x^2 + 9 = 0$ 16. $x^2 + 144 = 0$

17. $4x^2 + 81 = 0$ 18. $25x^2 + 49 = 0$ 19. $36x^2 - 121 = 0$ 20. $9x^2 - 225 = 0$

21. $5x^2 - 8 = 0$ 22. $3x^2 - 4 = 0$ 23. $7x^2 - 2 = 0$ 24. $8x^2 - 7 = 0$

25. $9x^2 + 1 = 5$

26. $25x^2 - 7 = 74$

27. $4x^2 + 8 = 4$

28. $x^2 - 5 = 7$

29. $2x^2 - 3 = 8$

30. $3x^2 + 2 = 1$

31. $(x - 5)^2 = 9$

32. $(x + 3)^2 = 4$

33. $(2x - 7)^2 = 1$

34. $\left(5x + \dfrac{1}{2}\right)^2 = 100$

35. $(2x - 1)^2 = 0$

36. $(3x + 4)^2 = 7$

37. $(4x - 5)^2 = 6$

38. $(x - 3)^2 = -4$

39. $(2x + 5)^2 = -49$

40. $(2x - 7)^2 = -100$

41. $(x - 1)^2 = -75$

42. $\left(3x - \dfrac{1}{2}\right)^2 = \dfrac{1}{2}$

43. $\left(5x + \dfrac{2}{3}\right)^2 = \dfrac{5}{9}$

44. $(2x + 3)^2 - 1 = 8$

45. $(4x - 1)^2 + 6 = 10$

46. $(2x - 4)^2 - 1 = 4$

47. $(3x + 1)^2 + 1 = 8$

48. $(x - 5)^2 + 7 = 3$

49. $(2x + 3)^2 - 1 = -8$

50. $(3x + 4)^2 + 10 = 2$

Find the solutions of the following quadratic equations by completing the square.

51. $x^2 - 6x - 16 = 0$

52. $x^2 + 8x + 15 = 0$

53. $x^2 + 10x + 21 = 0$

54. $x^2 - 2x - 63 = 0$

55. $4x^2 + 12x = 0$

56. $4x^2 - 16x - 84 = 0$

57. $9x^2 - 42x - 32 = 0$

58. $16x^2 + 24x - 7 = 0$

59. $x^2 + 4x - 1 = 0$

60. $x^2 - 8x + 6 = 0$

61. $x^2 - 2x - 7 = 0$

62. $x^2 + 8x - 11 = 0$

63. $4x^2 + 4x - 2 = 0$

64. $9x^2 - 12x - 3 = 0$

65. $9x^2 + 24x - 16 = 0$

66. $4x^2 - 4x - 49 = 0$

67. $x^2 + 4x + 29 = 0$

68. $x^2 + 8x + 80 = 0$

69. $x^2 - 2x + 37 = 0$

70. $x^2 - 6x + 18 = 0$

71. $4x^2 - 4x + 17 = 0$

72. $4x^2 - 12x + 45 = 0$

73. $9x^2 + 12x + 5 = 0$

74. $16x^2 + 8x + 10 = 0$

75. $x^2 - x - 2 = 0$

76. $x^2 + 3x - 13 = 0$

77. $x^2 + 5x + 6 = 0$

78. $3x^2 - 4x - 7 = 0$

79. $2x^2 + x - 6 = 0$

80. $3x^2 - 2x - 1 = 0$

81. $2x^2 - 3x - 2 = 0$

82. $3x^2 + 8x - 16 = 0$

83. $4x^2 + 12x + 7 = 0$

84. $4x^2 - 4x - 17 = 0$

85. $9x^2 - 12x + 1 = 0$

86. $9x^2 + 6x - 11 = 0$

87. $16x^2 - 8x - 7 = 0$

88. $18x^2 - 24x - 19 = 0$

89. $16x^2 + 24x - 63 = 0$

90. $12x^2 + 12x - 61 = 0$

91. $2x^2 - 10x + 37 = 0$

92. $2x^2 + 2x + 5 = 0$

93. $9x^2 + 6x + 5 = 0$

94. $9x^2 - 12x + 20 = 0$

95. $12x^2 - 12x + 7 = 0$

96. $16x^2 + 8x + 41 = 0$

97. $36x^2 + 96x + 73 = 0$

98. $36x^2 - 108x + 85 = 0$

7.4 *The Quadratic Formula*

The previous section demonstrated how to find the solutions of any quadratic equation by completing the square. Unfortunately, completing the square is rather tedious. In this section a formula will be developed, by completing the square, that will enable us to find the solutions of any quadratic equation.

Consider the general quadratic equation in standard form:

$$ax^2 + bx + c = 0, \qquad a \neq 0$$

Now solve for x by completing the square:

$$ax^2 + bx = -c \qquad \textit{Isolate the x terms.}$$

$$x^2 + \left(\frac{b}{a}\right)x = \frac{-c}{a} \qquad \begin{array}{l}\textit{Make the leading coefficient 1 by}\\ \textit{dividing both sides of the equation by a.}\end{array}$$

$$x^2 + \left(\frac{b}{a}\right)x + \underline{} = \frac{-c}{a} \qquad \underline{} = \left(\frac{1}{2}\cdot\frac{b}{a}\right)^2 = \left(\frac{b}{2a}\right)^2 = \frac{b^2}{4a^2}$$

$$x^2 + \left(\frac{b}{a}\right)x + \frac{b^2}{4a^2} = \frac{b^2}{4a^2} - \frac{c}{a}$$

$$\left(x + \frac{b}{2a}\right)^2 = \frac{b^2}{4a^2} - \frac{c}{a} \qquad \textit{Factor the left-hand side.}$$

$$\left(x + \frac{b}{2a}\right)^2 = \frac{b^2}{4a^2} - \frac{4ac}{4a^2} \qquad \begin{array}{l}\textit{Obtain the LCD on the}\\ \textit{right-hand side.}\end{array}$$

$$\left(x + \frac{b}{2a}\right)^2 = \frac{b^2 - 4ac}{4a^2}$$

$$x + \frac{b}{2a} = \pm\sqrt{\frac{b^2 - 4ac}{4a^2}} \qquad \begin{array}{l}\textit{By the Extraction of Roots}\\ \textit{theorem}\end{array}$$

$$x + \frac{b}{2a} = \pm\frac{\sqrt{b^2 - 4ac}}{\sqrt{4a^2}}$$

NOTE ▶ $\sqrt{4a^2} = |2a|$, so $\sqrt{4a^2} = 2a$ if $a > 0$ and $\sqrt{4a^2} = -2a$ if $a < 0$.

Thus,

$$\pm\frac{\sqrt{b^2 - 4ac}}{\sqrt{4a^2}} = \pm\frac{\sqrt{b^2 - 4ac}}{|2a|} = \pm\frac{\sqrt{b^2 - 4ac}}{2a}$$

whether a is positive or negative.

$$x + \frac{b}{2a} = \pm\frac{\sqrt{b^2 - 4ac}}{2a}$$

$$x = -\frac{b}{2a} \pm \frac{\sqrt{b^2 - 4ac}}{2a}$$

$$x = \frac{-b \pm \sqrt{b^2 - 4ac}}{2a}$$

This significant result is summarized in the following theorem.

Theorem 7.4.1   **Quadratic formula:** If $ax^2 + bx + c = 0$, where a, b, and c are real numbers with $a \neq 0$, then the solutions of the equation are

$$x = \frac{-b \pm \sqrt{b^2 - 4ac}}{2a}$$

REMARK   To apply the quadratic formula, the quadratic equation must be in standard form and a, b, and c must be properly identified.

E X A M P L E   7 . 4 . 1   Use the quadratic formula to find the solutions of the following quadratic equations.

1. $2x^2 + 5x - 3 = 0$

To apply the quadratic formula, we must identify a, b, and c. In this equation $a = 2$, $b = 5$, and $c = -3$. Thus,

$$x = \frac{-5 \pm \sqrt{5^2 - 4 \cdot 2 \cdot (-3)}}{2 \cdot 2}$$

$$= \frac{-5 \pm \sqrt{25 - (-24)}}{4}$$

$$= \frac{-5 \pm \sqrt{49}}{4}$$

$$= \frac{-5 \pm 7}{4}$$

so

$$x = \frac{-5 + 7}{4} \quad \text{or} \quad x = \frac{-5 - 7}{4}$$

$$x = \frac{2}{4} \qquad\qquad x = \frac{-12}{4}$$

$$x = \frac{1}{2} \qquad\qquad x = -3$$

The solutions of the equation could have been determined by using the factoring method of Chapter 3; that is,

$$2x^2 + 5x - 3 = 0$$

$$(2x - 1)(x + 3) = 0$$

$$2x - 1 = 0 \quad \text{or} \quad x + 3 = 0$$

$$2x = 1 \qquad\qquad x = -3$$

$$x = \frac{1}{2}$$

2. $3x^2 + 5 = 9x$

Write the equation in standard form:

$$3x^2 - 9x + 5 = 0$$

Now a, b, and c can be identified: $a = 3$, $b = -9$, and $c = 5$. So

$$x = \frac{-(-9) \pm \sqrt{(-9)^2 - 4 \cdot 3 \cdot 5}}{2 \cdot 3}$$

$$= \frac{9 \pm \sqrt{81 - 60}}{6}$$

$$= \frac{9 \pm \sqrt{21}}{6}$$

The solutions of this equation and the two that follow could not be determined by the factoring method.

3. $x^2 - 4x + 13 = 0$

$$a = 1, b = -4, c = 13$$

$$x = \frac{-(-4) \pm \sqrt{(-4)^2 - 4 \cdot 1 \cdot 13}}{2 \cdot 1}$$

$$= \frac{4 \pm \sqrt{16 - 52}}{2}$$

$$= \frac{4 \pm \sqrt{-36}}{2}$$

$$= \frac{4 \pm 6i}{2}$$

$$= \frac{2(2 \pm 3i)}{2}$$

$$= 2 \pm 3i$$

Try to avoid this mistake:

Incorrect	Correct
$x = \dfrac{\cancel{2} \pm 10i}{\cancel{2}}$ $= 1 \pm 10i$	$x = \dfrac{2 \pm 10i}{2}$ $= \dfrac{\cancel{2}(1 \pm 5i)}{\cancel{2}}$ $= 1 \pm 5i$ Remember that you can divide out only common *factors*, not terms.

4. $3 - \dfrac{2}{x} - \dfrac{4}{x^2} = 0$

Recall how to find the solutions of fractional equations. Multiply both sides of the equation by the LCD. Here the LCD is x^2.

$$\frac{x^2}{1} \cdot \left(3 - \frac{2}{x} - \frac{4}{x^2}\right) = 0 \cdot x^2$$

$$\frac{x^2}{1} \cdot 3 - \frac{x^2}{1} \cdot \frac{2}{x} - \frac{x^2}{1} \cdot \frac{4}{x^2} = 0 \cdot x^2$$

$$3x^2 - 2x - 4 = 0$$

Now $a = 3$, $b = -2$, and $c = -4$.

$$
\begin{aligned}
x &= \frac{-(-2) \pm \sqrt{(-2)^2 - 4 \cdot 3 \cdot (-4)}}{2 \cdot 3} \\[2mm]
&= \frac{2 \pm \sqrt{4 - (-48)}}{6} \\[2mm]
&= \frac{2 \pm \sqrt{52}}{6} \\[2mm]
&= \frac{2 \pm 2\sqrt{13}}{6} \\[2mm]
&= \frac{2(1 \pm \sqrt{13})}{6} \\[2mm]
&= \frac{1 \pm \sqrt{13}}{3}
\end{aligned}
$$

Neither solution is extraneous.

Try to avoid this mistake:

Incorrect	Correct
$x = -(-2) \pm \dfrac{\sqrt{(-2)^2 - 4 \cdot 3 \cdot (-4)}}{2 \cdot 3}$	$x = \dfrac{-(-2) \pm \sqrt{(-2)^2 - 4 \cdot 3 \cdot (-4)}}{2 \cdot 3}$
$x = 2 \pm \dfrac{\sqrt{52}}{6}$	$x = \dfrac{2 \pm \sqrt{52}}{6}$
$x = 2 \pm \dfrac{2\sqrt{13}}{6}$	$x = \dfrac{2 \pm 2\sqrt{13}}{6}$
$x = 2 \pm \dfrac{\sqrt{13}}{3}$	$x = \dfrac{1 \pm \sqrt{13}}{3}$
Remember that the fraction line goes all the way across.	

Solutions of quadratic equations may be rational numbers, irrational numbers, or nonreal numbers. The type of solution of a particular quadratic equation is determined by the radicand, $b^2 - 4ac$, in the quadratic formula. The expression

$b^2 - 4ac$ is called the **discriminant.** The following rules show how the discriminant characterizes the solutions of a quadratic equation.

If $ax^2 + bx + c = 0$, where a, b, and c are *rational* numbers with $a \neq 0$ and $b^2 - 4ac$ is

zero,	then there is one rational solution, called a double root.
a positive perfect square,	then there are two rational solutions.
positive, but not a perfect square,	then there are two irrational solutions.
negative,	then there are two nonreal solutions.

REMARK   If the discriminant is a perfect square, the quadratic is factorable.

EXAMPLE 7.4.2   Use the discriminant to characterize the solutions of the following quadratic equations.

1.
$$(x - 4)^2 + x^2 = 3x + 11$$
$$x^2 - 8x + 16 + x^2 = 3x + 11$$
$$2x^2 - 8x + 16 = 3x + 11$$
$$2x^2 - 11x + 5 = 0$$
$$a = 2, b = -11, c = 5$$
$$b^2 - 4ac = (-11)^2 - 4 \cdot 2 \cdot 5$$
$$= 121 - 40$$
$$= 81 \qquad A \ perfect \ square$$

This quadratic equation has two rational solutions.

2. $x^2 - 6x + 4 = 0$
$$a = 1, b = -6, c = 4$$
$$b^2 - 4ac = (-6)^2 - 4 \cdot 1 \cdot 4$$
$$= 36 - 16$$
$$= 20$$

This quadratic equation has two irrational solutions.

3. $2x^2 = 3x - 7$

$2x^2 - 3x + 7 = 0$

$a = 2, b = -3, c = 7$

$b^2 - 4ac = (-3)^2 - 4 \cdot 2 \cdot 7$

$= 9 - 56$

$= -47$

This quadratic equation has two nonreal solutions.

EXAMPLE 7.4.3   Determine the value of m so that the following quadratic equation will have one rational solution.

$$3x^2 - 4x + m = 0$$

$$a = 3, b = -4, c = m$$

$$b^2 - 4ac = (-4)^2 - 4 \cdot 3 \cdot m$$

$$= 16 - 12m$$

For the equation to have one rational solution, the discriminant must be zero. Thus

$$16 - 12m = 0$$

$$-12m = -16$$

$$m = \frac{-16}{-12}$$

$$m = \frac{4}{3}$$

EXAMPLE 7.4.4   Using a graphing calculator, find an approximation for the solutions of the following quadratic equation

$$2x^2 + 3x - 6 = 0$$

SOLUTION   Let's first graph the equation $y = 2x^2 + 3x - 6$. Your graph should look something like the one in Figure 7.4.1.

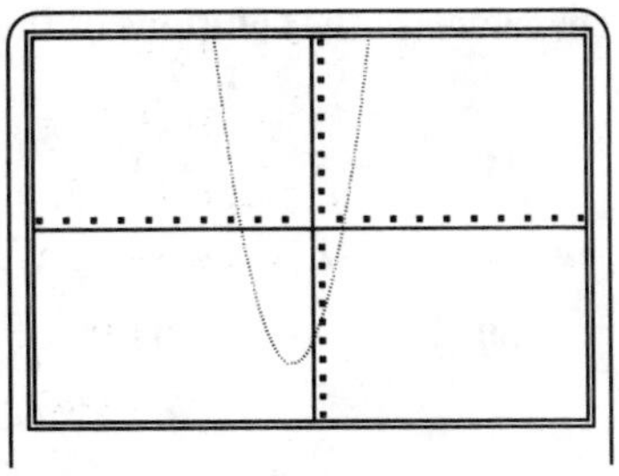

Figure 7.4.1

To find an approximation for the solutions of the equation $2x^2 + 3x - 6 = 0$, all we need to do is to find the x-intercepts of our parabola. (Recall: At the x-intercepts, $y = 0$.) Using the Zoom and Trace keys, we find the solutions to be

$$x \doteq -2.63 \quad \text{and} \quad x \doteq 1.16$$

E xercises 7.4

Use the quadratic formula to find the solutions of the following quadratic equations.

1. $2x^2 - 11x + 5 = 0$

2. $6x^2 + 11x + 4 = 0$

3. $4x^2 - 12x + 9 = 0$

4. $9x^2 + 6x + 1 = 0$

5. $x^2 - 9 = 0$

6. $4x^2 - 25 = 0$

7. $2x^2 - 8x = 0$

8. $3x^2 + 15x = 0$

9. $x^2 + 50 = 15x - 6$

10. $x^2 + 9x + 7 = 3x + 2$

11. $x(6x + 17) = -12$

12. $2x(x + 5) = 28$

13. $(3x + 5)(x - 3) = 5$

14. $(2x + 3)(x - 4) = 6$

15. $3x^2 - 7x + 1 = 0$

16. $5x^2 - 9x + 2 = 0$

17. $x(2x + 3) = 7$

18. $2(x^2 + 2) = 7x$

19. $3x^2 + 9x = 2x + 1$

20. $x^2 + 4x + 2 = 0$

21. $x^2 - 6x + 7 = 0$

22. $5x^2 - 8x - 1 = 0$

23. $2x(x + 3) = -1$

24. $3x(x + 2) = 2x + 5$

25. $2x^2 + 5x + 7 = 0$

26. $x^2 - 3x + 4 = 0$

27. $x(3x - 1) = -1$

28. $x^2 + 4 = 0$

29. $9x^2 - 2x + 28 = 3 - 2x$

30. $x^2 - 4x + 13 = 0$

31. $x^2 + 2x + 17 = 0$

32. $2x^2 - 2x + 1 = 0$

33. $9x^2 + 6x + 5 = 0$

34. $2x(x + 2) = -9$

35. $1 + \dfrac{1}{6x} - \dfrac{1}{3x^2} = 0$

36. $\dfrac{3}{2} + \dfrac{17}{2x} + \dfrac{6}{x^2} = 0$

37. $\dfrac{1}{2} - \dfrac{1}{x} - \dfrac{1}{x^2} = 0$

38. $1 - \dfrac{1}{x} - \dfrac{1}{4x^2} = 0$

39. $\dfrac{1}{4} + \dfrac{2}{x} + \dfrac{5}{x^2} = 0$

40. $\dfrac{3}{5} - \dfrac{4}{5x} + \dfrac{1}{3x^2} = 0$

Use the discriminant to characterize the solutions of the following quadratic equations.

41. $2x^2 - x - 6 = 0$ 　　　**42.** $3x^2 - 5x + 1 = 0$ 　　　**43.** $4x^2 - 12x + 9 = 0$ 　　　**44.** $x^2 + 3x + 1 = 0$

45. $2x^2 - 4x + 5 = 0$ 　　　**46.** $2x^2 + x + 6 = 0$ 　　　**47.** $4x^2 - 25x - 21 = 0$ 　　　**48.** $x^2 + 2x + 2 = 0$

49. $25x^2 + 20x + 4 = 0$ 　　**50.** $x^2 + 6x + 4 = 0$

Determine the value(s) of m so that the following quadratic equations will have one rational solution.

51. $4x^2 - 12x + m = 0$ 　　　**52.** $5x^2 - 2x + m = 0$ 　　　**53.** $mx^2 - 8x + 1 = 0$ 　　　**54.** $mx^2 + 6x - 3 = 0$

55. $9x^2 + mx + 1 = 0$ 　　　**56.** $9x^2 - mx + 4 = 0$

Determine the real values of m so that the following quadratic equations will have two nonreal solutions.

57. $x^2 + 2x + m = 0$ 　　　**58.** $5x^2 - 4x - m = 0$ 　　　**59.** $mx^2 + 2x - 4 = 0$ 　　　**60.** $mx^2 + 8x - 12 = 0$

61. Show that the sum of the two solutions of the quadratic equation $ax^2 + bx + c = 0$, where $a \neq 0$, is $-\dfrac{b}{a}$.

62. Show that the product of the two solutions of the quadratic equation $ax^2 + bx + c = 0$, where $a \neq 0$, is $\dfrac{c}{a}$.

Write Algebra

63. Describe the types of quadratic equations that are easier to solve by completing the square rather than by using the Quadratic Formula.

Use the Zoom and Trace keys of a graphing calculator to find approximations (to 2 decimal places) for the solutions of the following quadratic equations.

64. $3x^2 - 4x - 7 = 0$ 　　　　　　　　　**65.** $-2x^2 + 5x + 4 = 0$

66. $-1.67x^2 - 3x + 2.94 = 0$ 　　　　　**67.** $0.58x^2 + 4.04x + 3.007 = 0$

The following example illustrates how we can use the quadratic formula to solve for x in terms of y.

XAMPLE

Solve the following equation for x.

$$4x^2 + 3xy + 5 = 0$$

Use $a = 4$, $b = 3y$, and $c = 5$ in the quadratic formula.

$$x = \frac{-3y \pm \sqrt{(3y)^2 - 4 \cdot 4 \cdot 5}}{2 \cdot 4}$$

$$x = \frac{-3y \pm \sqrt{9y^2 - 80}}{8}$$

In exercises 68–75, solve each equation for x.

68. $4x^2 + xy + 1 = 0$

69. $9x^2 + xy + 4 = 0$

70. $x^2 - 5xy + 1 = 0$

71. $x^2 - 3xy + 16 = 0$

72. $5x^2 - 7xy + 5 = 0$

73. $4x^2 - 9xy + 4 = 0$

74. $x^2 + 2xy + y + 12 = 0$

75. $x^2 + 2xy + 3y + 10 = 0$

7.5 *Quadratic Equations Summary*

We have investigated several different methods for finding the solutions of quadratic equations. The following chart gives some hints for matching a quadratic equation with the method that most efficiently finds its solutions.

Form	Method	Example
1. $ax^2 + c = 0$ $a \neq 0, c \neq 0$	Use the Extraction of Roots theorem.	$9x^2 - 5 = 0$ $9x^2 = 5$ $x^2 = \dfrac{5}{9}$ $x = \pm\sqrt{\dfrac{5}{9}}$ $x = \pm\dfrac{\sqrt{5}}{3}$
2. $ax^2 + bx = 0$ $a \neq 0, b \neq 0$	Use the factoring method.	$6x^2 + 12x = 0$ $6x(x + 2) = 0$ $6x = 0$ or $x + 2 = 0$ $x = 0$ or $x = -2$
3. $(mx + n)^2 = p$ $m \neq 0$	Use the Extraction of Roots theorem.	$(2x - 1)^2 = 5$ $2x - 1 = \pm\sqrt{5}$ $2x = 1 \pm \sqrt{5}$ $x = \dfrac{1 \pm \sqrt{5}}{2}$
4. $ax^2 + bx + c = 0$ $a \neq 0, b \neq 0, c \neq 0$	**a.** The left-hand side is factorable. Use the factoring method. **b.** The left-hand side cannot be factored using integer coefficients, or you cannot determine the factorization. Use the quadratic formula.	**a.** $x^2 - x - 12 = 0$ $(x - 4)(x + 3) = 0$ $x - 4 = 0$ or $x + 3 = 0$ $x = 4$ or $x = -3$ **b.** $x^2 + x - 3 = 0$ $a = 1, b = 1, c = -3$ $x = \dfrac{-1 \pm \sqrt{1^2 - 4 \cdot 1 \cdot (-3)}}{2 \cdot 1}$ $x = \dfrac{-1 \pm \sqrt{1 - (-12)}}{2}$ $x = \dfrac{-1 \pm \sqrt{13}}{2}$
5. $x^2 + bx + c = 0$ b is even.	Use the completing the square method.	$x^2 - 6x - 12 = 0$ $x^2 - 6x = 12$ $x^2 - 6x + 9 = 12 + 9$ $(x - 3)^2 = 21$ $x - 3 = \pm\sqrt{21}$ $x = 3 \pm \sqrt{21}$

Exercises 7.5

Using the most efficient method, find the solution set of each quadratic equation.

1. $(2x + 1)(2x - 9) = -16x$

2. $(2x + 1)(4x + 3) = 10$

3. $x^2 - 7x + 2 = 0$

4. $(3x + 1)(x - 2) = 0$

5. $5x^2 - 2 = 0$

6. $x^2 - 12x + 37 = 0$

7. $(2x - 3)^2 - 11 = 14$

8. $(4x + 1)(4x + 3) = 2(8x + 1)$

9. $(4x + 1)(x - 7) = -7x - 36$

10. $(5x + 2)^2 = 10$

11. $(5x + 3)(x - 2) = (2x - 1)(x - 3)$

12. $x^2 - 5x + 3 = 0$

13. $6x^2 - 10x = 0$

14. $(3x - 2)^2 - 5 = 4$

15. $(2x + 5)(x - 3) = 0$

16. $3x^2 - 7 = 0$

17. $(3x + 2)(6x + 1) = 9$

18. $(4x + 1)(x - 3) = x - 16$

19. $(2x + 1)(2x + 5) = 4(3x + 1)$

20. $2x^2 + 5x - 1 = 0$

21. $(4x + 1)^2 = 13$

22. $2x^2 - 10x + 11 = 0$

23. $2x^2 + 3x - 3 = 0$

24. $6x^2 - 21x = 0$

25. $2x^2 - 6x + 3 = 0$

26. $3x^2 + 28x + 9 = 0$

27. $(3x - 1)(3x - 2) = 3x$

28. $(6x + 1)(x - 2) = (4x - 7)(x - 1)$

29. $(2x + 4)(x - 3) = (x + 3)(x - 1)$

30. $(9x - 5)(x + 1) = 4x - 3$

31. $6x^2 + 19x + 8 = 0$

32. $(3x - 2)(x - 1) = x$

33. $(x + 3)^2 + 5 = 1$

34. $(2x + 6)(x - 1) = (x + 1)(x - 4)$

35. $(6x - 9)(x + 2) = -18$

36. $8x^2 + 4x - 1 = 0$

37. $(2x + 1)(x - 1) = (x + 7)(x + 8)$

38. $(2x + 3)(x - 6) = (x + 4)(x + 6)$

39. $16x^2 + 8x - 5 = 0$

40. $(x + 7)^2 + 17 = 1$

41. $(4x - 1)(x + 1) = 3x + 4$

42. $(4x + 2)(x + 1) = 2$

43. $x^2 - 8x + 17 = 0$

44. $(3x + 4)(3x - 2) = 6x - 7$

7.6 *Applications*

In Chapters 3 and 4 application problems were investigated. In both chapters most of the application problems could be solved by finding the solutions of a quadratic equation.

The quadratic equations in Chapters 3 and 4 could be solved using the factoring method. Now we are able to examine application problems that lead to quadratic equations where the factoring method cannot be applied.

XAMPLE 7.6.1   The sum of a number and its reciprocal is 3. What is the number?

SOLUTION   Let x = number. Then

$$x + \frac{1}{x} = 3$$

$$\frac{x}{1}\left(x + \frac{1}{x}\right) = 3 \cdot \frac{x}{1} \qquad \textit{Multiply by the LCD, x.}$$

$$\frac{x}{1} \cdot x + \frac{x}{1} \cdot \frac{1}{x} = 3 \cdot \frac{x}{1}$$

$$x^2 + 1 = 3x$$

$$x^2 - 3x + 1 = 0$$

Note that the factoring method cannot be applied. Consequently, we will use the quadratic formula to find the solutions of this quadratic equation. Here $a = 1$, $b = -3$, and $c = 1$. Thus

$$x = \frac{-(-3) \pm \sqrt{(-3)^2 - 4 \cdot 1 \cdot 1}}{2 \cdot 1}$$

$$x = \frac{3 \pm \sqrt{9 - 4}}{2}$$

$$x = \frac{3 \pm \sqrt{5}}{2}$$

The two solutions are

$$\frac{3 + \sqrt{5}}{2} \quad \text{and} \quad \frac{3 - \sqrt{5}}{2}$$

Check:

$$\frac{3 + \sqrt{5}}{2} + \frac{2}{3 + \sqrt{5}} \stackrel{?}{=} 3$$

$$\frac{3 + \sqrt{5}}{2} + \frac{2}{3 + \sqrt{5}} \cdot \frac{(3 - \sqrt{5})}{(3 - \sqrt{5})}$$

$$\frac{3 + \sqrt{5}}{2} + \frac{2(3 - \sqrt{5})}{9 - 5}$$

$$\frac{3 + \sqrt{5}}{2} + \frac{\overset{1}{2}(3 - \sqrt{5})}{\underset{2}{4}}$$

$$\frac{3 + \sqrt{5}}{2} + \frac{3 - \sqrt{5}}{2}$$

$$\frac{6}{2}$$

$$3$$

The check for $\left(\dfrac{3 - \sqrt{5}}{2}\right)$ is similar.

XAMPLE 7.6.2 Bonnie and Bruce have a rectangular swimming pool that is 6 ft wide and 10 ft long. They are going to build a tile border of uniform width around the pool. They have 100 sq ft of tile. How wide should the border be?

SOLUTION Let x = width of the border. By examining Figure 7.6.1, we can construct a quadratic equation that will enable us to find x.

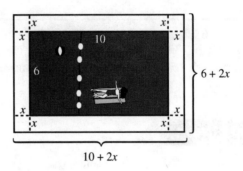

Figure 7.6.1

(area of large rectangle) − (area of swimming pool) = (area of border)

$$(10 + 2x)(6 + 2x) \quad - \quad\quad 60 \quad\quad = \quad\quad 100$$

$$60 + 20x + 12x + 4x^2 - 60 = 100$$

$$4x^2 + 32x = 100$$

$$4x^2 + 32x - 100 = 0$$

$$x^2 + 8x - 25 = 0 \quad\quad \textit{Divide both sides by 4.}$$

In the quadratic formula, $a = 1$, $b = 8$, and $c = -25$.

$$x = \frac{-8 \pm \sqrt{8^2 - 4 \cdot 1 \cdot (-25)}}{2 \cdot 1}$$

$$= \frac{-8 \pm \sqrt{64 + 100}}{2}$$

$$= \frac{-8 \pm \sqrt{164}}{2}$$

$$= \frac{-8 \pm 2\sqrt{41}}{2}$$

$$= \frac{2(-4 \pm \sqrt{41})}{2}$$

$$= -4 \pm \sqrt{41}$$

We reject the negative solution, $-4 - \sqrt{41}$, and conclude that $x = -4 + \sqrt{41}$. To obtain a decimal approximation for the width of the border, refer to Table II in Appendix A or use a calculator to find $\sqrt{41} \doteq 6.4$:

$$x \doteq -4 + 6.4$$

$$x \doteq 2.4$$

Thus the border should be approximately 2.4 ft wide.

E X A M P L E 7 . 6 . 3

Working together, Farmer Odell and her son can brand their cattle in 2 days. Working alone, Farmer Odell's son takes 3 more days to brand the cattle than his mother does. Working alone, how long does it take Farmer Odell to brand the cattle?

SOLUTION Let

$x =$ number of days it takes Farmer Odell to brand the cattle

$x + 3 =$ number of days it takes Farmer Odell's son to brand the cattle

$\dfrac{1}{x} =$ part of the herd branded by Farmer Odell in 1 day

$\dfrac{1}{x + 3} =$ part of the herd branded by Farmer Odell's son in 1 day

$\dfrac{1}{2} =$ part of the herd branded by Farmer Odell and her son in 1 day

part branded by Farmer Odell in 1 day	+	part branded by Farmer Odell's son in 1 day	=	part branded by both in 1 day
$\dfrac{1}{x}$	$+$	$\dfrac{1}{x + 3}$	$=$	$\dfrac{1}{2}$

$$\frac{2x(x+3)}{1}\left(\frac{1}{x}+\frac{1}{x+3}\right)=\frac{1}{2}\cdot\frac{2x(x+3)}{1}$$

Multiply both sides by the LCD, 2x(x + 3).

$$\frac{2x(x+3)}{1}\cdot\frac{1}{x}+\frac{2x(x+3)}{1}\cdot\frac{1}{x+3}=\frac{1}{2}\cdot\frac{2x(x+3)}{1}$$

$$2(x+3)+2x=x(x+3)$$

$$2x+6+2x=x^2+3x$$

$$4x+6=x^2+3x$$

$$0=x^2-x-6$$

This quadratic equation *can* be solved by factoring:

$$0=(x-3)(x+2)$$

$$x-3=0 \quad\text{or}\quad x+2=0$$

$$x=3 \qquad\qquad x=-2$$

We reject −2. Working alone it takes Farmer Odell 3 days to brand the herd.

EXAMPLE 7.6.4   Gail paddles her canoe to her favorite fishing spot 2 mi downstream. The trip downstream and back takes 5 hr. If the speed of the current is 1 mi/hr, what is the speed of Gail's canoe in still water?

SOLUTION   Let x = Gail's speed in still water. Use the table developed in Chapter 4.

	Distance	÷	Rate	=	Time
Downstream	2		$x+1$		$\dfrac{2}{x+1}$
Upstream	2		$x-1$		$\dfrac{2}{x-1}$

$$(\text{time downstream}) + (\text{time upstream}) = (\text{total time})$$

$$\frac{2}{x+1} \quad + \quad \frac{2}{x-1} \quad = \quad 5$$

$$\frac{(x + 1)(x - 1)}{1}\left(\frac{2}{x + 1} + \frac{2}{x - 1}\right) = 5 \cdot \frac{(x + 1)(x - 1)}{1} \qquad \textit{Multiply by the LCD,}\ (x + 1)(x - 1).$$

$$\frac{(x + 1)(x - 1)}{1} \cdot \frac{2}{x + 1} + \frac{(x + 1)(x - 1)}{1} \cdot \frac{2}{x - 1} = 5 \cdot \frac{(x + 1)(x - 1)}{1}$$

$$2(x - 1) + 2(x + 1) = 5(x^2 - 1)$$

$$2x - 2 + 2x + 2 = 5x^2 - 5$$

$$4x = 5x^2 - 5$$

$$0 = 5x^2 - 4x - 5$$

In the quadratic formula, $a = 5$, $b = -4$, and $c = -5$.

$$x = \frac{-(-4) \pm \sqrt{(-4)^2 - 4 \cdot 5 \cdot (-5)}}{2 \cdot 5}$$

$$= \frac{4 \pm \sqrt{16 - (-100)}}{10}$$

$$= \frac{4 \pm \sqrt{116}}{10}$$

$$= \frac{4 \pm 2\sqrt{29}}{10}$$

$$= \frac{2(2 \pm \sqrt{29})}{10}$$

$$= \frac{2 \pm \sqrt{29}}{5}$$

We reject $x = \dfrac{2 - \sqrt{29}}{5}$, since it is negative, and conclude that $x = \dfrac{2 + \sqrt{29}}{5}$. Using the decimal approximation $\sqrt{29} \doteq 5.4$,

$$x \doteq \frac{2 + 5.4}{5}$$

$$\doteq \frac{7.4}{5}$$

$$\doteq 1.5$$

Thus Gail's speed in still water is approximately 1.5 mi/hr.

EXAMPLE 7.6.5

An arrow is shot upward from the roof of a 96-ft-tall building with a speed of 48 ft/sec. The equation that gives the arrow's height above ground level is $h = -16t^2 + 48t + 96$, where h is the height (measured in feet) and t is the time (measured in seconds) after the arrow was shot. When does the arrow hit the ground?

SOLUTION The arrow hits the ground when $h = 0$. Thus we must solve the equation

$$0 = -16t^2 + 48t + 96$$
$$0 = -t^2 + 3t + 6 \qquad \text{Divide both sides by 16.}$$

In the quadratic formula, $a = -1$, $b = 3$, and $c = 6$.

$$t = \frac{-3 \pm \sqrt{3^2 - 4 \cdot (-1) \cdot 6}}{2 \cdot (-1)}$$

$$= \frac{-3 \pm \sqrt{9 - (-24)}}{-2}$$

$$= \frac{-3 \pm \sqrt{33}}{-2}$$

We reject $t = \dfrac{-3 + \sqrt{33}}{-2}$, since it is negative, and conclude that $t = \dfrac{-3 - \sqrt{33}}{-2}$. Using the decimal approximation $\sqrt{33} \doteq 5.7$,

$$t \doteq \frac{-3 - 5.7}{-2}$$

$$\doteq \frac{-8.7}{-2}$$

$$\doteq 4.4$$

Approximately 4.4 sec after the arrow is shot, it strikes the ground.

Exercises 7.6

 (Review over)

Find the solutions of the following problems. Beginning with exercise 9, determine a decimal approximation for any solution that is irrational.

1. Find two consecutive positive integers whose product is 42.

2. Find two consecutive positive integers whose product is 132.

3. The sum of a number and its reciprocal is 4. What is the number?

4. The sum of a number and its reciprocal is 6. What is the number?

5. The sum of the squares of two consecutive positive integers is 61. Find the integers.

6. The sum of the squares of two consecutive even integers is 100. Find the integers.

7. The square of a number minus 2 times the number is 14. What is the number?

8. Three times the square of a number is 6 more than twice the number. What is the number?

9. One leg of a right triangle is 2 in. longer than the other leg. The hypotenuse is 6 in. long. What are the lengths of the legs of the triangle? (*Hint:* Use the Pythagorean theorem.)

10. The hypotenuse of an isosceles right triangle is 20 in. long. What are the lengths of the legs of the triangle? (*Hint:* In an isosceles right triangle the legs have equal length.)

11. The base of a triangle is 1 in. less than twice the height. Find the base and the height if the area of the triangle is 6 sq in. (*Hint:* The area of a triangle $= \frac{1}{2} \cdot$ base $\cdot$ height.)

12. The height of a triangle is 2 in. more than 3 times the base. Find the base and the height if the area of the triangle is 5 sq in.

13. Megan and Michael have a rectangular baby swimming pool that is 5 ft wide and 9 ft long. They are going to build a tile border of uniform width around the pool. They have 40 sq ft of tile. How wide should the border be?

14. Minnie and Steve have a rectangular swimming pool that is 12 ft wide and 20 ft long. They are going to build a tile border of uniform width around the pool. They have 160 sq ft of tile. How wide should the border be?

15. A painting and its frame cover 108 sq in. The frame is $\frac{3}{2}$ in. wide. The length of the painting is 3 in. more than the width. What are the dimensions of the painting?

16. A painting and its frame cover 144 sq in. The frame is 3 in. wide at top and bottom and 2 in. wide at the sides. If the length of the painting is 2 in. more than the width, what are the dimensions of the painting? See Figure 7.6.2.

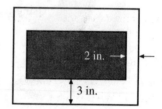

2 in. ⟶

3 in.

Figure 7.6.2

17. Patricia and Joe have a rectangular swimming pool that is 10 ft wide and 16 ft long. They are going to build a tile border of uniform width around three sides of the pool as shown in Figure 7.6.3. They have 90 sq ft of tile. How wide should the border be?

18. Patti and Frank have a *square* hot tub that measures 8 ft on a side. They are going to build a tile border of uniform width around three sides of the tub. They have 70 sq ft of tile. How wide should the border be?

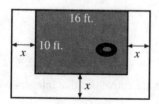

16 ft.

10 ft.

x x

x

Figure 7.6.3

19. One pipe can fill a reservoir 1 hr faster than another pipe. Together they fill the reservoir in 4 hr. How long does it take each pipe to fill the reservoir?

20. One pipe can fill a reservoir 2 hr faster than another pipe can. Together they fill the reservoir in 5 hr. How long does it take each pipe to fill the reservoir?

21. The Moss County Water Department's new computer can process the county's water bills 2 days faster than the old computer can. Working together the two computers process the county's water bills in 6 days. Working alone how long does it take each computer to process the water bills?

22. Working together Bud and Don can clean Moss County Park in 2.5 hr. Working alone Bud takes 1 more hour to clean the park than Don does. Working alone how long does it take Don to clean the park?

23. Kaori hikes up a 4-mi mountain trail. Her speed going down the trail is 1 mi/hr faster than her speed going up the trail. She spends a total of 5 hr hiking. What is Kaori's hiking speed up the trail?

24. Mona takes her motorboat to her favorite fishing spot 7 mi downstream. The trip downstream and back takes 3 hr. If the speed of the current is 2 mi/hr, what is the speed of Mona's boat in still water?

25. Daniel walks from his home to school, a distance of 3 mi. When he gets to school, he finds his brother's bicycle and rides it home. His bicycling speed is 5 mi/hr faster than his walking speed. His total walking and bicycling time is 1 hr. What is Daniel's walking speed?

26. Larry walks from his home to Al's Auto Repair, a distance of 4 mi. When he gets to Al's, he finds his car and drives it home. His driving speed is 20 mi/hr faster than his walking speed. His total walking and driving time is 2 hr. What is Larry's walking speed?

27. A ball is thrown upward from the roof of an 80-ft-tall building with a speed of 32 ft/sec. The equation that gives the ball's height above ground level is $h = -16t^2 + 32t + 80$. When does the ball hit the ground?

28. An astronaut on the surface of the moon throws a ball upward from an 80-ft-tall cliff with a speed of 32 ft/sec. The equation that gives the ball's height above ground level is $h = -2.6t^2 + 32t + 80$. When does the ball hit the ground?

29. A cannonball is fired upward from a 104-ft tall cliff with a speed of 96 ft/sec. The equation that gives the cannonball's height above ground level is $h = -16t^2 + 96t + 104$. When does the cannonball hit the ground? When does the cannonball reach its maximum height? What is the maximum height?

30. A cannonball is fired upward from a 44-ft-tall cliff with a speed of 128 ft/sec. The equation that gives the cannonball's height above ground level is $h = -16t^2 + 128t + 44$. When does the cannonball hit the ground? When does the cannonball reach its maximum height? What is the maximum height?

7.7 *Equations in Quadratic Form and Higher-Degree Equations*

A quadratic equation in the variable x is any equation that can be written in the form $ax^2 + bx + c = 0$, where a, b, and c are real numbers with $a \neq 0$. The equation $x^4 - 10x^2 + 9 = 0$ is not a quadratic equation. However,

$$x^4 - 10x^2 + 9 = 0$$

is equivalent to

$$(x^2)^2 - 10(x^2) + 9 = 0$$

If we let $w = x^2$, then $(x^2)^2 - 10(x^2) + 9 = 0$ becomes $w^2 - 10w + 9 = 0$, a quadratic equation in the variable w. Any equation that can be converted to the form $aw^2 + bw + c = 0$, where a, b, and c are real numbers with $a \neq 0$, is called an **equation in quadratic form.** The following example illustrates how to find the solutions of equations in quadratic form.

EXAMPLE 7.7.1 Find the solutions of the following equations.

1. $x^4 - 10x^2 + 9 = 0$

Let $w = x^2$; then $w^2 = x^4$. Substituting this into the given equation, we obtain

$$w^2 - 10w + 9 = 0$$

This equation can be solved by the factoring method.

$$(w - 9)(w - 1) = 0$$
$$w - 9 = 0 \quad \text{or} \quad w - 1 = 0$$
$$w = 9 \qquad\qquad w = 1$$

NOTE ▶ *Don't stop here.* We have found only w and still need to determine x. Since $w = x^2$,

$$x^2 = 9 \qquad \text{or} \quad x^2 = 1$$
$$x = \pm\sqrt{9} \qquad x = \pm\sqrt{1} \qquad \textit{Extracting the roots}$$
$$x = \pm 3 \qquad\quad x = \pm 1$$

2. $x^4 + 7x^2 - 18 = 0$

Let $w = x^2$; then $w^2 = x^4$. Substituting this into the given equation, we obtain

$$w^2 + 7w - 18 = 0$$
$$(w + 9)(w - 2) = 0$$
$$w + 9 = 0 \quad \text{or} \quad w - 2 = 0$$
$$w = -9 \qquad\qquad w = 2$$

Since $w = x^2$,

$$x^2 = -9 \qquad \text{or} \quad x^2 = 2$$
$$x = \pm\sqrt{-9} \qquad x = \pm\sqrt{2} \qquad \textit{Extracting the roots}$$
$$x = \pm 3i$$

3. $3x - 5\sqrt{x} - 2 = 0$

Let $w = \sqrt{x}$; then $w^2 = x$. Substituting this into the given equation, we obtain

$$3w^2 - 5w - 2 = 0$$
$$(w - 2)(3w + 1) = 0$$
$$w - 2 = 0 \quad \text{or} \quad 3w + 1 = 0$$
$$w = 2 \qquad\qquad 3w = -1$$
$$w = -\frac{1}{3}$$

Since $w = \sqrt{x}$,

$$\sqrt{x} = 2 \quad \text{or} \quad \sqrt{x} = -\frac{1}{3}$$

$$(\sqrt{x})^2 = 2^2 \qquad \textit{Discard } -\tfrac{1}{3}, \textit{ since } \sqrt{x}$$

$$x = 4 \qquad \textit{cannot be negative.}$$

This equation has only one solution, $x = 4$. (Remember to check your solutions to be sure that no extraneous solutions were generated. Since both sides were squared, we must check for extraneous solutions.)

NOTE ▶   When you have an equation of the form $a(f(x))^2 + b(f(x)) + c = 0$, make the substitution $w = f(x)$ to obtain a quadratic equation in the variable w.

4. $5(2x + 3)^2 - 29(2x + 3) - 6 = 0$

Let $w = 2x + 3$; then $w^2 = (2x + 3)^2$. Substituting this into the given equation, we obtain

$$5w^2 - 29w - 6 = 0$$

$$(5w + 1)(w - 6) = 0$$

$$5w + 1 = 0 \quad \text{or} \quad w - 6 = 0$$

$$5w = -1 \qquad\qquad w = 6$$

$$w = -\frac{1}{5}$$

Since $w = 2x + 3$,

$$2x + 3 = -\frac{1}{5} \quad \text{or} \quad 2x + 3 = 6$$

$$2x = -\frac{16}{5} \qquad\qquad 2x = 3$$

$$x = -\frac{8}{5} \qquad\qquad x = \frac{3}{2}$$

5. $2x^{-2} + 7x^{-1} - 4 = 0$

Let $w = x^{-1}$; then $w^2 = x^{-2}$. Substituting this into the given equation, we obtain

$$2w^2 + 7w - 4 = 0$$

$$(2w - 1)(w + 4) = 0$$

$$2w - 1 = 0 \quad \text{or} \quad w + 4 = 0$$

$$2w = 1 \qquad\qquad w = -4$$

$$w = \frac{1}{2}$$

Since $w = x^{-1}$.

$$x^{-1} = \frac{1}{2} \quad \text{or} \quad x^{-1} = -4$$

$$\frac{1}{x} = \frac{1}{2} \qquad \frac{1}{x} = -4$$

$$x = 2 \qquad x = -\frac{1}{4}$$

6. $5x^{2/3} + 2x^{1/3} - 7 = 0$

Let $w = x^{1/3}$; then $w^2 = x^{2/3}$. Substituting this into the given equation, we obtain

$$5w^2 + 2w - 7 = 0$$
$$(5w + 7)(w - 1) = 0$$
$$5w + 7 = 0 \quad \text{or} \quad w - 1 = 0$$
$$5w = -7 \qquad w = 1$$
$$w = -\frac{7}{5}$$

Since $w = x^{1/3}$,

$$x^{1/3} = -\frac{7}{5} \quad \text{or} \quad x^{1/3} = 1$$

$$(x^{1/3})^3 = \left(-\frac{7}{5}\right)^3 \qquad (x^{1/3})^3 = 1^3$$

$$x = -\frac{343}{125} \qquad x = 1$$

7. $\left(\frac{2x}{x + 1}\right)^2 - \left(\frac{2x}{x + 1}\right) - 12 = 0$

Let

$$w = \frac{2x}{x + 1}$$

Then

$$w^2 = \left(\frac{2x}{x + 1}\right)^2$$

Substituting this into the given equation, we obtain

$$w^2 - w - 12 = 0$$
$$(w - 4)(w + 3) = 0$$
$$w - 4 = 0 \quad \text{or} \quad w + 3 = 0$$
$$w = 4 \qquad\qquad w = -3$$

Since $w = \dfrac{2x}{x + 1}$,

$$\dfrac{2x}{x + 1} = 4 \qquad \text{or} \qquad \dfrac{2x}{x + 1} = -3$$

$$\dfrac{x + 1}{1} \cdot \dfrac{2x}{x + 1} = 4 \cdot \dfrac{x + 1}{1} \qquad \dfrac{x + 1}{1} \cdot \dfrac{2x}{x + 1} = -3 \cdot \dfrac{x + 1}{1} \qquad \textit{Multiply both sides by the LCD, x + 1.}$$

$$2x = 4x + 4 \qquad\qquad 2x = -3x - 3$$
$$-2x = 4 \qquad\qquad 5x = -3$$
$$x = -2 \qquad\qquad x = -\dfrac{3}{5}$$

NOTE ▶ Remember to check your solutions to be sure that no extraneous solutions were generated.

In summary, if an equation can be written as

$$a \cdot [P(x)]^2 + b \cdot [P(x)] + c = 0$$

where $P(x)$ is an algebraic expression and where a, b, and c are real numbers with $a \neq 0$, then it is an equation in quadratic form. To find the solutions of an equation in quadratic form, make the substitution

$$w = P(x)$$

The first part of this section demonstrated how to use substitution and factoring to find the solutions of equations in quadratic form. In the remainder of this section, factoring by grouping will be used to find the solutions of more complicated equations.

EXAMPLE 7.7.2 Use factoring by grouping to find the solutions of the following equations.

1. $2x^3 + 3x^2 - 8x - 12 = 0$

$x^2(2x + 3) - 4(2x + 3) = 0$ *A common factor of 2x + 3 has been generated.*

$(2x + 3)(x^2 - 4) = 0$

$(2x + 3)(x + 2)(x - 2) = 0$

By extending the Zero Product theorem.

$$2x + 3 = 0 \quad \text{or} \quad x + 2 = 0 \quad \text{or} \quad x - 2 = 0$$

$$2x = -3 \qquad\qquad x = -2 \qquad\qquad x = 2$$

$$x = -\frac{3}{2}$$

2.
$$5x^4 + 3x^3 - 40x - 24 = 0$$

$$x^3(5x + 3) - 8(5x + 3) = 0$$

$$(5x + 3)(x^3 - 8) = 0 \qquad \textit{Recall that } x^3 - 8 \textit{ is the difference of}$$

$$(5x + 3)(x - 2)(x^2 + 2x + 4) = 0 \qquad \textit{two cubes.}$$

$$5x + 3 = 0 \quad \text{or} \quad x - 2 = 0 \quad \text{or} \quad x^2 + 2x + 4 = 0$$

$$5x = -3 \qquad\qquad x = 2 \qquad a = 1, b = 2, c = 4$$

$$x = -\frac{3}{5} \qquad\qquad\qquad \textit{Use the quadratic formula:}$$

$$x = \frac{-2 \pm \sqrt{2^2 - 4 \cdot 1 \cdot 4}}{2 \cdot 1}$$

$$= \frac{-2 \pm \sqrt{4 - 16}}{2}$$

$$= \frac{-2 \pm \sqrt{-12}}{2}$$

$$= \frac{-2 \pm 2i\sqrt{3}}{2}$$

$$= \frac{\cancel{2}(-1 \pm i\sqrt{3})}{\cancel{2}}$$

$$= -1 \pm i\sqrt{3}$$

Exercises 7.7

Find the solutions of the following equations.

1. $x^4 - 5x^2 + 4 = 0$

2. $x^4 - 17x^2 + 16 = 0$

3. $4x^4 - 37x^2 + 9 = 0$

4. $4x^4 - 29x^2 + 25 = 0$

5. $x^4 + x^2 - 12 = 0$

6. $x^4 - 7x^2 - 8 = 0$

7. $2x^4 - 11x^2 + 5 = 0$

8. $3x^4 - 10x^2 + 8 = 0$

9. $3x^4 + 4x^2 + 1 = 0$

10. $4x^4 + 12x^2 + 9 = 0$

11. $x - 5\sqrt{x} + 4 = 0$

12. $x - 10\sqrt{x} + 9 = 0$

13. $2x - 11\sqrt{x} + 5 = 0$

14. $6x - 7\sqrt{x} + 2 = 0$

15. $2x - \sqrt{x} - 15 = 0$

16. $3x + 11\sqrt{x} - 4 = 0$

17. $2x + 7\sqrt{x} + 3 = 0$

18. $3x + 5\sqrt{x} + 2 = 0$

19. $(x + 1)^2 + (x + 1) - 20 = 0$

20. $(x - 6)^2 + 5(x - 6) + 6 = 0$

21. $(x - 3)^2 - 4 = 0$

22. $(2x + 7)^2 - 1 = 0$

23. $4(x - 1)^2 - 3(x - 1) - 10 = 0$

24. $3(x + 3)^2 + 11(x + 3) + 6 = 0$

25. $4(2x - 3)^2 + 8(2x - 3) + 3 = 0$

26. $6(2x + 1)^2 + 13(2x + 1) - 5 = 0$

27. $(3x - 4)^2 - 4(3x - 4) + 4 = 0$

28. $(2x + 1)^2 + 6(2x + 1) + 9 = 0$

29. $x^{-2} + x^{-1} - 6 = 0$

30. $x^{-2} - 6x.^{-1} + 5 = 0$

31. $x^{-2} + 6x^{-1} + 5 = 0$

32. $x^{-2} - 4x^{-1} + 3 = 0$

33. $2x^{-2} + 13x^{-1} - 7 = 0$

34. $4x^{-2} + 13x^{-1} + 3 = 0$

35. $6x^{-2} - 7x^{-1} + 2 = 0$

36. $10x^{-2} + 9x^{-1} - 9 = 0$

37. $x^{2/3} - x^{1/3} - 30 = 0$

38. $x^{2/3} - 3x^{1/3} - 4 = 0$

39. $3x^{2/3} - 7x^{1/3} + 2 = 0$

40. $4x^{2/3} + 16x^{1/3} + 15 = 0$

41. $6x^{2/3} - 5x^{1/3} - 4 = 0$

42. $4x^{2/3} - 12x^{1/3} + 5 = 0$

43. $\left(\dfrac{x}{x + 1}\right)^2 - 2\left(\dfrac{x}{x + 1}\right) - 15 = 0$

44. $\left(\dfrac{x}{2x - 1}\right)^2 - 7\left(\dfrac{x}{2x - 1}\right) + 12 = 0$

45. $\left(\dfrac{3x + 1}{x}\right)^2 + 6\left(\dfrac{3x + 1}{x}\right) + 8 = 0$

46. $\left(\dfrac{2x - 1}{2x}\right)^2 + 5\left(\dfrac{2x - 1}{2x}\right) - 6 = 0$

47. $6 \cdot \left(\dfrac{2x}{x - 3}\right)^2 - 5\left(\dfrac{2x}{x - 3}\right) - 6 = 0$

48. $4 \cdot \left(\dfrac{x}{x - 5}\right)^2 + 13\left(\dfrac{x}{x - 5}\right) + 3 = 0$

49. $4 \cdot \left(\dfrac{x + 1}{x - 1}\right)^2 - 8\left(\dfrac{x + 1}{x - 1}\right) + 3 = 0$

50. $3 \cdot \left(\dfrac{2x - 1}{x - 1}\right)^2 + 13\left(\dfrac{2x - 1}{x - 1}\right) + 4 = 0$

51. $x^4 - 8x^2 + 3 = 0$ (*Hint:* Use the quadratic formula.)

52. $x^4 + 6x^2 - 2 = 0$ (*Hint:* Use the quadratic formula.)

Use factoring by grouping to find the solutions of the following equations.

53. $x^3 + 7x^2 - x - 7 = 0$

54. $x^3 - 3x^2 - 9x + 27 = 0$

55. $3x^3 + 2x^2 - 12x - 8 = 0$

56. $2x^3 + x^2 - 50x - 25 = 0$

57. $4x^3 - 12x^2 - 9x + 27 = 0$

58. $4x^3 + 8x^2 - x - 2 = 0$

59. $9x^3 + 9x^2 - x - 1 = 0$

60. $9x^3 - 18x^2 - 4x + 8 = 0$

61. $8x^3 - 12x^2 - 2x + 3 = 0$

62. $12x^3 + 8x^2 - 3x - 2 = 0$

63. $5x^3 - 2x^2 + 5x - 2 = 0$

64. $3x^3 - 4x^2 + 12x - 16 = 0$

65. $8x^3 - 12x^2 + 2x - 3 = 0$

66. $9x^3 + 27x^2 + 4x + 12 = 0$

67. $5x^4 - 3x^3 - 5x + 3 = 0$

68. $x^4 + 2x^3 - 27x - 54 = 0$

69. $x^4 + 3x^3 + 27x + 81 = 0$

70. $2x^4 - x^3 + 16x - 8 = 0$

71. $x^5 - 5x^4 - x + 5 = 0$

72. $x^5 + x^4 - 16x - 16 = 0$

73. $x^7 + 3x^6 - x - 3 = 0$

74. $x^7 - x^6 - 64x + 64 = 0$

Write Algebra

75. Describe how you recognize an equation in quadratic form.

G R O U P   A C T I V I T Y

The following example illustrates how we can use some of the laws of exponents to solve some equations in which the unknown is contained in the base and in the exponent. Assume that x represents a *positive* real number.

EXAMPLE

Find the rational solution of the following equation.

$$x^x = (3x)^{2x}$$
$$(x^x)^{1/x} = \left[(3x)^{2x}\right]^{1/x} \qquad \textit{Raise both sides to the } \tfrac{1}{x} \textit{ power.}$$
$$x = (3x)^2$$
$$x = 9x^2$$
$$0 = 9x^2 - x$$
$$0 = x(9x - 1)$$
$$x = 0 \quad \text{or} \quad 9x - 1 = 0$$
$$x = \frac{1}{9}$$

But since $x \neq 0$, $x = \frac{1}{9}$ is the only possible solution. After checking, we conclude that the solution set is $\left\{\frac{1}{9}\right\}$.

In exercises 76–83, find the positive rational solutions of each equation.

76. $x^x = (4x)^{2x}$

77. $x^x = (5x)^{2x}$

78. $(5x)^x = (4x)^{2x}$

79. $(3x)^x = (4x)^{2x}$

80. $x^{3x} = (2x)^{4x}$

81. $x^{2x} = (4x)^{3x}$

82. $(8x)^x = x^{4x}$

83. $(27x)^x = x^{4x}$

84. The preceding example used the fact that $(x^x)^{1/x} = x$ if x is positive. Show, by example, that this equation is not true if x is negative.

7.8 Polynomial and Rational Inequalities

In Chapter 2 we demonstrated how to find the solutions of linear inequalities. In Chapter 3 and in this chapter we illustrated how to find the solutions of quadratic equations. This section will unite the reasoning behind these two types of problems. A quadratic inequality is just like a quadratic equation except that the equality symbol has been replaced with an inequality symbol. Thus a quadratic inequality is defined as follows.

Definition 7.8.1

Any inequality that can be written in the form

$$ax^2 + bx + c < 0 \quad \text{or} \quad ax^2 + bx + c \leq 0$$

where a, b, and c are real numbers with $a \neq 0$, is called a **quadratic inequality** in the variable x.

REMARK

As with quadratic equations, the form $ax^2 + bx + c < 0$ is called **standard form.** (The definition could have been stated with $>$ or $\geq$.)

Some examples of quadratic inequalities are

$$x^2 - 4x + 1 > 0$$
$$x^2 - 3x < 10$$
$$x(x + 3) \leq 1$$
$$x^2 \geq 4$$

Let's try to develop a technique to find the solutions of these inequalities.

Consider the quadratic inequality $x^2 \geq 4$. Some positive numbers that satisfy this inequality are 2, $2\frac{1}{2}$, 3, 5, and 10. In fact, any number greater than or equal to 2 is a solution of this inequality. Some negative numbers that satisfy this inequality are -2, $-2\frac{1}{2}$, -3, and -10. Any number less than or equal to -2 is a solution of this inequality. Thus the graph of the solution set is indicated by the shaded regions on the following number line:

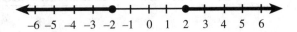

The solution set is

$$\{x : x \leq -2 \quad \text{or} \quad x \geq 2\}$$

REMARK

The values 2 and -2 are called **boundary points** of the solution because they are points where the solution set begins or ends.

There is a more precise method than picking numbers for determining what numbers satisfy a given inequality. The following steps will enable us to find the solutions of any quadratic inequality, provided that the left-hand side is factorable once the quadratic inequality has been written in standard form. Later in this section we will extend this method to cover the case in which the quadratic is not factorable.

Solving a Quadratic Inequality

1. Place the quadratic inequality in standard form.

2. Factor the left-hand side of the quadratic inequality. Use the Zero Product theorem and set each factor equal to zero. These values will be the boundary points of the solution set.

3. The boundary points will *usually* partition the number line into three distinct regions. Pick a number from each region, not including the boundary points, and determine whether it satisfies the inequality. It can be shown that if one number in a region satisfies the inequality, then all the numbers in that region must satisfy the inequality. Likewise, if the number does not satisfy the inequality, then all the numbers in the region fail to satisfy the inequality.

Let's try this procedure on $x^2 \geq 4$:

$$x^2 \geq 4$$
$$x^2 - 4 \geq 0 \qquad \textit{Place in standard form.}$$
$$(x + 2)(x - 2) \geq 0 \qquad \textit{Factor.}$$
$$x + 2 = 0 \quad \text{or} \quad x - 2 = 0 \qquad \textit{Set each factor equal to zero.}$$
$$x = -2 \qquad\qquad x = 2 \qquad \textit{These are the boundary points.}$$

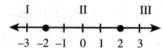

Graph the boundary points.

Now pick a number from each region and determine whether or not it satisfies the inequality. This step is shown in the following table:

Region	x	$x^2 - 4$	$x^2 - 4 \overset{?}{\geq} 0$
I	-3	$(-3)^2 - 4 = 5$	Yes
II	0	$0^2 - 4 = -4$	No
III	3	$3^2 - 4 = 5$	Yes

Test the regions.

Consequently, regions I and III make up the solution set

$$\{x : x \leq -2 \text{ or } x \geq 2\}$$

and the graph is

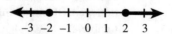

REMARKS

1. The boundary points -2 and 2 are shaded because they satisfy the inequality.

2. If the problem had been $x^2 > 4$, the boundary points -2 and 2 would have been denoted with open circles.

EXAMPLE 7.8.1

Find the solution of each of the following quadratic inequalities, and graph each solution set on a number line.

1.
$$x^2 - 3x < 10$$

$$x^2 - 3x - 10 < 0 \qquad \textit{Place in standard form.}$$

$$(x + 2)(x - 5) < 0 \qquad \textit{Factor.}$$

$$x + 2 = 0 \quad \text{or} \quad x - 5 = 0 \qquad \textit{Set each factor equal to zero.}$$

$$x = -2 \qquad\qquad x = 5 \qquad \textit{These are the boundary points and they are denoted with open circles.}$$

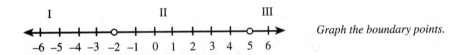

Graph the boundary points.

Region	x	$x^2 - 3x - 10$	$x^2 - 3x - 10 \overset{?}{<} 0$
I	-3	$(-3)^2 - 3(-3) - 10 = 8$	No
II	0	$0^2 - 3 \cdot 0 - 10 = -10$	Yes
III	6	$6^2 - 3 \cdot 6 - 10 = 8$	No

We thus obtain the solution set $\{x: -2 < x < 5\}$ and its graph

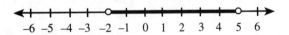

The solution is sometimes indicated by simply giving the inequality $-2 < x < 5$, or, using interval notation, as $(-2, 5)$.

2.
$$x^2 - 10x + 25 > 0$$

$$(x - 5)(x - 5) > 0 \qquad \textit{Factor.}$$

$$x - 5 = 0 \quad \text{or} \quad x - 5 = 0 \qquad \textit{Set each factor equal to zero.}$$

$$x = 5 \qquad\qquad x = 5 \qquad \textit{Here we obtain only one boundary point at } x = 5.$$

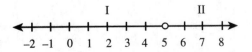

Graph the boundary point.

Region	x	$x^2 - 10x + 25$	$x^2 - 10x + 25 \overset{?}{>} 0$
I	2	$2^2 - 10 \cdot 2 + 25 = 9$	Yes
II	10	$10^2 - 10 \cdot 10 + 25 = 25$	Yes

The graph of the solution set is the colored region on the following number line:

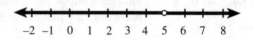

The graph indicates that every number but 5 satisfies the inequality $x^2 - 10x + 25 > 0$. The solution can be stated as either

$$x < 5 \quad \text{or} \quad x > 5,$$

using interval notation as $(-\infty, 5) \cup (5, \infty)$, or

$$x \neq 5$$

Try to avoid this mistake:

Incorrect	Correct
$2x^2 - 5x - 3 \geq 0$	$2x^2 - 5x - 3 \geq 0$
$(2x + 1)(x - 3) \geq 0$	$(2x + 1)(x - 3) \geq 0$ *Factor.*
$2x + 1 \geq 0 \quad$ or $\quad x - 3 \geq 0$	$2x + 1 = 0 \quad$ or $\quad x - 3 = 0$
$\qquad 2x \geq -1 \qquad\qquad x \geq 3$	$\qquad 2x = -1 \qquad\qquad x = 3$
$\qquad x \geq -\frac{1}{2}$	$\qquad x = -\frac{1}{2}$

Incorrect (continued): This problem was treated like a linear inequality, not a quadratic inequality. The inequality $x \geq -\frac{1}{2}$ indicates that numbers like 0, 1, $\frac{3}{2}$, and 2.4 are solutions, but they are not. Also, all the solutions less than $-\frac{1}{2}$ were left out.

Correct (continued):

Regions I and III check. So the solutions are $x \leq -\frac{1}{2}$ or $x \geq 3$ and the graph of the solution set is

Suppose we are given a quadratic inequality for which we cannot use the factoring method to find the boundary points. In this case we will use the quadratic formula in step 2 to find the boundary points. The following examples illustrate this process.

XAMPLE 7.8.2   Find the solution of the following quadratic inequality, and graph the solution on a number line.

$$x^2 - 4x + 1 > 0$$

Since the left-hand side is not factorable, the quadratic formula will be used to determine the boundary points from the equation $x^2 - 4x + 1 = 0$. In the quadratic formula, $a = 1$, $b = -4$, and $c = 1$:

$$x = \frac{-(-4) \pm \sqrt{(-4)^2 - 4 \cdot 1 \cdot 1}}{2 \cdot 1}$$

$$= \frac{4 \pm \sqrt{16 - 4}}{2}$$

$$= \frac{4 \pm \sqrt{12}}{2}$$

$$= \frac{4 \pm 2\sqrt{3}}{2}$$

$$= \frac{\cancel{2}(2 \pm \sqrt{3})}{\cancel{2}}$$

$$= 2 \pm \sqrt{3}$$

The boundary points are $x = 2 - \sqrt{3} \doteq 0.3$ and $x = 2 + \sqrt{3} \doteq 3.7$.

$$
\begin{array}{ccc}
\text{I} & \text{II} & \text{III}
\end{array}
$$

$$\xleftarrow{\hspace{0.5cm}} \underset{-1 \quad 0 \quad 1 \quad 2 \quad 3 \quad 4 \quad 5}{\longmapsto} \xrightarrow{\hspace{0.5cm}}$$

NOTE ▶   We use the decimal approximation of the boundary points to aid in plotting the boundary points on the number line.

Region	x	$x^2 - 4x + 1$	$x^2 - 4x + 1 \overset{?}{>} 0$
I	0	$0^2 - 4 \cdot 0 + 1 = 1$	Yes
II	1	$1^2 - 4 \cdot 1 + 1 = -2$	No
III	4	$4^2 - 4 \cdot 4 + 1 = 1$	Yes

We obtain the solutions $x < 2 - \sqrt{3}$ or $x > 2 + \sqrt{3}$ and the graph

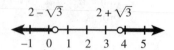

In interval notation the solution is $(-\infty, 2 - \sqrt{3}) \cup (2 + \sqrt{3}, \infty)$.

We have determined the solutions of quadratic inequalities in which the boundary points are rational or irrational numbers. In the next example we will find the solution of a quadratic inequality in which the boundary points are nonreal numbers.

EXAMPLE 7.8.3

Find the solution of the following quadratic inequality, and graph the solution on a number line:

$$x^2 + 4 \geq 0$$

Use the quadratic formula to determine the boundary points from the equation $x^2 + 4 = 0$. In the quadratic formula, $a = 1$, $b = 0$, and $c = 4$:

$$x = \frac{-0 \pm \sqrt{0^2 - 4 \cdot 1 \cdot 4}}{2 \cdot 1}$$

$$= \frac{0 \pm \sqrt{0 - 16}}{2}$$

$$= \frac{0 \pm \sqrt{-16}}{2}$$

$$= \frac{\pm 4i}{2}$$

$$= \pm 2i$$

The boundary points are nonreal numbers.

REMARK

When the boundary points are nonreal numbers, it can be shown that either all the real numbers satisfy the given inequality or none of the real numbers satisfy the inequality.

There is only one region, so pick any real number. Try $x = 7$:

$$7^2 + 4 \overset{?}{\geq} 0$$

$$49 + 4 \overset{?}{\geq} 0$$

$$53 \geq 0 \checkmark$$

Since the boundary points are nonreal and since $x = 7$ satisfies the inequality, all real numbers satisfy the inequality. The solution set is the set of all real numbers, and its graph is

$$-2 \quad -1 \quad 0 \quad 1 \quad 2$$

NOTE ▶ Since $x^2 \geq 0$ for all real numbers, it should be clear that $x^2 + 4 \geq 0$ for all real numbers.

The steps used to find the solutions of any quadratic inequality are summarized here.

---

1. Place the quadratic inequality in standard form.

2. Determine the boundary points by using either the factoring method or the quadratic formula.

3. If the boundary points are real numbers, they will partition the number line into distinct regions. Determine which regions satisfy the inequality by testing a number in each region.

4. If the boundary points are nonreal numbers, then there is only one region and either all the real numbers or none of the real numbers satisfy the given inequality. Pick any real number and determine the solution set.

E X A M P L E   7 . 8 . 4

Using a graphing calculator, find an approximation for the solutions of the following inequality and graph the solution on a number line:

$$x^2 - 4x - 6 < 0$$

SOLUTION

Let's first graph the equation $y = x^2 - 4x - 6$. Your graph should look something like the one in Figure 7.8.1.

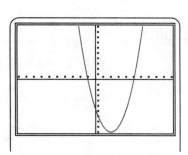

Figure 7.8.1

To find an approximation for the solutions of the inequality

$$x^2 - 4x - 6 < 0$$

all we need to do is to determine where the y values of the parabola are negative. From the graph it is clear that the y values are negative between the x-intercepts. Using the graphing calculator we find x-intercepts of approximately -1.16 and 5.16. Thus the solution of the inequality is approximately $-1.16 < x < 5.16$, and the graph of the solution is

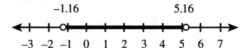

In the remainder of this section, we will find the solutions of **rational inequalities** and **polynomial inequalities.** The process is similar to that of solving quadratic inequalities.

A **rational inequality** is one that involves rational expressions such as

$$\frac{x + 5}{2x - 1} < 0.$$

This inequality is solved in the next example.

E X A M P L E   7 . 8 . 5   Find the solutions of the following rational inequalities, and graph the solutions on a number line.

1. $\dfrac{x + 5}{2x - 1} < 0$

Set the numerator and denominator equal to zero.

$$\begin{array}{ll} x + 5 = 0 & 2x - 1 = 0 \\ x = -5 & 2x = 1 \\ & x = \dfrac{1}{2} \end{array}$$

These are the boundary points.

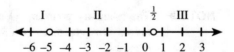

Region	x	$\dfrac{x+5}{2x-1}$	$\dfrac{x+5}{2x-1} \gtrless 0$
I	-6	$\dfrac{-6+5}{2(-6)-1} = \dfrac{-1}{-13} = \dfrac{1}{13}$	No
II	0	$\dfrac{0+5}{2(0)-1} = \dfrac{5}{-1} = -5$	Yes
III	1	$\dfrac{1+5}{2(1)-1} = \dfrac{6}{1} = 6$	No

Thus we obtain the solution $-5 < x < \frac{1}{2}$ and the graph

2. $\dfrac{2x+3}{x+4} \geq 1$

$\dfrac{2x+3}{x+4} - 1 \geq 0$ *Make one side of the inequality zero.*

$\dfrac{2x+3}{x+4} - \dfrac{x+4}{x+4} \geq 0$ *Combine the terms into one fraction.*

$\dfrac{2x+3-x-4}{x+4} \geq 0$

$\dfrac{x-1}{x+4} \geq 0$

$x - 1 = 0 \qquad x + 4 = 0$ *Set the numerator and the denominator equal to zero.*

$x = 1 \qquad\quad x = -4$ *These are the boundary points.*

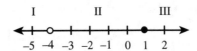

NOTE ▶   The boundary point at -4 is drawn with an open circle because -4 is not a solution. Observe that -4 produces a zero in the denominator in the original inequality.

Region	x	$\dfrac{x-1}{x+4}$	$\dfrac{x-1}{x+4} \overset{?}{\geq} 0$
I	-5	$\dfrac{-5-1}{-5+4} = \dfrac{-6}{-1} = 6$	Yes
II	0	$\dfrac{0-1}{0+4} = \dfrac{-1}{4}$	No
III	2	$\dfrac{2-1}{2+4} = \dfrac{1}{6}$	Yes

Thus the solution is $x < -4$ or $x \geq 1$ and the graph is

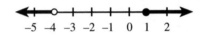

Try to avoid this mistake:

Incorrect	Correct

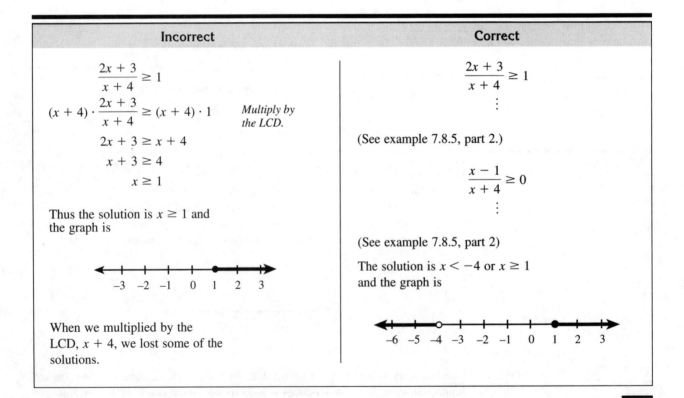

Incorrect

$$\frac{2x+3}{x+4} \geq 1$$

$$(x+4) \cdot \frac{2x+3}{x+4} \geq (x+4) \cdot 1 \qquad \textit{Multiply by the LCD.}$$

$$2x + 3 \geq x + 4$$

$$x + 3 \geq 4$$

$$x \geq 1$$

Thus the solution is $x \geq 1$ and the graph is

When we multiplied by the LCD, $x + 4$, we lost some of the solutions.

Correct

$$\frac{2x+3}{x+4} \geq 1$$

$$\vdots$$

(See example 7.8.5, part 2.)

$$\frac{x-1}{x+4} \geq 0$$

$$\vdots$$

(See example 7.8.5, part 2)

The solution is $x < -4$ or $x \geq 1$ and the graph is

The steps used to find the solutions of any rational inequality are summarized here.

1. Make one side of the inequality zero.

2. Combine all the expressions on the nonzero side into one fraction.

3. Set the numerator and the denominator of the fraction found in step 2 equal to zero. These values will be the boundary points of the solution set. The boundary points from the denominator are indicated with open circles.

4. The boundary points partition the number line into several distinct regions. Pick a number from each region and determine whether it satisfies the inequality.

A **polynomial inequality** is one that involves a polynomial. The quadratic polynomials considered earlier were polynomial inequalities that contained polynomials of degree 2. We now consider one containing a polynomial of degree 3. The polynomial is given in factored form, but if it were multiplied out, we would have

$$(x - 2)(x + 4)(x - 1) \geq 0$$
$$(x^2 + 2x - 8)(x - 1) \geq 0$$
$$x^3 + x^2 - 10x + 8 \geq 0$$

 XAMPLE 7.8.6

Find the solution of the following inequality, and graph the solution on a number line:

$$(x - 2)(x + 4)(x - 1) \geq 0$$

Since the left-hand side is factored and the right-hand side is zero, set each factor equal to zero to determine the boundary points:

$$x - 2 = 0 \qquad x + 4 = 0 \qquad x - 1 = 0$$
$$x = 2 \qquad\quad x = -4 \qquad\quad x = 1$$

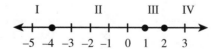

Region	x	$(x - 2)(x + 4)(x - 1)$	$(x - 2)(x + 4)(x - 1) \overset{?}{\geq} 0$
I	-5	$\begin{aligned}(-5 - 2) \cdot (-5 + 4) \cdot (-5 - 1) \\ = (-7) \cdot (-1) \cdot (-6) = -42\end{aligned}$	No
II	0	$\begin{aligned}(0 - 2) \cdot (0 + 4) \cdot (0 - 1) \\ = (-2) \cdot (4) \cdot (-1) = 8\end{aligned}$	Yes
III	$\dfrac{3}{2}$	$\begin{aligned}\left(\dfrac{3}{2} - 2\right) \cdot \left(\dfrac{3}{2} + 4\right) \cdot \left(\dfrac{3}{2} - 1\right) \\ = \left(-\dfrac{1}{2}\right) \cdot \left(\dfrac{11}{2}\right) \cdot \left(\dfrac{1}{2}\right) = -\dfrac{11}{8}\end{aligned}$	No
IV	3	$\begin{aligned}(3 - 2) \cdot (3 + 4) \cdot (3 - 1) \\ = (1) \cdot (7) \cdot (2) = 14\end{aligned}$	Yes

The solution is $-4 \leq x \leq 1$ or $x \geq 2$ and the graph is

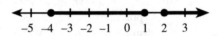

This set may be written using interval notation as $[-4, -1] \cup [2, \infty)$.

Exercises 7.8

Find the solutions of the following inequalities, and graph each solution on a number line. Express your answers using interval notation.

1. $x^2 + x - 20 < 0$ 2. $x^2 + 6x - 7 < 0$ 3. $x^2 + 9x + 18 > 0$

4. $x^2 - 4x \leq 0$ 5. $2x^2 + 3x > 0$ 6. $5x^2 - 9x \geq 0$

7. $x^2 + 10x + 16 \geq 0$ 8. $x^2 - 13x + 36 > 0$ 9. $x^2 - 7x + 6 \leq 0$

10. $x^2 - 9 \leq 0$ 11. $x^2 - 49 > 0$ 12. $x^2 - 25 > 0$

13. $2x^2 - x - 6 < 0$ 14. $3x^2 - 11x - 4 > 0$ 15. $8x^2 + 22x + 9 > 0$

16. $10x^2 - 41x + 21 < 0$ 17. $x^2 + 2x + 1 \geq 0$ 18. $x^2 - 8x + 16 > 0$

19. $4x^2 - 20x + 25 < 0$ 20. $9x^2 - 12x + 4 \leq 0$ 21. $x^2 - 6x + 7 > 0$

22. $x^2 + 4x - 1 \geq 0$ 23. $x^2 + 2x - 5 \leq 0$ 24. $x^2 - 2x - 4 < 0$

25. $x^2 - x - 6 < 0$ 26. $x^2 + 4x - 5 \leq 0$ 27. $2x^2 + 11x + 12 \geq 0$

28. $3x^2 + 7x + 2 > 0$ 29. $4x^2 - 4x + 1 < 0$ 30. $9x^2 - 6x - 2 > 0$

31. $16x^2 - 1 > 0$ 32. $8 + 2x - x^2 \leq 0$ 33. $-24 + 11x - x^2 \leq 0$

34. $30 + 7x - 2x^2 \geq 0$ 35. $-9 + 18x - 8x^2 > 0$ 36. $2 - 9x - 5x^2 > 0$

37. $7 + 12x - 4x^2 \leq 0$ 38. $-12 + 35x - 8x^2 \geq 0$ 39. $x^2 - 3x + 1 \geq 0$

40. $9x^2 - 12x - 1 \le 0$

41. $4x^2 - 1 > 0$

42. $9x^2 - 4 < 0$

43. $x^2 - 2x + 1 < 0$

44. $x^2 + 8x + 16 > 0$

45. $9x^2 + 30x + 25 \ge 0$

46. $4x^2 - 12x + 9 \le 0$

47. $x^2 + 1 > 0$

48. $4x^2 + 9 < 0$

49. $x^2 - 2x + 5 \le 0$

50. $x^2 + 6x + 25 > 0$

51. $x^2 + x < 6$

52. $x^2 + 12x \le -35$

53. $6x^2 - 7x \ge 3$

54. $2x^2 + 8x > -6$

55. $9x^2 > 16$

56. $x^2 + 9x + 1 < 5x + 6$

57. $x^2 + 11x + 7 < 3x - 5$

58. $2x^2 - 9x + 5 > 9 - 2x$

59. $3x^2 - 4x - 5 \ge x^2 + 4x + 1$

60. $3x^2 - 12x + 1 \ge 1 - 4x$

61. $3x^2 + 3x > x^2 - 1$

62. $2x^2 + 7x > 2x - 3$

63. $x(x - 8) \le -15$

64. $x(x - 7) \ge -6$

65. $x(2x + 1) \ge -6$

66. $x(x + 5) \le -9$

67. $2x(x + 4) \le x - 5$

68. $3x(3x + 1) < 3x + 16$

69. $4x(2x - 1) > 5x^2 - 1$

70. $2(x^2 + 4x + 5) \ge 10 - 4x$

71. $2x(3x + 2) < 5x^2 - 4$

72. $6x(2x - 3) > 5(x - 1)$

73. $5x(5x + 2) > 1$

74. $3x(3x + 2) \ge 5$

75. $(6x - 7)(x + 1) \le 5$

76. $(3x + 2)(2x + 1) \le 7$

77. $(3x - 1)(x - 1) < 3$

78. $(4x - 8)(x - 1) \le -3$

79. $(x + 9)(x - 1) < 6(x + 1)$

80. $x(4x - 7) \le 7(7 - x)$

81. $(3x + 1)(x - 1) < -2$

82. $(2x - 7)(x + 2) > -21$

83. $(2x - 3)(x - 1) < 3(x + 1)$

84. $(4x - 1)(x + 2) \ge 3(x - 1)$

85. $\dfrac{x - 4}{x + 1} < 0$

86. $\dfrac{x + 5}{x - 3} > 0$

87. $\dfrac{2x - 3}{x - 7} \ge 0$

88. $\dfrac{3x + 5}{x + 6} < 0$

89. $\dfrac{2x + 3}{x + 1} \ge 0$

90. $\dfrac{3x - 4}{x - 1} \le 0$

91. $\dfrac{2x}{x - 4} > 1$

92. $\dfrac{3x}{2x + 1} \ge 1$

93. $\dfrac{2x + 3}{x - 1} \le -1$

94. $\dfrac{2x - 1}{x - 2} < -1$

95. $\dfrac{x + 1}{x - 1} < \dfrac{1}{2}$

96. $\dfrac{x + 2}{x - 3} < \dfrac{1}{3}$

97. $(x - 1)(x - 2)(x + 3) > 0$

98. $(x + 3)(x - 4)(x + 6) > 0$

99. $(2x - 1)(x - 3)(x + 1) \le 0$

100. $(2x + 1)(2x + 3)(x - 5) < 0$

101. $(x - 1)^2(x + 3) \ge 0$

102. $(2x - 3)^2(x - 5) \le 0$

Write Algebra

103. In a polynomial inequality, if the boundary points are nonreal numbers, then all the real numbers or none of the real numbers satisfy the given inequality. Why?

104. In problems 97–100 notice that the solution regions alternate, but in problems 101 and 102, they do not. Can you explain why?

105. In the rational inequality $\frac{x - a}{x - b} \ge 0$, why is b never included as part of the solution?

🖩 **Use the Zoom and Trace keys of a graphing calculator to find approximate solutions of the following inequalities. Graph the solutions on a number line.**

106. $1.5x^2 + 6x + 1 < 0$

107. $0.9x^2 - 1.1x - 8 > 0$

108. $-1.34x^2 + 0.25x + 6.3 > 0$

109. $-2.05x^2 - 1.13x + 7.24 < 0$

GROUP ACTIVITY

In exercises 110–125, find and graph the solution set of each inequality. Describe the process that you use to find the solution sets.

110. $\sqrt{x} < 3x$

111. $\sqrt{x} < 2x$

112. $\sqrt{x} + 2 \geq x$

113. $\sqrt{x} + 6 \geq x$

114. $\sqrt{x-1} + 3 < x$

115. $\sqrt{x+1} + 1 < x$

116. $\sqrt{x+7} - x \geq 1$

117. $\sqrt{x+5} - x \geq 3$

118. $|x| + |x-4| < 6$

119. $|x-1| + |x+2| < 5$

120. $|x+2| - |x-3| \leq -1$

121. $|x+3| - |x-1| \leq 1$

122. $|x+3| + |x-4| < -6$

123. $|x-4| - |x+1| \leq 7$

124. $|x-3| - |x-5| \leq 8$

125. $|x-2| + |x+5| < 2$

Applied Algebra

The amount of water pressure that is lost as water passes through a fire hose is called friction loss. For a hose of given length and diameter, the friction loss is a function of the flow rate of the water through the hose. If a hose is 100 ft long and $2\frac{1}{2}$ in. in diameter, then the equation determining the friction loss is

$$F_L = 2Q^2 + Q$$

In this equation F_L measures the friction loss in pounds per square inch and Q represents the flow rate in 100 gal/min.

1. Find the friction loss for a 100-ft-long hose discharging 400 gal/min.

2. Find the flow rate in a 100-ft-long hose when the friction loss is 48 lb/sq in.

SOLUTION

1. Recall that Q represents that flow rate in 100 gal/min, so $Q = 4$.

$$
\begin{aligned}
F_L &= 2Q^2 + Q \\
&= 2 \cdot 4^2 + 4 \qquad \text{\textit{Substitute 4 for Q.}} \\
&= 2 \cdot 16 + 4 \\
&= 36 \text{ lb/sq in.}
\end{aligned}
$$

2. This time we substitute 48 for F_L in the equation:

$$F_L = 2Q^2 + Q$$
$$48 = 2Q^2 + Q$$
$$0 = 2Q^2 + Q - 48$$

Since the right-hand side of this equation cannot be factored, we will use the quadratic formula.

$$a = 2, b = 1, c = -48$$
$$Q = \frac{-1 \pm \sqrt{1^2 - 4 \cdot 2 \cdot (-48)}}{2 \cdot 2}$$
$$= \frac{-1 \pm \sqrt{1 + 384}}{4}$$
$$= \frac{-1 \pm \sqrt{385}}{4}$$
$$Q \doteq \frac{-1 \pm 19.6214}{4}$$

$$Q \doteq \frac{-1 + 19.6214}{4} \quad \text{or} \quad Q \doteq \frac{-1 - 19.6214}{4}$$
$$Q \doteq 4.66 \qquad\qquad\qquad \doteq -5.16$$

We reject the negative answer for the flow rate, since it is not feasible. Therefore the answer is approximately 4.66×100 gal/min = 466 gal/min.

The graph of this function is drawn in the accompanying figure, with the portion of the graph where $Q < 0$ shown as a dashed curve.

$$F_L = 2Q^2 + Q$$
$$= 2\left(Q^2 + \frac{1}{2}Q\right)$$
$$= 2\left(Q^2 + \frac{1}{2}Q + \frac{1}{16}\right) - \frac{1}{8}$$
$$= 2\left(Q + \frac{1}{4}\right)^2 - \frac{1}{8}$$
$$\text{vertex} = \left(-\frac{1}{4}, -\frac{1}{8}\right)$$

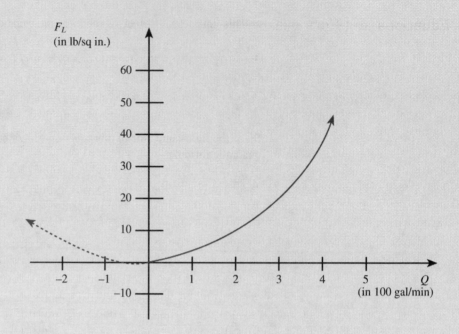

Your Turn

If a hose is 150 ft long and $2\frac{1}{2}$ in. in diameter, then the equation determining the friction loss is

$$F_L = 3Q^2 + 1.5Q$$

1. Find the friction loss when the flow rate is 400 gal/min.

2. Find the friction loss when the flow rate is 250 gal/min.

3. Find the flow rate when the friction loss is 31.5 lb/sq in.

4. Find the flow rate when the friction loss is 24 lb/sq in.

Chapter 7 Review

Terms to Remember

Parabola	p. 392	Quadratic formula	p. 419
Vertex	p. 392	Discriminant	p. 423
Axis	p. 392	Quadratic form	p. 437
Quadratic function	p. 392	Quadratic inequality	p. 445
Standard form of a quadratic function	p. 400	Standard form of a quadratic inequality	p. 445
Perfect square trinomial	p. 400	Boundary points	p. 445
Completing the square	p. 400	Rational inequality	p. 452
Extraction of roots	p. 413	Polynomial inequality	p. 452

Equations and Formulas

■ *Quadratic function:* Given an equation of the form

$$y = ax^2 + bx + c$$

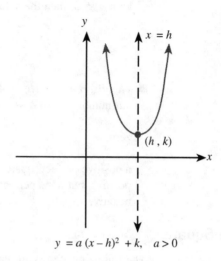

$$y = a(x-h)^2 + k, \quad a > 0$$

it can be changed to the form

$$y = a(x - h)^2 + k$$

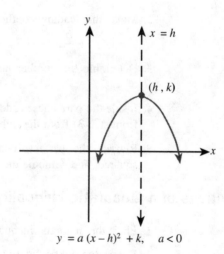

$$y = a(x-h)^2 + k, \quad a < 0$$

by completing the square in x.

1. The graph of the function is a parabola.

2. The graph opens upward if $a > 0$ or downward if $a < 0$.

3. Compared with the graph of $y = x^2$, the parabola is thinner if $|a| > 1$ or wider if $|a| < 1$.

4. The vertex is the point (h, k).

5. The graph is symmetric with respect to its axis, the vertical line $x = h$.

■ *Extraction of roots:* If $x^2 = a$, where a is any real number, then $x = \pm\sqrt{a}$.

■ *The quadratic formula:* If $ax^2 + bx + c = 0$, where a, b, and c are real numbers with $a \neq 0$, then the solutions of the equation are

$$x = \frac{-b \pm \sqrt{b^2 - 4ac}}{2a}$$

■ *The discriminant $(b^2 - 4ac)$:* If $ax^2 + bx + c = 0$, where a, b, and c are *rational* numbers with $a \neq 0$ and $b^2 - 4ac$ is

zero,	then there is one rational solution, called a double root.
a positive perfect square,	then there are two rational solutions.
positive, but not a perfect square,	then there are two irrational solutions.
negative,	then there are two nonreal solutions.

Completing the Square

The following steps will place any quadratic function in standard form.

1. Place the quadratic function in the form

$$y = ax^2 + bx + c$$

2. Make the leading coefficient 1 by factoring a out of the x terms—that is, $y = a\left[x^2 + \dfrac{b}{a}x\right] + c$.

3. Determine the number needed to complete the square—that is, $(\frac{1}{2} \cdot$ coefficient of $x)^2$.

4. Inside the parentheses add the number from step 3, and subtract $a \cdot$ (the number from step 3) from the right-hand side.

5. Rewrite the perfect square trinomial inside the parentheses as a binomial squared, and combine the numbers outside the parentheses.

Finding the Solutions of a Quadratic Inequality

1. Place the quadratic inequality in standard form.

2. Determine the boundary points by using either the factoring method or the quadratic formula.

3. If the boundary points are real numbers, they will partition the number line into distinct regions. Determine which regions satisfy the inequality by testing a number in each region.

4. If the boundary points are imaginary numbers, then either all the real numbers or none of the real numbers satisfy the given inequality. Pick any real number and determine the solution set.

Finding the Solutions of a Rational Inequality

1. Make one side of the inequality zero.

2. Combine all of the expressions on the nonzero side into one fraction.

3. Set the numerator and the denominator of the fraction found in step 2 equal to zero. These values will be the boundary points of the solution set. The boundary points from the denominator are indicated with open circles.

4. The boundary points partition the number line into several distinct regions. Pick a number from each region and determine whether it satisfies the inequality.

Review Exercises

7.1   Sketch the graph and determine the vertex of each of the following quadratic functions.

1. $y = 2x^2$

2. $y = -3x^2$

3. $y = \frac{1}{3}x^2$

4. $y = x^2 + 4$

5. $y = -2x^2 - 6$

6. $y = \frac{1}{2}x^2 - \frac{5}{2}$

7. $y = 4(x - 1)^2$

8. $y = -(x + 3)^2$

9. $y = \frac{1}{2}\left(x - \frac{3}{4}\right)^2$

10. $y = 2(x - 1)^2 + 7$

11. $y = -3\left(x - \frac{3}{2}\right)^2 + 5$

12. $y = \frac{1}{4}(x - 2)^2 - 8$

7.2   Place the quadratic functions in problems 13–24 in standard form. Also find the vertex and sketch the graph.

13. $y = x^2 - 6x + 14$

14. $y = x^2 + 8x + 15$

15. $y = -x^2 - 14x - 47$

16. $y = 2x^2 + 12x + 13$

17. $y = -3x^2 + 6x + 4$

18. $y = -4x^2 + 16x - 24$

19. $y = x^2 - \frac{4}{3}x + 3$

20. $y = -x^2 + \frac{1}{2}x + \frac{1}{2}$

21. $y = 2x^2 + 2x + \frac{7}{2}$

22. $y = 3x^2 - \frac{3}{2}x - \frac{13}{16}$

23. $y = -2x^2 + 10x - 6$

24. $y = -3x^2 - 4x - 3$

25. The Badday Tire Company has determined that the profit the company makes on steel-belted radial tires is given by the equation

$$y = -2x^2 + 360x - 14{,}060$$

where y is the profit measured in dollars per day and x is the number of steel-belted radial tires. Find the number of tires that will maximize the company's profit. What is the maximum profit?

26. The Moss County Agricultural Agent finds that 20 chickens in a chicken coop lay 30 eggs each per week. For every additional chicken in the coop, each chicken lays 1 egg fewer per week. The equation that determines the total number of eggs, y, that are laid each week by x chickens is

$$y = 50x - x^2$$

Find the number of chickens that will maximize the number of eggs laid per week. What is the maximum number of eggs laid?

7.3 Use the Extraction of Roots theorem to find the solutions of the following quadratic equations.

27. $x^2 = 16$

28. $x^2 + 81 = 0$

29. $x^2 - 121 = 0$

30. $25x^2 - 9 = 0$

31. $9x^2 + 16 = 0$

32. $5x^2 - 7 = 0$

33. $3x^2 + 1 = 0$

34. $(2x - 7)^2 = 25$

35. $(3x + 1)^2 = 13$

36. $(5x - 4)^2 + 6 = 8$

Find the solutions of the following quadratic equations by completing the square.

37. $x^2 - 6x + 5 = 0$

38. $x^2 + 10x + 26 = 0$

39. $x^2 - 2x - 1 = 0$

40. $4x^2 + 4x - 15 = 0$

41. $9x^2 - 12x + 20 = 0$

42. $4x^2 - 12x + 9 = 0$

43. $3x^2 + 8x + 3 = 0$

44. $9x^2 + 12x - 2 = 0$

45. $2x^2 - 2x - 1 = 0$

46. $25x^2 + 30x - 16 = 0$

7.4 Use the quadratic formula to find the solutions of the following quadratic equations.

47. $3x^2 - 5x - 2 = 0$

48. $4x^2 - 4x + 1 = 0$

49. $5x^2 - 4 = 0$

50. $2x^2 + 3x = 0$

51. $(x + 3)(2x - 1) = 15$

52. $x^2 + 9x + 9 = 2x - 3$

53. $x^2 + 6x + 7 = 0$

54. $2x^2 - 2x - 1 = 0$

55. $x^2 - 6x + 10 = 0$

56. $9x^2 - 30x + 29 = 0$

57. $3 - \dfrac{11}{x} - \dfrac{4}{x^2} = 0$

58. $\dfrac{1}{4} + \dfrac{9}{8x} + \dfrac{5}{4x^2} = 0$

59. $\dfrac{1}{13} - \dfrac{4}{13x} + \dfrac{1}{x^2} = 0$

60. $1 + \dfrac{1}{x} + \dfrac{5}{16x^2} = 0$

Use the discriminant to characterize the solutions of the following quadratic equations.

61. $9x^2 - 12x + 4 = 0$

62. $2x^2 - 7x - 4 = 0$

63. $3x^2 - 5x + 1 = 0$

64. $5x^2 - x + 6 = 0$

7.6 Find the solutions of the following problems. For the problems that have irrational solutions, determine a decimal approximation of the solution.

65. The sum of two numbers is 2 and the sum of their squares is $\frac{5}{2}$. Find the two numbers.

66. A number minus its square is -30. Find the number.

67. A building casts a shadow that is twice as long as the height of the building. The distance from the tip of the shadow to the top of the building is 300 ft. Find the height of the building.

68. The height of a triangle is 5 cm less than twice the base. The area of the triangle is 21 sq cm. Find the height and base of the triangle.

69. Val and Rosemary have a rectangular swimming pool that is 8 ft wide and 12 ft long. They are going to build a tile border of uniform width around the pool. They have 124 sq ft of tile. How wide is the border?

70. One pipe can fill a tank 3 hr faster than another pipe. Together they fill the tank in 5 hr. How long does it take each pipe to fill the tank?

71. Chief Running Bear shoots an arrow upward from an 80-ft-tall cliff with a speed of 48 ft/sec. The equation that gives the arrow's height above ground level is

$$h = -16t^2 + 48t + 80$$

When does the arrow hit the ground?

72. Corye and Kiki ran around their subdivision, a distance of $\frac{1}{2}$ mi. Corye ran 4 mi/hr slower than Kiki and took 2 min longer to finish. How fast did each run? (*Hint:* Change minutes to hours.)

7.7   Find the solutions of the following equations.

73. $x^4 + 5x^2 - 36 = 0$

74. $4x^4 - 7x^2 + 3 = 0$

75. $x - 4\sqrt{x} + 3 = 0$

76. $4x - 4\sqrt{x} - 3 = 0$

77. $4(x + 2)^2 + 9(x + 2) - 9 = 0$

78. $4(x - 4)^2 - 12(x - 4) + 5 = 0$

79. $x^{-2} + 9x^{-1} + 20 = 0$

80. $3x^{-2} - 11x^{-1} + 8 = 0$

81. $x^{2/3} - 3x^{1/3} - 10 = 0$

82. $6x^{2/3} + x^{1/3} - 12 = 0$

83. $\left(\dfrac{x}{x-1}\right)^2 - 4\left(\dfrac{x}{x-1}\right) - 5 = 0$

84. $\left(\dfrac{2x}{x+1}\right)^2 - 10\left(\dfrac{2x}{x+1}\right) + 21 = 0$

Use factoring by grouping to find the solutions of the following equations.

85. $3x^3 + 4x^2 - 27x - 36 = 0$

86. $5x^3 - 3x^2 + 5x - 3 = 0$

87. $4x^3 + 12x^2 - x - 3 = 0$

88. $18x^3 - 9x^2 + 8x - 4 = 0$

89. $7x^4 - 6x^3 - 7x + 6 = 0$

90. $2x^4 - 3x^3 + 16x - 24 = 0$

7.8   Find the solutions of the following inequalities, and graph each solution on a number line.

91. $2x^2 + 3x - 5 \geq 0$

92. $9x^2 - 8x - 17 < 0$

93. $x^2 - 4x + 13 \geq 0$

94. $4x^2 - 4x - 27 > 0$

95. $(x + 10)(x + 1) < 18x$

96. $2(x + 1)(x - 1) > x^2 + 6x - 11$

97. $(x + 2)(x + 3) > 3x + 10$

98. $2x(x + 4) > x + 4$

99. $(x - 7)(x - 1) < 2x$

100. $(2x + 3)^2 \leq x^2 - 5x - 1$

101. $5(2x + 3)(x + 1) \leq x(x + 1) - 1$

102. $(2x + 1)(2x + 3) < 4x - 2$

103. $\dfrac{2x - 3}{x + 4} \geq 0$

104. $\dfrac{x + 2}{x - 7} < 0$

105. $\dfrac{4x}{x - 2} < 1$

106. $\dfrac{3x + 2}{x + 4} \geq 2$

107. $(x + 3)(x - 4)(x - 7) \geq 0$

108. $(x - 2)^2(x - 4) \leq 0$

Chapter 7 Test

(You should be able to complete this test in 60 minutes.)

I. Place the following quadratic functions in standard form, if necessary. Also find the vertex and sketch the graph of each.

1. $y = -\dfrac{1}{2}(x + 1)^2 + 4$

2. $y = 2x^2 - 12x + 10$

II. Find the solutions of the following quadratic equations using the indicated method.

3. $(2x + 7)^2 = 16$, extraction of roots

4. $x^2 + 5x + 6 = 0$, completing the square

5. $4x^2 + 15x = 3x - 1$, quadratic formula

6. $16x^2 - 24x + 13 = 0$, quadratic formula

III. Find the solutions of the following equations.

7. $2x^4 - 7x^2 - 72 = 0$

8. $3x^{-2} - 10x^{-1} - 8 = 0$

9. $27x^3 + 45x^2 - 3x - 5 = 0$

IV. Find the solutions of the following inequalities, and graph each solution on a number line. Express your answers using interval notation.

10. $x^2 - 8x + 14 < 0$

11. $2x^2 - 5x - 12 \geq 0$

12. $\dfrac{x + 3}{x - 7} < 0$

13. $\dfrac{1}{x - 4} \geq 1$

V. Find the solutions of the following problems. For the problems that have irrational solutions, use a decimal approximation of the solution.

14. Noah needs to build a rectangular corral for two sheep up against one wall of the ark. He will run a fence around three sides but not along the side by the wall. If he has 32 cubits of the fencing material, what should the dimensions of the corral be to maximize its area?

15. The Duke of Earl has a rectangular castle that is 80 yd wide and 120 yd long. He is going to build a moat of uniform width around the castle. He has enough crocodiles to cover 1600 sq yd of water. How wide is the moat?

16. Captain Spalding, the African explorer, takes a barge 10 mi down the Congo River to photograph an underdeveloped tribe. The trip downstream and back takes $2\frac{1}{2}$ hr. If the speed of the current is 5 mi/hr, what is the speed of the barge in still water?

TEST YOUR MEMORY

These problems review Chapters 1–7.

I. Simplify each of the following. Assume that all variables represent positive numbers.

1. $\dfrac{(2^{-1}x^3y^{-4})^2}{(4^0x^{-3}y^{-1})^{-3}}$

2. $\left(\dfrac{125x^6y^{-4}}{8x^{-1}y^{1/2}}\right)^{2/3}$

II. Perform the indicated operations.

3. $\dfrac{4x^3 - 28x^2 + 24x}{5x^3 + 8x^2 - 5x - 8} \cdot \dfrac{x^2 + 7x + 6}{12x^5 - 432x^3}$

4. $\dfrac{2x - 3}{x^2 - 6x + 9} - \dfrac{x + 1}{x^2 + 2x - 15}$

5. $\dfrac{2 + x^{-1} - 15x^{-2}}{1 - 4x^{-1} - 21x^{-2}}$

III. Simplify each of the following. Assume that all variables represent positive numbers.

6. $2\sqrt{96x} + 3\sqrt{24x} - \sqrt{54x}$

7. $\sqrt[3]{135} - 5\sqrt[3]{40}$

8. $\sqrt[3]{\dfrac{2}{3x^2}}$

9. $\dfrac{\sqrt{x} + \sqrt{y}}{\sqrt{x} - \sqrt{y}}$

IV. Place the following quadratic functions in standard form, if necessary. Also find the vertex and sketch the graph of each.

10. $y = -(x - 1)^2 + 2$

11. $y = 2x^2 + 4x - 1$

V. Let $f(x) = 3x + 2$ and $g(x) = 2x^2 - 7$. Find the following.

12. $f(0)$

13. $g(-1)$

14. $[f(-2) + g(-1)]^2$

15. $8f(1) - 2g(2)$

16. $g(x) - f(x)$

VI. Find the equations of the following lines. Write the equations in slope-intercept form, if possible.

17. Through $(6, -2)$ and $(-9, 4)$

18. Through $(3, -4)$ and perpendicular to $y = 2x + 8$

19. Through $(-1, -6)$ and parallel to $2x + 3y = 6$

20. Through $(5, 7)$ and perpendicular to the y-axis

VII. Find the solutions of the following equations and check your solutions.

21. $\dfrac{2}{x + 3} = \dfrac{5}{x - 1}$

22. $\dfrac{x + 2}{2x + 1} - \dfrac{5}{2x + 3} = \dfrac{6x + 6}{4x^2 + 8x + 3}$

23. $\dfrac{x + 3}{x - 1} + \dfrac{3}{x + 2} = \dfrac{5x + 13}{x^2 + x - 2}$

24. $2 + \dfrac{5x - 12}{x - 4} = \dfrac{3x - 4}{x - 4}$

25. $\sqrt{3x - 2} + 4 = 0$

26. $\sqrt[3]{4x - 1} + 1 = 0$

27. $\sqrt{x + 2} - \sqrt{x - 3} = 1$

28. $2\sqrt{x + 7} = x - 1$

29. $2\left(\dfrac{x}{x+1}\right)^2 - 5\left(\dfrac{x}{x+1}\right) - 3 = 0$

30. $(x^2 + x)^2 - 8(x^2 + x) + 12 = 0$

31. $9x^4 - 37x^2 + 4 = 0$

32. $8x^3 + 12x^2 - 2x - 3 = 0$

VIII. Find the solutions of the following quadratic equations using the indicated method.

33. $(3x - 1)^2 = 25$, extraction of roots

34. $x^2 + 3x - 4 = 0$, completing the square

35. $x^2 - 4x - 2 = 0$, quadratic formula

36. $2x^2 - 6x + 5 = 0$, quadratic formula

IX. Find the solutions of the following inequalities and graph each solution set on a number line. Express your answers using interval notation.

37. $2 - 3x \geq -10$

38. $|2x + 3| < 5$

39. $x^2 - 2x - 8 < 0$

40. $x^2 - 6x - 1 \geq 0$

41. $\dfrac{x-2}{x+6} > 0$

42. $\dfrac{2}{x+1} \leq 1$

X. Graph the solution sets of the following inequalities.

43. $2x - y \leq 0$

44. $\begin{cases} y < 2x + 2 \\ x + 2y < 4 \end{cases}$

XI. Use algebraic expressions to find the solutions of the following problems. For the problems that have irrational solutions, use decimal approximations of the solutions.

45. Pedro has $3.20 in nickels, dimes, and quarters. The number of nickels is twice the number of quarters. He has five more dimes than quarters. How many of each type of coin does Pedro have?

46. One leg of a right triangle is 2 cm more than twice the other leg. The hypotenuse is 13 cm long. What are the lengths of the legs of the triangle?

47. The surface area of a cube varies directly as the square of an edge. The surface area is 54 sq in. when an edge is 3 in. Find the surface area when an edge is 6 in.

48. Tony takes twice as long as Maria to deliver the morning edition of the *Moss County Gazette*. Together they can deliver the newspapers in 4 hr. How long does it take each of them to deliver the newspapers?

49. Louise and Leroy have a rectangular swimming pool that is 8 ft wide and 12 ft long. They are going to build a tile border of uniform width around the pool. They have 120 sq ft of tile. How wide is the border?

50. Ahmad takes his old motorboat to his favorite fishing spot 8 mi downstream. The trip downstream and back takes 2 hr. If the speed of Ahmad's boat in still water is 10 mi/hr, what is the speed of the current?

C H A P T E R

Conic Sections

Ed makes greenhouses that have rectangular bases 10 ft wide and 12 ft long (see Figure A). The frame consists of PVC pipes, which are squeezed into fittings in the base and along the top. Find a mathematical model for the curve of the PVC pipes given that the height is $91\frac{3}{4}$ in. (see Figure B). Does your model fit the values shown in Figure C? In this chapter, we will study curves called conic sections. We will see that one of these curves fits the data of Ed's greenhouse better than the others. We can then answer questions such as, How close to the side of the greenhouse can a 6-ft-tall potted avocado tree be placed?

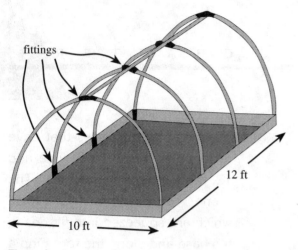

fittings

12 ft

10 ft

Figure A

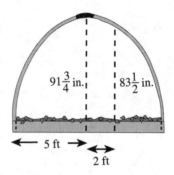

fittings

$91\frac{3}{4}$ in.

10 ft

Figure B

$91\frac{3}{4}$ in. $\quad$ $83\frac{1}{2}$ in.

5 ft

2 ft

Figure C

8.1 *Horizontal Parabolas*

In Chapter 7 we studied quadratic functions whose equations could be written in the form $y = ax^2 + bx + c$ and whose graphs were parabolas opening upward or downward. These parabolas are examples of **conic sections.** A conic section is a curve that comes from the intersection of a plane with a cone, as shown in Figure 8.1.1. We will continue the study of parabolas in this chapter, and then study other conic sections.

In the previous chapter we examined parabolas that were symmetric with respect to a vertical line passing through their vertex. We called this line the **axis of symmetry**—or, simply, **axis**—of the parabola. The axis of a parabola that comes from a quadratic function is vertical. A parabola in the Cartesian coordinate system can also have a horizontal axis. We call this type of parabola a **horizontal parabola.** This is the type we will consider in this section.

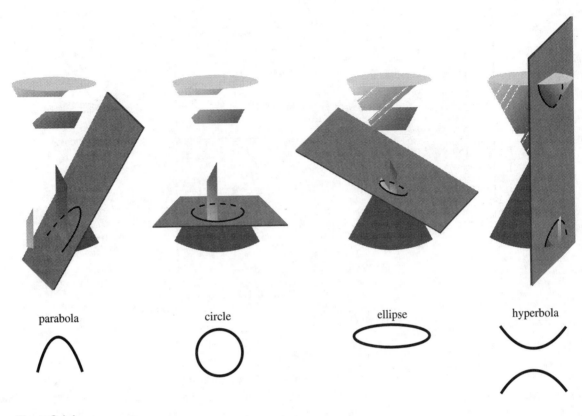

parabola   circle   ellipse   hyperbola

Figure 8.1.1

XAMPLE 8.1.1   Sketch the graphs of the following equations.

1. $x = y^2$

Some of the ordered pairs that satisfy the equation $x = y^2$ are given in the chart accompanying Figure 8.1.2. It is easier to obtain these ordered pairs by substituting values for y and solving for x. Plot these points and smooth the curve as in Figure 8.1.2.

REMARKS

a. The equation $x = y^2$ does *not* define a function. For one value of x, such as $x = 4$, there are two values of y, -2, and 2. Note also that the graph does not pass the vertical line test.

b. The shape of the parabola is the same as that of the graph of $y = x^2$, except that the axis is horizontal. In this case the axis of the parabola is the x-axis.

c. The vertex, $(0, 0)$, is the leftmost point of the parabola. Thus it is the point with the smallest x-coordinate.

x	y
4	-2
1	-1
$\dfrac{1}{4}$	$-\dfrac{1}{2}$
0	0
$\dfrac{1}{4}$	$\dfrac{1}{2}$
1	1
4	2

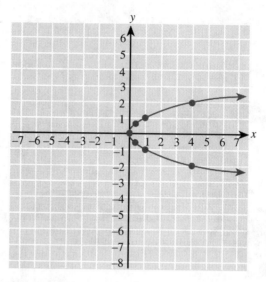

Figure 8.1.2

2. $x = (y - 2)^2 - 3$

Since $(y - 2)^2 \geq 0$, the smallest that x can be is $x = 0 - 3 = -3$, which occurs when $y - 2 = 0$ or $y = 2$. Thus the vertex is $(-3, 2)$. We now choose values of y on either side of $y = 2$ and obtain the ordered pairs in the chart accompanying Figure 8.1.3. This parabola has the same shape as the parabola in part 1, except that it is shifted to the left 3 units and up 2 units (see Figure 8.1.3).

x	y
1	0
-2	1
-3	2
-2	3
1	4

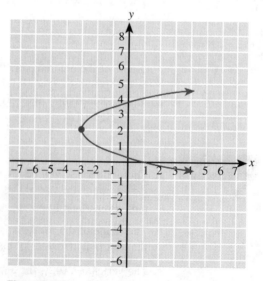

Figure 8.1.3

3. $x = -\dfrac{1}{2}(y - 1)^2 + 2$

Since $-\frac{1}{2}(y - 1)^2 \leq 0$, the largest that x can be is $x = 0 + 2 = 2$, which occurs when $y - 1 = 0$ or $y = 1$. Thus the vertex is $(2, 1)$. We now choose values of y on either side of $y = 1$ and obtain the ordered pairs in the chart. This parabola is wider than the two previous parabolas because of the $-\frac{1}{2}$ factor on $(y - 1)^2$. It also opens to the left (the negative direction for x) because of the negative sign on $-\frac{1}{2}$ (see Figure 8.1.4).

x	y
$-\dfrac{5}{2}$	-2
0	-1
$\dfrac{3}{2}$	0
2	1
$\dfrac{3}{2}$	2
0	3
$-\dfrac{5}{2}$	4

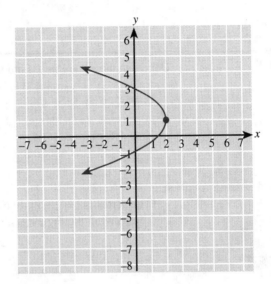

Figure 8.1.4

The techniques used in the previous example to sketch the parabolas are similar to those used in the previous chapter to graph quadratic functions. We summarize this information.

Given an equation in the form

$$x = a(y - k)^2 + h \qquad a \neq 0$$

1. The graph is a horizontal parabola opening to the right if $a > 0$, or to the left if $a < 0$.

2. Compared with the graph of $x = y^2$, the parabola is thinner if $|a| > 1$, or wider if $|a| < 1$.

3. The vertex is the point (h, k).

4. The graph is symmetric with respect to its axis, the horizontal line $y = k$.

Given an equation in the form

$$x = ay^2 + by + c, \qquad a \neq 0$$

it can be changed to the form

$$x = a(y - k)^2 + h$$

by *completing the square in y.* The following example illustrates this process.

E XAMPLE 8.1.2   Complete the square in y. Find the vertex and sketch the graph.

1. $x = y^2 + 4y + 3$

$x = (y^2 + 4y) + 3$

$x = (y^2 + 4y + \underline{\quad}) + 3 \qquad \underline{\quad} = (\frac{1}{2} \cdot 4)^2 = 2^2 = 4$

$x = (y^2 + 4y + 4) + 3 - 4 \qquad$ *We must subtract 4 to balance the 4 that is added*

$x = (y + 2)^2 - 1 \qquad\qquad$ *inside the parentheses.*

The vertex is $(-1, -2)$. Some points on the parabola are shown in the chart accompanying the Figure 8.1.5. The graph is shown in Figure 8.1.5.

x	y
3	−4
0	−3
−1	−2
0	−1
3	0

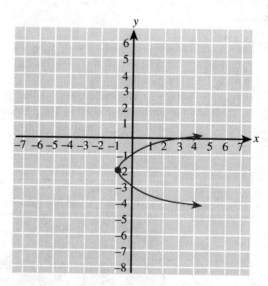

Figure 8.1.5

2. $x = \dfrac{1}{3}y^2 - 2y + 1$

$x = \dfrac{1}{3}(y^2 - 6y) + 1 \qquad$ *Make the leading coefficient 1 by factoring $\frac{1}{3}$ out of the y terms.*

$$x = \frac{1}{3}(y^2 - 6y + \underline{}) + 1 \qquad \underline{} = [\tfrac{1}{2} \cdot (-6)]^2 = (-3)^2 = 9$$

$$x = \frac{1}{3}(y^2 - 6y + 9) + 1 - 3 \qquad \textit{We must subtract 3, since } \tfrac{1}{3} \cdot 9 \textit{ is added to the right-hand side.}$$

$$x = \frac{1}{3}(y - 3)^2 - 2$$

The vertex is $(-2, 3)$. Some points on the parabola are shown in the chart. The graph is shown in Figure 8.1.6.

x	y
$\dfrac{10}{3}$	-1
1	0
$-\dfrac{2}{3}$	1
$-\dfrac{5}{3}$	2
-2	3
$-\dfrac{5}{3}$	4
$-\dfrac{2}{3}$	5
1	6
$\dfrac{10}{3}$	7

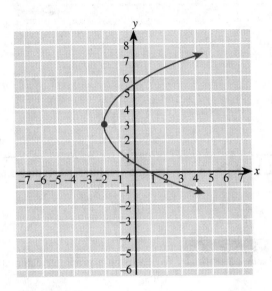

Figure 8.1.6

3. $x = -2y^2 - 6y$

    $x = -2(y^2 + 3y)$ *Make the leading coefficient 1 by factoring -2 out of the y terms.*

    $x = -2(y^2 + 3y + \underline{})$ $\underline{} = \left(\tfrac{1}{2} \cdot 3\right)^2 = \left(\tfrac{3}{2}\right)^2 = \tfrac{9}{4}$

    $x = -2\left(y^2 + 3y + \dfrac{9}{4}\right) + \dfrac{9}{2}$ *We must add $\tfrac{9}{2}$, since $-2 \cdot \tfrac{9}{4}$ is added to the right-hand side.*

    $x = -2\left(y + \dfrac{3}{2}\right)^2 + \dfrac{9}{2}$

The vertex is $\left(\tfrac{9}{2}, -\tfrac{3}{2}\right)$. Some points on the parabola are shown in the chart. The graph is shown in Figure 8.1.7.

x	y
-8	-4
0	-3
4	-2
$\dfrac{9}{2}$	$-\dfrac{3}{2}$
4	-1
0	0
-8	1

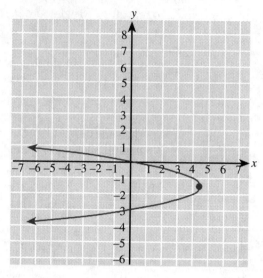

Figure 8.1.7

EXAMPLE 8.1.3

Using a graphing calculator, graph the equation

$$x = y^2$$

SOLUTION

When using a graphing calculator to graph an equation, we must first solve for y. Thus,

$$y^2 = x$$
$$y = \pm\sqrt{x} \qquad \textit{By the Extraction of Roots theorem}$$

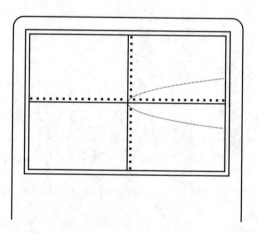

Figure 8.1.8

To graph this equation, we must enter two equations

$$y_1 = \sqrt{x} \quad \text{and} \quad y_2 = -\sqrt{x}$$

Your graph should look something like the one in Figure 8.1.8.

NOTE ▶   You might want to square up the grid to make the graph look more like a parabola.

REMARK   Every equation that can be written in the form $x = ay^2 + by + c$, where $a \neq 0$, does *not* define a function. The reader should observe that in each of the preceding examples, a single value of x usually corresponds to two values of y. Also, a horizontal parabola does not pass the vertical line test.

Exercises 8.1

Sketch the graphs of the following equations.

1. $x = 2y^2 - 3$

2. $x = 5y^2 - 7$

3. $x = 2 - y^2$

4. $x = 5 - 3y^2$

5. $x = \left(y + \dfrac{5}{2}\right)^2$

6. $x = \left(y - \dfrac{4}{3}\right)^2$

7. $x = (y + 3)^2 - 1$

8. $x = (y - 1)^2 + 3$

9. $x = 2(y + 1)^2 - 5$

10. $x = 3(y - 2)^2 - 4$

11. $x = -(y + 4)^2 - 2$

12. $x = -3\left(y - \dfrac{2}{3}\right)^2 + \dfrac{10}{3}$

13. $x = -\dfrac{3}{4}(y + 3)^2 + 1$

14. $x = \dfrac{1}{3}(y + 1)^2 - 3$

Complete the square in y. Find the vertex and sketch the graph of each equation.

15. $x = y^2 - 8y + 12$

16. $x = y^2 + 4y + 6$

17. $x = y^2 - 3y + 1$

18. $x = y^2 + y + 1$

19. $x = 4y^2 - 24y + 32$

20. $x = 2y^2 + 16y + 33$

21. $x = 2y^2 - y + 1$

22. $x = 3y^2 + 2y - 1$

23. $x = -y^2 + 10y - 24$

24. $x = -2y^2 + 2y + 2$

25. $3x = -2y^2 - 8y + 1$

26. $2x = y^2 - 6y + 5$

Write Algebra

27. Explain why any equation whose graph is a horizontal parabola does not define a function.

Use a graphing calculator to graph the following equations.

28. $y^2 = 4x$

29. $y^2 = -9x$

30. $(y - 5)^2 = x$

31. $(y + 3)^2 = -4x$

8.2 Circles

The conic section we will study now is the circle. You are probably more familiar with circles than with the other conic sections. In geometry much can be said about circles. However, we are primarily interested in studying circles in the Cartesian coordinate system and in the equations that determine which points lie on these circles.

We need to start with the definition of a circle.

Definition 8.2.1	A **circle** is a set of points in a plane that is located a fixed distance, called the **radius,** from a given point in the plane, called the **center** (see Figure 8.2.1).

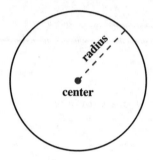

Figure 8.2.1

To determine the equation of a circle, let us call the radius r and the center $C(h, k)$. Recall the Distance Formula from Section 6.6, which is used to find the distance from a point $A(x_1, y_1)$ to a point $B(x_2, y_2)$:

$$d(A, B) = \sqrt{(x_2 - x_1)^2 + (y_2 - y_1)^2}$$

A point $P(x, y)$ is on the circle if it is a distance r from the center (see Figure 8.2.2).

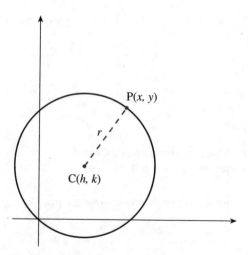

Figure 8.2.2

Thus

$$d(C, P) = \sqrt{(x - h)^2 + (y - k)^2} = r$$

Since r is always positive, we can square both sides of this equation and obtain the following equivalent equation.

quation of a Circle

$$(x - h)^2 + (y - k)^2 = r^2$$

where the center is $C(h, k)$ and the radius is $r > 0$.

XAMPLE 8.2.1   Find the equations of the following circles, and graph the equations.

1. Center $= C(2, -5)$, radius $= 3$

   To find the equation, we substitute 2 for h, -5 for k, and 3 for r:

   $$(x - 2)^2 + (y - (-5))^2 = 3^2$$
   $$(x - 2)^2 + (y + 5)^2 = 9$$

   We will *not* construct a table of ordered pairs in order to draw the circle. Instead, we will locate the center and draw a curve around it, staying 3 units away from the center. This is most easily done with a compass. If you don't have a compass, you can still roughly sketch the graph of the circle. Locate the points 3 units directly to the left and right of the center. Also locate the points 3 units directly above and below the center. Now smooth in a circle passing through these four points. You should obtain the circle shown in Figure 8.2.3.

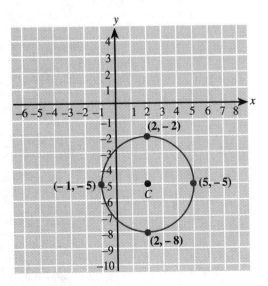

Figure 8.2.3

2. Center = $C(-3, 1)$, radius = $\sqrt{5}$

Substituting -3 for h, 1 for k, and $\sqrt{5}$ for r, we obtain

$$(x - (-3))^2 + (y - 1)^2 = (\sqrt{5})^2$$
$$(x + 3)^2 + (y - 1)^2 = 5$$

To draw the circle as in Figure 8.2.4, we use 2.2 as an approximation for $\sqrt{5}$.

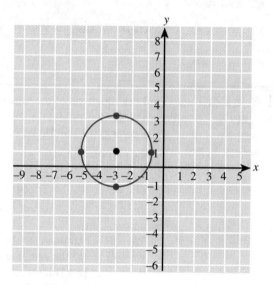

Figure 8.2.4

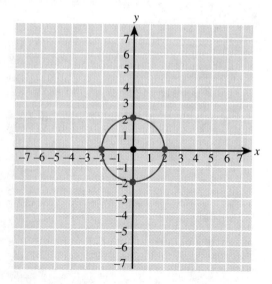

Figure 8.2.5

3. Center = $C(0, 0)$, radius = 2

Substituting 0 for h and k, and 2 for r, we obtain

$$(x - 0)^2 + (y - 0)^2 = 2^2$$
$$x^2 + y^2 = 4$$

The circle is graphed in Figure 8.2.5.

REMARKS

1. The center $(0, 0)$ is not a point on the circle; that is, it does not satisfy the equation $x^2 + y^2 = 4$. The center is used only to define the circle.

2. For $x = 0$ there are two values of y, -2 and 2. Thus the equation $x^2 + y^2 = 4$ does *not* define a function. Note also that the graph of *any* circle fails the vertical line test.

In the special case when the *center is at the origin* and the radius is r, the equation of the circle can be written in the form

$$x^2 + y^2 = r^2$$

So far we have obtained the equation and the graph from certain given information about the circle. Now we will start with the equation and obtain the center, radius, and graph.

EXAMPLE 8.2.2   Find the center and radius of the circle with the given equation; then draw the circle.

1. $(x - 4)^2 + (y + 2)^2 = 16$

Therefore,

$$x - 4 = x - h \qquad y + 2 = y - k \qquad 16 = r^2$$
$$h = 4 \qquad\qquad k = -2 \qquad\qquad r = 4$$

Thus, the center is $(4, -2)$ and the radius is 4 (see Figure 8.2.6).

2. $x^2 + y^2 = 3$

This equation is in the form

$$x^2 + y^2 = r^2$$

so the center is $(0, 0)$ (why ?) and

$$r^2 = 3$$
$$r = \sqrt{3} \doteq 1.7$$

The graph is shown in Figure 8.2.7.

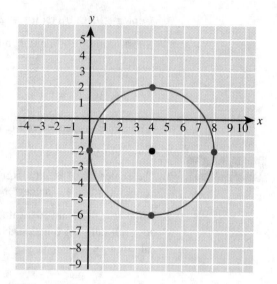

Figure 8.2.6

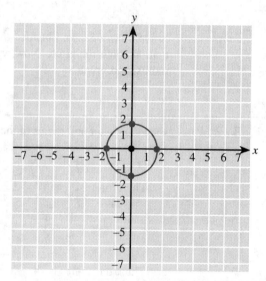

Figure 8.2.7

3. $x^2 + y^2 + 6x + 2y + 9 = 0$

This equation is starting out in a different form. By completing the square in x and in y, we can change it to the form $(x - h)^2 + (y - k)^2 = r^2$:

$$x^2 + y^2 + 6x + 2y + 9 = 0$$
$$x^2 + 6x + y^2 + 2y = -9 \qquad \text{\textit{Place the x terms together and the y terms together and subtract 9 from both sides.}}$$

$$(x^2 + 6x + \underline{}) + (y^2 + 2y + \underline{}) = -9$$

For the x's,
$$\underline{} = \left(\tfrac{1}{2} \cdot 6\right)^2 = 3^2 = 9.$$
For the y's,
$$\underline{} = \left(\tfrac{1}{2} \cdot 2\right)^2 = 1^2 = 1.$$

$$(x^2 + 6x + 9) + (y^2 + 2y + 1) = -9 + 9 + 1$$ *Add 9 and 1 to both sides.*

$$(x + 3)^2 + (y + 1)^2 = 1$$

Therefore,

$$x + 3 = x - h \qquad y + 1 = y - k \qquad 1 = r^2$$
$$h = -3 \qquad\qquad k = -1 \qquad\qquad 1 = r$$

Thus, the center is $(-3, -1)$ and the radius is 1 (see Figure 8.2.8).

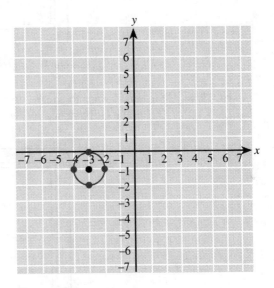

Figure 8.2.8

EXAMPLE 8.2.3

Using a graphing calculator graph the equation

$$x^2 + y^2 = 16$$

SOLUTION

Recall that when using a graphing calculator to graph an equation, we must first solve for *y*. Thus,

$$x^2 + y^2 = 16$$
$$y^2 = 16 - x^2$$
$$y = \pm\sqrt{16 - x^2} \qquad \textit{By the Extraction of Roots theorem}$$

To graph this equation we must enter two equations

$$y_1 = \sqrt{16 - x^2} \quad \text{and} \quad y_2 = -\sqrt{16 - x^2}$$

Your graph should look something like Figure 8.2.9.

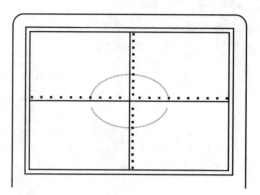

Figure 8.2.9

NOTE ▶   You might want to square up the grid to make the graph look more like a circle. This graph is illustrated in Figure 8.2.10.

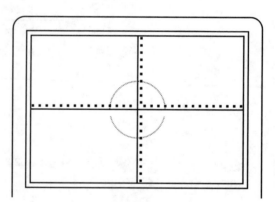

Figure 8.2.10

Exercises 8.2

Find the equations of the following circles, and graph the equations.

1. Center = $C(3, 4)$, radius = 2

2. Center = $C(-5, 1)$, radius = 1

3. Center = $C\left(\dfrac{1}{2}, -2\right)$, radius = 3

4. Center = $C\left(0, -\dfrac{4}{3}\right)$, radius = 4

5. Center = $C(-2, -3)$, radius = $\sqrt{3}$

6. Center = $C(4, 0)$, radius = $\dfrac{\sqrt{7}}{4}$

7. Center $= C(0, 0)$, radius $= 5$

8. Center $= C(0, 0)$, radius $= \dfrac{3}{2}$

9. Center $= C(0, 0)$, passing through $\left(\dfrac{3}{4}, -1\right)$

10. Center $= C(-3, 1)$, passing through $\left(3, \dfrac{7}{2}\right)$

11. Endpoints of a diameter are $(-1, -5)$ and $(5, 3)$.

12. Endpoints of a diameter are $(-12, 5)$ and $(-4, -11)$.

Find the center and radius of each circle with the given equation; then draw each circle.

13. $(x - 1)^2 + (y - 3)^2 = 9$

14. $(x + 6)^2 + (y - 2)^2 = 1$

15. $(x - 4)^2 + (y + 2)^2 = 4$

16. $(x + 3)^2 + (y + 1)^2 = 16$

17. $x^2 + \left(y - \dfrac{3}{2}\right)^2 = \dfrac{1}{4}$

18. $\left(x + \dfrac{5}{3}\right)^2 + y^2 = \dfrac{4}{9}$

19. $x^2 + y^2 = 6$

20. $x^2 + y^2 = 2$

21. $x^2 + y^2 = 1$

22. $x^2 + y^2 = 12$

23. $x^2 + y^2 + 4x - 5y - 2 = 0$

24. $x^2 + y^2 - 10x + 8y + 37 = 0$

25. $x^2 + y^2 + 2x - 12y + 36 = 0$

26. $x^2 + y^2 - x + 6y - 11 = 0$

27. $x^2 + y^2 + \dfrac{2}{3}y = 0$

28. $x^2 + y^2 - \dfrac{4}{5}x = 0$

29. $4x^2 + 4y^2 - 24x + 8y + 31 = 0$

30. $9x^2 + 9y^2 + 54x + 36y + 92 = 0$

Write Algebra

31. Given an equation of the form $(x - h)^2 + (y - k)^2 = r^2$:
    a. If $r^2 > 0$, what is the graph?
    b. If $r^2 = 0$, what is the graph?
    c. If $r^2 < 0$, what is the graph?

32. Find the area and the circumference of the circle in exercise 23.

Use a graphing calculator to graph the following equations.

33. $x^2 + y^2 = 25$

34. $x^2 + (y - 4)^2 = 11$

35. $(x - 3)^2 + y^2 = 16$

36. $(x + 1.5)^2 + (y - 3.6)^2 = 18.9$

G R O U P   A C T I V I T Y

In exercises 37–44, find the area of each shaded region.

37.

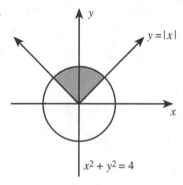

$y = |x|$

$x^2 + y^2 = 4$

38.

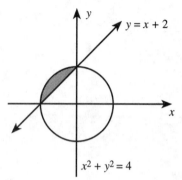

$y = x + 2$

$x^2 + y^2 = 4$

39.

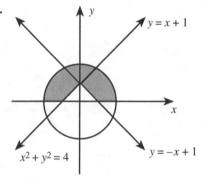

$y = x + 1$

$x^2 + y^2 = 4$

$y = -x + 1$

40.

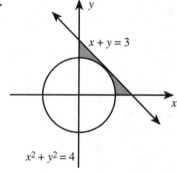

$x + y = 3$

$x^2 + y^2 = 4$

41.

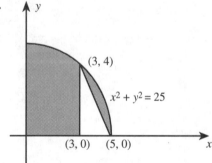

$(3, 4)$

$x^2 + y^2 = 25$

$(3, 0)$ $(5, 0)$

42.

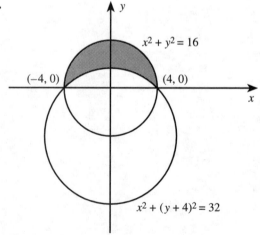

$x^2 + y^2 = 16$

$(-4, 0)$ $(4, 0)$

$x^2 + (y + 4)^2 = 32$

Continued

43.

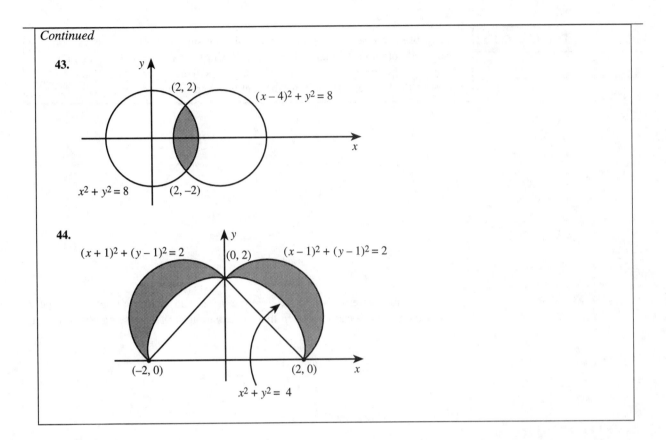

44.

8.3 *Ellipses*

The path that the earth takes as it orbits the sun is almost circular. It looks like a "flattened" circle. The earth does not stay a constant distance from the sun; that is, there are times when the earth is closer to the sun that at other times. There is a fixed point in space such that the sum of the distances from the center of the earth to this fixed point and to the center of the sun is constant (see Figure 8.3.1). The earth's orbit is an example of an ellipse.

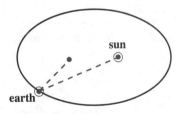

Figure 8.3.1

Definition 8.3.1	An **ellipse** is the set of all points in a plane, the sum of whose distances from two fixed points in the plane is constant. Each of the fixed points is called a **focus** of the ellipse. Together they are called **foci** (see Figure 8.3.2). The midpoint of the line segment joining the foci is called the **center** of the ellipse.

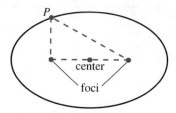

Figure 8.3.2

When the center of an ellipse is at the origin and the foci are on either the x-axis or the y-axis, it can be shown that the equation of the ellipse can be written in the form

$$\frac{x^2}{a^2} + \frac{y^2}{b^2} = 1$$

The x-intercepts are found by setting $y = 0$:

$$\frac{x^2}{a^2} + \frac{0}{b^2} = 1$$

$$\frac{x^2}{a^2} + 0 = 1$$

$$\frac{x^2}{a^2} = 1$$

$$x^2 = a^2$$

$$x = -a \quad \text{or} \quad x = a$$

In a similar fashion we can show that the y-intercepts are $y = -b$ and $y = b$. This type of ellipse is symmetric with respect to both the x-axis and the y-axis.

E XAMPLE 8.3.1   Graph the following equations.

1. $\dfrac{x^2}{25} + \dfrac{y^2}{9} = 1$

The x-intercepts are $x = -5$ and $x = 5$, and the y-intercepts are $y = -3$ and $y = 3$. These are the only points we use to graph the equation in Figure 8.3.3.

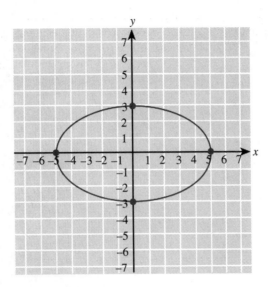

Figure 8.3.3

The foci of this ellipse are on the x-axis. We will not concern ourselves with the locations of the foci in order to graph the ellipse.

2. $\dfrac{x^2}{4} + \dfrac{y^2}{10} = 1$

The x-intercepts are $x = -2$ and $x = 2$, and the y-intercepts are

$$y = -\sqrt{10} \doteq -3.2$$

and

$$y = \sqrt{10} \doteq 3.2$$

The ellipse is graphed in Figure 8.3.4.

3. $4x^2 + y^2 = 4$

To change this equation to the form $\dfrac{x^2}{a^2} + \dfrac{y^2}{b^2} = 1$, we divide both sides by 4 in order to leave 1 on the right side of the equation. We write a 1 under x^2 to help us determine the x-intercepts.

$$\frac{4x^2}{4} + \frac{y^2}{4} = \frac{4}{4}$$

$$\frac{x^2}{1} + \frac{y^2}{4} = 1$$

The x-intercepts are $x = -1$ and $x = 1$, and the y-intercepts are $y = -2$ and $y = 2$. The ellipse is graphed in Figure 8.3.5.

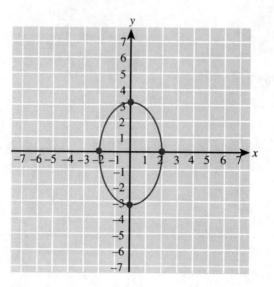

Figure 8.3.4

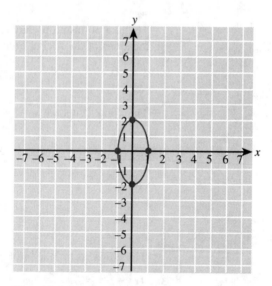

Figure 8.3.5

The line segments that pass through the center and join the intercepts are called the axes. The longer axis is called the **major axis,** and the shorter one is called the **minor axis** (see Figure 8.3.6). Their lengths are $2a$ and $2b$.

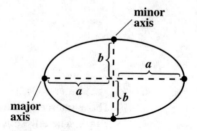

Figure 8.3.6

These lengths can be used to help us draw an ellipse when the center is not at the origin. The points at the end of the major axis are called **vertices.**

If the center of the ellipse is *(h, k)*, the equation of the ellipse can be written in the following form:

$$\frac{(x - h)^2}{a^2} + \frac{(y - k)^2}{b^2} = 1$$

XAMPLE 8.3.2   Graph the following equations.

1. $\dfrac{(x - 2)^2}{16} + \dfrac{(y + 3)^2}{4} = 1$

The center is $(2, -3)$. We plot points $a = 4$ units to the left and right of the center. We also plot points $b = 2$ units above and below the center. The ellipse is graphed in Figure 8.3.7.

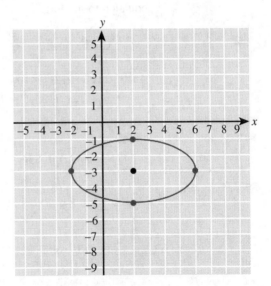

Figure 8.3.7

2. $\dfrac{(x + 4)^2}{3} + \dfrac{(y + 1)^2}{9} = 1$

The center is $(-4, -1)$. We plot points $a = \sqrt{3} \doteq 1.7$ units to the left and right of the center. We also plot points $b = 3$ units above and below the center. The ellipse is graphed in Figure 8.3.8.

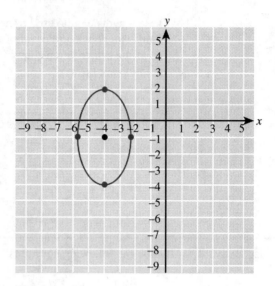

Figure 8.3.8

3. $4x^2 + 25y^2 + 12x - 100y + 9 = 0$

We must complete the square in x and complete the square in y to change this equation to the form

$$\frac{(x-h)^2}{a^2} + \frac{(y-k)^2}{b^2} = 1$$

$$4x^2 + 25y^2 + 12x - 100y + 9 = 0$$

$$4x^2 + 12x + 25y^2 - 100y = -9 \qquad \text{\textit{Group the x terms together and the y terms together and subtract 9 from both sides.}}$$

$$4(x^2 + 3x) + 25(y^2 - 4y) = -9 \qquad \text{\textit{Factor 4 out of the x terms and 25 out of the y terms.}}$$

$$4(x^2 + 3x + \underline{\ \ }) + 25(y^2 - 4y + \underline{\ \ }) = -9 \qquad \begin{array}{l} \textit{For x, } \underline{\ \ } = \left(\frac{1}{2} \cdot 3\right)^2 \\ \quad = \left(\frac{3}{2}\right)^2 = \frac{9}{4} \\ \textit{For y, } \underline{\ \ } = \left[\frac{1}{2} \cdot (-4)\right]^2 \\ \quad = (-2)^2 = 4. \end{array}$$

$$4\left(x^2 + 3x + \frac{9}{4}\right) + 25(y^2 - 4y + 4) = -9 + 9 + 100 \qquad \text{\textit{Add } } 4 \cdot \frac{9}{4} = 9 \text{ \textit{and}} \\ 25 \cdot 4 = 100 \text{ \textit{to both sides.}}$$

$$4\left(x + \frac{3}{2}\right)^2 + 25(y - 2)^2 = 100$$

$$\frac{4\left(x + \frac{3}{2}\right)^2}{100} + \frac{25(y - 2)^2}{100} = \frac{100}{100} \qquad \text{\textit{Divide both sides by 100 to get 1 on the right.}}$$

$$\frac{\left(x + \frac{3}{2}\right)^2}{25} + \frac{(y - 2)^2}{4} = 1$$

The center is $\left(-\frac{3}{2}, 2\right)$. We plot points $a = 5$ units to the left and right of the center. We also plot points $b = 2$ units above and below the center. The ellipse is graphed in Figure 8.3.9.

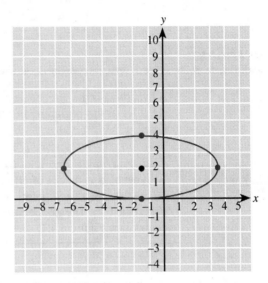

Figure 8.3.9

E xercises 8.3

Graph the following equations.

1. $\dfrac{x^2}{16} + \dfrac{y^2}{9} = 1$

2. $\dfrac{x^2}{36} + \dfrac{y^2}{16} = 1$

3. $\dfrac{x^2}{9} + \dfrac{y^2}{1} = 1$

4. $\dfrac{x^2}{1} + \dfrac{y^2}{16} = 1$

5. $\dfrac{x^2}{8} + \dfrac{y^2}{9} = 1$

6. $\dfrac{x^2}{25} + \dfrac{y^2}{24} = 1$

7. $8x^2 + 3y^2 = 48$

8. $2x^2 + 9y^2 = 18$

9. $12x^2 + 5y^2 = 60$

10. $20x^2 + 3y^2 = 60$

11. $\dfrac{(x - 4)^2}{4} + \dfrac{(y - 2)^2}{25} = 1$

12. $\dfrac{(x + 1)^2}{27} + \dfrac{(y - 5)^2}{1} = 1$

13. $\dfrac{\left(x + \frac{1}{2}\right)^2}{9} + \dfrac{(y + 4)^2}{5} = 1$

14. $\dfrac{(x + 2)^2}{3} + \dfrac{\left(y + \frac{5}{2}\right)^2}{16} = 1$

15. $\dfrac{x^2}{49} + \dfrac{(y - 2)^2}{20} = 1$

16. $\dfrac{(x + 5)^2}{1} + \dfrac{y^2}{40} = 1$

17. $9x^2 + 25y^2 - 18x - 150y + 9 = 0$

18. $9x^2 + 4y^2 + 54x - 8y - 59 = 0$

19. $9x^2 + 4y^2 - 24x + 8y - 16 = 0$

20. $25x^2 + 18y^2 - 200x + 12y - 48 = 0$

21. $64x^2 + 225y^2 - 448x - 300y + 484 = 0$

22. $9x^2 + 24y^2 + 30x + 24y + 30 = 0$

Write Algebra

23. Given an equation of the form $\dfrac{(x - h)^2}{a^2} + \dfrac{(y - k)^2}{b^2} = 1$,

as a gets closer to b, what happens to the shape of the graph?

8.4 *Hyperbolas*

A hyperbola is the only conic section that consists of two separate curves. (See Figure 8.1.1 at the beginning of the chapter). Its definition is similar to that of an ellipse.

Definition 8.4.1

A **hyperbola** is the set of all points in a plane, the difference of whose distances from two fixed points in the plane is constant. The fixed points are called the **foci**. The midpoint of the line segment joining the foci is called the **center** (see Figure 8.4.1).

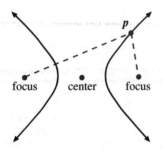

Figure 8.4.1

When the center of a hyperbola is at the origin and the foci are on the x-axis, it can be shown that the equation of the hyperbola can be written in the form

$$\frac{x^2}{a^2} - \frac{y^2}{b^2} = 1$$

This equation differs from that of an ellipse only by the minus sign. This minus sign causes this type of hyperbola not to have y-intercepts. If $x = 0$,

$$0 - \frac{y^2}{b^2} = 1$$

$$\frac{y^2}{b^2} = -1$$

$$y^2 = -b^2$$

$$y = \pm bi$$

Thus, there are no y-intercepts. But if we set $y = 0$,

$$\frac{x^2}{a^2} - 0 = 1$$

$$x^2 = a^2$$

$$x = -a \quad \text{or} \quad x = a$$

Therefore, the x-intercepts are $x = -a$ and $x = a$.

EXAMPLE 8.4.1

Graph the equation $\dfrac{x^2}{16} - \dfrac{y^2}{4} = 1$.

The x-intercepts can be found by taking the square roots of the number under x^2. Thus the x-intercepts are $x = -4$ and $x = 4$. These points are called the **vertices** of the hyperbola. Although the square roots of the number under y^2 do not yield y-intercepts, they still serve a purpose. Construct a rectangle whose sides pass vertically through ± 4 on the x-axis and horizontally through ± 2 on the y-axis. Draw the diagonals of the rectangle that pass through the origin, and then extend them in each direction as in Figure 8.4.2. Each half of the hyperbola is a curve that passes

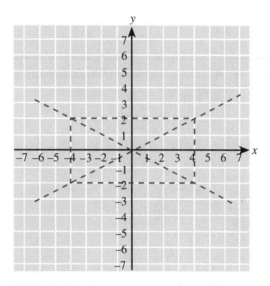

Figure 8.4.2

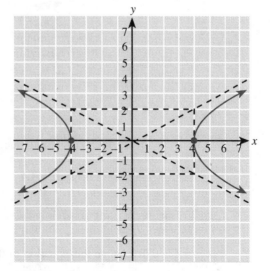

Figure 8.4.3

through a vertex and gets closer to these extended diagonals as it moves away from the origin (see Figure 8.4.3). The extended diagonals are called **asymptotes.** An asymptote is a line to which a graph gets closer and closer, usually without ever reaching it.

In general, to graph $\dfrac{x^2}{a^2} - \dfrac{y^2}{b^2} = 1$, we draw the rectangle that passes through $\pm a$ on the x-axis and through $\pm b$ on the y-axis. We then draw the asymptotes as dashed lines passing through the corners of the rectangles. The only points we plot are the vertices. We then draw the hyperbola as two curves passing through the vertices and approaching the asymptotes, without ever touching them. The graph is symmetric with respect to the x-axis and y-axis.

E X A M P L E   8 . 4 . 2   Graph the following equations.

1. $\dfrac{x^2}{12} - \dfrac{y^2}{1} = 1$

   Using $a = \sqrt{12} = 2\sqrt{3}$ and $b = 1$, we get the vertices at $x = -2\sqrt{3}$ and $x = 2\sqrt{3}$ and the asymptotes as shown in Figure 8.4.4.

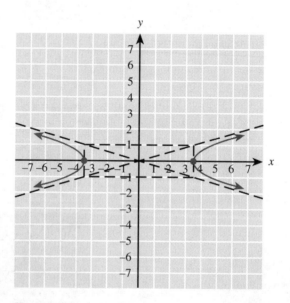

   Figure 8.4.4

2. $5x^2 - 4y^2 = 20$

   To change this equation to the form $\dfrac{x^2}{a^2} - \dfrac{y^2}{b^2} = 1$, we divide both sides by 20:

$$\frac{5x^2}{20} - \frac{4y^2}{20} = \frac{20}{20}$$

$$\frac{x^2}{4} - \frac{y^2}{5} = 1$$

Using $a = 2$ and $b = \sqrt{5}$, we get the vertices at $x = -2$ and $x = 2$, and also the asymptotes needed to graph the hyperbola in Figure 8.4.5.

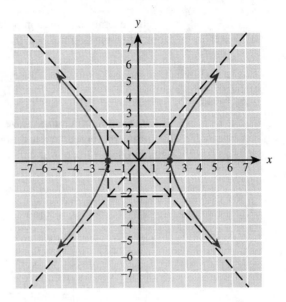

Figure 8.4.5

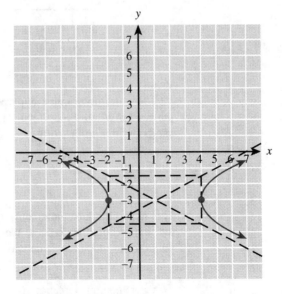

Figure 8.4.6

3. $\dfrac{(x - 1)^2}{9} - \dfrac{(y + 3)^2}{2} = 1$

The center of this hyperbola is shifted from the origin to $(1, -3)$. The vertices are located $a = 3$ units to the left and right to the center, at $(-2, -3)$ and $(4, -3)$. With the vertices and the points $b = \sqrt{2} \doteq 1.4$ units above and below the center, we can draw the asymptotes. In this case the foci are located on the horizontal line $y = -3$ passing through the center and vertices. The hyperbola is graphed in Figure 8.4.6.   ▬

   In general, the equation of the hyperbola with center (h, k) and foci on the horizontal line passing through the center can be written in the form

$$\dfrac{(x - h)^2}{a^2} - \dfrac{(y - k)^2}{b^2} = 1$$

The line segment joining the vertices V_1 and V_2 is called the **transverse axis.** The line segment perpendicular to the transverse axis and passing through the center b units in either direction is called the **conjugate axis** (see Figure 8.4.7).

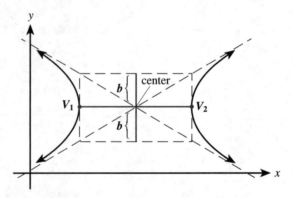

Figure 8.4.7

   Now let us examine hyperbolas with foci located vertically above and below the center. When the center is at the origin, the equation of this type of hyperbola can be written in the form

$$\dfrac{y^2}{a^2} - \dfrac{x^2}{b^2} = 1$$

This hyperbola has y-intercepts at $y = -a$, and $y = a$, but no x-intercepts. We call the y-intercepts the vertices. We will construct a rectangle to help us draw the asymptotes in much the same way as before.

EXAMPLE 8.4.3

Graph the following equations.

1. $\dfrac{y^2}{25} - \dfrac{x^2}{4} = 1$

Using $a = 5$ and $b = 2$, we get the vertices at $y = -5$ and $y = 5$, and also the asymptotes needed to graph the hyperbola in Figure 8.4.8.

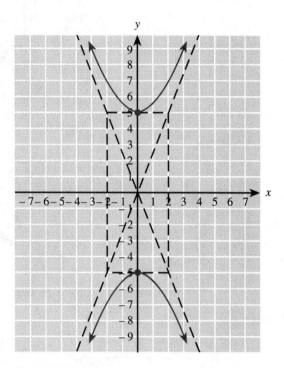

Figure 8.4.8

REMARK   The equation $\dfrac{x^2}{4} - \dfrac{y^2}{25} = 1$ would have the same asymptotes. However, because the y^2 term follows the minus sign, this latter hyperbola would have x-intercepts instead of y-intercepts, and it would open left and right instead of up and down.

2. $\dfrac{(y + 2)^2}{9} - \dfrac{(x + 1)^2}{9} = 1$

The center of this hyperbola is the point $(-1, -2)$. The vertices are located $a = 3$ units above and below the center, at $(-1, 1)$ and $(-1, -5)$. With the vertices and the points $b = 3$ units to the right and left of the center, we can draw the asymptotes. The foci are located on the vertical line $x = -1$ passing through the center and vertices (see Figure 8.4.9).

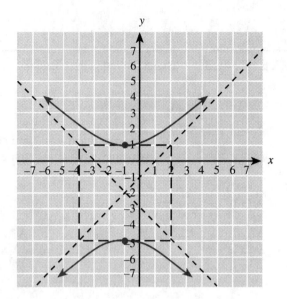

Figure 8.4.9

In general, the equation of the hyperbola with center (h, k) and foci on the vertical line passing through the center can be written in the form

$$\frac{(y - k)^2}{a^2} - \frac{(x - h)^2}{b^2} = 1$$

3. $9x^2 - 3y^2 - 6x + 18y - 17 = 0$

We must complete the square in x and in y in order to change this equation into the form

$$\frac{(x - h)^2}{a^2} - \frac{(y - k)^2}{b^2} = 1 \quad \text{or} \quad \frac{(y - k)^2}{a^2} - \frac{(x - h)^2}{b^2} = 1$$

$$9x^2 - 3y^2 - 6x + 18y - 17 = 0$$

$$9x^2 - 6x - 3y^2 + 18y = 17 \qquad \text{\textit{Group the x terms and the y terms together and add 17 to both sides.}}$$

$$9\left(x^2 - \frac{2}{3}x\right) - 3(y^2 - 6y) = 17 \qquad \text{\textit{Factor 9 out of the x terms and -3 out of the y terms.}}$$

$$9\left(x^2 - \frac{2}{3}x + \underline{\quad}\right) - 3(y^2 - 6y + \underline{\quad}) = 17 \qquad \text{\textit{For x, }} \underline{\quad} = [\tfrac{1}{2} \cdot (-\tfrac{2}{3})]^2 \\ = (-\tfrac{1}{3})^2 = \tfrac{1}{9}. \\ \text{\textit{For y, }} \underline{\quad} = [\tfrac{1}{2} \cdot (-6)]^2 \\ = (-3)^2 = 9.$$

$$9\left(x^2 - \frac{2}{3}x + \frac{1}{9}\right) - 3(y^2 - 6y + 9) = 17 + 1 - 27 \qquad \text{\textit{Add $9 \cdot \frac{1}{9} = 1$ and $-3 \cdot 9 = -27$ to both sides.}}$$

$$9\left(x - \frac{1}{3}\right)^2 - 3(y - 3)^2 = -9$$

$$\frac{9\left(x - \frac{1}{3}\right)^2}{-9} - \frac{3(y - 3)^2}{-9} = \frac{-9}{-9}$$

Divide both sides by -9 to get 1 on the right.

$$-\frac{\left(x - \frac{1}{3}\right)^2}{1} + \frac{(y - 3)^2}{3} = 1$$

$$\frac{(y - 3)^2}{3} - \frac{\left(x - \frac{1}{3}\right)^2}{1} = 1$$

The center is $\left(\frac{1}{3}, 3\right)$. The vertices are $a = \sqrt{3} \doteq 1.7$ units above and below the center, at $\left(\frac{1}{3}, 3 + \sqrt{3}\right)$ and $\left(\frac{1}{3}, 3 - \sqrt{3}\right)$. With the vertices and the points $b = 1$ unit to the right and left of the center, we can draw the asymptotes for the hyperbola in Figure 8.4.10.

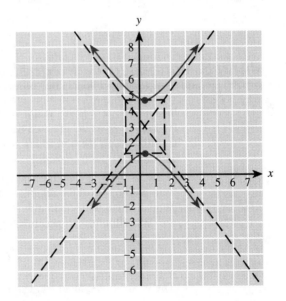

Figure 8.4.10

Exercises 8.4

Graph the following equations.

1. $\dfrac{x^2}{1} - \dfrac{y^2}{1} = 1$

2. $\dfrac{y^2}{1} - \dfrac{x^2}{1} = 1$

3. $\dfrac{y^2}{25} - \dfrac{x^2}{6} = 1$

4. $\dfrac{y^2}{10} - \dfrac{x^2}{3} = 1$

5. $\dfrac{y^2}{\frac{1}{4}} - \dfrac{x^2}{\frac{9}{16}} = 1$

6. $\dfrac{y^2}{1} - \dfrac{x^2}{\frac{1}{9}} = 1$

7. $x^2 - 9y^2 = 36$

8. $x^2 - 16y^2 = 16$

9. $9y^2 - 2x^2 = -18$

10. $y^2 - 2x^2 = -8$

11. $y^2 - x^2 = 4$

12. $x^2 - y^2 = 25$

13. $\dfrac{(x-3)^2}{16} - \dfrac{(y+2)^2}{4} = 1$

14. $\dfrac{(x+2)^2}{9} - \dfrac{(y-1)^2}{16} = 1$

15. $\dfrac{(x+4)^2}{6} - \dfrac{y^2}{2} = 1$

16. $\dfrac{(y+5)^2}{5} - \dfrac{x^2}{10} = 1$

17. $\dfrac{(y+1)^2}{\frac{25}{4}} - \dfrac{(x-2)^2}{\frac{16}{9}} = 1$

18. $x^2 - 5y^2 + 2x - 30y - 69 = 0$

19. $x^2 - 36y^2 - 2x + 288y - 539 = 0$

20. $4x^2 - 8y^2 - 4x - 24y - 49 = 0$

21. $9x^2 - 16y^2 - 24x + 24y + 151 = 0$

22. $4x^2 - 18y^2 + 24x + 24y + 37 = 0$

Write Algebra

23. Explain why an equation of the form $\dfrac{x^2}{a^2} - \dfrac{y^2}{b^2} = 1$ does not define a function.

24. Describe how to determine whether a hyperbola opens up and down or right and left.

8.5 *Graphing Second-Degree Inequalities*

In this section we will graph second-degree inequalities in two variables in much the same way as we graphed linear inequalities in two variables in Section 6.7. We will first graph the **boundary curve,** which is a conic section. The boundary curve will separate the Cartesian plane into two or three regions. We will then determine which of these regions contains the points that satisfy the inequality by testing a point within one of the regions. To indicate that all the points in a region satisfy the inequality, we will shade it in.

We will consider **second-degree inequalities** that can be written in the form

$$Ax^2 + Cy^2 + Dx + Ey + F < 0$$

(The inequality symbol, $<$, can be replaced by $\leq$, $>$, or $\geq$.)

XAMPLE 8.5.1   Graph the following inequalities.

1. $y > x^2 - 4x + 1$

Step 1: We obtain the equation of the boundary curve by replacing the inequality symbol with an equals sign:

$$y = x^2 - 4x + 1$$

This is the equation of a parabola that opens upward. By completing the square, we find the vertex to be $(2, -3)$.

$$y = (x^2 - 4x) + 1$$
$$y = (x^2 - 4x + 4) + 1 - 4$$
$$y = (x - 2)^2 - 3$$

When we graph the parabola, as in Figure 8.5.1, we draw a *dashed* curve to signify that the points on the boundary curve do *not* satisfy the inequality $y > x^2 - 4x + 1$.

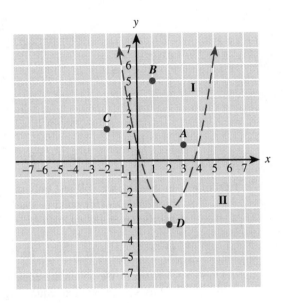

Figure 8.5.1

Step 2: We now need to determine which of the two regions contains the points that satisfy the inequality. For this first example only, we will test two points in each region.

$A(3, 1)$: $\quad 1 \overset{?}{>} 3^2 - 4 \cdot 3 + 1$

$\qquad\qquad 1 \overset{?}{>} 9 - 12 + 1$

$\qquad\qquad 1 > -2$

Point A checks.

$C(-2, 2)$: $\quad 2 \overset{?}{>} (-2)^2 - 4(-2) + 1$

$\qquad\qquad 2 \overset{?}{>} 4 + 8 + 1$

$\qquad\qquad 2 \not> 13$

Point C does not check.

$B(1, 5)$: $\quad 5 \overset{?}{>} 1^2 - 4 \cdot 1 + 1$

$\qquad\qquad 5 \overset{?}{>} 1 - 4 + 1$

$\qquad\qquad 5 > -2$

Point B checks.

$D(2, -4)$: $\quad -4 \overset{?}{>} 2^2 - 4 \cdot 2 + 1$

$\qquad\qquad -4 \overset{?}{>} 4 - 8 + 1$

$\qquad\qquad -4 \not> -3$

Point D does not check.

Region I, which consists of all the points "inside," or above, the parabola, is shaded in Figure 8.5.2 because the points tested from region I satisfy the inequality. We conclude that *all* the points in region I satisfy the inequality. Region II, which consists of all the points "outside," or below, the parabola, is not shaded, since the points tested from region II do not satisfy the inequality. We conclude that *none* of the points in region II satisfy the inequality.

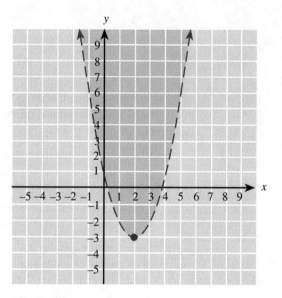

Figure 8.5.2

NOTE ▶ In general, we will test only one point. *If it satisfies the inequality, then the region that contains this point is shaded. If the point does not satisfy the inequality, then the region on the opposite side of the boundary curve is shaded.*

2. $x^2 + y^2 - 4x - 8y + 16 \geq 0$

Step 1: We obtain the equation of the boundary curve by replacing the inequality symbol with an equals sign:

$$x^2 + y^2 - 4x - 8y + 16 = 0$$

By completing the square in x and in y, we determine that this is the equation of a circle with center $C(2, 4)$ and radius $r = 2$.

$$(x^2 - 4x) + (y^2 - 8y) = -16$$
$$(x^2 - 4x + 4) + (y^2 - 8y + 16) = -16 + 4 + 16$$
$$(x - 2)^2 + (y - 4)^2 = 4$$

When we graph the circle in Figure 8.5.3, we draw a *solid* curve to signify that the points of the boundary curve satisfy the inequality.

Step 2: Let us test the origin, $(0, 0)$, which lies outside the circle:

$$0^2 + 0^2 - 4 \cdot 0 - 8 \cdot 0 + 16 \overset{?}{\geq} 0$$
$$0 + 0 - 0 - 0 + 16 \overset{?}{\geq} 0$$
$$16 \geq 0$$

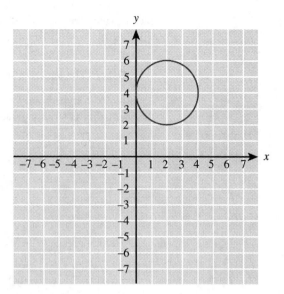

Figure 8.5.3

The origin checks. We therefore shade the region of all points outside the circle in Figure 8.5.4.

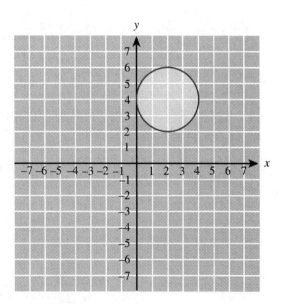

Figure 8.5.4

We can write the inequality in the form $(x - 2)^2 + (y - 4)^2 \geq 4$. The points that satisfy this inequality are those on the circle, which are exactly 2 units from the center, and those outside the circle, which are more than 2 units from the center of the circle.

3. $\dfrac{x^2}{9} - \dfrac{y^2}{4} > 1$

Step 1: The equation of the boundary curve is $\dfrac{x^2}{9} - \dfrac{y^2}{4} = 1$, which is the equation of a hyperbola. We draw the hyperbola dashed in Figure 8.5.5, since the inequality is strictly greater than.

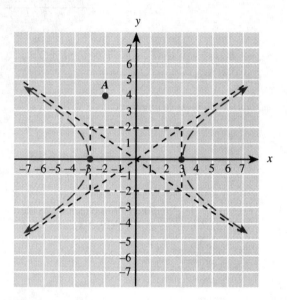

Figure 8.5.5

Step 2: Let us test the point $A(-2, 4)$, which lies between the two curves of the hyperbola:

$$\frac{(-2)^2}{9} - \frac{4^2}{4} \overset{?}{>} 1$$

$$\frac{4}{9} - \frac{16}{4} \overset{?}{>} 1$$

$$\frac{4}{9} - 4 \overset{?}{>} 1$$

$$\frac{-32}{9} \not> 1$$

Since point A does not check, we shade the region on the opposite side of the boundary curve in Figure 8.5.6. This region consists of two parts: the points to the right of the right half of the hyperbola and the points to the left of the left half of the hyperbola.

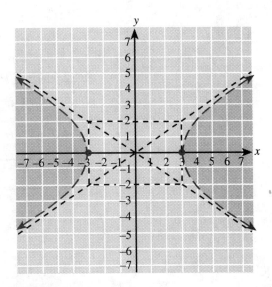

Figure 8.5.6

In the next example we will graph a **system of inequalities,** which is two (or more) inequalities considered at the same time. We will graph a system by graphing each inequality and then finding the intersection.

XAMPLE 8.5.2   Graph the following system of inequalities.

$$\begin{cases} 9x^2 + y^2 \le 9 \\ x \le y^2 - 4y \end{cases}$$

For the inequality $9x^2 + y^2 \le 9$, the equation of the boundary is

$$9x^2 + y^2 = 9$$

$$\frac{9x^2}{9} + \frac{y^2}{9} = \frac{9}{9}$$

$$\frac{x^2}{1} + \frac{y^2}{9} = 1$$

This is the equation of an ellipse, which we draw with a solid curve. By testing the point (0, 0):

$$9 \cdot 0^2 + 0^2 \overset{?}{\le} 9$$

$$0 \le 9$$

we obtain the shaded region in Figure 8.5.7.

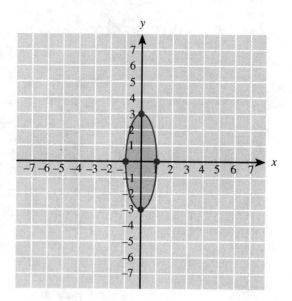

Figure 8.5.7

For the inequality $x \leq y^2 - 4y$, the equation of the boundary is

$$x = y^2 - 4y$$
$$x = (y^2 - 4y + 4) - 4$$
$$x = (y - 2)^2 - 4$$

This is the equation of a horizontal parabola, which we draw with a solid curve. By testing the point (3, 2):

$$3 \overset{?}{\leq} 2^2 - 4 \cdot 2$$
$$3 \overset{?}{\leq} 4 - 8$$
$$3 \nleq -4$$

we obtain the shaded region outside the parabola, as shown in Figure 8.5.8. The graph of the system is where the two shaded regions intersect.

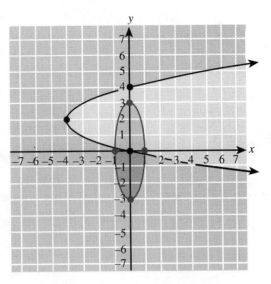

Figure 8.5.8

Exercises 8.5

Graph the following inequalities.

1. $x > y^2 - 1$

2. $\dfrac{x^2}{9} - \dfrac{y^2}{25} > 1$

3. $x^2 + y^2 < 4$

4. $y \le (x - 2)^2$

5. $\dfrac{x^2}{16} + \dfrac{y^2}{5} < 1$

6. $x \ge 3 - 2y^2$

7. $y < x^2 + 3$

8. $x^2 + y^2 \ge 8$

9. $5y^2 - 16x^2 < 80$

10. $25x^2 + 12y^2 > 300$

11. $y > 2x^2 + 4x - 2$

12. $x \le y^2 + 6y + 8$

13. $x^2 + y^2 - 10x + 6y + 33 < 0$

14. $3x^2 - 2y^2 - 8y - 26 \le 0$

15. $x^2 + 4y^2 + 6x + 40y + 105 \le 0$

16. $x^2 + 6x + 2y + 7 > 0$

17. $12x^2 - y^2 - 8y - 4 \le 0$

18. $4x^2 + y^2 - 16x - 20 \ge 0$

19. $3y^2 - x - 4y - 4 > 0$

20. $x^2 + y^2 + 2x - 2y - 23 > 0$

Graph the following systems of inequalities.

21. $\begin{cases} x^2 + (y - 3)^2 \le 8 \\ y < 4 - x^2 \end{cases}$

22. $\begin{cases} y \ge x^2 - 2x + 1 \\ x + y \le 3 \end{cases}$

23. $\begin{cases} 4y^2 - 9x^2 \le 36 \\ 3x + y \le 1 \end{cases}$

24. $\begin{cases} x^2 + 4y^2 - 16y \le 0 \\ x^2 + y^2 - 8x - 4y + 16 \ge 0 \end{cases}$

25. $\begin{cases} x < -2y^2 - 8y - 5 \\ 2x < y^2 - 2y - 5 \end{cases}$

26. $\begin{cases} 2x^2 + 5y^2 < 20 \\ x < 2y^2 - 2 \\ y > 2x \end{cases}$

27. $\begin{cases} y \le x^2 + 3 \\ y \ge 2x + 2 \\ x \ge -1 \end{cases}$

28. $\begin{cases} (x - 1)^2 + (y + 2)^2 \le 9 \\ (x + 2)^2 + (y + 2)^2 \le 4 \\ (x + 1)^2 + (y + 2)^2 \ge 1 \end{cases}$

Applied Algebra

Ed makes greenhouses that have rectangular bases 10 ft wide and 12 ft long. The frame consists of PVC pipes that are squeezed into fittings in the base and along the top. Find a mathematical model for the curve of the PVC pipes, given that the height is $91\frac{3}{4}$ in. Does your model fit the data in Figure D? Tom bought one of Ed's greenhouses to protect his potted avocado tree during the winter. If the tree is 6 ft tall, how close to the side of the greenhouse can he place the tree?

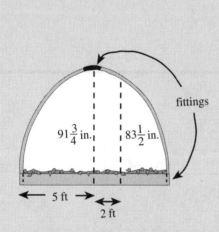

Figure D

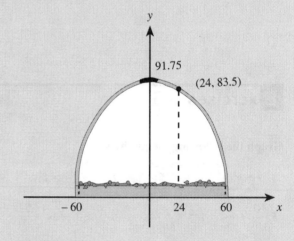

Figure E

SOLUTION

We will place the arch of the PVC pipes in the Cartesian coordinate system, as shown in Figure E, using inches as the units. Since the distances from the origin to the x- and y-intercepts are not the same, we know the curve cannot be a semicircle. Our two most obvious choices are a parabola and an ellipse. If the curve is a parabola with the vertex on the y-axis, opening down, the equation would have the form $y = -ax^2 + 91.75$. Since an x-intercept is $(60, 0)$, we can substitute 60 for x and 0 for y and solve for a.

$$0 = -a \cdot 60^2 + 91.75$$
$$0 = -3600a + 91.75$$
$$3600a = 91.75$$
$$a = \frac{91.75}{3600}$$
$$a \doteq 0.02549$$

Thus, the equation would be $y = -0.02549x^2 + 91.75$. Let us check to see if the point (24, 83.5) would lie on the graph of this equation. Substituting $x = 24$, we get

$$y = -0.02549(24)^2 + 91.75 \doteq 77.07$$

which is about 6.43 in. too small. Let us now try using an ellipse. Since the center is at the origin, the equation would have the form

$$\frac{x^2}{60^2} + \frac{y^2}{91.75^2} = 1$$

How close does the point (24, 83.5) come to the graph of this equation? We substitute 24 for x:

$$\frac{24^2}{60^2} + \frac{y^2}{91.75^2} = 1$$

$$0.16 + \frac{y^2}{91.75^2} = 1$$

$$\frac{y^2}{91.75^2} = 0.84$$

$$y^2 \doteq 7071$$

$$y \doteq \pm 84.09$$

Choosing $y = 84.09$, we obtain an answer that is only 0.59 in. too large, which is a better approximation than the first one. Using the equation for an ellipse, let us approximate how close to the side of the greenhouse Tom can place his potted avocado tree. This time we substitute 72 (6 ft = 72 in.) for y.

$$\frac{x^2}{60^2} + \frac{72^2}{91.75^2} = 1$$

$$\frac{x^2}{3600} + 0.6158 \doteq 1$$

$$\frac{x^2}{3600} \doteq 0.3842$$

$$x^2 \doteq 1383$$

$$x \doteq \pm 37.19$$

Choosing the positive value for x, we determine that the pot can be approximately 37 in. from the middle, or 60 in. − 37 in. = 23 in. from the side of the greenhouse.

Your Turn:

Ed also makes a larger greenhouse that is 15 ft wide and 9 ft $1\frac{1}{2}$ in. high. Find a mathematical model for the curve of the PVC pipes. If Ed hangs a hanging basket 2 ft over from the center and the basket hangs down 26 in., how high is the bottom of the basket? See the accompanying figure.

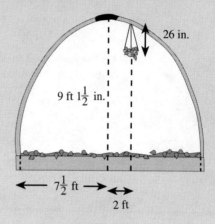

9 ft $1\frac{1}{2}$ in.

26 in.

$7\frac{1}{2}$ ft

2 ft

Chapter 8 Review

Terms to Remember

Conic section	p. 470	Hyperbola	p. 494
Axis of symmetry	p. 470	Foci of a hyperbola	p. 494
Horizontal parabola	p. 470	Center of a hyperbola	p. 494
Circle	p. 478	Vertices of a hyperbola	p. 495
Radius	p. 478	Asymptote	p. 496
Center of a circle	p. 478	Transverse axis	p. 498
Ellipse	p. 488	Conjugate axis	p. 498
Foci of an ellipse	p. 488	Boundary curve	p. 502
Center of an ellipse	p. 488	Second-degree inequalities	p. 502
Major axis	p. 490	System of inequalities	p. 507
Minor axis	p. 490		
Vertices of an ellipse	p. 491		

Equations and Properties of Conic Sections

■ *Horizontal parabola*

Given an equation in the form

$$x = ay^2 + by + c$$

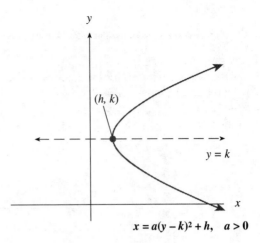

$$x = a(y - k)^2 + h, \quad a > 0$$

it can be changed to the form

$$x = a(y - k)^2 + h$$

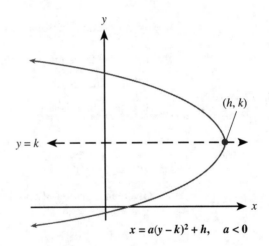

$$x = a(y - k)^2 + h, \quad a < 0$$

by completing the square in y.

1. The graph opens to the right if $a > 0$, or to the left if $a < 0$.

2. Compared with the graph of $x = y^2$, the parabola is thinner if $|a| > 1$ or wider if $|a| < 1$.

3. The vertex is the point (h, k).

4. The graph is symmetric with respect to its axis, the horizontal line $y = k$.

■ *Circle*

Given an equation of a circle in the form

$$x^2 + y^2 + dx + ey + f = 0$$

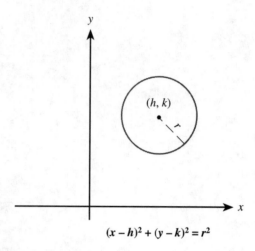

$$(x - h)^2 + (y - k)^2 = r^2$$

it can be changed to the form

$$(x - h)^2 + (y - k)^2 = r^2$$

by completing the square in x and in y.

1. The center is the point (h, k).

2. The radius is r.

3. If the center is at the origin, the equation simplifies to

$$x^2 + y^2 = r^2$$

■ *Ellipse*

Given an equation of an ellipse in the form

$$Ax^2 + Cy^2 + Dx + Ey + F = 0$$

it can be changed to the form

$$\frac{(x - h)^2}{a^2} + \frac{(y - k)^2}{b^2} = 1$$

by completing the square in x and in y, and making the constant on the right side equal to 1.

1. The center is the point (h, k).

2. If $a > b$, the ellipse is elongated horizontally.

3. If $a < b$, the ellipse is elongated vertically.

4. The points $(h + a, k)$ and $(h - a, k)$, which are a units to the right and left of the center, and the points $(h, k + b)$ and $(h, k - b)$, which are b units above and below the center, are used to draw the ellipse.

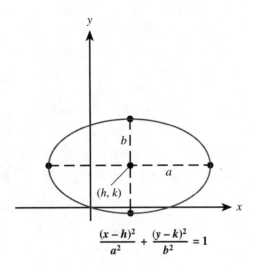

$$\frac{(x-h)^2}{a^2} + \frac{(y-k)^2}{b^2} = 1$$

5. The graph is symmetric with respect to the major axis and minor axis, which pass through the center and connect the points in part 4.

6. If the center is at the origin, the equation simplifies to

$$\frac{x^2}{a^2} + \frac{y^2}{b^2} = 1$$

■ *Hyperbola*

Given an equation of a hyperbola in the form

$$Ax^2 + Cy^2 + Dx + Ey + F = 0$$

it can be changed to the form

$$\frac{(x-h)^2}{a^2} - \frac{(y-k)^2}{b^2} = 1 \quad \text{or} \quad \frac{(y-k)^2}{a^2} - \frac{(x-h)^2}{b^2} = 1$$

by completing the square in x and in y and making the constant on the right side equal to 1.

1. The center is the point (h, k).

2. The graph is symmetric with respect to the transverse axis that joins the vertices and the conjugate axis that lies on the perpendicular bisector of the transverse axis.

3. Let $\dfrac{(x-h)^2}{a^2}$ be the positive term.

    a. The hyperbola opens left and right, approaching the asymptotes and passing through the vertices $(h - a, k)$ and $(h + a, k)$.

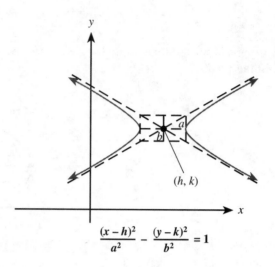

$$\frac{(x-h)^2}{a^2} - \frac{(y-k)^2}{b^2} = 1$$

b. The asymptotes are found by extending the diagonals of the rectangle that passes through the vertices and the points $(h, k - b)$ and $(h, k + b)$.

4. Let $\dfrac{(y - k)^2}{a^2}$ be the positive term.

a. The hyperbola opens upward and downward, approaching the asymptotes and passing through the vertices $(h, k - a)$ and $(h, k + a)$.

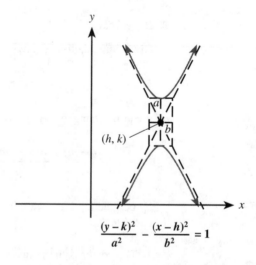

$$\frac{(y-k)^2}{a^2} - \frac{(x-h)^2}{b^2} = 1$$

b. The asymptotes are found by extending the diagonals of the rectangle that passes through the vertices and the points $(h - b, k)$ and $(h + b, k)$.

5. If the center is at the origin, the equation simplifies to

$$\frac{x^2}{a^2} - \frac{y^2}{b^2} = 1 \quad \text{or} \quad \frac{y^2}{a^2} - \frac{x^2}{b^2} = 1$$

Review Exercises

8.1 Sketch the graphs of the following equations. If necessary, complete the square in y to find the vertex.

1. $x = \frac{1}{2}y^2 + 1$

2. $x = 4 - 2y^2$

3. $x = (y + 4)^2 + 1$

4. $x = \frac{1}{3}(y - 3)^2 + 2$

5. $x = -3(y - 2)^2 - 1$

6. $x = -\left(y + \frac{5}{2}\right)^2 + \frac{9}{4}$

7. $x = y^2 - 2y - 1$

8. $x = y^2 - 8y + 17$

9. $x = \frac{3}{2}y^2 + 6y + 3$

10. $4x = -y^2 - 2y + 4$

8.2 Find the equations of the following circles, and graph the equations.

11. Center $= C(-2, 5)$, radius $= 1$

12. Center $= C\left(\frac{3}{2}, 4\right)$, radius $= \frac{5}{2}$

13. Center $= C(0, 0)$, radius $= \sqrt{6}$

14. Center $= C(1, 3)$, radius $2\sqrt{2}$

Find the center and radius of each circle with the given equation; then draw the circles.

15. $(x + 2)^2 + (y - 4)^2 = 5$

16. $\left(x + \frac{1}{2}\right)^2 + (y + 3)^2 = 4$

17. $x^2 + y^2 = 9$

18. $3x^2 + 3y^2 = \frac{1}{3}$

19. $x^2 + y^2 - 6x - 2y + 8 = 0$

20. $x^2 + y^2 - 8x + 10y + 32 = 0$

8.3 Graph the following equations.

21. $\frac{x^2}{4} + \frac{y^2}{20} = 1$

22. $\frac{x^2}{25} + \frac{y^2}{9} = 1$

23. $4x^2 + 8y^2 = 32$

24. $6x^2 + 2y^2 = 6$

25. $\frac{x^2}{36} + \frac{(y + 2)^2}{4} = 1$

26. $\frac{(x - 3)^2}{3} + \frac{y^2}{16} = 1$

27. $\frac{(x - 1)^2}{10} + \frac{(y - 4)^2}{1} = 1$

28. $\frac{\left(x + \frac{1}{2}\right)^2}{9} + \frac{(y - 3)^2}{4} = 1$

29. $25x^2 + 4y^2 + 150x + 8y + 129 = 0$

30. $4x^2 + y^2 - 40x + 10y + 121 = 0$

8.4 Graph the following equations.

31. $\frac{x^2}{9} - \frac{y^2}{16} = 1$

32. $\frac{y^2}{6} - \frac{x^2}{6} = 1$

33. $y^2 - 3x^2 = 12$

34. $9x^2 - y^2 = 9$

35. $\frac{(x - 2)^2}{8} - \frac{y^2}{3} = 1$

36. $\frac{(y - 4)^2}{24} - \frac{x^2}{9} = 1$

37. $\frac{\left(y + \frac{1}{2}\right)^2}{36} - \frac{\left(x + \frac{3}{2}\right)^2}{9} = 1$

38. $\frac{\left(x + \frac{4}{3}\right)^2}{25} - \frac{\left(y - \frac{7}{3}\right)^2}{9} = 1$

39. $x^2 - 8y^2 + 6x + 32y - 39 = 0$

40. $8x^2 - y^2 - 56x + 2y + 129 = 0$

8.5 **Graph the following inequalities.**

41. $\dfrac{x^2}{25} + \dfrac{y^2}{10} < 1$

42. $x < y^2 - 3$

43. $x^2 + y^2 \geq 16$

44. $\dfrac{y^2}{7} - \dfrac{x^2}{1} < 1$

45. $y > 4 - x^2$

46. $5x^2 + 2y^2 > 20$

47. $6x^2 - 4y^2 \geq 24$

48. $x \geq -2y^2 - 8y - 3$

49. $4x^2 + 4y^2 + 12x - 16y - 39 < 0$

50. $9x^2 - 2y^2 - 12y < 0$

Graph the following systems of inequalities.

51. $\begin{cases} x^2 + y^2 > 4 \\ \dfrac{x^2}{20} - \dfrac{y^2}{16} \leq 1 \end{cases}$

52. $\begin{cases} y < x^2 + 3 \\ y \geq 1 \end{cases}$

53. $\begin{cases} \dfrac{x^2}{25} + \dfrac{y^2}{4} < 1 \\ \dfrac{x^2}{9} + \dfrac{y^2}{36} > 1 \end{cases}$

54. $\begin{cases} (x - 2)^2 + (y - 1)^2 \leq 16 \\ x > -\dfrac{1}{2}(y + 1)^2 + 2 \\ 3x + 2y < 9 \end{cases}$

Chapter 8 Test

(You should be able to complete this test in 60 min.)

I. Graph the following equations.

1. $\dfrac{y^2}{14} - \dfrac{x^2}{4} = 1$

2. $x = \dfrac{3}{2}(y - 2)^2 - 6$

3. $x^2 + y^2 = 10$

4. $3x^2 + 2y^2 = 18$

5. $x = -y^2 - 8y - 14$

6. $x^2 + y^2 + 12x - 6y + 44 = 0$

7. $2x^2 - 8y^2 - 20y + \dfrac{39}{2} = 0$

8. $5x^2 + 36y^2 - 10x - 360y + 725 = 0$

II. Find the equation of the following circle.

9. Center $= C(-3, -1)$ and passing through $(1, 5)$

III. Graph the following inequality.

10. $x < y^2 + y$

IV. Graph the following system of inequalities.

11. $\begin{cases} \dfrac{x^2}{25} + \dfrac{y^2}{9} > 1 \\ \dfrac{x^2}{4} - \dfrac{y^2}{1} > 1 \end{cases}$

TEST YOUR MEMORY

These problems review Chapters 1–8.

I. Simplify each of the following. Assume that all variables represent positive numbers.

1. $\dfrac{(6^0 x^{-2} y)^4}{(3x^4 y^{-2})^{-3}}$

2. $\dfrac{(125xy^{-4})^{2/3}}{(9x^{-1} y^2)^{3/2}}$

II. Perform the indicated operations.

3. $\dfrac{3ax + 3bx + ay + by}{2ax + 2bx - ay - by} \div \dfrac{9x^2 - y^2}{6x^2 - 5xy + y^2}$

4. $\dfrac{x - 4}{x^2 - 4} - \dfrac{x - 4}{2x^2 - 4x}$

III. Simplify each of the following. Assume that all variables represent positive numbers.

5. $4x\sqrt{12xy} - 6\sqrt{3x^3 y} - \sqrt{27x^3 y}$

6. $\sqrt[3]{32} - 2\sqrt[3]{500}$

7. $\sqrt{\dfrac{5x}{9y^3}}$

8. $\dfrac{\sqrt{2} + \sqrt{3}}{\sqrt{6} - 1}$

IV. Let $f(x) = \dfrac{5}{x - 1}$ and $g(x) = x^3 + 1$. Find the following.

9. $f(3)$

10. $g(-2)$

11. $[g(2)]^2$

12. $g(3a)$

V. Find the equations of the following lines. Write the equations in slope-intercept form, if possible.

13. Through $(2, 6)$ and $(-3, -9)$

14. Through $(0, -2)$ and parallel to $y = 3x + 7$

15. Through $(-1, 3)$ and perpendicular to $2x - 5y = 10$

16. Through $(5, 4)$ and perpendicular to $y = -1$

VI. Find the equations of the following circles.

17. Center $= C(2, -3)$, radius $= 5$

18. Center $= C(-1, -4)$ and passing through $(2, 3)$

VII. Graph the following equations.

19. $2x - y = 6$

20. $y = x^2 + 2x - 2$

21. $x^2 + y^2 = 9$

22. $9y^2 - 4x^2 = 36$

23. $9x^2 + y^2 - 18x + 6y + 9 = 0$

24. $x = 1 - 4y^2$

VIII. Find the solutions of the following equations and check your solutions.

25. $x^2 - 6x + 13 = 0$

26. $4x^2 - 8x + 1 = 0$

27. $\dfrac{4x + 1}{x - 2} = \dfrac{5}{x + 5}$

28. $\dfrac{9x + 2}{2x + 1} = \dfrac{4}{x + 1}$

29. $\dfrac{1}{4}(2x - 1) + \dfrac{1}{3}(x + 4) = 1$

30. $x^4 - 26x^2 + 25 = 0$

31. $|5x - 1| = 13$

32. $\dfrac{1}{2}\left(4x + \dfrac{1}{3}\right) - \dfrac{2}{3}(x + 2) = 1$

33. $x^4 - 20x^2 + 64 = 0$

34. $\sqrt{2x + 2} - \sqrt{x - 3} = 2$

35. $\dfrac{x - 4}{x + 2} + \dfrac{3}{x - 3} = \dfrac{5x}{x^2 - x - 6}$

36. $\dfrac{x + 1}{2x + 1} - \dfrac{2}{x - 1} = \dfrac{2}{2x^2 - x - 1}$

37. $\sqrt{2x + 1} + \sqrt{x + 4} = 3$

38. $|2x - 1| = |3x + 2|$

IX. Find the solutions of the following inequalities and graph each solution set on a number line. Express your answers using interval notation.

39. $2x + 1 \le -3$

40. $x^2 - 2x - 6 < 0$

41. $2x^2 - 5x - 3 > 0$

42. $\dfrac{x - 4}{x - 1} \le 0$

X. Graph the following systems of inequalities.

43. $\begin{cases} 3x + 2y < 6 \\ y < 3 \end{cases}$

44. $\begin{cases} x^2 + y^2 \le 4 \\ x \ge 0 \end{cases}$

XI. Use algebraic expressions to find the solutions of the following problems. For problems that have irrational solutions, use a decimal approximation of the solution.

45. The area of a triangle is 22 sq in. The base is 9 in. more than $\frac{1}{2}$ of the height. Find the base and the height of the triangle.

46. Maria has \$4.40 in nickels and dimes. She has a total of 50 coins. How many nickels and how many dimes does she have?

47. Biff leaves the Organic Food Mart at 1:30 P.M. traveling southward at a rate of 48 mi/hr. At 2:00 P.M. Buffy leaves the Organic Food Mart and follows the same route as Biff, but at a rate of 60 mi/hr. How long will it take Buffy to catch Biff?

48. One pipe can fill a tank 15 min faster than another pipe can. Together they fill the tank in 18 min. How long does it take each pipe to fill the tank?

49. Darren throws a ball upward from the roof of a 16-ft-tall building with a speed of 96 ft/sec. The equation that gives the ball's height above ground level is $h = -16t^2 + 96t + 16$. When does the ball hit the ground?

50. In Moss County the sales tax is directly proportional to the amount purchased. A television set that sells for \$200 has a sales tax of \$16. What is the sales tax rate in Moss County?

Systems of Equations

Tanya, owner of Tanya's T-Shirts, sells customized T-shirts at the Moss County Mall. Tanya buys plain T-shirts from a wholesaler for $5 a shirt. She sells the customized T-shirts for $9 a shirt. Each month Tanya has fixed costs of $1200. Her fixed costs include wages, rent, utilities, and miscellaneous expenses. How many T-shirts must Tanya sell just to break even? (*Note:* When Tanya breaks even, her revenue equals her costs.) How many T-shirts must Tanya sell to generate a monthly profit of $2000?

In this chapter we will study systems of equations. We will see how systems of equations can be used to determine Tanya's break-even point. In addition, we will use the material developed in Chapter 6 to find Tanya's cost function, revenue function, and profit function.

9.1   *Linear Systems in Two Variables*

Two or more equations considered together form a **system of equations.** The equations

$$x + y = 4$$

and

$$2x - y = 5$$

are an example of a system of equations. Each of these equations is a linear equation. (Recall that the graph of a linear equation is a straight line.) Thus this system is called a **linear system** of equations in the two variables x and y.

The **solutions** of a system of equations are the ordered pairs that satisfy *every* equation in the system. One method for determining the solutions of a linear system of two equations is to sketch the graph of each equation in the system (all on the same axes) and find the points of intersection. Since the points of intersection satisfy each equation in the system, the points of intersection identify the solutions. This method is called the **graphing method.**

 XAMPLE 9.1.1

Using the graphing method, determine the solutions of the following linear systems of equations.

1. $x + y = 4$
     $2x - y = 5$

From Figure 9.1.1 the solution to this system appears to be the ordered pair $(3, 1)$. If $(3, 1)$ is the solution, $(3, 1)$ must satisfy both equations.

In the first equation,	In the second equation,
$3 + 1 \overset{?}{=} 4$	$2(3) - 1 \overset{?}{=} 5$
$4 = 4 \ \checkmark$	$5 = 5 \ \checkmark$

Thus, $(3, 1)$ is the solution of this linear system of equations.

REMARK

A linear system of equations that has one intersection point (i.e., one solution) is called **independent.**

2. $x - y = -2$
     $3x - 3y = 5$

The lines in Figure 9.1.2 are parallel; that is, they have no points in common. Therefore the solution set is $\varnothing$.

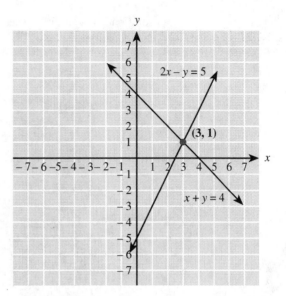

Figure 9.1.1

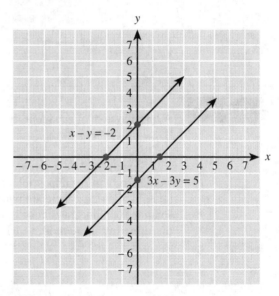

Figure 9.1.2

REMARK   A linear system of equations that has no intersection points (i.e., has no solution) is called **inconsistent.**

3. $x + 2y = 4$
$2x + 4y = 8$

The lines in Figure 9.1.3 coincide; that is, they have the same graph. Any ordered pair that satisfies one equation also satisfies the other equation. Thus the solution set is an infinite set of ordered pairs that satisfy either one of the original equations.

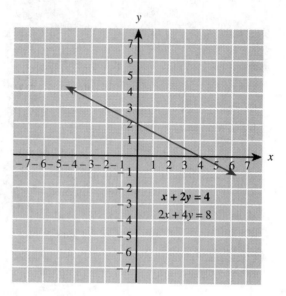

Figure 9.1.3

REMARK

A linear system of equations that has infinitely many solutions is called **dependent.**

Suppose we were given a linear system in which the solution is $\left(\frac{17}{8}, -\frac{9}{13}\right)$. It would be almost impossible to construct a graph that could *exactly locate* $\left(\frac{17}{8}, -\frac{9}{13}\right)$. However, there are two common algebraic methods that precisely find the solutions of a linear system of equations. The first method we will examine is called the **substitution method,** which consists of the following steps:

1. Solve one of the equations for one of the variables.

2. *Substitute* the expression obtained in step 1 in the other equation.

3. You now have an equation with one variable. Solve this equation. (*Note:* At this step only one coordinate of the solution has been obtained.)

4. Substitute the value obtained in step 3 into the equation of step 1, and solve this equation. You now have both coordinates of the solution.

5. Check the solution in both equations of the system.

EXAMPLE 9.1.2

Using the substitution method, determine the solutions of the following linear systems of equations.

1. $5x - 2y = 7$

$$y = x + 3$$

The second equation already has y in terms of x. Substitute $x + 3$ for y in the first equation:

$$5x - 2(x + 3) = 7$$
$$5x - 2x - 6 = 7$$
$$3x - 6 = 7$$
$$3x = 13$$
$$x = \frac{13}{3}$$

Since $y = x + 3$ and $x = \dfrac{13}{3}$,

$$y = \frac{13}{3} + 3$$

$$y = \frac{22}{3}$$

Thus, the solution is $\left(\frac{13}{3}, \frac{22}{3}\right)$. The check is left to the reader.

2. $x + y = 4$

$2x - y = 5$

Let's solve for x in the first equation:

$$x + y = 4$$
$$x = 4 - y$$

Substitute $4 - y$ for x in the second equation:

$$2(4 - y) - y = 5$$
$$8 - 2y - y = 5$$
$$8 - 3y = 5$$
$$-3y = -3$$
$$y = 1$$

Since $x = 4 - y$ and $y = 1$,

$$x = 4 - 1$$
$$x = 3$$

The solution is (3, 1). Note that this is the same system as in Example 9.1.1, part 1.

NOTE ▶   We could just as well have solved the second equation for y and proceeded from that point.

3. $x - y = -2$
$3x - 3y = 5$

Solve for x in the first equation:

$$x - y = -2$$
$$x = y - 2$$

Substitute $y - 2$ for x in the second equation:

$$3(y - 2) - 3y = 5$$
$$3y - 6 - 3y = 5$$
$$-6 = 5 \qquad \textit{A false statement}$$

This is the same system as in Example 9.1.1, part 2. The graphing method showed that this system was inconsistent. This observation leads to the following conclusion: *If all the variables are eliminated and the resulting statement is false, then the given system is inconsistent; that is, the solution set is $\varnothing$.*

4. $x + 2y = 4$
$2x + 4y = 8$

Solve for x in the first equation:

$$x + 2y = 4$$
$$x = 4 - 2y$$

Substitute $4 - 2y$ for x in the second equation:

$$2(4 - 2y) + 4y = 8$$
$$8 - 4y + 4y = 8$$
$$8 = 8 \qquad \textit{A true statement}$$

This is the same system as in Example 9.1.1, part 3. The graphing method showed that this system was *dependent*. This observation leads to another conclusion: *If all the variables are eliminated and the resulting statement is true, then the given system is dependent.*

The second common algebraic method for finding solutions of linear systems of equations is called the addition or **elimination method.** The elimination method makes use of the additive property of equality and consists of the following steps:

1. Eliminate any fractions and express each equation in the form
   $ax + by = c$.

2. Multiply each equation by some nonzero number so that the sum of the
   coefficients of either x or y is zero.

3. Add the equations obtained in step 2.

4. You now have an equation with one variable. Solve this equation.

5. Substitute the value determined in step 4 into one of the equations that
   contain both variables, and solve this equation. You now have both coordinates of the solution.

6. Check the solution in both equations.

EXAMPLE 9.1.3   Using the elimination method, determine the solutions of the following linear systems of equations.

1. $x + y = 4$
     $2x - y = 5$

The sum of the coefficients of the y's is zero. Thus, when we add the equations, the y's will be eliminated:

$$
\begin{array}{ll}
x + y = 4 & \\
\underline{2x - y = 5} & \\
3x \quad\;\; = 9 & \textit{Solve this equation.} \\
x \quad\;\; = 3 &
\end{array}
$$

Don't stop here! We have found only the x-coordinate of the solution. Substitute 3 for x in the first equation:

$$3 + y = 4$$
$$y = 1$$

The solution is $(3, 1)$.

2. $2x - 5y = 1$
    $3x - 4y = 2$

We can choose to eliminate either variable. If we multiply the first equation by -3 and multiply the second equation by 2, the sum of the coefficients of x will be zero:

$$2x - 5y = 1 \xrightarrow[\text{sides by } -3.]{\textit{Multiply both}} -6x + 15y = -3$$

$$3x - 4y = 2 \xrightarrow[\text{sides by } 2.]{\textit{Multiply both}} \underline{6x - 8y = 4}$$

$$7y = 1$$

$$y = \frac{1}{7}$$

Substitute $\frac{1}{7}$ for y in the first equation:

$$2x - 5\left(\frac{1}{7}\right) = 1$$

$$2x - \frac{5}{7} = 1$$

$$2x = \frac{12}{7}$$

$$x = \frac{6}{7}$$

The solution is $\left(\frac{6}{7}, \frac{1}{7}\right)$.

Try to avoid this mistake:

Incorrect	**Correct**
$4x - 3y = 2$	$4x - 3y = 2$
$3x - 5y = -2$	$3x - 5y = -2$
$12x - 9y = 2$	$12x - 9y = 6$
$-12x + 20y = -2$	$-12x + 20y = 8$

Here, we forgot to multiply the right side of the equation.

3. $\quad x - y = -2$

$\quad 3x - 3y = 5$

Multiply the first equation by -3 to eliminate x:

$$x - y = -2 \xrightarrow[\quad]{\substack{\textit{Multiply both} \\ \textit{sides by } -3.}} -3x + 3y = 6$$

$$3x - 3y = 5 \qquad\qquad\qquad \underline{3x - 3y = 5}$$

$$0 = 11$$

$$\textit{A false statement}$$

If all the variables are eliminated and the resulting statement is false, the given system is inconsistent; that is, the solution set is $\varnothing$. (This is the same system as in Example 9.1.1, part 2.)

If all the variables are eliminated and the resulting statement is true, the given system is dependent.

4. $\dfrac{3}{4}x - \dfrac{1}{2}y = \dfrac{3}{4}$

$\dfrac{2}{3}x + \dfrac{1}{9}y = \dfrac{1}{3}$

Multiply the top equation by 4 and the bottom equation by 9 to eliminate all fractions:

$$\begin{array}{lll}
\dfrac{3}{4}x - \dfrac{1}{2}y = \dfrac{3}{4} & \text{\textit{Multiply both}} & 3x - 2y = 3 \\
 & \text{\textit{sides by 4.}} & \\
\dfrac{2}{3}x + \dfrac{1}{9}y = \dfrac{1}{3} & \text{\textit{Multiply both}} & 6x + \ \ y = 3 \\
 & \text{\textit{sides by 9.}} &
\end{array}$$

Multiply the second equation by 2 to eliminate y:

$$\begin{array}{lll}
3x - 2y = 3 & & 3x - 2y = 3 \\
6x + \ \ y = 3 & \text{\textit{Multiply both}} & 12x + 2y = 6 \\
 & \text{\textit{sides by 2.}} & \overline{15x \qquad \ \ = 9} \\
 & & x = \dfrac{9}{15} \\
 & & x = \dfrac{3}{5}
\end{array}$$

To find y, substitute $\frac{3}{5}$ for x in the equation $6x + y = 3$:

$$6\left(\dfrac{3}{5}\right) + y = 3$$

$$\dfrac{18}{5} + y = 3$$

$$y = -\dfrac{3}{5}$$

The solution is $\left(\frac{3}{5}, -\frac{3}{5}\right)$.

We have examined three methods for finding the solutions of linear systems of equations. The graphing method does not necessarily produce exact answers, since graphs have limited accuracy. The substitution method works best when one of the variables has a coefficient of 1. Usually the elimination method is the superior method for finding the solutions of a linear system of equations.

Exercises 9.1

Using the graphing method, determine the solutions of the following linear systems of equations. Identify the system as independent, inconsistent, or dependent.

1. $x + 3y = 11$
$2x - y = -6$

2. $2x + y = 3$
$3x - y = -3$

3. $-2x + 3y = 4$
$x - 4y = -2$

4. $3x - 4y = -1$
$5x + 2y = -19$

5. $x + 4y = 6$
$x + 4y = 0$

6. $2x - 10y = 5$
$3x - 15y = -10$

7. $3x - y = 1$
$-6x + 2y = -2$

8. $2x - 4y = 10$
$3x - 6y = 15$

Using the substitution method, determine the solutions of the following linear systems of equations. Identify the system as independent, inconsistent, or dependent.

9. $2x - 3y = 7$
$y = x - 2$

10. $3x - y = 8$
$y = 2x - 6$

11. $-2x + 5y = 11$
$x = y - 2$

12. $x + 2y = 7$
$y = 3 - 2x$

13. $x - 2y = 5$
$x = 2y + 11$

14. $2x - y = 7$
$y = 2x - 7$

15. $x - y = 11$
$2x + 3y = 5$

16. $2x - y = -1$
$x + 6y = 4$

17. $3x - 5y = 8$
$2x - y = -1$

18. $x + 2y = 5$
$-2x - 4y = 7$

19. $2x - 3y = 4$
$-3x + 4y = -1$

20. $3x + 2y = 0$
$2x - 5y = 2$

21. $2x - 6y = 14$
$5x - 15y = 35$

22. $-5x + 2y = -3$
$-2x + 4y = 1$

Using the elimination method, determine the solutions of the following linear systems of equations. Identify the system as independent, inconsistent, or dependent.

23. $-3x + y = 7$
$3x + 2y = 1$

24. $x - 2y = 9$
$3x + 2y = -2$

25. $x - 5y = -2$
$3x - 5y = 3$

26. $-x + 3y = 7$
$2x + 3y = 0$

27. $2x - 3y = -1$
$x - 2y = 1$

28. $x + 2y = 7$
$2x - 3y = 9$

29. $2x - y = 7$
$6x - 3y = 11$

30. $-x + 2y = 3$
$3x - 6y = -9$

31. $2x + 3y = 7$
$-4x + 2y = 1$

32. $6x + 3y = 4$
$3x - 4y = -3$

33. $-5x - 2y = 6$
$2x - 5y = -2$

34. $3x + 4y = 0$
$-2x + 7y = 1$

35. $2x - 4y = 10$
$-3x + 6y = -15$

36. $10x + 5y = -3$
$6x + 3y = 0$

37. $\dfrac{1}{8}x - \dfrac{3}{4}y = -\dfrac{13}{4}$
$\dfrac{2}{3}x + \dfrac{1}{3}y = 0$

38. $\dfrac{2}{3}x - \dfrac{5}{6}y = \dfrac{20}{3}$
$-\dfrac{3}{2}x + \dfrac{1}{4}y = -8$

39. $\dfrac{1}{3}x - \dfrac{1}{2}y = \dfrac{2}{3}$
$\dfrac{1}{4}x - \dfrac{3}{8}y = \dfrac{1}{2}$

40. $\dfrac{1}{2}x + \dfrac{2}{3}y = -1$
$\dfrac{3}{8}x + \dfrac{1}{2}y = \dfrac{1}{4}$

41. $\dfrac{1}{2}x - \dfrac{1}{8}y = 3$

$\dfrac{2}{3}x + \dfrac{1}{6}y = -\dfrac{1}{2}$

42. $\dfrac{2}{3}x - \dfrac{3}{4}y = \dfrac{5}{6}$

$-\dfrac{1}{6}x - \dfrac{1}{12}y = -\dfrac{3}{4}$

Using the elimination or substitution method, determine the solutions of the following systems of equations. (Hint: Make the substitution $u = \frac{1}{x}$ and $v = \frac{1}{y}$.)

43. $\dfrac{1}{x} + \dfrac{2}{y} = 5$

$\dfrac{1}{x} - \dfrac{2}{y} = 1$

44. $\dfrac{1}{x} - \dfrac{3}{y} = 13$

$\dfrac{2}{x} + \dfrac{3}{y} = -19$

45. $-\dfrac{3}{x} + \dfrac{2}{y} = \dfrac{1}{2}$

$\dfrac{2}{x} - \dfrac{1}{y} = -\dfrac{5}{12}$

46. $\dfrac{1}{x} - \dfrac{5}{y} = -\dfrac{8}{3}$

$\dfrac{4}{x} + \dfrac{2}{y} = -\dfrac{10}{3}$

47. $\dfrac{2}{x} - \dfrac{3}{y} = -2$

$-\dfrac{3}{x} + \dfrac{4}{y} = -5$

48. $\dfrac{5}{x} - \dfrac{2}{y} = -7$

$\dfrac{2}{x} - \dfrac{3}{y} = 6$

Write Algebra

49. What must be true about the solution set of a system of two linear equations if the system is (a) dependent, (b) inconsistent, and (c) independent?

50. In a linear system of two equations, what can you conclude if one equation is a multiple of the other?

The following exercises require the use of a graphing calculator. Use the Trace and Zoom keys of a graphing calculator to approximate the solution (the point of intersection) of each system.

51. $x + y = 4$
$2x - y = 5$

NOTE ▶ This is the same system as Example 9.1.2 (part 2). We found the exact solution to be (3, 1). Can you explain why your first approximation is slightly off? You will have to zoom in several times before you get the exact answer.

52. $y = 3.154x + 7.8$
$2.5x + 3.9y = 6.4$

53. $-3x + y = 7$
$3x + 2y = 1$

54. $y = \sqrt{2}x - 3$
$x = -\sqrt{3}y + 1$

55. $2.3x + 1.5y = -4.8$
$4.7x - 3.1y = -20.2$

9.2 *Linear Systems in Three Variables*

Section 9.1 examined linear systems of two equations in two variables. In this section we will extend the ideas of Section 9.1. However, we must first generalize some of the concepts of Chapter 6.

A *solution* to an equation in three variables, such as $x + 3y - 4z = 8$, is an **ordered triple.** The ordered triple has the property that the substitutions of the first coordinate for x, the second coordinate for y, and the third coordinate for z in the equation yield a true statement. For example, (1, 5, 2) is a solution of the equation $x + 3y - 4z = 8$, since $1 + 3(5) - 4(2) = 8$ is a true statement.

An equation of the form $ax + by + cz = d$, where a, b, c, and d are real numbers with a, b, and c not all equal to zero, is called a **linear equation in three variables.** To graph a linear equation in three variables, a coordinate system with three axes must be constructed. It can be shown that the graph of the linear equation $ax + by + cz = d$, where a, b, c, and d are real numbers not all zero, is a flat surface called a plane.

In the previous section the solutions of a system of equations in two variables were defined to be the ordered pairs that satisfy every equation in the system—that is, the points of intersection. Likewise, in a system of equations in three variables, the solutions are the ordered triples that satisfy every equation in the system, again the points of intersection.

An example of a linear system of three equations in three variables is:

$$2x + y - 3z = -11$$
$$x - 2y + z = -3$$
$$-x + 3y - 4z = -6$$

The graphing method for determining the solutions in three dimensions is impractical. The substitution method is almost always inferior to the elimination method. Therefore, in this section we will use only the elimination method.

We first pick one pair of equations and eliminate one variable. Then we eliminate the same variable from another pair of equations. We now have a linear system of two equations in two variables. This process is illustrated with the system given earlier.

If we multiply the second equation by -2 and add the result to the first equation, the sum of the coefficients of x will be zero.

$$
\begin{array}{l}
2x + y - 3z = -11 \\
x - 2y + z = -3
\end{array}
\xrightarrow[\text{sides by } -2.]{\text{Multiply both}}
\begin{array}{l}
2x + y - 3z = -11 \\
\underline{-2x + 4y - 2z = 6} \\
 5y - 5z = -5
\end{array}
$$

Now we combine the *third* equation with either the first or second equation and eliminate x again. If we simply add the second and third equations, the sum of the coefficients of x will be zero.

$$
\begin{array}{l}
x - 2y + z = -3 \\
\underline{-x + 3y - 4z = -6} \\
 y - 3z = -9
\end{array}
$$

Now let us solve this linear system in two variables.

$$
\begin{array}{l}
5y - 5z = -5 \\
y - 3z = -9
\end{array}
\xrightarrow[\text{sides by } -5.]{\text{Multiply both}}
\begin{array}{l}
5y - 5z = -5 \\
\underline{-5y + 15z = 45} \\
 10z = 40 \\
z = 4
\end{array}
$$

To find y, substitute 4 for z in the equation $y - 3z = -9$.

$$y - 3(4) = -9$$
$$y - 12 = -9$$
$$y = 3$$

To find x, substitute 4 for z and 3 for y in any of the equations of the original system. Choosing the second equation,

$$x - 2y + z = -3$$
$$x - 2(3) + 4 = -3$$
$$x - 6 + 4 = -3$$
$$x - 2 = -3$$
$$x = -1$$

Thus the solution is $(-1, 3, 4)$. The check is left to the reader.

The elimination method for finding the solutions of a linear system of three equations in three variables is summarized here.

1. Express each equation in the form $ax + by + cz = d$.

2. Multiply *any* two of the equations by some nonzero number so that the sum of the coefficients of the same variable, x, y, or z, in those two equations is zero.

3. Add the two equations from step 2. You now have an equation with two variables.

4. Combine multiples of the *remaining* equation and one of the equations used in step 2. Pick multiples so that the variable that was eliminated in step 3 is eliminated here.

5. Add the two equations from step 4. You now have another equation with the same two variables as the result of step 3.

6. The equations from step 3 and step 5 form a linear system in two variables. Use the techniques of the previous section to find the solutions of this system.

7. Substitute the values determined in step 6 into one of the equations that contains the remaining variable, and solve this equation. You now have all three coordinates of the solution.

8. Check the solution in *every* equation in the system.

As previously mentioned, the graph of a linear equation in three variables is a plane. Figure 9.2.1 shows how a linear system of three equations in three variables might appear. Figure 9.2.1(a) shows an *independent system*, in which there is one point common to all three planes. Figure 9.2.1(b) shows *inconsistent systems*, in which there are no points common to all three planes. Figure 9.2.1(c) shows *dependent systems.* On the left there is a line common to all three planes; that is, there are an infinite number of points in common. On the right the planes coincide; again there are an infinite number of points in common.

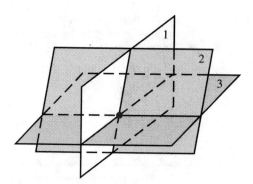

Figure 9.2.1(a)

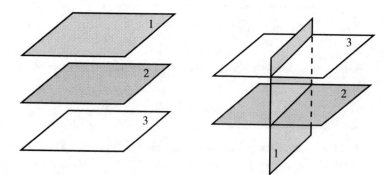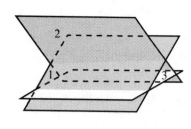

Figure 9.2.1(b)

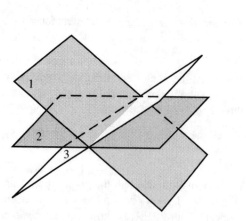

Figure 9.2.1(c)

XAMPLE 9.2.1

Determine whether the following linear systems are independent, inconsistent, or dependent. If the system is independent, find the solution.

1. $$x - 2y + z = -2$$
    $$3x + y + 2z = -2$$
    $$-2x + 3y - z = 0$$

If we multiply the second equation by 2 and add the result to the first equation, the sum of the coefficients of y will be zero:

$$x - 2y + z = -2 \xrightarrow{\hspace{2cm}} x - 2y + z = -2$$
$$3x + y + 2z = -2 \xrightarrow[\text{sides by 2.}]{\text{Multiply both}} 6x + 2y + 4z = -4$$
$$\overline{\hspace{2cm} 7x \hspace{1cm} + 5z = -6}$$

Now we must combine the *third* equation with either the first or second equation and eliminate y. If we multiply the second equation by -3 and add the result to the third equation, the sum of the coefficients of y will be zero:

$$3x + y + 2z = -2 \xrightarrow[\text{sides by} -3.]{\text{Multiply both}} -9x - 3y - 6z = 6$$
$$-2x + 3y - z = 0 \xrightarrow{\hspace{2cm}} -2x + 3y - z = 0$$
$$\overline{\hspace{2cm} -11x \hspace{1cm} - 7z = 6}$$

We now have a linear system in two variables:

$$7x + 5z = -6 \xrightarrow[\text{sides by 11.}]{\text{Multiply both}} 77x + 55z = -66$$
$$-11x - 7z = 6 \xrightarrow[\text{sides by 7.}]{\text{Multiply both}} -77x - 49z = 42$$
$$\overline{\hspace{2cm} 6z = -24}$$
$$z = -4$$

To find x, substitute -4 for z in the equation $7x + 5z = -6$:

$$7x + 5(-4) = -6$$
$$7x - 20 = -6$$
$$7x = 14$$
$$x = 2$$

To find y, substitute -4 for z and 2 for x in any of the equations of the original system. Choosing the first equation, we get

$$x - 2y + z = -2$$
$$2 - 2y + (-4) = -2$$
$$-2y - 2 = -2$$
$$-2y = 0$$
$$y = 0$$

Thus the system is independent and the solution is $(2, 0, -4)$. The check is left to the reader.

2.
$$x - 3y + z = -7$$
$$2x - 6y + 2z = 1$$
$$-3x + y + 4z = 3$$

If we multiply the first equation by -2 and add the result to the second equation, the sum of the coefficients of x will be zero:

$$x - 3y + z = -7 \xrightarrow[\text{sides by } -2.]{\text{Multiply both}} -2x + 6y - 2z = 14$$
$$2x - 6y + 2z = 1 \xrightarrow{\hspace{3cm}} \underline{\hspace{0.3cm} 2x - 6y + 2z = 1}$$
$$0 = 15$$

A false statement, so there is no solution.

Thus we have an inconsistent system.

NOTE ▶ Remember, when the elimination method yields a false statement, the given system is inconsistent.

3.
$$-3x + y + z = 1$$
$$4x - 3y + 2z = 2$$
$$5x - 3y + z = 1$$

If we multiply the first equation by -2 and add the result to the second equation, the sum of the coefficients of z will be zero:

$$-3x + y + z = 1 \xrightarrow[\text{sides by } -2.]{\text{Multiply both}} 6x - 2y - 2z = -2$$
$$4x - 3y + 2z = 2 \xrightarrow{\hspace{3cm}} \underline{\hspace{0.3cm} 4x - 3y + 2z = 2}$$
$$10x - 5y = 0$$

Now we must combine the third equation with either the first or second equation and eliminate z. If we multiply the first equation by -1 and add the result to the third equation, the sum of the coefficients of z will be zero:

$$-3x + y + z = 1 \xrightarrow[\text{sides by } -1.]{\text{Multiply both}} 3x - y - z = -1$$
$$5x - 3y + z = 1 \xrightarrow{\hspace{3cm}} \underline{\hspace{0.3cm} 5x - 3y + z = 1}$$
$$8x - 4y = 0$$

We now have a linear system in two variables:

$$10x - 5y = 0 \xrightarrow[\text{sides by } -4.]{\text{Multiply both}} -40x + 20y = 0$$
$$8x - 4y = 0 \xrightarrow[\text{sides by } 5.]{\text{Multiply both}} \underline{\hspace{0.3cm} 40x - 20y = 0}$$
$$0 = 0$$

A true statement, so there are infinitely many solutions.

Thus, we have a *dependent system*.

NOTE ▶   Remember, when the elimination method yields a true statement, the given system is dependent.

The elimination method can be used to find the solutions of linear systems of more than three equations in more than three variables. See exercises 31 and 32.

E xercises 9.2

Determine whether the following linear systems are independent, inconsistent, or dependent. If the system is independent, find the solution.

1. $2x + y - 3z = -4$
$x - y + 4z = 13$
$-x + 3y + z = 4$

2. $x - 2y - 5z = -34$
$2x + y + 3z = 22$
$x + 3y - z = 11$

3. $5x + y + z = 6$
$-2x + 3y - 4z = -17$
$4x - 2y - z = -7$

4. $-3x + 2y + 4z = 6$
$7x - y + 3z = 23$
$2x + 3y + z = 7$

5. $2x + 3y - z = -12$
$4x - 5y + z = 10$
$-6x + 7y + z = -4$

6. $4x + y + 5z = 16$
$-3x - y + 3z = -5$
$2x + y - 4z = 1$

7. $-2x + 3y + 2z = 16$
$3x - 3y + z = -14$
$3x - 2y - 5z = -16$

8. $-4x + 5y + z = 2$
$-2x - 3y + 3z = 20$
$3x + 2y - 4z = -29$

9. $2x - 3y - 2z = -12$
$3x + 4y + 4z = 20$
$2x + 5y - 2z = 4$

10. $-3x + 4y + 2z = 10$
$5x - 2y + 3z = 15$
$3x - 3y + 2z = 10$

11. $5x + y - z = -2$
$x + 3y + 4z = 8$
$3x - 3y - z = -7$

12. $-2x + 3y + 4z = -17$
$3x - y - z = 11$
$x + 2y - 5z = 6$

13. $x - 2y + 2z = 0$
$-2x + y + 4z = 2$
$7x - 2y - 2z = 0$

14. $2x - 4y + z = 4$
$x + 4y + 5z = 0$
$-x + 2y - 3z = -2$

15. $2x - 4y + 3z = 11$
$x + 3y - 5z = -22$
$2y - z = -3$

16. $3x - 2y - 5z = -11$
$x + 3y = 3$
$2x - y + 3z = 18$

17. $4x - 5y - 4z = 0$
$2x + 3y = -1$
$-y + z = 2$

18. $-3x + 3y - 2z = -4$
$6x - z = 3$
$x + 2y = -3$

19. $2x - y = 0$
$3x + 4z = -15$
$3y - 2z = 0$

20. $-x + 3y = 10$
$4x - 5z = -22$
$-2y + 3z = 10$

21. $3x - 4y + z = -11$
$3y - 4z = -7$
$z = 4$

22. $-2x + 3y + 2z = -22$
$y + 5z = -19$
$z = -3$

23. $2x - 4y + 6z = -2$
$x - 2y + 3z = -1$
$-3x + 6y - 9z = 3$

24. $-8x + 16y - 4z = -28$
$6x - 12y + 3z = 21$
$2x - 4y + z = 7$

25. $x + 5y + 3z = 11$
$-x + 6y + 8z = 22$
$2x + 7y + 3z = 13$

26. $x + y - 3z = -2$
$-3x + 5y + z = -2$
$x - 6y + 4z = 5$

27. $-6x + 3y - 9z = -1$
$2x - y + 3z = 7$
$4x - 2y + 6z = 0$

28. $-x - 4y + 2z = -8$
$2x + 8y - 4z = 9$
$x + 4y - 2z = 3$

29. $x + 2y - 2z = 7$
$9x - 6y + 3z = 7$
$3x - 2y + z = 1$

30. $-2x + y + 2z = 0$
$x + 2y + 3z = 7$
$2x - y - 2z = -4$

31. $x - 2y + z + w = 4$
$2x + y - z + 2w = -1$
$-x + 3y + 2z - w = 5$
$3x - y + 4z - 3w = 3$

32. $2x - y + 4z - w = 1$
$-x + y - 2z + 2w = 5$
$3x + 2y + z - w = 0$
$x - 2y - z - 2w = -4$

Write Algebra

33. Consider a linear system of three equations with three variables x, y, and z. If the variable x is missing from one of the equations, what would be your next step in finding the solution of the system?

34. Describe how you would use the elimination method to find the solution of a linear system of four equations with four variables.

35. Describe the simplest method for solving

$$-2x + 3y + 2z = -22$$
$$y + 5z = -19$$
$$z = -3$$

G R O U P   A C T I V I T Y

36. Working together Abe and Ben can do a job in two days. Ben and Cato can do the job in $2\frac{2}{5}$ days, while Abe and Cato can do the job in $1\frac{5}{7}$ days. Find the number of days it will take Abe to do the job working alone.

37. Working together Alice and Berta can do a job in two days. Berta and Carol can do the job in $2\frac{2}{11}$ days, while Alice and Carol can do the job in $3\frac{3}{7}$ days. Find the number of days it will take Alice to do the job working alone.

38. Working together Ahmad, Barry, and Clark can do a job in $1\frac{1}{11}$ hours. Ahmad and Clark can do the job in $2\frac{2}{5}$ hours, while Barry and Clark can do the job in $1\frac{1}{2}$ hours. Find the number of hours it will take Ahmad to do the job working alone.

39. Working together Ann, Bernice, and Carlota can do a job in $1\frac{5}{19}$ hours. Ann and Carlota can do the job in $1\frac{1}{2}$ hours, while Bernice and Carlota can do the job in $3\frac{3}{7}$ hours. Find the number of hours it will take Ann to do the job working alone.

I N T E R N E T   C O N N E C T I O N

The table on page 542 contains statistics on U.S. foreign trade. These statistics were prepared by the Department of Commerce and can be found on the Internet. The Web address is http://www.clark.net/pub/lschank/highlights-trade.txt. In the following example we will use some of the information in this table to construct a model to predict future levels of the total U.S. trade balance in goods and services.

Continued

XAMPLE

Use the equation $y = a(x - 1987)^2 + b(x - 1987) + c$ to create a quadratic model that predicts the total U.S. trade balance in goods and services for the years after 1987.

SOLUTION

Given the equation

$$y = a(x - 1987)^2 + b(x - 1987) + c$$

we let

$$x = \text{the year (after 1987)}$$

and

$$y = \text{the total trade balance (measured in \$ billion)}$$

Since our model contains three unknowns, a, b, and c, we need to use three data points to determine a, b, and c. Let's choose 1988, 1991, and 1994. Thus the three data points for our model are (1988, −114.8), (1991, −27.9), and (1994, −108.1).

Using the data point (1988, − 114.8) in our model, $y = a(x - 1987)^2 + b(x - 1987) + c$, we obtain

$$-114.8 = a(1988 - 1987)^2 + b(1988 - 1987) + c$$
$$-114.8 = a + b + c \qquad (1)$$

Using the data point (1991, −27.9) we obtain

$$-27.9 = a(1991 - 1987)^2 + b(1991 - 1987) + c$$
$$-27.9 = 16a + 4b + c \qquad (2)$$

Using the data point (1994, −108.1) we obtain

$$-108.1 = a(1994 - 1987)^2 + b(1994 - 1987) + c$$
$$-108.1 = 49a + 7b + c \qquad (3)$$

We now have a linear system of equations

$$a + b + c = -114.8$$
$$16a + 4b + c = -27.9$$
$$49a + 7b + c = -108.1$$

Continued

From equation (1),

$$a = -114.8 - b - c$$

We now "back" substitute this value for a into equations (2) and (3). Substituting into equation (2) we obtain

$$16(-114.8 - b - c) + 4b + c = -27.9$$
$$-1836.8 - 16b - 16c + 4b + c = -27.9$$

or

$$-12b - 15c = 1808.9$$

Likewise, substituting into equation (3) we obtain

$$49(-114.8 - b - c) + 7b + c = -108.1$$
$$-5625.2 - 49b - 49c + 7b + c = -108.1$$
$$-42b - 48c = 5517.1$$

We now have the "reduced" system

$$-12b - 15c = 1808.9$$
$$-42b - 48c = 5517.1$$

Sparing the reader the details, it can be shown that $a \doteq -9.3$, $b \doteq 75.3$, and $c \doteq -180.8$. Hence, our model is given by the equation

$$y = -9.3(x - 1987)^2 + 75.3(x - 1987) - 180.8$$

Using this model let's predict the U.S. balance of trade for the year 2001.

$$y = -9.3(2001 - 1987)^2 + 75.3(2001 - 1987) - 180.8$$
$$= -9.3(196) + 75.3(14) - 180.8$$
$$= -949.4$$

Thus, *from our model* the predicted trade deficit for the year 2001 will be $949.4 billion. It is hoped that trends will change, and our model will not be very accurate.

Continued

Exercises

1. Construct a model using the equation

$$y = a(x - 1987)^2 + b(x - 1987) + c$$

   to predict the total U.S. trade balance in goods and services for the years after 1987.

2. Construct a model using the equation

$$y = a(x - 1987)^2 + b(x - 1987) + c$$

   to predict the U.S. trade balance in goods only for the years after 1987.

3. Using your models from exercises 1 and 2 predict the total trade balance, and the trade balance in goods only for the year 2001.

Group Activity

4. Working in groups use different data points to generate other quadratic models. Try to determine which models yield more accurate predictions for particular years.

5. Discuss the reasons why a particular model might yield more accurate results for a given year than another model.

Further Exploration

Surf the Internet to see if you can obtain any data on the U.S. balance of trade for the years after 1994. If you are successful, test the accuracy of your models from exercises 1 and 2.

Continued

Table 1. U.S. Trade in Goods and Services ($ Billion)

Year	EXPORTS			IMPORTS			TRADE BALANCE		
	Total	Goods	Services	Total	Goods	Services	Total	Goods	Services
1960	25.9	19.7	6.3	22.4	14.8	7.7	3.5	4.9	−1.4
1961	26.4	20.1	6.3	22.2	14.5	7.7	4.2	5.6	−1.4
1962	27.7	20.8	6.9	24.4	16.3	8.1	3.4	4.5	−1.2
1963	29.6	22.3	7.3	25.4	17.0	8.4	4.2	5.2	−1.0
1964	33.3	25.5	7.8	27.3	18.7	8.6	6.0	6.8	−0.8
1965	35.3	26.5	8.8	30.6	21.5	9.1	4.7	5.0	−0.3
1966	38.9	29.3	9.6	36.0	25.5	10.5	2.9	3.8	−0.9
1967	41.3	30.7	10.7	38.7	26.9	11.9	2.6	3.8	−1.2
1968	45.5	33.6	11.9	45.3	33.0	12.3	0.3	0.6	−0.4
1969	49.2	36.4	12.8	49.1	35.8	13.3	0.1	0.6	−0.5
1970	56.6	42.5	14.2	54.4	39.9	14.5	2.3	2.6	−0.3
1971	59.7	43.3	16.4	61.0	45.6	15.4	−1.3	−2.3	1.0
1972	67.2	49.4	17.8	72.7	55.8	16.9	−5.4	−6.4	1.0
1973	91.2	71.4	19.8	89.3	70.5	18.8	1.9	0.9	1.0
1974	120.9	98.3	22.6	125.2	103.8	21.4	−4.3	−5.5	1.2
1975	132.6	107.1	25.5	120.2	98.2	22.0	12.4	8.9	3.5
1976	142.7	114.7	28.0	148.8	124.2	24.6	−6.1	−9.5	3.4
1977	152.3	120.8	31.5	179.5	151.9	27.6	−27.2	−31.1	3.8
1978	178.4	142.1	36.4	208.2	176.0	32.2	−29.8	−33.9	4.2
1979	224.1	184.4	39.7	248.7	212.0	36.7	−24.6	−27.6	3.0
1980	271.8	224.3	47.6	291.2	249.8	41.5	−19.4	−25.5	6.1
1981	294.4	237.0	57.4	310.6	265.1	45.5	−16.2	−28.0	11.9
1982	275.2	211.2	64.1	299.4	247.6	51.7	−24.2	−36.5	12.3
1983	266.0	201.8	64.2	323.8	268.9	54.9	−57.8	−67.1	9.3
1984	290.9	219.9	71.0	400.1	332.4	67.7	−109.2	−112.5	3.3
1985	288.8	215.9	72.9	410.9	338.1	72.8	−122.1	−122.2	0.1
1986	309.5	223.3	86.1	448.3	368.4	79.8	−138.8	−145.1	6.3
1987	348.0	250.2	97.8	500.0	409.8	90.2	−152.0	−159.6	7.6
1988	430.2	320.2	110.0	545.0	447.2	97.9	−114.8	−127.0	12.1
1989	489.0	362.1	126.8	579.3	477.4	101.9	−90.3	−115.2	24.9
1990	537.6	389.3	148.3	616.0	498.3	117.7	−78.4	−109.0	30.7
1991	581.2	416.9	164.3	609.1	490.7	118.4	−27.9	−73.8	45.9
1992	616.9	440.4	176.6	657.3	536.5	120.9	−40.4	−96.1	55.7
1993	641.7	456.9	184.8	717.4	589.4	128.0	−75.7	−132.6	56.9
1994	696.4	502.8	193.6	804.5	669.1	135.4	−108.1	−166.3	58.2

SOURCE: The data were downloaded and modified from the Department of Commerce's U.S. Global Trade Outlook CD-ROM.

NOTE: Balance of payments basis for goods reflects adjustments for timing, coverage, and valuation to the data compiled by the Census Bureau. The major adjustments concern military trade of U.S. defense agencies, additional nonmonetary gold transactions, and inland freight in Canada and Mexico.

Goods valuation: f.a.s. for exports and customs value for imports.

Data reflect all revisions through June 1994.

9.3 *Applications*

Having discussed techniques for finding the solutions of linear systems of equations, we are now prepared to examine application problems where linear systems of equations can be used. Consider the following examples.

XAMPLE 9.3.1   The sum of two numbers is 18. The larger number is 9 more than twice the smaller number. What are the numbers?

SOLUTION   Let

$$x = \text{larger number}$$
$$y = \text{smaller number}$$

Using the information given in the problem, we obtain the linear system of equations

$$x + y = 18$$
$$x = 2y + 9$$

Let's find the solution of this system using the substitution method. Since the second equation has x in terms of y, substitute $2y + 9$ for x into the first equation:

$$2y + 9 + y = 18$$
$$3y + 9 = 18$$
$$3y = 9$$
$$y = 3$$

Since $x = 2y + 9$ and $y = 3$,

$$x = 2(3) + 9$$
$$x = 6 + 9$$
$$x = 15$$

The two numbers are 15 and 3.

XAMPLE 9.3.2   Jerry has $3.20 in nickels and dimes. He has a total of 44 coins. How many nickels and how many dimes does Jerry have?

SOLUTION   Let

$$x = \text{number of nickels}$$
$$y = \text{number of dimes}$$

So

$$5x = \text{value of the nickels in cents}$$
$$10y = \text{value of the dimes in cents}$$

The following chart organizes the information given in the problem:

Coins	Nickels	Dimes	Total
Number	x	y	44
Value	$5x$	$10y$	320

Using the table, we obtain the linear system of equations:

$$x + y = 44$$
$$5x + 10y = 320$$

Using the elimination method, we get

$$x + y = 44 \xrightarrow[\text{sides by } -5.]{\text{Multiply both}} -5x - 5y = -220$$
$$5x + 10y = 320 \longrightarrow \underline{5x + 10y = 320}$$
$$5y = 100$$
$$y = 20$$

Substitute 20 for y in the first equation:

$$x + 20 = 44$$
$$x = 24$$

Jerry has 24 nickels and 20 dimes. The check is left to the reader.

XAMPLE 9.3.3   Charlie's Chocolate City has two types of chocolate: Beggar's Delight, which sells for $2 a pound, and Millionaire's Dream, which sells for $12 a pound. How many

pounds of each type of chocolate must Charlie mix to create 50 lb of Worker's Fantasy, which will sell for $5 a pound?

SOLUTION Let

$$x = \text{number of pounds of Beggar's Delight in the mix}$$
$$y = \text{number of pounds of Millionaire's Dream in the mix}$$

So

$$2x = \text{cost of Beggar's Delight}$$
$$12y = \text{cost of Millionaire's Dream}$$

The following chart organizes the information given in the problem:

Chocolate	Beggar's Delight	Millionaire's Dream	Worker's Fantasy
Amount	x	y	50
Cost	$2x$	$12y$	$5 \cdot 50 = 250$

Using the table, we obtain the linear system of equations:

$$x + y = 50$$
$$2x + 12y = 250$$

Using the elimination method, we get

$$x + y = 50 \xrightarrow[\text{sides by } -2.]{\textit{Multiply both}} -2x - 2y = -100$$
$$2x + 12y = 250 \xrightarrow{} \underline{2x + 12y = 250}$$
$$10y = 150$$
$$y = 15$$

Substitute 15 for y into the first equation:

$$x + 15 = 50$$
$$x = 35$$

Therefore, Charlie uses 35 lb of Beggar's Delight and 15 lb of Millionaire's Dream to create 50 lb of Worker's Fantasy.

EXAMPLE 9.3.4   Ronda invested $9000 at Moss County Savings and Loan. She invested part of her money in an account paying 12% simple interest, and she invested the rest of her money in an account paying 8% simple interest. After 1 yr her interest income was $920. How much did Ronda invest in each account?

SOLUTION

Let

$$x = \text{amount invested at } 12\%$$
$$y = \text{amount invested at } 8\%$$

So

$$0.12x = \text{interest earned in the account paying } 12\%$$
$$0.08y = \text{interest earned in the account paying } 8\%$$

The following chart organizes the information given in the problem:

Account	12% Account	8% Account	Total
Amount invested	x	y	9000
Interest earned	$0.12x$	$0.08y$	920

Using the table, we obtain the linear system of equations:

$$x + y = 9000$$
$$0.12x + 0.08y = 920$$

First multiply both sides of the second equation by 100 to clear the decimals:

$$x + y = 9000 \longrightarrow x + y = 9000$$
$$0.12x + 0.08y = 920 \xrightarrow{\text{Multiply both sides by 100.}} 12x + 8y = 92{,}000$$

$$x + y = 9000 \xrightarrow{\text{Multiply both sides by } -8.} -8x - 8y = -72{,}000$$
$$12x + 8y = 92{,}000 \longrightarrow \underline{12x + 8y = 92{,}000}$$
$$4x = 20{,}000$$
$$x = 5000$$

Substitute 5000 for x in the first equation:

$$5000 + y = 9000$$
$$y = 4000$$

Ronda invested $5000 in the 12% account and $4000 in the 8% account.

XAMPLE 9.3.5

Mrs. Walton bought 2 small, 1 medium, and 1 large orange drink at the Sunshine Orange Palace for $1.75. Later that day she bought 1 small, 1 medium, and 2 large orange drinks for $2.00. Finally, just before closing time, she purchased 1 small, 2 medium, and 3 large orange drinks for $3.05. What is the cost of each small, medium, and large orange drink?

SOLUTION   Let

$$x = \text{cost of 1 small orange drink}$$
$$y = \text{cost of 1 medium orange drink}$$
$$z = \text{cost of 1 large orange drink}$$

Using the information given in the problem, we obtain the linear system of equations:

$$2x + y + z = 1.75$$
$$x + y + 2z = 2.00$$
$$x + 2y + 3z = 3.05$$

Eliminating y in the first two equations yields:

$$2x + y + z = 1.75 \longrightarrow 2x + y + z = 1.75$$
$$x + y + 2z = 2.00 \xrightarrow{\text{Multiply both sides by } -1.} -x - y - 2z = -2.00$$
$$\overline{x - z = -0.25}$$

Eliminating y in the last two equations yields:

$$x + y + 2z = 2.00 \xrightarrow{\text{Multiply both sides by } -2.} -2x - 2y - 4z = -4.00$$
$$x + 2y + 3z = 3.05 \longrightarrow x + 2y + 3z = 3.05$$
$$\overline{-x - z = -0.95}$$

Adding the two equations in x and z we obtain:

$$
\begin{aligned}
x - z &= -0.25 \\
\underline{-x - z} &= \underline{-0.95} \\
-2z &= -1.20 \\
z &= 0.60
\end{aligned}
$$

Substitute 0.60 for z in the equation $x - z = -0.25$.

$$
\begin{aligned}
x - 0.60 &= -0.25 \\
x &= 0.35
\end{aligned}
$$

Substitute 0.60 for z and 0.35 for x in the equation $2x + y + z = 1.75$.

$$
\begin{aligned}
2(0.35) + y + 0.60 &= 1.75 \\
0.70 + y + 0.60 &= 1.75 \\
y + 1.30 &= 1.75 \\
y &= 0.45
\end{aligned}
$$

Thus, each small drink costs $0.35, each medium drink costs $0.45, and each large drink costs $0.60.

NOTE ▶   In solving application problems of this type, you need as many equations as un-knowns in order to get a unique solution.

Exercises 9.3

1. The sum of two numbers is 25. The larger number is 1 more than 5 times the smaller number. What are the numbers?

2. The sum of two numbers is 26. The larger number is 5 more than twice the smaller number. What are the numbers?

3. The sum of two numbers is 6, and their difference is 10. What are the numbers?

4. The sum of two numbers is 17, and their difference is 25. What are the numbers?

5. The difference of two numbers is 10. Twice the larger number minus the smaller number is 14. What are the numbers?

6. The difference of two numbers is 12. Twice the larger number plus the smaller number is 18. What are the numbers?

7. Maggie has $2.80 in nickels and dimes. She has a total of 40 coins. How many nickels and how many dimes does Maggie have?

8. Jan has $5.90 in dimes and quarters. She has a total of 32 coins. How many dimes and how many quarters does Jan have?

9. Sue spent $5.28 on 20¢ and 32¢ stamps. She bought a total of 21 stamps. How many 20¢ stamps and how many 32¢ stamps did she buy?

10. Elaine spent $5.96 on 12¢ and 26¢ stamps. She bought a total of 31 stamps. How many 12¢ stamps and how many 26¢ stamps did she buy?

11. Slim sold 37 tickets to the Moss County Rodeo. Adult tickets cost $3.50 and children's tickets cost $1.25. He collected $73.25 from the sale of the tickets. How many adult tickets and how many children's tickets did he sell?

12. Chauncey sold 35 tickets to the Moss County Opera's production of *The Barber of Seville*. Adult tickets cost $5.50 and children's tickets cost $2.75. He collected $162.25 from the sale of the tickets. How many adult tickets and how many children's tickets did he sell?

13. Irene has $4.50 in nickels, dimes, and quarters. She has a total of 34 coins, and she has 2 more nickels than quarters. How many nickels, dimes, and quarters does she have?

14. Maude has $5.80 in nickels, dimes, and quarters. She has a total of 39 coins, and she has 3 more quarters than dimes. How many nickels, dimes, and quarters does she have?

15. Sonoki bought 4 pencils and 1 eraser for $0.62 at the Moss County Drugstore. The next week she bought 3 pencils and 7 erasers for $0.84. What is the cost of each pencil and eraser?

16. Otis bought 3 double-scoop and 2 single-scoop ice cream cones at the Ice Cream Palace for $3.95. Later that day he bought 4 double-scoop and 3 single-scoop ice cream cones for $5.45. What is the cost of each kind of ice cream cone?

17. Coach Reeves buys 3 basketballs and 2 footballs for $60 at Athletic World. Coach Johnson buys 2 basketballs and 5 footballs for $73. What is the cost of each football and basketball?

18. Sheila bought 2 cans of paint and 3 paintbrushes for $53 at the Moss County Hardware Store. Edith bought 3 cans of paint and 1 paintbrush for $55. What is the cost of each can of paint and each paintbrush?

19. Larry bought 1 saw, 1 hammer, and 1 screwdriver for $36 at the Moss County Hardware Store. Moe bought 2 saws, 2 hammers, and 3 screwdrivers for $76. Curley bought 1 saw, 2 hammers, and 11 screwdrivers for $86. What is the cost of each saw, hammer, and screwdriver?

20. Andy bought 5 roses, 3 tulips, and 1 daisy for $15.25 at Floral World. Omar bought 2 roses, 2 tulips, and 1 daisy for $7.75. Bob bought 1 rose, 1 tulip, and 6 daisies for $8.00. What is the cost of each kind of flower?

21. Jimmy's Peanut Palace has two types of peanut butter: Mr. Peanut, which sells for $2 a pound, and King Peanut, which sells for $7 a pound. How many pounds of each type of peanut butter must Jimmy mix to create 20 lb of Sir Peanut, which will sell for $4 a pound?

22. Colombian coffee beans are selling for $2.50 a pound, and Moss County coffee beans are selling for $0.75 a pound on the New York Coffee Exchange. How many pounds of each type of coffee bean must be purchased to create 28 lb of coffee that will sell for $1.50 per pound?

23. Patti invested $7000 at her local savings and loan. She invested part of her money in an account paying 9% simple interest, and she invested the rest of her money in an account paying 11% simple interest. After 1 yr her interest income was $750. How much did Patti invest in each account?

24. Barbara invested $15,000 at her local savings and loan. She invested part of her money in an account paying 7% simple interest, and she invested the rest of her money in an account paying 12% simple interest. After 1 yr her interest income was $1500. How much did Barbara invest in each account?

25. Lone Star Antifreeze contains 74% ethylene glycol. Big Time Oil Antifreeze contains 58% ethylene glycol. How many gallons of each antifreeze must be used to make 10 gal of antifreeze that is 68% ethylene glycol?

26. Marvin's Glass Cleaner contains 64% ammonia. Betty's Glass Cleaner contains 48% ammonia. How many ounces of each glass cleaner must be used to make 40 oz of glass cleaner that is 58% ammonia?

27. Determine the values of a, b, and c so that the graph of the function

$$y = ax^2 + bx + c$$

contains the points $(1, 5)$, $(-2, 14)$, and $(-1, 7)$.

28. Determine the values of a, b, and c so that the graph of the function

$$y = ax^2 + bx + c$$

contains the points $(4, -2)$, $(2, 0)$, and $(-1, -12)$.

29. The sum of the digits of a two-digit number is 9. If the digits are reversed, the new number is 9 larger than the original number. What is the original number? (*Hint:* Let t = tens digit of the original number and u = units digit of the original number; then $10t + u$ = original number.)

30. The sum of the digits of a two-digit number is 11. If the digits are reversed, the new number is 63 larger than the original number. What is the original number? (Use the hint in exercise 29.)

31. Find the equation of the circle passing through the points $(-3, -1)$, $(-1, 3)$, and $(5, -5)$. *Hint:* These ordered pairs must satisfy the equation

$$x^2 + y^2 + Dx + Ey + F = 0$$

G R O U P   A C T I V I T Y

In exercises 32–35, use a system of equations to solve each of the following problems.

32. The area of a rectangle remains unchanged when its length is decreased by 4 ft and its width is increased by 2 ft. Likewise, its area remains unchanged when its length is increased by 4 ft and its width is decreased by 1 ft. What are the dimensions of the original rectangle?

33. The area of a rectangle remains unchanged when its length is increased by 9 ft and its width is decreased by 4 ft. Likewise, its area remains unchanged when its length is increased by 3 ft and its width is decreased by 2 ft. What are the dimensions of the original rectangle?

34. At the normal current rate, Sue's motorboat can travel 10 miles downstream in the same time that it can travel 6 miles upstream. If the speed of the current doubles, the 10-mile trip downstream takes only $\frac{5}{6}$ of an hour. What is the speed of Sue's boat in still water? What is the normal rate of the current?

35. At the normal current rate, Thieu's motorboat can travel 14 miles downstream in the same time that it can travel 6 miles upstream. If the speed of the current doubles, the 14-mile trip downstream takes only $\frac{7}{9}$ of an hour. What is the speed of Thieu's boat in still water? What is the normal rate of the current?

9.4 *Nonlinear Systems of Equations*

A system of equations that contains at least one equation that is not linear is called a **nonlinear system of equations.** The graphing, substitution, and elimination methods were used in Section 9.1 to find the solutions of linear systems of equations. In this section we will learn how to apply these methods to find the solutions of nonlinear systems of equations.

EXAMPLE 9.4.1   Determine the real solutions of the following systems of equations.

1. $x^2 + y^2 = 25$
   $x - 2y = -5$

Solve for x in the second equation:

$$x - 2y = -5$$
$$x = 2y - 5$$

Substitute $2y - 5$ for x in the first equation:

$$(2y - 5)^2 + y^2 = 25$$
$$4y^2 - 20y + 25 + y^2 = 25$$
$$5y^2 - 20y + 25 = 25$$
$$5y^2 - 20y = 0$$
$$y^2 - 4y = 0$$
$$y(y - 4) = 0$$
$$y = 0 \quad \text{or} \quad y - 4 = 0$$
$$y = 4$$

Since $x = 2y - 5$, when $y = 0$, $x = 2 \cdot 0 - 5 = -5$, and when $y = 4$, $x = 2 \cdot 4 - 5 = 3$. Thus the solution set of the system is $\{(-5, 0), (3, 4)\}$. Figure 9.4.1 illustrates the graph of this system.

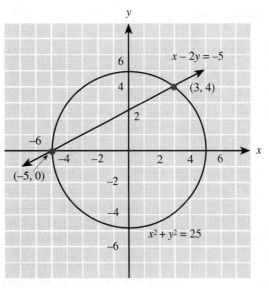

Figure 9.4.1

REMARK

When a nonlinear system of two equations in two variables contains one linear equation, use the substitution method and the following steps to find the solutions of the system:

1. Solve the *linear equation* for one of the variables and *substitute* that value into the nonlinear equation.

2. You will now have an equation with one variable. Solve this equation.

3. Substitute the value determined in step 2 into the linear equation. You now have both coordinates of the solutions.

4. Check the solutions in both equations.

2. $x + 3y = 7$

$\qquad xy = 2$

Since the first equation is linear, we will solve for one of the variables in the first equation. Solve for x:

$$x + 3y = 7$$
$$x = -3y + 7$$

Substitute $-3y + 7$ for x in the second equation:

$$(-3y + 7)y = 2$$
$$-3y^2 + 7y = 2$$
$$0 = 3y^2 - 7y + 2$$
$$0 = (3y - 1)(y - 2)$$
$$3y - 1 = 0 \quad \text{or} \quad y - 2 = 0$$
$$3y = 1 \qquad\qquad y = 2$$
$$y = \frac{1}{3}$$

Since $x = -3y + 7$, when $y = \frac{1}{3}$, $x = -3(\frac{1}{3}) + 7 = 6$, and when $y = 2$, $x = -3(2) + 7 = 1$. Thus the solution set of the system is $\{(6, \frac{1}{3}),(1, 2)\}$. Figure 9.4.2 illustrates the graph of this system.

3. $x^2 - y = 0$

$2x \ - y = 5$

Since the second equation is linear, we will solve for one of the variables in the second equation. Solve for y:

$$2x - y = 5$$
$$-y = -2x + 5$$
$$y = 2x - 5$$

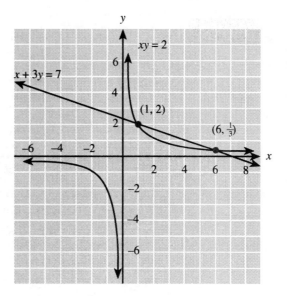

Figure 9.4.2

Substitute $2x - 5$ for y in the first equation:

$$x^2 - (2x - 5) = 0$$
$$x^2 - 2x + 5 = 0$$

Using the quadratic formula to solve for x, we obtain

$$x = \frac{-(-2) \pm \sqrt{(-2)^2 - 4 \cdot 1 \cdot 5}}{2 \cdot 1}$$
$$= \frac{2 \pm \sqrt{4 - 20}}{2}$$
$$= \frac{2 \pm \sqrt{-16}}{2}$$
$$= \frac{2 \pm 4i}{2} = 1 \pm 2i$$

Thus, there are no real solutions to the system. Figure 9.4.3 illustrates the graph of this system.

REMARK

In Section 9.1 we saw that the graphing method had limited precision in locating ordered pairs. This problem shows that the graphing method cannot find nonreal solutions.

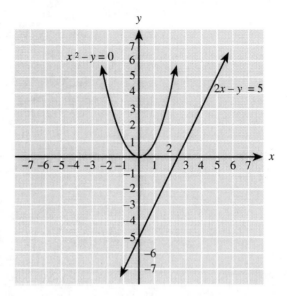

Figure 9.4.3

XAMPLE 9.4.2   Determine the real solutions of the following systems of equations.

1. $9x^2 + 25y^2 = 100$
    $x^2 + y^2 = 9$

Since both equations contain only x^2 and y^2 terms, we can use the elimination method to find the solutions of this system.

$$9x^2 + 25y^2 = 100 \longrightarrow 9x^2 + 25y^2 = 100$$

$$x^2 + y^2 = 9 \quad \xrightarrow[\text{sides by } -9.]{\textit{Multiply both}} \quad \underline{-9x^2 - 9y^2 = -81}$$

$$16y^2 = 19$$

$$y^2 = \frac{19}{16}$$

$$y = \pm\sqrt{\frac{19}{16}}$$

$$y = \pm\frac{\sqrt{19}}{4}$$

Substitute both $\dfrac{\sqrt{19}}{4}$ and $-\dfrac{\sqrt{19}}{4}$ for y into either equation. Let's use the equation $x^2 + y^2 = 9$. When $y = \dfrac{\sqrt{19}}{4}$, $x^2 + y^2 = 9$ becomes

$$x^2 + \left(\frac{\sqrt{19}}{4}\right)^2 = 9$$

$$x^2 + \frac{19}{16} = 9$$

$$x^2 = \frac{125}{16}$$

$$x = \pm\sqrt{\frac{125}{16}}$$

$$x = \pm\frac{5\sqrt{5}}{4}$$

So $\left(\dfrac{5\sqrt{5}}{4}, \dfrac{\sqrt{19}}{4}\right)$ and $\left(-\dfrac{5\sqrt{5}}{4}, \dfrac{\sqrt{19}}{4}\right)$ are solutions. Similarly, when $y = -\dfrac{\sqrt{19}}{4}$, $x^2 + y^2 = 9$ becomes

$$x^2 + \left(\frac{-\sqrt{19}}{4}\right)^2 = 9$$

$$x^2 + \frac{19}{16} = 9$$

$$x^2 = \frac{125}{16}$$

$$x = \pm\sqrt{\frac{125}{16}}$$

$$x = \pm\frac{5\sqrt{5}}{4}$$

So $\left(\dfrac{5\sqrt{5}}{4}, -\dfrac{\sqrt{19}}{4}\right)$ and $\left(-\dfrac{5\sqrt{5}}{4}, -\dfrac{\sqrt{19}}{4}\right)$ are solutions. Thus the solution set is $\left\{\left(\dfrac{5\sqrt{5}}{4}, \dfrac{\sqrt{19}}{4}\right), \left(-\dfrac{5\sqrt{5}}{4}, \dfrac{\sqrt{19}}{4}\right), \left(\dfrac{5\sqrt{5}}{4}, -\dfrac{\sqrt{19}}{4}\right), \left(-\dfrac{5\sqrt{5}}{4}, -\dfrac{\sqrt{19}}{4}\right)\right\}$.

Figure 9.4.4 illustrates the graph of this system.

REMARK   If both equations in the system contain only second-degree terms, the elimination method is *usually* the superior method for finding the solutions of the system.

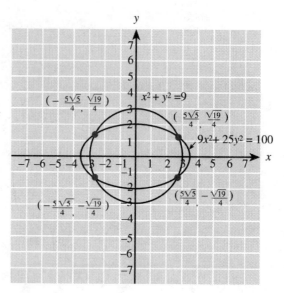

Figure 9.4.4

2. $x^2 - y^2 = 1$
   $x^2 + y\ \ = 7$

   Using the elimination method, we have

$$x^2 - y^2 = 1 \xrightarrow[\text{sides by } -1.]{\textit{Multiply both}} -x^2 + y^2 = -1$$

$$x^2 + y\ = 7 \longrightarrow \underline{\quad x^2 + y\ =\ \ \ 7\quad}$$

$$y^2 + y = 6$$
$$y^2 + y - 6 = 0$$
$$(y + 3)(y - 2) = 0$$
$$y + 3 = 0 \quad \text{or} \quad y - 2 = 0$$
$$y = -3 \qquad\qquad y = 2$$

Substitute -3 for y into either equation. Let's use the equation $x^2 + y = 7$:

$$x^2 + (-3) = 7$$
$$x^2 = 10$$
$$x = \pm\sqrt{10}$$

So $(\sqrt{10}, -3)$ and $(-\sqrt{10}, -3)$ are solutions. Similarly, when $y = 2$, $x^2 + y = 7$ becomes

$$x^2 + 2 = 7$$
$$x^2 = 5$$
$$x = \pm\sqrt{5}$$

So $(\sqrt{5}, 2)$ and $(-\sqrt{5}, 2)$ are solutions. Thus the solution set is $\{(\sqrt{10}, -3), (-\sqrt{10}, -3), (\sqrt{5}, 2), (-\sqrt{5}, 2)\}$. Figure 9.4.5 illustrates the graph of this system.

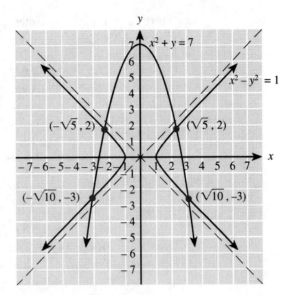

Figure 9.4.5

To find the solutions of the next example, we will use both the elimination and the substitution methods.

XAMPLE 9.4.3   Determine the real solutions of the following system of equations:

$$x^2 + y^2 = 13$$
$$x^2 + 4xy + y^2 = 37$$

Using the elimination method, we obtain

$$x^2 + y^2 = 13 \xrightarrow[\text{sides by } -1.]{\text{Multiply both}} -x^2 \qquad - y^2 = -13$$
$$x^2 + 4xy + y^2 = 37 \xrightarrow{\hspace{2cm}} \underline{x^2 + 4xy + y^2 = \quad 37}$$
$$4xy \quad = \quad 24$$

Now solve the equation $4xy = 24$ for x or y. Solving for x, we have

$$4xy = 24$$
$$x = \frac{24}{4y}$$
$$x = \frac{6}{y}$$

Substitute $\frac{6}{y}$ for x into either equation of the system. Let's use the equation $x^2 + y^2 = 13$:

$$\left(\frac{6}{y}\right)^2 + y^2 = 13$$

$$\frac{36}{y^2} + y^2 = 13$$

$$\frac{y^2}{1}\left(\frac{36}{y^2} + y^2\right) = 13 \cdot y^2 \qquad \textit{Multiply by the LCD, } y^2.$$

$$36 + y^4 = 13y^2$$

$$y^4 - 13y^2 + 36 = 0$$

$$(y^2 - 9)(y^2 - 4) = 0$$

$$(y + 3)(y - 3)(y + 2)(y - 2) = 0$$

$$\begin{array}{cccc} y + 3 = 0 & y - 3 = 0 & y + 2 = 0 & y - 2 = 0 \\ y = -3 & y = 3 & y = -2 & y = 2 \end{array}$$

To find x, substitute these values of y into the equation $x = \frac{6}{y}$.

- When $y = -3$, $x = \frac{6}{-3} = -2$.
- When $y = 3$, $x = \frac{6}{3} = 2$.
- When $y = -2$, $x = \frac{6}{-2} = -3$.
- When $y = 2$, $x = \frac{6}{2} = 3$.

Thus, the solution set is $\{(-2, -3), (2, 3), (-3, -2), (3, 2)\}$. Figure 9.4.6 illustrates the graph of this system.

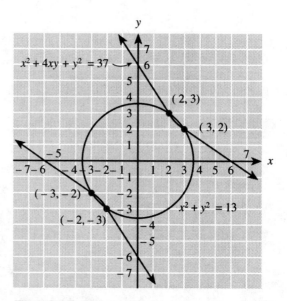

Figure 9.4.6

To graph the equation $x^2 + 4xy + y^2 = 37$ requires methods beyond the scope of this text.

E xercises 9.4

Determine the real solutions of the following systems of equations.

1. $x^2 + y^2 = 4$
$x + y = 2$

2. $x^2 + y^2 = 10$
$2x - y = -5$

3. $x^2 + y^2 = 4$
$x - 2y = -2$

4. $x^2 + y^2 = 1$
$x - y = -1$

5. $x^2 - y = 0$
$4x - y = 4$

6. $x^2 - y = -1$
$-x + y = 3$

7. $x^2 - y = 1$
$4x - y = 6$

8. $x^2 + y = 3$
$3x - 2y = -4$

9. $-x + y^2 = 0$
$x - 4y = -4$

10. $2x + y = 3$
$x - y^2 = 1$

11. $2x + y = -6$
$xy = 4$

12. $3x + y = 6$
$xy = 3$

13. $xy = 1$
$3x - 2y = 5$

14. $x + y = 0$
$xy = 9$

15. $x^2 + 25y^2 = 25$
$x - 5y = -1$

16. $x^2 + 3y^2 = 52$
$x + y = -2$

17. $2x^2 + 3y^2 = 4$
$x - y = -2$

18. $4x^2 + y^2 = 13$
$2x + y = 5$

19. $x^2 - 2y^2 = 17$
$x - 2y = -1$

20. $y^2 - 2x^2 = 2$
$x - y = -1$

21. $4y^2 - x^2 = 4$
$3x - 2y = 2$

22. $3x^2 - y^2 = 1$
$2x - y = 0$

23. $x^2 + y^2 = 12$
$3x^2 - 4y^2 = 8$

24. $x^2 - 3y^2 = 17$
$x^2 + y^2 = 1$

25. $5x^2 + 2y^2 = 7$
$x^2 + y^2 = 2$

26. $9x^2 + 4y^2 = 36$
$x^2 + y^2 = 9$

27. $-7x^2 + 2y^2 = 13$
$2x^2 + 3y^2 = 7$

28. $3x^2 + 5y^2 = 12$
$2x^2 - 3y^2 = 8$

29. $x^2 + y^2 = 2$
$3x - 2y^2 = -5$

30. $x^2 + 4y^2 = 20$
$x^2 - y = 6$

31. $x^2 + 4y^2 = 4$
$x^2 - y = -1$

32. $x^2 - y^2 = 4$
$x^2 - y = 4$

33. $x^2 - 4y^2 = -16$
$2x - y^2 = 0$

34. $x^2 + y^2 = 14$
$x^2 + y = 2$

35. $x^2 + y^2 = 5$
$x^2 + 2xy + y^2 = 9$

36. $x^2 + y^2 = 29$
$x^2 + 3xy + y^2 = 59$

37. $4x^2 + y^2 = 37$
$4x^2 - 2xy + y^2 = 31$

38. $x^2 - y^2 = 12$
$x^2 - 2xy - y^2 = -4$

39. $y = \sqrt{2x + 1}$
$2x + y = 5$

40. $y = \dfrac{3}{x^2}$

$y = \dfrac{x + 2}{x}$

Solve the following problems.

41. The sum of two numbers is 11. The sum of the squares of the numbers is 65. What are the numbers?

42. The difference of two numbers is 1. The sum of the squares of the numbers is 145. What are the numbers?

43. The perimeter of a rectangle is 32 ft. The rectangle has an area of 48 sq ft. What are the dimensions of the rectangle?

44. The perimeter of a rectangle is 44 ft. The rectangle has an area of 112 sq ft. What are the dimensions of the rectangle?

45. The hypotenuse of a right triangle is 5 ft long. The area of the triangle is 6 sq ft. What are the lengths of the legs of the triangle?

46. The hypotenuse of a right triangle is $\sqrt{13}$ ft long. The area of the triangle is 3 sq ft. What are the lengths of the legs of the triangle?

47. Find the equation of the ellipse with center at the origin and passing through the points $(2\sqrt{3}, -1)$ and $(2, \sqrt{3})$. (*Hint:* These ordered pairs must satisfy the equation $\dfrac{x^2}{a^2} + \dfrac{y^2}{b^2} = 1$.)

48. Find the equation of the circle with center on the y-axis and passing through the points $(-2, 0)$ and $(0, 6)$. [*Hint:* These ordered pairs must satisfy the equation $x^2 + (y - k)^2 = r^2$.]

Write Algebra

49. Describe the types of nonlinear systems that are more easily solved using the substitution method.

50. Given a system containing a linear and a quadratic function, why can the solution set contain at most two ordered pairs?

The following exercises require the use of a graphing calculator. Use the Trace and Zoom keys of a graphing calculator to approximate the solution (the point of intersection) of the following systems.

51. $3x + 2y = 4$
$\quad\quad y = x^2$

52. $x^2 + y^2 = 4$
$\quad\quad x + y = 2$

53. $y = x^2 + 3x - 5$
$\quad\quad y = -0.5x^2 - 2x + 6$

54. $x^2 + y^2 = 45$
$\quad\quad y = 0.8x^2 + 3x - 6.4$

G R O U P   A C T I V I T Y

In exercises 55–60, find the values of a so that the given system of equations has only one ordered pair solution.

55. $x + y = a$
$\quad\quad x^2 + y^2 = 25$

56. $x + y = a$
$\quad\quad y = x^2 + 3$

57. $x - y = a$
$\quad\quad y = \sqrt{x}$

58. $x + y = a$
$\quad\quad y = \dfrac{1}{x}$

59. $x = y^2 + a$
$\quad\quad x - y = 4$

60. $x^2 + y^2 = 4$
$\quad\quad (x - 5)^2 + (y - 1)^2 = a^2$

Applied Algebra

Tanya, owner of Tanya's T-Shirts, sells customized T-shirts at the Moss County Mall. Tanya buys plain T-shirts from a wholesaler for $5 a shirt. She sells the customized T-shirts for $9 a shirt. Each month Tanya has fixed costs of $1200. Her fixed costs include wages, rent, utilities, and miscellaneous expenses. How many T-shirts must Tanya sell just to break even? How many T-shirts must Tanya sell to generate a monthly profit of $2000?

SOLUTION

When Tanya breaks even, her monthly revenue will equal her monthly costs. Let's first find equations that calculate Tanya's cost and revenue.

If Tanya sells 20 customized T-shirts, her revenue will be $180:

$$(\text{price}) \cdot (\text{number sold}) = 9 \cdot 20 = 180$$

So if we let

$$x = \text{the number of T-shirts sold}$$

and

$$R = \text{the revenue when } x \text{ T-shirts are sold}$$

Then

$$\text{revenue} = (\text{price}) \cdot (\text{number sold})$$
$$R = 9x$$

Since R is a function of x, we sometimes use function notation:

$$R(x) = 9x$$

Now let's find Tanya's cost equation. Again suppose that Tanya sells 20 customized T-shirts. Then Tanya's monthly cost will be

$$\begin{aligned}
\text{total cost} &= \text{cost of the T-shirts} + \text{fixed costs} \\
&= 5 \cdot 20 \qquad\qquad\quad + 1200 \\
&= 100 \qquad\qquad\qquad + 1200 \\
&= \$1300
\end{aligned}$$

In general, we say

$$\text{total cost} = \text{variable costs} + \text{fixed costs}$$

NOTE ▶ The variable cost depends upon how many T-shirts Tanya sells.

Again, let

$$x = \text{the number of T-shirts sold}$$

and

$$C = \text{the total cost when } x \text{ T-shirts are sold}$$

Then

$$\text{total cost} = \text{variable costs} + \text{fixed costs}$$
$$C = \qquad 5x \qquad + \qquad 1200$$

Since C is a function of x, we again use function notation:

$$C(x) = 5x + 1200.$$

Both the revenue and cost functions are graphed in the accompanying figure.

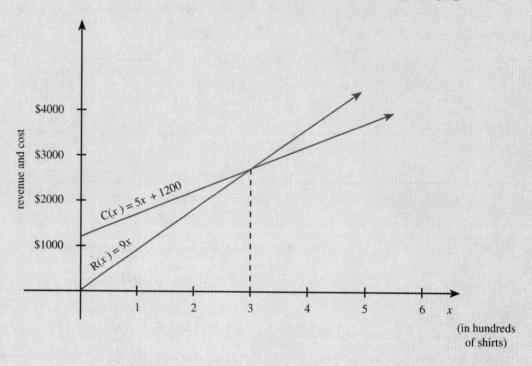

The break-even point is the point at which the revenue equals the cost. From our graph the break-even point (the point of intersection) appears to occur when $x = 300$. Let's verify this fact algebraically.

Recall that

$$R(x) = 9x$$

and

$$C(x) = 5x + 1200$$

At the break-even point

$$R(x) = C(x)$$

In our case,

$$9x = 5x + 1200$$
$$4x = 1200$$
$$x = 300$$

Thus, the break-even point occurs when Tanya sells 300 T-shirts.

To determine the profit, we use the fact that

$$\text{profit} = \text{revenue} - \text{cost}$$

So in our case,

$$\begin{aligned} \text{profit} &= 9x - (5x + 1200) \\ &= 9x - 5x - 1200 \\ &= 4x - 1200 \end{aligned}$$

Let's agree to use $P(x)$ to indicate the profit when x T-shirts are sold. Thus,

$$P(x) = 4x - 1200$$

We want to find how many T-shirts Tanya must sell to generate a monthly profit of $2000. In other words, we want to find x when $P(x) = 2000$:

$$\begin{aligned} 2000 &= 4x - 1200 \\ 3200 &= 4x \\ 800 &= x \end{aligned}$$

Hence Tanya must sell 800 T-shirts to generate a monthly profit of $2000.

Your Turn

Weber, the owner of Weber's Fresh Produce, sells watermelons at his watermelon stand. Weber buys watermelons from the farmers in Moss County for 20¢ a pound. He sells the watermelons at his watermelon stand for 45¢ a pound. Each day Weber has fixed costs of $5.

1. Find Weber's daily revenue and cost functions.

2. Sketch the graph of Weber's daily revenue and cost functions.

3. What is Weber's break-even point?

4. Find Weber's daily profit function.

5. How many pounds of watermelons must Weber sell to generate a daily profit of $40?

Chapter 9 Review

Terms to Remember

System of equations	p. 522	Dependent system	p. 524
Linear system of equations	p. 522	Substitution method	p. 524
Solutions of a system of equations	p. 522	Elimination method	p. 526
		Ordered triple	p. 531

Graphing method   **p. 522**   Linear equation in three vari-   **p. 532**
Independent system   **p. 522**   ables
Inconsistent system   **p. 523**   Nonlinear system of equations   **p. 550**

Methods for Finding the Solutions of a Linear System of Equations

■ *Substitution Method* (two equations in two variables)

1. Solve one of the equations for one of the variables.

2. *Substitute* the expression obtained in step 1 into the other equation.

3. You now have an equation with one variable. Solve this equation. (At this step you have obtained only one coordinate of the solution.)

4. Substitute the value obtained in step 3 into the equation of step 1, and solve this equation. You now have both coordinates of the solution.

5. Check your solution in both equations of the system.

■ *Elimination Method* (two equations in two variables)

1. Express each equation in the form $ax + by = c$.

2. Multiply each equation by some nonzero number so that the sum of the coefficients of either x or y is zero.

3. Add the equations obtained in step 2.

4. You now have an equation with one variable. Solve this equation.

5. Substitute the value determined in step 4 into one of the equations that contains both variables, and solve this equation. You now have both coordinates of the solution.

6. Check the solution in both equations.

■ *Elimination Method* (three equations in three variables)

1. Express each equation in the form $ax + by + cz = d$.

2. Multiply *any* two of the equations by some nonzero number so that the sum of the coefficients of x, y, or z in those two equations is zero.

3. Add the two equations from step 2. You now have an equation with two variables.

4. Combine multiples of the *remaining* equation and one of the equations used in step 2. Pick multiples so that the variable that was eliminated in step 3 is eliminated here.

5. Add the equations from step 4. You now have another equation with two variables.

6. The equations from step 3 and step 5 form a linear system in two variables. Find the solution of this system of equations.

7. Substitute the values determined in step 6 into one of the equations that contains the remaining variable, and solve this equation. You now have all three coordinates of the solution.

8. Check the solution in every equation in the system.

Review Exercises

9.1 Using graphs, determine the solutions of the following linear systems of equations. Identify the system as independent, inconsistent, or dependent.

1. $2x - y = 5$
   $x + 3y = -8$

2. $x + y = -6$
   $3x - y = -2$

3. $x - 2y = 4$
   $2x - 4y = 0$

4. $6x - 2y = 10$
   $3x - y = 5$

Using the substitution method, determine the solutions of the following linear systems of equations. Identify the system as independent, inconsistent, or dependent.

5. $3x + 2y = 19$
   $y = x + 2$

6. $2x - 6y = 13$
   $x = 3y - 1$

7. $3x - 9y = -21$
   $x = 3y - 7$

8. $x - 3y = 5$
   $2x - 6y = 1$

9. $2x + 4y = -11$
   $x + y = -3$

10. $6x - 2y = 10$
    $9x - 3y = 15$

Using the elimination method, determine the solutions of the following linear systems of equations. Identify the system as independent, inconsistent, or dependent.

11. $2x + y = 8$
    $3x - y = 12$

12. $2x - 3y = 4$
    $-6x + 9y = -1$

13. $2x + 8y = 7$
    $x + 4y = 1$

14. $2x - 5y = 13$
    $-3x + 6y = -16$

15. $-\frac{1}{2}x + \frac{1}{3}y = \frac{1}{6}$
    $\frac{1}{4}x - \frac{1}{2}y = \frac{1}{4}$

16. $\frac{1}{4}x + \frac{1}{3}y = 1$
    $\frac{1}{2}x + \frac{2}{3}y = \frac{3}{4}$

9.2 Using the elimination method, determine whether the following linear systems are independent, inconsistent, or dependent. If the system is independent, find the solution.

17. $x - 3y + z = 8$
    $2x + y - z = 0$
    $-x - y + 4z = 11$

18. $3x + 4y + 2z = -2$
    $-x + 3y - z = -1$
    $2x + y + z = 3$

19. $x - 3y + 2z = 10$
    $5x + 2y + z = 3$
    $-2x - y + 3z = 1$

20. $-2x - y + 4z = 2$
    $3x - y - z = 0$
    $x + 2y - 5z = -3$

21. $-2x - 6y + 2z = 8$
    $3x + 9y - 3z = -12$
    $x + 3y - z = -4$

22. $2x + y - z = 4$
    $x - 2y - 5z = 0$
    $6x + 3y - 3z = -5$

23. $x + 4y - z = 0$
    $-x - y + 2z = -4$
    $3x + y + 3z = 5$

24. $2x - y + 2z = -1$
    $3x + 4y - z = -5$
    $-x + 2y + 2z = 5$

25. $-3x + 2y + z = 5$
    $x - 2y + z = 1$
    $2x - y - z = -4$

26. $3x + 2y = 7$
    $y - 4z = -10$
    $-x + 3z = 8$

9.3 Find the solutions of the following problems.

27. The sum of two numbers is 19. The larger number is 5 more than the smaller number. What are the numbers?

28. Nancy has $3.20 in nickels and dimes. She has a total of 41 coins. How many nickels and how many dimes does Nancy have?

29. Betty Sue sold 32 tickets to the Moss County Community Theater's production of *Oklahoma*. Adult tickets cost $4.00 and children's tickets cost $1.50. She collected $103 from the sale of the tickets. How many adult tickets and how many children's tickets did she sell?

30. Kay bought 3 ballpoint pens and 1 spiral notebook for $1.70 at the Moss County Drugstore. Four days later she bought 2 ballpoint pens and 2 spiral notebooks for $2.36. What is the cost of each ballpoint pen and each spiral notebook?

31. Andy's Apple Emporium has two types of apple juice: "Apple of Your Eye" apple juice, which sells for $1.50 a quart, and "Generic" apple juice, which sells for $0.70 a quart. How many quarts of each type of apple juice must Andy blend to make 40 qt of "Andy's Homestyle" apple juice, which will sell for $1.20 a quart?

32. Sidney invested $6000 at his local bank. He invested part of his money in an account paying 11% simple interest, and he invested the rest of his money in an account paying 5% simple interest. After 1 yr his interest income was $570. How much did Sidney invest in each account?

33. Big D toothpaste contains 42% fluoride. Big G toothpaste contains 24% fluoride. How many ounces of each toothpaste must be used to make 12 oz of toothpaste that contains 36% fluoride?

34. Esther has $4.05 in nickels, dimes, and quarters. She has a total of 41 coins, and she has 4 more dimes than nickels. How many nickels, dimes, and quarters does she have?

35. Coach Williams buys 1 football, 1 basketball, and 1 soccer ball for $47 at Athletic World. Coach Doyle buys 2 footballs, 1 basketball, and 2 soccer balls for $80. Coach Sharp buys 2 footballs, 3 basketballs, and 1 soccer ball for $93. What is the cost of each football, each basketball, and each soccer ball?

9.4 Determine the solutions of the following systems of equations.

36. $x^2 + y^2 = 13$
$x - y = 5$

37. $x^2 + y = 3$
$x - y = 3$

38. $x - y^2 = 5$
$x - 2y = 5$

39. $2x - y = 3$
$xy = 20$

40. $x^2 + 4y^2 = 20$
$x - 2y = -6$

41. $y^2 - 9x^2 = 9$
$x - y = 3$

42. $x^2 + 2y^2 = 16$
$3x^2 + y^2 = 18$

43. $2x^2 - 3y^2 = 8$
$3x^2 + 4y^2 = 29$

44. $x^2 + y^2 = 9$
$x + y^2 = -3$

45. $x^2 - 4y^2 = 5$
$x^2 - y = 8$

46. $x^2 + y^2 = 10$
$x^2 + 8xy + y^2 = 34$

47. $x^2 + y^2 = 20$
$x^2 + 5xy + y^2 = 60$

Solve the following problems.

48. The sum of two numbers is 7. The sum of the squares of the numbers is 25. What are the numbers?

49. The perimeter of a rectangle is 36 ft. The rectangle has an area of 72 sq ft. What are the dimensions of the rectangle?

Chapter 9 Test

(You should be able to complete this test in 60 minutes.)

I. Using the substitution or elimination method, determine the solutions of the following linear systems of equations. Identify the system as independent, inconsistent, or dependent.

1. $2x - 3y = 5$
       $y = 2x + 2$

2. $2x + 3y = 13$
       $3x - 4y = 11$

3. $3x - 6y = 11$
       $x - 2y = 5$

4. $6x + 12y = -4$
       $9x + 18y = -6$

5. $x - 2y + 3z = 7$
         $2x + y + z = 4$
       $-3x + 2y + 3z = 19$

6. $2x + 3y - z = -7$
          $x + 2y + 5z = 0$
        $-2x + y + 4z = -6$

II. Determine the solutions of the following systems of equations.

7. $x^2 + y^2 = 1$
       $x - 2y = -1$

8. $2x^2 + y^2 = 3$
       $3x^2 - 2y^2 = 1$

9. $x^2 + y^2 = 29$
       $x^2 + 14xy + y^2 = 169$

III. Find the solutions of the following problems.

10. Beth bought 3 bath towels and 2 hand towels for $31 at Louie's Dry Goods. The next week she bought 2 bath towels and 5 hand towels for $39. What is the cost of each kind of towel?

11. Fred's Nursery has two types of fertilizer: "Let Your Garden Grow" fertilizer, which sells for $1.70 a pound, and "This Might Work" fertilizer, which sells for $0.50 a pound. How many pounds of each type of fertilizer must Fred blend to obtain 30 lb of "Fred's Famous Fertilizer," which will sell for $0.90 a pound?

12. The perimeter of a rectangle is 40 ft. The rectangle has an area of 84 sq ft. What are the dimensions of the rectangle?

TEST YOUR MEMORY

These problems review Chapters 1–9.

I. Simplify each of the following. Assume that all variables represent positive numbers.

1. $\left(\dfrac{10x^{-9}y^4}{6x^{-2}y^{10}}\right)^2$

2. $\dfrac{(4x^3)^{-1/2}x^2}{x^{1/3}}$

3. $x\sqrt{48x} - 3\sqrt{12x^3} - \sqrt{300x^3}$

4. $\sqrt[3]{\dfrac{3}{4x^2}}$

II. Perform the indicated operations.

5. $\dfrac{x-4}{x^2 - 8x + 12} + \dfrac{x-11}{x^2 - 2x - 24}$

6. $\dfrac{2 - 11x^{-1} + 12x^{-2}}{1 - 7x^{-1} + 12x^{-2}}$

III. Let $f(x) = \dfrac{x}{x-2}$ **and** $g(x) = \dfrac{3}{x-1}$. **Find the following.**

7. $f(0)$

8. $3f(4) - 5g(3)$

9. $[g(-4)]^3$

10. $f(x) - g(x)$

IV. Find the equations of the following lines. Write the equations in slope-intercept form, if possible.

11. Through $(10, 1)$ and $(-2, 5)$

12. Through $(-2, -3)$ and perpendicular to $3x + 4y = 1$

V. Find the equations of the following circles.

13. Center $= C(0, -4)$, radius $= 6$

14. Center $= C(-1, 3)$ and passing through $(1, -5)$

VI. Graph the following equations.

15. $x = y^2 + 1$

16. $4x^2 + 25y^2 = 100$

17. $4x^2 - y^2 = 4$

18. $y = 2x - 1$

19. $y = x^2 - 6x + 10$

20. $x^2 + y^2 - 4x + 2y + 1 = 0$

VII. Find the solutions of the following equations and check your solutions.

21. $3x^3 + x^2 - 12x - 4 = 0$

22. $2x^3 + 7x^2 - 2x - 7 = 0$

23. $1 - \dfrac{3x + 14}{x + 3} = \dfrac{4x + 7}{x + 3}$

24. $\dfrac{2x + 1}{x + 1} - \dfrac{5}{x + 3} = \dfrac{7x}{x^2 + 4x + 3}$

25. $\sqrt[3]{2x - 1} + 3 = 1$

26. $\sqrt{3x + 1} + 5 = 1$

27. $2x^2 - 2x + 5 = 0$

28. $9x^2 - 6x - 1 = 0$

29. $\sqrt{2x + 11} + \sqrt{x + 6} = 2$

30. $\sqrt{2x} + \sqrt{4x - 1} = 2$

VIII. **Using the substitution or elimination method, determine the solutions of the following linear systems of equations. Identify each system as independent, inconsistent, or dependent.**

31. $4y - 6x = -11$
$\quad\quad x = y + 2$

32. $6x - 3y = 15$
$\quad\quad\quad y = 2x - 5$

33. $2x + 3y = -5$
$\quad\quad 3x - 4y = -16$

34. $6x - 3y = 1$
$\quad\quad 4x - 2y = -3$

35. $\quad x + 2y - z = -4$
$\quad\quad 2x - y + 3z = 7$
$\quad\quad -x + y - 4z = -13$

IX. **Determine the solutions of the following systems of equations.**

36. $\quad\quad y = x^2 - 4x + 7$
$\quad\quad 2x + y = 10$

37. $\quad\quad x^2 + y^2 = 17$
$\quad\quad x^2 + 3xy + y^2 = 29$

X. **Find the solutions of the following inequalities and graph each solution set on a number line. Express your answers using interval notation.**

38. $1 - 3x \le -5$

39. $x^2 - 4x - 5 < 0$

40. $\dfrac{2x}{x+3} > 1$

XI. **Graph the solution sets of the following systems of inequalities.**

41. $\begin{cases} y < 2x \\ 2x + 3y < 8 \end{cases}$

42. $\begin{cases} y \le 4 - x^2 \\ y \ge x + 2 \end{cases}$

XII. **Use algebraic expressions to find the solutions of the following problems.**

43. Karen has $2.50 in quarters and nickels. She has a total of 22 coins. How many quarters and how many nickels does she have?

44. One leg of a right triangle is 7 yd longer than the other leg. The hypotenuse is 17 yd long. What are the lengths of the legs of the triangle?

45. The temperature varies directly as the height of a column of mercury in a thermometer. If a column of mercury is 2 in. high when the temperature is 60° F, what will the temperature be when the column is 3 in. high?

46. Jesse wants to plant a rectangular garden next to a straight river. He has 80 ft of fencing and will not fence the side of the garden by the river. What dimensions will produce the largest garden?

47. Miri's motorboat can travel 18 mi down the Red River in the same time that it can travel 12 mi up the Red River. If the speed of the current in the Red River is 1 mi/hr, what is the speed of Miri's boat in still water?

48. Margie and Charles have a square hot tub that measures 9 ft on a side. They are going to build a tile border of uniform width around the tub. They have 115 sq ft of tile. How wide should the border be?

49. Sean bought 3 ballpoint pens and 2 pencils for $1.71 at the Moss County Community College Bookstore. Two days later he bought 5 ballpoint pens and 1 pencil for $2.57. What is the cost of 1 ballpoint pen and 1 pencil?

50. Moss County "Silver Dollars" contain 4% pure silver. Bernal County "Silver Dollars" contain 54% pure silver. How many ounces of each silver dollar must be melted together to produce a 40-oz metal bar that is 24% pure silver?

Exponential and Logarithmic Functions

The population of most animals in captivity is small in comparison with the animal population in the wild. This concerns zookeepers for two reasons. The inbreeding among a particular species may cause the loss of genetic variation and may introduce undesirable genetic traits. To understand this problem, scientists study the change in heterozygosity (genetic variation) from one generation to the next. A formula that predicts this change is

$$H_t = H_0\left(1 - \frac{1}{2N}\right)^t$$

where H_0 is the initial heterozygosity, H_t is the heterozygosity after t generations, and N is the number of animals.

This equation is an example of an *exponential equation*. After we have studied exponential and logarithmic equations in this chapter, we will be able to answer questions such as these:

1. What will the heterozygosity be for New Guinea singing dogs after 50 years?

2. When will the heterozygosity for New Guinea singing dogs drop to 50% of its initial heterozygosity?

10.1 *Inverse Functions*

Given a function, many times we are interested in a function or relation that behaves in an opposite or inverse manner. In this chapter we will study the exponential function and its inverse, the logarithmic function. First we must develop a working knowledge of what we mean by the inverse of a function.

Let us start with a simple example. The equation $y = 2x$ defines a function that takes a value of x and doubles it to get the corresponding value of y. The diagram in Figure 10.1.1 illustrates how values of x in the domain are doubled to get values for y in the range. These ordered pairs are listed in the chart accompanying Figure 10.1.1:

x	y
1	2
3	6
$\dfrac{1}{2}$	1
-2	-4

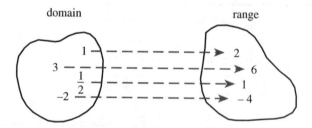

Figure 10.1.1

Suppose we wanted an equation that does the opposite or "undoes" the results of the original function. The equation would be $y = \frac{1}{2}x$. This equation defines a function that takes a value for x and halves it to get the corresponding value for y. The diagram in Figure 10.1.2 illustrates the halving behavior of this function. The table lists the ordered pairs from the diagram:

x	y
2	1
6	3
1	$\dfrac{1}{2}$
-4	-2

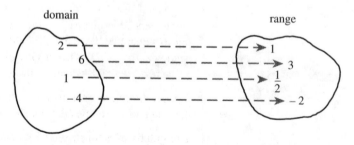

Figure 10.1.2

REMARKS

1. Starting with the ordered pairs of the original function, we can generate ordered pairs for the second function by interchanging the x- and y-coordinates. This is done here.

$$y = 2x \qquad\qquad y = \frac{1}{2}x$$

$$(1, 2) \longrightarrow (2, 1)$$
$$(3, 6) \longrightarrow (6, 3)$$
$$\left(\frac{1}{2}, 1\right) \longrightarrow \left(1, \frac{1}{2}\right)$$
$$(-2, -4) \longrightarrow (-4, -2)$$

When the coordinates are interchanged, the new relation is called the **inverse** of the original function. Thus $y = \frac{1}{2}x$ is the inverse of $y = 2x$. This reasoning motivates the following definition.

Definition 10.1.1 Given a relation R, the **inverse** of R, which is denoted by R^{-1}, is the set $\{(y,\ x):(x,\ y) \in R\}$.

NOTE ▶   The notation $\{(y, x):(x, y) \in R\}$ is read "the set of all ordered pairs (y, x) such that (x, y) is an element of R."

2. Since the x- and y-coordinates are interchanged in the inverse, we can derive the equation of the inverse by interchanging the variables x and y in the original equation and then solving for y:

$$y = 2x \xrightarrow[\substack{\text{x and y.}}]{\text{Interchanging}} x = 2y$$

$$\left(\frac{1}{2}\right) \cdot x = \left(\frac{1}{2}\right) \cdot 2y$$

$$\frac{1}{2}x = y$$

3. In general, *the inverse of a function does not have to be a function.* In this example, since the inverse is a function, we can use functional notation when writing its equation. If we use f for the original function, then the original equation becomes $f(x) = 2x$. The inverse of f is denoted by f^{-1}. Its equation becomes $f^{-1}(x) = \frac{1}{2}x$, which is read "f inverse of x is one-half x."

4. Since x and y are interchanged, the domain of f becomes the range of f^{-1}, and the range of f becomes the domain of f^{-1}.

5. The graphs of a function and its inverse are related in an interesting way. Both f and f^{-1} are *linear* functions. Their graphs are sketched in Figure 10.1.3 using

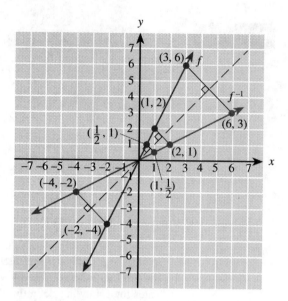

Figure 10.1.3

the points from the tables used for Figures 10.1.1 and 10.1.2. The diagonal, $y = x$, is drawn as a dashed line. The graph of f^{-1} is said to be "the reflection about the diagonal" of the graph of f. If line segments are drawn between $(3, 6)$ and $(6, 3)$, $(1, 2)$ and $(2, 1)$, $(\frac{1}{2}, 1)$ and $(1, \frac{1}{2})$, and $(-2, -4)$ and $(-4, -2)$, then the diagonal is the perpendicular bisector of these segments.

In general, the graph of the inverse can be roughly sketched by reflecting the graph of the original function about the diagonal. If the two graphs cross, they will do so on the diagonal itself (see Figure 10.1.4).

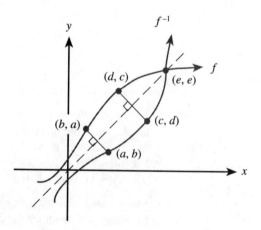

Figure 10.1.4

EXAMPLE 10.1.1   Find an equation that defines the inverse. If the inverse is a function, write it in the form $y = f^{-1}(x)$. Also graph both equations on the same coordinate plane.

1. $f(x) = \dfrac{2}{3}x - 2$

To find an equation for the inverse, interchange x and y.

$$y = \frac{2}{3}x - 2 \rightarrow x = \frac{2}{3}y - 2$$

$$x + 2 = \frac{2}{3}y \qquad \textit{Solve for y.}$$

$$\frac{3}{2} \cdot (x + 2) = \left(\frac{3}{2}\right) \cdot \left(\frac{2}{3}\right)y$$

$$\frac{3}{2}x + 3 = y$$

In this case, the inverse of f is a function. Thus, $f^{-1}(x) = \frac{3}{2}x + 3$. f and f^{-1} are graphed in Figure 10.1.5 using the techniques developed in Section 6.5. The y-intercept for f, $(0, -2)$, becomes the x-intercept for f^{-1}, $(-2, 0)$. Similarly, the x-intercept for f, $(3, 0)$, becomes the y-intercept for f^{-1}, $(0, 3)$.

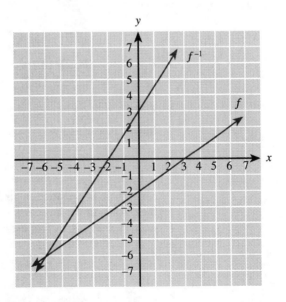

Figure 10.1.5

f	
x	*y*
-2	6
-1	3
$-\frac{1}{2}$	$2\frac{1}{4}$
0	2
$\frac{1}{2}$	$2\frac{1}{4}$
1	3
2	6

Inverse of f

x	*y*
6	-2
3	-1
$2\frac{1}{4}$	$-\frac{1}{2}$
2	0
$2\frac{1}{4}$	$\frac{1}{2}$
3	1
6	2

2. $f(x) = x^2 + 2$

$$y = x^2 + 2 \xrightarrow[\text{x and y.}]{\textit{Interchange}} x = y^2 + 2 \qquad \textit{Solve for y.}$$

$$x - 2 = y^2$$

$$y = +\sqrt{x-2} \quad \text{or} \quad y = -\sqrt{x-2}$$

The inverse is not a function, since for one value for *x*, there are two values for *y* (positive and negative). The two equations correspond to the top half and bottom half of a horizontal parabola.

For the graphs of *f* and the inverse of *f*, we use the points given in the charts. The graphs are shown in Figure 10.1.6.

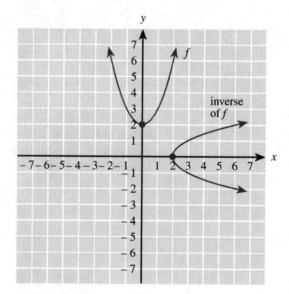

Figure 10.1.6

3. $f(x) = \sqrt{3-x}$

Interchange *x* and *y*.

$$y = \sqrt{3-x} \rightarrow x = \sqrt{3-y}$$

Since there is a restriction on the range of *f*, $y \geq 0$, there must be a restriction on the domain of the inverse, $x \geq 0$. Similarly, since there is a restriction on the domain of *f*, $x \leq 3$, there must be a restriction on the range of the inverse, $y \leq 3$.

$$x^2 = (\sqrt{3-y})^2 \qquad \textit{Solve for y.}$$

$$x^2 = 3 - y \qquad x \geq 0$$

$$y + x^2 = 3 \qquad x \geq 0$$

$$y = -x^2 + 3 \qquad x \geq 0$$

The inverse is a function, so we write it as

$$f^{-1}(x) = -x^2 + 3 \qquad x \geq 0$$

For the graphs we use the points given in the charts. The graphs are sketched in Figure 10.1.7.

	f	
x		y
3		0
2		1
−1		2
−6		3

	f⁻¹	
x		y
0		3
1		2
2		−1
3		−6

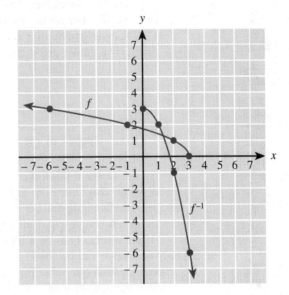

Figure 10.1.7

4. $f(x) = |x + 3|$

Interchange x and y. Since there is a restriction on the range of f, $y \geq 0$, there must be a restriction on the domain of the inverse, $x \geq 0$.

$$y = |x + 3| \quad\longrightarrow\quad x = |y + 3|, \qquad x \geq 0$$
$$y + 3 = x \qquad \text{or} \quad y + 3 = -x \qquad \textit{Solve for y.}$$
$$y = x - 3, x \geq 0 \qquad\qquad y = -x - 3, x \geq 0$$

The inverse is not a function, since there are two y values for one x value. The two equations correspond to the top half and bottom half of a horizontal "V." For the graph of f and the inverse of f, we use the points given in the charts. The graphs are sketched in Figure 10.1.8.

f	
x	**y**
−6	3
−4	1
−3	0
−2	1
0	3
2	5
3	6

Inverse of f

x	y
3	−6
1	−4
0	−3
1	−2
3	0
5	2
6	3

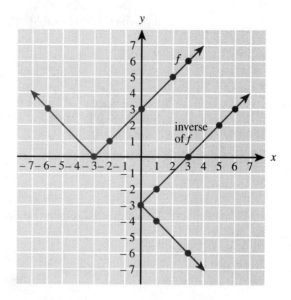

Figure 10.1.8

In the first and third examples, we started with a function and found the inverse, which also turned out to be a function. But this was not the case in the second and fourth examples. Let us determine a condition that will guarantee that the inverse of a function will itself be a function.

If the inverse is to be a function, then when one number is substituted for x, only one value can be found for y. Interchanging x and y, this means that in the original function one value for y results in only one value for x. This condition determines the definition of a special type of function.

Definition 10.1.2	A function f is said to be a **one-to-one function** (or just a **1–1 function**) if each value for y corresponds to only one value for x.

REMARK This definition is sometimes written more formally as follows: f is a one-to-one function if $f(a) = f(b)$ implies that $a = b$; that is, if two y values are equal, then their x values must be equal.

NOTE ▶ *The inverse of a function* f *is itself a function if and only if* f *is a one-to-one function.*

XAMPLE 10.1.2 Determine whether the following functions are one-to-one. If they are, find the inverse function.

1. $f(x) = x^4$

Suppose that $f(a) = f(b)$. Then

$$a^4 = b^4$$
$$a^4 - b^4 = 0$$
$$(a^2 - b^2)(a^2 + b^2) = 0$$
$$(a - b)(a + b)(a^2 + b^2) = 0$$
$$a - b = 0 \quad \text{or} \quad a + b = 0 \quad \text{or} \quad a^2 + b^2 = 0$$
$$a = b \qquad\qquad a = -b \qquad\qquad a^2 = -b^2 \qquad \text{\textit{This has}}$$

imaginary solutions.

Thus, f is *not* a one-to-one function, since a is not necessarily equal to b.

2. $f(x) = \dfrac{3}{x - 1}$

Suppose that $f(a) = f(b)$. Then

$$\frac{3}{a - 1} = \frac{3}{b - 1}$$
$$3(b - 1) = 3(a - 1)$$
$$3b - 3 = 3a - 3$$
$$3b = 3a$$
$$b = a$$

Thus, f is a one-to-one function. To find the inverse, interchange x and y:

$$y = \frac{3}{x - 1} \longrightarrow x = \frac{3}{y - 1}$$
$$x \cdot (y - 1) = 3 \qquad \text{\textit{Solve for y.}}$$
$$xy - x = 3$$
$$xy = x + 3$$
$$y = \frac{x + 3}{x}$$

So $f^{-1}(x) = \dfrac{x + 3}{x}$.

REMARK Consider the graphs of the functions $y = x^2$ and $y = x$ shown in Figures 10.1.9 and 10.1.10. The function $y = x^2$ is not one-to-one, since there are ordered pairs with the same second coordinate and different first coordinates—for example, $(-2, 4)$ and $(2, 4)$. This is indicated on the graph by the fact that these two points lie on the same horizontal line. The function $y = x$ is one-to-one, since every second coordinate has only one first coordinate. This is indicated on the graph by the fact that no two points lie on the same horizontal line. These observations motivate the **horizontal line test.** The horizontal line test states that if every horizontal line crosses the graph of a given function only once, then that function is one-to-one.

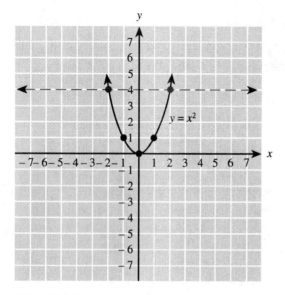

Figure 10.1.9

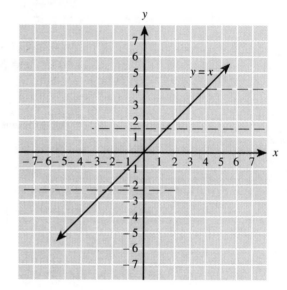

Figure 10.1.10

In Example 6.3.3 in Chapter 6 we saw how two functions could be combined using the operations of addition, subtraction, multiplication, and division to obtain a new function. Now we will briefly examine another type of operation that can be performed on two (or more) functions. We call this operation **composition.**

Let us start with the two functions $f(x) = 2x + 1$ and $g(x) = x^2$. Let us also go back to the notion of a function machine, which was discussed in Section 6.3. In Figure 10.1.11 we have placed the two machines for f and g next to each other so that when a number is outputted from f, it is immediately inputted into g. In this

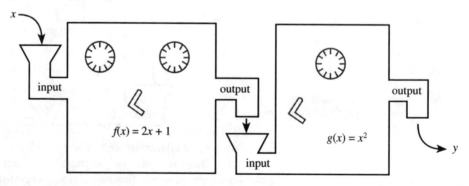

Figure 10.1.11

manner we have constructed a new function, which we call the *composition of g with f.* Notationally, it is written as $g \circ f$, and its defining equation is

$$
\begin{aligned}
(g \circ f)(x) &= g[f(x)] \\
&= g[2x + 1] \qquad \textit{Substitute 2x + 1 for f(x).} \\
&= (2x + 1)^2
\end{aligned}
$$

Let us start with an input of $x = 3$ for f and follow it through the two function machines f and g. Letting f operate on 3 as in Figure 10.1.12, we get

$$
\begin{aligned}
f(3) &= 2 \cdot 3 + 1 \\
&= 6 + 1 \\
&= 7
\end{aligned}
$$

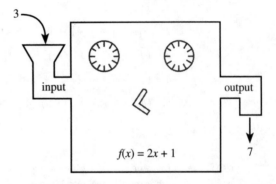

Figure 10.1.12

Now we will take this output from f and input it into g, as in Figure 10.1.13:

$$
\begin{aligned}
g(7) &= 7^2 \\
&= 49
\end{aligned}
$$

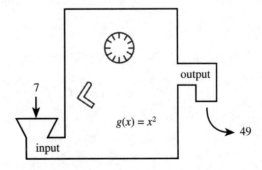

Figure 10.1.13

Thus, $x = 3$ is sent to $y = 49$.

Let us now compare this result with the value of the defining equation $(g \circ f)(x) = (2x + 1)^2$ at $x = 3$:

$$(g \circ f)(3) = (2 \cdot 3 + 1)^2$$
$$= (6 + 1)^2$$
$$= (7)^2$$
$$= 49$$

 XAMPLE 10.1.3 For $f(x) = 2x + 1$ and $g(x) = x^2$, find the following.

1. $g(2) = 2^2 = 4$

2. $f[g(2)] = f[4] = 2 \cdot 4 + 1 = 8 + 1 = 9$

3. $(f \circ g)(x) = f[g(x)] = f[x^2] = 2x^2 + 1$

4. $(f \circ g)(2) = 2 \cdot 2^2 + 1 = 2 \cdot 4 + 1 = 8 + 1 = 9$

5. $(f \circ g)(-5) = 2(-5)^2 + 1 = 2 \cdot 25 + 1$
$$= 50 + 1 = 51$$

NOTE ▶ In general, $(f \circ g)(x) \neq (g \circ f)(x)$. For example, we found $(g \circ f)(x) = (2x + 1)^2$, whereas $(f \circ g)(x) = 2x^2 + 1$.

We will now consider the special case of the composition of a function with its inverse. Consider the function $f(x) = \frac{2}{3}x - 2$, whose inverse we found to be $f^{-1}(x) = \frac{3}{2}x + 3$ in Example 10.1.1, part 1. First let us find $(f \circ f^{-1})(x)$:

$$(f \circ f^{-1})(x) = f[f^{-1}(x)]$$
$$= f\left[\frac{3}{2}x + 3\right]$$
$$= \frac{2}{3}\left[\frac{3}{2}x + 3\right] - 2$$
$$= x + 2 - 2 = x$$

Now let us find $(f^{-1} \circ f)(x)$:

$$(f^{-1} \circ f)(x) = f^{-1}[f(x)]$$
$$= f^{-1}\left[\frac{2}{3}x - 2\right]$$
$$= \frac{3}{2}\left[\frac{2}{3}x - 2\right] + 3$$
$$= x - 3 + 3 = x$$

Thus, $(f \circ f^{-1})(x) = x$ and $(f^{-1} \circ f)(x) = x$. It turns out that for any function f whose inverse is also a function, we have the property

$$(f \circ f^{-1})(x) = x$$

and

$$(f^{-1} \circ f)(x) = x$$

This property should not be surprising, since the inverse is the function that does the opposite of, or "undoes," the results of the original function. Whatever value f sends x to, f^{-1} should send this value back to x! (See Figure 10.1.14.)

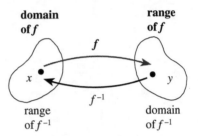

Figure 10.1.14

NOTE ▶ It can be shown that, given two functions f and g, if $(f \circ g)(x) = x$ and $(g \circ f)(x) = x$, then f and g are inverse functions.

XAMPLE 10.1.4   Determine whether the following pairs of functions are inverses by finding $(f \circ g)(x)$ and $(g \circ f)(x)$.

1. $f(x) = 3x - 1$, $g(x) = \dfrac{x + 1}{3}$

$$(f \circ g)(x) = f[g(x)] \qquad\qquad (g \circ f)(x) = g[f(x)]$$

$$= f\left[\frac{x + 1}{3}\right] \qquad\qquad\quad = g[3x - 1]$$

$$= 3\left[\frac{x + 1}{3}\right] - 1 \qquad\qquad = \frac{3x - 1 + 1}{3}$$

$$= x + 1 - 1 \qquad\qquad\qquad = \frac{3x}{3}$$

$$= x \qquad\qquad\qquad\qquad\quad = x$$

Thus, f and g are inverses.

2. $f(x) = \dfrac{2x}{x - 1}$, $g(x) = \dfrac{x}{x + 2}$

$$(f \circ g)(x) = f[g(x)]$$

$$= f\left[\dfrac{x}{x + 2}\right]$$

$$= \dfrac{2\left[\dfrac{x}{x + 2}\right]}{\dfrac{x}{x + 2} - 1}$$

$$= \dfrac{\dfrac{2x}{x + 2}}{\dfrac{x}{x + 2} - \dfrac{x + 2}{x + 2}}$$

$$= \dfrac{\dfrac{2x}{x + 2}}{\dfrac{-2}{x + 2}}$$

$$= \dfrac{2x}{x + 2} \cdot \dfrac{x + 2}{-2}$$

$$= -x$$

Thus, f and g are not inverses.

Exercises 10.1

Find an equation that defines the inverse. If the inverse is a function, write it in the form $y = f^{-1}(x)$. Also graph both equations on the same coordinate plane.

1. $f(x) = 3x + 1$

2. $f(x) = 2x - 4$

3. $f(x) = 3 - 2x$

4. $f(x) = 2 - 4x$

5. $f(x) = \dfrac{x + 3}{4}$

6. $f(x) = \dfrac{x + 1}{2}$

7. $f(x) = \dfrac{2x - 5}{3}$

8. $f(x) = \dfrac{4x - 1}{3}$

9. $f(x) = x^2 - 3$

10. $f(x) = x^2 - 5$

11. $f(x) = \dfrac{1}{2}x^2 + 2$

12. $f(x) = \dfrac{1}{3}x^2 + 4$

13. $f(x) = 5 - 2x^2$

14. $f(x) = 7 - 3x^2$

15. $f(x) = \dfrac{6 - x^2}{4}$

16. $f(x) = \dfrac{2 - x^2}{3}$

17. $f(x) = \sqrt{x + 2}$

18. $f(x) = \sqrt{x + 5}$

19. $f(x) = \sqrt{x - 1}$

20. $f(x) = \sqrt{x - \dfrac{7}{4}}$

21. $f(x) = \sqrt{5 - 2x}$

22. $f(x) = \sqrt{4 - x}$

23. $f(x) = \sqrt{x} + 1$

24. $f(x) = \sqrt{x} - 1$

25. $f(x) = |x + 1|$

26. $f(x) = |x + 4|$

27. $f(x) = |x - 2|$

28. $f(x) = |x - 3|$

29. $f(x) = |x| + 2$ **30.** $f(x) = |x| - 2$ **31.** $f(x) = x^3$ **32.** $f(x) = x^3 + 2$

33. $f(x) = \sqrt[3]{x + 1}$ **34.** $f(x) = \sqrt[3]{x - 2}$

Determine whether the following functions are one-to-one. If they are, find the inverse functions.

35. $f(x) = x^5$ **36.** $f(x) = x^6$ **37.** $f(x) = \dfrac{3}{x + 2}$ **38.** $f(x) = \dfrac{6}{x - 3}$

39. $f(x) = \dfrac{x}{x - 4}$ **40.** $f(x) = \dfrac{x}{x + 5}$ **41.** $f(x) = \dfrac{1}{x^2 - 4}$ **42.** $f(x) = \dfrac{3}{x^2 - 1}$

43. $f(x) = |2x + 3|$ **44.** $f(x) = |5x - 2|$ **45.** $f(x) = x^2 + 4$ **46.** $f(x) = x^2 - 3$

47. $f(x) = \sqrt{x + 2}$ **48.** $f(x) = \sqrt{x - 4}$ **49.** $f(x) = \sqrt[3]{x - 5}$ **50.** $f(x) = \sqrt[3]{x + 7}$

For each pair of functions find the following: (a) $(f \circ g)(2)$, (b) $(f \circ g)(-3)$, (c) $(f \circ g)(x)$, (d) $(g \circ f)(3)$, (e) $(g \circ f)(-1)$, and (f) $(g \circ f)(x)$.

51. $f(x) = 2x + 1$, $g(x) = 3 - x$ **52.** $f(x) = 3x - 4$, $g(x) = 2 - x$ **53.** $f(x) = x^2 + 1$, $g(x) = x - 5$

54. $f(x) = x^2 - 4$, $g(x) = x + 3$ **55.** $f(x) = 4$, $g(x) = 2x$ **56.** $f(x) = 3x$, $g(x) = 5$

Determine whether the following pairs of functions are inverses by finding $(f \circ g)(x)$ and $(g \circ f)(x)$.

57. $f(x) = 3x + 5$, $g(x) = x - \dfrac{5}{3}$ **58.** $f(x) = 2x - 7$, $g(x) = -2x + 7$

59. $f(x) = \dfrac{2x + 3}{4}$, $g(x) = \dfrac{4x - 3}{2}$ **60.** $f(x) = \dfrac{6x - 2}{5}$, $g(x) = \dfrac{5x + 2}{6}$

61. $f(x) = \dfrac{2}{x + 3}$, $g(x) = \dfrac{2 - 3x}{x}$ **62.** $f(x) = \dfrac{2x}{x + 5}$, $g(x) = \dfrac{5x}{2 - x}$

63. $f(x) = \dfrac{1}{x - 2}$, $g(x) = \dfrac{2 + x}{x}$ **64.** $f(x) = \dfrac{x}{x - 1}$, $g(x) = \dfrac{x + 1}{x}$

Write Algebra

65. Give an example of a linear function that does not have an inverse function.

66. Explain the relation between the domain and range of a function and those of its inverse.

67. Explain how you can use the horizontal line test to determine if a graph is the graph of a one-to-one function.

⌨ **The following exercises require the use of a graphing calculator. In each of the exercises, graph the functions $f(x), g(x)$, and $y = x$. By examining the graphs, determine if $g(x) = f^{-1}(x)$. (*Hint:* Square up the grid.)**

68. $f(x) = 2x + 5$; $g(x) = -\frac{1}{2}x + 6$ **69.** $f(x) = \frac{4}{3}x - 2$; $g(x) = \frac{3}{4}x + \frac{3}{2}$

70. $f(x) = x^3 + 2$; $g(x) = \sqrt[3]{x - 2}$ **71.** $f(x) = x^2 + 3$; $g(x) = \sqrt{x - 3}$

In exercises 72–77, find the value of a so that $f(f(x)) = x$.

72. $f(x) = \dfrac{1}{x + a}$

73. $f(x) = \dfrac{x}{x + a}$

74. $f(x) = \dfrac{ax}{x - 3}$

75. $f(x) = \dfrac{ax}{x + 2}$

76. $f(x) = \dfrac{x}{3x + a}$

77. $f(x) = \dfrac{x}{2x + a}$

10.2   *Exponential Functions and Equations*

For the remainder of this chapter we will study two types of functions that have widespread applications in many different fields. We will examine such diverse applications as computing compound interest, measuring population growth, and measuring the decay of radioactive elements. The two functions that will help us in these different areas are the exponential function and the logarithmic function.

We will begin by studying exponential functions. An **exponential function** is one in which the variable appears in an exponent. For example, consider the function defined by the equation $f(x) = 2^x$. In order to work with this function, we have to know the domain; that is, what is the largest set of numbers that can be substituted for x so that $f(x)$ is a real number?

In Chapter 1 we gave a meaning to exponents that are natural numbers. For example, we know that

$$2^4 = 2 \cdot 2 \cdot 2 \cdot 2 = 16 \quad \text{and} \quad 2^1 = 2$$

So the domain contains the set of natural numbers. In Chapter 4 we gave definitions for exponents that are zero or negative integers. For example, we know that

$$2^0 = 1 \quad \text{and} \quad 2^{-3} = \frac{1}{2^3} = \frac{1}{8}$$

So the domain contains the set of integers. In Chapter 5 we gave a definition for exponents that are fractions. For example, we know that

$$2^{1/2} = \sqrt{2} \quad \text{and} \quad 2^{-2/3} = \frac{1}{\sqrt[3]{2^2}}$$

So the domain contains the set of rational numbers.

If the domain is to be the set of all real numbers, then we must have a meaning for exponents that are irrational numbers. For example, what would 2^π be? We could approximate 2^π by approximating π with a rational number and then computing 2 to this power. The better the approximation for π, the better is the approximation for 2^π. The table gives approximations for π and 2^π that are successively more and more accurate.

n	2^n
3	8
3.1	8.5741877
3.14	8.8152409
3.142	8.8274699
3.1416	8.8250228
3.14159	8.8249616
3.141593	8.8249799
3.1415927	8.8249781
3.14159265	8.8249778

As n gets closer and closer to the actual value of π, it is apparent that 2^n is getting closer and closer to some number (8.8249778 . . .). This number is 2^π. We cannot write the exact value for 2^π any more than we can write the exact value for π. Actually there is more to this argument than we are prepared to go into in a course of this level. We are interested only in the fact that a value for 2^x exists for any irrational number x. Thus the domain of the function $f(x) = 2^x$ includes the set of irrational numbers, and therefore the domain of f is the set of all real numbers.

The graph of f is shown in Figure 10.2.1. To arrive at it we plotted points only for integer values of x and then drew the curve. This is how we will sketch the graph of any exponential function.

x	y
-3	$\frac{1}{8}$
-2	$\frac{1}{4}$
-1	$\frac{1}{2}$
0	1
1	2
2	4
3	8

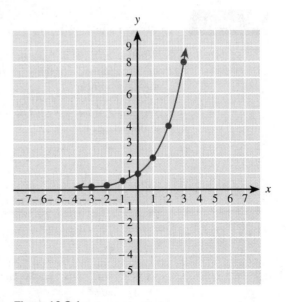

Figure 10.2.1

In general, the graph of the exponential function $f(x) = a^x$ has the properties illustrated in Figure 10.2.2. If $a > 1$, then the graph rises to the right and gets closer to the x-axis as x gets smaller in the negative direction [see Figure 10.2.2(a)]. Thus, the negative half of the x-axis is a horizontal asymptote for the graph. Recall that an asymptote is a line to which a graph gets closer and closer, usually without ever reaching it. If $a = 1$, then the graph is a horizontal line $y = 1$ [see Figure 10.2.2(b)]. If $0 < a < 1$, then the graph falls to the right and gets closer to the x-axis as x gets larger [see Figure 10.2.2(c)]. Thus, the positive half of the x-axis is a horizontal asymptote for the graph.

(a)
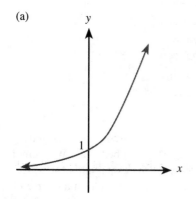

(b)
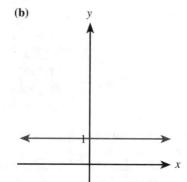

(c)
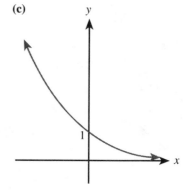

Figure 10.2.2

REMARKS

1. We will not consider the case where $a \leq 0$.

2. When $a > 0$, but $a \neq 1$, the *exponential function $f(x) = a^x$ is a one-to-one function and therefore its inverse is a function.* We will study this inverse function in the next section.

3. When $a > 0$ but $a \neq 1$, the domain is the set of all real numbers, and the range is the set $\{y : y > 0\}$.

XAMPLE 10.2.1   Graph the following functions.

1. $f(x) = 5^x$

    Notice that this graph in Figure 10.2.3 is steeper than the graph of $f(x) = 2^x$.

x	y
-2	$\frac{1}{25}$
-1	$\frac{1}{5}$
0	1
1	5
2	25

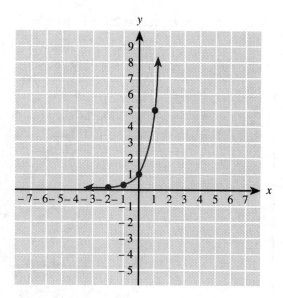

Figure 10.2.3

2. $f(x) = \left(\dfrac{1}{2}\right)^x$

This graph in Figure 10.2.4 is the "mirror image" of the graph of $f(x) = 2^x$ with respect to the y-axis.

$f(x) = (\frac{1}{2})^x$

x	y
-3	8
-2	4
-1	2
0	1
1	$\frac{1}{2}$
2	$\frac{1}{4}$
3	$\frac{1}{8}$

$f(x) = 2^{-x}$

x	y
-3	8
-2	4
-1	2
0	1
1	$\frac{1}{2}$
2	$\frac{1}{4}$
3	$\frac{1}{8}$

Figure 10.2.4

3. $f(x) = 2^{-x}$

This function generates the same ordered pairs as in part 2, so the graph is the same. The reason for this is that $2^{-x} = (2^{-1})^x = (\frac{1}{2})^x$.

4. $f(x) = 2^{x+2}$

This graph in Figure 10.2.5 has the same shape as the graph of $f(x) = 2^x$ but is shifted 2 units to the left.

x	y
-5	$\dfrac{1}{8}$
-4	$\dfrac{1}{4}$
-3	$\dfrac{1}{2}$
-2	1
-1	2
0	4
1	8

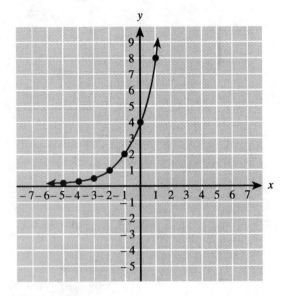

Figure 10.2.5

5. $f(x) = 10(2^{-x/4})$

It is important to note that the base is 2 and that we multiply by 10 *after* evaluating the exponential expression $2^{-x/4}$. The graph is shown in Figure 10.2.6. The graph looks flat due to the difference in scales on the x- and y-axes.

x	y
-8	40
-6	28.28
-4	20
-2	14.14
0	10
2	7.07
4	5
6	3.54
8	2.5
10	1.77
12	1.25

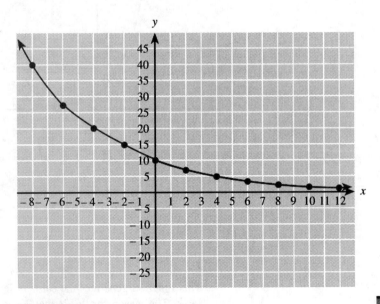

Figure 10.2.6

In this last example and in later functions that we will consider, it is helpful to be able to evaluate exponential expressions with the aid of a calculator that has a $\boxed{y^x}$ key. This can be done with calculators that employ algebraic logic by using the following steps:

1. Enter the base.

2. Press the $\boxed{y^x}$ key.

3. Enter the exponent.

4. Press the $\boxed{=}$ key.

5. Read the answer from the display.

In Example 10.2.1, part 5, when $x = -6$, this exponent is

$$-\left(-\frac{6}{4}\right) = \frac{3}{2}, \quad \text{or} \quad 1.5$$

The corresponding y value is found as follows:

$$2 \boxed{y^x} 1.5 \boxed{=} 2.8284271 \boxed{\times} 10 \boxed{=} 28.284271$$

NOTE ▶ A very important number in engineering and the sciences is

$$e = 2.718281828459045\ldots$$

This number is an irrational number. The exponential function with base e is therefore a significant function. To graph it we need to raise e to various powers, which can be done with a calculator (see exercise 47 in Exercises 10.3).

EXAMPLE 10.2.2 Sketch the graph of $f(x) = e^x$.

See Figure 10.2.7.

x	y
−3	0.05
−2	0.14
−1	0.37
0	1
1	2.72
2	7.39
3	20.09

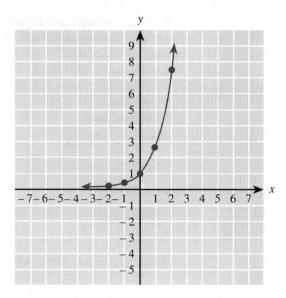

Figure 10.2.7

Population growth is just one of many applications of exponential functions. In the following example and in the problems at the end of this section, we will examine some of these applications.

XAMPLE 10.2.3 The population of Moss County can be described by the equation $P = 60,000(1.015)^t$, where t is the number of years after 1950. What was the population of Moss County in 1950? 1960? 1970?

SOLUTION The population of Moss County in 1950, which is called the initial population, can be found by letting $t = 0$. Thus,

$$P = 60,000(1.015)^0 = 60,000 \cdot 1 = 60,000$$

In 1960, $t = 10$, so

$$P = 60,000(1.015)^{10}$$
$$= 60,000(1.16054)$$
$$\doteq 69,632$$

In 1970, $t = 20$, so

$$P = 60,000(1.015)^{20}$$
$$= 60,000(1.346855)$$
$$\doteq 80,811$$

We round to the nearest whole number, since we are dealing with numbers of people.

In many applications we need to solve equations of the form $y = a^x$, where the values of only two of the variables y, a, or x are known. Depending on which variable is unknown, these equations can be solved using one of the three methods illustrated in examples 10.2.4–10.2.6. If y is unknown, the exponential expression a^x needs to be evaluated.

EXAMPLE 10.2.4   Solve the equation.

$$y = 4^{-3}$$

$$= \frac{1}{4^3} \qquad \textit{Simplify the right side.}$$

$$= \frac{1}{64}$$

If a is unknown, the equation must be solved for a by making the exponent of a equal to one. This is done by raising both sides to the power equal to the reciprocal of the power on a. (Since a is a base, we will be interested in only positive real solutions for a.)

EXAMPLE 10.2.5   Solve the equation.

$$9 = a^{2/3}$$
$$(9)^{3/2} = (a^{2/3})^{3/2} \qquad \textit{Raise both sides to the power } \tfrac{3}{2}.$$
$$3^3 = a^1$$
$$27 = a$$

If x is unknown, the equation $y = a^x$ is called an **exponential equation.** It can sometimes be solved by writing both sides as exponential expressions with the same base. If the expressions are equal and the bases are equal, the exponents must be equal. This method can be used because the exponential function is a one-to-one function.

One-to-One Property of Exponential Functions

Let a be a positive real number, with $a \neq 1$, and let x and z be real numbers. If $a^x = a^z$, then $x = z$.

XAMPLE 10.2.6   Solve the following equations.

1. $125 = 5^x$

    $5^3 = 5^x$ *Write 125 as a power of 5.*

    $3 = x$ *Equate the exponents.*

2. $\dfrac{1}{27} = 9^{x+2}$

    $3^{-3} = (3^2)^{x+2}$ *Write both sides as exponential expressions with base 3.*

    $3^{-3} = 3^{2x+4}$

    $-3 = 2x + 4$ *Equate the exponents.*

    $-7 = 2x$

    $-\dfrac{7}{2} = x$

Exercises 10.2

Sketch the graphs of the following functions.

1. $f(x) = 2^{x+3}$

2. $f(x) = 2^{x-1}$

3. $f(x) = 2^{3-x}$

4. $f(x) = 2^{x/2}$

5. $f(x) = 2^{-x/2}$

6. $f(x) = 10(2^{x/3})$

7. $f(x) = 3 + 2^x$

8. $f(x) = -5 + 2^x$

9. $f(x) = 5 - 2^x$

10. $f(x) = 3^x$

11. $f(x) = 4^x$

12. $f(x) = \left(\dfrac{1}{3}\right)^x$

13. $f(x) = \left(\dfrac{1}{4}\right)^x$

14. $f(x) = \left(\dfrac{3}{4}\right)^x$

15. $f(x) = \left(\dfrac{6}{5}\right)^x$

16. $f(x) = 3^{-2x}$

17. $f(x) = 16(4^{x/8})$

18. $f(x) = 1.02^x$

19. $f(x) = 100(1.05)^{4x}$

20. $f(x) = 48(3^{-x/200})$

Solve the following equations. Let a represent a positive real number.

21. $y = 5^{-2}$

22. $y = 16^{-1/2}$

23. $y = 27^{2/3}$

24. $y = 16^{3/4}$

25. $y = 25^{-3/2}$

26. $y = 4^{-5/2}$

27. $16 = a^2$

28. $81 = a^4$

29. $3 = a^{1/3}$

30. $2 = a^{1/4}$

31. $36 = a^{2/3}$

32. $27 = a^{3/4}$

33. $5 = a^{-1/3}$

34. $\dfrac{2}{7} = a^{-1/2}$

35. $\dfrac{1}{16} = a^{-4/3}$

36. $9 = a^{-2/5}$

37. $8 = 2^x$

38. $9 = 3^x$

39. $\dfrac{1}{4} = 2^x$

40. $\dfrac{1}{27} = 3^x$

41. $\dfrac{4}{9} = \left(\dfrac{2}{3}\right)^x$

42. $\dfrac{9}{25} = \left(\dfrac{3}{5}\right)^x$

43. $\dfrac{1}{4} = 8^x$

44. $\dfrac{1}{81} = 27^x$

45. $\dfrac{1}{25} = 5^{x-1}$

46. $16 = 2^{3x+1}$ **47.** $8 = 4^{x/3}$ **48.** $125 = 5^{-x/2}$ **49.** $\dfrac{1}{9} = 4(6^{3x})$ **50.** $96 = 3(4^{-0.5x})$

The following applications require a calculator.

51. The population of Texas since 1940 can be roughly approximated by the function

$$P = 6.4(1.02)^t$$

where P is measured in millions and t is the number of years after 1940. According to this function, what was the population in these years: 1940, 1950, 1960, 1970, and 1980? You can compare your answers with the table of actual values given here. If this growth rate continues, what will the population be in 2000?

52. Mary had her apartment sprayed for roaches on December 1. All but 5 of the roaches were annihilated. The number of roaches in her apartment can now be described by the equation $R = 5(1.15)^t$, where t represents the number of days that have passed since December 1. How many roaches will there be on December 10? December 25?

Year	Actual Values	Function Values
1940	6,414,824	
1950	7,711,194	
1960	9,579,677	
1970	11,196,730	
1980	14,228,000	
2000	?	

53. If $5000 is deposited in an account paying 8% annual interest, compounded quarterly, the amount of money, P, in the account t years later can be described by the equation $P = 5000(1.02)^{4t}$. How much money will be in the account after 1 yr? 5 yr? In approximately how many years will the money in the account be at least twice as much as was deposited?

54. If $800 is deposited in an account paying 6% annual interest, compounded monthly, the amount of money, P, in the account t years later can be described by the equation $P = 800(1.005)^{12t}$. How much money will be in the account after 1 yr? 5 yr? In approximately how many years will the money in the account be at least twice as much as was deposited?

55. If every month $100 is deposited in an account paying 6% annual interest, compounded monthly, the amount of money, P, in the account t years later can be described by the equation $P = 20,100(1.005^{12t} - 1)$. How much money will be in the account after 1 yr? 5 yr?

56. The radioactive element radium decays, turning into lead, as time passes. Starting with an initial quantity of 10 g, the amount of radium, A, left after t years is given by the equation $A = 10(2^{(-1/1620)t})$. How much radium is left after 810 yr? 1620 yr? 3240 yr?

57. Suppose an object that starts at an initial temperature of T_0 (at time $t = 0$) is placed in a room with the surrounding air at temperature A. Newton's law of cooling (or heating) states that the temperature of the object, T, at time t is given by the equation

$$T = A + (T_0 - A)e^{kt}$$

where k is a constant that depends on the object. A bologna sandwich is removed from a refrigerator at 7:00 A.M., where it had been kept at 40°F, and is placed in a room where the air temperature is 70°F. The equation that gives its temperature is

$$T = 70 + (40 - 70)e^{-0.015t}$$

where t is measured in minutes and T in degrees Fahrenheit. What will the temperature of the sandwich be at 8:00 A.M.? 9:00 A.M.?

Write Algebra

58. If $1^x = 1^y$, explain why we cannot conclude that $x = y$.

59. Explain why we cannot easily apply the one-to-one property of exponential functions to solve the equation $3^x = 7$.

60. Given the exponential function $f(x) = b^x$, describe the graph if   **a.** $b > 1$ or   **b.** $0 < b < 1$.

The following exercises require the use of a graphing calculator.

61. Graph $y = 2^x$, $y = 1.5^x$, and $y = 1.2^x$. Predict the graph of $y = 1.1^x$. Now graph $y = 1.1^x$. Describe the pattern you see in the graphs.

62. Graph $y = 2^x$, $y = 2^{x-2}$, and $y = 2^{x-4}$. Predict the graph of $y = 2^{x-6}$. Now graph $y = 2^{x-6}$. Describe the pattern you see in the graphs.

63. Graph $y = 2^x$, $y = 2^x + 2$, and $y = 2^x + 4$. Describe the pattern you see in the graphs. Determine the horizontal asymptotes.

64. Graph $y = (\frac{1}{2})^x - 1$, $y = (\frac{1}{2})^x - 3$, and $y = (\frac{1}{2})^x - 5$. Describe the pattern you see in the graphs. Determine the horizontal asymptotes.

G R O U P   A C T I V I T Y

In exercises 65–72, solve the equations.

65. $2^{3x+1} \cdot 4^{x-5} = 8$

66. $2^{4x-5} \cdot 4^{x+3} = 8$

67. $2^{3x+1} \cdot 4^{2x} = 32^{x-1}$

68. $2^{4x-3} \cdot 4^{x+3} = 8^{x-2}$

69. $2^{x+1} \cdot 4^{x-3} = 8^{x+2}$

70. $2^{2x-1} \cdot 4^{x+3} = 16^{x-2}$

71. $2^{2x-4} \cdot 4^{2x-1} = 64^{x-1}$

72. $2^{x+4} \cdot 4^{2x+3} = 32^{x+2}$

The table on pages 600 and 601 contains statistics obtained from the Internet on cigarette smoking by persons 18 years of age and over in the United States. The Web address of this table is

http://www.cdc.gov/nchs/datawh/statab/pubd/hus-t62h.htm

In the following example we will use some of the information in this table to construct a model to predict future smoking levels.

XAMPLE

Construct an exponential model to predict the percentage of cigarette smokers among the population of all persons 18 years and over in the United States (age adjusted).

SOLUTION

Since we are *assuming* an exponential model, we will use the equation

$$y = a \cdot b^{x-1900}$$

In this equation we will let

x = the year

y = the percentage of the population that smokes cigarettes

where a and b are constants that must be determined. Since our model contains two unknowns (a and b), we need to use two data points to determine a and b. We can choose any two years to generate these points. Let's choose 1965 and 1985. Thus, the two data points are (1965, 42.3) and (1985, 30.0). Given our model $y = a \cdot b^{x-1900}$ and using the data point (1965, 42.3), we obtain

$$42.3 = a \cdot b^{1965-1900}$$
$$42.3 = a \cdot b^{65}$$

or

$$\frac{42.3}{b^{65}} = a \qquad (1)$$

Using the data point (1985, 30.0), we obtain

$$30 = a \cdot b^{1985-1900}$$
$$30 = a \cdot b^{85}$$

Continued

or

$$\frac{30}{b^{85}} = a \tag{2}$$

We now have a system of equations

$$a = \frac{42.3}{b^{65}}$$

$$a = \frac{30}{b^{85}}$$

Solving this system using the substitution method we obtain

$$\frac{30}{b^{85}} = \frac{42.3}{b^{65}}$$

$$42.3b^{85} = 30b^{65}$$

$$\frac{b^{85}}{b^{65}} = \frac{30}{42.3}$$

$$b^{20} = \frac{30}{42.3}$$

$$b = \left(\frac{30}{42.3}\right)^{1/20}$$

$$b \doteq 0.983$$

From equation (2)

$$a = \frac{30}{b^{85}}$$

$$a \doteq \frac{30}{(0.983)^{85}}$$

$$a \doteq 129$$

Hence, our model is given by the equation

$$y = 129 \cdot (0.983)^{x-1900}$$

Using this model let's predict the percent cigarette smoking level for all persons 18 years of age and over (age adjusted) for the year 2001.

$$y = 129 \cdot (0.983)^{2001-1900}$$

$$= 129 \cdot (0.983)^{101}$$

$$\doteq 22.8$$

Continued

Thus, using our model the predicted percentage of all persons 18 years of age and over that will be cigarette smokers in the United States in the year 2001 will be approximately 22.8%.

Exercises

I. In the following problems construct an *exponential model* to predict the percentage of cigarette smokers for the indicated category. Use the equation $y = a \cdot b^{x-1900}$, and let x = the year, and y = percentage of the category that smokes cigarettes.

1. All males 45–64 years of age.

2. White males 18–24 years of age.

3. Black males 35–44 years of age.

4. All females 18–24 years of age.

5. White females 25–34 years of age.

6. Black females 25–34 years of age.

II. Predict the percentage of cigarette smokers for all the categories from the exercises in part I for the year 2004.

Group Activity

1. Working in groups use different data points to generate other exponential models (other values for a and b). Try to determine which models yield more accurate predictions for particular years.

2. Discuss the reasons why a particular model might yield more accurate results for a given year than another model.

Further Exploration

Surf the Internet to see if you can find any data on cigarette smoking for the years after 1993. If you are successful, test the accuracy of your models from part I.

Continued

Current Cigarette Smoking by Persons 18 Years of Age and Over, According to Sex, Race, and Age: United States, Selected Years 1965–93

[Data are based on household interviews of a sample of the civilian noninstitutionalized population.]

Sex, race, and age	1965	1974	1979	1983	1985	1987	1988	1990	1991	1992	1993
All persons		Percent of persons 18 years of age and over									
18 years and over, age adjusted	42.3	37.2	33.5	32.2	30.0	28.7	27.9	25.4	25.4	26.4	25.0
18 years and over, crude	42.4	37.1	33.5	32.1	30.1	28.8	28.1	25.5	25.6	26.5	25.0
All males											
18 years and over, age adjusted	51.6	42.9	37.2	34.7	32.1	31.0	30.1	28.0	27.5	28.2	27.5
18 years and over, crude	51.9	43.1	37.5	35.1	32.6	31.2	30.8	28.4	28.1	28.6	27.7
18–24 years	54.1	42.1	35.0	32.9	28.0	28.2	25.5	26.6	23.5	28.0	28.8
25–34 years	60.7	50.5	43.9	38.8	38.2	34.8	36.2	31.6	32.8	32.8	30.2
35–44 years	58.2	51.0	41.8	41.0	37.6	36.6	36.5	34.5	33.1	32.9	32.0
45–64 years	51.9	42.6	39.3	35.9	33.4	33.5	31.3	29.3	29.3	28.6	29.2
65 years and over	28.5	24.8	20.9	22.0	19.6	17.2	18.0	14.6	15.1	16.1	13.5
White:											
18 years and over, age adjusted	50.8	41.7	36.5	34.1	31.3	30.4	29.5	27.6	27.0	28.0	27.0
18 years and over, crude	51.1	41.9	36.8	34.5	31.7	30.5	30.1	28.0	27.4	28.2	27.0
18–24 years	53.0	40.8	34.3	32.5	28.4	29.2	26.7	27.4	25.1	30.0	30.4
25–34 years	60.1	49.5	43.6	38.6	37.3	33.8	35.4	31.6	32.1	33.5	29.9
35–44 years	57.3	50.1	41.3	40.8	36.6	36.2	35.8	33.5	32.1	30.9	31.2
45–64 years	51.3	41.2	38.3	35.0	32.1	32.4	30.0	28.7	28.0	28.1	27.8
65 years and over	27.7	24.3	20.5	20.6	18.9	16.0	16.9	13.7	14.2	14.9	12.5
Black:											
18 years and over, age adjusted	59.2	54.0	44.1	41.3	39.9	39.0	36.5	32.2	34.7	32.0	33.2
18 years and over, crude	60.4	54.3	44.1	40.6	39.9	39.0	36.5	32.5	35.0	32.2	32.7
18–24 years	62.8	54.9	40.2	34.2	27.2	24.9	18.6	21.3	15.0	16.2	19.9
25–34 years	68.4	58.5	47.5	39.9	45.6	44.9	41.6	33.8	39.4	29.5	30.7
35–44 years	67.3	61.5	48.6	45.5	45.0	44.0	42.5	42.0	44.4	47.5	36.9

Continued

45–64 years	57.9	57.8	50.0	44.8	46.1	44.3	43.2	36.7	42.0	35.4	42.4
65 years and over	36.4	29.7	26.2	38.9	27.7	30.3	29.8	21.5	24.3	28.3	27.9

All females

18 years and over, age adjusted	34.0	32.5	30.3	29.9	28.2	26.7	26.0	23.1	23.6	24.8	22.7
18 years and over, crude	33.9	32.1	29.9	29.5	27.9	26.5	25.7	22.8	23.5	24.6	22.5
18–24 years	38.1	34.1	33.8	35.5	30.4	26.1	26.3	22.5	22.4	24.9	22.9
25–34 years	43.7	38.8	33.7	32.6	32.0	31.8	31.3	28.2	28.4	30.1	27.3
35–44 years	43.7	39.8	37.0	33.8	31.5	29.6	27.8	24.8	27.6	27.3	27.4
45–64 years	32.0	33.4	30.7	31.0	29.9	28.6	27.7	24.8	24.6	26.1	23.0
65 years and over	9.6	12.0	13.2	13.1	13.5	13.7	12.8	11.5	12.0	12.4	10.5

White:

18 years and over, age adjusted	34.3	32.3	30.6	30.1	28.3	27.2	26.2	23.9	24.2	25.7	23.7
18 years and over, crude	34.0	31.7	30.1	29.4	27.7	26.7	25.7	23.4	23.7	25.1	23.1
18–24 years	38.4	34.0	34.5	36.5	31.8	27.8	27.5	25.4	25.1	28.5	26.8
25–34 years	43.4	38.6	34.1	32.2	32.0	31.9	31.0	28.5	28.4	31.5	28.4
35–44 years	43.9	39.3	37.2	34.8	31.0	29.2	28.3	25.0	27.0	27.6	27.3
45–64 years	32.7	33.0	30.6	30.6	29.7	29.0	27.7	25.4	25.3	25.8	23.4
65 years and over	9.8	12.3	13.8	13.2	13.3	13.9	12.6	11.5	12.1	12.6	10.5

Black:

18 years and over, age adjusted	32.1	35.9	30.8	31.8	30.7	27.2	27.1	20.4	23.1	23.9	19.8
18 years and over, crude	33.7	36.4	31.1	32.2	31.0	28.0	27.8	21.2	24.4	24.2	20.8
18–24 years	37.1	35.6	31.8	32.0	23.7	20.4	21.8	10.0	11.8	10.3	8.2
25–34 years	47.8	42.2	35.2	38.0	36.2	35.8	37.2	29.1	32.4	26.9	24.7
35–44 years	42.8	46.4	37.7	32.7	40.2	35.3	27.6	25.5	35.3	32.4	31.5
45–64 years	25.7	38.9	34.2	36.3	33.4	28.4	29.5	22.6	23.4	30.9	21.3
65 years and over	7.1	8.9	8.5	13.1	14.5	11.7	14.8	11.1	9.6	11.1	10.2

Source: Centers for Disease Control and Prevention, National Center for Health Statistics, Division of Health Interview Statistics: Data from the National Health Interview Survey; data computed by the Division of Health and Utilization Analysis from data compiled by the Division of Health Interview Statistics.

National Center for Health Statistics. *Health, United States, 1995,* Hyattsville, Md.: Public Health Service, p. 173. 1996. (The NCHS Web site contains instructions to download the Acrobat® reader required to view this publication.)

Notes: Estimates for 1992 and beyond are not strictly comparable with those for earlier years, and estimates for 1992 and 1993 are not strictly comparable with each other due to a change in the definition of current smoker in 1992 and the use of a split sample in 1992. See discussion of current smoker in Appendix II.

10.3 *Logarithmic Functions*

In the previous section we noted that the exponential function is one-to-one and therefore its inverse is a function. In this section we will begin our study of this inverse function—the logarithmic function.

Let us start with a particular example $y = 2^x$. By interchanging x and y, we obtain an equation defining the inverse function:

$$y = 2^x \xrightarrow{\textit{Inverse}} x = 2^y$$

Normally we solve for y at this point, but this cannot be done using algebraic manipulations. Instead we must introduce new notation for this function. We write $x = 2^y$ as $y = \log_2 x$, read "y equals the log base two of x." The "log" is an abbreviation for logarithm. This example leads to the following definition.

Definition 10.3.1 | The equation $y = \log_a x$, which defines a **logarithmic function,** is equivalent to the equation $a^y = x$. Recall, $a > 0$ and $a \neq 1$.

REMARK

1. A logarithm is just a fancy way of saying that y is the exponent placed on a to obtain x.

2. $x = 2^y$ and $y = \log_2 x$ are two forms of the equation defining the same function. They are called, respectively, the **exponential form** and the **logarithmic form.**

3. We generally use the exponential form of the defining equation to generate ordered pairs:

$$x = 2^y$$

   By substituting values for y in $x = 2^y$, we can easily obtain the ordered pairs in the chart on page 603.

4. Using the ordered pairs, we sketch the graphs of the two functions in Figure 10.3.1 and observe the relationship between them, which was discussed in Section 10.1. Note that in the equation $y = \log_2 x$, as y gets smaller, x gets closer to zero; that is, the negative half of the y-axis is a vertical asymptote for the graph.

In general, the graph of $f(x) = \log_a x$ has the properties illustrated in Figure 10.3.2. If $a > 1$, the graph rises to the right and gets closer and closer to the y-axis to the left, without ever reaching it [see Figure 10.3.2(a)]. If $0 < a < 1$, the graph falls to the right and gets closer and closer to the y-axis to the left, without ever reaching it [see Figure 10.3.2(b)]. From the graphs we note that x must be greater than zero. *Thus we can never take the log of a negative number or zero, no matter what the base.* For example, if

$$\log_{10}(-5) = y$$

then $10^y = -5$, which is impossible!

$y = \log_2 x$	
x	y
$\dfrac{1}{8}$	-3
$\dfrac{1}{4}$	-2
$\dfrac{1}{2}$	-1
1	0
2	1
4	2
8	3

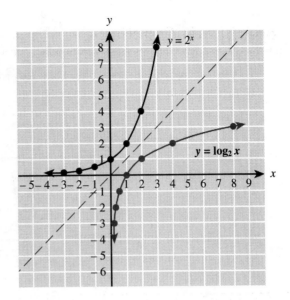

Figure 10.3.1

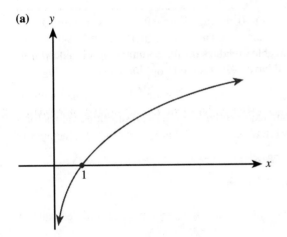

(a)

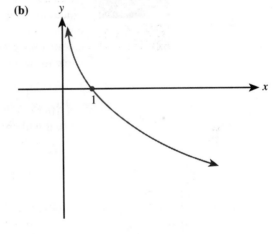

(b)

Figure 10.3.2

XAMPLE 10.3.1   Sketch the graphs of the following functions.

1. $f(x) = \log_{10} x$

   We first rewrite this equation in exponential form:

   $$y = \log_{10} x \longrightarrow 10^y = x$$

   Picking values for y and solving for x, we generate the ordered pairs in the table accompanying Figure 10.3.3. Note that this graph in Figure 10.3.3 rises much more slowly than did the graph of $f(x) = \log_2 x$.

x	y
$\dfrac{1}{100}$	-2
$\dfrac{1}{10}$	-1
1	0
10	1
100	2

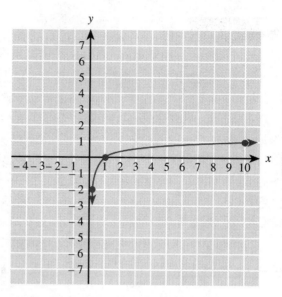

Figure 10.3.3

REMARK Since our number system has a base of 10, the logarithmic function with base 10 is an important function. It has a special name, the **common logarithm,** and a special notation, $\log_{10} x = \log x$. There is a table of values for the common log in Appendix A. Also, any scientific calculator will have a key marked $\boxed{\log}$ for the common log.

2. $f(x) = \log_e x$

After rewriting this equation in exponential form, we can substitute values for y and find approximate values for x.

$$y = \log_e x \longrightarrow x = e^y$$

The graph is sketched in Figure 10.3.4.

x	y
0.05	-3
0.14	-2
0.37	-1
1	0
2.72	1
7.39	2

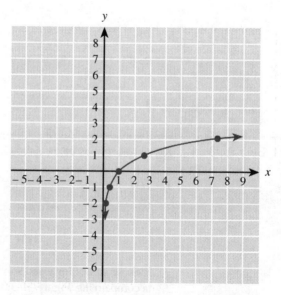

Figure 10.3.4

REMARK   The previous example is called the **natural logarithm** function, and it has its own special notation, $\log_e x = \ln x$. Any scientific calculator will have a key marked $\boxed{\ln}$, which can be used to find $\ln x$ for any $x > 0$.

In the next example we will look at the graph of a more complicated function than $f(x) = \log_a x$. We will observe the relationship between this graph and those of the previous examples.

XAMPLE 10.3.2   Sketch the graph of the following function.

$$f(x) = \log_2(x - 3)$$

Let us first rewrite this equation in exponential form and solve for x:

$$y = \log_2(x - 3) \longrightarrow x - 3 = 2^y$$
$$x = 2^y + 3$$

This graph in Figure 10.3.5 has the same shape as the graph of $f(x) = \log_2 x$, but it is shifted 3 units to the right.

x	y
$\dfrac{25}{8}$	-3
$\dfrac{13}{4}$	-2
$\dfrac{7}{2}$	-1
4	0
5	1
7	2
11	3

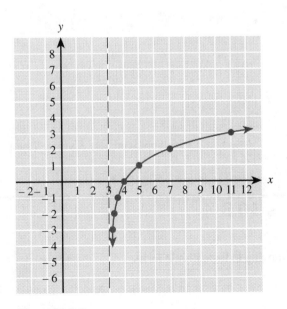

Figure 10.3.5

We conclude the section with an application of the natural logarithmic function.

EXAMPLE 10.3.3 If a certain amount of money is deposited in an account paying an annual interest rate of r, compounded continuously, the time it will take for the money to double is

$$t = \frac{\ln 2}{r}$$

where t is measured in years. How long will it take to double the money if $r = 6\%$? $r = 8\%$?

SOLUTION When $r = 6\%$, use $6\% = 0.06$ to find

$$t = \frac{\ln 2}{0.06}$$
$$\doteq \frac{0.69314718}{0.06}$$
$$\doteq 11.55245301 \text{ years}$$

or approximately 11 years, 6 months, and 19 days. When $r = 8\%$, use $8\% = 0.08$ to find

$$t = \frac{\ln 2}{0.08}$$
$$\doteq \frac{0.69314718}{0.08}$$
$$\doteq 8.664339757 \text{ years}$$

or approximately 8 years, 7 months, and 29 days.

Exercises 10.3

Rewrite the following in logarithmic form.

1. $5^{-3} = \frac{1}{125}$
2. $4^2 = 16$
3. $10^4 = 10000$
4. $7^0 = 1$
5. $6^1 = 6$
6. $49^{1/2} = 7$
7. $y = 3^x$
8. $y = 4^x$
9. $y = 3^{x+2}$
10. $y = 4^{x-2}$
11. $P = e^{2t}$
12. $A = e^{t/4}$
13. $M = (1.02)^{12t}$
14. $N = (1.5)^{6t}$
15. $Q = Q_0 a^{kt}$
16. $P = P_0(1 + r)^t$

Rewrite the following in exponential form.

17. $\log_8 64 = 2$

18. $\log_5 625 = 4$

19. $\log_9 3 = \dfrac{1}{2}$

20. $\log_2 \dfrac{1}{8} = -3$

21. $\log_{10} 10 = 1$

22. $\log_4 1 = 0$

23. $y = \log_5 x$

24. $y = \log_7 x$

25. $y = \log_5(2x + 1)$

26. $y = \log_7(x^2 - 1)$

27. $\log_{27} 9 = x$

28. $\log_4 8 = x$

29. $4 = \log_3(1 - 3x)$

30. $\log_8(x^2 + x) = 1$

31. $\log_x 16 = 2$

32. $\log_x 5 = \dfrac{1}{2}$

33. $\log(x + 1) = 4$

34. $y = \log(x - 2)$

35. $y = \ln x^2$

36. $\ln(3x + 2) = -1$

Sketch the graphs of the following functions.

37. a. $f(x) = \log_3 x$ **b.** $f(x) = \log_3(x + 4)$

38. a. $f(x) = \log_4 x$ **b.** $f(x) = \log_4(x + 3)$

39. a. $f(x) = \log_6 x$ **b.** $f(x) = \log_6(x - 2)$

40. a. $f(x) = \log_8 x$ **b.** $f(x) = \log_8(x - 1)$

41. $f(x) = \log_2(2x)$

42. $f(x) = \log_2\left(\dfrac{x}{2}\right)$

43. $f(x) = \log_2(2x + 3)$

44. $f(x) = \log_2(3x - 2)$

45. $f(x) = \log_2 x^2$

46. $f(x) = \log_2 \sqrt{x}$

47. Many calculators have either an $\boxed{e^x}$ key or an $\boxed{\ln}$ key, but not both. If you have a calculator that has only an $\boxed{\ln}$ key, you can still use it to evaluate both functions. To evaluate powers of e with an $\boxed{\ln}$ key requires the use of an $\boxed{\text{INV}}$ or $\boxed{\text{2nd}}$ key, both of which cause the calculator to perform the inverse function. Of course the inverse of $y = \ln x$ is $y = e^x$. So to evaluate e^x, enter the exponent and press the keys $\boxed{\text{INV}}$ and $\boxed{\ln}$. Use this method to evaluate the following powers of e:

a. $e^{2.534}$ **b.** e^{142} **c.** $e^{0.00385}$

d. $e^{-12.56}$ (*Hint:* You may need to evaluate $e^{-12.56}$

     as $\dfrac{1}{e^{12.56}}$.)

48. In a telegraph cable, the speed, s, of the signal is determined by the equation

$$s = x^2 \ln\left(\dfrac{1}{x}\right)$$

where x is the ratio of the radius of the core of the cable to the thickness of the cable's insulation. Find the speed when $x = \frac{1}{2}$.

49. A natural number is prime if it is greater than 1 and divisible by only itself and 1. According to the *prime number theorem*, the number of primes less than x, where x is large, is approximately $\frac{x}{\ln x}$. By this theorem, how many primes are less than 100? There are actually 25 primes less than 100. Is 100 "large" enough for this theorem?

Write Algebra

50. Explain why 1 is not permitted as a base for a logarithm.

51. Explain why we cannot find the logarithm of a negative number.

52. Given the logarithmic function $f(x) = \log_b x$, describe the graph if   **a.** $b > 1$ or   **b.** $0 < b < 1$.

The following exercises require the use of a graphing calculator.

53. Graph $y = \ln x$, $y = \ln(6x)$, and $y = \ln(\frac{x}{6})$. Describe the pattern you see in the graphs.

54. Graph $f(x) = \ln(x^2)$ and $g(x) = (\ln x)^2$. The graphs illustrate that $\ln(x^2) \neq (\ln x)^2$. Determine the domain of f and the domain of g.

55. Graph $f(x) = \ln(x^2)$ and $g(x) = 2 \ln x$. What do the graphs illustrate about the functions?

56. Graph $f(x) = \ln(6x)$ and $g(x) = \ln 6 + \ln x$. What do the graphs illustrate about the functions?

G R O U P   A C T I V I T Y

In exercises 57–64, find the indicated function values for

$$f(x) = \log(20x - 17) - \log(2x + 9)$$

57. $f(1)$

58. $f(10)$

59. $f(50)$

60. $f(100)$

61. $f(200)$

62. $f(1000)$

63. $f(5000)$

64. $f(10,000)$

65. Predict the value that $f(x)$ approaches as x takes on larger and larger values. Explain your conjecture.

10.4 *Properties of Logarithms*

The importance of the logarithmic function is based on the properties that this function has. In this section we will study these properties and show how they correspond to the properties of exponents (because they *are* exponents!) that we studied in Chapter 4. In the following properties, b, M, and N represent positive real numbers and $b \neq 1$.

Product Property

$$\log_b MN = \log_b M + \log_b N$$

In words, this property states that the log of a product of factors is equal to the sum of the logs of the individual factors. To prove this property, we let

$$\log_b M = m \quad \text{and} \quad \log_b N = n$$

We then rewrite these equations in exponential form:

$$b^m = M \quad \text{and} \quad b^n = N$$

Now

$$MN = b^m \cdot b^n = b^{m+n}$$

If we rewrite $MN = b^{m+n}$ in logarithmic form, we get

$$\log_b MN = \quad m \quad + \quad n$$
$$= \log_b M + \log_b N$$

Quotient Property

$$\log_b \frac{M}{N} = \log_b M - \log_b N$$

In words, this property states that the log of a quotient is equal to the log of the numerator minus the log of the denominator. The proof of this property is left to the reader as exercise 60 at the end of this section.

Power Property

$$\log_b N^r = r \cdot \log_b N$$

In words, this property states that the log of an exponential expression is equal to the exponent times the log of the base. The proof of this property is left to the reader as exercise 61 at the end of this section.

EXAMPLE 10.4.1

Use the properties of logarithms to express each of the following as sums or differences of logarithms.

1. $\log_b x^3 y^2 = \log_b x^3 + \log_b y^2$ *Product property*
$$= 3 \log_b x + 2 \log_b y \qquad \textit{Power property}$$

2. $\log_b \dfrac{5x^4}{yz} = \log_b 5x^4 - \log_b yz$ *Quotient property*

$$= \log_b 5 + \log_b x^4 - (\log_b y + \log_b z) \qquad \textit{Product property}$$
$$= \log_b 5 + 4 \log_b x - \log_b y - \log_b z \qquad \textit{Power property}$$

3. $\log_b \sqrt[3]{\dfrac{x^2}{y + z}} = \log_b\left(\dfrac{x^2}{y + z}\right)^{1/3}$

$$= \log_b \dfrac{x^{2/3}}{(y + z)^{1/3}}$$

$$= \log_b x^{2/3} - \log_b(y + z)^{1/3} \qquad \textit{Quotient property}$$

$$= \dfrac{2}{3} \log_b x - \dfrac{1}{3} \log_b(y + z) \qquad \textit{Power property}$$

Stop here! $\log_b(y + z)$ cannot be expressed as $\log_b y + \log_b z$.

Try to avoid these mistakes:

Incorrect	Correct
1. $\log_b MN = \log_b M \cdot \log_b N$	$\log_b MN = \log_b M + \log_b N$
2. $\log_b\left(\dfrac{M}{N}\right) = \dfrac{\log_b M}{\log_b N}$	$\log_b\left(\dfrac{M}{N}\right) = \log_b M - \log_b N$
3. $\log_b(M + N) = \log_b M + \log_b N$	$\log_b(M + N)$ and $\log_b(M - N)$
4. $\log_b(M - N) = \log_b M - \log_b N$	cannot be changed using the properties of logarithms.

EXAMPLE 10.4.2   Use the properties of logarithms to express each of the following as a single logarithm.

1. $\log_b x + 3 \log_b y - \dfrac{1}{2} \log_b z = \log_b x + \log_b y^3 - \log_b z^{1/2} \qquad \textit{Power property}$

$$= \log_b xy^3 - \log_b \sqrt{z} \qquad \textit{Product property}$$

$$= \log_b \dfrac{xy^3}{\sqrt{z}} \qquad \textit{Quotient property}$$

2. $\dfrac{3}{4} \log_b x - \dfrac{1}{4} \log_b y - \dfrac{1}{2} \log_b z = \log_b x^{3/4} - \log_b y^{1/4} - \log_b z^{1/2} \qquad \textit{Power property}$

$$= \log_b x^{3/4} - (\log_b y^{1/4} + \log_b z^{1/2})$$

$$= \log_b x^{3/4} - \log_b(y^{1/4} \cdot z^{1/2}) \qquad \textit{Product property}$$

$$= \log_b \dfrac{x^{3/4}}{y^{1/4}z^{1/2}} \qquad \textit{Quotient property}$$

or

$$\log_b\left(\frac{x^3}{yz^2}\right)^{1/4} = \log_b \sqrt[4]{\frac{x^3}{yz^2}}$$

The next property allows you to take a log to any base and rewrite it as a quotient of two logs to an entirely different base.

Change of Base Property

$$\log_b N = \frac{\log_a N}{\log_a b}$$

To prove this property, let us suppose that

$$\log_b N = x \qquad \log_a N = y \qquad \log_a b = z$$

Then

$$b^x = N \qquad a^y = N \qquad a^z = b$$

Now $b^x = N = a^y$. But since $b = a^z$, we can substitute and get

$$(a^z)^x = a^y$$

or

$$a^{zx} = a^y$$

So $z \cdot x = y$, since the bases are the same. This means

$$\log_a b \cdot \log_b N = \log_a N$$

or

$$\log_b N = \frac{\log_a N}{\log_a b}$$

With common or natural logarithms and the change of base property, we can evaluate logs to any base.

EXAMPLE 10.4.3   Use the change of base property to evaluate the following.

$$\log_4 17 = \frac{\log 17}{\log 4} \qquad \textit{Change of base property}$$

$$\doteq \frac{1.23044892}{0.60205999} \qquad \begin{array}{l} \textit{Follow the sequence} \\ 17 \boxed{\log} \boxed{\div} 4 \boxed{\log} \boxed{=} \\ \textit{on your calculator.} \end{array}$$

$$\doteq 2.04373142$$

NOTE ▶   Using natural logarithms, we get the same answer.

$$\log_4 17 = \frac{\ln 17}{\ln 4} \qquad \textit{Change of base property}$$

$$\doteq \frac{2.83321334}{1.38629436}$$

$$\doteq 2.04373142$$

There are two more properties of logarithms that result directly from the fact that $y = a^x$ and $y = \log_a x$ are inverse functions. We know that the composition of a function with its inverse yields a function that sends x back to x. Thus if $f(x) = a^x$ and $g(x) = \log_a x$, then

$$\begin{aligned} (f \circ g)(x) &= f[g(x)] \\ &= f[\log_a x] \\ &= a^{\log_a x} \\ &= x \end{aligned}$$

Figure 10.4.1 illustrates this property.

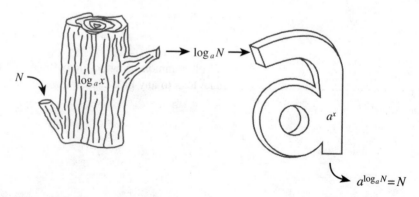

Figure 10.4.1

Since N should be sent back to N, we must have that

$$a^{\log_a N} = N$$

Similarly,

$$
\begin{aligned}
(g \circ f)(x) &= g[f(x)] \\
&= g[a^x] \\
&= \log_a a^x = x
\end{aligned}
$$

Figure 10.4.2 illustrates this property. In this case we must have that

$$\log_a a^N = N$$

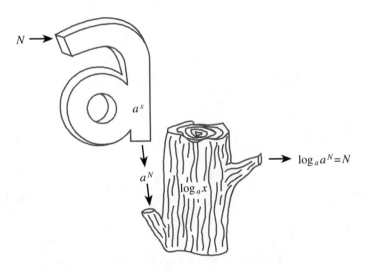

Figure 10.4.2

E X A M P L E   1 0 . 4 . 4   Simplify each of the following.

1. $5^{\log_5 7} = 7$

2. $3^{2 \cdot \log_3 11} = 3^{\log_3 11^2}$
$$= 11^2, \quad \text{or} \quad 121$$

3. $\log_2 2^5 = 5$

4. $\log_6 \sqrt{6} = \log_6 6^{1/2}$

$$= \frac{1}{2}$$

5. $\log_3 9 = \log_3 3^2$

$$= 2$$

XAMPLE 10.4.5   A decibel is the smallest increase in intensity of sound that can be detected by the human ear. A sound of intensity I is rated at a certain number of decibels (dB), D, by the equation

$$D = 10 \log\left(\frac{I}{10^{-16}}\right)$$

Find the decibel ratings of the following sounds with the given intensities:

whisper, $I = 10^{-14}$: $D = 10 \log\left(\frac{10^{-14}}{10^{-16}}\right)$

$$= 10 \log 10^2 = 10 \cdot 2 = 20 \text{ dB}$$

ordinary speech, $I = 10^{-10}$: $D = 10 \log\left(\frac{10^{-10}}{10^{-16}}\right)$

$$= 10 \log 10^6 = 10 \cdot 6 = 60 \text{ dB}$$

car horn, $I = 10^{-7}$: $D = 10 \log\left(\frac{10^{-7}}{10^{-16}}\right)$

$$= 10 \log 10^9 = 10 \cdot 9 = 90 \text{ dB}$$

jet plane, $I = 10^{-4}$: $D = 10 \log\left(\frac{10^{-4}}{10^{-16}}\right)$

$$= 10 \log 10^{12} = 10 \cdot 12 = 120 \text{ dB}$$

NOTE ▶   The following two facts can come in handy:

1. $\log_a a = 1$, since $a^1 = a$.

2. $\log_a 1 = 0$, since $a^0 = 1$.

XAMPLE 10.4.6   Given that $\log_b 3 = 1.4$, $\log_b 5 = 2.0$, and $\log_b 7 = 2.4$, find each of the following.

1. $\log_b 21 = \log_b 3 \cdot 7$

$$= \log_b 3 + \log_b 7$$

$$= 1.4 + 2.4 = 3.8$$

2. $\log_b 0.6 = \log_b \dfrac{3}{5}$

$= \log_b 3 - \log_b 5$

$= 1.4 - 2.0 = -0.6$

3. $\log_b 81 = \log_b 3^4$

$= 4 \cdot \log_b 3$

$= 4(1.4) = 5.6$

4. $\log_5 b = \dfrac{\log_b b}{\log_b 5}$

$= \dfrac{1}{2}$ or 0.5

Exercises 10.4

Use the properties of logarithms to express each of the following as sums or differences of logarithms.

1. $\log_b x^2 y^5$

2. $\log_b xy^4 z^3$

3. $\log_b \dfrac{(xy)^3}{\sqrt{z}}$

4. $\log_b \dfrac{\sqrt{xy}}{z^2}$

5. $\log_b \dfrac{x^5}{y^2 z^4}$

6. $\log_b \dfrac{2x^3}{yz^4}$

7. $\log_b \sqrt{\dfrac{x}{y^3}}$

8. $\log_b \sqrt[3]{\dfrac{x^2}{y^5}}$

9. $\log_b \dfrac{\sqrt{x}}{\sqrt[3]{yz^2}}$

10. $\log_b \dfrac{x\sqrt[3]{y}}{\sqrt[4]{z^2 w^7}}$

11. $\log_b x(y^{2t})$

12. $\log_b P(a^{-kt})$

Use the properties of logarithms to express each of the following as a single logarithm.

13. $2\log_b x + \dfrac{1}{2}\log_b y + \log_b z$

14. $\dfrac{2}{3}\log_b x + 3\log_b y + 4\log_b z$

15. $\dfrac{1}{4}\log_b x + \dfrac{3}{4}\log_b y - \log_b z$

16. $\dfrac{4}{3}\log_b x + \log_b y - \dfrac{1}{3}\log_b z$

17. $\dfrac{1}{2}\log_b x - 2\log_b y - \dfrac{2}{3}\log_b z$

18. $3\log_b x - \dfrac{1}{3}\log_b y - \dfrac{3}{2}\log_b z$

Use the change of base property to evaluate the following.

19. $\log_6 35$

20. $\log_2 15$

21. $\log_7 0.0419$

22. $\log_{12} 0.00052$

23. $\log_{1/3} 4$

24. $\log_{0.1} 18$

Simplify each of the following.

25. $6^{\log_6 2}$

26. $8^{\log_8 7}$

27. $4^{3\log_4 5}$

28. $7^{2\log_7 9}$

29. $\sqrt{5^{\log_5 4}}$

30. $\sqrt[3]{3^{\log_3 8}}$

31. $\log_9 9^3$

32. $\log_5 5^4$

33. $\log_7 \sqrt[3]{7}$

34. $\log_3 \sqrt{3}$

35. $\log_2 8$

36. $\log_5 125$

37. $\log_3 \dfrac{1}{9}$

38. $\log_4 \dfrac{1}{64}$

39. $\log_4 8$

40. $\log_9 27$

41. $\log_{16} \dfrac{1}{8}$

42. $\log_{25} \dfrac{1}{125}$

Given that $\log_b 3 = 1.4$, $\log_b 5 = 2.0$, and $\log_b 7 = 2.4$, find each of the following.

43. $\log_b 15$

44. $\log_b 35$

45. $\log_b 45$

46. $\log_b 75$

47. $\log_b 49$

48. $\log_b 125$

49. $\log_b \sqrt{3}$

50. $\log_b \sqrt[3]{7}$

51. $\log_b \dfrac{7}{5}$

52. $\log_b \dfrac{3}{7}$

53. $\log_b \dfrac{9}{5}$

54. $\log_b \dfrac{25}{27}$

55. $\log_b \sqrt[3]{\dfrac{15}{7}}$

56. $\log_b \sqrt{\dfrac{21}{5}}$

57. $\log_b \dfrac{5\sqrt[4]{75}}{63}$

58. In chemistry, the pH of a substance is defined by the equation

$$pH = -\log[H^+]$$

where $[H^+]$ is the hydrogen ion concentration, measured in moles per liter. A substance is acidic if its pH is less than 7 and basic if its pH is greater than 7. Find the pH of the following substances and determine which are acidic and which are basic.

Substance	$[H^+]$
Human blood	$(3.8)10^{-8}$
Gastric juice	$(1.0)10^{-2}$
Pure water	$(1.0)10^{-7}$
Cow's milk	$(2.5)10^{-7}$
Lemon	$(5.0)10^{-3}$
Crackers	$(6.3)10^{-9}$

59. The Richter scale measures the magnitude, M, of the shock waves of an earthquake in terms of the energy released, E:

$$M = \frac{\log E - 11.4}{1.5}$$

How much did the following earthquakes measure on the Richter scale?

Earthquake	E
Alaska, 1964	$(1.41)10^{24}$
San Francisco, 1906	$(5.96)10^{23}$

60. Prove the quotient property of logarithms. [*Hint:* This proof is similar to the proof of the product property.]

61. Prove the power property of logarithms. [*Hint:* This proof is similar to the proof of the product property.]

62. Show that $-\ln|\sqrt{1 + x^2} - x| = \ln|\sqrt{1 + x^2} + x|$.
   a. Use the power property to eliminate the minus sign.
   b. Multiply top and bottom by $|\sqrt{1 + x^2} + x|$.

63. Express as a single logarithm:

$$2 \ln |x + 3| + \frac{1}{2} \ln |2x - 1| - \ln |x^2 - x - 3|$$

64. Solve for t: $A = A_0 e^{kt}$.

Write Algebra

65. Describe the process you would use to express $5 \log_b x + 2 \log_b y - 3 \log_b z$ as a single logarithm.

66. Describe the difference between each pair of values.
   a. $\log_2 4^3$ and $(\log_2 4)^3$
   b. $\dfrac{\log_3 27}{\log_3 3}$ and $\log_3 \dfrac{27}{3}$
   c. $\log_3(x + y)$ and $(\log_3 x + \log_3 y)$

In exercises 67–74, find the values of m and n in each pair of equations.

67. $\log_2 4 = m$
$\log_4 2 = n$

68. $\log_5 25 = m$
$\log_{25} 5 = n$

69. $\log_3 \frac{1}{9} = m$
$\log_{1/9} 3 = n$

70. $\log_4 \frac{1}{16} = m$
$\log_{1/16} 4 = n$

71. $\log_3 27 = m$
$\log_{27} 3 = n$

72. $\log_4 64 = m$
$\log_{64} 4 = n$

73. $\log_5 \frac{1}{125} = m$
$\log_{1/125} 5 = n$

74. $\log_2 \frac{1}{8} = m$
$\log_{1/8} 2 = n$

75. Predict the value of $\log_a b \cdot \log_b a$; $a > 0$, $b > 0$, $a \neq 1$, $b \neq 1$. Verify your conjecture.

10.5 *Logarithmic and More Exponential Equations*

In Section 10.2 we solved exponential equations in which both sides could be written as exponential expressions with the same base. In this section we will solve exponential equations where this cannot be done. We will find that this type of equation arises frequently in applications.

EXAMPLE 10.5.1   Solve the following equations.

1.

$$15 = 2^x$$

$\ln 15 = \ln 2^x$ *Take the natural log of both sides.*

$\ln 15 = x \cdot \ln 2$ *By the power property of logs*

$$\frac{\ln 15}{\ln 2} = x$$

$$\frac{2.7080502}{0.693147} \doteq x$$

$$3.9068906 \doteq x$$

NOTE ▶   We took the log of both sides so that we could use the power property of logarithms to "move" x out of the exponent. We used the natural log because calculators have natural log keys. For the same reason we can also use the common log, as in the next example.

2.

$$5^{2-x} = 9^{x/4}$$

$$\log 5^{2-x} = \log 9^{x/4} \qquad \textit{Take the common log of both sides.}$$

$$(2 - x)\log 5 = \frac{x}{4}\log 9$$

$$2\log 5 - x\log 5 = x \cdot \frac{1}{4}\log 9$$

$$2\log 5 = x\log 5 + x \cdot \frac{1}{4}\log 9$$

$$2\log 5 = x\left(\log 5 + \frac{1}{4}\log 9\right)$$

$$\frac{2\log 5}{\log 5 + \dfrac{1}{4}\log 9} = x$$

$$\frac{2(0.6989700)}{0.6989700 + \dfrac{1}{4}(0.9542425)} \doteq x$$

$$\frac{1.3979400}{0.6989700 + 0.2385606} \doteq x$$

$$\frac{1.3979400}{0.9375306} \doteq x$$

$$1.4910873 \doteq x$$

When an initial deposit of D is placed in an account paying an annual interest rate of r, compounded n times a year, then the total amount of money, M, in the account t years later is given by the equation

$$M = D\left(1 + \frac{r}{n}\right)^{nt}$$

XAMPLE 10.5.2(a)   Suppose $500 is deposited in an account paying 8% annual interest, compounded quarterly. When will there be $600 in the account?

SOLUTION   In this situation, $D = 500$, $r = 0.08$, and $n = 4$. So the equation giving the value of the account is

$$M = 500\left(1 + \frac{0.08}{4}\right)^{4t}$$

$$= 500(1 + 0.02)^{4t}$$

$$= 500(1.02)^{4t}$$

The question is, for what value of t will $M = 600$?

$$600 = 500(1.02)^{4t}$$

$$\frac{600}{500} = (1.02)^{4t} \qquad \textit{Divide both sides by 500 to isolate the exponential expression.}$$

$$1.2 = (1.02)^{4t}$$

$$\log 1.2 = \log(1.02)^{4t}$$

$$\log 1.2 = 4t \cdot \log(1.02)$$

$$\frac{\log 1.2}{4 \log(1.02)} = t$$

$$\frac{0.0791812}{4(0.0086002)} \doteq t$$

$$2.3017344 \; yr \doteq t$$

This is approximately 2 yr 3.6 mo. But since interest is paid only every 3 mo (quarterly), the account won't actually reach $600 until after 6 mo of the third year. In other words, it will take $2\frac{1}{2}$ yr for the account to reach $600. ▬

XAMPLE 10.5.2(b)   Use a graphing calculator to find an approximation for the solution of the equation in Example 10.5.2(a).

SOLUTION

$$1.2 = (1.02)^{4t}$$

$$0 = (1.02)^{4t} - 1.2 \qquad \textit{Subtract 1.2 from both sides to obtain zero on one side.}$$

$$y = (1.02)^{4t} - 1.2 \qquad \textit{Set the right side equal to y.}$$

Now graph $y = (1.02)^{4x} - 1.2$. Your graph should look something like Figure 10.5.1.

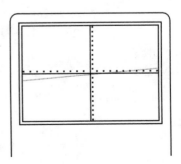

Figure 10.5.1

To find the solution of the equation

$$0 = (1.02)^{4t} - 1.2$$

all we need to do is find the x-intercept of the graph. Using the Zoom and Trace keys, we find the solution to be

$$t \doteq 2.30$$

Let us now turn our attention to **logarithmic equations** of the form

$$\log_b N = n$$

where the unknown x is in the position of b, N, or n. We will solve equations of this form by rewriting them in exponential form:

$$b^n = N$$

and then using the techniques covered earlier to solve them.

XAMPLE 10.5.3   Solve the following equations.

1. $\log_x 9 = \dfrac{2}{3}$

$x^{2/3} = 9$

$(x^{2/3})^{3/2} = 9^{3/2}$ *Change the power of x to 1.*

$x^1 = 3^3$

$x = 27$

2. $\log_4 \dfrac{1}{8} = x$

$4^x = \dfrac{1}{8}$

$(2^2)^x = \dfrac{1}{2^3}$ *Rewrite with the same base.*

$2^{2x} = 2^{-3}$

$2x = -3$ *Equate powers.*

$x = -\dfrac{3}{2}$

3. $\log_5 x = -2$

$5^{-2} = x$

$\dfrac{1}{5^2} = x$ *Simplify.*

$\dfrac{1}{25} = x$

4. $\log_2 x + \log_2(x - 2) = 3$

First, we will use the product property of logarithms to rewrite the left side of the equation as a single logarithm:

$$\log_2 x(x - 2) = 3$$
$$\log_2(x^2 - 2x) = 3$$
$$x^2 - 2x = 2^3 \qquad \textit{Change from logarithmic}$$
$$\textit{to exponential form.}$$
$$x^2 - 2x = 8$$
$$x^2 - 2x - 8 = 0$$
$$(x - 4)(x + 2) = 0$$
$$x - 4 = 0 \quad \text{or} \quad x + 2 = 0$$
$$x = 4 \qquad\qquad x = -2$$

Let us check the solutions.

$x = 4$:

$$\log_2 4 + \log_2(4 - 2) \stackrel{?}{=} 3$$
$$\log_2 4 + \log_2 2$$
$$\log_2 2^2 + \log_2 2$$
$$2 + 1$$
$$3 \ \checkmark$$

$x = -2$:

$$\log_2 (-2) + \log_2(-2 - 2) \stackrel{?}{=} 3$$

The log of a negative number is undefined. Thus $x = -2$ is an extraneous solution.

NOTE ▶   Using one of the properties of logarithms to find the solutions of a logarithmic equation might generate extraneous solutions. Thus, if you use a property of logarithms, check your solutions.

If both sides of the equation contain only logarithms, we take a different approach. For these equations we need the following property:

◯ne-to-One Property of Logarithms

Let b, M, and N be positive real numbers, with $b \neq 1$.
If $\log_b M = \log_b N$, then $M = N$.

Like the exponential function, the logarithmic function

$$f(x) = \log_b x$$

is one-to-one. Thus, if $f(M) = f(N)$, which means $\log_b M = \log_b N$, then $M = N$. The process of deducing from $\log_b M = \log_b N$ that $M = N$ is called "dropping the logs."

EXAMPLE 10.5.4   Solve the following equation:

$$\log_5 2x + \log_5(x + 1) = \log_5(3x + 10)$$

Again, we use the product property of logarithms to rewrite the left side of the equation as a single logarithm:

$$\log_5 [2x \cdot (x + 1)] = \log_5(3x + 10)$$
$$2x \cdot (x + 1) = 3x + 10 \qquad \textit{Drop the logs.}$$
$$2x^2 + 2x = 3x + 10$$
$$2x^2 - x - 10 = 0$$
$$(2x - 5)(x + 2) = 0$$
$$2x - 5 = 0 \quad \text{or} \quad x + 2 = 0$$
$$2x = 5 \qquad\qquad x = -2$$
$$x = \frac{5}{2}$$

Check: $x = \frac{5}{2}$.

$$\log_5 2 \cdot \frac{5}{2} + \log_5\left(\frac{5}{2} + 1\right) \overset{?}{=} \log_5\left(3 \cdot \frac{5}{2} + 10\right)$$

$$\log_5 5 + \log_5\left(\frac{7}{2}\right) \qquad\qquad \log_5\left(\frac{15}{2} + 10\right)$$

$$\log_5 5 \cdot \frac{7}{2} \qquad\qquad \log_5\left(\frac{35}{2}\right)$$

$$\log_5 \frac{35}{2} \quad \checkmark$$

$x = -2$:

$$\log_5 2(-2) + \log_5(-2 + 1) \overset{?}{=} \log_5(3(-2) + 10)$$
$$\log_5(-4) + \log_5(-1) \neq \log_5 4$$

Thus, the only solution is $x = \frac{5}{2}$, since we can't take the log of a negative number.

Exercises 10.5

Solve the following equations. Use a calculator to approximate the solutions to exercises 1–14.

1. $5 = 11^x$

2. $7 = 16^x$

3. $25 = 4^x$

4. $14 = 3^x$

5. $6^{x+3} = 2.5$

6. $3^{x-2} = 6.3$

7. $2^{5x-2} = 3.5^x$

8. $8^{3x+1} = 4.1^x$

9. $2.54^{3-x} = 7^{x+2}$

10. $8.05^{4-x} = 9^{x+5}$

11. $3 = 4 \cdot (5.2)^t$

12. $9 = 5 \cdot (3.1)^t$

13. $200 = 125 \cdot (2.45)^{2t}$

14. $75 = 500 \cdot (6.91)^{3t}$

15. $\log_x 9 = -2$

16. $\log_x \left(\dfrac{1}{16} \right) = -4$

17. $\log_x 4 = \dfrac{2}{5}$

18. $\log_x 27 = \dfrac{3}{2}$

19. $\log_3 \left(\dfrac{1}{27} \right) = x$

20. $\log_5 \left(\dfrac{1}{25} \right) = x$

21. $\log_8 16 = x$

22. $\log_{27} 9 = x$

23. $\log_6 x = -3$

24. $\log_2 x = -4$

25. $\log x = 5$

26. $\ln x = 2$

27. $\log_3 x + \log_3 (x + 6) = 3$

28. $\log_4(x - 3) + \log_4(x + 3) = 2$

29. $\log_4(x + 12) - \log_4(x - 3) = 2$

30. $\log 5x - \log(x - 2) = 0$

31. $2 \log_4 x = 3$

32. $3 \log_8(x - 4) = 2$

33. $\log_2 5 + \log_2(x - 3) = \log_2(x + 5)$

34. $\log_7 2 + \log_7(x + 4) = \log_7(5 - x)$

35. $\log(x + 1) = \log(x + 9) - \log x$

36. $\log x^2 - \log 2 = \log(x + 4)$

37. $\dfrac{1}{2} \log_4(2x + 1) = \log_4 3$

38. $\dfrac{1}{2} \log_5(x + 4) = \log_5(x - 2)$

39. Suppose $10,000 is deposited in an account paying 10% annual interest, compounded semi-annually. When will there be $15,000 in the account? $20,000?

40. Suppose $400 is deposited in an account paying 6% annual interest, compounded monthly. When will there be $800 in the account? $1000?

41. The population of Texas since 1940 was given in exercise 51 of section 10.2 as

$$P = 6.4(1.02)^t$$

where P is measured in millions and t is the number of years after 1940. According to this function, when will the population reach 25 million?

42. Professor Ben Zene at Moss County University has discovered a new radioactive element, bolognium, which decays as time passes, turning into rubber. Starting with an initial quantity of 25 g, the amount of bolognium, A, left after t hours is given by the equation

$$A = 25e^{-0.04t}$$

After how many hours will the amount of bolognium be reduced to 10 g?

Write Algebra

43. Explain why we check for extraneous solutions when we solve logarithmic equations.

44. Given $\log_b x = \log_b y$ with $b \neq 1$, explain why we can conclude $x = y$.

45. Given an equation of the form $b^x = c$ with $b > 1$ and $c > 0$, explain why this equation has exactly one solution.

Use the Zoom and Trace keys of a graphing calculator to find an approximation for the solution of the following exponential equations.

46. $5 = 11^x$ **47.** $5.4 = (8.92)^x$ **48.** $\sqrt{7} = (9.7)^x$

G R O U P   A C T I V I T Y

In exercises 49–60, solve the equations.

49. $\log(x - 4) + \log(x - 2) = \log(x^2 - 6x + 8)$

50. $\log(x + 5) + \log(x - 1) = \log(x^2 + 4x - 5)$

51. $\log(x + 2) + \log(x + 6) = \log(x^2 + 8x + 12)$

52. $\log(x + 4) + \log(x + 3) = \log(x^2 + 7x + 12)$

53. $\log(x - 1) - \log(x - 6) = \log\left(\dfrac{x - 1}{x - 6}\right)$

54. $\log(x - 5) - \log(x - 2) = \log\left(\dfrac{x - 5}{x - 2}\right)$

55. $\log(x + 4)^3 = 3 \log(x + 4)$

56. $\log(x - 1)^2 = 2 \log(x - 1)$

57. $\log(x^2 + 1) + \log 5 = \log(5x^2 + 5)$

58. $\log(x + 2) = \log x + \log 2$

59. $\log(5x - 2) = \log 5x - \log 2$

60. $\log(x + 1) + \log(x - 6) = \log(x^2 + 5x - 6)$

61. Explain why the solution set of the equations in exercises 49–56 is not the set of all real numbers.

Applied Algebra

The change in heterozygosity from one generation to the next in a particular species of animals in captivity is given by the formula

$$H_t = H_0\left(1 - \frac{1}{2N}\right)^t$$

where H_0 is the initial heterozygosity, H_t is the heterozygosity after t generations, and N is the number of animals.

1. What will the heterozygosity be for New Guinea singing dogs after 50 years?

2. When will the heterozygosity for New Guinea singing dogs drop to 50% of its initial heterozygosity?

SOLUTION

New Guinea singing dogs (named for their unique vocalization-like yodeling) have been in the zoos of the United States for only about 15 years. In 1992 there were just 100 in captivity in the United States. Their generation length is about 2.5 years. Therefore, in 50 years, $\frac{50}{2.5} = 20$ generations will occur. Substituting $t = 20$ and $N = 100$ into the equation yields

$$H_{20} = H_0\left(1 - \frac{1}{2 \cdot 100}\right)^{20}$$
$$= H_0(1 - 0.005)^{20}$$
$$= H_0(0.995)^{20}$$
$$= H_0(0.90461)$$

So after 50 years, or 20 generations, the heterozygosity will be about 90% of the initial heterozygosity.

To answer the second question, we substitute $N = 100$ and $H_t = 50\%$ of $H_0 = 0.5H_0$ into the equation and solve for t.

$$0.5H_0 = H_0\left(1 - \frac{1}{200}\right)^{t}$$
$$0.5 = (0.995)^{t} \qquad \textit{Divide both sides by } H_0 \textit{ and simplify inside the parentheses.}$$
$$\log 0.5 = \log(0.995)^{t} \qquad \textit{Take the log of both sides.}$$
$$\log 0.5 = t \cdot \log(0.995)$$
$$\frac{\log 0.5}{\log 0.995} = t$$
$$t \doteq 138 \text{ generations}$$

or

$$138 \cdot (2.5) = 345 \text{ years}$$

Your Turn

The Bali mynah, one of the rarest birds in the world, numbered only about 70 in the wild in 1992. At present there are about 500 in North American zoos. Their generation length is about 2 years.

1. What will the heterozygosity for the 500 Bali mynahs be after 50 years?

2. When will the heterozygosity for Bali mynahs drop to 50% of its initial heterozygosity?

Chapter 10 Review

Terms to Remember

Inverse of a relation or function	**p. 573**	Logarithmic function	**p. 602**
		Exponential form	**p. 602**
One-to-one function	**p. 578**	Logarithmic form	**p. 602**
Horizontal line test	**p. 580**	Common logarithm	**p. 604**
Composition of functions	**p. 580**	Natural logarithm	**p. 605**
Exponential function	**p. 586**	Logarithmic equation	**p. 620**
Exponential equation	**p. 593**		

Notation

R^{-1}	The inverse of a relation R
$y = f^{-1}(x)$	When the inverse of a function $y = f(x)$ is itself a function, we denote it by $y = f^{-1}(x)$.
1–1	Shorthand for one-to-one function
$f \circ g$	Composition of functions, f of g
$\log x$	Common logarithm, $\log x = \log_{10} x$
$\ln x$	Natural logarithm, $\ln x = \log_e x$

Properties

Inverse function:

$$(f \circ f^{-1})(x) = x$$
$$(f^{-1} \circ f)(x) = x$$

For the remaining properties, let a, b, M, and N be positive real numbers with $a \neq 1$ and $b \neq 1$. Also let r, x, and z be real numbers.

- *One-to-one property of exponential functions:* If $a^x = a^z$, then $x = z$.
- *One-to-one property of logarithmic functions:* If $\log_b M = \log_b N$, then $M = N$.
- *Product:* $\log_b MN = \log_b M + \log_b N$
- *Quotient:* $\log_b \left(\frac{M}{N}\right) = \log_b M - \log_b N$
- *Power:* $\log_b N^r = r \log_b N$
- *Change of base:* $\log_b N = \dfrac{\log_a N}{\log_a b}$
- *Inverse:* $a^{\log_a N} = N$, $\quad \log_a a^N = N$

Formulas

Compound interest:

$$M = D\left(1 + \frac{r}{n}\right)^{nt}$$

M is the amount of money in an account, after t years, from an initial deposit of D, earning an annual interest rate of r, compounded n times a year.

Review Exercises

10.1 **Find an equation that defines the inverse. If the inverse is a function, write it in the form $y = f^{-1}(x)$. Also graph both equations on the same coordinate plane.**

1. $f(x) = 3x - 4$ **2.** $f(x) = \frac{5}{2}x + 1$ **3.** $f(x) = \frac{2x + 4}{3}$ **4.** $f(x) = x^2 + 2$

5. $f(x) = 3 - \frac{1}{2}x^2$ **6.** $f(x) = \sqrt{x + \frac{3}{2}}$ **7.** $f(x) = \sqrt{3 - x}$ **8.** $f(x) = |x + 3|$

Determine whether the following functions are one-to-one. If they are, find the inverse function.

9. $f(x) = \frac{2}{x + 1}$ **10.** $f(x) = \frac{x}{x - 2}$ **11.** $f(x) = |x + 4|$

12. $f(x) = 2x^2 - 3$ **13.** $f(x) = \sqrt{3x + 1}$ **14.** $f(x) = \sqrt[3]{2 - x}$

For each pair of functions find the following: (a) $(f \circ g)(-2)$, (b) $(f \circ g)(x)$, (c) $(g \circ f)(4)$, and (d) $(g \circ f)(x)$.

15. $f(x) = 4x - 7$, $g(x) = x + 5$ **16.** $f(x) = 2x^2 + 1$, $g(x) = 3 - x$

Determine whether the following pairs of functions are inverses by finding $(f \circ g)(x)$ and $(g \circ f)(x)$.

17. $f(x) = \frac{1}{4}x + 3$, $g(x) = 4x - 12$ **18.** $f(x) = \frac{3x - 2}{5}$, $g(x) = \frac{2 - 5x}{3}$

19. $f(x) = \frac{x}{x + 4}$, $g(x) = \frac{4x}{1 - x}$ **20.** $f(x) = \frac{5}{2x + 1}$, $g(x) = \frac{5 - x}{2x}$

10.2 **Sketch the graphs of the following functions.**

21. $f(x) = 2^{x+4}$ **22.** $f(x) = 2^{x/4}$ **23.** $f(x) = 4 + 2^x$

24. $f(x) = 4(2^{x/4})$ **25.** $f(x) = \left(\frac{3}{2}\right)^x$ **26.** $f(x) = \left(\frac{2}{3}\right)^x$

Solve the following equations.

27. $y = 8^{-2/3}$ **28.** $y = 9^{5/2}$ **29.** $64 = a^{3/4}$ **30.** $7 = a^{-1/2}$

31. $\frac{1}{9} = 3^x$ **32.** $8 = 16^{x+1}$ **33.** $\frac{3}{4} = \left(\frac{16}{9}\right)^{3x+1}$ **34.** $25 = \left(\frac{1}{5}\right)^{2-x}$

▦ **The following application exercises require a calculator.**

35. On April 1 Allen sprayed his garden for aphids, annihilating all but 10 of the little varmints. The number of aphids in his garden can now be described by the equation $A = 10(1.12)^t$, where t represents the number of days that have passed since April 1. How many aphids will there be in Allen's garden on April 15? May 1?

36. If $1000 is deposited in an account paying 6% annual interest, compounded semiannually, the amount of money, M, in the account t years later can be described by the equation $M = 1000(1.03)^{2t}$. How much money will be in the account after 3 yr? 10 yr? Approximately how long will it take for there to be $2000 in the account?

10.3   **Rewrite the following in logarithmic form.**

37. $4^{1/2} = 2$

38. $9^0 = 1$

39. $y = 2^{x-5}$

40. $y = 10^{3x}$

41. $B = e^{kt}$

42. $Q = Q_0(2^{-t/500})$

Rewrite the following in exponential form.

43. $\log_3 3 = 1$

44. $\log 100 = 2$

45. $y = \log_3(x + 2)$

46. $y = 2 \log_8(3x)$

47. $2t = \log_7 100$

48. $k = \ln\left(\dfrac{P}{P_0}\right)$

Sketch the graphs of the functions in exercises 49–52.

49. $y = \log_2 4x$

50. $y = 4 \log_2 x$

51. $y = \log_2(x + 4)$

52. $y = 4 + \log_2 x$

53. The time that it takes an instructor to learn the names of her students depends on the number of students in the class. The number of days, t, that it takes an instructor to learn N names may be given by the equation

$$t = 2 - 12 \ln\left(1 - \frac{N}{40}\right) \qquad N < 40$$

How long will it take the instructor to learn 25 names? 35 names?

10.4   **Use the properties of logarithms to express each of the following as sums or differences of logarithms.**

54. $\log_b x^4 \sqrt[3]{y}$

55. $\log_b \sqrt{\dfrac{x^3}{y^5}}$

56. $\log_b \dfrac{a}{yz^3}$

57. $\log_b D(e^{rt})$

Use the properties of logarithms to express each of the following as a single logarithm.

58. $3 \log_b x + \dfrac{1}{3} \log_b y - \dfrac{3}{4} \log_b z$

59. $\dfrac{3}{2} \log_b x - \dfrac{1}{2} \log_b y - \log_b z$

Use the change of base property to evaluate the following.

60. $\log_{1/2} 11$

61. $\log_5 24$

Simplify each of the following.

62. $3^{\log_3 7}$

63. $\log_5 \sqrt[3]{5}$

64. $\log_6 36$

65. $\log_2 \dfrac{1}{8}$

66. $\log_4 32$

Given that $\log_b 3 = 1.4$, $\log_b 5 = 2.0$, and $\log_b 7 = 2.4$, find each of the following.

67. $\log_b 63$

68. $\log_b \dfrac{7}{3}$

69. $\log_b \sqrt{7}$

70. $\log_b \sqrt{\dfrac{21}{25}}$

71. Refer to exercise 58 in Exercises 10.4. Find the hydrogen ion concentration of the following substances with the given pH:

Substance	pH
Vinegar	2.5
Human spinal fluid	7.4
Egg white	8.0

10.5   Solve the following equations. Use a calculator to approximate the solutions.

72. $9 = 2^x$

73. $7.5 = 3^{x-4}$

74. $18^x = 7.2^{2x-1}$

75. $250 = 100(2)^{0.03t}$

Solve the following equations.

76. $\log_x 8 = -\dfrac{3}{2}$

77. $\log_9 \dfrac{1}{81} = x$

78. $\log_4 8 = x$

79. $\log_7 x = -2$

80. $\log_2(3x - 1) + \log_2 x = 2$

81. $\log_3(4x + 1) - \log_3(x - 4) = 1$

82. $\dfrac{1}{2} \log_9(x - 2) = \dfrac{3}{4}$

83. $\log_5 3 + \log_5(x + 7) = \log_5(13 - 5x)$

84. $\log(9x - 4) - \log x = \log(x + 5)$

85. $2 \log_6(5x + 1) = \log_6(13 - 3x)$

86. Suppose $5000 is deposited in an account paying $6\frac{1}{4}\%$ annual interest, compounded monthly. When will there be $50,000 in the account?

Chapter 10 Test

(You should be able to complete this test in 60 minutes.)

I. Inverse functions and composition of functions

1. Find the equation that defines the inverse of $f(x) = 3x - 6$, and write it in the form $y = f^{-1}(x)$. Also graph both equations on the same coordinate plane.

2. Determine whether $f(x) = \dfrac{x}{2x + 1}$ and $g(x) = \dfrac{-x}{2x - 1}$ are inverses by finding $(f \circ g)(x)$ and $(g \circ f)(x)$.

II. Graph the following functions.

3. $f(x) = 3^{2-x}$

4. $f(x) = \log_3(x + 2)$

III. Change to the indicated form.

5. $A = A_0 e^{rt}$,   logarithmic

6. $y = \log_3(2x - 5)$,   exponential

IV. Simplify the following.

7. $2^{\log_2 9}$

8. $\log_7 \sqrt{7}$

9. $\log_8 \frac{1}{4}$

10. Use the properties of logarithms to express $\frac{1}{3}\log_b x + 2 \log_b y - \frac{3}{2}\log_b z$ as a single logarithm.

V. Solve the following equations.

11. $9 = a^{-1/2}$

12. $64 = 2^{3x+5}$

13. $7^{3-x} = 1.5$

14. $\log_2(x + 3) + \log_2(x + 9) = 4$

15. $\log 7 - \log(2x - 3) = \log(x - 4)$

VI. Application

16. On July 4 Gilbert sprayed his gardenias for whiteflies, chasing away all but 750 of the nasty critters. The number of whiteflies on the gardenias can now be given by the equation $W = 750(e^{0.09t})$, where t represents the number of days that have passed since July 4. How many whiteflies will there be on July 18?

TEST YOUR MEMORY

These problems review Chapters 1–10.

I. Simplify each of the following. Assume that all variables represent positive numbers.

1. $\left(\dfrac{8x^{-4}y^5}{12x^8y}\right)^{-3}$

2. $\dfrac{(64x^4)^{-2/3}x^2}{(16x^5)^{1/2}}$

3. $4x\sqrt[3]{48} - \sqrt[3]{162x^3}$

4. $\dfrac{\sqrt{2}}{1+\sqrt{2}}$

II. Perform the indicated operations.

5. $\dfrac{2ax + 2bx - 3a - 3b}{2x^2 - 11x + 12} \cdot \dfrac{x^2 - 16}{a^3 + b^3}$

6. $\dfrac{2x}{x+3} + \dfrac{11x+3}{x^2+x-6} - \dfrac{6}{x-2}$

III. Let $f(x) = 2x - 5$ and $g(x) = x^2 + 4$. Find the following.

7. $f(-2)$

8. $g(x) - f(x)$

9. $f^{-1}(x)$

10. $(f \circ g)(x)$

11. $(g \circ f)(x)$

IV. Find the equations of the following lines. Write the equations in slope-intercept form, if possible.

12. Through $(-1, 8)$ and $(2, -7)$

13. Through $(4, -2)$ and perpendicular to $x - 2y = 13$

V. Find the equations of the following circles.

14. Center $= C(-2, -5)$, radius $= 4$

15. Center $= C(1, -7)$ and passing through $(4, -3)$

VI. Graph the following equations.

16. $x^2 + y^2 = 4$

17. $x = 4 - y^2$

18. $25x^2 - 4y^2 = 100$

19. $9x^2 + 4y^2 - 36x + 8y + 4 = 0$

20. $y = 2x^2 + 4x + 4$

21. $f(x) = 2^{x+1}$

22. $f(x) = \log_3(x - 2)$

VII. Simplify the following.

23. $\log_3 \sqrt[3]{3}$

24. $2^{\log_2 16}$

VIII. Find the solutions of the following equations and check your solutions.

25. $2\left(\dfrac{2x}{x+1}\right)^2 + 7\left(\dfrac{2x}{x+1}\right) - 4 = 0$

26. $3\left(\dfrac{3x}{x-2}\right)^2 + 14\left(\dfrac{3x}{x-2}\right) - 5 = 0$

27. $\sqrt{3x} - \sqrt{x+1} = 1$

28. $\sqrt{5x} - \sqrt{x+4} = 2$

29. $4x^4 - 29x^2 + 25 = 0$

30. $36x^4 - 13x^2 + 1 = 0$

31. $5x^2 - 6x + 2 = 0$

32. $x^2 - 2x + 17 = 0$

33. $\dfrac{1}{125} = 25^{x+3}$

34. $8 = 4^{2x+1}$

35. $\log(x + 5) + \log(x - 1) = \log(6x + 19)$

36. $\log(3x - 9) - \log(x + 5) = \log(x - 6)$

37. $\log_2 3x - \log_2(x - 1) = 2$

38. $\log_2 x + \log_2(x - 4) = 5$

IX. Using the substitution or elimination method, determine the solutions of the following linear systems of equations. Identify each system as independent, inconsistent, or dependent.

39. $2x + 3y = -1$
$3x + 7y = -2$

40. $4x - 6y = 3$
$6x - 9y = -2$

41. $3x - 2y = 1$
$y = 3x + 4$

42. $2x + y - 3z = 14$
$x - y - z = 5$
$-x + 2y + 2z = -9$

X. Determine the solutions of the following systems of equations.

43. $x^2 + y^2 = 25$
$x^2 - y = 13$

44. $2x^2 + 3y^2 = 15$
$x^2 - y^2 = 10$

XI. Use algebraic expressions to find the solutions of the following problems.

45. A rectangle has a length that is 3 ft less than 3 times the width. The area is 90 sq ft. What are the dimensions of the rectangle?

46. A cannonball is fired upward from a 336-ft-tall cliff with a speed of 64 ft/sec. The equation that gives the cannonball's height above ground level is $h = -16t^2 + 64t + 336$. When does the cannonball hit the ground? When does the cannonball reach its maximum height? What is the maximum height?

47. Vanna takes her motorboat to her favorite fishing spot 6 mi downstream. The trip downstream and back takes $1\frac{1}{2}$ hr. If the speed of the current is 3 mi/hr, what is the speed of Vanna's boat in still water?

48. Jevey bought 2 apples and 5 pears for $2.80 at the Moss County Farmer's Market. The next day she returned and bought 4 apples and 3 pears for $2.94. What is the cost of 1 apple and 1 pear?

49. Pete's Popcorn Palace has two types of popcorn: Jumbo Deluxe, which sells for $3 a pound, and Small Fry, which sells for $0.50 a pound. How many pounds of each type of popcorn must Pete mix to create 10 pounds of Pete's Plain Popcorn, which will sell for $2 a pound?

50. On July 1 Dottie finds 20 g of bernallium (a heretofore undiscovered radioactive substance). The amount of bernallium that Dottie has is given by the equation $A = 20(e^{-.007t})$, where A is the amount of bernallium (measured in grams) and t represents the number of days that have passed since July 1. How much of the bernallium will remain on July 31?

Sequences and Series

In a codicil to his will Benjamin Franklin bequeathed $5000 (1000 pounds sterling) to the "inhabitants of the city of Philadelphia." The money was to be loaned to "young married artificers" at 5% interest compounded annually. The city of Philadelphia received the $5000 in 1791. If the terms of Franklin's will were satisfied, how much money did he expect the Benjamin Franklin Fund to have in 1891?

When we finish this chapter, we will (with the aid of a calculator) be able to determine quickly the amount of money Franklin projected that the city of Philadelphia would receive. (*Note:* The terms of the will could not be carried out. However, by 1907 the Benjamin Franklin Fund in Philadelphia amounted to $172,350).

11.1 Sequences and Series

Suppose that Nancy and Patti are playing in the Moss County Invitational Golf Tournament. The rules of the tournament state that the winner of the first hole will collect $2, and then the winnings double from hole to hole (i.e., the winner of the second hole will collect $4, the winner of the third hole will collect $8, and so on). The equation that expresses the winnings on each hole is $a(n) = 2^n$, where $n = 1$, 2, 3, ..., 18. Note that the equation $a(n) = 2^n$ defines a function, and that a restriction has been placed on the domain of this function. This example leads to the following definition.

Definition 11.1.1	A **finite sequence** is a function whose domain is $\{1, 2, 3, ..., k\}$ for some natural number k. An **infinite sequence** is a function whose domain is the set of natural numbers. The elements of the range of the function are called **terms.** The list of terms stated in order is called a **sequence.**

In the example, the terms 2, 4, 8, ..., 262,144 form a sequence. The equation $a(n) = 2^n$ generates the *terms of this sequence:*

$$a(1) = 2^1 = 1$$
$$a(2) = 2^2 = 4$$
$$a(3) = 2^3 = 8$$
$$\vdots$$
$$a(18) = 2^{18} = 262{,}144$$

When dealing with sequences, it is common practice to use subscript notation instead of function notation. Thus,

$$a_1 = 2$$
$$a_2 = 4$$
$$a_3 = 8$$
$$\vdots$$
$$a_{18} = 262{,}144$$

The equation $a(n) = 2^n$ is equivalent to $a_n = 2^n$. a_n determines the nth term of the sequence and is called the **general term** of the sequence. The general term generates the sequence.

EXAMPLE 11.1.1   Write the first five terms of the sequence with general term $a_n = 3n - 1$:

$$a_1 = 3(1) - 1 = 2$$
$$a_2 = 3(2) - 1 = 5$$

$$a_3 = 3(3) - 1 = 8$$
$$a_4 = 3(4) - 1 = 11$$
$$a_5 = 3(5) - 1 = 14$$

Thus, the first five terms of the sequence are

$$2, \quad 5, \quad 8, \quad 11, \quad 14$$

EXAMPLE 11.1.2   Find the ninth term of the sequence with general term $a_n = (-1)^n(2n + \frac{3}{n})$.

SOLUTION   To find the ninth term, we must determine a_9:

$$a_9 = (-1)^9\left(2 \cdot 9 + \frac{3}{9}\right)$$

$$= -1\left(18 + \frac{1}{3}\right)$$

$$= -\frac{55}{3}$$

Suppose that Patti won only the first four holes of the Moss County Invitational Golf Tournament. Her winnings would be $2 + $4 + $8 + $16 = $30. Notice that Patti's winnings are the sum of consecutive terms of a sequence. This example illustrates the next definition of this section.

Definition 11.1.2   A sum of consecutive terms of a sequence is called a **series.** If the sequence is infinite, the series is called an **infinite series.**

REMARKS   1. The notation S_n denotes the sum of the first n terms of a sequence. Thus, given the finite sequence $a_1, a_2, a_3, ..., a_n,$

$$S_1 = a_1$$
$$S_2 = a_1 + a_2$$
$$S_3 = a_1 + a_2 + a_3$$
$$S_n = a_1 + a_2 + a_3 + \cdots + a_n$$

2. $S_n = S_{n-1} + a_n$

3. S_∞ denotes the sum of all terms of an infinite sequence. Thus, given the infinite sequence $b_1, b_2, b_3, ..., b_n, ...,$

$$S_\infty = b_1 + b_2 + b_3 + \cdots + b_n + \cdots$$

Believe it or not, in Section 11.3 we will see a case where S_∞ can be calculated.

E XAMPLE 11.1.3   Find S_5 when $a_n = 3n - 1$.

SOLUTION

$$S_5 = a_1 + a_2 + a_3 + a_4 + a_5$$
$$= 2 + 5 + 8 + 11 + 14 \qquad \textit{From Example 11.1.1}$$
$$= 40$$

The series $S_n = a_1 + a_2 + a_3 + \cdots + a_n$ can be expressed in a condensed form by using **summation notation.** Summation notation uses the Greek letter Σ (sigma) in the following manner:

$$\sum_{i=1}^{n} a_i = a_1 + a_2 + a_3 + \cdots + a_n \qquad (1)$$

REMARKS

1. The right-hand side of equation (1) is called the *expanded form* of the series.

2. The left-hand side of equation (1) is the series written in *summation notation.*

The symbol $\Sigma_{i=1}^{n} a_i$ is read "the summation of a_i as i goes from 1 to n." The letter i is called the **index of summation.** Any letter can be used as the index of summation; i, j, and k are the most common choices. The expression immediately to the right of Σ is always the general term expressed in terms of the index of summation.

The Σ notation implies sum. The following example demonstrates how to use the Σ notation.

Evaluate $\Sigma_{i=1}^{4} 2^i$.

$$\sum_{i=1}^{4} 2^i = 2^1 + 2^2 + 2^3 + 2^4$$
$$= 2 + 4 + 8 + 16$$
$$= 30$$

NOTE ▶   To evaluate a summation, express the summation in expanded form. The expanded form is found by successively substituting for i (the index) the consecutive integers beginning with the number at the base of Σ and ending with the number at the top of Σ.

EXAMPLE 11.1.4   Write the following series in expanded form and evaluate.

$$1. \quad \sum_{i=1}^{3} (2i + 1) = (2(1) + 1) + (2(2) + 1) + (2(3) + 1)$$

$$= 3 + 5 + 7$$

$$= 15$$

$$2. \quad \sum_{i=3}^{7} (4i - 5) = (4(3) - 5) + (4(4) - 5) + (4(5) - 5) + (4(6) - 5) + (4(7) - 5)$$

$$= 7 + 11 + 15 + 19 + 23$$

$$= 75$$

In this example the summation did not begin with $i = 1$. In fact, the number that begins the summation can be any integer.

Given a series in expanded form, it is sometimes useful to write that series in summation notation. The next example shows how to make this conversion.

EXAMPLE 11.1.5   Write the following series in summation notation.

1. $1 + 3 + 5 + 7 + 9$

By examining the series, we can observe that the general term is given by $a_i = 2i - 1$. Thus $1 + 3 + 5 + 7 + 9 = \sum_{i=1}^{5} (2i - 1)$.

REMARKS

a. Since the summation does not have to start with $i = 1$, the solution is not unique. Another solution is

$$1 + 3 + 5 + 7 + 9 = \sum_{i=0}^{4} (2i + 1)$$

b. The method used to find the general term is trial and error.

2. $\dfrac{1}{2} + \dfrac{2}{3} + \dfrac{3}{4} + \dfrac{4}{5} + \cdots$

Here we have an infinite series. The general term is $a_i = \frac{i}{i+1}$. The infinite series expressed in summation notation is

$$\frac{1}{2} + \frac{2}{3} + \frac{3}{4} + \frac{4}{5} + \cdots = \sum_{i=1}^{\infty} \frac{i}{i + 1}$$

Exercises 11.1

Write the first five terms of the sequences with the following general terms.

1. $a_n = 3n + 2$ 　　　　　**2.** $a_n = 5n + 6$ 　　　　　**3.** $a_n = 4n - 1$ 　　　　　**4.** $a_n = -3n - 1$

5. $a_n = 3^n$ 　　　　　**6.** $a_n = 4^{n-1}$ 　　　　　**7.** $a_n = \left(\dfrac{2}{3}\right)^n$ 　　　　　**8.** $a_n = \left(\dfrac{3}{4}\right)^n$

9. $a_n = \dfrac{1}{n}$ 　　　　　**10.** $a_n = \dfrac{2}{n^2}$ 　　　　　**11.** $a_n = n^2 - 4$ 　　　　　**12.** $a_n = 2n^2 + 1$

13. $a_n = n^2 + n + 1$ 　　　　　**14.** $a_n = n^2 - n - 1$ 　　　　　**15.** $a_n = (-1)^n(2n + 1)$ 　　　　　**16.** $a_n = (-1)^n(2n)$

17. $a_n = \dfrac{(-1)^{n+1}}{n}$ 　　　　　**18.** $a_n = \dfrac{(-1)^{n+1}}{n^2}$

Find the indicated term of the sequence with the following general terms.

19. $a_n = 5n + 7;\quad a_6$ 　　　　　**20.** $a_n = -4n + 12;\quad a_8$ 　　　　　**21.** $a_n = \left(\dfrac{9}{5}\right)^n;\quad a_3$

22. $a_n = 4\left(\dfrac{2}{3}\right)^n;\quad a_4$ 　　　　　**23.** $a_n = \dfrac{2n}{n^3 + 1};\quad a_9$ 　　　　　**24.** $a_n = (2n + 1)(3n - 4);\quad a_7$

25. $a_n = (-1)^n\left(3n + \dfrac{n^2}{n + 1}\right);\quad a_5$ 　　　　　**26.** $a_n = (-1)^n(n + n^2 + n^3);\quad a_6$

Find S_5 for the sequences with the following general terms. (*Hint:* Use the results of exercises 1–18.)

27. $a_n = 3n + 2$ 　　　　　**28.** $a_n = 5n + 6$ 　　　　　**29.** $a_n = 4n - 1$ 　　　　　**30.** $a_n = -3n - 1$

31. $a_n = 3^n$ 　　　　　**32.** $a_n = 4^{n-1}$ 　　　　　**33.** $a_n = \left(\dfrac{2}{3}\right)^n$ 　　　　　**34.** $a_n = \left(\dfrac{3}{4}\right)^n$

35. $a_n = \dfrac{1}{n}$ 　　　　　**36.** $a_n = \dfrac{2}{n^2}$ 　　　　　**37.** $a_n = n^2 - 4$ 　　　　　**38.** $a_n = 2n^2 + 1$

39. $a_n = n^2 + n + 1$ 　　　　　**40.** $a_n = n^2 - n - 1$ 　　　　　**41.** $a_n = (-1)^n(2n + 1)$ 　　　　　**42.** $a_n = (-1)^n(2n)$

43. $a_n = \dfrac{(-1)^{n+1}}{n}$ 　　　　　**44.** $a_n = \dfrac{(-1)^{n+1}}{n^2}$

Write the following series in expanded form and evaluate.

45. $\sum_{i=1}^{8}(3i - 7)$ 　　　　　**46.** $\sum_{i=1}^{6}(2i + 5)$ 　　　　　**47.** $\sum_{i=1}^{3}(2i^2 + i + 3)$ 　　　　　**48.** $\sum_{i=1}^{5}(5i^2 - 4)$

49. $\sum_{i=1}^{4}\left(\dfrac{2}{3}\right)^i$ 　　　　　**50.** $\sum_{i=1}^{5}\left(\dfrac{3}{2}\right)^i$ 　　　　　**51.** $\sum_{i=3}^{7}2i(i + 4)$ 　　　　　**52.** $\sum_{i=0}^{3}4i(i^2 + 1)$

53. $\sum_{i=2}^{6}(-1)^i(i)$ 　　　　　**54.** $\sum_{i=2}^{5}(-1)^{i+1}(3i^2)$ 　　　　　**55.** $\sum_{i=6}^{7}(-1)^{i+1}\left(\dfrac{1}{2i}\right)$ 　　　　　**56.** $\sum_{i=8}^{10}(-1)^i\left(\dfrac{1}{i - 4}\right)$

Write the following series in summation notation.

57. $1 + 2 + 3 + 4 + 5 + 6 + 7$ 　　　　　**58.** $6 + 8 + 10 + 12$ 　　　　　**59.** $7 + 9 + 11 + 13 + 15 + 17$

60. $-3 + (-1) + 1 + 3$ 　　　　　**61.** $3 + 9 + 27 + 81 + 243$ 　　　　　**62.** $2 - 4 + 8 - 16 + 32$

63. $80 + 40 + 20 + 10$

64. $1 + 4 + 9 + 16 + 25 + 36$

65. $1 - 8 + 27 - 64 + 125$

66. $\dfrac{1}{4} + \dfrac{4}{9} + \dfrac{9}{16} + \dfrac{16}{25}$

67. $\dfrac{1}{3} + \dfrac{1}{9} + \dfrac{1}{27} + \cdots$

68. $10 + 20 + 30 + \cdots$

69. $2 + 5 + 8 + \cdots$

70. $1 - \dfrac{1}{2} + \dfrac{1}{4} - \dfrac{1}{8} + \cdots$

Find the solutions of the following problems.

71. What is the prize money for the eighth hole of the Moss County Invitational Golf Tournament, mentioned at the beginning of the section?

72. Bennie is on the "Cookie-Free" diet. The "Cookie-Free" diet says that every day you must cut your cookie consumption in half. On the first day of the diet Bennie eats 32 cookies. Therefore, he can eat only 16 cookies the second day, 8 cookies the third day, and so on. How many cookies can Bennie eat during the first week of his diet?

Write Algebra

73. Explain the difference between a sequence and a series.

74. Given the series $\sum_{i=2}^{10} (-1)^i (3i + 5)$, what is the effect of the factor $(-1)^i$ on the terms of the series?

11.2 *Arithmetic Sequences and Series*

Suppose you are given the sequence with general term $a_n = 4n - 3$. The first five terms of the sequence are 1, 5, 9, 13, 17. Note that each term after the first term can be found by adding 4 to the previous term. The sequence 1, 5, 9, 13, 17 is an example of a particular type of sequence called an arithmetic sequence.

Definition 11.2.1	An **arithmetic sequence** is a sequence in which each term, after the first term, can be found by adding a fixed constant to the preceding term. This fixed constant is called the **common difference** and is denoted by the letter d.

REMARKS

1. To find the common difference of any arithmetic sequence, simply subtract from *any* term (other than a_1) the term that precedes it.

2. If the equation that defines the general term, a_n, is first degree in n, then the sequence defined by a_n is an arithmetic sequence.

 EXAMPLE 11.2.1 Find the common difference of the following arithmetic sequence:

$$7, 5, 3, 1, -1, \ldots$$

SOLUTION

One way to find the common difference is as follows:

$$\text{common difference} = \text{fourth term} - \text{third term}$$
$$d = a_4 - a_3$$
$$d = 1 - 3$$
$$d = -2$$

NOTE ▶   We can use any two consecutive terms to find the common difference.

EXAMPLE 11.2.2   Write the first five terms of an arithmetic sequence with first term 7 and common difference 3.

SOLUTION

Since we have an arithmetic sequence with common difference 3, each term can be found by adding 3 to the preceding term:

$$a_1 = 7$$
$$a_2 = a_1 + 3 = 7 + 3 = 10$$
$$a_3 = a_2 + 3 = 10 + 3 = 13$$
$$a_4 = a_3 + 3 = 13 + 3 = 16$$
$$a_5 = a_4 + 3 = 16 + 3 = 19$$

Thus, the first five terms of the sequence are

$$7, \quad 10, \quad 13, \quad 16, \quad 19$$

Suppose the sequence $a_1, a_2, a_3, a_4, \ldots, a_n$ is an arithmetic sequence with common difference d; then

$$a_1 = a_1$$
$$a_2 = a_1 + d$$
$$a_3 = a_2 + d = (a_1 + d) + d = a_1 + 2d$$
$$a_4 = a_3 + d = (a_1 + 2d) + d = a_1 + 3d$$

These results are generalized in the following formula.

General Term of an Arithmetic Sequence

Given an arithmetic sequence with first term a_1 and common difference d, the general term, a_n, is determined by the formula

$$a_n = a_1 + (n - 1)d$$

EXAMPLE 11.2.3   Find a_{19} for the arithmetic sequence with $a_1 = 24$ and $d = -\frac{1}{2}$.

SOLUTION   From the formula,

$$
\begin{aligned}
a_{19} &= a_1 + (19 - 1)d \\
&= a_1 + 18d \\
&= 24 + 18\left(-\frac{1}{2}\right) \\
&= 24 - 9 \\
&= 15
\end{aligned}
$$

EXAMPLE 11.2.4   Find the general term, a_n, for the arithmetic sequence 8, 11, 14, 17, ...

SOLUTION   To find a_n, we must know a_1 and d. Clearly, $a_1 = 8$. We can find d by subtracting from any term (other than a_1) the term that precedes it. So

$$
\begin{aligned}
d &= a_2 - a_1 \\
&= 11 - 8 \\
&= 3
\end{aligned}
$$

Now

$$
\begin{aligned}
a_n &= a_1 + (n - 1)d \\
&= 8 + (n - 1)3 \\
&= 8 + 3n - 3 \\
&= 3n + 5
\end{aligned}
$$

The equation defining a_n is first degree in n.

EXAMPLE 11.2.5   Given an arithmetic sequence with $a_8 = 1$ and $a_{14} = 3$, find a_1 and d.

SOLUTION   From the general term formula,

$$
\begin{aligned}
a_8 &= a_1 + (8 - 1)d \quad \text{so} \quad a_8 = a_1 + 7d \\
a_{14} &= a_1 + (14 - 1)d \quad \text{so} \quad a_{14} = a_1 + 13d
\end{aligned}
$$

Substituting 1 for a_8 and 3 for a_{14} yields the following system of equations:

$$1 = a_1 + 7d \xrightarrow[\text{sides by } -1.]{\textit{Multiply both}} -1 = -a_1 - 7d$$

$$3 = a_1 + 13d \xrightarrow{\hspace{3cm}} \underline{\quad 3 = \quad a_1 + 13d}$$

$$2 = \quad\quad 6d$$

$$\frac{2}{6} = d$$

$$\frac{1}{3} = d$$

Substitute $\frac{1}{3}$ for d into the first equation of the system:

$$1 = a_1 + 7\left(\frac{1}{3}\right)$$

$$1 = a_1 + \frac{7}{3}$$

$$1 - \frac{7}{3} = a_1$$

$$-\frac{4}{3} = a_1$$

Thus, $a_1 = -\frac{4}{3}$ and $d = \frac{1}{3}$.

Recall from Section 11.1 that a series is a sum of consecutive terms of a sequence. In the next definition we examine a particular type of series.

Definition 11.2.2	A sum of consecutive terms of an arithmetic sequence is called an **arithmetic series.**

In the previous section we associated with the sequence $a_1, a_2, a_3, \ldots, a_n$ the series $S_n = a_1 + a_2 + a_3 + \cdots + a_n$. Now suppose that the sequence $a_1, a_2, a_3, \ldots, a_n$ is an arithmetic sequence. Then S_n can be expressed as

$$S_n = a_1 + (a_1 + d) + (a_1 + 2d) + \cdots + (a_1 + (n-1)d) \qquad (1)$$

or

$$S_n = a_n + (a_n - d) + (a_n - 2d) + \cdots + (a_n - (n-1)d) \qquad (2)$$

[Equation (2) is formed by reversing the terms of the series and *subtracting* the common difference from each term to generate the next term.]

Add equations (1) and (2):

$$2S_n = \frac{(a_1 + a_n) + (a_1 + a_n) + (a_1 + a_n) + \cdots (a_1 + a_n)}{n \text{ times}}$$

$$2S_n = n(a_1 + a_n)$$

so that

$$S_n = \frac{n}{2}(a_1 + a_n) \tag{3}$$

Because

$$a_n = a_1 + (n - 1)d$$

then

$$S_n = \frac{n}{2}(a_1 + a_1 + (n - 1)d)$$

$$S_n = \frac{n}{2}(2a_1 + (n - 1)d) \tag{4}$$

These two results are summarized in the following formulas.

Sum of the First n Terms of an Arithmetic Sequence

Given an arithmetic sequence with first term a_1, common difference d, and general term a_n, the sum of the first n terms of the sequence, denoted by S_n, is determined by the formulas

$$S_n = \frac{n}{2}(a_1 + a_n) \tag{3}$$

$$S_n = \frac{n}{2}(2a_1 + (n - 1)d) \tag{4}$$

REMARKS

1. To use equation (3) you must know a_1, a_n, and n; that is, you must know the first term, the nth term, and the number of terms in the sequence.

2. To use equation (4) you must know the first term, the number of terms, and the common difference.

EXAMPLE 11.2.6 Find the sum of the first eight terms (i.e., find S_8) of the arithmetic sequence with first term 2 and eighth term 11.

SOLUTION Since we are given n, a_1, and a_n, we will use equation (3). In this case $n = 8$, $a_1 = 2$, and $a_8 = 11$.

$$S_8 = \frac{8}{2}(a_1 + a_8)$$
$$= \frac{8}{2}(2 + 11)$$
$$= 4(13)$$
$$= 52$$

EXAMPLE 11.2.7 Find S_{10} of the arithmetic sequence with first term -3 and common difference $\frac{1}{2}$.

SOLUTION Since we are given n, a_1, and d, we will use equation (4). In this case $n = 10$, $a_1 = -3$, and $d = \frac{1}{2}$.

$$S_{10} = \frac{10}{2}(2a_1 + (10 - 1)d)$$
$$= \frac{10}{2}\left(2(-3) + (10 - 1)\frac{1}{2}\right)$$
$$= 5\left(-6 + \frac{9}{2}\right)$$
$$= 5\left(-\frac{3}{2}\right)$$
$$= -\frac{15}{2}$$

EXAMPLE 11.2.8 Evaluate the series

$$\sum_{i=1}^{20} (4i - 3)$$

SOLUTION   The general term, $4i - 3$, is first degree in i. Therefore, $\sum_{i=1}^{20} (4i - 3)$ is an arithmetic series. Let's write the series in expanded form listing only the first three terms and the last term:

$$\sum_{i=1}^{20} (4i - 3) = (4(1) - 3) + (4(2) - 3) + (4(3) - 3) + \cdots + (4(20) - 3)$$

$$= 1 + 5 + 9 + \cdots + 77$$

Since this is an arithmetic series and since we know $n = 20$, $a_1 = 1$, and $a_{20} = 77$, we will use equation (3):

$$S_{20} = \frac{20}{2}(a_1 + a_{20})$$

$$= \frac{20}{2}(1 + 77)$$

$$= 10(78)$$

$$= 780$$

Exercises 11.2

Find the common difference of each of the following arithmetic sequences.

1. 6, 8, 10, 12, 14, ...

2. $-2, -5, -8, -11, -14, \ldots$

3. $\frac{1}{2}, 0, -\frac{1}{2}, -1, -\frac{3}{2}, \ldots$

4. $\frac{5}{3}, \frac{7}{3}, 3, \frac{11}{3}, \frac{13}{3}, \ldots$

5. 7, 7, 7, 7, 7, ...

6. $-1.1, -1.2, -1.3, -1.4, -1.5, \ldots$

Find the indicated term of each of the following arithmetic sequences.

7. $a_1 = 8$ and $d = 7$; a_9

8. $a_1 = -3$ and $d = 10$; a_7

9. 6, 8, 10, 12, 14, ...; a_{18}

10. $-2, -5, -8, -11, -14, \ldots$; a_{25}

11. $\frac{1}{2}, 0, -\frac{1}{2}, -1, -\frac{3}{2}, \ldots$; a_{30}

12. $\frac{5}{3}, \frac{7}{3}, 3, \frac{11}{3}, \frac{13}{3}, \ldots$; a_{41}

13. 7, 7, 7, 7, 7, ...; a_{99}

14. $-1.1, -1.2, -1.3, -1.4, -1.5, \ldots$; a_{12}

Find the general term, a_n, of each of the following arithmetic sequences.

15. 7, 9, 11, 13, ...

16. 10, 12, 14, 16, ...

17. $-2, -5, -8, -11, \ldots$

18. $-1, -4, -7, -10, \ldots$

19. $-\frac{5}{2}, -2, -\frac{3}{2}, -1, \ldots$

20. $-\frac{5}{4}, -\frac{1}{2}, \frac{1}{4}, 1, \ldots$

21. 1, $-3, -7, -11, \ldots$

22. $-7, -12, -17, -22, \ldots$

23. $-\frac{7}{4}, -\frac{13}{4}, -\frac{19}{4}, -\frac{25}{4}, \ldots$

24. $-\dfrac{7}{3}, -3, -\dfrac{11}{3}, -\dfrac{13}{3}, \ldots$

Find a_1 and d in the arithmetic sequences with the following terms.

25. $a_7 = 22$ and $a_{12} = 37$

26. $a_6 = 17$ and $a_{10} = 33$

27. $a_9 = -\dfrac{37}{2}$ and $a_{16} = -\dfrac{65}{2}$

28. $a_4 = -\dfrac{17}{3}$ and $a_{10} = -\dfrac{32}{3}$

29. $a_5 = \dfrac{23}{6}$ and $a_9 = \dfrac{13}{2}$

30. $a_4 = -\dfrac{8}{3}$ and $a_{16} = -\dfrac{35}{3}$

Find the indicated sums of the following arithmetic sequences.

31. $a_1 = 1$ and $a_{10} = 27$; S_{10}

32. $a_1 = 2$ and $a_{18} = 53$; S_{18}

33. $-7, -2, 3, 8, \ldots$; S_{19}

34. $-5, -1, 3, 7, \ldots$; S_{14}

35. $-\dfrac{8}{3}, -\dfrac{10}{3}, -4, -\dfrac{14}{3}, \ldots$; S_{25}

36. $\dfrac{1}{2}, 2, \dfrac{7}{2}, 5, \ldots$; S_{29}

37. $a_1 = -\dfrac{15}{2}$ and $d = -\dfrac{1}{2}$; S_{23}

38. $a_1 = \dfrac{11}{4}$ and $d = \dfrac{3}{4}$; S_{20}

Evaluate the following series.

39. $\sum_{i=1}^{15} (2i + 7)$

40. $\sum_{i=1}^{24} (3i - 5)$

41. $\sum_{i=1}^{42} (7 - 3i)$

42. $\sum_{i=1}^{18} (6 - 9i)$

43. $\sum_{i=1}^{48} \left(\dfrac{1}{2}i - \dfrac{3}{4} \right)$

44. $\sum_{i=1}^{57} \left(\dfrac{2}{3}i + 7 \right)$

45. $\sum_{i=1}^{11} \left(\dfrac{1}{5}i - \dfrac{2}{3} \right)$

46. $\sum_{i=1}^{7} \left(\dfrac{3}{4} - \dfrac{1}{8}i \right)$

Write Algebra

47. Given a_3 and a_8 in an arithmetic sequence, describe how you would find the common difference, d.

48. If the terms of an arithmetic sequence are getting smaller, what can you conclude about the common difference?

11.3 Geometric Sequences and Series

Suppose you are given the sequence with general term $a_n = 3(2^n)$. The first five terms of the sequence are 6, 12, 24, 48, 96. Note that after the first term, each term can be found by multiplying the preceding term by 2. The sequence 6, 12, 24, 48, 96 is an example of a particular type of sequence called a geometric sequence.

Definition 11.3.1	A **geometric sequence** is a sequence in which each term after the first term can be found by multiplying the preceding term by a fixed constant. This fixed constant is called the **common ratio** and is denoted by the letter r.

REMARK To find the common ratio of any geometric sequence, simply divide any term (other than a_1) by the term that precedes it.

XAMPLE 11.3.1   Find the common ratio of the following geometric sequence:

$$3, \quad 12, \quad 48, \quad 192, \ldots$$

SOLUTION   One way to find the common ratio is as follows:

$$\text{common ratio} = \text{third term} \div \text{second term}$$
$$r = a_3 \div a_2$$
$$= 48 \div 12$$
$$= 4$$

NOTE ▶   We can use any two consecutive terms to find the common ratio.

XAMPLE 11.3.2   Write the first five terms of a geometric sequence with first term 32 and common ratio $\frac{1}{2}$.

SOLUTION   Since we have a geometric sequence with common ratio $\frac{1}{2}$, each term can be found by multiplying the preceding term by $\frac{1}{2}$:

$$a_1 = 32$$
$$a_2 = a_1 \cdot \frac{1}{2} = 32 \cdot \frac{1}{2} = 16$$
$$a_3 = a_2 \cdot \frac{1}{2} = 16 \cdot \frac{1}{2} = 8$$
$$a_4 = a_3 \cdot \frac{1}{2} = 8 \cdot \frac{1}{2} = 4$$
$$a_5 = a_4 \cdot \frac{1}{2} = 4 \cdot \frac{1}{2} = 2$$

Thus the first five terms of the sequence are

$$32, \quad 16, \quad 8, \quad 4, \quad 2$$

Suppose that the sequence $a_1, a_2, a_3, a_4, \ldots, a_n$ is a geometric sequence with common ratio r; then

$$a_1 = a_1$$
$$a_2 = a_1 r$$
$$a_3 = a_2 \cdot r = (a_1 r) \cdot r = a_1 r^2$$
$$a_4 = a_3 \cdot r = (a_1 r^2) \cdot r = a_1 r^3$$

These results are generalized in the following formula.

General Term of a Geometric Sequence

Given a geometric sequence with first term a_1 and common ratio r, the general term, a_n, is determined by the formula

$$a_n = a_1 r^{n-1}$$

EXAMPLE 11.3.3   Find a_5 for the geometric sequence with $a_1 = 90$ and $r = -\frac{1}{3}$.

SOLUTION   From the general-term formula,

$$
\begin{aligned}
a_5 &= a_1 r^{5-1} \\
&= a_1 r^4 \\
&= 90\left(-\frac{1}{3}\right)^4 \\
&= 90\left(\frac{1}{81}\right) \\
&= \frac{90}{81} \\
&= \frac{10}{9}
\end{aligned}
$$

EXAMPLE 11.3.4   Find the general term, a_n, for the geometric sequence $54, -36, 24, -16, \ldots$

SOLUTION   To find a_n, we must know a_1 and r. Clearly $a_1 = 54$. We can find r by dividing any term (other than a_1) by the term that precedes it. So

$$
\begin{aligned}
r &= a_2 \div a_1 \\
&= -36 \div 54 \\
&= -\frac{2}{3}
\end{aligned}
$$

Now

$$
\begin{aligned}
a_n &= a_1 r^{n-1} \\
&= 54\left(-\frac{2}{3}\right)^{n-1}
\end{aligned}
$$

Definition 11.3.2	A sum of consecutive terms of a geometric sequence is called a **geometric series.**

In the first section of this chapter, we associated with the sequence $a_1, a_2, a_3, ..., a_n$ the series $S_n = a_1 + a_2 + a_3 + \cdots + a_n$. Suppose that the sequence $a_1, a_2, a_3, ..., a_n$ is a geometric sequence. Then S_n can be expressed as

$$S_n = a_1 + a_1r + a_1r^2 + \cdots + a_1r^{n-1} \tag{1}$$

so

$$-rS_n = -a_1r - a_1r^2 - \cdots - a_1r^{n-1} - a_1r^n \tag{2}$$

[Equation (2) is formed by multiplying both sides of equation (1) by $-r$.]
    Add equations (1) and (2):

$$S_n - rS_n = a_1 - a_1r^n$$

so that

$$(1 - r)S_n = a_1 - a_1r^n$$

$$S_n = \frac{a_1 - a_1r^n}{1 - r} \qquad \text{provided } r \neq 1$$

This result is summarized in the following formula.

Sum of the First n Terms of a Geometric Sequence

Given a geometric sequence with first term a_1 and common ratio r, the sum of the first n terms of the sequence, denoted by S_n, is determined by the formula

$$S_n = \frac{a_1 - a_1r^n}{1 - r} \qquad \text{provided } r \neq 1$$

REMARK   To use this formula you must know a_1, r, and n; that is, you must know the first term, the common ratio, and the number of terms in the sequence.

 XAMPLE 11.3.5   Find the sum of the first five terms (i.e., find S_5) of the geometric sequence

$$6, \quad 12, \quad 24, \quad 48, ...$$

SOLUTION   From the formula,

$$S_5 = \frac{a_1 - a_1 r^5}{1 - r}$$

Clearly, $a_1 = 6$. We can find r in the following fashion:

$$r = a_2 \div a_1$$
$$= 12 \div 6$$
$$= 2$$

Thus,

$$S_5 = \frac{6 - 6(2^5)}{1 - 2}$$
$$= \frac{6 - 6(32)}{1 - 2}$$
$$= \frac{6 - 192}{-1}$$
$$= 186$$

EXAMPLE 11.3.6   Evaluate the series

$$\sum_{i=1}^{7} 3^{i-1}$$

SOLUTION   Let's write this series in expanded form listing only the first three terms and the last term:

$$\sum_{i=1}^{7} 3^{i-1} = 3^{1-1} + 3^{2-1} + 3^{3-1} + \cdots + 3^{7-1}$$
$$= 3^0 + 3^1 + 3^2 + \cdots + 3^6$$
$$= 1 + 3 + 9 + \cdots + 729$$

This is a geometric series. Therefore, we can apply the formula for the sum of the first n terms of a geometric sequence. In this case $a_1 = 1$, $r = 3$, and $n = 7$.

$$S_7 = \frac{1 - 1 \cdot 3^7}{1 - 3}$$
$$= \frac{1 - 1 \cdot (2187)}{1 - 3}$$

$$= \frac{-2186}{-2}$$

$$= 1093$$

Hence $\Sigma_{i=1}^{7} 3^{i-1} = 1093$.

In the first section of this chapter, it was stated that we would find a case where the sum of the terms of an infinite sequence could be determined. We are now ready to examine that case. Consider the following infinite geometric sequence:

$$\frac{1}{2}, \frac{1}{4}, \frac{1}{8}, \frac{1}{16}, \frac{1}{32}, \cdots \tag{3}$$

Let's investigate the sum of the first n terms of this sequence for increasingly large values of n:

$$S_1 = \frac{1}{2} = 0.5$$

$$S_2 = \frac{1}{2} + \frac{1}{4} = \frac{3}{4} = 0.75$$

$$S_3 = \frac{1}{2} + \frac{1}{4} + \frac{1}{8} = \frac{7}{8} = 0.875$$

$$S_4 = \frac{1}{2} + \frac{1}{4} + \frac{1}{8} + \frac{1}{16} = \frac{15}{16} = 0.9375$$

$$S_5 = \frac{1}{2} + \frac{1}{4} + \frac{1}{8} + \frac{1}{16} + \frac{1}{32} = \frac{31}{32} = 0.96875$$

Notice that as n becomes larger, S_n is apparently becoming closer to 1. We can extend this argument by examining the formula for the sum of the first n terms of a geometric sequence:

$$S_n = \frac{a_1 - a_1 r^n}{1 - r}$$

If $|r| < 1$, then as n becomes larger, r^n approaches zero and so $a_1 \cdot r^n$ approaches zero, as is illustrated in sequence (3). This reasoning leads to the following formula:

Sum of the Terms of an Infinite Geometric Sequence

Given an infinite geometric sequence with first term a_1 and common ratio r, the sum of the terms of the sequence, denoted by S_∞, is determined by the formula

$$S_\infty = \frac{a_1}{1 - r} \qquad |r| < 1$$

REMARK

We can determine the sum of the terms of an infinite geometric sequence only when $|r| < 1$.

EXAMPLE 11.3.7   Find the sum of the terms of the following infinite geometric sequences (i.e., find S_∞), if possible.

1. $\dfrac{1}{2}, \dfrac{1}{4}, \dfrac{1}{8}, \dfrac{1}{16}, \ldots$

    To find the sum of the terms of an infinite geometric sequence, we must know a_1 and r. Clearly, $a_1 = \frac{1}{2}$. We can find r in the following fashion:

    $$r = a_2 \div a_1$$
    $$= \frac{1}{4} \div \frac{1}{2}$$
    $$= \frac{1}{2}$$

    Since $|r| < 1$, we can determine the sum of this sequence. From the formula,

    $$S_\infty = \frac{a_1}{1 - r}$$
    $$= \frac{\dfrac{1}{2}}{1 - \dfrac{1}{2}}$$
    $$= \frac{\dfrac{1}{2}}{\dfrac{1}{2}}$$
    $$= 1$$

2. 6, 12, 24, 48, …

    Here $a_1 = 6$, so

    $$r = a_2 \div a_1$$
    $$= 12 \div 6$$
    $$= 2$$

    Since $|r| \geq 1$, we cannot find the sum of the terms of this infinite geometric sequence.

EXAMPLE 11.3.8   Evaluate the series

$$\sum_{i=1}^{\infty} \left(\frac{2}{3}\right)^{i-1}$$

SOLUTION   Let's write this series in expanded form listing only the first four terms:

$$\sum_{i=1}^{\infty} \left(\frac{2}{3}\right)^{i-1} = \left(\frac{2}{3}\right)^{1-1} + \left(\frac{2}{3}\right)^{2-1} + \left(\frac{2}{3}\right)^{3-1} + \left(\frac{2}{3}\right)^{4-1} + \cdots$$

$$= \left(\frac{2}{3}\right)^{0} + \left(\frac{2}{3}\right)^{1} + \left(\frac{2}{3}\right)^{2} + \left(\frac{2}{3}\right)^{3} + \cdots$$

$$= 1 + \frac{2}{3} + \frac{4}{9} + \frac{8}{27} + \cdots$$

This is an infinite geometric series with $a_1 = 1$ and $r = \frac{2}{3}$. Therefore, we can apply the formula for the sum of the terms of an infinite geometric sequence:

$$S_{\infty} = \frac{1}{1 - \frac{2}{3}}$$

$$= \frac{1}{\frac{1}{3}}$$

$$= 3$$

Hence $\sum_{i=1}^{\infty} \left(\frac{2}{3}\right)^{i-1} = 3$.

In the next example we will apply the formula for the sum of the terms of an infinite geometric sequence to write a repeating decimal as a quotient of two integers.

EXAMPLE 11.3.9   Using an infinite geometric series, write the following repeating decimal as a quotient of two integers:

$$0.\overline{7} = 0.7777\ldots$$

NOTE ▶   Recall that the bar over the 7 means that the 7 repeats indefinitely:

$$0.\overline{7} = 0.7777\ldots$$

$$= \frac{7}{10} + \frac{7}{100} + \frac{7}{1000} + \frac{7}{10,000} + \cdots$$

The right-hand side of the equation is an infinite geometric series with $a_1 = \frac{7}{10}$ and $r = \frac{1}{10}$. Thus the sum of the right-hand side is

$$S_\infty = \frac{\dfrac{7}{10}}{1 - \dfrac{1}{10}}$$

$$= \frac{\dfrac{7}{10}}{\dfrac{9}{10}}$$

$$= \frac{7}{9}$$

Hence $0.\overline{7} = \frac{7}{9}$.

Exercises 11.3

Find the common ratio of each of the following geometric sequences.

1. 4, 12, 36, 108, ...

2. $\frac{1}{2}$, 2, 8, 32, ...

3. $9, 6, 4, \frac{8}{3}, \ldots$

4. $\frac{9}{10}, \frac{9}{100}, \frac{9}{1000}, \frac{9}{10,000}, \cdots$

5. $-\frac{1}{2}, \frac{3}{4}, -\frac{9}{8}, \frac{27}{16}, \cdots$

6. $1, -\frac{3}{4}, \frac{9}{16}, -\frac{27}{64}, \cdots$

Find the indicated term of each of the following geometric sequences.

7. $a_1 = 6$ and $r = 2$; a_5

8. $a_1 = 12$ and $r = \frac{1}{3}$; a_6

9. 4, 12, 36, 108, ...; a_6

10. $\frac{1}{2}$, 2, 8, 32, ...; a_7

11. $9, 6, 4, \frac{8}{3}, \ldots$; a_7

12. $\frac{9}{10}, \frac{9}{100}, \frac{9}{1000}, \frac{9}{10,000}, \cdots$; a_6

13. $-\frac{1}{2}, \frac{3}{4}, -\frac{9}{8}, \frac{27}{16}, \cdots$; a_8

14. $1, -\frac{3}{4}, \frac{9}{16}, -\frac{27}{64}, \cdots$; a_7

Find the general term, a_n, of each of the following geometric sequences.

15. 2, 4, 8, 16, ...

16. 1, 5, 25, 125, ...

17. $\frac{25}{4}, \frac{125}{8}, \frac{625}{16}, \frac{3125}{32}, \cdots$

18. $\frac{4}{9}, \frac{8}{27}, \frac{16}{81}, \frac{32}{243}, \cdots$

19. $-6, 18, -54, 162, \ldots$

20. $12, -48, 192, -768, \ldots$

21. $16, 12, 9, \frac{27}{4}, \ldots$

22. $1, \frac{5}{2}, \frac{25}{4}, \frac{125}{8}, \cdots$

23. $\frac{7}{8}, \frac{49}{16}, \frac{343}{32}, \frac{2401}{64}, \cdots$

24. $-\frac{1}{2}, \frac{3}{8}, -\frac{9}{32}, \frac{27}{128}, \cdots$

Find the indicated sums of the following geometric sequences, if possible.

25. $a_1 = -3$ and $r = 2$; S_6

26. $a_1 = 5$ and $r = 3$; S_4

27. 8, 12, 18, 27, ...; S_5

28. 27, -18, 12, -8, ...; S_6

29. $\dfrac{5}{2}, -\dfrac{25}{4}, \dfrac{125}{8}, -\dfrac{625}{16}, \dots;$ S_5

30. $\dfrac{3}{16}, \dfrac{3}{8}, \dfrac{3}{4}, \dfrac{3}{2}, \dots;$ S_6

31. $9, 6, 4, \dfrac{8}{3}, \dots;$ S_∞

32. 64, 48, 36, 27, ...; S_∞

33. $\dfrac{1}{3}, -\dfrac{1}{9}, \dfrac{1}{27}, -\dfrac{1}{81}, \dots;$ S_∞

34. $1, -\dfrac{4}{5}, \dfrac{16}{25}, -\dfrac{64}{125}, \dots;$ S_∞

35. 2, 4, 8, 16, ...; S_∞

36. $\dfrac{1}{3}, \dfrac{1}{2}, \dfrac{3}{4}, \dfrac{9}{8}, \dots;$ S_∞

Evaluate the following series, if possible.

37. $\sum_{i=1}^{6} 2^i$

38. $\sum_{i=1}^{4} 5^{i-1}$

39. $\sum_{i=1}^{5} 6\left(\dfrac{2}{3}\right)^{i-1}$

40. $\sum_{i=1}^{7} 8\left(-\dfrac{3}{2}\right)^{i-2}$

41. $\sum_{i=1}^{4} \dfrac{1}{10}\left(\dfrac{1}{2}\right)^{i}$

42. $\sum_{i=1}^{6} \dfrac{2}{3}\left(\dfrac{3}{2}\right)^{i-1}$

43. $\sum_{i=1}^{\infty} 3^i$

44. $\sum_{i=1}^{\infty} 7^{i-1}$

45. $\sum_{i=1}^{\infty} \left(\dfrac{2}{3}\right)^{i+1}$

46. $\sum_{i=1}^{\infty} \left(\dfrac{3}{4}\right)^{i}$

47. $\sum_{i=1}^{\infty} 8\left(-\dfrac{1}{2}\right)^{i+2}$

48. $\sum_{i=1}^{\infty} \dfrac{3}{4}\left(-\dfrac{5}{8}\right)^{i-1}$

Using an infinite geometric series, write the following repeating decimals as a quotient of two integers.

49. $0.\overline{8}$

50. $0.\overline{4}$

51. $0.\overline{56}$

52. $0.\overline{23}$

53. $0.\overline{123}$

54. $0.\overline{351}$

55. $0.5\overline{2}$

56. $4.7\overline{1}$

57. $1.8\overline{1}$

58. $5.\overline{72}$

59. $2.9\overline{34}$

60. $3.0\overline{25}$

Find the solutions of the following problems.

61. A ball is dropped from a height of 80 ft. On each bounce it will rebound three-fourths of its previous height. What is the total distance the ball will travel before coming to rest?

62. A ball is dropped from a height of 60 ft. On each bounce it will rebound two-thirds of its previous height. What is the total distance the ball will travel before coming to rest?

63. Home prices in Moss County are depreciating at a rate of 10% per year. If a home today is worth $60,000, how much will it be worth after 6 yr?

64. An old coin is currently selling for $500. Spike the coin dealer predicts that the coin will appreciate in value at a rate of 20% per year. If Spike is correct, how much will the coin be worth after 5 yr?

65. Suppose Patti wins the first 10 holes of the Moss County Invitational Golf Tournament. What are Patti's total winnings? (*Hint:* Recall that the equation $a_n = 2^n$ expresses the winnings on each hole. Thus, Patti won $2 for the first hole, $4 for the second hole, and so on.)

Write Algebra

66. Given a_2 and a_5 in a geometric sequence, describe how you would find the common ratio, r.

67. Describe the process for converting repeating decimals to fractions by using infinite geometric series.

G R O U P   A C T I V I T Y

68. Given a geometric series with $a_5 - a_4 = -4$ and $a_2 - a_1 = -32$, find S_5.

69. Given a geometric series with $a_5 - a_4 = -8$ and $a_2 - a_1 = -27$, find S_5.

70. (i) Find the sum of the geometric series $1 + 3 + 9 + 27 + 81$.

(ii) Find the sum of the geometric series $1 + \frac{1}{3} + \frac{1}{9} + \frac{1}{27} + \frac{1}{81}$.

71. (i) Find S_n for the geometric series $1 + r + r^2 + r^3 + \cdots + r^{n-1}$

(ii) Find S_n for the geometric series $1 + \frac{1}{r} + \frac{1}{r^2} + \frac{1}{r^3} + \cdots + \frac{1}{r^{n-1}}$

(iii) What is the relationship between the sum found in part (i) and the sum found in part (ii)?

72. Find the sum.
$\frac{1}{5} + \frac{2}{5^2} + \frac{1}{5^3} + \frac{2}{5^4} + \cdots$

73. Find the sum.
$\frac{1}{9} + \frac{2}{9^2} + \frac{1}{9^3} + \frac{2}{9^4} + \cdots$

11.4 *The Binomial Theorem and Pascal's Triangle*

In Chapter 3 we learned how to expand binomials raised to natural number powers. However, the natural numbers we examined were quite small, usually 1, 2, or 3. Rather than use the multiplication techniques of Chapter 3, in this section we will find a quicker way of expanding binomials raised to relatively large natural number powers.

The following list shows the expansion of the binomial $(a + b)^n$ when $n = 1, 2, 3, 4,$ and 5:

$$(a + b)^1 = a + b$$
$$(a + b)^2 = a^2 + 2ab + b^2$$
$$(a + b)^3 = a^3 + 3a^2b + 3ab^2 + b^3$$
$$(a + b)^4 = a^4 + 4a^3b + 6a^2b^2 + 4ab^3 + b^4$$
$$(a + b)^5 = a^5 + 5a^4b + 10a^3b^2 + 10a^2b^3 + 5ab^4 + b^5$$

These expansions yield the following observations given $(a + b)^n$, where n is a natural number:

1. The expansion has $n + 1$ terms.

2. The first term in the expansion is a^n and the last term is b^n.

3. The degree of each term is n; that is, the sum of the exponents of a and b in each term is n.

4. In successive terms the exponent of a is decreased by 1 and the exponent of b is increased by 1.

With these observations we can predict the form of the expansion of $(a + b)^6$:

$$(a + b)^6 = a^6 + \underline{\quad} a^5b + \underline{\quad} a^4b^2 + \underline{\quad} a^3b^3 + \underline{\quad} a^2b^4 + \underline{\quad} ab^5 + b^6$$

The problem here is to fill in the blanks—that is, find the coefficients of all the terms other than the first and last term. Let's return to the expansions of $(a + b)^n$ and examine only the coefficients in each expansion.

In Figure 11.4.1 the coefficients have been arranged in a triangular array. Each row of the array begins and ends with 1. *Note that the numbers between the ones can be determined by finding the sum of the two nearest numbers in the row above.* A schematic method for finding the numbers between the 1's is as follows: Find the sum of the numbers at the top of the V. This sum is the number in the next row at the base of the V. The triangular array in the figure is called **Pascal's triangle.**

$(a + b)^1$

$(a + b)^2$

$(a + b)^3$

$(a + b)^4$

$(a + b)^5$

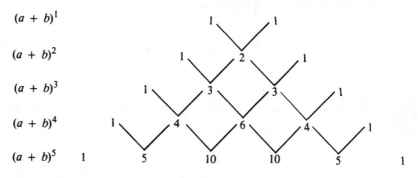

Figure 11.4.1

To find the coefficients of the expansion of $(a + b)^6$, we will simply add one more row to the triangle in Figure 11.4.1 to make Figure 11.4.2. Hence

$$(a + b)^6 = a^6 + 6a^5b + 15a^4b^2 + 20a^3b^3 + 15a^2b^4 + 6ab^5 + b^6$$

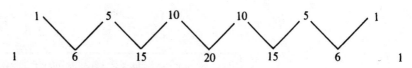

Figure 11.4.2

XAMPLE 11.4.1   Use Pascal's triangle to expand $(x + 2)^4$.

SOLUTION   From the observations at the start of this section, we know that

$$(x + 2)^4 = x^4 + \underline{\hspace{1cm}} x^3(2) + \underline{\hspace{1cm}} x^2(2)^2 + \underline{\hspace{1cm}} x(2)^3 + 2^4$$

We can now fill in the blanks using the fourth row of Pascal's triangle:

$$(x + 2)^4 = x^4 + 4x^3(2) + 6x^2(2)^2 + 4x(2)^3 + 2^4$$
$$(x + 2)^4 = x^4 + 8x^3 + 24x^2 + 32x + 16$$

Pascal's triangle makes the expansion of binomials raised to natural number powers relatively simple. However, if we wanted the expansion of a binomial to a large natural number power, say 25, then we would have to construct Pascal's triangle all the way to the 25th row. Fortunately, a theorem has been developed that somewhat lessens this problem. First some new terminology needs to be introduced.

Definition 11.4.1   The product of the first n natural numbers, denoted by the symbol $n!$, is called **n factorial:**

$$n! = n \cdot (n - 1) \cdot (n - 2) \cdots 3 \cdot 2 \cdot 1$$

XAMPLE 11.4.2   Evaluate each of the following.

1. $4! = 4 \cdot 3 \cdot 2 \cdot 1$
    $= 24$

2. $6! = 6 \cdot 5 \cdot 4 \cdot 3 \cdot 2 \cdot 1$
    $= 720$

Since $n! = n \cdot (n - 1) \cdot (n - 2) \cdots 3 \cdot 2 \cdot 1$, then

$$(n - 1)! = (n - 1) \cdot (n - 2) \cdot (n - 3) \cdots 3 \cdot 2 \cdot 1 \tag{1}$$

Multiply both sides of equation (1) by n:

$$n \cdot (n - 1)! = n \cdot (n - 1) \cdot (n - 2) \cdot (n - 3) \cdots 3 \cdot 2 \cdot 1$$

$$n \cdot (n - 1)! = n! \tag{2}$$

Equation (2) illustrates the fact that

$$7 \cdot 6! = 7!$$
$$12 \cdot 11! = 12!$$

However, when $n = 1$, equation (2) becomes

$$1 \cdot (0)! = 1!$$
$$1 \cdot (0)! = 1$$

This observation leads to the following definition.

Definition 11.4.2	$0! = 1$

Let's return to the expansion of $(a + b)^4$. We know that $(a + b)^4 = a^4 + 4a^3b + 6a^2b^2 + 4ab^3 + b^4$. Note the following:

1. The coefficient of the second term can be expressed as

$$4 = \frac{4}{1!} = \frac{n}{1!}$$

where n is the exponent of $a + b$.

2. Likewise, the coefficient of the third term can be expressed as

$$6 = \frac{4 \cdot 3}{2!} = \frac{n(n - 1)}{2!}$$

3. Likewise, the coefficient of the fourth term can be expressed as

$$4 = \frac{4 \cdot 3 \cdot 2}{3!} = \frac{n(n - 1)(n - 2)}{3!}$$

The pattern on the right-hand side of these equations is generalized in the following theorem, which is given without proof.

Binomial Theorem

$$(a + b)^n = a^n + \frac{n}{1!} a^{n-1}b$$

$$+ \frac{n(n - 1)}{2!} a^{n-2}b^2 + \frac{n(n - 1)(n - 2)}{3!} a^{n-3}b^3 + \cdots$$

$$+ \frac{n(n - 1)(n - 2) \cdots (n - r + 2)}{(r - 1)!} a^{n-r+1}b^{r-1} + \cdots + b^n$$

where n is any natural number.

XAMPLE 11.4.3 Use the binomial theorem to expand the following.

1. $(x + y)^5 = x^5 + \dfrac{5}{1!}x^4y + \dfrac{5 \cdot 4}{2!}x^3y^2 + \dfrac{5 \cdot 4 \cdot 3}{3!}x^2y^3 + \dfrac{5 \cdot 4 \cdot 3 \cdot 2}{4!}xy^4 + y^5$

   $= x^5 + 5x^4y + 10x^3y^2 + 10x^2y^3 + 5xy^4 + y^5$

2. $(a - 2b)^4 = (a + (-2b))^4$ *Rewrite the binomial as a sum.*

   $= a^4 + \dfrac{4}{1!}a^3(-2b) + \dfrac{4 \cdot 3}{2!}a^2(-2b)^2 + \dfrac{4 \cdot 3 \cdot 2}{3!}a(-2b)^3 + (-2b)^4$

   $= a^4 - 8a^3b + 24a^2b^2 - 32ab^3 + 16b^4$

NOTE ▶ When a difference is raised to a power, the signs always *alternate*.

E xercises 11.4

Use Pascal's triangle to expand the following binomials.

1. $(x + 2y)^6$ **2.** $(3a + b)^5$ **3.** $(x - 3)^5$ **4.** $\left(2x - \dfrac{1}{3}\right)^6$

5. $\left(\dfrac{5}{2} - 2a\right)^4$ **6.** $\left(\dfrac{1}{2}x + \dfrac{2}{3}y\right)^5$

Evaluate each of the following.

7. $5!$ **8.** $8!$ **9.** $\dfrac{4!}{8!}$ **10.** $\dfrac{5!}{10!}$

11. $\dfrac{10!}{4!}$ **12.** $\dfrac{0!}{0!}$ **13.** $\dfrac{4 \cdot 3 \cdot 2}{3!}$ **14.** $\dfrac{6 \cdot 5 \cdot 4 \cdot 3}{4!}$

Use the binomial theorem to expand the following binomials.

15. $(a + 3)^4$ **16.** $(2x + 5)^4$ **17.** $(3x + 4y)^4$ **18.** $(2x + y)^7$

19. $\left(a - \dfrac{3}{2}b\right)^5$ **20.** $\left(\dfrac{1}{3}x - 2y\right)^6$ **21.** $(x - 2y)^7$ **22.** $(x - 3y)^5$

23. $\left(\dfrac{1}{2}x + y\right)^6$ **24.** $\left(\dfrac{1}{2}a + \dfrac{1}{3}b\right)^6$ **25.** $(x + 2y^2)^5$ **26.** $(3a^2 + 4y)^4$

Write Algebra

27. Describe how to simplify the fraction $\frac{8!}{10!}$.

Applied Algebra

In a codicil to his will Benjamin Franklin bequeathed $5000 (1000 pounds sterling) to the "inhabitants of the city of Philadelphia." The money was to be loaned to "young married artificers" at 5% interest compounded annually. The city of Philadelphia received the $5000 in 1791. If the terms of Franklin's will were satisfied, how much money did he expect the Benjamin Franklin Fund to have in 1891?

SOLUTION   Recall the simple interest formula

$$I = PRT$$

In this formula I represents the interest, P represents the principal, R represents the rate, and T is the time measured in years.

Thus, the interest earned in 1792 would be determined as follows:

$$
\begin{aligned}
I &= PRT \\
&= (5000)(0.05)(1) \\
&= 0.05(5000) \qquad \text{\textit{We will see in a moment why we}} \\
&\qquad\qquad\qquad\quad \text{\textit{use this form for the interest.}}
\end{aligned}
$$

Hence the amount in the Franklin Fund

after 1 year (i.e., 1792):
$$
\begin{aligned}
\text{principal} + \text{interest} &= 5000 + 0.05(5000) \\
&= 5000(1 + 0.05) \\
&= 5000(1.05)
\end{aligned}
$$

after 2 years (i.e., 1793):
$$
\begin{aligned}
\text{principal} + \text{interest} &= 5000(1.05) + 0.05(5000)(1.05) \\
&= 5000(1.05)(1 + 0.05) \\
&= 5000(1.05)^2
\end{aligned}
$$

after 3 years (i.e., 1794):
$$
\begin{aligned}
\text{principal} + \text{interest} &= 5000(1.05)^2 + 0.05(5000)(1.05)^2 \\
&= 5000(1.05)^2(1 + 0.05) \\
&= 5000(1.05)^3
\end{aligned}
$$

NOTE ▶   The amounts in the Franklin Fund form a geometric sequence

$$5000(1.05), \quad 5000(1.05)^2, \quad 5000(1.05)^3, \ldots$$

In this sequence $a_1 = 5000(1.05)$ and $r = 1.05$.
Since we want to determine the value of the fund in 1891 we need to find a_{100}.

$$
\begin{aligned}
a_{100} &= a_1 r^{99} \\
&= 5000(1.05) \cdot (1.05)^{99} \\
&= 5000(1.05)^{100} \\
&\doteq 657{,}506
\end{aligned}
$$

Thus, Benjamin Franklin expected the value of his fund to be approximately $657,506 100 years after his death.

Your Turn

1. If Franklin's plan were to operate for 200 years, how much money would the Franklin Fund have in 1991?

2. If the interest rate was 6% compounded annually, how much money would the Franklin Fund have in 1891?

3. If the interest rate was 5% compounded annually and Franklin bequeathed $6000, how much money would the Franklin Fund have in 1891?

Chapter 11 Review

Terms to Remember

Finite sequence	**p. 634**	Arithmetic sequence	**p. 639**
Infinite sequence	**p. 634**	Common difference	**p. 639**
Term of a sequence	**p. 634**	Arithmetic series	**p. 642**
Sequence	**p. 634**	Geometric sequence	**p. 646**
General term	**p. 634**	Common ratio	**p. 646**
Series	**p. 635**	Geometric series	**p. 649**
Infinite series	**p. 635**	Pascal's triangle	**p. 657**
Summation notation	**p. 636**	*n* factorial	**p. 658**
Index of summation	**p. 636**		

Key Definitions

A *finite sequence* is a function whose domain is $\{1, 2, 3, ..., k\}$ for some natural number k. An *infinite sequence* is a function whose domain is the set of natural numbers. The elements of the range of the function are called *terms*. The list of terms stated in order is called a *sequence*.

A sum of consecutive terms of a sequence is called a *series*. If the sequence is infinite, the series is called an *infinite series*.

Given the finite sequence $a_1, a_2, a_3, ..., a_n$, then

$$S_n = a_1 + a_2 + a_3 + \cdots + a_n$$

Given the infinite sequence $b_1, b_2, b_3, ..., b_n, ...$, then

$$S_\infty = b_1 + b_2 + b_3 + \cdots + b_n + \cdots$$

$$\sum_{i=1}^{n} a_i = a_1 + a_2 + a_3 + \cdots + a_n$$

An *arithmetic sequence* is a sequence in which each term after the first term can be found by adding a fixed constant to the preceding term. This fixed constant is called the *common difference*.

A sum of consecutive terms of an arithmetic sequence is called an *arithmetic series.*

A *geometric sequence* is a sequence in which each term after the first term can be found by multiplying the preceding term by a fixed constant. This fixed constant is called the *common ratio.*

A sum of consecutive terms of a geometric sequence is called a *geometric series.*

The product of the first n natural numbers, denoted by the symbol $n!$, is called *n factorial:*

$$n! = n(n - 1) \cdot (n - 2) \cdot \cdots \cdot 3 \cdot 2 \cdot 1$$

$$n \cdot (n - 1)! = n!$$

$$0! = 1$$

Formulas

■ *General term of an arithmetic sequence:* Given an arithmetic sequence with first term a_1 and common difference d, the general term, a_n, is determined by the formula

$$a_n = a_1 + (n - 1)d$$

■ *Sum of the first n terms of an arithmetic sequence:* Given an arithmetic sequence with first term a_1, common difference d, and general term a_n, the sum of the first n terms of the sequence, denoted by S_n, is determined by the formulas

$$S_n = \frac{n}{2}(a_1 + a_n)$$

$$S_n = \frac{n}{2}(2a_1 + (n - 1)d)$$

■ *General term of a geometric sequence:* Given a geometric sequence with first term a_1 and common ratio r, the general term, a_n, is determined by the formula

$$a_n = a_1 r^{n-1}$$

■ *Sum of the first n terms of a geometric sequence:* Given a geometric sequence with first term a_1 and common ratio r, the sum of the first n terms of the sequence, denoted by S_n, is determined by the formula

$$S_n = \frac{a_1 - a_1 r^n}{1 - r} \qquad \text{provided } r \neq 1$$

■ *Sum of the terms of an infinite geometric sequence:* Given an infinite geometric sequence with first term a_1 and common ratio r, the sum of the terms of the sequence, denoted by S_∞, is determined by the formula

$$S_\infty = \frac{a_1}{1 - r} \qquad \text{where } |r| < 1$$

■ *Binomial theorem:*

$$(a + b)^n = a^n + \frac{n}{1!} a^{n-1}b + \frac{n(n-1)}{2!} a^{n-2}b^2 + \frac{n(n-1)(n-2)}{3!} a^{n-3}b^3 + \cdots$$

$$+ \frac{n(n-1)(n-2)\cdots(n-r+2)}{(r-1)!} a^{n-r+1}b^{r-1} + \cdots + b^n$$

where n is any natural number.

Review Exercises

11.1 Write the first five terms of the sequences with the following general terms.

1. $a_n = 4n + 2$

2. $a_n = 3^{n-1}$

3. $a_n = 2n^2 - 3$

4. $a_n = \left(\frac{1}{2}\right)^n + 1$

5. $a_n = (-1)^n(3n + 2)$

6. $a_n = \frac{1}{2n}$

Find each indicated term of the sequences with the following general terms.

7. $a_n = 6n - 5$; a_4

8. $a_n = n^2 - n - 12$; a_6

9. $a_n = 8 \cdot \left(\frac{3}{2}\right)^n$; a_4

Find S_5 for the sequences with the following general terms. (*Hint:* Use the results of exercises 1–6.)

10. $a_n = 4n + 2$

11. $a_n = 3^{n-1}$

12. $a_n = 2n^2 - 3$

13. $a_n = \left(\frac{1}{2}\right)^n + 1$

14. $a_n = (-1)^n(3n + 2)$

15. $a_n = \frac{1}{2n}$

Write the following series in expanded form and evaluate.

16. $\Sigma_{i=1}^{7} (2i + 4)$

17. $\Sigma_{i=1}^{3} \left(\frac{2}{5}\right)^i$

18. $\Sigma_{i=1}^{4} (3^i - 2i)$

19. $\Sigma_{i=0}^{6} (i^2 + i + 1)$

20. $\Sigma_{i=2}^{6} \frac{60}{i}$

21. $\Sigma_{i=1}^{4} (-1)^i(2i + 1)$

Write the following series in summation notation.

22. $3 + 5 + 7 + 9 + 11 + 13$

23. $2 + 4 + 8 + 16 + 32 + 64$

24. $1 + 8 + 27 + 64 + 125$

25. $\frac{3}{2} + \frac{4}{3} + \frac{5}{4} + \frac{6}{5} + \frac{7}{6} + \frac{8}{7}$

26. $1 - 4 + 9 - 16 + 25 - 36$

27. Gus is taking a typing course. At the end of the first week Gus can type 28 words per minute. His speed then increases weekly at a rate of 12 words per minute. How fast will Gus be typing at the end of the fourth week?

11.2   Find the common difference of each of the following arithmetic sequences.

28. $5, 9, 13, 17, 21, \ldots$

29. $2, \dfrac{3}{2}, 1, \dfrac{1}{2}, 0, \ldots$

Find the indicated term of each of the following arithmetic sequences.

30. $a_1 = 7$ and $d = 2$; a_{10}

31. $4, 2, 0, -2, -4, \ldots$; a_{24}

32. $2, \dfrac{8}{3}, \dfrac{10}{3}, 4, \dfrac{14}{3}, \ldots$; a_{31}

Find the general term, a_n, of the following arithmetic sequences.

33. $5, 8, 11, 14, \ldots$

34. $3, 1, -1, -3, \ldots$

35. $4, \dfrac{9}{2}, 5, \dfrac{11}{2}, \ldots$

36. $-1, -\dfrac{5}{3}, -\dfrac{7}{3}, -3, \ldots$

Find a_1 and d in the arithmetic sequences with the following terms.

37. $a_8 = 41$ and $a_{13} = 61$

38. $a_{10} = 27$ and $a_{24} = 62$

Find the indicated sums of the following arithmetic sequences.

39. $a_1 = 8$ and $a_{15} = -14$; S_{15}

40. $3, 7, 11, 15, \ldots$; S_{22}

41. $a_1 = \dfrac{5}{3}$ and $d = -\dfrac{2}{3}$; S_{12}

Evaluate the following series.

42. $\Sigma_{i=1}^{17} (3i + 7)$

43. $\Sigma_{i=1}^{24} \left(\dfrac{3}{8}i - 2 \right)$

44. $\Sigma_{i=1}^{8} \left(\dfrac{5}{2} - \dfrac{3}{2}i \right)$

Find the solutions of the following problems.

45. Find the sum of the even integers between 1 and 201.

46. On the day of its grand opening, Beth's Burger City sells 5 hamburgers. Each succeeding day, Beth's Burger City sells 3 more hamburgers than on the previous day. How many hamburgers does Beth sell in her first week?

11.3   Find the common ratio of each of the following geometric sequences.

47. $3, -9, 27, -81, \ldots$

48. $24, 36, 54, 81, \ldots$

Find the indicated term of each of the following geometric sequences.

49. $a_1 = 7$ and $r = 3$; a_5

50. $27, 18, 12, 8, \ldots$; a_6

51. $40, -20, 10, -5, \ldots$; a_8

Find the general term, a_n, of each of the following geometric sequences.

52. $1, 4, 16, 64, \ldots$

53. $-15, 45, -135, 405, \ldots$

54. $\dfrac{9}{4}, \dfrac{27}{8}, \dfrac{81}{16}, \dfrac{243}{32}, \ldots$

55. $\dfrac{1}{6}, \dfrac{1}{10}, \dfrac{3}{50}, \dfrac{9}{250}, \ldots$

Find the indicated sums of the following geometric sequences, if possible.

56. $a_1 = 4$ and $r = -3$; S_5

57. $16, 20, 25, \dfrac{125}{4}, \ldots$; S_5

58. $54, -36, 24, -16, \ldots$; S_∞

59. $16, 24, 36, 54, \ldots$; S_∞

Evaluate the following series, if possible.

60. $\sum_{i=1}^{5} 5(2)^i$

61. $\sum_{i=1}^{6} 16 \cdot \left(-\dfrac{3}{2}\right)^{i-2}$

62. $\sum_{i=1}^{\infty} \dfrac{5}{2} \cdot \left(\dfrac{7}{5}\right)^{i-1}$

63. $\sum_{i=1}^{\infty} 4 \cdot \left(\dfrac{1}{2}\right)^i$

Using an infinite geometric series, write each of the following repeating decimals as a quotient of two integers.

64. $0.\overline{2}$ **65.** $0.6\overline{1}$ **66.** $0.\overline{34}$ **67.** $0.\overline{285}$ **68.** $1.\overline{52}$

69. A ball is dropped from a height of 100 ft. On each bounce it will rebound to one-fourth of its previous height. What is the total distance the ball will travel before coming to rest?

70. On the day of its grand opening, Bud's Burger City sells 5 hamburgers. Each succeeding day, Bud's Burger City sells twice as many hamburgers as on the previous day. How many burgers does Bud sell in his first week?

11.4   Use Pascal's triangle to expand the following binomials.

71. $(3a + b)^6$

72. $(2x - 3)^5$

Evaluate each of the following.

73. $7!$

74. $\dfrac{9!}{6!}$

Use the binomial theorem to expand the following binomials.

75. $(x + 4)^4$

76. $(3x - 2y)^6$

77. $\left(\dfrac{1}{2}x - \dfrac{2}{3}y\right)^5$

78. $(2x^2 + 5)^4$

Chapter 11 Test

(You should be able to complete this test in 60 minutes.)

I. In the following arithmetic sequence find the common difference, d, the general term, a_n, and S_{24}.

1. $4, \dfrac{11}{2}, 7, \dfrac{17}{2}, \ldots$

II. In the following geometric sequence find the common ratio, r, the general term, a_n, and S_5.

2. $32, 48, 72, 108, \ldots$

III. Evaluate the following series, if possible.

3. $\sum_{i=1}^{5}(2i^2 - 3i + 7)$

4. $\sum_{i=2}^{6}(-1)^i\left(\left(\frac{i}{2}\right)^2 + 5\right)$

5. $\sum_{i=1}^{20}(5i - 7)$

6. $\sum_{i=1}^{7} 3 \cdot (2^i)$

7. $\sum_{i=1}^{\infty}\left(\frac{9}{8}\right)^i$

8. $\sum_{i=1}^{\infty} 3 \cdot \left(\frac{2}{3}\right)^{i-1}$

IV. Using an infinite geometric series, write the following repeating decimal as a quotient of two integers.

9. $0.\overline{56}$

V. Evaluate the following.

10. $5!$

11. $\dfrac{12!}{9!}$

VI. Use the binomial theorem to expand the following binomial.

12. $(3x + 4y)^5$

VII. Applications

13. On the day of its grand opening, Bob's Burger City sells 64 hamburgers. Each succeeding day, Bob's Burger City sells half as many hamburgers as on the previous day. How many hamburgers does Bob sell in his first week?

14. A ball is dropped from a height of 90 ft. On each bounce it will rebound to two-thirds of its previous height. What is the total distance that the ball will travel before coming to rest?

TEST YOUR MEMORY

These problems review Chapters 1–11.

I. Simplify each of the following. Assume that all variables represent positive numbers.

1. $\left(\dfrac{250x^4y}{16x^{-2}y^8}\right)^{-2/3}$

2. $\dfrac{(3xy^{-3})^2}{(2x^{1/2}y^4)^{-3}}$

3. $4\sqrt{20x} - \sqrt{180x} - 3\sqrt{45x}$

4. $\dfrac{\sqrt{x} + \sqrt{3}}{\sqrt{x} + \sqrt{12}}$

II. Perform the indicated operations.

5. $\dfrac{3x^3 + 4x^2 - 12x - 16}{15x^4 - 30x^3 - 120x^2} \cdot \dfrac{3x^2 - 12x}{x^3 - 8}$

6. $\dfrac{4x - 14}{x^2 + 2x - 8} - \dfrac{x - 6}{x^2 - 4}$

III. Let $f(x) = 3x + 2$ and $g(x) = 2x^2 - 1$. Find the following.

7. $g(-3)$

8. $3f(1) - 4g(-4)$

9. $f^{-1}(x)$

10. $(f \circ g)(x)$

IV. Find the following equations.

11. The line passing through $(9, 1)$ and $(-6, 7)$: Write the equation in slope-intercept form.

12. The circle with center $= C(-3, 7)$, radius $= 6$

V. Graph the following equations.

13. $x = 2y^2 + 3$

14. $25x^2 + 4y^2 = 100$

15. $y^2 - 4x^2 = 16$

16. $y = 3x - 2$

17. $y = -x^2 + 6x - 10$

18. $x^2 + y^2 - 6x + 4y + 9 = 0$

19. $f(x) = \log_2(x + 2)$

20. $f(x) = 3^{x-1}$

VI. Evaluate the following.

21. $4^0 + 4^{-1} - 4^{1/2}$

22. $\log_5 \sqrt{5}$

23. $\dfrac{10!}{6!}$

VII. Evaluate the following series, if possible.

24. $\displaystyle\sum_{i=1}^{30} (4i + 3)$

25. $\displaystyle\sum_{i=1}^{6} 7(3^i)$

VIII. Find the solutions of the following equations and check your solutions.

26. $2x^2 - 6x + 5 = 0$

27. $4x^2 - 4x + 17 = 0$

28. $8x^3 + 20x^2 - 2x - 5 = 0$

29. $27x^3 + 36x^2 - 3x - 4 = 0$

30. $|3x - 2| = 7$

31. $|5x - 1| = 8$

32. $27 = 9^{2x-1}$

33. $\dfrac{1}{8} = 16^{3x+1}$

34. $\dfrac{x + 10}{x - 5} + \dfrac{2}{x - 4} = \dfrac{5x - 22}{x^2 - 9x + 20}$

35. $\dfrac{x + 6}{x - 8} + \dfrac{2}{x - 3} = \dfrac{3x - 19}{x^2 - 11x + 24}$

36. $\sqrt{2x - 1} - \sqrt{x + 3} = 1$

37. $\sqrt{2x + 5} - \sqrt{x + 3} = 2$

38. $\log_2(3x + 1) + \log_2(x + 1) = 3$

39. $\log_2(2x + 1) + \log_2(2x - 2) = 2$

40. $\log(2x + 13) - \log(x - 1) = \log(2x - 1)$

41. $\log(4x - 1) - \log(x - 2) = \log(3x + 2)$

IX. Determine the solutions of the following systems of equations.

42. $3x - 6y = -1$
$5x - 10y = 4$

43. $x^2 + y^2 = 25$
$x - 7y = -25$

44. $2x - y + 3z = -1$
$x + y - 2z = 9$
$-x + y + 5z = -4$

X. Use algebraic expressions to find the solutions of the following problems.

45. One leg of a right triangle is 10 mm less than 3 times the other leg. The hypotenuse is 10 mm long. What are the lengths of the legs of the triangle?

46. Jake's motorboat can travel 12 mi down Spring Creek in the same time that it can travel 8 mi up Spring Creek. If the speed of the current in Spring Creek is 3 mi/hr, what is the speed of Jake's boat in still water?

47. Jane and Bob have a rectangular swimming pool that is 9 ft wide and 12 ft long. They are going to build a tile border of uniform width around the pool. They have 100 sq ft of tile. How wide is the border?

48. Rita invested $9000 at the Peoples Bank of Moss County. She invested part of her money in a super-savers account paying $6\frac{1}{2}\%$ simple interest, and she invested the rest of her money in a Christmas Club account paying 8% simple interest. After 1 yr her interest income was $660. How much did Rita invest in each account?

49. In Miss Grundy's algebra classes it has been determined that the number of A's she assigns varies inversely as the square root of the number of apples she receives during the semester. If Miss Grundy receives 4 apples and assigns 12 A's, how many A's will Miss Grundy assign when she receives 9 apples?

50. A ball is dropped from a height of 40 ft. On each bounce it will rebound to one-half of its previous height. What is the total distance that the ball will travel before coming to rest?

Tables

I. Powers and Corresponding Roots of Some Natural Numbers

n	n^3	n^4	n^5	n^6	n^7
1	1	1	1	1	1
2	8	16	32	64	128
3	27	81	243	729	2187
4	64	256	1024	4096	
5	125	625	3125		
6	216	1296			
7	343	2401			
8	512	4096			
9	729	6561			
10	1000	10,000	100,000	1,000,000	10,000,000
$\sqrt[3]{n}$	n				
$\sqrt[4]{n}$		n			
$\sqrt[5]{n}$			n		
$\sqrt[6]{n}$				n	
$\sqrt[7]{n}$					n

II. Squares and Square Roots of Natural Numbers from 1 to 100

n	n^2	$\sqrt{n}$	$\sqrt{10n}$	n	n^2	$\sqrt{n}$	$\sqrt{10n}$
1	1	1.000	3.162	34	1156	5.831	18.439
2	4	1.414	4.472	35	1225	5.916	18.708
3	9	1.732	5.477	36	1296	6.000	18.974
4	16	2.000	6.325	37	1369	6.083	19.235
5	25	2.236	7.071	38	1444	6.164	19.494
6	36	2.449	7.746	39	1521	6.245	19.748
7	49	2.646	8.367	40	1600	6.325	20.000
8	64	2.828	8.944	41	1681	6.403	20.248
9	81	3.000	9.487	42	1764	6.481	20.494
10	100	3.162	10.000	43	1849	6.557	20.736
11	121	3.317	10.488	44	1936	6.633	20.976
12	144	3.464	10.954	45	2025	6.708	21.213
13	169	3.606	11.402	46	2116	6.782	21.448
14	196	3.742	11.832	47	2209	6.856	21.679
15	225	3.873	12.247	48	2304	6.928	21.909
16	256	4.000	12.649	49	2401	7.000	22.136
17	289	4.123	13.038	50	2500	7.071	22.361
18	324	4.243	13.416	51	2601	7.141	22.583
19	361	4.359	13.784	52	2704	7.211	22.804
20	400	4.472	14.142	53	2809	7.280	23.022
21	441	4.583	14.491	54	2916	7.348	23.238
22	484	4.690	14.832	55	3025	7.416	23.452
23	529	4.796	15.166	56	3136	7.483	23.664
24	576	4.899	15.492	57	3249	7.550	23.875
25	625	5.000	15.811	58	3364	7.616	24.083
26	676	5.099	16.125	59	3481	7.681	24.290
27	729	5.196	16.432	60	3600	7.746	24.495
28	784	5.292	16.733	61	3721	7.810	24.698
29	841	5.385	17.029	62	3844	7.874	24.900
30	900	5.477	17.321	63	3969	7.937	25.100
31	961	5.568	17.607	64	4096	8.000	25.298
32	1024	5.657	17.889	65	4225	8.062	25.495
33	1089	5.745	18.166	66	4356	8.124	25.690

II. Squares and Square Roots *(Continued)*

n	n^2	$\sqrt{n}$	$\sqrt{10n}$	n	n^2	$\sqrt{n}$	$\sqrt{10n}$
67	4489	8.185	25.884	84	7056	9.165	28.983
68	4624	8.246	26.077	85	7225	9.220	29.155
69	4761	8.307	26.268	86	7396	9.274	29.326
70	4900	8.367	26.458	87	7569	9.327	29.496
71	5041	8.426	26.646	88	7744	9.381	29.665
72	5184	8.485	26.833	89	7921	9.434	29.833
73	5329	8.544	27.019	90	8100	9.487	30.000
74	5476	8.602	27.203	91	8281	9.539	30.166
75	5625	8.660	27.386	92	8464	9.592	30.332
76	5776	8.718	27.568	93	8649	9.644	30.496
77	5929	8.775	27.749	94	8836	9.695	30.659
78	6084	8.832	27.928	95	9025	9.747	30.822
79	6241	8.888	28.107	96	9216	9.798	30.984
80	6400	8.944	28.284	97	9409	9.849	31.145
81	6561	9.000	28.460	98	9604	9.899	31.305
82	6724	9.055	28.636	99	9801	9.950	31.464
83	6889	9.110	28.810	100	10000	10.000	31.623

III. Powers of e

x	e^x	e^{-x}	x	e^x	e^{-x}
0.00	1.0000	1.0000	0.16	1.1735	0.8521
0.01	1.0101	0.9901	0.17	1.1853	0.8437
0.02	1.0202	0.9802	0.18	1.1972	0.8353
0.03	1.0305	0.9705	0.19	1.2092	0.8270
0.04	1.0408	0.9608	0.20	1.2214	0.8187
0.05	1.0513	0.9512	0.21	1.2337	0.8106
0.06	1.0618	0.9418	0.22	1.2461	0.8025
0.07	1.0725	0.9324	0.23	1.2586	0.7945
0.08	1.0833	0.9331	0.24	1.2712	0.7866
0.09	1.0942	0.9139	0.25	1.2840	0.7788
0.10	1.1052	0.9048	0.30	1.3499	0.7408
0.11	1.1163	0.8958	0.35	1.4191	0.7047
0.12	1.1275	0.8869	0.40	1.4918	0.6703
0.13	1.1388	0.8781	0.45	1.5683	0.6376
0.14	1.1503	0.8694	0.50	1.6487	0.6065
0.15	1.1618	0.8607	0.55	1.7333	0.5769

III. Powers of *e (Continued)*

x	e^x	e^{-x}	x	e^x	e^{-x}
0.60	1.8221	0.5488	3.1	22.198	0.0450
0.65	1.9155	0.5220	3.2	24.533	0.0408
0.70	2.0138	0.4966	3.3	27.113	0.0369
0.75	2.1170	0.4724	3.4	29.964	0.0334
0.80	2.2255	0.4493	3.5	33.115	0.0302
0.85	2.3396	0.4274	3.6	36.598	0.0273
0.90	2.4596	0.4066	3.7	40.447	0.0247
0.95	2.5857	0.3867	3.8	44.701	0.0224
1.0	2.7183	0.3679	3.9	49.402	0.0202
1.1	3.0042	0.3329	4.0	54.598	0.0183
1.2	3.3201	0.3012	4.1	60.340	0.0166
1.3	3.6693	0.2725	4.2	66.686	0.0150
1.4	4.0552	0.2466	4.3	73.700	0.0136
1.5	4.4817	0.2231	4.4	81.451	0.0123
1.6	4.9530	0.2019	4.5	90.017	0.0111
1.7	5.4739	0.1827	4.6	99.484	0.0101
1.8	6.0496	0.1653	4.7	109.95	0.0091
1.9	6.6859	0.1496	4.8	121.51	0.0082
2.0	7.3891	0.1353	4.9	134.29	0.0074
2.1	8.1662	0.1225	5.0	148.41	0.0067
2.2	9.0250	0.1108	5.5	244.69	0.0041
2.3	9.9742	0.1003	6.0	403.43	0.0025
2.4	11.023	0.0907	6.5	665.14	0.0015
2.5	12.182	0.0821	7.0	1096.6	0.0009
2.6	13.464	0.0743	7.5	1808.0	0.0006
2.7	14.880	0.0672	8.0	2981.0	0.0003
2.8	16.445	0.0608	8.5	4914.8	0.0002
2.9	18.174	0.0550	9.0	8103.1	0.0001
3.0	20.086	0.0498	10.00	22026	0.00005

IV. Common Logarithms

x	0	1	2	3	4	5	6	7	8	9
1.0	0.0000	0.004321	0.008600	0.01284	0.01703	0.02119	0.02531	0.02938	0.03342	0.03743
1.1	0.04139	0.04532	0.04922	0.05308	0.05690	0.06070	0.06446	0.06819	0.07188	0.07555
1.2	0.07918	0.08279	0.08636	0.08991	0.09342	0.09691	0.1004	0.1038	0.1072	0.1106
1.3	0.1139	0.1173	0.1206	0.1239	0.1271	0.1303	0.1335	0.1367	0.1399	0.1430
1.4	0.1461	0.1492	0.1523	0.1553	0.1584	0.1614	0.1644	0.1673	0.1703	0.1732

IV. Common Logarithms *(Continued)*

x	0	1	2	3	4	5	6	7	8	9
1.5	0.1761	0.1790	0.1818	0.1847	0.1875	0.1903	0.1931	0.1959	0.1987	0.2014
1.6	0.2041	0.2068	0.2095	0.2122	0.2148	0.2175	0.2201	0.2227	0.2253	0.2279
1.7	0.2304	0.2330	0.2355	0.2380	0.2405	0.2430	0.2455	0.2480	0.2504	0.2529
1.8	0.2553	0.2577	0.2601	0.2625	0.2648	0.2673	0.2695	0.2718	0.2742	0.2765
1.9	0.2788	0.2810	0.2833	0.2856	0.2878	0.2900	0.2923	0.2945	0.2967	0.2989
2.0	0.3010	0.3032	0.3054	0.3075	0.3096	0.3118	0.3139	0.3160	0.3181	0.3201
2.1	0.3222	0.3243	0.3263	0.3284	0.3304	0.3324	0.3345	0.3365	0.3385	0.3404
2.2	0.3424	0.3444	0.3464	0.3483	0.3502	0.3522	0.3541	0.3560	0.3579	0.3598
2.3	0.3617	0.3636	0.3655	0.3674	0.3692	0.3711	0.3729	0.3747	0.3766	0.3784
2.4	0.3802	0.3820	0.3838	0.3856	0.3874	0.3892	0.3909	0.3927	0.3945	0.3962
2.5	0.3979	0.3997	0.4014	0.4031	0.4048	0.4065	0.4082	0.4099	0.4116	0.4133
2.6	0.4150	0.4166	0.4183	0.4200	0.4216	0.4232	0.4249	0.4265	0.4281	0.4298
2.7	0.4314	0.4330	0.4346	0.4362	0.4378	0.4393	0.4409	0.4425	0.4440	0.4456
2.8	0.4472	0.4487	0.4502	0.4518	0.4533	0.4548	0.4564	0.4579	0.4594	0.4609
2.9	0.4624	0.4639	0.4654	0.4669	0.4683	0.4698	0.4713	0.4728	0.4742	0.4757
3.0	0.4771	0.4786	0.4800	0.4814	0.4829	0.4843	0.4857	0.4871	0.4886	0.4900
3.1	0.4914	0.4928	0.4942	0.4955	0.4969	0.4983	0.4997	0.5011	0.5024	0.5038
3.2	0.5051	0.5065	0.5079	0.5092	0.5105	0.5119	0.5132	0.5145	0.5159	0.5172
3.3	0.5185	0.5198	0.5211	0.5224	0.5237	0.5250	0.5263	0.5276	0.5289	0.5302
3.4	0.5315	0.5328	0.5340	0.5353	0.5366	0.5378	0.5391	0.5403	0.5416	0.5428
3.5	0.5441	0.5453	0.5465	0.5478	0.5490	0.5502	0.5514	0.5527	0.5539	0.5551
3.6	0.5563	0.5575	0.5587	0.5599	0.5611	0.5623	0.5635	0.5647	0.5658	0.5670
3.7	0.5682	0.5694	0.5705	0.5717	0.5729	0.5740	0.5752	0.5763	0.5775	0.5786
3.8	0.5798	0.5809	0.5821	0.5832	0.5843	0.5855	0.5866	0.5877	0.5888	0.5899
3.9	0.5911	0.5922	0.5933	0.5944	0.5955	0.5966	0.5977	0.5988	0.5999	0.6010
4.0	0.6021	0.6031	0.6042	0.6053	0.6064	0.6075	0.6085	0.6096	0.6107	0.6117
4.1	0.6128	0.6138	0.6149	0.6160	0.6170	0.6180	0.6191	0.6201	0.6212	0.6222
4.2	0.6232	0.6243	0.6253	0.6263	0.6274	0.6284	0.6294	0.6304	0.6314	0.6325
4.3	0.6335	0.6345	0.6355	0.6365	0.6375	0.6385	0.6395	0.6405	0.6415	0.6425
4.4	0.6435	0.6444	0.6454	0.6464	0.6474	0.6484	0.6493	0.6503	0.6513	0.6522
4.5	0.6532	0.6542	0.6551	0.6561	0.6571	0.6580	0.6590	0.6599	0.6609	0.6618
4.6	0.6628	0.6637	0.6646	0.6656	0.6665	0.6675	0.6684	0.6693	0.6702	0.6712
4.7	0.6721	0.6730	0.6739	0.6749	0.6758	0.6767	0.6776	0.6785	0.6794	0.6803

IV. Common Logarithms *(Continued)*

x	0	1	2	3	4	5	6	7	8	9
4.8	0.6812	0.6821	0.6830	0.6839	0.6848	0.6857	0.6866	0.6875	0.6884	0.6893
4.9	0.6902	0.6911	0.6920	0.6928	0.6937	0.6946	0.6955	0.6964	0.6972	0.6981
5.0	0.6990	0.6998	0.7007	0.7016	0.7024	0.7033	0.7042	0.7050	0.7059	0.7067
5.1	0.7076	0.7084	0.7093	0.7101	0.7110	0.7118	0.7126	0.7135	0.7143	0.7152
5.2	0.7160	0.7168	0.7177	0.7185	0.7193	0.7202	0.7210	0.7218	0.7226	0.7235
5.3	0.7243	0.7251	0.7259	0.7267	0.7275	0.7284	0.7292	0.7300	0.7308	0.7316
5.4	0.7324	0.7332	0.7340	0.7348	0.7356	0.7364	0.7372	0.7380	0.7388	0.7396
5.5	0.7404	0.7412	0.7419	0.7427	0.7435	0.7443	0.7451	0.7459	0.7466	0.7474
5.6	0.7482	0.7490	0.7497	0.7505	0.7513	0.7520	0.7528	0.7536	0.7543	0.7551
5.7	0.7559	0.7566	0.7574	0.7582	0.7589	0.7597	0.7604	0.7612	0.7619	0.7627
5.8	0.7634	0.7642	0.7649	0.7657	0.7664	0.7672	0.7679	0.7686	0.7694	0.7701
5.9	0.7709	0.7716	0.7723	0.7731	0.7738	0.7745	0.7752	0.7760	0.7767	0.7774
6.0	0.7782	0.7789	0.7796	0.7803	0.7810	0.7818	0.7825	0.7832	0.7839	0.7846
6.1	0.7853	0.7860	0.7868	0.7875	0.7882	0.7889	0.7896	0.7903	0.7910	0.7917
6.2	0.7924	0.7931	0.7938	0.7945	0.7952	0.7959	0.7966	0.7973	0.7980	0.7987
6.3	0.7993	0.8000	0.8007	0.8014	0.8021	0.8028	0.8035	0.8041	0.8048	0.8055
6.4	0.8062	0.8069	0.8075	0.8082	0.8089	0.8096	0.8102	0.8109	0.8116	0.8122
6.5	0.8129	0.8136	0.8142	0.8149	0.8156	0.8162	0.8169	0.8176	0.8182	0.8189
6.6	0.8195	0.8202	0.8209	0.8215	0.8222	0.8228	0.8235	0.8241	0.8248	0.8254
6.7	0.8261	0.8267	0.8274	0.8280	0.8287	0.8293	0.8299	0.8306	0.8312	0.8319
6.8	0.8325	0.8331	0.8338	0.8344	0.8351	0.8357	0.8363	0.8370	0.8376	0.8382
6.9	0.8388	0.8395	0.8401	0.8407	0.8414	0.8420	0.8426	0.8432	0.8439	0.8445
7.0	0.8451	0.8457	0.8463	0.8470	0.8476	0.8482	0.8488	0.8494	0.8500	0.8506
7.1	0.8513	0.8519	0.8525	0.8531	0.8537	0.8543	0.8549	0.8555	0.8561	0.8567
7.2	0.8573	0.8579	0.8585	0.8591	0.8597	0.8603	0.8609	0.8615	0.8621	0.8627
7.3	0.8633	0.8639	0.8645	0.8651	0.8657	0.8663	0.8669	0.8675	0.8681	0.8686
7.4	0.8692	0.8698	0.8704	0.8710	0.8716	0.8722	0.8727	0.8733	0.8739	0.8745
7.5	0.8751	0.8756	0.8762	0.8768	0.8774	0.8779	0.8785	0.8791	0.8797	0.8802
7.6	0.8808	0.8814	0.8820	0.8825	0.8831	0.8837	0.8842	0.8848	0.8854	0.8859
7.7	0.8865	0.8871	0.8876	0.8882	0.8887	0.8893	0.8899	0.8904	0.8910	0.8915
7.8	0.8921	0.8927	0.8932	0.8938	0.8943	0.8949	0.8954	0.8960	0.8965	0.8971
7.9	0.8976	0.8982	0.8987	0.8993	0.8998	0.9004	0.9009	0.9015	0.9020	0.9025
8.0	0.9031	0.9036	0.9042	0.9047	0.9053	0.9058	0.9063	0.9069	0.9074	0.9079
8.1	0.9085	0.9090	0.9096	0.9101	0.9106	0.9112	0.9117	0.9122	0.9128	0.9133
8.2	0.9138	0.9143	0.9149	0.9154	0.9159	0.9165	0.9170	0.9175	0.9180	0.9186
8.3	0.9191	0.9196	0.9201	0.9206	0.9212	0.9217	0.9222	0.9227	0.9232	0.9238

IV. Common Logarithms *(Continued)*

x	0	1	2	3	4	5	6	7	8	9
8.4	0.9243	0.9248	0.9253	0.9258	0.9263	0.9269	0.9274	0.9279	0.9284	0.9289
8.5	0.9294	0.9299	0.9304	0.9309	0.9315	0.9320	0.9325	0.9330	0.9335	0.9340
8.6	0.9345	0.9350	0.9355	0.9360	0.9365	0.9370	0.9375	0.9380	0.9385	0.9390
8.7	0.9395	0.9400	0.9405	0.9410	0.9415	0.9420	0.9425	0.9430	0.9435	0.9440
8.8	0.9445	0.9450	0.9455	0.9460	0.9465	0.9469	0.9474	0.9479	0.9484	0.9489
8.9	0.9494	0.9499	0.9504	0.9509	0.9513	0.9518	0.9523	0.9528	0.9533	0.9538
9.0	0.9542	0.9547	0.9552	0.9557	0.9562	0.9566	0.9571	0.9576	0.9581	0.9586
9.1	0.9590	0.9595	0.9600	0.9605	0.9609	0.9614	0.9619	0.9624	0.9628	0.9633
9.2	0.9638	0.9643	0.9647	0.9652	0.9657	0.9661	0.9666	0.9671	0.9675	0.9680
9.3	0.9685	0.9689	0.9694	0.9699	0.9703	0.9708	0.9713	0.9717	0.9722	0.9727
9.4	0.9731	0.9736	0.9741	0.9745	0.9750	0.9754	0.9759	0.9763	0.9768	0.9773
9.5	0.9777	0.9782	0.9786	0.9791	0.9795	0.9800	0.9805	0.9809	0.9814	0.9818
9.6	0.9823	0.9827	0.9832	0.9836	0.9841	0.9845	0.9850	0.9854	0.9859	0.9863
9.7	0.9868	0.9872	0.9877	0.9881	0.9886	0.9890	0.9894	0.9899	0.9903	0.9908
9.8	0.9912	0.9917	0.9921	0.9926	0.9930	0.9934	0.9939	0.9943	0.9948	0.9952
9.9	0.9956	0.9961	0.9965	0.9969	0.9974	0.9978	0.9983	0.9987	0.9991	0.9996

Determinants and Cramer's Rule

B.1 Determinants

In this section we will introduce some new mathematical tools. We will discuss the application of these tools in the next section. A square array of numbers enclosed by a pair of vertical lines represents a **determinant.**

The following represent determinants:

$$\begin{vmatrix} 5 & 9 \\ 3 & 8 \end{vmatrix}, \quad \begin{vmatrix} 1 & 2 & 3 \\ 4 & 5 & 6 \\ 7 & 8 & 9 \end{vmatrix}, \quad \begin{vmatrix} -1 & 0 & 2 & 3 \\ 0 & 4 & 0 & 0 \\ 2 & 1 & 0 & -1 \\ 0 & 4 & 5 & 1 \end{vmatrix}$$

The numbers in the array are called **elements.** The **rows** of the array are the horizontal lines of numbers. The **columns** of the array are the vertical lines of numbers. The array

$$\begin{vmatrix} 5 & 9 \\ 3 & 8 \end{vmatrix}$$

has two rows and two columns, and represents a **second-order** determinant. Similarly,

$$\begin{vmatrix} 1 & 2 & 3 \\ 4 & 5 & 6 \\ 7 & 8 & 9 \end{vmatrix}$$

represents a **third-order** determinant.

A *determinant is a number.* The following definition shows how to calculate the value of any second-order determinant.

Definition B.1.1	$\begin{vmatrix} a_1 & b_1 \\ a_2 & b_2 \end{vmatrix} = a_1 b_2 - a_2 b_1$

EXAMPLE B.1.1

Calculate the value of the following determinant:

$$\begin{vmatrix} 5 & 9 \\ 3 & 8 \end{vmatrix}$$

In this case $a_1 = 5$, $a_2 = 3$, $b_1 = 9$, and $b_2 = 8$. Thus

$$\begin{vmatrix} 5 & 9 \\ 3 & 8 \end{vmatrix} = 5 \cdot 8 - 3 \cdot 9$$
$$= 40 - 27$$
$$= 13$$

NOTE ▶ A schematic method of calculating the value of a second-order determinant is as follows:

$$= 5 \cdot 8 - 3 \cdot 9$$
$$= 40 - 27$$
$$= 13$$

Subtract the product of the numbers on the diagonal going up from the product of the numbers on the diagonal going down.

The definition for finding the value of a second-order determinant is relatively simple. Unfortunately, the calculation of higher-order determinants is more complicated. The following definition states how to find the value of any third-order determinant.

Definition B.1.2

$$\begin{vmatrix} a_1 & b_1 & c_1 \\ a_2 & b_2 & c_2 \\ a_3 & b_3 & c_3 \end{vmatrix} = a_1b_2c_3 - a_1b_3c_2 - a_2b_1c_3 + a_2b_3c_1 + a_3b_1c_2 - a_3b_2c_1$$

Obviously this definition is very cumbersome. The next definition will lead to another method of calculating the value of any third-order determinant.

Definition B.1.3

The *minor* of an element in a third-order determinant is the second-order determinant formed by deleting the row and column in which the element appears.

XAMPLE B.1.2   In the determinant

$$\begin{vmatrix} 1 & 0 & 3 \\ 5 & -1 & 6 \\ 2 & 8 & -7 \end{vmatrix}$$

find the minor of 8. Delete 8's row and column:

$$\begin{vmatrix} 1 & 3 \\ 5 & 6 \end{vmatrix}$$

Thus, the minor of 8 is the determinant

$$\begin{vmatrix} 1 & 3 \\ 5 & 6 \end{vmatrix} = 1 \cdot 6 - 5 \cdot 3$$
$$= 6 - 15$$
$$= -9$$

The following reasoning shows how minors can be used to find the value of any third-order determinant:

$$\begin{vmatrix} a_1 & b_1 & c_1 \\ a_2 & b_2 & c_2 \\ a_3 & b_3 & c_3 \end{vmatrix} = a_1b_2c_3 - a_1b_3c_2 - a_2b_1c_3 + a_2b_3c_1 + a_3b_1c_2 - a_3b_2c_1$$

$$= a_1(b_2c_3 - b_3c_2) - a_2(b_1c_3 - b_3c_1) + a_3(b_1c_2 - b_2c_1)$$

$$= a_1 \cdot \begin{vmatrix} b_2 & c_2 \\ b_3 & c_3 \end{vmatrix} - a_2 \cdot \begin{vmatrix} b_1 & c_1 \\ b_3 & c_3 \end{vmatrix} + a_3 \cdot \begin{vmatrix} b_1 & c_1 \\ b_2 & c_2 \end{vmatrix}$$

$$= a_1 \cdot (\text{minor of } a_1) - a_2 \cdot (\text{minor of } a_2) + a_3 \cdot (\text{minor of } a_3)$$

This equation is called the expansion of the determinant by minors about the first column. By using similar algebraic procedures, it can be shown that

$$\begin{vmatrix} a_1 & b_1 & c_1 \\ a_2 & b_2 & c_2 \\ a_3 & b_3 & c_3 \end{vmatrix} = -a_2 \cdot (\text{minor of } a_2) + b_2 \cdot (\text{minor of } b_2) - c_2 \cdot (\text{minor of } c_2)$$

This equation is called the expansion of the determinant by minors about the second row. Notice the alternating signs in both of these expansions. These two expansions are examples of *the evaluation property of third-order determinants:* A third-order determinant can be evaluated by expanding the determinant by minors about any row or column. A + or − sign is placed before the elements in the expansion according to the position of the elements in the *sign diagram:*

$$\begin{matrix} + & - & + \\ - & + & - \\ + & - & + \end{matrix}$$

If an element is in a + position, a + sign is placed before the element in the expansion. Similarly, if an element is in a − position, a − sign is placed before the element in the expansion. The sign diagram is easy to remember. Simply start with a + sign in the upper left-hand corner, and alternate signs as you move either horizontally or vertically.

NOTE ▶   Higher-order determinants are calculated using an extension of this property.

XAMPLE B.1.3

Using minors, calculate the value of the following determinant.

$$\begin{vmatrix} 1 & 0 & 3 \\ 5 & -1 & 6 \\ 2 & 8 & -7 \end{vmatrix}$$

The evaluation property of third-order determinants states that the determinant can be evaluated by expanding the determinant by minors about any row or column. Let's choose the first column:

$$\begin{vmatrix} 1 & 0 & 3 \\ 5 & -1 & 6 \\ 2 & 8 & -7 \end{vmatrix} = +1 \cdot \begin{vmatrix} -1 & 6 \\ 8 & -7 \end{vmatrix} - 5 \cdot \begin{vmatrix} 0 & 3 \\ 8 & -7 \end{vmatrix} + 2 \cdot \begin{vmatrix} 0 & 3 \\ -1 & 6 \end{vmatrix}$$

$$= 1 \cdot (7 - 48) - 5 \cdot (0 - 24) + 2 \cdot (0 - (-3))$$
$$= 1(-41) - 5(-24) + 2(3)$$
$$= -41 + 120 + 6$$
$$= 85$$

In Example B.1.1 a schematic method for calculating the value of a second-order determinant was presented. There is a similar schematic method for calculating the value of a third-order determinant. This method will be illustrated by calculating the determinant of the previous example:

$$\begin{vmatrix} 1 & 0 & 3 \\ 5 & -1 & 6 \\ 2 & 8 & -7 \end{vmatrix} = \begin{vmatrix} 1 & 0 & 3 \\ 5 & -1 & 6 \\ 2 & 8 & -7 \end{vmatrix} \begin{matrix} 1 & 0 \\ 5 & -1 \\ 2 & 8 \end{matrix}$$

$$= (7 + 0 + 120) - (-6 + 48 + 0)$$
$$= 127 - 42$$
$$= 85$$

Copy the first two columns immediately to the right of the determinant. Construct the diagonals as shown in the preceding scheme. Calculate the products of the numbers on each diagonal. Subtract the sum of the products on the diagonals going up from the sum of the products on the diagonals going down.

REMARK   A word of caution: The *diagonal method* does not correctly calculate fourth- or higher-order determinants. The diagonal method can be applied to only second- and third-order determinants.

Exercises B.1

Calculate the value of each of the following determinants.

1. $\begin{vmatrix} 5 & 4 \\ 3 & 8 \end{vmatrix}$

2. $\begin{vmatrix} 9 & 3 \\ 6 & 1 \end{vmatrix}$

3. $\begin{vmatrix} -2 & 8 \\ -1 & 7 \end{vmatrix}$

4. $\begin{vmatrix} 0 & 5 \\ -2 & -9 \end{vmatrix}$

5. $\begin{vmatrix} 1 & -2 \\ 3 & -6 \end{vmatrix}$

6. $\begin{vmatrix} 5 & -3 \\ 10 & -6 \end{vmatrix}$

7. $\begin{vmatrix} -3 & 1 \\ 4 & 2 \end{vmatrix}$

8. $\begin{vmatrix} 4 & -7 \\ 1 & 9 \end{vmatrix}$

9. $\begin{vmatrix} 0 & 7 \\ 0 & 5 \end{vmatrix}$

10. $\begin{vmatrix} 0 & 0 \\ -8 & 2 \end{vmatrix}$

11. $\begin{vmatrix} -3 & -4 \\ -5 & -6 \end{vmatrix}$

12. $\begin{vmatrix} -6 & -9 \\ -2 & -4 \end{vmatrix}$

Using *minors*, calculate the value of each of the following determinants.

13. $\begin{vmatrix} 1 & 2 & 3 \\ 4 & 5 & 6 \\ 7 & 8 & 9 \end{vmatrix}$

14. $\begin{vmatrix} 4 & 1 & 1 \\ 7 & 3 & 2 \\ 2 & 5 & 6 \end{vmatrix}$

15. $\begin{vmatrix} 3 & -1 & 2 \\ 0 & 8 & 5 \\ -2 & 4 & 6 \end{vmatrix}$

16. $\begin{vmatrix} -5 & 0 & 1 \\ 2 & 3 & -4 \\ 1 & -1 & 6 \end{vmatrix}$

17. $\begin{vmatrix} 2 & -1 & 0 \\ 0 & 5 & 0 \\ 3 & 4 & -7 \end{vmatrix}$

18. $\begin{vmatrix} -7 & 8 & 1 \\ 3 & 0 & 0 \\ 0 & 4 & -9 \end{vmatrix}$

19. $\begin{vmatrix} 1 & -2 & 5 \\ 3 & 0 & -1 \\ 2 & -4 & 10 \end{vmatrix}$

20. $\begin{vmatrix} 3 & -2 & 6 \\ 1 & 5 & 2 \\ -4 & 0 & -8 \end{vmatrix}$

21. $\begin{vmatrix} 1 & 2 & -4 \\ 0 & 7 & 3 \\ 2 & 0 & 6 \end{vmatrix}$

22. $\begin{vmatrix} 5 & -9 & 0 \\ 1 & 3 & -2 \\ 0 & 9 & 4 \end{vmatrix}$

23. $\begin{vmatrix} 1 & 2 & -4 \\ 2 & 0 & 6 \\ 0 & 7 & 3 \end{vmatrix}$

24. $\begin{vmatrix} -9 & 5 & 0 \\ 3 & 1 & -2 \\ 9 & 0 & 4 \end{vmatrix}$

25. $\begin{vmatrix} 4 & 7 & -2 \\ 0 & 0 & 0 \\ 3 & -1 & 5 \end{vmatrix}$

26. $\begin{vmatrix} 2 & -1 & 0 \\ 6 & 4 & 0 \\ -3 & 2 & 0 \end{vmatrix}$

27. $\begin{vmatrix} -1 & 0 & 2 & 3 \\ 0 & 4 & 0 & 0 \\ 2 & 1 & 0 & -1 \\ 0 & 4 & 5 & 1 \end{vmatrix}$

28. $\begin{vmatrix} 5 & 6 & 0 & 1 \\ 0 & 0 & 2 & -3 \\ -2 & 1 & 0 & 8 \\ 3 & 0 & 0 & 4 \end{vmatrix}$

Using the *diagonal method*, calculate the value of each of the following third-order determinants.

29. $\begin{vmatrix} 1 & -3 & 4 \\ 2 & 5 & 3 \\ 4 & 1 & -7 \end{vmatrix}$

30. $\begin{vmatrix} 5 & -3 & 4 \\ -1 & -2 & 7 \\ 8 & 3 & -4 \end{vmatrix}$

31. $\begin{vmatrix} 0 & 5 & 3 \\ -1 & 9 & 2 \\ 6 & -2 & 4 \end{vmatrix}$

32. $\begin{vmatrix} 4 & -2 & 5 \\ 3 & 0 & -1 \\ -2 & -6 & 1 \end{vmatrix}$

33. $\begin{vmatrix} 5 & 4 & -6 \\ 0 & 2 & 4 \\ 3 & -1 & 0 \end{vmatrix}$ **34.** $\begin{vmatrix} 2 & 1 & 0 \\ -1 & 0 & 5 \\ 3 & 4 & 0 \end{vmatrix}$ **35.** $\begin{vmatrix} -2 & -4 & 1 \\ 0 & 1 & 5 \\ -6 & -8 & 2 \end{vmatrix}$ **36.** $\begin{vmatrix} 0 & 1 & 0 \\ -2 & 4 & 4 \\ 3 & -5 & -6 \end{vmatrix}$

37. $\begin{vmatrix} 2 & 0 & 0 \\ 8 & 4 & 0 \\ -1 & 2 & 3 \end{vmatrix}$ **38.** $\begin{vmatrix} 5 & 6 & 7 \\ 0 & -3 & 4 \\ 0 & 0 & 2 \end{vmatrix}$

Verify the following equations.

39. $\begin{vmatrix} a_1 & b_1 & c_1 \\ a_2 & b_2 & c_2 \\ a_3 & b_3 & c_3 \end{vmatrix} = - \begin{vmatrix} a_1 & b_1 & c_1 \\ a_3 & b_3 & c_3 \\ a_2 & b_2 & c_2 \end{vmatrix}$

40. $\begin{vmatrix} a_1 & b_1 & c_1 \\ a_2 & b_2 & c_2 \\ a_3 & b_3 & c_3 \end{vmatrix} = - \begin{vmatrix} a_1 & c_1 & b_1 \\ a_2 & c_2 & b_2 \\ a_3 & c_3 & b_3 \end{vmatrix}$

41. $\begin{vmatrix} ka_1 & kb_1 & kc_1 \\ a_2 & b_2 & c_2 \\ a_3 & b_3 & c_3 \end{vmatrix} = k \cdot \begin{vmatrix} a_1 & b_1 & c_1 \\ a_2 & b_2 & c_2 \\ a_3 & b_3 & c_3 \end{vmatrix}$

42. $\begin{vmatrix} a_1 & b_1 & c_1 \\ ka_1 & kb_1 & kc_1 \\ a_3 & b_3 & c_3 \end{vmatrix} = 0$

🖩 **The following exercises require the use of a graphing calculator. Calculate the value of each determinant.**

43. $\begin{vmatrix} -4.5 & -2 & 5 \\ 3 & 8.3 & 1.9 \\ -4 & 6.2 & 3.8 \end{vmatrix}$

44. $\begin{vmatrix} 3 & 8 & 1 & -2 & 4 & 5 \\ -2 & 0 & -1 & 6 & 2 & 2 \\ 4 & -3 & 3 & -1 & -1 & -5 \\ 0 & -1 & 4 & -4 & 2 & -1 \\ 1 & 6 & 5 & -9 & -6 & 3 \\ -2 & 7 & -2 & 3 & 0 & 4 \end{vmatrix}$

B.2 *Cramer's Rule*

In the previous section we learned how to evaluate determinants. In this section we will learn how to use determinants to find the solutions of linear systems of equations. Consider the linear system of equations:

$$a_1x + b_1y = c_1$$
$$a_2x + b_2y = c_2$$

where a_1, a_2, b_1, b_2, c_1, and c_2 are real numbers. Let's find the solution of this system by using the elimination method:

$$a_1x + b_1y = c_1 \quad \overset{\text{\textit{Multiply both sides by } } b_2.}{\underset{\text{\textit{Multiply both sides by } } -b_1.}{\longrightarrow}}$$

$$a_2x + b_2y = c_2$$

$$\begin{aligned} a_1b_2x + b_1b_2y &= c_1b_2 \\ \underline{-a_2b_1x - b_2b_1y} &= \underline{-c_2b_1} \\ (a_1b_2 - a_2b_1)x &= c_1b_2 - c_2b_1 \end{aligned}$$

$$x = \frac{c_1b_2 - c_2b_1}{a_1b_2 - a_2b_1}$$

Notice that both the numerator and the denominator can be expressed as second-order determinants; that is,

$$x = \frac{c_1 b_2 - c_2 b_1}{a_1 b_2 - a_2 b_1} = \frac{\begin{vmatrix} c_1 & b_1 \\ c_2 & b_2 \end{vmatrix}}{\begin{vmatrix} a_1 & b_1 \\ a_2 & b_2 \end{vmatrix}} \qquad a_1 b_2 - a_2 b_1 \neq 0$$

In a similar fashion it can be shown that

$$y = \frac{a_1 c_2 - a_2 c_1}{a_1 b_2 - a_2 b_1} = \frac{\begin{vmatrix} a_1 & c_1 \\ a_2 & c_2 \end{vmatrix}}{\begin{vmatrix} a_1 & b_1 \\ a_2 & b_2 \end{vmatrix}} \qquad a_1 b_2 - a_2 b_1 \neq 0$$

Using determinants to find the solution of a linear system of equations is known as Cramer's rule.

Cramer's Rule for Two Linear Equations in Two Variables

The solution (x, y) of the linear system of equations

$$a_1 x + b_1 y = c_1$$
$$a_2 x + b_2 y = c_2$$

is given by the formula

$$x = \frac{D_x}{D} \quad \text{and} \quad y = \frac{D_y}{D}$$

where

$$D = \begin{vmatrix} a_1 & b_1 \\ a_2 & b_2 \end{vmatrix}, \qquad D_x = \begin{vmatrix} c_1 & b_1 \\ c_2 & b_2 \end{vmatrix}, \qquad D_y = \begin{vmatrix} a_1 & c_1 \\ a_2 & c_2 \end{vmatrix}$$

provided $D \neq 0$.

REMARKS

1. If $D = 0$, the system is either dependent or inconsistent. Use the elimination or substitution method to determine whether the system is dependent or inconsistent.

2. D is the determinant formed by the coefficients of x and y.

3. To obtain D_x replace the first column in D (i.e., the column of coefficients of x) with the column of constants. To obtain D_y replace the second column in D (i.e., the column of coefficients of y) with the column of constants.

XAMPLE B.2.1

Using Cramer's rule, determine the solution of the following linear system of equations.

1. $x + y = 4$
    $2x - y = 5$

$$D = \begin{vmatrix} 1 & 1 \\ 2 & -1 \end{vmatrix} = -3$$

$$D_x = \begin{vmatrix} 4 & 1 \\ 5 & -1 \end{vmatrix} = -9$$

$$D_y = \begin{vmatrix} 1 & 4 \\ 2 & 5 \end{vmatrix} = -3$$

By Cramer's rule,

$$x = \frac{D_x}{D} \qquad y = \frac{D_y}{D}$$

$$= \frac{-9}{-3} \qquad = \frac{-3}{-3}$$

$$= 3 \qquad = 1$$

Thus, the solution is (3, 1). Note that this is the same system as in Example 9.1.1, part 1.

2. $x + 3y = 4$
    $2x + 6y = -1$

$$D = \begin{vmatrix} 1 & 3 \\ 2 & 6 \end{vmatrix} = 0$$

Thus, this system must be inconsistent or dependent. Using any of the methods of Section 9.1, it can be shown that this system is inconsistent. ▬▬▬

As shown next, Cramer's rule can be extended to linear systems in three variables.

Cramer's Rule for Three Linear Equations in Three Variables

The solution (x, y, z) of the linear system of equations

$$a_1x + b_1y + c_1z = d_1$$
$$a_2x + b_2y + c_2z = d_2$$
$$a_3x + b_3y + c_3z = d_3$$

is given by the formula

$$x = \frac{D_x}{D}, \qquad y = \frac{D_y}{D}, \qquad z = \frac{D_z}{D}$$

where

$$D = \begin{vmatrix} a_1 & b_1 & c_1 \\ a_2 & b_2 & c_2 \\ a_3 & b_3 & c_3 \end{vmatrix}, \qquad D_x = \begin{vmatrix} d_1 & b_1 & c_1 \\ d_2 & b_2 & c_2 \\ d_3 & b_3 & c_3 \end{vmatrix}, \qquad D_y = \begin{vmatrix} a_1 & d_1 & c_1 \\ a_2 & d_2 & c_2 \\ a_3 & d_3 & c_3 \end{vmatrix}$$

$$D_z = \begin{vmatrix} a_1 & b_1 & d_1 \\ a_2 & b_2 & d_2 \\ a_3 & b_3 & d_3 \end{vmatrix}$$

provided $D \neq 0$.

REMARKS

1. Again, if $D = 0$, the system is either dependent or inconsistent.

2. The determinants D, D_x, D_y, and D_z are constructed in a manner similar to a linear system of two equations in two variables.

EXAMPLE B.2.2   Using Cramer's rule, determine the solution of the following linear system of equations.

$$3x + y - z = 4$$
$$-x + y + 3z = 0$$
$$x + 2y + z = 1$$

$$D = \begin{vmatrix} 3 & 1 & -1 \\ -1 & 1 & 3 \\ 1 & 2 & 1 \end{vmatrix} = -8$$

$$D_x = \begin{vmatrix} 4 & 1 & -1 \\ 0 & 1 & 3 \\ 1 & 2 & 1 \end{vmatrix} = -16$$

$$D_y = \begin{vmatrix} 3 & 4 & -1 \\ -1 & 0 & 3 \\ 1 & 1 & 1 \end{vmatrix} = 8$$

$$D_z = \begin{vmatrix} 3 & 1 & 4 \\ -1 & 1 & 0 \\ 1 & 2 & 1 \end{vmatrix} = -8$$

By Cramer's rule,

$$x = \frac{D_x}{D} \qquad y = \frac{D_y}{D} \qquad z = \frac{D_z}{D}$$

$$= \frac{-16}{-8} \qquad = \frac{8}{-8} \qquad = \frac{-8}{-8}$$

$$= 2 \qquad\qquad = -1 \qquad = 1$$

Thus, the solution is $(2, -1, 1)$.

Exercises B.2

Using Cramer's rule, determine the solutions of the following linear systems of equations.

1. $2x + 3y = 7$
   $5x - y = 26$

2. $x - 7y = 7$
   $3x + 4y = -29$

3. $-4x + 9y = -3$
   $-6x - 7y = -25$

4. $8x - 3y = 4$
   $-2x + 5y = 16$

5. $-x + 9y = -2$
   $-2x + 11y = -4$

6. $3x - 5y = 15$
   $6x - 5y = 15$

7. $-x + 8y = 7$
   $3x - 2y = -10$

8. $9x + 2y = 14$
   $3x - y = -2$

9. $2x + 15y = -8$
   $4x - 3y = -5$

10. $-4x + 3y = 4$
    $2x - 9y = -7$

11. $3x - y = 1$
    $6x - 2y = -5$

12. $-x + 3y = 7$
    $2x - 6y = 11$

13. $6x + 3y = -9$
    $10x + 5y = -15$

14. $2x - 4y = 12$
    $3x - 6y = 18$

15. $x - 2y + z = 7$
    $3x + y - 2z = -1$
    $-x + 3y + z = -2$

16. $2x - y - z = -1$
    $-3x + y + 2z = 0$
    $x + 4y - z = -3$

17. $5x + 2y - z = 12$
    $-x - 3y + 2z = -6$
    $x - y - 4z = 10$

18. $-3x + y + 2z = -4$
    $2x - 4y + 3z = 5$
    $x + 5y - z = -9$

19. $x - 2y + 3z = 4$
$-x + 4y + z = 5$
$5x + 2y - z = -6$

20. $3x + 4y - z = 0$
$-5x + 2y + z = -11$
$-x - 2y + 3z = -7$

21. $x + y + 3z = 1$
$-x + 4y - 6z = 2$
$-2x + 3y - 3z = 5$

22. $3x - y + 2z = 1$
$5x + y - z = -3$
$2x + 4y + z = 7$

23. $x + 2y - 3z = 6$
$-x + 4y + z = -7$
$3x - 3y + 5z = -8$

24. $x + y - 4z = -1$
$2x - 6y + 5z = -6$
$3x - y + 2z = -5$

25. $x + y - 3z = -2$
$-3x + 5y + z = -2$
$x - 6y + 4z = 5$

26. $-2x + 4y - 2z = -8$
$x - 2y + z = 4$
$3x - 6y + 3z = 12$

27. $2x + 3y - z = 4$
$4x + 6y - 2z = 7$
$-x + 5y - 3z = -1$

28. $x - 5y + 3z = 8$
$x - 5y + 3z = -3$
$2x - 10y + 6z = -9$

29. $x - 2y + z + w = 4$
$2x + y - z + 2w = -1$
$-x + 3y + 2z - w = 5$
$3x - y + 4z - 3w = 3$

30. $2x - y + 4z - w = 1$
$-x + y - 2z + 2w = 5$
$3x + 2y + z - w = 0$
$x - 2y - z - 2w = -4$

The following exercises require the use of a graphing calculator. Using Cramer's rule, determine the solutions of the following linear systems of equations.

31. $6x + 2.5y + 2.4z - w = -9.8$
$-1.2x + 4y - 8z - 3w = 17.6$
$3.4x - y + 2z + 1.5w = 0.7$
$2.4x + 2y - 4z - 2w = 13$

32. $4v - w + 3x - 2y - 7z = 6$
$v + 2w + 5x - y - 2z = 7$
$2v - 3w - 2x + 4y + 3z = 0$
$-3v - w + x + 5y + 4z = -12$
$5v - 4w - 3x + 3y + z = 8$

GROUP ACTIVITY

In exercises 33–38, determine the value(s) of a so that each system will not be independent.

33. $x + 2y - 3z = 7$
$ax - 3y - z = 4$
$2x + y + z = 8$

34. $ax + 3y - z = 1$
$2x - y - 4z = 3$
$x + 2y + 5z = -1$

35. $2x - y + z = 4$
$x + y + az = -1$
$-3x - 2y - z = 2$

36. $x + y + 3z = 4$
$2x + ay - z = 5$
$-x - y + 2z = -1$

37. $ax + y - 3z = 4$
$-x + ay - z = 1$
$-3x + 2y + z = 2$

38. $ax + 2y - z = 4$
$6x + ay + 3z = 0$
$2x - y + z = -1$

39. Describe your process for finding the values of a.

Counting, Permutations, Combinations, and Probability

C.1 Counting, Permutations, and Combinations

The ability to count objects in a set is essential in mathematics. In this section we will develop some techniques for counting that will be useful in our discussion of probability in the next section.

Suppose three roads connect the towns of Red Creek and Fair Haven and two roads connect the towns of Fair Haven and Oswego. How many ways are there of traveling from Red Creek to Oswego that pass through Fair Haven? This situation may be represented by a **tree diagram,** as in Figure C.1.1.

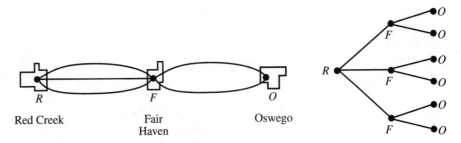

Figure C.1.1

Since we may take any one of three roads from Red Creek to Fair Haven followed by either of two roads from Fair Haven to Oswego, there are $3 \cdot 2 = 6$ ways to travel from Red Creek to Oswego through Fair Haven. This illustrates the following principle.

Fundamental Principle of Counting

> If an event E_1 can occur in m ways and an event E_2 can occur in n ways, then E_1 followed by E_2 can occur in $m \cdot n$ ways.

The fundamental principle of counting generalizes further: If E_1 can occur in n_1 ways, E_2, in n_2 ways, ..., and E_k, in n_k ways, then the events can occur in order in $n_1 \cdot n_2 \cdots n_k$ ways. The next example illustrates that this principle works with three events.

EXAMPLE C.1.1   A college student scheduling her classes notes that she can take any one of three English classes, any one of four mathematics classes, and either of two history classes. Among how many different schedules must she choose if she is required to take one class in each area?

SOLUTION   Let's denote the three possible English classes by E_1, E_2, and E_3, the math classes by M_1, M_2, M_3, and M_4, and the history classes by H_1 and H_2. Instead of drawing the entire tree diagram for this example, we will draw just one branch of it. Figure C.1.2 illustrates that once an English class is chosen, there are four choices for a math class. Figure C.1.3 illustrates that once a math class is chosen, there are two choices for a history class. Thus there are $4 \cdot 2 = 8$ ways of choosing a math and history class, once an English class is chosen. (See Figure C.1.4.)

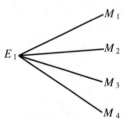

Figure C.1.2

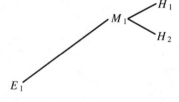

Figure C.1.3

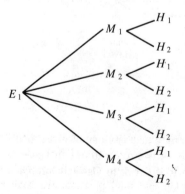

Figure C.1.4

Since there are three English classes from which to choose, we would have similar branches for E_2 and E_3. Thus, there are $3 \cdot 4 \cdot 2 = 24$ different schedules from which this student may choose.

NOTE ▶   It is not necessary to draw a tree diagram for every problem once we realize how to apply the fundamental principle of counting.

EXAMPLE C.1.2   Five children line up in a row for a photograph. In how many different ways can these children be arranged?

SOLUTION   Any one of the five children may occupy the first position.

$$\underline{\quad 5 \quad} \cdot \underline{\qquad} \cdot \underline{\qquad} \cdot \underline{\qquad} \cdot \underline{\qquad}$$

Any one of the remaining four children could be next.

$$\underline{\quad 5 \quad} \cdot \underline{\quad 4 \quad} \cdot \underline{\qquad} \cdot \underline{\qquad} \cdot \underline{\qquad}$$

Continuing, we see that there are $5 \cdot 4 \cdot 3 \cdot 2 \cdot 1 = 120$ ways to arrange 5 children in order.

Recall from Section 11.4 that this product can be expressed using factorial notation as $5!$.

An arrangement or ordering of n objects is called a *permutation* of the n objects. As example C.1.2 illustrates, there are $n!$ permutations of a set of n objects.

EXAMPLE C.1.3   Eight horses compete in a race. How many different finishing orders are there if there are no ties?

SOLUTION   Since a particular finishing order is an arrangement of the eight horses in sequence, this question asks for the number of permutations of 8 objects. There are $8! = 8 \cdot 7 \cdot 6 \cdot 5 \cdot 4 \cdot 3 \cdot 2 \cdot 1 = 40{,}320$ different finishing orders.

Suppose that we wish to count the number of different arrangements of k objects taken from a set of n objects ($1 \leq k \leq n$). As before, we have n choices for the first position, $n - 1$ choices for the second, $n - 2$ choices for the third, and so on. To fill in kth and final position, we have $n - k + 1$ objects from which to choose.

$$\frac{n}{1st}\ \frac{n-1}{2nd}\ \frac{n-2}{3rd}\ \cdots\ \frac{n-k+1}{kth}$$

By the fundamental principle of counting, there are $n \cdot (n-1) \cdot (n-2) \cdots$ $(n-k+1)$ different arrangements of k objects chosen from a set of n objects. Let us denote this number by $_nP_k$. Then, if $1 \leq k < n$,

$$_nP_k = n \cdot (n-1) \cdot (n-2) \cdots (n-k+1)$$

$$= n \cdot (n-1) \cdot (n-2) \cdots (n-k+1) \cdot \frac{(n-k) \cdot (n-k-1) \cdots 3 \cdot 2 \cdot 1}{(n-k) \cdot (n-k-1) \cdots 3 \cdot 2 \cdot 1}$$

$$= \frac{n \cdot (n-1) \cdot (n-2) \cdots 3 \cdot 2 \cdot 1}{(n-k) \cdot (n-k-1) \cdots 3 \cdot 2 \cdot 1}$$

$$= \frac{n!}{(n-k)!}$$

Recall that we define $0! = 1$. With this definition, the relation $_nP_k = \dfrac{n!}{(n-k)!}$ is correct for $k = n$ as well as $k = 1, 2, 3, ..., n-1$.

> The number of **permutations** (arrangements) of n objects taken k at a time is
>
> $$_nP_k = \frac{n!}{(n-k)!}, \qquad 1 \leq k \leq n$$

EXAMPLE C.1.4   Which number is larger, $_6P_2$ or $_6P_4$?

$$_6P_2 = \frac{6!}{(6-2)!} = \frac{6!}{4!} = \frac{6 \cdot 5 \cdot 4!}{4!} = 6 \cdot 5 = 30$$

$$_6P_4 = \frac{6!}{(6-4)!} = \frac{6!}{2!} = \frac{6 \cdot 5 \cdot 4 \cdot 3 \cdot 2!}{2!} = 6 \cdot 5 \cdot 4 \cdot 3 = 360$$

$_6P_4$ is (much) larger than $_6P_2$.

EXAMPLE C.1.5   How many different ways are there to arrange three letters chosen from $\{a, b, c, d, e\}$?

SOLUTION   There are

$$_5P_3 = \frac{5!}{(5-3)!} = \frac{5!}{2!} = 5 \cdot 4 \cdot 3 = 60$$

different arrangements. Some of these are (a, c, e), (b, c, d), (c, a, b), and (e, b, d).

NOTE ▶   We used ordered triple notation to express the different orderings. For example (a, c, e) means a first, c second, and e last.

Suppose instead of counting the different ways of arranging three letters we are simply interested in counting the number of three-element *subsets* of $\{a, b, c, d, e\}$. Remember that two sets are considered equal if they contain the same elements, regardless of order.

XAMPLE C.1.6   List all possible three-letter subsets of $\{a, b, c, d, e\}$. The three-element subsets of $\{a, b, c, d, e\}$ are:

$$\{a, b, c\}, \quad \{a, b, d\}, \quad \{a, b, e\}, \quad \{a, c, d\}, \quad \{a, c, e\}$$
$$\{a, d, e\}, \quad \{b, c, d\}, \quad \{b, c, e\}, \quad \{b, d, e\}, \quad \{c, d, e\}$$

Example C.1.6 illustrates that there are 10 ways of choosing three elements from a set containing five elements. Each three-element subset is called a *combination* of five objects taken three at a time. In general, an r-element subset chosen from a set of n objects is called a *combination of n objects taken r at a time*, which is denoted by $_nC_r$. (Another common notation is $\binom{n}{r}$, which reads "the number of ways of choosing r objects from a set of n objects," or, more succinctly, "n choose r.") For a combination the order or arrangement of the r objects is not important. Two r-element subsets are considered equal if they contain the same elements regardless of order.

What is the relationship between the number of combinations of n objects taken r at a time, $_nC_r$, and the number of permutations of n objects taken r at a time, $_nP_r$? In Example C.1.6, the combination $\{a, b, c\}$ can be arranged or ordered in $3! = 6$ ways: (a, b, c), (a, c, b), (b, a, c), (b, c, a), (c, a, b), and (c, b, a). In the same way each of the remaining nine combinations corresponds to $3! = 6$ permutations. Thus, $_5P_3 = (3!) \cdot (_5C_3)$, or

$$_5C_3 = \frac{_5P_3}{3!} = \frac{\dfrac{5!}{(5-3)!}}{3!} = \frac{5!}{3!2!} = \frac{5 \cdot 4 \cdot 3 \cdot 2 \cdot 1}{(3 \cdot 2 \cdot 1 \cdot) \cdot (2 \cdot 1)} = 10$$

In general, since r elements in an r-element subset can be arranged in $r!$ ways, each combination corresponds to $r!$ permutations. Thus, $_nP_r = (r!) \cdot (_nC_r)$, or

$$_nC_r = \frac{_nP_r}{r!} = \frac{\dfrac{n!}{(n-r)!}}{r!} = \frac{n!}{r!(n-r)!}$$

The number of **combinations** of n objects taken r at a time is

$$_nC_r = \frac{n!}{r!(n-r)!}, \qquad where\ 0 \le r \le n$$

EXAMPLE C.1.7

A class of 20 students is selecting two representatives to serve on a school committee. In how many different ways can this be done?

SOLUTION

This question asks for the number of 2-element subsets in a set of 20 elements. The number of such subsets is

$$_{20}C_2 = \frac{20!}{2!(20-2)!} = \frac{20!}{2!18!} = \frac{20 \cdot 19 \cdot 18!}{2!18!} = \frac{20 \cdot 19}{2 \cdot 1} = 190$$

There are 190 different ways to select 2 students from a class of 20.

EXAMPLE C.1.8

A class of 20 students wishes to elect a president and vice president. In how many ways can this be done?

SOLUTION

This question asks for the number of different ways of *arranging* 2 elements from a set of 20. (If a and b are students, imagine that (a, b) denotes "a is chosen president and b is chosen vice president." Then $(a, b) \ne (b, a)$; in other words, *order is important.*) We need to compute the number of *permutations* of 20 objects taken 2 at a time:

$$_{20}P_2 = \frac{20!}{(20-2)!} = \frac{20!}{18!} = \frac{20 \cdot 19 \cdot 18!}{18!} = 20 \cdot 19 = 380$$

There are 380 ways of electing officers.

E XAMPLE C.1.9   A committee of 2 men and 3 women is to be formed from a group of 8 men and 10 women. How many different committees are possible?

SOLUTION   The number of ways of choosing 2 men from a group of 8 is

$$_8C_2 = \frac{8!}{2!\,6!} = 28$$

The number of ways of selecting 3 women from 10 is

$$_{10}C_3 = \frac{10!}{3!\,7!} = 120$$

By the fundamental counting principle, there are $28 \cdot 120 = 3360$ possible committees.

E XAMPLE C.1.10   How many 5-card hands drawn from a deck of 52 cards contain at least one ace?

SOLUTION   There are

$$_{52}C_5 = \frac{52!}{5!\,47!} = 2,598,960$$

different 5-card hands that can be drawn from a deck of 52 cards. The number of hands that contain *no* aces is

$$_{48}C_5 = \frac{48!}{5!43!} = 1,712,304$$

(We are counting the number of ways of choosing 5 cards from a "reduced" deck of 48 cards that contains no aces.) The number of 5-card hands that contain at least one ace is $_{52}C_5 - {}_{48}C_5 = 886,656$.

E xercises C.1

Use the fundamental principle of counting to find the solutions in exercises 1–8.

1. Pepe's Pizza serves round or square pizzas with two types of crust and eight possible toppings. How many different ways can you order pizza if your coupon allows for only one topping?

2. Ernie's Ice Cream stand serves three flavors of ice cream on two types of cones with or without sprinkles. How many different ice cream cones can be ordered? (Assume one flavor per cone.)

3. A coin is flipped once and a die is rolled once. The results are recorded using ordered-pair notation [for example, $(H, 5)$]. How many different outcomes are possible?

4. Two cards are chosen in order from a 52-card deck. In how many different ways can this be done in each case?
   a. The first card must be a spade and the second card may be a heart or a diamond.
   b. Both cards must be spades.

5. How many three-digit numbers can be formed using 2, 3, and 5 in each case?
   a. The are no restrictions.
   b. No repetition of digits is allowed.
   c. The number must be even.
   d. The number must be less than 350.

6. In how many ways can a 10-question multiple-choice exam be answered if each question has four possible responses?

7. Telephone area codes consist of three digits. The first digit must not be 0 or 1, the middle digit must be 0 or 1, and the last digit must not be 0. How many different area codes are possible?

8. In a certain state, automobile license plates consist of two letters followed by three digits. How many different license plates are possible in each case?
   a. There are no restrictions on letters or numbers.
   b. The first letter cannot be O.

In exercises 9–16, evaluate the given expression.

9. $\dfrac{n!}{(n-1)!}$

10. $\dfrac{n!}{(n-2)!}$

11. $_7P_5$

12. $_8P_3$

13. $_4P_1$

14. $_9P_1$

15. $_{11}P_{11}$

16. $_5P_5$

In exercises 17–22, use permutations to solve each problem.

17. In how many ways can nine faculty be assigned to teach nine courses? (Assume one teacher per course.)

18. Ten bands march in a parade. How many different orders are possible if the host band is to march first?

19. Eight horses run in a race. In how many ways can first-, second-, and third-place winners be determined if a tie is not possible?

20. Ten contestants vie for the title of Miss Moss County at the Moss County Beauty Pageant. In how many ways can Miss Moss County, first runner-up, and second runner-up be determined?

21. Six children are playing musical chairs. In how many different ways can a row of five chairs be occupied by these children?

22. How many ways are there to elect a president, vice president, and secretary in a class of 24 students?

If the n objects in a set are not all distinguishable, then the number of distinguishable permutations is

$$\frac{n!}{(n_1! \cdot n_2! \cdots n_k!)}$$

where there are n_1 objects of the first type, n_2 objects of the second type, and so on for k types.

23. In how many different ways can the letters in the word COMMITTEE be arranged?

24. Find the number of distinguishable permutations of the letters in the word BANANA.

25. In how many distinguishable ways can the product a^4bc^3 be written without exponents? (For example, $a \cdot a \cdot c \cdot c \cdot c \cdot a \cdot b \cdot a$ is one such way.)

26. Aunt Bea is entering the pie contest at the Moss County fair. She has baked 4 apple pies, 3 cherry pies and 2 rhubarb pies.
   a. In how many ways can she arrange them on the window sill if pies of the same type are grouped?
   b. In how many distinguishable ways can the pies be arranged if pies of the same type need not be grouped together?

In exercises 27–34, find the number.

27. $_7C_2$

28. $_5C_4$

29. $_4C_1$

30. $_6C_0$

31. $_7C_7$

32. $_nC_{n-1}$

33. $_nC_1$

34. $_nC_n$

35. List all three-element permutations of $\{a, b, c, d\}$. List all three-element combinations.

36. List all three-element permutations of $\{x, y, z, w\}$. List all three-element combinations.

37. A student must answer 10 of 12 questions on a mathematics exam. How many different ways are there to accomplish this?

38. Forty people enroll in a tennis camp. In how many different ways may players be paired together?

39. Twenty golfers compete in a tournament. How many different foursomes are possible?

40. Suppose six points are arranged in the plane such that no three are collinear. How many different lines are determined?

41. Suppose six points are arranged in a plane such that no three are collinear. How many different triangles are determined?

42. In how many ways can 13 cards be selected from a 52-card deck?

43. An examination consists of 10 short-answer questions and 4 essay questions. A student must answer 8 short-answer questions and 3 essay questions. In how many ways can these 11 questions be selected?

44. A club has 30 members.
   a. How many ways are there to choose four members to serve on the social committee?
   b. How many ways are there to choose a president, vice president, secretary, and treasurer?

45. The mathematics department at Moss County College has five full professors, seven associate professors, and four assistant professors.
   a. In how many ways can a committee of six be formed if rank is not considered?
   b. In how many ways can a committee of six be formed if all three ranks must have equal representation?

46. A class contains 12 boys and 15 girls. A team of 4 students is to be chosen to represent the class in academic competition.
   a. How many different teams are possible?
   b. How many different all-girl teams are possible?
   c. How many teams containing at least one boy are possible?

47. If n and r are nonnegative integers with $r \leq n$, prove that $_nC_r = {}_nC_{n-r}$.

Write Algebra

48. Explain the difference between a permutation and a combination.

C.2 *Probability*

If two coins are flipped, what are the chances of obtaining two heads? How likely is it that a 5-card poker hand drawn from a well-shuffled deck of 52 cards will contain 4 aces? The mathematical theory of probability was begun in the seventeenth century by French mathematicians Pierre Fermat and Blaise Pascal as a means of analyzing similar games of chance.

In probability theory an **experiment** is any process whose outcome is not known in advance with certainty. Tossing a coin and rolling a pair of dice are examples of experiments. The result of an experiment is an **outcome.** If our experiment consists of tossing two coins in succession, then one possible outcome is *HT* (the first coin is heads, the second, tails.) Other outcomes for this experiment are *HH, TH,* and *TT.* We will always assume that *all outcomes are equally likely.* The **sample space** *S* of an experiment is the set of all possible outcomes of the experiment.

EXAMPLE C.2.1   A single die (singular of dice) is rolled. Find the sample space of this experiment.

Since 1, 2, 3, 4, 5, or 6 may be rolled, the sample space is $S = \{1, 2, 3, 4, 5, 6\}$.

An **event** *E* is any subset of the sample space *S*. It is worth noting that the entire sample space *S* and the empty set $\varnothing$ are events. Each individual outcome is also an event (sometimes called a simple event).

EXAMPLE C.2.2   In the experiment of rolling a single die, $E_1 = \{2, 4, 6\}$ is the event of rolling an even number, whereas $E_2 = \{2, 3, 5\}$ is the event of rolling a prime number.

As the examples considered thus far suggest, in our discussion of probability we will consider only sample spaces with a finite number of elements. In this setting we have the following reasonable definition of the probability of an event.

Definition C.2.1

Let *E* be a subset of a finite sample space *S*. The **probability of event** *E,* denoted by $P(E)$, is

$$P(E) = \frac{n(E)}{n(S)}$$

where $n(E)$ is the number of elements (outcomes) in *E* and $n(S)$ is the number of elements (outcomes) in *S*.

If E is any event, clearly $0 \leq n(E) \leq n(S)$. Dividing by $n(S)$ yields

$$0 \leq \frac{n(E)}{n(S)} \leq 1$$

Thus, $0 \leq P(E) \leq 1$; that is, the probability of any event is always at least 0 and never exceeds 1. You should convince yourself that $P(E) = 1$ if and only if $E = S$. In this case we say event E *is a* **certain event**. Also, $P(E) = 0$ if and only if $E = \emptyset$. In this case we say E is an **impossible event**, or it is not possible for E to occur.

EXAMPLE C.2.3   A single die is rolled. Find the probability of each event.

1. Rolling an 8

2. Rolling a number divisible by 3

3. Rolling a number less than 10

SOLUTION   In this experiment, $S = \{1, 2, 3, 4, 5, 6\}$. Let E_1, E_2, and E_3 be the three events.

1. The first event is impossible (8 is not in the sample space), so $P(E_1) = 0$.

2. $E_2 = \{3, 6\}$; thus,

$$P(E_2) = \frac{n(E_2)}{n(S)} = \frac{2}{6} = \frac{1}{3}$$

The probability of rolling a number divisible by 3 is $\frac{1}{3}$.

3. For the third event, every outcome in the sample space is less than 10, so $E_3 = S$ and $P(E_3) = 1$. Thus, E_3 is a certain event.

EXAMPLE C.2.4   Two balanced dice, one red and one green, are rolled. What is the probability that the sum of the two numbers showing is 7?

SOLUTION   Since the result of rolling either die can be any one of six numbers, our sample space S consists of 36 possible outcomes:

(1, 1)	(1, 2)	(1, 3)	(1, 4)	(1, 5)	(1, 6)
(2, 1)	(2, 2)	(2, 3)	(2, 4)	(2, 5)	(2, 6)
(3, 1)	(3, 2)	(3, 3)	(3, 4)	(3, 5)	(3, 6)
(4, 1)	(4, 2)	(4, 3)	(4, 4)	(4, 5)	(4, 6)
(5, 1)	(5, 2)	(5, 3)	(5, 4)	(5, 5)	(5, 6)
(6, 1)	(6, 2)	(6, 3)	(6, 4)	(6, 5)	(6, 6)

where (5, 3), for example, indicates "5 on the red die, 3 on the green die." Since the sum of the numbers is to be 7,

$$E = \{(1, 6), (2, 5), (3, 4), (4, 3), (5, 2), (6, 1)\}$$

Thus,

$$P(E) = \frac{n(E)}{n(S)} = \frac{6}{36} = \frac{1}{6}$$

If we roll a pair of dice 600 times, for example, we would expect to roll a 7 about 100 times.

REMARK

In Example C.2.4 you may wonder why we did not choose {2, 3, 4, ..., 12} as our sample space, corresponding to the eleven possible sums. The reason is simple: These outcomes *are not equally likely*. (The probability of rolling 2 is $\frac{1}{36}$, as you may check, whereas the probability of rolling 7 is $\frac{1}{6}$.) Remember that *all outcomes must be equally likely*.

XAMPLE C.2.5

Three fair coins are tossed. What is the probability that all three faces will be the same?

SOLUTION

The tree diagram in Figure C.2.1 illustrates that there are eight outcomes in the sample space.

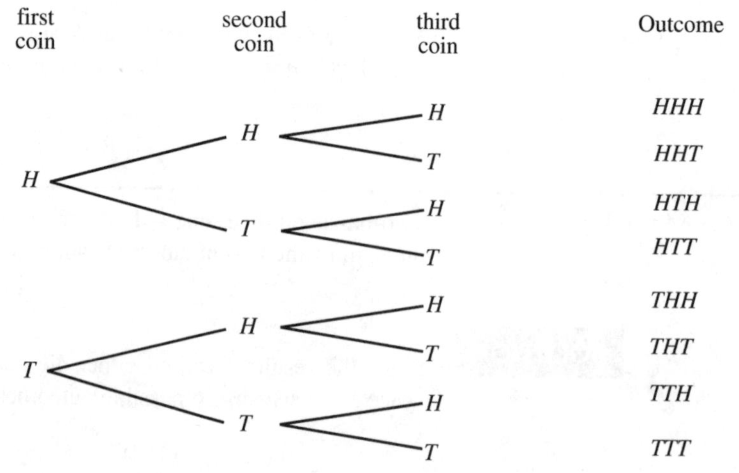

first coin	second coin	third coin	Outcome
H	H	H	HHH
		T	HHT
	T	H	HTH
		T	HTT
T	H	H	THH
		T	THT
	T	H	TTH
		T	TTT

Figure C.2.1

$$S = \{HHH, HHT, HTH, HTT, THH, THT, TTH, TTT\}$$

Since $E = \{HHH, TTT\}$,

$$P(E) = \frac{n(E)}{n(S)} = \frac{2}{8} = \frac{1}{4}$$

XAMPLE C.2.6   Five cards are dealt from a 52-card deck. Find the probability that all 5 cards are black.

SOLUTION   There are $_{52}C_5$ different ways of being dealt a 5-card hand from a deck of 52 cards. Thus our sample space S contains $_{52}C_5$ outcomes (an outcome is a particular 5-card hand). A 52-card deck contains 26 black cards. There are $_{26}C_5$ different 5-card hands where all cards are black. Thus, if E represents the event of drawing a hand consisting of 5 black cards, then

$$P(E) = \frac{n(E)}{n(S)} = \frac{_{26}C_5}{_{52}C_5} = \frac{\dfrac{26!}{5!\,21!}}{\dfrac{52!}{5!\,47!}} = \frac{65,780}{2,598,960} \approx 0.0253$$

Suppose S is a finite sample space and E_1 and E_2 are disjoint events (E_1 and E_2 have no elements in common). Then the probability of E_1 or E_2 is

$$P(E_1 \cup E_2) = \frac{n(E_1 \cup E_2)}{n(S)}$$
$$= \frac{n(E_1) + n(E_2)}{n(S)}$$
$$= \frac{n(E_1)}{n(S)} + \frac{n(E_2)}{n(S)}$$
$$= P(E_1) + P(E_2)$$

Thus, if E_1 and E_2 are disjoint events, then $P(E_1 \cup E_2) = P(E_1) + P(E_2)$. In general, if $E_1, E_2, \ldots, E_n$ are mutually disjoint events (E_i and E_j have no element in common if $i \neq j$) associated with the same experiment, then

$$P(E_1 \cup E_2 \cup \cdots \cup E_n) = P(E_1) + P(E_2) + \cdots + P(E_n)$$

XAMPLE C.2.7   A jar contains 5 red balls, 3 black balls, and 2 white balls. Find the probability of drawing a black or a white ball.

SOLUTION

Let E_1 denote the event of drawing a black ball and E_2 denote the event of drawing a white ball. The sample space S consists of 10 outcomes corresponding to the 10 balls.

$$P(E_1) = \frac{n(E_1)}{n(S)} = \frac{3}{10} \quad \text{and} \quad P(E_2) = \frac{n(E_2)}{n(S)} = \frac{2}{10}$$

Therefore,

$$P(E_1 \cup E_2) = P(E_1) + P(E_2) = \frac{3}{10} + \frac{2}{10} = \frac{5}{10} = \frac{1}{2}$$

Let E' denote the set of outcomes in S that are not in E. For example, in the experiment of rolling a die, where $S = \{1, 2, 3, 4, 5, 6\}$, if $E = \{3, 6\}$, then $E' = \{1, 2, 4, 5\}$. E' is called the **complement** of E. Clearly $S = E \cup E'$, and E and E' are disjoint. Thus,

$$P(S) = 1$$
$$P(E \cup E') = 1$$
$$P(E) + P(E') = 1$$

which implies that

$$P(E) = 1 - P(E')$$

Example C.2.8 illustrates that it is sometimes easier to determine $P(E)$ by finding $P(E')$ and using the preceding relation than to find $P(E)$ directly.

 XAMPLE C.2.8

What is the probability that a family with three children has at least one girl? Assume boy-girl births are equally likely.

SOLUTION

Since each child can be either a boy or a girl, the sample space contains $2 \cdot 2 \cdot 2 = 8$ outcomes. If E is the event of having at least one girl, then E' is the event of having no girls—that is, of having exactly three boys. So,

$$P(E') = \frac{1}{8} \quad \text{and} \quad P(E) = 1 - P(E') = 1 - \frac{1}{8} = \frac{7}{8}$$

Thus the probability of having at least one girl in a family of three children is $\frac{7}{8}$.

Suppose that someone deals you a card and you catch a glimpse of it. If you notice that it is a red face card, what is the probability that it is a king? This question illustrates **conditional probability.** We have some additional information that allows us to eliminate some outcomes from the sample space. Since there are 3 face cards in diamonds and 3 face cards in hearts, there are 6 possible red face cards. Thus the probability that it is a king is $\frac{2}{6}$, or $\frac{1}{3}$. If we let R and K be the events

$$R = \text{the card is a red face card}$$
$$K = \text{the card is a king}$$

then we denote the probability that the card is a king *given that* it is a red face card by $P(K|R)$. Figure C.2.2 illustrates the events R and K and their relationship. From this figure we note that

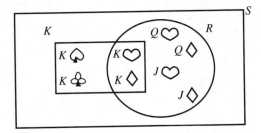

Figure C.2.2

$$P(K|R) = \frac{n(K \cap R)}{n(R)}$$

This formula is true in general.

Conditional Probability

If A and B are events in a sample space S, then the probability of A given that B occurs is denoted by $P(A|B)$ and is calculated by

$$P(A|B) = \frac{n(A \cap B)}{n(B)}$$

Suppose that we are dealt two cards from a deck of 52 cards. What is the probability that they are both red face cards? Let R_1 and R_2 be the events

$$R_1 = \text{first card is a red face card}$$
$$R_2 = \text{second card is a red face card}$$

For the first event, we can easily see that

$$P(R_1) = \frac{n(R_1)}{n(S)} = \frac{6}{52} = \frac{3}{26}$$

But for R_2, S has only 51 outcomes and $n(R_2)$ depends on whether the first card was a red face card or not. *Given that* the first card was a red face card, then we can say that there are 5 red face cards left, so

$$P(R_2|R_1) = \frac{5}{51}$$

Now what? How do we combine these two probabilities to figure out the probability that *both* cards are red face cards? What we need to determine is the probability that R_1 occurs *and* that R_2 occurs, *given that* R_1 has already occurred. So we use conditional probability:

$$P(R_2|R_1) = \frac{n(R_1 \cap R_2)}{n(R_1)} \qquad \textit{Definition of conditional probability}$$

$$= \frac{n(R_1 \cap R_2)}{n(R_1)} \cdot \frac{\dfrac{1}{n(S)}}{\dfrac{1}{n(S)}} \qquad \textit{Divide top and bottom by n(S).}$$

$$= \frac{\dfrac{n(R_1 \cap R_2)}{n(S)}}{\dfrac{n(R_1)}{n(S)}}$$

$$= \frac{P(R_1 \cap R_2)}{P(R_1)}$$

$$P(R_1) \cdot P(R_2|R_1) = P(R_1 \cap R_2) \qquad \textit{Multiply both sides by P(R}_1\textit{).}$$

Thus, the probability that the first card *and* the second card are red face cards is

$$P(R_1 \cap R_2) = P(R_1) \cdot P(R_2|R_1)$$

$$= \frac{3}{26} \cdot \frac{5}{51}$$

$$= \frac{1}{26} \cdot \frac{5}{17}$$

$$= \frac{5}{442}$$

$$\doteq 0.0113$$

This result is called the **product rule** and is generalized as follows.

*P*roduct Rule

If A and B are events in a sample space S, then the probability of A and B is given by

$$P(A \cap B) = P(A) \cdot P(B|A)$$

EXAMPLE C.2.9   A jar contains nine red and three green marbles. If two marbles are pulled out, what is the probability that they are both red?

SOLUTION

Let us assume that the marbles are pulled out one at a time. We are looking for the probability that the first marble is red and the second marble is red. Let us call these events R_1 and R_2, respectively. We can illustrate this experiment with the tree diagram in Figure C.2.3.

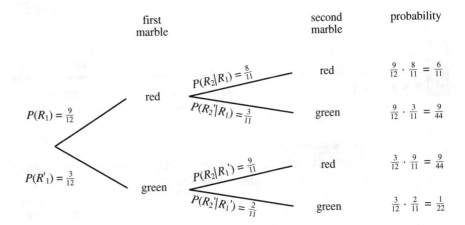

Figure C.2.3

Since all the possible cases were covered, the sum of the probabilities should be 1. Checking:

$$\frac{6}{11} + \frac{9}{44} + \frac{9}{44} + \frac{1}{22} = \frac{24}{44} + \frac{9}{44} + \frac{9}{44} + \frac{2}{44}$$

$$= \frac{44}{44}$$

$$= 1$$

The probability that both are red is illustrated in the top branch where we have the probabilities of the events R_1, and R_2 given that R_1 has occurred. This probability is $\frac{6}{11}$.

What if we performed the experiment in Example C.2.9, but after selecting the first marble we replaced it in the jar before selecting the second marble? Then the occurrence of the first event does not affect the probability of the second event. In this case, we say the events are **independent.** For independent events, $P(R_2|R_1) = P(R_2)$, so

$$P(R_1 \cap R_2) = P(R_1) \cdot P(R_2|R_1)$$
$$= P(R_1) \cdot P(R_2)$$

In general, we have the following rule.

Independent Events

If A and B are independent events in a sample space S, then

$$P(A \cap B) = P(A) \cdot P(B)$$

EXAMPLE C.2.10   A jar contains nine red and three green marbles. A marble is pulled out and then replaced. Then a second marble is selected. What is the probability that both are red?

SOLUTION   Since the first marble is replaced before pulling the second marble, the events are independent. Thus,

$$
\begin{aligned}
P(R_1 \cap R_2) &= P(R_1) \cdot P(R_2) \\
&= \frac{9}{12} \cdot \frac{9}{12} \\
&= \frac{3}{4} \cdot \frac{3}{4} \\
&= \frac{9}{16}
\end{aligned}
$$

Exercises C.2

Find the indicated probabilities in the following experiments.

1. A single die is rolled. What is the probability of rolling
 a. 4?
 b. 4 or 1?
 c. An odd number?
 d. A number less than 3?

2. Two coins are tossed in succession. What is the probability of getting
 a. Two heads?
 b. At least one head?
 c. No heads?

3. A single card is drawn from a deck of 52 cards. What is the probability of drawing
 a. The ace of spades?
 b. An ace?
 c. A spade?
 d. An ace or a spade?

4. A single card is drawn from a deck of 52 cards. What is the probability of drawing
 a. A seven?
 b. A red card?
 c. A heart?
 d. A face card (J, Q, or K of any suit)?

5. A pair of balanced dice are rolled. Find the probability that the sum of the two numbers showing is
   a. 10.
   b. 1.
   c. Less than or equal to 4.

6. A pair of balanced dice are rolled. Find the probability that the sum of the two numbers showing is
   a. Even.
   b. Prime.

7. A pair of balanced dice are rolled. Find the probability that the numbers showing are
   a. Equal.
   b. Unequal.

8. Three balanced dice are rolled. Find the probability that
   a. All three numbers are equal.
   b. At least two numbers differ.

9. Three coins are tossed in succession.
   a. Find the probability of obtaining exactly two heads.
   b. Find the probability of obtaining at least two heads.

10. Three coins are tossed in succession. Find the probability of obtaining exactly two heads or exactly two tails.

11. If two balanced dice are rolled, what is the probability that the difference between the two numbers that appear will be less than 3?

12. A fair coin is tossed once and a single die is rolled once. What is the probability that a tail is obtained on the coin toss and an even number is obtained on the die roll?

13. A family has four children. What is the probability that (a) exactly two of the children are boys? (b) At least two are boys?

14. A family has five children. What is the probability that (a) exactly three of the children are girls? (b) At least one child is a girl?

15. In a certain state lottery, six different numbers are chosen at random from the numbers 1 through 40. Winning combinations do not depend on the order in which the numbers are chosen.
   a. How many different lottery outcomes are possible?
   b. What is the probability of winning the lottery?

16. In a certain class there are 8 boys and 12 girls. The teacher intends to form a reading group of 6 students. What is the probability that the group will consist of 3 boys and 3 girls?

17. Find the probability that a 5-card hand drawn from a deck of 52 cards contains 3 red cards and 2 black cards.

18. Five cards are drawn from a 52-card deck. Find the probability of obtaining four of a kind (for example, four aces or four tens).

19. Five cards are drawn from a 52-card deck. Find the probability of obtaining 3 aces and 2 kings.

20. Five cards are drawn from a 52-card deck. Find the probability of obtaining at least 1 heart.

21. A coin is tossed four times in succession. Find the probability of obtaining at least one head.

22. A monkey is given the letters *t*, *p*, *s*, and *o* to arrange to form a four-letter "word." What is the probability that the monkey will form an actual four-letter word?

23. Suppose eight people are seated in a random manner in a row of eight seats. What is the probability that two specified individuals will be seated next to each other?

24. Find the probability of picking an ace of spades from a standard deck given that the card is black.

25. Find the probability of picking a queen from a standard deck given that the card is a face card.

26. A pair of dice are rolled. What is the probability that the sum of the two numbers is 7 given that it is odd?

27. A pair of dice are rolled. What is the probability that the sum of the two numbers is 6 given that both numbers are the same?

28. A drawer contains 12 black socks and 6 brown socks. Two socks are pulled out (in the dark). What is the probability that they are both black?

29. A drawer contains 10 blue socks and 15 gray socks (the "borrowers" took one). Two socks are pulled out. What is the probability that they are both blue?

30. A class contains 15 male students and 18 female students. One student is chosen as president and another is chosen as secretary. What is the probability that both are female?

31. A class contains 16 male students and 12 female students. One student is chosen as "most popular" and another is chosen as "most likely to succeed." What is the probability that both are male?

32. A box contains 10 green balls and 8 red balls. A ball is selected and replaced; then a second ball is pulled. What is the probability that both balls are green?

33. A box contains 17 green balls and 5 red balls. A ball is selected and replaced; then a second ball is pulled out. What is the probability that both balls are red?

34. A pond contains 25 bass and 40 perch. Jim Bob catches a fish and releases it back into the pond. He later catches another fish. What is the probability that both fish were bass?

35. A pond contains 50 perch and 30 catfish. Audrey catches a fish and releases it back into the pond. She later catches another fish. What is the probability that both fish were catfish?

Graphing Calculators

Advances in technology can often eliminate dull and repetitive tasks. In addition, technological advances can enable us to find patterns and relationships that we might not have otherwise observed. This appendix gives a brief introduction to the graphing calculator (specifically, the TI-81 from Texas Instruments) and illustrates some of its many uses.

One of the most important features of the graphing calculator is the viewing screen. The viewing screen of the TI-81 is a grid of picture elements called *pixels*.

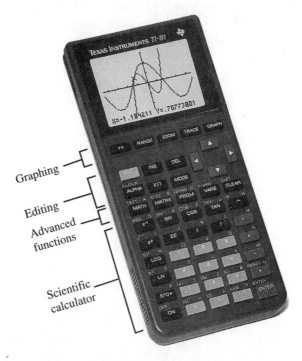

Graphing

Editing

Advanced functions

Scientific calculator

Figure D.1 *Courtesy of Texas Instruments, Inc.*

This screen has 64 rows and 96 columns of pixels. A pixel will light up if it contains a plotted point on the graph of a curve. ·

Figure D.1 illustrates the keyboard on the TI-81. In addition, this figure shows the four main groupings of the keys. We will devote most of our attention to the graphing keys.

XAMPLE D.1   Use a graphing calculator to sketch the graph of $y = 2x + 3$.

SOLUTION

There are two common sequences of key strokes that can be used to sketch the graph of $y = 2x + 3$. Let's first determine if our calculator is set on the standard viewing screen. Press $\boxed{\text{RANGE}}$ and determine if your calculator shows the screen in Figure D.2.

If your screen has different values than those illustrated in Figure D.2, press $\boxed{\text{ZOOM}}$ $\boxed{6}$ and press $\boxed{\text{RANGE}}$ again. Your calculator will now be set for the standard viewing screen.

NOTE ▶   In all the problems in this appendix, we will begin graphing on the standard viewing screen.

Now let's graph $y = 2x + 3$ on the standard viewing screen. Press $\boxed{Y=}$ (the cursor should be flashing to the right of $y_1 =$). Now press $\boxed{2}$ $\boxed{\times}$ $\boxed{\text{X|T}}$ $\boxed{+}$ $\boxed{3}$. We have entered the desired equation in the calculator. We can obtain the graph by either of the following sequences of key strokes: Press $\boxed{\text{GRAPH}}$ or $\boxed{\text{ZOOM}}$ $\boxed{6}$. The graph is illustrated in Figure D.3.

RANGE
X min = –10
X max = 10
X scl = 1
Y min = –10
Y max = 10
Y scl = 1
X res = 1

Figure D.2

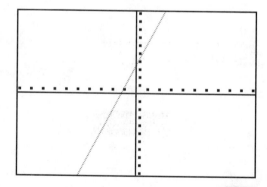

Figure D.3

One final note about the viewing screen. Suppose we want to sketch the graph of the equation $y = x^2 + 11$. If we try to sketch the graph of this equation in the standard viewing screen, no pixel will light up. This result is due to the fact that the largest y value in the standard viewing screen is 10, and the smallest y value of the equation $y = x^2 + 11$ is 11. Thus, to obtain a graph for $y = x^2 + 11$, we must change the viewing rectangle. Let's set the RANGE as indicated in Figure D.4. Now press $\boxed{Y =}$ and then press $\boxed{CLEAR}$ to eliminate the previous equation. (The cursor should be flashing to the right of $y_1 =$). Next press $\boxed{X|T}$ $\boxed{X^2}$ $\boxed{+}$ $\boxed{1}$ $\boxed{1}$. Now press $\boxed{GRAPH}$. The screen should display the graph in Figure D.5.

RANGE
X min = –10
X max = 10
X scl = 1
Y min = –2
Y max = 30
Y scl = 1
X res = 1

Figure D.4

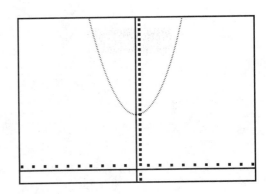

Figure D.5

NOTE ▶   You might try different range settings in order to investigate different parts of the graph.

The graphing calculator can be used to find approximate solutions of many types of equations. Consider the following example.

XAMPLE D.2   Find the solutions of $x^2 - 5 = x + 1$. First use algebraic methods, and then find an approximation for the solutions by using the graphing calculator.

SOLUTION   First, use standard algebraic methods.

$$x^2 - 5 = x + 1 \qquad \textit{A quadratic equation}$$
$$x^2 - x - 6 = 0 \qquad \textit{Place the equation in standard form.}$$
$$(x - 3)(x + 2) = 0 \qquad \textit{Factor.}$$
$$x - 3 = 0 \quad \text{or} \quad x + 2 = 0 \qquad \textit{Use the zero product theorem.}$$
$$x = 3 \quad \text{or} \qquad x = -2$$

Thus, the solution set is $\{3, -2\}$.

Using the graphing calculator, let's make sure we have the standard viewing screen by pressing $\boxed{\text{ZOOM}}$ $\boxed{6}$. Now press $\boxed{y =}$. The cursor should be flashing to the right of $y_1 =$ (press $\boxed{\text{CLEAR}}$ if an old equation is present). Press $\boxed{\text{X|T}}$ $\boxed{x^2}$ $\boxed{-}$ $\boxed{5}$ (we have entered the left-hand side of the equation as y_1). Now press $\boxed{\blacktriangledown}$. The cursor should be flashing to the right of $y_2 =$. Press $\boxed{\text{X|T}}$ $\boxed{+}$ $\boxed{1}$ (we have entered the right-hand side of the equation as y_2). Now press $\boxed{\text{GRAPH}}$. Your screen should look like Figure D.6.

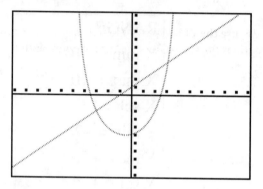

Figure D.6

To obtain an approximation for the solutions of the equation $x^2 - 5 = x + 1$, press $\boxed{\text{TRACE}}$. A flashing point will appear near the vertex of the parabola. Next press $\boxed{\blacktriangleright}$ until the flashing point appears to be at the right-hand intersection point of the line and the parabola. Note the x value at the bottom of the screen. This value is one of the approximate solutions of the equation; it is $x \doteq 3.053$. Next press $\boxed{\blacktriangleleft}$ until the flashing point appears at the left-hand point of the intersection. This happens when $x = -2$ (the exact solution we obtained earlier). Thus, the values we obtained with the graphing calculator are $x = -2$ and $x \doteq 3.053$.   ▬

NOTE ▶   You can improve the approximation $x \doteq 3.053$ if you know more about how to use the ZOOM key.

The graphing calculator can be a very powerful and helpful tool. This appendix only hints at some of the uses of the graphing calculator.

E xercises D

In problems 1–20, use a graphing calculator to sketch the graph of the given equation. Sketch the graph in the standard viewing screen.

1. $y = 3x - 2$

2. $y = x + 2$

3. $5x + 2y = 3$

4. $3x + 2y = 1$

5. $y = \frac{1}{2}x^2 - 3$

6. $y = \frac{1}{3}x^2 - 2$

7. $y = x^2 - 4x + 3$

8. $y = x^2 - 6x + 1$

9. $y = -x^2 + 6x - 2$

10. $y = -x^2 + 8x - 9$

11. $y = \sqrt{x - 2} + 3$

12. $y = \sqrt{x + 3} - 4$

13. $y = \sqrt{25 - x^2}$

14. $y = \sqrt{36 - x^2}$

15. $y = |x + 2| - 4$

16. $y = |x - 3| - 3$

17. $y = x^3 - 6x - 3$

18. $y = x^3 - 4x + 1$

19. $y = x^3 - 5x^2 + 3x + 9$

20. $y = x^4 - 5x^2 + 4$

In problems 21–36, use the ZOOM and TRACE keys of a graphing calculator to find approximations for the solutions of the given equation.

21. $x + 2 = 7 - 2x$

22. $4x + 1 = 9 - x$

23. $5 - 2(x + 7) = 2(2x + 1) + 6$

24. $8 - 3(x + 3) = 2(2x + 5) - 1$

25. $x^2 - 7 = 2x - 1$

26. $5 - x^2 = 3 - \frac{1}{2}x$

27. $x^2 - 2x - 5 = -x^2 + 6x - 1$

28. $2x^2 - x - 4 = -x^2 + x + 5$

29. $\sqrt{36 - x^2} = x^2 - 4x - 5$

30. $\sqrt{49 - x^2} = |x + 2| - 1$

31. $\sqrt{x + 7} - x = 5$

32. $\sqrt{x + 3} - x = -3$

33. $|x - 2| = 8 - \frac{1}{2}x^2$

34. $x + |x| + 1 = 4 - x^2$

35. $x^3 - 6x + 2 = x + 5$

36. $x^3 - 4x - 3 = x - 1$

Answers to Odd-Numbered Problems

CHAPTER 1   Exercises 1.1 (p. 6)

1. False   **3.** True   **5.** False   **7.** True   **9.** False   **11.** True   **13.** True   **15.** True   **17.** True   **19.** False
21. $\{1, 2, 3, \ldots, 10\}$ or M **23.** $\{2, 4, 6, 8, 10\}$ or P **25.** $\{1, 2, 3, \ldots, 10\}$ or M **27.** $\{1, 2, 3, 4, \ldots\}$ or N **29.** $\varnothing$
31. $\{1, 2, 3, 4, \ldots\}$ or N **33.** $\{1, 2, 3, \ldots, 10\}$ or M **35.** $\{1, 2, 3, \ldots, 10\}$ or M **37.** $\{1, 2, 3, \ldots, 10\}$ or M
39. $\{2, 4, 6, 8, 10\}$ or P **41.** $\varnothing, \{1\}, \{2\}, \{3\}, \{1, 2\}, \{1, 3\}, \{2, 3\}, \{1, 2, 3\}$

Exercises 1.2 (p. 13)

1. 25   **3.** -59 **5.** -480 **7.** 144   **9.** -23 **11.** 22   **13.** -13 **15.** -34 **17.** -1 **19.** 600
21. 42   **23.** 0   **25.** -49 **27.** -8 **29.** -625 **31.** 0   **33.** -16 **35.** -24 **37.** 101   **39.** 420
41. Undefined   **43.** Indeterminant   **45.** -384 **47.** $-1,000,000$ **49.** $-1,000,000$

Exercises 1.3 (p. 23)

1. $\dfrac{4}{3}$ **3.** $\dfrac{35}{18}$ **5.** $\dfrac{8}{9}$ **7.** $\dfrac{1}{5}$ **9.** $\dfrac{3}{2}$ **11.** $\dfrac{1}{6}$ **13.** $\dfrac{57}{40}$ **15.** $\dfrac{20}{9}$ **17.** $\dfrac{9}{25}$ **19.** $\dfrac{3}{2}$ **21.** $-\dfrac{81}{2}$ **23.** $\dfrac{103}{78}$

25. $-\dfrac{15}{2}$ **27.** $-\dfrac{25}{12}$ **29.** $-\dfrac{1}{2}$ **31.** 0.375   **33.** $-0.222 \ldots = -0.\overline{2}$ **35.** $0.\overline{142857}$ **37.** $\dfrac{13}{20}$ **39.** $\dfrac{31}{8}$

41. $\dfrac{1}{400}$ **43.** -24.853 **45.** 2.12   **47.** 13.945   **49.** -739.5 **51.** 20.003   **53.** -24.8 **55. (a)** 11 in.

(b) 10.67 m   **57. (a)** \$289.69   **(b)** $36\dfrac{1}{2}$ hr   **59.** $16\dfrac{3}{8}$ **61. (a)** $\dfrac{5}{9}$ **(b)** $\dfrac{5}{11}$

Exercises 1.4 (p. 29)

1. True   **3.** False   **5.** True   **7.** True   **9.** False   **11.** False   **13.** 4.24 mm   **15.** 5.39 m

17. 8.6 in. $= 8\dfrac{3}{5}$ in.   **19.** 24.04 in. $= 24\dfrac{1}{25}$ in.   **21.** 3.77 m   **23.** 38.47 cm^2 **25.** $9\dfrac{5}{8}$ in.^{2} **27.** 1.6180340

29. Φ is one unit larger than $\dfrac{1}{\Phi}$ **31. (a)** 1.4142136   **(b)** 1414.2136   **(c)** 0.2135623   **(d)** 1.4142135623   **(e)** Same

Exercises 1.5 (p. 38)

1. True   **3.** False   **5.** True   **7.** False   **9.** False   **11.** False   **13.** False   **15.** True   **17.** False

19.   **21.**   **23.**   **25.** $\frac{5}{4} < 1.4$ **27.** $-7 < 0$

29. $-6 < -2$ **31.** $\sqrt{6} > 2$ **33.** $\pi > 3$ **35.** $2.\overline{2} > 2.22$ **37.** Symmetric   **39.** Reflexive   **41.** Transitive property of inequality   **43.** Substitution   **45.** Identity property of multiplication   **47.** Commutative property of addition

49. Identity property of addition   **51.** Trichotomy   **53.** Additive inverse: 6, Multiplicative inverse: $-\frac{1}{6}$ **55.** Additive inverse: $\frac{3}{7}$, Multiplicative inverse: $-\frac{7}{3}$ **57.** Additive inverse: $\sqrt{8}$, Multiplicative inverse: $-\frac{1}{\sqrt{8}}$ **59.** $-0.375 = -\frac{3}{8}$, Additive inverse: 0.375, Multiplicative inverse: $-\frac{8}{3}$, or $-2.\overline{6}$ **61.** $0.333... = \frac{1}{3}$, Additive inverse: $-0.333...$, Multiplicative inverse: 3   **63.** $2 \cdot 5 + 2 \cdot 7 = 24$ **65.** $6x + 6y + 24$ **67.** $5x - 15$ **69.** $5(a + b)$ **71.** $(7 + 10 + 5)x = 22x$ **73.** $xy + x$ **75.** Associative property of multiplication   **77.** Commutative property of addition, associative property of addition   **79.** Commutative property of addition, associative property of addition, inverse property of addition, identity property of addition   **81.** Commutative property of multiplication, associative property of multiplication, inverse property of multiplication, identity property of multiplication   **83.** Distributive property, inverse property of addition, multiplication property of zero   **85.** x must be less than 2 $(x < 2)$

Exercises 1.6 (p. 46)

1. 10   **3.** 18   **5.** 45   **7.** $-\frac{3}{4}$ **9.** 1   **11.** $-\frac{4}{3}$ or $-1.\overline{3}$ **13.** -9 **15.** -5 **17.** 112   **19.** -15

21. 40   **23.** 8   **25.** $\frac{4}{3}$ or $1.\overline{3}$ **27.** -1 **29.**   **31.**

33.   **35.**   **37.**   **39.**

41. 7   **43.** 0   **45.** -4 **47.** -2π **49.** $<$ **51.** $<$ **53.** $=$ **55.** $>$ **57.** 13   **59.** -36 **61.** 11   **63.** 2   **65.** 1   **67.** 42   **69.** 6   **71.** $\frac{13}{6}$ **73.** 7   **75.** 130

APPLIED ALGEBRA—YOUR TURN (p. 48)

1. $T_{WC} = 3.5°$

CHAPTER 1   Review Exercises (p. 51)

1. True   **3.** False   **5.** True   **7.** True   **9.** False   **11.** {4, 8}   **13.** {2, 4, 6, 8} or W **15.** {4, 8, 12, ..., 100}   **17.** {2, 4, 6, 8, ..., 100} or Z **19.** {4, 8, 12} or X **21.** -37 **23.** -12 **25.** -13 **27.** -52 **29.** 8   **31.** -132 **33.** -120 **35.** 95   **37.** 121   **39.** 0   **41.** 5   **43.** Indeterminant   **45.** $\frac{2}{3}$ **47.** $\frac{1}{2}$ **49.** $\frac{12}{7}$ **51.** $\frac{5}{8}$ **53.** $\frac{19}{12}$ **55.** $\frac{1}{16}$ **57.** $-\frac{33}{35}$ **59.** $-\frac{14}{45}$ **61.** $\frac{3}{2}$ **63.** $\frac{7}{12}$ **65.** $-\frac{16}{33}$ **67.** $\frac{41}{8}$ or $5\frac{1}{8}$ **69.** $\frac{2}{3}$

71. 0.3125   **73.** $0.\overline{36}$ **75.** $\frac{6}{25}$ **77.** $\frac{13}{5}$ **79.** -18.172 **81.** 2.65   **83.** -34.545 **85.** 1216.875

87. (a) $\frac{49}{16}$ or $3\frac{1}{16}$ sq in.   **(b)** 3.735 cm^2 **(c)** 12.024 m^2 **(d)** $\frac{45}{8}$ or $5\frac{5}{8}$ ft^2 **89.** 8-oz bowl   **91.** True

93. True   **95.** False   **97.** 2.83 cm   **99.** $9.43 = 9\frac{43}{100}$, or $\frac{943}{100}$ in.   **101.** $\frac{77}{8} = 9\frac{5}{8}$ sq in.   **103.** 13.188 cm

105. True   **107.** True   **109.** True   **111.** True   **113.** False   **115.**

117. $<$

119. $>$ **121.** $>$ **123.** $>$ **125.** Symmetric   **127.** Transitive property of inequality   **129.** Distributive

131. Identity property of multiplication   **133.** Associative property of addition   **135.** Additive inverse: $\dfrac{11}{2}$, Multiplicative

inverse: $-\dfrac{2}{11}$ **137.** Additive inverse: -26, Multiplicative inverse: $\dfrac{1}{26}$ **139.** $5x + 35$ **141.** $7a + 7b + 21$

143. $(8 + 12)x = 20x$ **145.** Commutative property of addition, associative property of addition   **147.** Distributive, inverse

property of addition, multiplication property of zero   **149.** -12 **151.** 46   **153.** $\dfrac{3}{2}$ **155.** 3   **157.** -6.05

159. -9 **161.** 178   **163.**

165.

167. $>$ **169.** $>$ **171.** $<$

173. 36   **175.** -7 **177.** 3   **179.** $\dfrac{18}{13}$ **181.** $-\dfrac{11}{12}$

CHAPTER 1   Test (p. 55)

1. False   **3.** True   **5.** False   **7.** True   **9.** $\{2, 4\}$ **11.** $\{2, 4, 6, \dots\}$ or E **13.** $<$ **15.** $<$

17.

19. Additive inverse: $-\dfrac{5}{7}$, multiplicative inverse: $\dfrac{7}{5}$ **21.** $80 - 20 + 40 = 100$

23. Trichotomy   **25.** Associative property of multiplication, inverse property of multiplication, identity property of

multiplication   **27.** -6 **29.** -64 **31.** $-\dfrac{2}{75}$ **33.** 3.254   **35.** 1   **37.** $-\dfrac{1}{2}$ **39.** -1 **41.** 192   **43.** 26

45. 0.12   **47.** 38   **49.** 0

CHAPTER 2   Exercises 2.1 (p. 65)

1. 20   **3.** $-\dfrac{13}{3}$ **5.** 170   **7.** 36   **9.** $-\dfrac{3}{2}$ **11.** $\dfrac{8}{3}$ **13.** 11   **15.** $-\dfrac{8}{3}$ **17.** 0   **19.** $-\dfrac{21}{2}$ **21.** -6

23. 18.5   **25.** 1   **27.** -8 **29.** $-\dfrac{4}{15}$ **31.** $\varnothing$ **33.** $\dfrac{1}{3}$ **35.** $-\dfrac{5}{6}$ **37.** -12 **39.** 0   **41.** -7 **43.** $\dfrac{13}{12}$

45. All real numbers   **47.** -6 **49.** $\dfrac{6}{11}$ **51.** $-\dfrac{2}{3}$ **53.** $\varnothing$ **55.** -9 **57.** 0   **59.** 8   **61.** -5 **63.** 6

65. 39   **67.** $\dfrac{5}{8}$ **69.** 13.87

Exercises 2.2 (p. 74)

1. $x \geq 5$ $[5, \infty)$

3. $x \leq 8$ $(-\infty, 8]$

5. $x \leq 9$ $(-\infty, 9]$

7. $x > 0$ $(0, \infty)$

9. $x \geq -\dfrac{2}{3}$ $[-\dfrac{2}{3}, \infty)$

11. $x > -12$ $(-12, \infty)$

13. $x \leq -4$ $(-\infty, -4]$

15. $x > -2$ $(-2, \infty)$

17. $x < 9$ $(-\infty, 9)$

19. $x \geq -\dfrac{5}{6}$ $[-\dfrac{5}{6}, \infty)$

21. $x < 0.3$ $(-\infty, 0.3)$

23. $x \leq 4$ $(-\infty, 4]$

25. $x < \dfrac{5}{2}$ $(-\infty, \dfrac{5}{2})$

27. $x < -6$ $(-\infty, -6)$

29. $x \geq -12$ $[-12, \infty)$

31. $x > 7$ $(7, \infty)$

33. $x \le \frac{1}{4}$ $(-\infty, \frac{1}{4}]$

35. $x < -8$ $(-\infty, -8)$

37. $x \le \frac{5}{2}$ $(-\infty, \frac{5}{2}]$

39. $x < -2$ $(-\infty, -2)$

41. $x < -\frac{15}{2}$ $(-\infty, -\frac{15}{2})$

43. $x > 2$ $(2, \infty)$

45. $x \le -4$ $(-\infty, -4]$

47. $x < 0$ $(-\infty, 0)$

49. $x < \frac{3}{2}$ $(-\infty, \frac{3}{2})$

51. $3 < x < 10$ $(3, 10)$

53. $-3 < x < 5$ $(-3, 5)$

55. $-2 \le x \le 4$ $[-2, 4]$

57. $2 \le x \le 6$ $[2, 6]$

59. $1 < x \le 5$ $(1, 5]$

61. $-3 < x < -1$ $(-3, -1)$

63. $3 < x < 12$ $(3, 12)$

65. $-\frac{10}{3} < x < 10$ $(-\frac{10}{3}, 10)$

67. $\varnothing$

69. $x > -1$ $(-1, \infty)$

71. $x \le \frac{2}{3}$ or $x \ge \frac{5}{2}$ $(-\infty, \frac{2}{3}] \cup [\frac{5}{2}, \infty)$

73. $-2 < x < \frac{6}{5}$ $(-2, \frac{6}{5})$

75. All real numbers $(-\infty, \infty)$

Exercises 2.3 (p. 78)

1. $R = \frac{D}{T}$ **3.** $W = \frac{A}{L}$ **5.** $r = \frac{C}{2\pi}$ **7.** $h = \frac{3v}{\pi r^2}$ **9.** $G = \frac{aR^2}{m}$ **11.** $L = \frac{P - 2W}{2}$ or $\frac{P}{2} - W$

13. $z = 180 - x - y$ **15.** $m = \frac{y - b}{x}$ or $\frac{y}{x} - \frac{b}{x}$ **17.** $m = \frac{y - y_1}{x - x_1}$ **19.** $y = \frac{S - 2x^2}{4x}$ or $\frac{S}{4x} - \frac{x}{2}$ **21.** $x_2 = 2x - x_1$

23. $r = \frac{P - 2L}{\pi + 2}$ **25.** $w = \frac{S - 2LH}{2L + 2H}$ **27.** $m = \frac{2T}{v^2 + 2ad}$ **29.** $37° C$ **31.** 6 hr **33.** $9\frac{1}{4}\%$ **35.** 16 ft **37.** 17 m

39. 14 in. **41.** 4 in.

Exercises 2.4 (p. 87)

1. 21 **3.** -5 **5.** $\frac{5}{6}$ **7.** 3, 10 **9.** $-4, 9$ **11.** 10, 13 **13.** 25, 26, 27 **15.** 35, 37, 39 **17.** 12, 14, 16

19. 6 m by 21 m **21.** 4 in. by 9 in. **23.** 3 cm, 6 cm, 8 cm **25.** 8 nickels, 17 quarters **27.** 18 dimes, 12 quarters

29. 19 five-dollar, 21 ten-dollar, and 12 twenty-dollar bills **31.** $84,210.53 **33.** $11,290 **35.** 2 hr **37.** $2\frac{1}{2}$ hr

Exercises 2.5 (p. 94)

1. $\{2, -2\}$ **3.** $\{23, -23\}$ **5.** $\varnothing$ **7.** $\varnothing$ **9.** $\{0\}$ **11.** $\{7, -7\}$ **13.** $\varnothing$ **15.** $\left\{\frac{13}{5}, -\frac{13}{5}\right\}$ **17.** $\{0\}$

19. $\{4, -4\}$ **21.** $\{6, -6\}$ **23.** $\{9, -9\}$ **25.** $\{9, -9\}$ **27.** $\{4, -4\}$ **29.** $\varnothing$ **31.** $\varnothing$ **33.** $\{16, -2\}$

35. $\varnothing$ **37.** $\{6, -4\}$ **39.** $\left\{-\frac{9}{8}, \frac{21}{8}\right\}$ **41.** $\{3, -11\}$ **43.** $\left\{\frac{1}{3}\right\}$ **45.** $\left\{\frac{17}{2}, -\frac{9}{2}\right\}$ **47.** $\varnothing$ **49.** $\left\{\frac{1}{2}\right\}$

51. $\{3, -9\}$ **53.** $\left\{\frac{8}{3}, 0\right\}$ **55.** $\{3, 0\}$ **57.** $\varnothing$ **59.** $\{3.3, -0.3005291\}$ **61.** $\{-5, 1\}$ **63.** $\left\{\frac{3}{5}, -3\right\}$

65. $\left\{-8, \frac{10}{9}\right\}$ **67.** $\{1, 4\}$ **69.** $\{14, -2\}$ **71.** $\left\{\frac{88}{21}, \frac{80}{39}\right\}$ **73.** $\left\{-\frac{1}{2}\right\}$ **75.** $\left\{\frac{1}{12}\right\}$ **77.** $\left\{\frac{-b - c}{a}, \frac{-b + c}{a}\right\}$

81. $\left\{\dfrac{3}{2}, -\dfrac{5}{4}\right\}$ **83.** $\left\{-\dfrac{5}{4}, -\dfrac{3}{2}\right\}$ **85.** $\left\{4, -\dfrac{10}{3}\right\}$ **87.** The solutions of $|x - |ax + b|| = c$ are obtained by solving the equations $ax + b = x + c$ and $ax + b = -x - c$.

Exercises 2.6 (p. 100)

1. $-1 < x < 1$
$(-1, 1)$

3. $x < -2$ or $x > 2$
$(-\infty, -2) \cup (2, \infty)$

5. $-\dfrac{3}{2} \le x \le \dfrac{3}{2}$
$[-\dfrac{3}{2}, \dfrac{3}{2}]$

7. All real numbers
$(-\infty, \infty)$

9. $-5 < x < 5$
$(-5, 5)$

11. $x \le -1$ or $x \ge 1$
$(-\infty, -1] \cup [1, \infty)$

13. $-2 < x < 2$
$(-2, 2)$

15. $-3 < x < 3$
$(-3, 3)$

17. All real numbers
$(-\infty, \infty)$

19. $-6 \le x \le 4$
$[-6, 4]$

21. $x \le -4$ or $x \ge 6$
$(-\infty, -4] \cup [6, \infty)$

23. $\dfrac{5}{2} < x < \dfrac{7}{2}$
$(\dfrac{5}{2}, \dfrac{7}{2})$

25. $-1 \le x \le \dfrac{13}{5}$
$[-1, \dfrac{13}{5}]$

27. $x < -\dfrac{1}{2}$ or $x > \dfrac{5}{2}$
$(-\infty, -\dfrac{1}{2}) \cup (\dfrac{5}{2}, \infty)$

29. $-1 \le x \le 4$
$[-1, 4]$

31. $0 < x < \dfrac{2}{3}$
$(0, \dfrac{2}{3})$

33. $x < -\dfrac{7}{3}$ or $x > 1$
$(-\infty, -\dfrac{7}{3}) \cup (1, \infty)$

35. $-6 < x < 2$
$(-6, 2)$

37. $x \le -4$ or $x \ge -1$
$(-\infty, -4] \cup [-1, \infty)$

39. $1 < x < 7$
$(1, 7)$

41. All real numbers
$(-\infty, \infty)$

43. $-11 < x < -3$
$(-11, -3)$

45. $x \le -\dfrac{7}{6}$ or $x \ge \dfrac{11}{6}$
$(-\infty, -\dfrac{7}{6}] \cup [\dfrac{11}{6}, \infty)$

47. $-1 \le x \le 3$
$[-1, 3]$

49. $x < -5$ or $x > -2$
$(-\infty, -5) \cup (-2, \infty)$

51. $x \ne -1$
$(-\infty, -1) \cup (-1, \infty)$

53. $x = \dfrac{4}{3}$

55. $x \ne 3$
$(-\infty, 3) \cup (3, \infty)$

57. All real numbers
$(-\infty, \infty)$

59. $\varnothing$

65. $-7 \le x \le -3$ or $3 \le x \le 7$

67. $-5 < x < -2$ or $2 < x < 5$

69. $-6 \le x \le -2$ or $2 \le x \le 6$

71. $-4 \le x \le -2$ or $4 \le x \le 6$

73. $-4 < x < -1$ or $7 < x < 10$

75. $-6 \le x \le -3$ or $-1 \le x \le 2$

APPLIED ALGEBRA—YOUR TURN (p. 102)

1. 1.44 in.   **3.** 1.152 in.   **5.** 163.75° F   **7.** 49.1$\overline{6}$° F

CHAPTER 2   Review Exercises (p. 104)

1. 8   **3.** $\dfrac{16}{3}$ **5.** -6 **7.** 3   **9.** $\dfrac{5}{3}$ **11.** 0   **13.** $x < 4$ $(-\infty, 4)$ 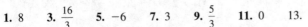   **15.** $x > -8$ $(-8, \infty)$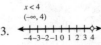

17. $x > \frac{9}{2}$, $(\frac{9}{2}, \infty)$

19. $x > 1$, $(1, \infty)$

21. $-5 \le x \le 3$, $[-5, 3]$

23. $-3 < x < 0$, $(-3, 0)$

25. $\varnothing$

27. $a = \frac{F}{m}$ **29.** $x = \frac{y - b}{m}$ **31.** $y = 4x + 2$ **33.** $z = \frac{2x + 3y - 5}{4}$ **35.** Width $= 6$ m **37.** $20°$ C **39.** 11

41. 11, 12, and 13 **43.** 23 nickels, 14 dimes, 7 quarters **45.** 4:40 P.M. **47.** $\left\{-\frac{3}{2}, \frac{3}{2}\right\}$ **49.** $\{-5, 13\}$

51. $\left\{-3, \frac{17}{5}\right\}$ **53.** $\varnothing$ **55.** $\{-7, -1\}$ **57.** $\left\{\frac{1}{3}\right\}$ **59.** $1 \le x \le \frac{5}{3}$, $[1, \frac{5}{3}]$ **61.** $0 < x < 4$, $(0, 4)$ **63.** $\varnothing$

65. $x < -3 \text{ or } x > \frac{9}{5}$, $(-\infty, -3) \cup (\frac{9}{5}, \infty)$

67. $x \le -4 \text{ or } x \ge 0$, $(-\infty, -4] \cup [0, \infty)$

69. All real numbers $(-\infty, \infty)$

CHAPTER 2 Test (p. 106)

1. 16 **3.** -27 **5.** $\varnothing$ **7.** 1 or 5 **9.** $-7 < x < -2$, $(-7, -2)$ **11.** $x < -\frac{7}{2} \text{ or } x > 2$, $(-\infty, -\frac{7}{2}) \cup (2, \infty)$ **13.** $-1 < x < \frac{8}{3}$, $(-1, \frac{8}{3})$

15. $a = \frac{2A - hb}{h}$ **17.** 50 fives, 25 tens, 8 twenties

CHAPTERS 1 and 2 Test Your Memory (p. 107)

1. False **3.** True **5.** $\{0, 1, 2, 3, 4, 5\}$ **7.** $\varnothing$ **9.** Commutative property of addition; associative property of addition; inverse property of addition; identity property of addition **11.** 16 **13.** $-\frac{3}{50}$ **15.** $-\frac{41}{60}$ **17.** 9 **19.** -5

21. $\frac{12}{25}$ **23.** 6 **25.** -36 **27.** $\left\{\frac{2}{3}\right\}$ **29.** $\left\{\frac{3}{2}\right\}$ **31.** $\left\{-\frac{1}{5}\right\}$ **33.** $\{-7\}$ **35.** $\left\{\frac{1}{3}, 4\right\}$

37. $x > -2$, $(-2, \infty)$ **39.** $-3 < x < 4$, $(-3, 4)$ **41.** $-1 \le x < 6$, $[-1, 6)$ **43.** $b = \frac{2A}{h}$ **45.** 4 **47.** \$52,083.33

49. 5 twenties, 12 fives

CHAPTER 3 Exercises 3.1 (p. 115)

In 1–15 the first answer lists the terms of the polynomial, the second answer lists the coefficients, the third answer states the degree, and the fourth answer specifies what type of polynomial.

1. (i) $9x^2, 4x, -7$ (ii) $9, 4, -7$ (iii) 2nd degree (iv) trinomial **3.** (i) $23abc$ (ii) 23 (iii) 3rd degree (iv) monomial **5.** (i) $27x^3, 8y^3$ (ii) $27, 8$ (iii) 3rd degree (iv) binomial **7.** (i) 19 (ii) 19

(iii) zero degree (iv) monomial **9.** (i) $\frac{1}{2}x^3yz^4$ (ii) $\frac{1}{2}$ (iii) 8th degree (iv) monomial

11. (i) $\frac{2}{3}x, \frac{1}{8}y, -.9z$ (ii) $\frac{2}{3}, \frac{1}{8}, -.9$ (iii) 1st degree (iv) trinomial **13.** (i) $\frac{3x}{2}, -\frac{8y^2}{5}$ (ii) $\frac{3}{2}, -\frac{8}{5}$

(iii) 2nd degree (iv) binomial **15.** (i) $12x^5, -9x^4, 2x^2, x, 61$ (ii) $12, -9, 2, 1, 61$ (iii) 5th degree

(iv) polynomial **17.** $-100z^2 + 81; -100$ **19.** $-4x^4 + 17x^2 + x + 12; -4$ **21.** $7; 7$ **23.** $2y^3 + y^2 - 8y - 4; 2$

25. $-16t^2 - 32t + 300; -16$ **27.** $-\frac{8x^5}{3} + \frac{2x^3}{7}; -\frac{8}{3}$ **29.** $-18a + 4b + 20$ **31.** $-2.84x + 3.68y + .4z$ **33.** $5x^3 +$

$\frac{47}{5}y^5 - \frac{1}{5}z^2$ **35.** $-19q - 15r + 101s + t^2$ **37.** 0   **39.** $-16j + 11k$ **41.** $-12k + 13m + 42n$ **43.** $\frac{5}{9}y - \frac{11}{4}z$

45. $x^3 + 2x^2 - 14x - 12$ **47.** $-\frac{7}{15}x^2y + \frac{1}{4}x$ **49.** $7x^3 + 4x^2 - 3x + 11$ **51.** $5x^2 - 7xy - 2y + 2x - 6y^2$

53. $-3x^3 + 3x^2 - 3x - 1$ **55.** $-13a^2 - 6a - 2b + b^2$ **57.** $2x^2 - 6x + 1$ **59.** $-2x^2 - x + 20$ **61.** $-a + 2b$

63. $-2x^2 + 2$ **65.** $16cd - 8$ **67.** $5x + 12y - 15z$ **69.** $-10x^2 - 6x + 2$ **71.** $3x^2y - 2xy^2 - 12x - 6z + 23$

73. $-4.86x^2 - 1.0502y^2$ **75.** $6m + 14n - 9j + 10k$ **77.** $\frac{10}{9}a + \frac{5}{12}b$ **79. (a)** Area $= yz - 4x^2$

(b) Volume $= x(y - 2x)(z - 2x)$ **81.** Cost = \$5.70; income = \$1.60; loss = \$4.10; profit or loss $= -2x - 5y - 120$

Exercises 3.2 (p. 124)

1. x^{13} **3.** y^{18} **5.** x^{10} **7.** y^{56} **9.** x^9y^9 **11.** x^8y^{36} **13.** $40x^7y^{10}$ **15.** $-42x^6y^4$ **17.** $90x^9y^{30}$

19. $9x^2y^6$ **21.** $16x^{12}y^{48}$ **23.** $216x^9y^{15}$ **25.** $-200x^{16}y^{12}$ **27.** $12a^3b - 60a^2b$ **29.** $50x^4 - 30x^3 + 40x^2$

31. $-56m^4n^7 + 63m^3n^9 - 105m^4n^{10}$ **33.** $72x^2 + 59x + 12$ **35.** $15x^2 - 38x + 11$ **37.** $5x^2 - 13xy - 6y^2$

39. $30m^2 + 21mn - 18n^2$ **41.** $16a^2 - 1$ **43.** $4x^2 - 81$ **45.** $x^4 - y^4$ **47.** $\frac{x^2}{4} - 4y^2$ **49.** $x^2 + 8x + 16$

51. $x^2 - 14x + 49$ **53.** $9x^2 + 24xy + 16y^2$ **55.** $\frac{1}{4}a^2 - 3ab + 9b^2$ **57.** $x^4 + 2x^2 + 1$ **59.** $x^3 - 8$

61. $2x^3 - 5x^2 + x + 2$ **63.** $45x^3 - 26x^2 - 61x + 42$ **65.** $3x^3 + x^2y - 7xy^2 - 5y^3$ **67.** $5x^4 + 6x^3 - 10x^2 + 10x - 3$

69. $40x^4 - 80x^3 - 166x^2 - 28x - 63$ **71.** $6x^5 - 34x^4 - 4x^3 + 16x^2$ **73.** $4y^3 - 31y - 15$ **75.** $2x^4 + 3x^3 - 16x^2$

$- 11x + 30$ **77.** x^{3n+5} **79.** x^{2n^2+10n} **81.** $x^{2n} - 2x^n - 8$ **83.** $x^{2n} + 6x^n + 9$

Exercises 3.3 (p. 130)

1. $5(2x^2 - 7)$ **3.** $5x + 11y - 10z$ **5.** $4x^2(2x + 8y - 5z)$ **7.** $3x^3y^3(1 - 5x^2 + 3xy^2)$ **9.** $-2x^2(3x^3 + 9x + 5)$

11. $(3x - 5)(x - 7)$ **13.** $(x - 3)(2x + 1)$ **15.** $(2i - 1)(h + 3k)$ **17.** $2(x - 2y)(2x - 3y)$ **19.** $5x^2(y + 8)(3x + 1)$

21. $2yz(1 - 5x)(4y - 3z^2)$ **23.** $3x(x + y)[x(x + y) + 4]$ **25.** $(2a + 3b)(x + y)$ **27.** $(3c - d)(4a - b)$

29. $2(3f - g)(c + d)$ **31.** $(x - 2y)(2x - 1)$ **33.** $(3n - 4)(5m - n)$ **35.** $3(2x + 5y)(x - 7)$ **37.** $xy(3 - 5y)(x - 4y)$

39. $(4x - 3)(3x^2 + 1)$ **41.** $(3y + 7)(y^2 + 6)$ **43.** $3x^3(2x - 1)(x^2 - 2)$ **45.** $4x(2x^2 + 3)(x^3 + 5)$

47. $x(3x^2 - y)(4x - 7)$ **49.** $3(2b^2 - a^3)(9b - 2a)$ **51.** $x^ny^n(x^ny^n - 1)$ **53.** $x^4(x^n + 1)$

Exercises 3.4 (p. 134)

1. $(x + 3)(x - 3)$ **3.** $(z + 8)(z - 8)$ **5.** $(6z + 7b)(6z - 7b)$ **7.** $2(x + 5)(x - 5)$ **9.** $3xy(3x + 2y)(3x - 2y)$

11. $4x^2 + 9$ is prime   **13.** $3(x^2 + 16)$ **15.** $(9y^2 + 4z^2)(3y + 2z)(3y - 2z)$ **17.** $(x + 4)(x - 10)$

19. $(x + 4 + 2y)(x + 4 - 2y)$ **21.** $(2x - 3y + 3z)(2x - 3y - 3z)$ **23.** $(y - 4)(y^2 + 4y + 16)$

25. $(2x + 1)(4x^2 - 2x + 1)$ **27.** $(2m + 3n)(4m^2 - 6mn + 9n^2)$ **29.** $(4x - 3y)(16x^2 + 12xy + 9y^2)$

31. $3(5y - 1)(25y^2 + 5y + 1)$ **33.** $2(3r + 4s)(9r^2 - 12rs + 16s^2)$ **35.** $(y - 5)(y^2 - y + 7)$

37. $(x - 6)(7x^2 + 15x + 21)$ **39.** $9x(x^2 - xy + y^2)$ **41.** $(x + y)(x - y)(x^2 - xy + y^2)(x^2 + xy + y^2)$

43. $(2x - 3)(x + 1)(x - 1)$ **45.** $(r - 6)(r - 2)(r^2 + 2r + 4)$ **47.** $(5x + 2)(x + 3)(x - 3)$

49. $(a + b)(a - b)(3a - 4b)$ **51.** $(3x - 5)(x - 1)(x^2 + x + 1)$ **53.** $(4y - 3)(y + 2)(y^2 - 2y + 4)$

55. $(2x + 1)(2x - 1)(x - 2)(x^2 + 2x + 4)$ **57.** $7x(x + 1)(x - 1)(x + 2y)(x - 2y)$

59. $5y(y + 1)(y^2 - y + 1)(x - 2y)(x^2 + 2xy + 4y^2)$ **61.** $(x + 2)(x + 2)(x + 2)$

63. $(x^n + 1)(x^n - 1)$ **65.** $(x^{2n} + 9)(x^n + 3)(x^n - 3)$ **67.** $(x^n + 3)(x + 5)$

69. $(x + 2)(x - 2)(3x + y)$ **71.** $3xy^2(x^2 + 2 - 5y)$ **73.** $3y^4(2x - 5)(x + 4y)$ **75.** $5x(x + 1)(x - 1)$

77. $(25 + m^2)(5 + m)(5 - m)$

Exercises 3.5 (p. 142)

1. $(x - 2)(x - 7)$ **3.** $(x - 7)(x + 5)$ **5.** $(x - 1)(x - 8)$ **7.** $(x + 10)(x - 3)$ **9.** $x^2 - x + 7$ is prime

11. $(x - 9)(x + 4)$ **13.** $(x - 3)^2$ **15.** $(x + 9)^2$ **17.** $3(x + 1)(x + 7)$ **19.** $-4(x - 7)(x + 4)$ **21.** $-2x(x + 4)^2$

23. $6x(x^2 + 3x - 1)$ **25.** $3x^2y^2(x^2 + 3x + 1)$ **27.** $(x - 3y)(x - 2y)$ **29.** $(x + y)(x + 5y)$ **31.** $(x + 10y)(x - 2y)$

33. $(3x + 1)(2x - 5)$ **35.** $2x^2 + 5x - 14$ is prime   **37.** $(4x - 1)(3x + 7)$ **39.** $6x^2 - 9x - 4$ is prime

41. $(3x - 1)(3x - 4)$ **43.** $(4x - 3)(2x + 9)$ **45.** $(3x + 5)^2$ **47.** $8x^2 - 7x + 20$ is prime   **49.** $(9x - 1)(x + 2)$

51. $(5x + 1)^2$ **53.** $(6x + 5)(x - 3)$ **55.** $2(8x + 3)(x - 2)$ **57.** $5(2x - 7)(2x + 1)$ **59.** $2x(5x^2 - 7x + 4)$
61. $3x(5x - 1)(x + 3)$ **63.** $3(3x + 4y)^2$ **65.** $(9x + 2y)(x - 2y)$ **67.** $2(4x - 3y)(x - y)$ **69.** $(x^2 + 3)(x^2 + 9)$
71. $(x + 2)(x - 2)(x^2 + 3)$ **73.** $(2x^2 + 5)(2x^2 - 3)$ **75.** $(3x - 4)(3x - 1)$ **77.** $(6x - 7)(4x + 1)$
79. $(3x + 6y - 8)(2x + 4y - 1)$ **81.** $(x + 3 + y)(x + 3 - y)$ **83.** $(2x + y + 7)(2x - y - 7)$
85. $(2x + y + z - 1)(2x + y - z + 1)$ **87.** $(x - 1 + y + z)(x - 1 - y - z)$ **89.** $(3x + 2 + y - 2z)(3x + 2 - y + 2z)$
91. $(x - 4)(x + 3 + 5y)$ **93.** $(2x - 1)(x - 5 - 3k)$ **95.** $(x + y)(x + 3y + z)$ **97.** $(x^n - 3)(x^n - 5)$
99. $(2x^n + 1)(x^n - 3)$ **101.** $(3x + 5)(x + 2)$ **103.** $(4x - 7)(x + 2)$ **105.** $(6x + 5)(x - 2)$ **107.** $(3x - 4)(3x + 1)$
109. $(4x + 1)(x + 10)$ **111.** $(4x + 1)(3x - 7)$ **113.** $(8x - 5)(x - 2)$ **115.** $(6x - 5)(2x - 3)$ **117.** $(12x + 5)(x - 2)$
119. $4(3 + y)(3 - y)$ **121.** $6(2x - 1)(x + 2)$ **123.** $(y + 4)(y^2 - 4y + 16)$ **125.** $(y + 8)(y + 1)(y - 1)$
127. $(3x - 4)(3x - 1)$ **131.** $(x^2 + 2x + 2)(x^2 - 2x + 2)$ **133.** $(8x^2 + 4x + 1)(8x^2 - 4x + 1)$
135. $(x^2 + x + 7)(x^2 - x + 7)$ **137.** $(x^2 + 2x - 5)(x^2 - 2x - 5)$ **139.** $(2x^4 + 2x^2y + y^2)(2x^4 - 2x^2y + y^2)$

Exercises 3.6 (p. 145)

1. $(x - 2)(x - 1)$ **3.** $(2x + 1)(x + 3)$ **5.** $(2x + 3y)(4x^2 - 6xy + 9y^2)$ **7.** $(x^2 + 3)(2x + 5)$ **9.** $2(6x + 1)(x - 3)$
11. $2x(x - 1)(x - 9)$ **13.** $(5x + 2)(x - 1)$ **15.** $(x - 4y)(x + 3y)$ **17.** $(5x + 10y + 1)(x + 2y + 6)$
19. $(x + 6)(x - 6)$ **21.** $2xy^3(3x + y)(3x - 2y)$ **23.** $(2x + 2y + 3)(2x - 2y - 3)$ **25.** $(x + 3 + 4y)(x + 3 - 4y)$
27. $(4x - 1)(16x^2 + 4x + 1)$ **29.** $(y + 4)(y - 2)$ **31.** $(2x - 2y - 1)(2x - 2y - 3)$ **33.** $(x + 3y)^2$
35. $(2x + 5)(2x - 5)(x^2 + 2)$ **37.** $(x - 2 + 3y)(x - 2 - 3y)$ **39.** $(5x + 4y)(5x - 4y)$ **41.** $(x + 1)(x - 1)(3x + y)$
43. $(x + 3 + 3y)(x + 3 - 3y)$ **45.** $(3x - 2y)(x - 2y)$ **47.** $6ab^3(3a^2 - 2b^2 - 1)$ **49.** $(4x + 3y)(2x - 3y)$
51. $4(x - 3)^2$ **53.** $(x + 1)(x - 1)(x + 3)(x - 3)$ **55.** $(x + y + m + 3)(x + y - m - 3)$
57. $(x + 4 + 5y)(x^2 + 8x + 16 - 5xy - 20y + 25y^2)$

Exercises 3.7 (p. 150)

The quadratic, linear, and constant terms are listed, in order, following the equation.
1. $5x^2 - 7x + 1 = 0; 5x^2, -7x, 1$ **3.** $-6x^2 + 5x - 1 = 0; -6x^2, 5x, -1$ **5.** $2x^2 + 14 = 0; 2x^2, 0x, 14$
7. $7x^2 + 2x + 4 = 0; 7x^2, 2x, 4$ **9.** $4x^2 - 25 = 0; 4x^2, 0x, -25$ **11.** $x^2 + 4x + 9 = 0; x^2, 4x, 9$
13. $x^2 - x - 12 = 0; x^2, -x, - 12$ **15.** $6x^2 - 40x - 14 = 0; 6x^2, -40x, -14$ **17.** $3x^2 - 19x + 18 = 0; 3x^2, -19x, 18$
19. $-3x^2 + 17x + 29 = 0; -3x^2, 17x, 29$ **21.** $\{-4, -2\}$ **23.** $\{7, -5\}$ **25.** $\{-8, 2\}$ **27.** $\{0, -2\}$ **29.** $\left\{0, \frac{1}{3}\right\}$
31. $\{5, -5\}$ **33.** $\left\{\frac{3}{2}, -\frac{3}{2}\right\}$ **35.** $\left\{\frac{1}{4}, -\frac{1}{4}\right\}$ **37.** $\left\{-\frac{1}{2}, 4\right\}$ **39.** $\left\{-\frac{3}{8}, -2\right\}$ **41.** $\left\{-\frac{1}{6}, 7\right\}$ **43.** $\left\{-\frac{3}{4}, \frac{1}{2}\right\}$
45. $\left\{\frac{9}{4}, \frac{1}{3}\right\}$ **47.** $\left\{\frac{2}{5}, 1\right\}$ **49.** $\{-3\}$ **51.** $\{4\}$ **53.** $\{-7, 2\}$ **55.** $\left\{-\frac{5}{2}, 3\right\}$ **57.** $\left\{-\frac{1}{4}, \frac{5}{2}\right\}$ **59.** $\left\{\frac{5}{3}, 8\right\}$
61. $\{-4\}$ **63.** $\left\{\frac{2}{3}, -\frac{2}{3}\right\}$ **65.** $\left\{\frac{1}{3}, -\frac{7}{2}\right\}$ **67.** $\{-8, 3\}$ **69.** $\{2, -5\}$ **71.** $\left\{\frac{3}{5}, -\frac{4}{3}\right\}$ **73.** $\{2, 4\}$ **75.** $\left\{-\frac{2}{3}, 9\right\}$
79. $x = 1$ or 8; sum 65 **81.** $x = 5$ or 6; sum 61 **83.** $x = 3$ or 4; sum 25 **85.** $x = 3$ or 5; sum 34

Exercises 3.8 (p. 155)

1. 8 and 9 **3.** 6 and 8; -8 and -6 **5.** 3 and 4 **7.** 3 and 7; $-\frac{7}{2}$ and -6 **9.** 4 and -3; 3 and -4
11. 3 ft $\times$ 14 ft **13.** 6 in. and 8 in. **15.** Base = 8 in., height = $\frac{7}{2}$ in. **17.** After 5 sec = 1520 ft; after 12 sec =
2304 ft; hits ground at 24 sec. **19.** After 4 sec **21.** After 4 sec **23.** $\frac{3}{2}$ ft **25.** 6 in. $\times$ 10 in.

APPLIED ALGEBRA—YOUR TURN (p. 157)

1. 80 mi/hr

CHAPTER 3   Review Exercises (p. 159)

In 1–5 the first answer lists the terms of the polynomial; the second answer lists the coefficients; the third answer specifies what type of polynomial; and the fourth answer states the degree.

1. (i) $5x^2, -x, 2$ (ii) $5, -1, 2$ (iii) trinomial   (iv) 2nd degree   **3.** (i) $\frac{y}{3}, x$ (ii) $\frac{1}{3}, 1$ (iii) binomial

(iv) 1st degree   **5.** (i) $y^3, \frac{3xy}{2}, -\frac{x^2y}{4}, \frac{2}{3}$ (ii) $1, \frac{3}{2}, -\frac{1}{4}, \frac{2}{3}$ (iii) polynomial   (iv) 3rd degree   **7.** $10x^2 - 1.5y$

9. $-y^3 - 5y^2 + 4y + 3$ **11.** $x^2 + 9xy - 9y^2 + 4y$ **13.** $-5.2a - 3.5b + .25c + 11.6$ **15.** $-3x^2 - 9xy + 7y^2 +$

$3x + 2y$ **17.** $a - \frac{3b}{2} - \frac{19}{6}$ **19.** $8x^2 - 7y^2$ **21.** x^{17} **23.** $28a^5b^9$ **25.** $-72x^{11}y^{14}$ **27.** $-10x^4 + 2x^3 - 6x^2$

29. $4x^2 + 27xy + 35y^2$ **31.** $4a^2 - 81$ **33.** $x^2 - 25$ **35.** $y^2 - 2y + 1$ **37.** $64x^3 + 1$ **39.** $y^4 + 4y^2 + 16$

41. $2x^2yz^3(5xyz - 4y^4 - 7x^2z^4)$ **43.** $3ax(2a^2 - ax + 1)$ **45.** $x(x - 9)$ **47.** $(x + 6)(x - 6)$ **49.** $y^2 + 16$ is prime

51. $2x(x + 3)(x - 3)$ **53.** $(y^2 + 9)(y + 3)(y - 3)$ **55.** $x^2 - 3x + 4$ is prime   **57.** $(x - 5y)(x + 2y)$

59. $(x - 7)(x - 7)$ **61.** $(3x - 4)(x - 3)$ **63.** $(6x + 1)(3x + 4)$ **65.** $3x(3x + 5)(2x - 3)$ **67.** $(x + 1)(x - 1)(x^2 + 5)$

69. $9x(x + 2)$ **71.** $(x + 3)(x^2 - 3x + 9)$ **73.** $(y - 1)(y^2 + y + 1)$ **75.** $4x(7x^2 - 12xy + 12y^2)$

77. $(5xy^2 + 1)(3x + y)$ **79.** $(2a + b)(4a^2 - 2ab + b^2)(a + 2b)$ **81.** $(4x + 3)(2y + 5)$ **83.** $(3y - 2)(2x + 3y)(2x - 3y)$

85. $(x - 4y - 5z)(x - 4y + 5z)$ **87.** $(x + 2)^3$ **89.** $(6x + 5)(2x - 3)$ **91.** $(8x - 9)(2x + 1)$

The quadratic, linear, and constant terms are listed, in order, following the equation.

93. $7x^2 - 5x + 1 = 0; 7x^2, -5x, 1$ **95.** $3x^2 - 7x - 40 = 0; 3x^2; -7x, -40$ **97.** $10x^2 - 20x + 3 = 0; 10x^2, -20x, 3$

99. $3x^2 + 7x - 37 = 0; 3x^2, 7x, -37$ **101.** $\{-2, 13\}$ **103.** $\{2, 7\}$ **105.** $\{0, 4\}$ **107.** $\{-6, 6\}$ **109.** $\{-2, 2\}$

111. $\{4\}$ **113.** $\left\{\frac{8}{3}, 3\right\}$ **115.** $\left\{-\frac{7}{3}, -\frac{1}{2}\right\}$ **117.** $\{-9, 1\}$ **119.** $\left\{-1, -\frac{5}{6}\right\}$ **121.** $\left\{-\frac{3}{2}, 9\right\}$ **123.** $\left\{-\frac{6}{7}, 2\right\}$

125. $\{-6, 10\}$ **127.** $\{-2, -1\}$ **129.** 11 and 12   **131.** $\frac{9}{2}$ ft $\times$ 8 ft   **133.** base = 4 cm, height = 14 cm

135. After $\frac{1}{2}$ sec, 4 ft high; lands on ground after 1 sec   **137.** 2 yd

CHAPTER 3   Test (p. 162)

1. Terms: $x^3, 5x^2, -x, 7$; coefficients: $1, 5, -1, 7$; third-degree polynomial   **3.** $2a^2b - 2a - 2b - 5a^2 - 2ab^2 - 3ab$

5. $-10x^5y^6 + 15x^8y^5$ **7.** $6x^3 - 21x + 99$ **9.** $(3x + 4)(x + 2)$ **11.** $x^2 - 4x + 5$ is prime.

13. $(4x + 1)(4x - 1)(x - 3)$ **15.** $(3x - 2y - 5)(3x - 2y + 5)$ **17.** $5a(2a^3x^3 - ax + a^5)$ **19.** $(3x + 2)(3x - 2)$

21. $\{-2, 10\}$ **23.** $\left\{-\frac{4}{3}, -\frac{1}{3}\right\}$ **25.** After 2 sec

CHAPTERS 1–3   Test Your Memory (p.164)

1. True   **3.** $\{0, 5, 10, 15, 20, 30\}$ **5.** -8 **7.** $\frac{5}{12}$ **9.** 8   **11.** 9   **13.** $5x^3 - 5x^2 + 8x + 11$

15. $7x^2 - 19xy - 6y^2$ **17.** $(4x + 3)(2x - 5)$ **19.** $3a(5a + 2)(25a^2 - 10a + 4)$ **21.** $4x(2x^2 + 3xy - 1)$

23. $(2m - 3n + 2)(2m - 3n - 2)$ **25.** $\{2, 9\}$ **27.** $\{-2\}$ **29.** $\left\{\frac{1}{2}\right\}$ **31.** $\left\{-\frac{7}{2}\right\}$ **33.** $\{-1, 11\}$ **35.** $\left\{-\frac{1}{4}, \frac{1}{2}\right\}$

37. $\{0, 3\}$ **39.**
$$x \le -4$$
$$(-\infty, -4]$$
⟵+++++●++++⟶
$-8\,-7\,-6\,-5\,-4\,-3\,-2\,-1\ 0$
   41.
$$x < -1 \text{ or } x > 5$$
$$(-\infty, -1) \cup (5, \infty)$$
⟵+++○++++++○+⟶
$-3\,-2\,-1\ 0\ 1\ 2\ 3\ 4\ 5\ 6\ 7$
   43. 5 ft by 13 ft   **45.** 6 m and 8 m   **47.** $\frac{3}{2}$ ft

49. 18 nickels, 13 dimes

CHAPTER 4 Exercises 4.1 (p. 175)

1. 32 **3.** 25 **5.** $-\dfrac{1}{7}$ **7.** 256 **9.** 16 **11.** $\dfrac{1}{216}$ **13.** -64 **15.** 121 **17.** $\dfrac{1}{81}$ **19.** $-\dfrac{1}{64}$ **21.** $\dfrac{3}{8}$

23. $\dfrac{1}{2}$ **25.** $\dfrac{50}{23}$ **27.** $\dfrac{73}{8}$ **29.** $\dfrac{8}{27}$ **31.** 1 **33.** $\dfrac{8}{5}$ **35.** $\dfrac{9}{16}$ **37.** $\dfrac{27}{25}$ **39.** 225 **41.** $\dfrac{7}{2}x^2 y$ **43.** $\dfrac{8x^9}{y^{15}}$

45. $4x^{12}y^2$ **47.** $\dfrac{21}{x^3 y^5}$ **49.** $\dfrac{1}{48y^8}$ **51.** $\dfrac{8x^3}{y^6}$ **53.** $\dfrac{25x^2}{4y^6}$ **55.** $\dfrac{8x^{12}y^6}{125}$ **57.** $\dfrac{125y^9}{27x^{15}}$ **59.** $\dfrac{4x^9}{5y^7}$ **61.** $\dfrac{2}{x^2 y^2}$

63. $\dfrac{16x^{12}}{25y^{30}}$ **65.** $\dfrac{-432y^{33}}{x^4}$ **67.** $\dfrac{9y^2}{25x^8}$ **69.** $\dfrac{16x^4}{9y^8}$ **71.** $\dfrac{9}{4x^6 y^2}$ **73.** $\dfrac{1}{216x^{24}y^{12}}$ **75.** $\dfrac{8x^6}{27y^9}$ **77.** $2x - y$

79. $\dfrac{1}{x^2 + 6x + 9}$ **81.** $\dfrac{1}{x^2 + 2xy + y^2}$ **83.** $\dfrac{x^2 - y^2}{xy}$ **85.** x^{10m} **87.** $\dfrac{1}{x^{8m}}$ **89.** x^{4m} **91.** $(x - 7)^4 (2x - 3)^3 (18x - 71)$

93. $\dfrac{4(1 - x)}{(x + 1)^3}$ **95.** $\dfrac{-3}{(x - 2)^2}$

Exercises 4.2 (p. 181)

1. $\dfrac{4x^3}{3y}$ **3.** $5x^4 y^6$ **5.** $\dfrac{-9c^2}{4b}$ **7.** $\dfrac{7x^2}{4y^2 z^4}$ **9.** $-\dfrac{1}{4x^2 y^4 z^6}$ **11.** $\dfrac{x(3x + y)}{y^2(x - 2y)}$ **13.** $\dfrac{x + 2y}{x}$ **15.** $-\dfrac{2x}{y}$ **17.** $\dfrac{x + 2}{x - 7}$

19. $\dfrac{3x - 1}{2x - 3}$ **21.** $\dfrac{x^2 - 2x - 3}{x^2 + 6x + 8}$ **23.** $-\dfrac{x + 5}{x + 2}$ **25.** $\dfrac{2x - 3}{x - 4}$ **27.** $\dfrac{x}{2x - 1}$ **29.** $\dfrac{x^2 + x - 2}{2x(x + 5)}$ **31.** $-\dfrac{x + 4}{2x + 3}$

33. $\dfrac{x^2 - 2x + 4}{3x - 4}$ **35.** $\dfrac{9x^2 + 3x + 1}{2(x + 1)}$ **37.** $\dfrac{2x - y}{x + y}$ **39.** $-\dfrac{3a - 4}{a + 2}$ **41.** $\dfrac{x^2 + y}{x + y^2}$ **43. (a)** $\dfrac{x - 3}{x - 2}$ **(b)** $2, 0, \dfrac{5}{4}$

(c) $2, 0,$ undefined **(d)** No, the reduction eliminated a division by zero.

Exercises 4.3 (p. 186)

1. $\dfrac{10x^4 y^3}{21a^5 b}$ **3.** $\dfrac{9b}{2xy}$ **5.** $\dfrac{1}{2x}$ **7.** $\dfrac{1}{7ax}$ **9.** 1 **11.** $5xy^2$ **13.** $9x^2 y^4$ **15.** $\dfrac{5ax^5}{4y^3}$ **17.** $2axyz^2$ **19.** $\dfrac{2x}{7a^2 by^2}$

21. $\dfrac{4bx^2}{3a^5 y^3}$ **23.** $\dfrac{10b}{3a^2 x^2}$ **25.** $\dfrac{35b^3 x^3}{3a^4 y^2}$ **27.** $\dfrac{12a^2 b}{7c^2 d^4}$ **29.** $\dfrac{2}{5y}$ **31.** $-\dfrac{3}{4y}$ **33.** $-5xy^2$ **35.** $\dfrac{2x + 1}{x + 5}$ **37.** $\dfrac{2x - 3}{x + 8}$

39. $\dfrac{5x + 2}{3x - 2}$ **41.** $-\dfrac{x + 2}{5x}$ **43.** $-\dfrac{1}{4x}$ **45.** $-\dfrac{x + 3}{2x - 1}$ **47.** $-\dfrac{x - 2}{x + 4}$ **49.** $3x + 2$ **51.** $\dfrac{1}{(2x + 1)(x - 4)}$ **53.** $\dfrac{x - y}{x + y}$

55. $\dfrac{x^2 + 2x + 4}{x + 3}$ **57.** $\dfrac{x^2 - xy + y^2}{2x - y}$ **59.** $\dfrac{(a - 3b)(x - 3)}{(a - b)(x + 1)}$ **61.** 1 **63.** $-\dfrac{2x + 3}{3x^2(3x + 2)}$ **65.** $\dfrac{x^2 + x + 1}{4x^2}$

Exercises 4.4 (p. 192)

1. $\dfrac{5}{x}$ **3.** $\dfrac{2x}{3a}$ **5.** $-\dfrac{7k}{st}$ **7.** $\dfrac{12x}{2x + 3}$ **9.** $\dfrac{-2x - 5}{4x - 7}$ **11.** 2 **13.** $\dfrac{3x - 8}{2x^2 + x - 11}$ **15.** $\dfrac{1}{2x - 1}$ **17.** $x + 1$

19. $x + 4$ **21.** $\dfrac{16y + 15x}{10xy}$ **23.** $\dfrac{22y - 3cx}{10xy}$ **25.** $\dfrac{9x + 28y}{12x^2 y^2}$ **27.** $\dfrac{7x + 2y}{y}$ **29.** $\dfrac{7x + 19}{(x + 3)(x + 2)}$ **31.** $\dfrac{-7x - 34}{(2x + 5)(x - 3)}$

33. $\dfrac{9x^2 + 4x}{(x + 4)(3x - 4)}$ **35.** $\dfrac{-16x^2 - 18x}{(5x + 3)(2x - 3)}$ **37.** $\dfrac{-x^2 + 17x + 3}{(2x - 3)(x + 6)}$ **39.** $\dfrac{8x^2 + 6x - 7}{(2x + 5)(2x - 3)}$ **41.** $\dfrac{3x + 5}{x + 1}$ **43.** $\dfrac{4x - 5}{x - 4}$

45. $\dfrac{-4x + 15}{x(x - 5)(x + 5)}$ **47.** $\dfrac{x - 3}{x(x + 3)}$ **49.** $\dfrac{7x + 11}{(x + 1)(x + 3)(x - 5)}$ **51.** $\dfrac{4x^2 + 15x + 2}{(3x + 1)(x + 4)(x - 2)}$ **53.** $\dfrac{5x - 2}{(x - 4)(x + 2)}$

55. $\dfrac{3x^2 + 5x - 21}{(2x - 3)(x + 1)(x + 3)}$ **57.** $\dfrac{3x^2 - 8x - 17}{(x + 1)(x - 1)(x - 5)}$ **59.** $\dfrac{x^2 - 24x + 11}{(x - 3)^2(x + 1)}$ **61.** $\dfrac{5x + 6}{(3x + 5)(x + 4)}$ **63.** $\dfrac{x - 19}{(2x - 5)(x - 4)}$

65. $\dfrac{2x + 1}{(3x - 2)(x - 3)}$ **67.** $\dfrac{2x + 7}{(x + 3)(x + 2)}$ **69.** $\dfrac{-4x^2 - 25x - 24}{(4x + 1)(2x - 3)(2x + 5)}$ **71.** $\dfrac{x + 7}{(3x + 2)(2x - 5)}$ **73.** $\dfrac{3x^2 + 6xy}{(2x + y)(x + y)(x - y)}$

75. $\dfrac{x + 8y}{(x + 3y)(x + 4y)}$ **77.** $\dfrac{-x^2 - 19xy - 5y^2}{(2x + 3y)(x - y)^2}$ **79.** $\dfrac{-x + 57}{12(x + 1)}$ **81.** $\dfrac{5x^2 - 6x - 2}{(2x - 1)(x + 3)(x - 1)}$ **83.** $\dfrac{2x^2 + 4x + 47}{(x + 5)(x + 4)(x - 4)}$

85. $\dfrac{2x - 11}{(x + 2)(x - 2)}$

Exercises **4.5** (p. 200)

1. $\dfrac{16}{63}$ **3.** $\dfrac{5x}{4y}$ **5.** $\dfrac{6y}{5x^3}$ **7.** $\dfrac{8a^3b^2}{55x^4y^2}$ **9.** $\dfrac{(5x+3)(x-2)}{(2x-1)(x+4)}$ **11.** 2   **13.** $\dfrac{3y}{4x^2}$ **15.** -1 **17.** $\dfrac{2(x+7)}{x(x-1)}$

19. $-\dfrac{3xy^2}{2}$ **21.** 1   **23.** $\dfrac{(x+4)(x+5)}{2x^2(x+3)}$ **25.** $\dfrac{2x+3}{x-7}$ **27.** $\dfrac{2y+7x}{12xy}$ **29.** $\dfrac{xy(x+y)}{2}$ **31.** $\dfrac{30y-5xy}{20x-2xy}$ **33.** $\dfrac{xy}{x+y}$

35. $\dfrac{y^2-x^2}{x^2y^2}$ **37.** $\dfrac{2x+1}{4x-3}$ **39.** $\dfrac{xy}{x+y}$ **41.** xy **43.** $\dfrac{5y-2x}{xy(x-1)}$ **45.** $-\dfrac{3x+1}{9x^2}$ **47.** $-\dfrac{1}{4x}$ **49.** $-\dfrac{x+3}{3x^2}$

51. $-\dfrac{2}{2+h}$ **53.** $-\dfrac{5}{xy}$ **55.** $\dfrac{x-1}{-x+3}$ **57.** $\dfrac{4(x+3)(x-2)}{6x+13}$ **59.** $-\dfrac{x+5}{4}$ **61.** 3   **63.** $\dfrac{6x-2}{-2x-1}$

65. $\dfrac{2x^2+9x-18}{x^2+x-2}$ **67.** $\dfrac{3x-3}{4x-5}$ **69.** $\dfrac{x-2}{2x+1}$ **71.** $\dfrac{xy}{y+x}$ **73.** $\dfrac{1}{x(x+5)}$ **75.** $\dfrac{5}{2x}$ **77.** $\dfrac{x+3}{x+2}$

Exercises **4.6** (p. 206)

1. $5x^3-7x^2+4$ **3.** x^2+2x-8 **5.** $5x^2-7x-1$ **7.** $5x^2-8xy-y^2$ **9.** $-7xy+\dfrac{2x}{y^2}+\dfrac{1}{xy}$

11. $\dfrac{3y^2}{5x}+\dfrac{9x}{2y^2}-\dfrac{7}{10xy}$ **13.** $\dfrac{3}{8xyz}-\dfrac{5x^2}{4yz}-\dfrac{9z^3}{x^2y}$ **15.** $\dfrac{5z}{2x^2y}+\dfrac{7x^2}{6y^3z}+\dfrac{y^3}{3xz^4}$ **17.** $5x+4-\dfrac{2}{x-3}$ **19.** $6x-5+\dfrac{1}{4x-7}$

21. $8x-4+\dfrac{5}{5x-3}$ **23.** $9x-8$ **25.** $x^2-8x+7-\dfrac{2}{2x+3}$ **27.** $3x^2-2-\dfrac{5}{4x+7}$ **29.** $8x^2+3x+2+\dfrac{4}{5x+2}$

31. $5x+10+\dfrac{7}{x-2}$ **33.** $2x+1$ **35.** $3x+2-\dfrac{1}{3x-2}$ **37.** $5x-2$ **39.** $2x^2+3x+4+\dfrac{5}{2x-3}$

41. $3x^2+x-3-\dfrac{1}{2x+6}$ **43.** $2x^2-x+4+\dfrac{2x-1}{x^2+x-3}$ **45.** $x^2+2x+3+\dfrac{5x-1}{2x^2-x-1}$ **47.** x^2+2x+4

49. $9x^2-12x+16$ **51.** x^3-2x^2+4x-8 **53.** $3x-5+\dfrac{2}{x^2+2x+1}$ **55.** $4x-7$ **57.** $x^2-6x+2+\dfrac{3x+5}{2x^2-4}$

59. $x^4+3x^2+4+\dfrac{5x+1}{2x^2-3}$

Exercises **4.7** (p. 211)

1. $x-1+\dfrac{3}{x-2}$ **3.** $2x-4-\dfrac{1}{x+2}$ **5.** $9x+\dfrac{4}{x-\frac{1}{3}}$ **7.** $2x^2-x-4-\dfrac{6}{x-3}$ **9.** $x^2-2x-3+\dfrac{2}{x+2}$

11. $3x^2-x-1+\dfrac{4}{x+4}$ **13.** $4x^2-2x-4+\dfrac{3}{x+\frac{1}{2}}$ **15.** $6x^2+9-\dfrac{2}{x-\frac{2}{3}}$ **17.** $x^3-4x^2+3x-5+\dfrac{15}{x+2}$

19. $2x^3-x^2+7+\dfrac{2}{x-1}$ **21.** $16x^3-12x^2+4-\dfrac{5}{x+\frac{3}{4}}$ **23.** $x^2-5x+25$ **25.** $x^5+2x^4+4x^3+8x^2+16x+32$

27. $x+\sqrt{5}$ **29.** $x^3-\sqrt{3}x^2+3x-3\sqrt{3}$

33. -20 **35.** -108 **37.** -5

$(x+1)(x-1)(2x-5)$ $4x(x+3)(x-3)$ $5x(x+1)(x-1)$

$(x-4)(x+1)(x-1)(2x-5)$ $4x(x+3)^2(x-3)$ $5x(x+1)(x-1)(x+2)$

39. 45   **41.** 2

$(x+1)(3x+1)(3x-1)$ $(x+3)(x-3)(2x-1)$

$(x+4)(x+1)(3x+1)(3x-1)$ $(x-2)(x+3)(x-3)(2x-1)$

Exercises **4.8** (p. 215)

1. $\left\{-\dfrac{4}{3}\right\}$ **3.** $\{2\}$ **5.** $\varnothing$ **7.** $\{-7\}$ **9.** $\{-3\}$ **11.** $\{4\}$ **13.** $\left\{\dfrac{3}{2}\right\}$ **15.** All real numbers except $x=3$

17. $\varnothing$ **19.** $\left\{\dfrac{3}{4}\right\}$ **21.** $\{-3,2\}$ **23.** $\left\{-\dfrac{5}{3},\dfrac{3}{2}\right\}$ **25.** $\{3\}$ **27.** $\left\{\dfrac{7}{2}\right\}$ **29.** $\left\{\dfrac{-4}{3}\right\}$ **31.** $\left\{-\dfrac{5}{3},-1\right\}$ **33.** $\{-1\}$

35. $\left\{-\frac{5}{3}, \frac{4}{3}\right\}$ **37.** $\{2\}$ **39.** $\varnothing$ **41.** $\left\{-\frac{5}{2}, 3\right\}$ **43.** $\left\{-1, \frac{16}{3}\right\}$ **45.** $\left\{-\frac{9}{7}, 4\right\}$ **47.** $\{-1, 18\}$ **49.** $\{-2, 10\}$

51. $\left\{-\frac{8}{7}, 1\right\}$ **53.** $\varnothing$ **55.** $\{-3, 2\}$ **57.** $\{1, 16\}$ **59.** $\dfrac{5x - 4}{10x(x - 2)}$ **61.** $\{-7, -1\}$ **63.** $\dfrac{(x + 1)(3x - 4)}{(2x - 3)(x + 2)}$

65. $\left\{-\frac{1}{3}\right\}$ **67.** $\{5\}$ **69.** $\{7\}$ **71.** $\dfrac{-x - 23}{(2x + 1)(x - 4)}$ **73.** $\left\{-\frac{5}{8}\right\}$ **75.** $\dfrac{7}{x - 2}$ **77.** $\left\{-\frac{18}{5}, 1\right\}$ **79.** $\dfrac{5(5x + 3)}{3x^3}$

81. $\dfrac{-2x + 27}{6x}$ **83.** $\dfrac{x^2 + 13x - 2}{(x + 1)(x - 1)(2x + 1)}$ **85.** $\dfrac{-4x^2}{x + 5}$ **87.** $\{-8, 3\}$

Exercises 4.9 (p. 224)

1. 3 **3.** -3 **5.** $\dfrac{1}{5}$ **7.** $\dfrac{3}{4}$ or $\dfrac{4}{3}$ **9.** 3 and 4 **11.** 6 and 8 **13.** $1\frac{1}{5}$ hr (1 hr 12 min) **15.** $\dfrac{2}{3}$ hr (40 min)

17. 20 hr **19.** Barbara—12 hr; Gary—24 hr **21.** 3 min **23.** 2 and 3 hr **25.** 2.5 mi/hr **27.** 10 mi/hr

29. 90 mi/hr **31.** $1\frac{1}{2}$ hr **33.** 2 mi **35.** 48 mi

APPLIED ALGEBRA—YOUR TURN (p. 227)

1. 810 bass

CHAPTER 4 Review Exercises (p. 228)

1. 625 **3.** 216 **5.** 64 **7.** 1 **9.** -81 **11.** 256 **13.** $-\dfrac{1}{125}$ **15.** $\dfrac{1}{4096}$ **17.** $\dfrac{3}{4}$ **19.** $-\dfrac{19}{16}$ **21.** $\dfrac{64}{125}$

23. $\dfrac{216}{125}$ **25.** 1 **27.** $\dfrac{25}{9}$ **29.** 256 **31.** $\dfrac{8x^{17}}{25y^{16}}$ **33.** $\dfrac{x^{10}}{36}$ **35.** $\dfrac{49x^4}{16y^2}$ **37.** $\dfrac{4}{25x^{10}y^8}$ **39.** $\dfrac{-125}{216x^6y^9}$

41. $\dfrac{3y^9}{2x^{16}}$ **43.** $-\dfrac{2y^5}{x^5}$ **45.** $\dfrac{4y^{18}}{x^6}$ **47.** $\dfrac{9x^8}{25y^4}$ **49.** $\dfrac{1}{y^2 + 8y + 16}$ **51.** x^{6m} **53.** x^{7m} **55.** $\dfrac{5}{(x + 3)^2}$ **57.** $\dfrac{4x^4}{5y^5}$

59. $-\dfrac{y^2z}{5}$ **61.** $\dfrac{x(3 - 2xy^4)}{2y(2y + 3x^3)}$ **63.** $-\dfrac{x - 3}{x}$ **65.** $\dfrac{2(x - 2)}{x - 5y}$ **67.** $\dfrac{x - 3}{2x - 3}$ **69.** $\dfrac{4a^2 - 2ab + b^2}{4a + b}$

71. $-\dfrac{(x + 2y)(x - 2y)}{x + y}$ **73.** $\dfrac{3y^5}{x^3}$ **75.** $25a^2c^9$ **77.** $\dfrac{14x^5}{by^3z^4}$ **79.** $\dfrac{2x(x + 3)}{3(x - 3)}$ **81.** $-\dfrac{2x + y}{3x + y}$ **83.** $\dfrac{1}{x + 2}$

85. $-\dfrac{a - 3b}{3ab}$ **87.** $\dfrac{1}{x + 3}$ **89.** $\dfrac{2(x - 5)}{x(x + 5)}$ **91.** $\dfrac{3}{xy}$ **93.** $\dfrac{10x^2}{5x - 1}$ **95.** $\dfrac{5}{x - 7}$ **97.** $\dfrac{1}{y - 4}$ **99.** $\dfrac{5x + 12}{3x^2}$

101. $\dfrac{2x - 3}{x - 5}$ **103.** $\dfrac{5x^2 + 30x - 2}{(5x - 1)(x + 4)}$ **105.** $\dfrac{4x - 1}{2x - 5}$ **107.** $\dfrac{14}{3(x - 2)}$ **109.** $\dfrac{7}{(x + 7)(x - 7)}$ **111.** $\dfrac{3x^2 - 63x + 72}{(x - 4)(x + 3)(x - 8)}$

113. $\dfrac{2x^2 + 8xy - 4y^2}{(2x - 3y)(x - 2y)(2x + y)}$ **115.** $\dfrac{x - 4}{(x - 1)(2x - 1)}$ **117.** $\dfrac{-11x + 14}{(2x + 1)(x + 3)(x - 3)}$ **119.** $\dfrac{2y^3}{3x^3}$ **121.** $-\dfrac{3x + 7}{x + 7}$

123. $\dfrac{x^2 + 3x + 9}{2x - 1}$ **125.** $\dfrac{1}{5(x - 4)}$ **127.** $\dfrac{4x + 3}{2 - 3x}$ **129.** $\dfrac{2xy(3y + 2x^2)}{5}$ **131.** $\dfrac{2x^2y}{y - 3x}$ **133.** $\dfrac{x^3y^3}{(y + x)(y^2 + x^2)}$

135. $\dfrac{-4}{3(x - 1)}$ **137.** $\dfrac{13x - 9}{-19x - 3}$ **139.** $-\dfrac{1}{3(2 + h)}$ **141.** $\dfrac{x - 4}{2(x - 1)}$ **143.** $\dfrac{1}{x(x + 3)}$ **145.** $4xy^2 - 6x^2y - 5x^3$

147. $4a^2b^2 + 5a + \dfrac{9}{2ab}$ **149.** $x - 5 + \dfrac{10}{2x + 1}$ **151.** $5x^2 + 2x + 6 + \dfrac{16}{x - 2}$ **153.** $6x - 15 + \dfrac{64}{2x + 5}$

155. $6x^2 + 4x - 2 + \dfrac{2}{3x - 2}$ **157.** $3x^3 - 6x^2 + 8x - 5$ **159.** $2x^2 - x - 5 + \dfrac{9x + 10}{3x^2 + x + 4}$ **161.** $x^3 + 3x^2 + 9x + 27$

163. $4x^2 + 3x - 2$ **165.** $x^3 + 2x - 4$ **167.** $3x - 9 + \dfrac{32}{x + 3}$ **169.** $2x^2 + 7x + 13 + \dfrac{49}{x - 4}$

171. $x^3 + 5x + 6 + \dfrac{12}{x - 2}$ **173.** $6x^3 - 12x^2 + 18x - 34 + \dfrac{48}{x + \frac{3}{2}}$ **175.** $x^4 + 2x^3 + 4x^2 + 8x + 16$ **177.** $\{-1\}$

179. $\varnothing$ **181.** $\{16\}$ **183.** $\left\{\dfrac{16}{5}\right\}$ **185.** $\left\{-5, \dfrac{5}{3}\right\}$ **187.** $\{1\}$ **189.** $\left\{-4, \dfrac{2}{3}\right\}$ **191.** $\left\{2, \dfrac{11}{4}\right\}$ **193.** $\dfrac{11}{5}$

195. 12 and 14 **197.** 6 days **199.** 30 min **201.** 360 mi/hr **203.** Up 12 mi/hr, down 30 mi/hr

CHAPTER 4   Test (p. 233)

1. $\dfrac{1}{9}$ **3.** $36a^6b^6$ **5.** $-\dfrac{9a^2 + 6ab + 4b^2}{4b^2 + 9a^2}$ **7.** $\dfrac{4x + 3}{x - 6}$ **9.** $\dfrac{10x + 1}{2x(x + 3)}$ **11.** $\dfrac{-1}{2x}$ **13.** $x^2 - 3x + 4$

15. $2x^4 - x^3 - 4x^2 - \dfrac{5}{x - 3}$ **17.** $\{-2\}$ **19.** 72 min

CHAPTERS 1–4   Test Your Memory (p. 235)

1. $-\dfrac{16}{3}$ **3.** $\dfrac{4}{9}$ **5.** $\dfrac{36y^6}{x^{10}}$ **7.** $(6x - 5y)(2x + 3y)$ **9.** $4(x^2 + 25y^2)$ **11.** $(5x - 4)(x + 2)(x - 2)$ **13.** $72x^8y^{22}$

15. $7x^3 - 29x^2 + 2x + 8$ **17.** $\dfrac{x^2 + 3x + 9}{(x + 6)(x - 2)}$ **19.** $\dfrac{3x - 11}{(x - 1)(x + 3)}$ **21.** $\dfrac{3x - 1}{x + 3}$ **23.** $\dfrac{7}{4x}$ **25.** $2x^2 + x - 6 +$

$\dfrac{3x - 2}{x^2 - 3x + 4}$ **27.** $\left\{-\dfrac{4}{3}, -\dfrac{1}{2}\right\}$ **29.** $\left\{-1, \dfrac{3}{2}\right\}$ **31.** $\left\{-\dfrac{5}{4}, \dfrac{9}{2}\right\}$ **33.** $\{-4\}$ **35.** $\left\{\dfrac{7}{3}\right\}$ **37.** $\{-3, 0\}$

39. $\{9\}$ **41.** $\begin{array}{c} x \geq -2 \\ [-2, \infty) \end{array}$ ├┼┼●┼┼┼┼┼┤   **43.** 4 ft by 18 ft   **45.** 6 fives, 32 tens   **47.** $\dfrac{3}{2}$ hr   **49.** 4 mi/hr

$-4\,-3\,-2\,-1\ 0\ 1\ 2\ 3\ 4$

CHAPTER 5   Exercises 5.1 (p. 242)

1. 6   **3.** 14   **5.** $\dfrac{5}{6}$ **7.** $\dfrac{12}{5}$ **9.** Not a real number   **11.** Not a real number   **13.** -2 **15.** 0   **17.** -1

19. 4   **21.** Not a real number   **23.** $\dfrac{2}{3}$ **25.** 4   **27.** -5 **29.** $-\dfrac{2}{5}$ **31.** 2   **33.** 3   **35.** $-\dfrac{1}{2}$ **37.** 32

39. Not a real number   **41.** -4 **43.** 8   **45.** $-\dfrac{1}{64}$ **47.** $\dfrac{8}{27}$ **49.** $\dfrac{1}{7}$ **51.** $\dfrac{1}{16}$ **53.** $\dfrac{1}{125}$ **55.** $\dfrac{5}{4}$ **57.** 9

59. $\dfrac{27}{8}$ **61.** 5   **63.** $\dfrac{1}{49}$ **65.** 729   **67.** 3   **69.** $5^{8/15}$ **71.** $\dfrac{1}{243}$ **73.** $6^{1/2}$ **75.** 6   **77.** $x^{5/6}$ **79.** $9x^{8/3}$

81. $16x^{11/6}$ **83.** $x^{7/6}$ **85.** $\dfrac{2}{3x^{11/12}}$ **87.** $x^{5/2}$ **89.** $-\dfrac{3x^{8/3}}{2y^{5/2}}$ **91.** $\dfrac{2y^{7/6}}{x^{3/4}}$ **93.** $\dfrac{2x^2}{3y^{1/6}}$ **95.** $\dfrac{5x^{1/6}y^{1/12}}{2}$ **97.** $\dfrac{2y^5}{3x^2}$

99. $\dfrac{y^{31/6}}{4x^{1/6}}$ **101.** $3x^{1/2} - 7x^{1/6}$ **103.** $\dfrac{10}{x^{1/2}} + \dfrac{25}{x}$ **105.** $\dfrac{21}{x^{5/6}} + 28x^{1/3}y^{3/4}$ **107.** $x^{17/4} - 3x + 1$ **109.** $x^{1/2} - 4$

111. $x - 10x^{1/2} + 25$ **113.** $x - 8$ **115.** x^m **117.** $x^{m/12}$ **119.** $\dfrac{1}{x^{9m}}$ **121.** $2x^{5m/8}$ **123.** $\dfrac{-1}{2x^{3/2}(x + 1)^{1/2}}$

125. $(2x + 1)^{1/2}(3x - 1)^{1/3}(17x + 1)$ **127.** $\dfrac{5x - 1}{(2x + 1)^{1/2}(3x - 2)^{2/3}}$

Internet Connection (p. 244) answers can be found at http://www.hbcollege.com

Exercises 5.2 (p. 250)

1. 10   **3.** $\dfrac{6}{5}$ **5.** 0   **7.** Not a real number   **9.** 3   **11.** -1 **13.** $\dfrac{2}{3}$ **15.** 4   **17.** 4   **19.** 4   **21.** -4

23. 0   **25.** 8   **27.** $\dfrac{2}{3}$ **29.** 6   **31.** -27 **33.** $\dfrac{10}{7}$ **35.** $\dfrac{5}{3}$ **37.** 3   **39.** $\sqrt[3]{49}$ **41.** $\sqrt[5]{-5}$ **43.** $\sqrt{\dfrac{3}{4}}$

45. $\sqrt{x^3}$ **47.** $\sqrt[4]{64x^3y^3}$ **49.** $\sqrt{x + 3y}$ **51.** $\sqrt{16x^2 - 25y^2}$ **53.** $7^{1/2}$ **55.** $4^{2/3}$ **57.** $x^{1/2}$

59. x^6 **61.** $t^{8/3}$ **63.** x^2 **65.** k **67.** a^5 **69.** $(5xy^2)^{2/5}$ **71.** $(x^3 + 1)^{1/3}$ **73.** $(7x^3y)^3 = 343x^9y^3$

75. $(-2xy^3)^2 = 4x^2y^6$ **77.** $x + 3y$ **79.** $x^2 + 2xy + y^2$ **81.** $x^{5/3}$ **83.** $5y^5$ **85.** $-3x^4$ **87.** $\dfrac{1}{x^{1/12}}$

89. x^2 **91.** x **93.** $x^{1/20}$ **95.** $x^{2/15}$ **97.** $x^{1/24}$ **99.** $9x^2y^4$ **101.** $10xy^5z^6$ **103.** $2x^2y^3$

105. $-4x^3y^{10}z$ **107.** $2xy^4z^6$ **109.** $2x + 1$ **111.** $x + 3y$

Exercises 5.3 (p. 259)

1. $\sqrt{10}$ **3.** $5\sqrt{6}$ **5.** 13   **7.** $2\sqrt[3]{3}$ **9.** $5\sqrt[3]{6}$ **11.** $\sqrt[4]{14}$ **13.** $4\sqrt{3}$ **15.** $5\sqrt{6}$ **17.** $4\sqrt{5}$ **19.** $6\sqrt{3}$
21. $2\sqrt[3]{2}$ **23.** $-2\sqrt[3]{5}$ **25.** $2\sqrt[4]{5}$ **27.** $4x\sqrt{2x}$ **29.** $4x^2\sqrt[3]{2x}$ **31.** $2x\sqrt[4]{4x^3}$ **33.** xy **35.** $3xy^4\sqrt{5xy}$

37. $4x^3y^2\sqrt{2x}$ **39.** $5x^2y^2z^3\sqrt[3]{3y}$ **41.** $12x^2y^3z^4\sqrt{xyz}$ **43.** $4xy^3\sqrt[3]{y}$ **45.** $3x^3z\sqrt[3]{3y^2z}$ **47.** $2xy\sqrt[4]{2xy^3}$

49. $\dfrac{\sqrt{3}}{3}$ **51.** $\dfrac{\sqrt{2}}{2}$ **53.** $\dfrac{1}{2}$ **55.** $\dfrac{\sqrt{6}}{2}$ **57.** $\dfrac{5\sqrt{x}}{x}$ **59.** $\dfrac{\sqrt{2x}}{x}$ **61.** $\dfrac{\sqrt[3]{2}}{2}$ **63.** $\dfrac{3x}{4y^2}$ **65.** $\dfrac{x}{2}$ **67.** $\dfrac{\sqrt{2x}}{3x}$

69. $\dfrac{\sqrt{14x}}{2x}$ **71.** $\dfrac{\sqrt{14xy}}{2y}$ **73.** $\dfrac{\sqrt{6xy}}{3y^2}$ **75.** $\dfrac{5x^2\sqrt{2y}}{4y^4}$ **77.** $\dfrac{2x^2\sqrt{10}}{5y}$ **79.** $\dfrac{2}{3x}$ **81.** $\dfrac{\sqrt[3]{7x^2}}{2x}$ **83.** $\dfrac{\sqrt[3]{10x}}{2x}$

85. $\dfrac{2\sqrt[3]{18x}}{3x}$ **87.** $\dfrac{\sqrt[3]{75x^2y^2}}{5x^2}$ **89.** $\dfrac{2}{x^2}$ **91.** $\dfrac{\sqrt[4]{375}}{5}$ **93.** $\dfrac{\sqrt[4]{56xy^3}}{2y}$ **95.** $\dfrac{\sqrt[4]{20y}}{2y}$ **97.** $\sqrt{x}$ **99.** $2\sqrt{x}$ **101.** x^4y^2

105. x^4 **107.** $|x^5|\sqrt{x}$ **109.** $4|x|$ **111.** $5x^2\sqrt{2x}$ **113.** $|y^3|\sqrt{xy}$ **115.** $10|y|\sqrt{2xy}$ **117.** $|x|\sqrt{1-x^6}$

119. $3|x|\sqrt{1+2x^4}$ **121.** $|x-7|$ **123.** $|3x+5y|$

Absolute value signs are unnecessary when a variable represents a nonnegative real number, or exponents are even.

Exercises 5.4 (p. 262)

1. $-10\sqrt{7}$ **3.** $-5\sqrt[3]{10}$ **5.** 0 **7.** $-7x\sqrt[3]{10y}$ **9.** $7\sqrt{3}$ **11.** $15\sqrt{2}$ **13.** $5\sqrt[3]{4}$ **15.** $\sqrt[4]{2}$ **17.** $7\sqrt{3x}$

19. $12x\sqrt{2x}$ **21.** $\sqrt[3]{4x}$ **23.** $5\sqrt[4]{2x}$ **25.** $7\sqrt{3x}-13\sqrt{6y}$ **27.** $-5x^2\sqrt{3}+6y^2\sqrt{2}$ **29.** $4\sqrt{7}+12\sqrt{5}$

31. $3\sqrt[3]{2}-2\sqrt[3]{5}$ **33.** $(-1+5x)\sqrt{10x}$ **35.** $(9y-7)\sqrt{2y}$ **37.** $(2x^2-3x+20)\sqrt{5}$ **39.** $2\sqrt[3]{2x^2y}$ **41.** $2x\sqrt[3]{3xy}$

43. $-\sqrt[4]{5xy}$ **45.** $y^2\sqrt[4]{xy}$ **47.** $\dfrac{11\sqrt{2}}{2}$ **49.** $\dfrac{23\sqrt{5}}{5}$ **51.** $-\dfrac{5\sqrt[3]{4}}{2}$ **53.** $\dfrac{14\sqrt{3}}{3}$ **55.** $\dfrac{29\sqrt{6}}{6}$ **57.** $\dfrac{11\sqrt{2x}}{2}$

59. $-\dfrac{3\sqrt[3]{2x}}{2}$ **61.** $\sqrt{2x}\left(\dfrac{2-x}{2x}\right)$ **63.** $\sqrt{2x}\left(\dfrac{8x+1}{2x}\right)$

Exercises 5.5 (p. 266)

1. $2\sqrt{3}+\sqrt{15}$ **3.** $14\sqrt{6}-2\sqrt{30}+4\sqrt{3}$ **5.** $3\sqrt{x}+x$ **7.** $8\sqrt{3y}-2y\sqrt{3y}+6\sqrt{2y}$ **9.** $4\sqrt{2}-6+2\sqrt{14}-3\sqrt{7}$

11. $5\sqrt{2}+\sqrt{10}+\sqrt{70}+\sqrt{14}$ **13.** $6\sqrt{6}-12\sqrt{21}+12\sqrt{10}-24\sqrt{35}$ **15.** $6x+5\sqrt{x}-4$

17. $8x+22\sqrt{xy}+15y$ **19.** $8-2\sqrt{15}$ **21.** $13+4\sqrt{3}$ **23.** $126-36\sqrt{6}$ **25.** $x-2\sqrt{2xy}+2y$

27. $4x+28\sqrt{x}+49$ **29.** $x+2\sqrt{x-1}$ **31.** $x+11-6\sqrt{x+2}$ **33.** $4x-2-2\sqrt{3x^2-2x}$ **35.** 2 **37.** 69

39. $x-49$ **41.** $2x-3y$ **43.** $\sqrt[3]{12}-5\sqrt[3]{3}+2\sqrt[3]{4}-10$ **45.** $\sqrt[3]{4}+2\sqrt[3]{6}+\sqrt[3]{9}$ **47.** $\sqrt[3]{4x^2}-1$ **49.** 2

51. $x-y$ **53.** $2-\sqrt{3}$ **55.** $\sqrt{5}-\sqrt{3}$ **57.** $\dfrac{x\sqrt{x}-x\sqrt{y}}{x-y}$ **59.** $7+4\sqrt{3}$ **61.** $\dfrac{x+9\sqrt{x}+20}{x-16}$

63. $\dfrac{3\sqrt{2}+3+2\sqrt{3}+\sqrt{6}}{3}$ **65.** $\dfrac{x+2\sqrt{xy}+y}{x-y}$ **67.** $-\dfrac{29+17\sqrt{3}}{2}$ **69.** $\dfrac{\sqrt{6}+6}{10}$ **71.** $\dfrac{6x-13\sqrt{xy}+6y}{9x-4y}$

73. $\dfrac{2x\sqrt{6}-4\sqrt{3xy}-15\sqrt{xy}+15y\sqrt{2}}{3x-6y}$ **75.** $\sqrt{2}$ **77.** $\sqrt{22}$ **79.** $2\sqrt{3}$

Exercises 5.6 (p. 274)

1. $\{4\}$ **3.** $\{-3,3\}$ **5.** $\varnothing$ **7.** $\{21\}$ **9.** $\left\{\dfrac{4}{3},2\right\}$ **11.** $\varnothing$ **13.** $\{2\}$ **15.** $\left\{\dfrac{4}{3}\right\}$ **17.** $\{-3,5\}$ **19.** $\left\{-\dfrac{9}{2}\right\}$

21. $\left\{\dfrac{11}{3}\right\}$ **23.** $\{8\}$ **25.** $\{1\}$ **27.** $\left\{\dfrac{2}{3}\right\}$ **29.** $\varnothing$ **31.** $\{-2\}$ **33.** $\{8\}$ **35.** $\left\{\dfrac{4}{3}\right\}$ **37.** $\{-1\}$ **39.** $\{5\}$

41. $\{25\}$ **43.** $\{1,9\}$ **45.** $\{5\}$ **47.** $\{-1\}$ **49.** $\{-2\}$ **51.** $\{2\}$ **53.** $\left\{-\dfrac{1}{9}\right\}$ **55.** $\{-2\}$ **57.** $\{-1\}$

59. $\{3\}$ **61.** $\{3,15\}$ **63.** $\{-1\}$

Exercises 5.7 (p. 282)

1. $9i$ **3.** $-2\sqrt{5}i$ **5.** $12\sqrt{3}i$ **7.** $7i$ **9.** $-11i$ **11.** $5\sqrt{3}i$ **13.** $9+3i$ **15.** $4\sqrt{3}-7\sqrt{2}i$

17. $9\sqrt{2}-4\sqrt{6}i$ **19.** -12 **21.** -13 **23.** 3 **25.** $-5i$ **27.** $12\sqrt{2}i$ **29.** $\dfrac{9}{2}$ **31.** -5 **33.** $\dfrac{3}{2}i$

35. $7-3i$ **37.** $-\dfrac{1}{12}+\dfrac{7}{24}i$ **39.** $-8-3i$ **41.** $-\dfrac{17}{12}+\dfrac{13}{36}i$ **43.** $21+63i$ **45.** $14+8i$ **47.** $22-21i$

49. $-2-26i$ **51.** 29 **53.** 4 **55.** $-5+12i$ **57.** $2+2\sqrt{3}i$ **59.** $-\dfrac{5}{3}i$ **61.** $3-5i$ **63.** $\dfrac{1}{2}-2i$

65. $\dfrac{1}{5} - \dfrac{1}{10}i$ **67.** $\dfrac{6}{13} + \dfrac{9}{13}i$ **69.** $\dfrac{3}{5} + \dfrac{1}{5}i$ **71.** $\dfrac{10}{17} + \dfrac{6}{17}i$ **73.** $\dfrac{1}{2} - \dfrac{3}{2}i$ **75.** $7 - 5i$ **77.** $\dfrac{18}{17} - \dfrac{13}{17}i$

79. $\dfrac{7}{41} - \dfrac{22}{41}i$ **81.** $-i$ **83.** i **85.** $-6i$ **87.** -2 **89.** i **91.** $-i$ **93.** -1 **95.** $2i$ **97.** -3

APPLIED ALGEBRA—YOUR TURN (p. 284)

1. 455 gal/min   **3.** 13.69 lb/sq in.

CHAPTER 5   Review Exercises (p. 286)

1. 13   **3.** $\dfrac{3}{5}$ **5.** -4 **7.** 1   **9.** -1 **11.** 3   **13.** $-\dfrac{3}{4}$ **15.** -3 **17.** -9 **19.** 81   **21.** 4   **23.** -8

25. $\dfrac{1}{16}$ **27.** $-\dfrac{1}{16}$ **29.** 2   **31.** $\dfrac{9}{4}$ **33.** -8 **35.** 2   **37.** 3125   **39.** $\dfrac{1}{4}$ **41.** $\dfrac{2}{5}$ **43.** $3x^{1/7}$ **45.** $\dfrac{1}{x^{7/24}}$

47. $x^{1/4}$ **49.** $\dfrac{3y^{19/12}}{x^{1/21}}$ **51.** $\dfrac{y^{3/4}}{8x^{9/14}}$ **53.** $\dfrac{2}{27x^{16/3}y^{1/2}}$ **55.** $6x^{1/3} + 3$ **57.** $x^{2/3} - 9$ **59.** $x + 27$ **61.** x^{3m}

63. $\dfrac{1}{x^{11m/4}}$ **65.** $25x^{3m/2}$ **67.** $\dfrac{5}{3}(6x - 1)^{3/2}(x + 3)^{2/3}(15x + 26)$ **69.** 21   **71.** Not a real number   **73.** -2 **75.** 2

77. 9   **79.** 3   **81.** $\dfrac{2}{3}$ **83.** 12   **85.** $\dfrac{1}{2}$ **87.** $\sqrt[5]{36}$ **89.** $\sqrt[4]{343x^3}$ **91.** $\sqrt{27x^3 + 135x^2 + 225x + 125}$ or

$(3x + 5)\sqrt{3x + 5}$ **93.** $11^{1/2}$ **95.** $a^{3/4}$ **97.** $m^{7/2}$ **99.** p^3 **101.** $(3x^2y)^{3/2}$ **103.** $a + b$ **105.** $y^{13/12}$

107. $x^{1/6}$ **109.** $x^{1/6}$ **111.** $4xy^2$ **113.** $3xy^2$ **115.** $x + 1$ **117.** $\sqrt{30}$ **119.** 7   **121.** $3\sqrt[3]{4}$ **123.** $5\sqrt{3}$

125. $3\sqrt{11}$ **127.** $3\sqrt[3]{3}$ **129.** $2\sqrt[4]{3}$ **131.** $xy^2\sqrt[4]{2y}$ **133.** $2x^2y^6\sqrt{6}$ **135.** $2xy^4\sqrt[3]{3x}$ **137.** $\dfrac{2\sqrt{x}}{x}$

139. $\dfrac{5\sqrt{3y}}{3y}$ **141.** $\dfrac{x^2\sqrt{6xy}}{y^2}$ **143.** $\dfrac{x\sqrt{35y}}{14y^2}$ **145.** $\dfrac{x\sqrt[3]{4}}{2}$ **147.** $\dfrac{\sqrt[3]{50}}{5}$ **149.** $\dfrac{x^2\sqrt[3]{175}}{10}$ **151.** $\dfrac{3ax\sqrt[4]{y^3}}{y}$

153. $x\sqrt{x}$ **155.** $y\sqrt[3]{x^2}\sqrt{y}$ **157.** $-4\sqrt[3]{6}$ **159.** $10x^2\sqrt{2y}$ **161.** $6\sqrt{2}$ **163.** $2\sqrt[3]{5} + 6\sqrt[3]{9}$

165. $-2x\sqrt{x} - 2x\sqrt{2x}$ **167.** $7p\sqrt[3]{2p} - 2\sqrt[3]{2p}$ **169.** $-19x^2y^3\sqrt{5x} + 2x^2y^2\sqrt{10y}$ **171.** $11x^3y\sqrt[4]{x}$

173. $-\dfrac{8\sqrt[3]{25}}{5}$ **175.** $\dfrac{17\sqrt{10}}{10}$ **177.** $-\dfrac{5\sqrt[3]{4x}}{2}$ **179.** $15\sqrt{2} + 6\sqrt{5} - 18$ **181.** $\sqrt{21} - \sqrt{3} + 5\sqrt{7} - 5$

183. $63 + 8\sqrt{5}$ **185.** $9x + 9\sqrt{xy} - 4y$ **187.** $4y + 4\sqrt{y} + 1$ **189.** $x - 4$ **191.** 6   **193.** $\dfrac{\sqrt{21} + \sqrt{3}}{6}$

195. $-2\sqrt{3} - 3\sqrt{2}$ **197.** $\dfrac{x\sqrt{15} - \sqrt{6xy} + \sqrt{5xy} - y\sqrt{2}}{5x - 2y}$ **199.** $\dfrac{6x\sqrt{2} + 8\sqrt{6xy} - 5\sqrt{3xy} - 20y}{9x - 48y}$

201. $4\sqrt{15x} + 8\sqrt{3x} - 7\sqrt{10y} - 14\sqrt{2y}$ **203.** $\{1\}$ **205.** $\{-2, 8\}$ **207.** $\{-4, 1\}$ **209.** $\{3\}$ **211.** $\left\{\dfrac{1}{4}\right\}$

213. $\{5\}$ **215.** $\left\{4, \dfrac{24}{5}\right\}$ **217.** $\{10\}$ **219.** $14i$ **221.** -42 **223.** 2   **225.** $5\sqrt{2}i$ **227.** $7i$ **229.** $13 + 5i$

231. $3 - 12i$ **233.** $30 + 48i$ **235.** $33 + 19i$ **237.** 61   **239.** $\dfrac{3}{2} - \dfrac{5}{2}i$ **241.** $\dfrac{2}{5} - \dfrac{6}{5}i$ **243.** $\dfrac{13}{17} - \dfrac{1}{17}i$

245. 1   **247.** $-i$ **249.** -1

CHAPTER 5   Test (p. 289)

1. $\dfrac{1}{3}$ **3.** $24x^{10}y^{13/2}$ **5.** $\dfrac{1}{125}$ **7.** $16x\sqrt{2y}$ **9.** $\dfrac{\sqrt[3]{2x^2}}{x}$ **11.** $\sqrt{3} - \sqrt{2}$ **13.** $2x + 5\sqrt{x} - 12$ **15.** $\{-2\}$

17. $5 + 6i$ **19.** $-\dfrac{3}{2} - 8i$ **21.** $-i$ **23.** $20 + 5i$ **25.** $6 - 2i$

CHAPTERS 1–5   Test Your Memory (p. 291)

1. $\dfrac{11}{9}$ **3.** $\dfrac{44}{9}$ **5.** $4x^6y^{12}$ **7.** $x^{1/6}$ **9.** $(1 - 5t)(1 + 5t + 25t^2)$ **11.** $(2x + y + 3)(2x + y + 4)$ **13.** $\dfrac{x - 5}{4x^2}$

15. $\dfrac{x^2 + 10x + 11}{(x - 2)^2(x + 3)}$ **17.** $\dfrac{x^3y^3}{(x^2 + y^2)(x + y)}$ **19.** $\dfrac{1}{36}$ **21.** $4x\sqrt{2y}$ **23.** $\dfrac{\sqrt{15xy}}{5y}$ **25.** $4x - 28\sqrt{x} + 49$ **27.** $8\sqrt{5}i$

29. $14 - 18i$ **31.** $\{13\}$ **33.** $\left\{-\dfrac{3}{2}, 4\right\}$ **35.** $\{3\}$ **37.** $\{1, 3\}$ **39.** $\left\{\dfrac{1}{2}\right\}$ **41.** $\{11\}$ **43.** $\{-6, -2\}$

45. 12 ft by 35 ft   **47.** 14 quarters, 8 dimes   **49.** 12 hr

CHAPTER 6   Exercises 6.1 (p. 301)

1. $A(4, 0)$, $B(1, 6)$, $C(0, 4)$, $D(0, 0)$, $E(-3, 2)$, $F\left(-\dfrac{13}{2}, 0\right)$, $G(-4, -5)$, $H(0, -2)$, $I(7, -6)$

3. 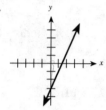   **5.** 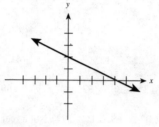   **7.**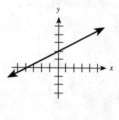

9. 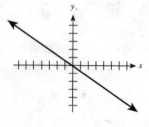   **11.** 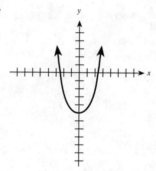   **13.**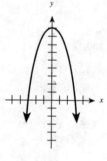

15. 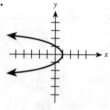   **17.** 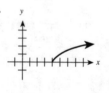   **19.** 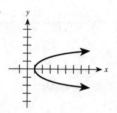   **21.**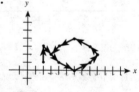

25.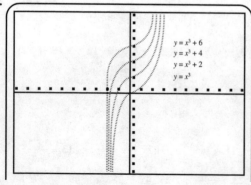

$y = x^3 + 6$
$y = x^3 + 4$
$y = x^3 + 2$
$y = x^3$

27.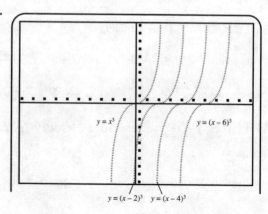

$y = x^3$ $y = (x - 6)^3$

$y = (x - 2)^3$ $y = (x - 4)^3$

29.

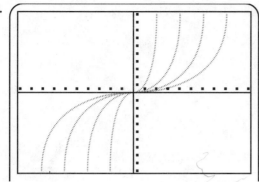

31. $y = 2x$ $y = x + \sqrt{x^2}$

x	y
-4	-8
-3	-6
-2	-4
-1	-2
0	0
1	2
2	4
3	6
4	8

x	y
-4	0
-3	0
-2	0
-1	0
0	0
1	2
2	4
3	6
4	8

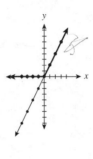

Graphs coincide for $x \geq 0$, since, if $x \geq 0$, $\sqrt{x^2} = x$ so $x + \sqrt{x^2} = x + x = 2x$.

33. $y = x + 1$ $y = \sqrt{x^2 + 2x + 1}$

x	y
-4	-3
-3	-2
-2	-1
-1	0
0	1
1	2
2	3
3	4
4	5

x	y
-4	3
-3	2
-2	1
-1	0
0	1
1	2
2	3
3	4
4	5

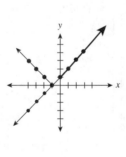

Graphs coincide for $x \geq -1$, since if $x \geq -1$, $\sqrt{x^2 + 2x + 1} = \sqrt{(x + 1)^2} = x + 1$.

Exercises 6.2 (p. 308)

1. $V = \left\{ (-5, 17), (0, 7), \left(\frac{3}{2}, 4 \right), (2.1, 2.8) \right\}$; domain of $V = \left\{ -5, 0, \frac{3}{2}, 2.1 \right\}$; range of $V = \{17, 7, 4, 2.8\}$

3. $X = \left\{ \left(\frac{1}{2}, 0 \right), (1, 1), (3, \sqrt{5}), (5, 3) \right\}$; domain of $X = \left\{ \frac{1}{2}, 1, 3, 5 \right\}$; range of $X = \{0, 1, \sqrt{5}, 3\}$

5. $Z = \{(-2, 5), (0, 3), (2, 5)\}$; domain of $Z = \{-2, 0, 2\}$; range of $Z = \{5, 3\}$ **7.** Domain = all real numbers; range = all real numbers; function   **9.** Domain = all real numbers; range = $\{y : y \geq 5\}$; function   **11.** Domain = $\{x : x \leq 2\}$; range = all real numbers; not a function   **13.** Domain = $\{x : x \geq -3\}$; range = $\{y : y \geq 0\}$; function

15. Domain = $\{x : x \geq 2\}$; range = $\{y : y \geq 0\}$; function   **17.** Domain = all real numbers; range = $\{4\}$; function

19. Domain = all real numbers; range = all real numbers   **21.** Domain = all real numbers; range = $\{y : y \geq 0\}$

23. Domain = $\{x : x \leq 1\}$; range = $\{y : y \leq 0\}$ **25.** Domain = $\{x : x \geq 4\}$; range = all real numbers

27. Domain = all real numbers; range = $\{4\}$ **29.** Function   **31.** Function   **33.** Not a function   **35.** Function

37. Not a function   **39. (a)** $\{(15620, 1978), (16550, 1979), (18105, 1980), (18782, 1981), (18105, 1982), (20750, 1983)\}$; not a function   **(b)** $\{(1978, 15620), (1979, 16550), (1980, 18105), (1981, 18782), (1982, 18105), (1983, 20750)\}$; function

45.

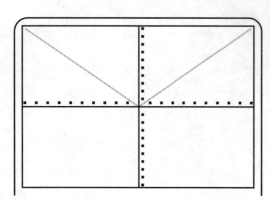

Function; domain = $(-\infty, \infty)$; range = $[0, \infty)$

47.

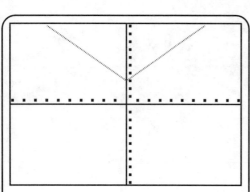

Function; domain = $(-\infty, \infty)$; range = $[3, \infty)$

49.

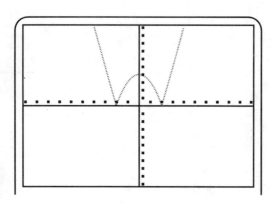

Function; domain = $(-\infty, \infty)$; range = $[0, \infty)$

Exercises 6.3 (p. 316)

1. $f(-2) = -7, f\left(-\dfrac{1}{2}\right) = -4, f(0) = -3, f\left(\dfrac{1}{3}\right) = -\dfrac{7}{3}, f(1) = -1, f(a) = 2a - 3, f(a + h) = 2a + 2h - 3$

3. $f(-2) = 2, f\left(-\dfrac{1}{2}\right) = -\dfrac{7}{4}, f(0) = -2, f\left(\dfrac{1}{3}\right) = -\dfrac{17}{9}, f(1) = -1, f(a) = a^2 - 2, f(a + h) = a^2 + 2ah + h^2 - 2$

5. $f(-2) = -16, f\left(-\dfrac{1}{2}\right) = -\dfrac{1}{4}, f(0) = 0, f\left(\dfrac{1}{3}\right) = \dfrac{2}{27}, f(1) = 2, f(a) = 2a^3, f(a + h) = 2a^3 + 6a^2h + 6ah^2 + 2h^3$

7. $f(-2) = -2, f\left(-\dfrac{1}{2}\right) = -2, f(0) = -2, f\left(\dfrac{1}{3}\right) = -2, f(1) = -2, f(a) = -2, f(a + h) = -2$

9. $(f + g)(x) = 2x^2 + 6x - 3, (f - g)(x) = 2x^2 + 4x + 1, (fg)(x) = 2x^3 + x^2 - 11x + 2, \left(\dfrac{f}{g}\right)(x) = \dfrac{2x^2 + 5x - 1}{x - 2}$

11. $(f + g)(x) = 2x^2 - 4$, $(f - g)(x) = 14$, $(fg)(x) = x^4 - 4x^2 - 45$, $\left(\dfrac{f}{g}\right)(x) = \dfrac{x^2 + 5}{x^2 - 9}$ **13.** $(f + g)(x) = 3x - 8$,

$(f - g)(x) = x + 14$, $(fg)(x) = 2x^2 - 19x - 33$, $\left(\dfrac{f}{g}\right)(x) = \dfrac{2x + 3}{x - 11}$ **15.** $(f + g)(x) = x^3 + 5x - 4$, $(f - g)(x) = x^3 + x - 6$,

$(fg)(x) = 2x^4 + x^3 + 6x^2 - 7x - 5$, $\left(\dfrac{f}{g}\right)(x) = \dfrac{x^3 + 3x - 5}{2x + 1}$ **17.** -7 **19.** 5 **21.** -199 **23.** -343 **25.** $\dfrac{25}{1681}$

27. 3 **29.** 0 **31.** $2x^2 + 5x + 3$ **33.** $\dfrac{x^2 - 4}{2x + 1}$ **35.** $12x^2 + 10x - 1$ **37.** $6x + 1$ **39.** $3a + 3h - 2$ **41.** 3

43. $2a + 2h + 2$ **45.** 2 **47.** $3a^2 + 5a + 3h^2 + 5h - 2$ **49.** $6a + 3h + 5$ **51.** 122 **53.** 14 **57.** 14

59. 6 **61.** -7 **63.** 10

Internet Connection (p. 319) answers can be found at http://www.hbcollege.com

Exercises 6.4 (p. 331)

1.

3.
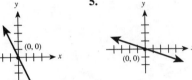

5.

7.
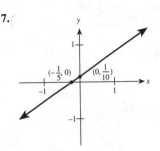

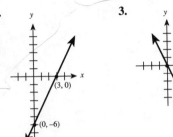

9.
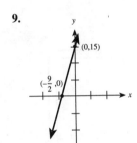

11.
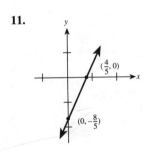

13.
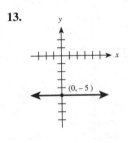

15.
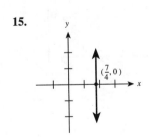

17. $\dfrac{1}{2}$ **19.** $-\dfrac{6}{7}$ **21.** 3 **23.** -18 **25.** $\dfrac{15}{16}$ **27.** 0 **29.**

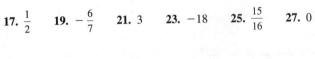

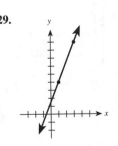

31.
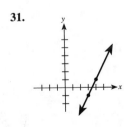

33.
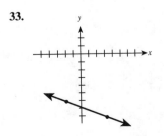

35.
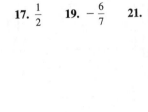

37. 2 **39.** -2 **41.** $-\dfrac{1}{3}$

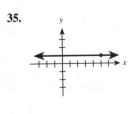

43. $\frac{1}{2}$ **45.** $\frac{10}{3}$ **47.** 2   **49.** 0   **51.** Undefined   **53.**

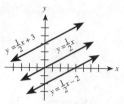

The lines are parallel.

55.

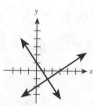

The lines are perpendicular.

63. (a) 1996–616, 2000–528, 2004–440   **(b)** $g(x) = 748 - 22(x - 1990)$

where x stands for the year after 1990.

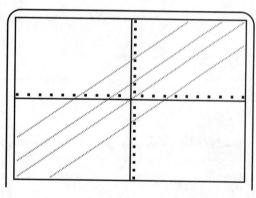

(c)

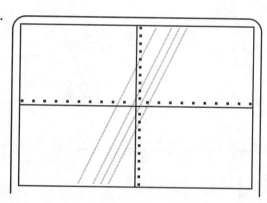

(d) $m = -22$ **(e)** 2024

69.

71.

73. -2 **75.** $\frac{3}{2}$ **77.** $\frac{A - 10}{2}$

Exercises 6.5 (p. 344)

1. $y = 5x + \frac{1}{2}$ **3.** $y = x - 9$ **5.** $y = -\frac{1}{2}x - 3$ **7.** $y = 11$ **9.** $m = 3; b = -4$ **11.** $m = \frac{3}{5}; b = -\frac{7}{5}$

13. $m = \frac{6}{5}; b = \frac{8}{5}$ **15.** $m = \frac{2}{7}; b = -\frac{4}{7}$ **17.** $m = 0; b = 3$ **19.** $y = 3x + 1$ **21.** $y = \frac{9}{2}x - \frac{11}{2}$

23. $y = \frac{4}{3}x - \frac{7}{6}$ **25.** $y = 5$ **27.** $y = -x + 2$ **29.** $y = -\frac{3}{4}x$ **31.** $y = -\frac{3}{2}x$ **33.** $y = -\frac{11}{12}x + \frac{41}{24}$

35. $y = -4$ **37.** $y = \frac{2}{3}x - 2$ **39.** $y = 3x - 5$ **41.** $y = -\frac{5}{4}x + \frac{59}{4}$ **43.** $y = \frac{3}{4}x - \frac{63}{80}$ **45.** $x = -8$

47. $y = -7$ **49.** $y = -\frac{1}{3}x - \frac{5}{3}$ **51.** $y = \frac{4}{5}x + \frac{43}{5}$ **53.** $y = -\frac{4}{3}x - \frac{4}{15}$ **55.** $y = -9$ **57.** $x = 3$

59. m of $\overline{AB} = \frac{7}{6}$, m of $\overline{BC} = -\frac{6}{7}$. Thus, $\overline{AB}$ is perpendicular to $\overline{BC}$.

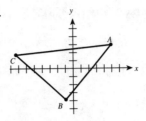

61. m of $\overline{AB} = \frac{1}{2}$, m of $\overline{BC} = -2$, m of $\overline{CD} = \frac{1}{2}$, m of $\overline{AD} = -2$. Thus, a rectangle is formed.

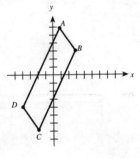

63. m of $\overline{AB} = -\frac{3}{2}$, m of $\overline{CD} = -\frac{3}{2}$, m of $\overline{AD} = 2$, m of $\overline{BC} = 2$. Thus, $\overline{AB}$ is parallel to $\overline{CD}$, and $\overline{AD}$ is parallel to $\overline{BC}$.

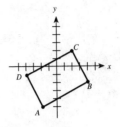

65. $C = \frac{5}{9}F - \frac{160}{9}$ **67. (a)** $y = b$ **(b)** $x = a$ **(c)** $x = a$ **(d)** $y = b$ **69.** x-axis: $y = 0$; y-axis: $x = 0$

71.

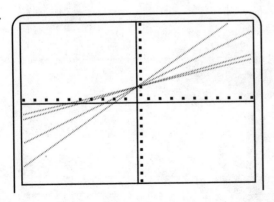

73.

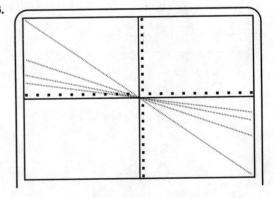

75. (a) $y = 4x + 4$ **(b)** $y = 14x + 14$ **(c)** $y = \frac{1}{2}x + \frac{1}{2}$ **(d)** $y = 2Ax + 2A$

77. (a) $y = -\frac{4}{3}x - 4$ **(b)** $y = -\frac{10}{3}x - 10$ **(c)** $y = -\frac{4}{9}x - \frac{4}{3}$ **(d)** $y = -\frac{2A}{9}x - \frac{2A}{3}$

Internet Connection (p. 348) answers can be found at http://www.hbcollege.com

Exercises 6.6 (p. 358)

1. $d = 8, M = (6, -1)$ **3.** $d = 10, M = (2, 7)$ **5.** $d = \sqrt{109}; M = \left(-\frac{1}{2}, 0\right)$ **7.** $d = 2\sqrt{41}, M = (-1, 4)$

9. $d = \sqrt{85}, M = \left(-1, -\frac{3}{2}\right)$ **11.** $d = 6\sqrt{2}; M = (1, 4)$ **13.** $d = \frac{25}{6}, M = \left(\frac{3}{4}, \frac{-4}{3}\right)$ **15.** $d = \sqrt{147.05} \doteq 12.126,$

$M = (-5.3, -0.85)$ **17.** $d = 2\sqrt{2}; M = \left(\frac{3}{2}\sqrt{3}, \frac{3}{2}\sqrt{5}\right)$ **19.** $B = (-1, -6)$ **21.** $\left(\frac{11}{2}, 6\right)$ **23.** $d(A, C) = \sqrt{65},$

$d(A, B) = \sqrt{20}, d(B, C) = \sqrt{45}; [d(A, C)]^2 = [d(A, B)]^2 + [d(B, C)]^2$ **25.** $d(A, C) = \sqrt{68} = d(B, C)$ **27.** $d(A, B) = \sqrt{5} =$

$d(C, D), d(A, D) = \sqrt{80} = d(B, C)$; slope of $\overline{AB} = -\frac{1}{2}$, slope of $\overline{BC} = 2$. Opposite sides have the same length and $\overline{AB} \perp \overline{BC}$.

29. $(8, -3)$ and $(-4, -3)$ **31.** $d(P, C) = d(Q, C) = d(R, C) = 5; (x - 2)^2 + (y + 1)^2 = 25$ **33.** $M = \left(\frac{b}{2}, \frac{c}{2}\right),$

$N = \left(\frac{a + b}{2}, \frac{c}{2}\right); AC = a, MN = \sqrt{\left(\frac{a}{2}\right)^2 + 0} = \frac{a}{2} = \frac{1}{2}(AC)$ **37.** $y = 2x + 3; 3\sqrt{5}$ **39.** $y = 3x - 2; 3\sqrt{10}$

41. $y = 4x + 1; 3\sqrt{17}$ **43.** $|x_2 - x_1|\sqrt{1 + m^2}$

Exercises 6.7 (p. 366)

1.
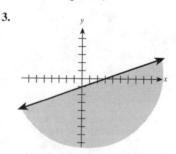

3.
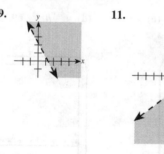

5.
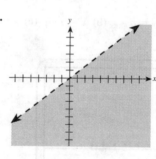

7.
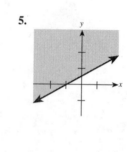

9.
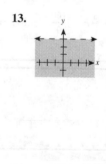

11.
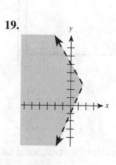

13.
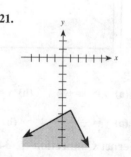

15.
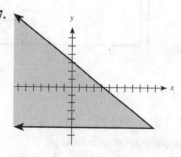

17.

19.

21.

23.

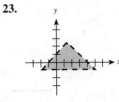

25.

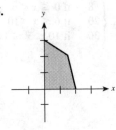

27. $\begin{cases} x \geq 0 \\ y \geq 0 \end{cases}$

Exercises 6.8 (p. 373)

1. $k = 8$ **3.** $k = \dfrac{1}{2}$ **5.** $k = 9$ **7.** $k = -2$ **9.** $k = \dfrac{3}{2}$ **11.** $C = kr; k = 2\pi$ **13.** $A = kE^2; k = 6$

15. $A = kH(B_1 + B_2); k = \dfrac{1}{2}$ **17.** $H = \dfrac{kV}{R^2}; k = \dfrac{3}{\pi}$ **19.** 108   **21.** $\dfrac{5}{2}$ **23.** 144   **25.** $\dfrac{48}{5}$ **27.** 6   **29.** 144 ft

31. 42 lb   **33.** \$90   **35.** 100%   **37.** 20 lb/sq in.   **39.** 9×10^{-11} Newtons

Exercises 6.9 (p. 381)

1.

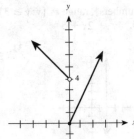

3.

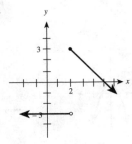

5.

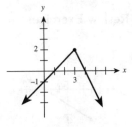

7.

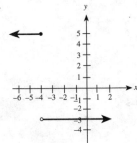

9.

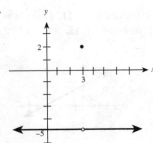

11.

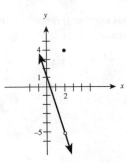

13.

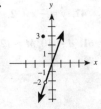

15.

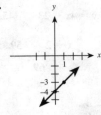

17.

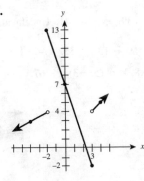

19.

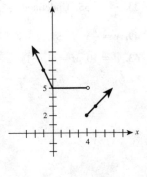

21.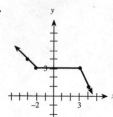

23.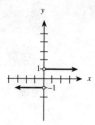

25. $f(x) = \begin{cases} 8 & \text{if } 0 \le x < 6 \\ 20 & \text{if } 6 \le x < 10 \\ 30 & \text{if } 10 \le x \le 14 \end{cases}$

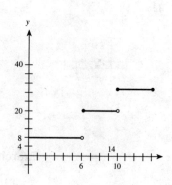

APPLIED ALGEBRA—YOUR TURN (p. 383)

1. $1\frac{3}{4}$ in.

CHAPTER 6 Review Exercises (p. 385)

1. 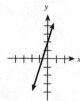 **3.** 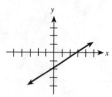 **5.** **7.**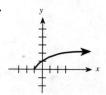

9. Domain = {all real numbers}, range = {all real numbers}; function
function **13.** Domain = {$x : x \ge 3$}, range = {$y : y \ge 0$}; function **11.** Domain = {all real numbers}, range = {$y : y \ge 3$};

15. 8 **17.** 2704 **19.** $4x^2 + 2x + 5$

21. $5a + 5h - 4$ **23.** 5 **25.** 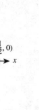 **27.** 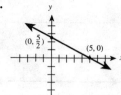 **29.** 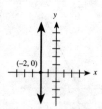 **31.** $\frac{2}{3}$

33. $\frac{7}{3}$ **35.** Undefined **37.** 2 **39.** $-\frac{1}{2}$ **41.** Undefined **43.** $y = 4x - 23$ **45.** $y = 1$ **47.** $y = \frac{2}{3}x + 2$

49. $y = \frac{7}{3}x$ **51.** $x = 2$ **53.** $y = 3x - 13$ **55.** $y = -1$ **57.** $y = -\frac{4}{5}x - 2$ **59.** $x = 4$ **61.** $d = 10, M = (2, 6)$

63. $d = 10, M = (1, -1)$ **65.** $d = 8, M = (5, -3)$ **67.** **69.**

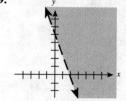

71.

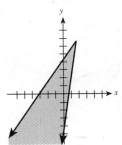

73. (a) $A = ks^2$ **(b)** $k = \dfrac{\sqrt{3}}{4}$ **75.** 3300   **77.**

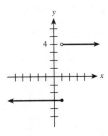

79.

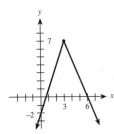

81.

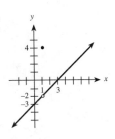

83. $f(x) = \begin{cases} 12.5 & \text{if } 0 \le x \le 5 \\ 11.5 & \text{if } 5 < x \le 20 \\ 14 & \text{if } 20 < x \le 25 \\ 4 & \text{if } 25 < x \le 26 \end{cases}$

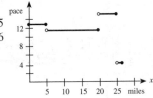

CHAPTER 6   Test (p. 387)

1.

domain = {all real numbers}
range = {y:y ≥ −3}

3. $-\dfrac{3}{5}$ **5.** 9   **7.** Domain = $\{x : x \ne 2\}$ **9.** $2a + h$ **11.**

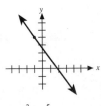

$y = -\dfrac{3}{2}x + \dfrac{5}{2}$

13. $y = \dfrac{1}{4}x + 2$ **15.** $d = \dfrac{13}{2}, M = \left(\dfrac{9}{4}, 1\right)$ **17.**

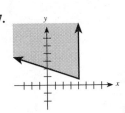

CHAPTERS 1–6   Test Your Memory (p. 389)

1. $-\dfrac{3}{4}$ **3.** $\dfrac{72x^8}{y^3}$ **5.** $\dfrac{x-3}{3x(x-1)}$ **7.** $x + 4$ **9.** $-2\sqrt{5x}$ **11.** $\dfrac{\sqrt{7xy}}{2y}$ **13.** $16x + 24\sqrt{x} + 9$ **15.** -12

17.

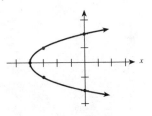

domain = [−4,∞)
range = (−∞,∞)

19. -14 **21.** 121   **23.** $4a^2 + 7$ **25.**

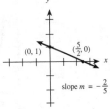

slope $m = -\dfrac{2}{5}$

27. $y = \dfrac{5}{4}x + \dfrac{21}{4}$ **29.** $y = -\dfrac{4}{3}x - \dfrac{37}{3}$ **31.**

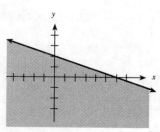

$x < \dfrac{5}{2}$

$\left(-\infty, \dfrac{5}{2}\right)$

33.

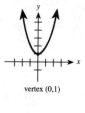

35. $\{26\}$ **37.** $\left\{-\dfrac{4}{5}, \dfrac{8}{5}\right\}$ **39.** $\left\{-\dfrac{4}{3}, 2\right\}$ **41.** $\left\{\dfrac{1}{2}\right\}$ **43.** $\varnothing$ **45.** Base is 18 in. Height is 7 in.   **47.** 40 hr

49. 400 ft

CHAPTER 7   Exercises 7.1 (p. 399)

1.

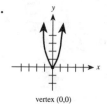

vertex $(0,0)$

3.

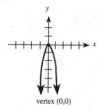

vertex $(0,0)$

5.

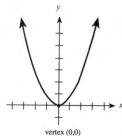

vertex $(0,0)$

7.

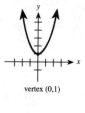

vertex $(0,1)$

9.

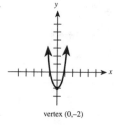

vertex $(0,-2)$

11.

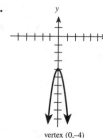

vertex $(0,-4)$

13.

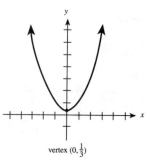

vertex $\left(0, \dfrac{1}{3}\right)$

15.

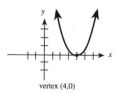

vertex $(4,0)$

17.

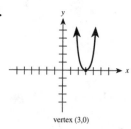

vertex $(3,0)$

19.

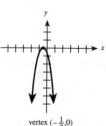

vertex $\left(-\dfrac{1}{2}, 0\right)$

21.

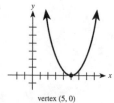

vertex $(5,0)$

23.

vertex $(4, 1)$

25.

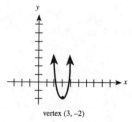

vertex $(3, -2)$

27.

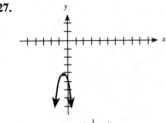

vertex $(-\frac{1}{2}, -4)$

29.

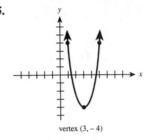

vertex $(5, \frac{1}{3})$

31.

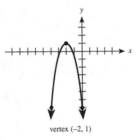

vertex $(-3, 1)$

33.

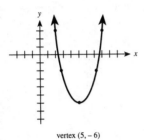

vertex $(5, -6)$

35.

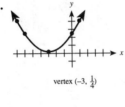

vertex $(3, -4)$

37.

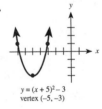

vertex $(-2, 1)$

39.

vertex $(\frac{1}{2}, -2)$

41.

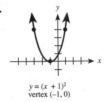

vertex $(-3, \frac{1}{4})$

Exercises 7.2 (p. 411)

1. 9   **3.** 121   **5.** $\dfrac{9}{4}$ **7.** $\dfrac{49}{4}$

9.

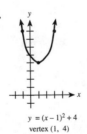

$y = (x - 1)^2 + 4$
vertex $(1, 4)$

11.

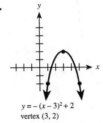

$y = (x + 1)^2$
vertex $(-1, 0)$

13.

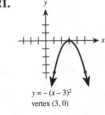

$y = (x + 5)^2 - 3$
vertex $(-5, -3)$

15.

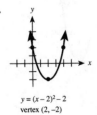

$y = (x - 2)^2 - 2$
vertex $(2, -2)$

17.

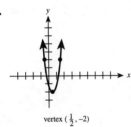

$y = -(x - 3)^2 + 2$
vertex $(3, 2)$

19.

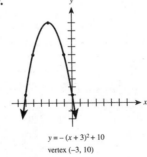

$y = -(x + 3)^2 + 10$
vertex $(-3, 10)$

21.

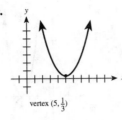

$y = -(x - 3)^2$
vertex $(3, 0)$

23.

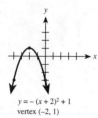

$y = -(x+2)^2 + 1$
vertex $(-2, 1)$

25.

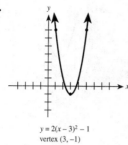

$y = 2(x-3)^2 - 1$
vertex $(3, -1)$

27.

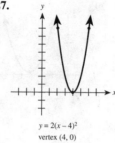

$y = 2(x-4)^2$
vertex $(4, 0)$

29.

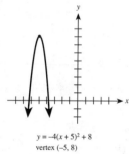

$y = 3(x-2)^2 + 5$
vertex $(2, 5)$

31.

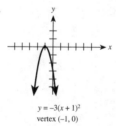

$y = 3(x+2)^2 - 7$
vertex $(-2, -7)$

33.

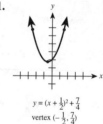

$y = -2(x-3)^2 - 5$
vertex $(3, -5)$

35.

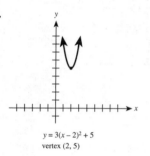

$y = -4(x+5)^2 + 8$
vertex $(-5, 8)$

37.

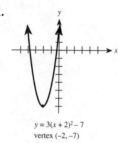

$y = -3(x+1)^2$
vertex $(-1, 0)$

39.

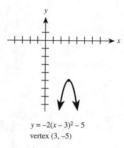

$y = (x - \frac{1}{2})^2 + \frac{3}{4}$
vertex $(\frac{1}{2}, \frac{3}{4})$

41.

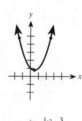

$y = (x + \frac{1}{2})^2 + \frac{7}{4}$
vertex $(-\frac{1}{2}, \frac{7}{4})$

43.

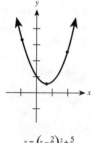

$y = (x - \frac{2}{3})^2 + \frac{5}{9}$
vertex $(\frac{2}{3}, \frac{5}{9})$

45.

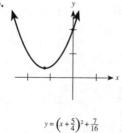

$y = (x + \frac{5}{4})^2 + \frac{7}{16}$
vertex $(-\frac{5}{4}, \frac{7}{16})$

47.

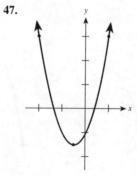

$y = 2(x + \frac{1}{2})^2 - \frac{3}{2}$
vertex $(-\frac{1}{2}, -\frac{3}{2})$

49.

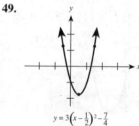

$y = 3(x - \frac{1}{2})^2 - \frac{7}{4}$
vertex $(\frac{1}{2}, -\frac{7}{4})$

51.

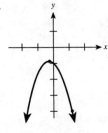

$$y = -2\left(x + \frac{1}{4}\right)^2 - \frac{7}{8}$$
vertex $\left(-\frac{1}{4}, -\frac{7}{8}\right)$

53.

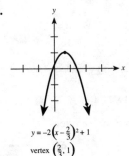

$$y = -2\left(x - \frac{2}{3}\right)^2 + 1$$
vertex $\left(\frac{2}{3}, 1\right)$

55.

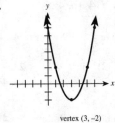

vertex $(3, -2)$

57.

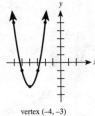

vertex $(-4, -3)$

59.

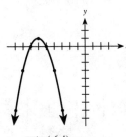

vertex $(-6, 1)$

61.

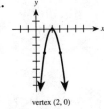

vertex $(2, 0)$

63.

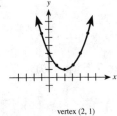

vertex $(2, 1)$

65. 120 dolls, $25,400

67. 30 ft $\times$ 60 ft   **69.** 25 ft $\times$ 50 ft   **71.** Max height at 12 sec, max height = 2304 ft, hits ground after 24 sec

75.

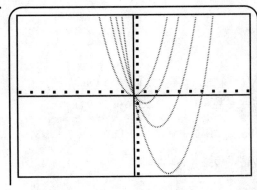

77. (a) $(-1.50, -7.25)$
(b) $(1.50, -1.50)$
(c) $(-3.67, 7.17)$
(d) $(1.35, 3.09)$

79. (a)

x	y
2	4
3	3
4	2
5	1
6	0
7	1
8	2
9	3
10	4

(b) 6

(c) $\sqrt{x^2 - 2(6x - 18)}$
$= \sqrt{x^2 - 12x + 36}$
$= \sqrt{(x - 6)^2}$, which is a minimum
when $x = 6$.

Exercises 7.3 (p. 417)

1. ± 5 **3.** ± 13 **5.** $\pm 4\sqrt{5}$ **7.** $\pm 7\sqrt{2}$ **9.** $\pm 14i$ **11.** $\pm 11i$ **13.** ± 8 **15.** $\pm 3i$ **17.** $\pm \frac{9}{2}i$ **19.** $\pm \frac{11}{6}$

21. $\pm \frac{2\sqrt{10}}{5}$ **23.** $\pm \frac{\sqrt{14}}{7}$ **25.** $\pm \frac{2}{3}$ **27.** $\pm i$ **29.** $\pm \frac{\sqrt{22}}{2}$ **31.** 8 or 2   **33.** 4 or 3   **35.** $\frac{1}{2}$ **37.** $\frac{5 \pm \sqrt{6}}{4}$

39. $\frac{-5 \pm 7i}{2}$ **41.** $1 \pm 5\sqrt{3}i$ **43.** $\frac{-2 \pm \sqrt{5}}{15}$ **45.** $\frac{3}{4}$ or $\frac{-1}{4}$ **47.** $\frac{-1 \pm \sqrt{7}}{3}$ **49.** $\frac{-3 \pm \sqrt{7}i}{2}$ **51.** -2 or 8

53. -7 or -3 **55.** 0 or -3 **57.** $-\frac{2}{3}$ or $\frac{16}{3}$ **59.** $-2 \pm \sqrt{5}$ **61.** $1 \pm 2\sqrt{2}$ **63.** $\frac{-1 \pm \sqrt{3}}{2}$ **65.** $\frac{-4 \pm 4\sqrt{2}}{3}$

67. $-2 \pm 5i$ **69.** $1 \pm 6i$ **71.** $\frac{1 \pm 4i}{2}$ **73.** $\frac{-2 \pm i}{3}$ **75.** -1 or 2   **77.** -3 or -2 **79.** -2 or $\frac{3}{2}$

81. $-\frac{1}{2}$ or 2   **83.** $\frac{-3 \pm \sqrt{2}}{2}$ **85.** $\frac{2 \pm \sqrt{3}}{3}$ **87.** $\frac{1 \pm 2\sqrt{2}}{4}$ **89.** $\frac{-3 \pm 6\sqrt{2}}{4}$ **91.** $\frac{5 \pm 7i}{2}$ **93.** $\frac{-1 \pm 2i}{3}$

95. $\frac{3 \pm 2\sqrt{3}i}{6}$ **97.** $\frac{-8 \pm 3i}{6}$

Exercises 7.4 (p. 425)

1. $\frac{1}{2}$ or 5   **3.** $\frac{3}{2}$ **5.** ± 3 **7.** 0 or 4   **9.** 7 or 8   **11.** $-\frac{4}{3}$ or $-\frac{3}{2}$ **13.** -2 or $\frac{10}{3}$ **15.** $\frac{7 \pm \sqrt{37}}{6}$

17. $\frac{-3 \pm \sqrt{65}}{4}$ **19.** $\frac{-7 \pm \sqrt{61}}{6}$ **21.** $3 \pm \sqrt{2}$ **23.** $\frac{-3 \pm \sqrt{7}}{2}$ **25.** $\frac{-5 \pm \sqrt{31}i}{4}$ **27.** $\frac{1 \pm \sqrt{11}i}{6}$ **29.** $\pm \frac{5}{3}i$

31. $-1 \pm 4i$ **33.** $\frac{-1 \pm 2i}{3}$ **35.** $-\frac{2}{3}$ or $\frac{1}{2}$ **37.** $1 \pm \sqrt{3}$ **39.** $-4 \pm 2i$ **41.** Two rational solutions

43. One rational solution   **45.** Two nonreal solutions   **47.** Two rational solutions   **49.** One rational solution

51. $m = 9$ **53.** $m = 16$ **55.** $m = \pm 6$ **57.** $m > 1$ **59.** $m < -\frac{1}{4}$ **61.** $\frac{-b + \sqrt{b^2 - 4ac}}{2a} + \frac{-b - \sqrt{b^2 - 4ac}}{2a} =$

$\frac{-2b}{2a} = -\frac{b}{a}$ **65.** $\{-0.64, 3.14\}$ **67.** $\{-6.12, -0.85\}$ **69.** $x = \frac{-y \pm \sqrt{y^2 - 144}}{18}$ **71.** $x = \frac{3y \pm \sqrt{9y^2 - 64}}{2}$

73. $x = \frac{9y \pm \sqrt{81y^2 - 64}}{8}$ **75.** $x = -y \pm \sqrt{y^2 - 3y - 10}$

Exercises 7.5 (p. 429)

1. $\left\{\pm \frac{3}{2}\right\}$ **3.** $\left\{\frac{7 \pm \sqrt{41}}{2}\right\}$ **5.** $\left\{\pm \frac{\sqrt{10}}{5}\right\}$ **7.** $\{4, -1\}$ **9.** $\left\{\frac{5}{2} \pm i\right\}$ **11.** $\{\pm \sqrt{3}\}$ **13.** $\left\{0, \frac{5}{3}\right\}$ **15.** $\left\{-\frac{5}{2}, 3\right\}$

17. $\left\{-\frac{7}{6}, \frac{1}{3}\right\}$ **19.** $\left\{\pm \frac{1}{2}i\right\}$ **21.** $\left\{\frac{-1 \pm \sqrt{13}}{4}\right\}$ **23.** $\left\{\frac{-3 \pm \sqrt{33}}{4}\right\}$ **25.** $\left\{\frac{3 \pm \sqrt{3}}{2}\right\}$ **27.** $\left\{\frac{2 \pm \sqrt{2}}{3}\right\}$

29. $\{2 \pm \sqrt{13}\}$ **31.** $\left\{-\frac{8}{3}, -\frac{1}{2}\right\}$ **33.** $\{-3 \pm 2i\}$ **35.** $\left\{0, -\frac{1}{2}\right\}$ **37.** $\{19, -3\}$ **39.** $\left\{\frac{-1 \pm \sqrt{6}}{4}\right\}$ **41.** $\left\{\pm \frac{\sqrt{5}}{2}\right\}$

43. $\{4 \pm i\}$

Exercises 7.6 (p. 435)

1. 6 and 7   **3.** $2 \pm \sqrt{3}$ **5.** 5 and 6   **7.** $1 \pm \sqrt{15}$ **9.** $-1 + \sqrt{17} \doteq 3.1$ in., $1 + \sqrt{17} \doteq 5.1$ in.

11. height $= \frac{1 + \sqrt{97}}{4} \doteq 2.7$ in., base $= \frac{-1 + \sqrt{97}}{2} \doteq 4.4$ in.   **13.** $\frac{-7 + \sqrt{89}}{2} \doteq 1.2$ ft   **15.** 6 in. by 9 in.

17. $-9 + 3\sqrt{14} \doteq 2.2$ ft   **19.** $\frac{7 + \sqrt{65}}{2} \doteq 7.53$ hr and $\frac{9 + \sqrt{65}}{2} \doteq 8.53$ hr   **21.** $5 + \sqrt{37} \doteq 11.1$ days and $7 + \sqrt{37} \doteq$

13.1 days   **23.** Up $= \frac{3 + \sqrt{89}}{10} \doteq 1.2$ mi/hr, down $= \frac{13 + \sqrt{89}}{10} \doteq 2.2$ mi/hr   **25.** $\frac{1 + \sqrt{61}}{2} \doteq 4.4$ mi/hr   **27.** $1 + \sqrt{6} \doteq$

3.4 sec   **29.** Hits ground at $\frac{6 + \sqrt{62}}{2} \doteq 6.95$ sec, max. height at 3 sec, max. height $= 248$ ft

Exercises 7.7 (p. 442)

1. $\pm 1, \pm 2$ **3.** $\pm \frac{1}{2}, \pm 3$ **5.** $\pm 2i, \pm \sqrt{3}$ **7.** $\pm \frac{\sqrt{2}}{2}, \pm \sqrt{5}$ **9.** $\pm \frac{\sqrt{3}}{3}i, \pm i$ **11.** 1, 16   **13.** $\frac{1}{4}, 25$ **15.** 9

17. $\varnothing$ **19.** $-6, 3$ **21.** 1, 5   **23.** $-\frac{1}{4}, 3$ **25.** $\frac{3}{4}, \frac{5}{4}$ **27.** 2   **29.** $-\frac{1}{3}, \frac{1}{2}$ **31.** $-1, -\frac{1}{5}$ **33.** $-\frac{1}{7}, 2$

35. $\frac{3}{2}, 2$ **37.** $-125, 216$ **39.** $\frac{1}{27}, 8$ **41.** $-\frac{1}{8}, \frac{64}{27}$ **43.** $-\frac{5}{4}, -\frac{3}{4}$ **45.** $-\frac{1}{5}, -\frac{1}{7}$ **47.** $-9, \frac{3}{4}$ **49.** $-3, 5$

51. $\pm \sqrt{4 \pm \sqrt{13}}$ **53.** $-7, \pm 1$ **55.** $-\frac{2}{3}, \pm 2$ **57.** $\pm \frac{3}{2}, 3$ **59.** $-1, \pm \frac{1}{3}$ **61.** $\pm \frac{1}{2}, \frac{3}{2}$ **63.** $\pm i, \frac{2}{5}$

65. $\pm \frac{1}{2}i, \frac{3}{2}$ **67.** $\frac{-1 \pm \sqrt{3}i}{2}, \frac{3}{5}, 1$ **69.** $\frac{3 \pm 3\sqrt{3}i}{2}, -3$ **71.** $\pm i, \pm 1, 5$ **73.** $\frac{1 \pm \sqrt{3}i}{2}, \frac{-1 \pm \sqrt{3}i}{2}, \pm 1, -3$

77. $\frac{1}{25}$ **79.** $\frac{3}{16}$ **81.** $\frac{1}{64}$ **83.** 3

Exercises 7.8 (p. 456)

1. $-5 < x < 4$
$(-5, 4)$

3. $x < -6$ or $x > -3$
$(-\infty, -6) \cup (-3, \infty)$

5. $x < -\frac{3}{2}$ or $x > 0$
$(-\infty, -\frac{3}{2}) \cup (0, \infty)$

7. $x \le -8$ or $x \ge -2$
$(-\infty, -8] \cup [-2, \infty)$

9. $1 \le x \le 6$
$[1, 6]$

11. $x < -7$ or $x > 7$
$(-\infty, -7) \cup (7, \infty)$

13. $-\frac{3}{2} < x < 2$
$(-\frac{3}{2}, 2)$

15. $x < -\frac{9}{4}$ or $x > -\frac{1}{2}$
$(-\infty, -\frac{9}{4}) \cup (-\frac{1}{2}, \infty)$

17. All real numbers
$(-\infty, \infty)$

19. $\varnothing$

21. $x < 3 - \sqrt{2}$ or $x > 3 + \sqrt{2}$
$(-\infty, 3 - \sqrt{2}) \cup (3 + \sqrt{2}, \infty)$

23. $-1 - \sqrt{6} \le x \le -1 + \sqrt{6}$
$[-1 - \sqrt{6}, -1 + \sqrt{6}]$

25. $-2 < x < 3$
$(-2, 3)$

27. $x \le -4$ or $x \ge -\frac{3}{2}$
$(-\infty, -4] \cup [-\frac{3}{2}, \infty)$

29. $\varnothing$

31. $x < -\frac{1}{4}$ or $x > \frac{1}{4}$
$(-\infty, -\frac{1}{4}) \cup (\frac{1}{4}, \infty)$

33. $x \le 3$ or $x \ge 8$
$(-\infty, 3] \cup [8, \infty)$

35. $\frac{3}{4} < x < \frac{3}{2}$
$(\frac{3}{4}, \frac{3}{2})$

37. $x \le -\frac{1}{2}$ or $x \ge \frac{7}{2}$
$(-\infty, -\frac{1}{2}] \cup [\frac{7}{2}, \infty)$

39. $x \le \frac{3 - \sqrt{5}}{2}$ or $x \ge \frac{3 + \sqrt{5}}{2}$
$(-\infty, \frac{3 - \sqrt{5}}{2}] \cup [\frac{3 + \sqrt{5}}{2}, \infty)$

41. $x < -\frac{1}{2}$ or $x > \frac{1}{2}$
$(-\infty, -\frac{1}{2}) \cup (\frac{1}{2}, \infty)$

43. $\varnothing$

45. All real numbers
$(-\infty, \infty)$

47. All real numbers
$(-\infty, \infty)$

49. $\varnothing$

51. $-3 < x < 2$
$(-3, 2)$

53. $x \le -\frac{1}{3}$ or $x \ge \frac{3}{2}$
$(-\infty, -\frac{1}{3}] \cup [\frac{3}{2}, \infty)$

55. $x < -\frac{4}{3}$ or $x > \frac{4}{3}$
$(-\infty, -\frac{4}{3}) \cup (\frac{4}{3}, \infty)$

57. $-6 < x < -2$
$(-6, -2)$

59. $x \le 2 - \sqrt{7}$ or $x \ge 2 + \sqrt{7}$
$(-\infty, 2 - \sqrt{7}] \cup [2 + \sqrt{7}, \infty)$

61. $x < -1$ or $x > -\frac{1}{2}$
$(-\infty, -1) \cup (-\frac{1}{2}, \infty)$

63. $3 \le x \le 5$
$[3, 5]$

65. All real numbers
$(-\infty, \infty)$

67. $-\frac{5}{2} \le x \le -1$
$[-\frac{5}{2}, -1]$

69. $x < \frac{1}{3}$ or $x > 1$
$(-\infty, \frac{1}{3}) \cup (1, \infty)$

$$x < \frac{-1 - \sqrt{2}}{5} \text{ or } x > \frac{-1 + \sqrt{2}}{5}$$

$$(-\infty, \frac{-1 - \sqrt{2}}{5}) \cup (\frac{-1 + \sqrt{2}}{5}, \infty)$$

71. ∅

73. ◄—○——○——►
$\frac{-1 - \sqrt{2}}{5}$ $\frac{-1 + \sqrt{2}}{5}$

$$-\frac{4}{3} \le x \le \frac{3}{2}$$
$$[-\frac{4}{3}, \frac{3}{2}]$$

75. ◄——●———●——►
$-\frac{4}{3}$ $\frac{3}{2}$

$$\frac{2 - \sqrt{10}}{3} < x < \frac{2 + \sqrt{10}}{3}$$
$$(\frac{2 - \sqrt{10}}{3}, \frac{2 + \sqrt{10}}{3})$$

77. ◄——○———○——►
$\frac{2 - \sqrt{10}}{3}$ $\frac{2 + \sqrt{10}}{3}$

$$-5 < x < 3$$
$$(-5, 3)$$

79. ◄—○││││││○—►
-5 3

81. ∅

$$0 < x < 4$$
$$(0, 4)$$

83. ◄—○│││○—►
0 4

$$-1 < x < 4$$
$$(-1, 4)$$

85. ◄—○│││││○—►
-1 4

$$x \le \frac{3}{2} \text{ or } x > 7$$
$$(-\infty, \frac{3}{2}] \cup (7, \infty)$$

87. ◄————●│││││○—►
$\frac{3}{2}$ 7

$$x \le -\frac{3}{2} \text{ or } x > -1$$
$$(-\infty, -\frac{3}{2}] \cup (-1, \infty)$$

89. ◄——●—○——►
$-\frac{3}{2}$ -1

$$x < -4 \text{ or } x > 4$$
$$(-\infty, -4) \cup (4, \infty)$$

91. ◄—○│││││││○—►
-4 4

$$-\frac{2}{3} \le x < 1$$
$$[-\frac{2}{3}, 1)$$

93. ◄——●———○——►
$-\frac{2}{3}$ 1

$$-3 < x < 1$$
$$(-3, 1)$$

95. ◄——○│││○——►
-3 1

$$-3 < x < 1 \text{ or } x > 2$$
$$(-3, 1) \cup (2, \infty)$$

97. ◄—○│││○—○—►
-3 $1\ 2$

$$x \le -1 \text{ or } \frac{1}{2} \le x \le 3$$
$$(-\infty, -1] \cup [\frac{1}{2}, 3]$$

99. ◄——●│○│●——►
-1 $\frac{1}{2}$ 3

$$x \ge -3$$
$$[-3, \infty)$$

101. ◄——●————►
-3

$$x < -2.43 \text{ or } x > 3.65$$
$$(-\infty, -2.43) \cup (3.65, \infty)$$

107. ◄—○│││││││○—►
-2.43 3.65

$$x < -2.17 \text{ or } x > 1.62$$
$$(-\infty, -2.17) \cup (1.62, \infty)$$

109. ◄—○│││││││○—►
-2.17 1.62

$$x > \frac{1}{4}$$

111. ◄│││○││││►
$\frac{1}{4}$

$$0 \le x \le 9$$

113. ◄│││●│││││││●│││►
0 9

$$x > 3$$

115. ◄│││○│││││►
3

$$-5 \le x \le -1$$

117. ◄│●│││●│││││►
-5 -1

$$-3 < x < 2$$

119. ◄│││○││││○││►
-3 2

$$x \le -\frac{1}{2}$$

121. ◄◄│││●│││││►
$-\frac{1}{2}$

123. All real numbers

125. ∅

Use the solutions to the corresponding equation to partition the number line into intervals. Check a point from each interval. In exercises 110–117 the domain is restricted.

APPLIED ALGEBRA—YOUR TURN (p. 460)

1. 54 lb/sq in.   **3.** 300 gal/min

CHAPTER 7   Review Exercises (p. 463)

1.

vertex (0, 0)

3.

vertex (0, 0)

5.

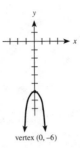

vertex (0, –6)

7.

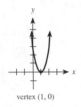

vertex (1, 0)

9.

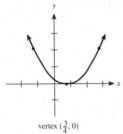

vertex $(\frac{3}{4}, 0)$

11.

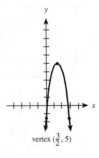

vertex $(\frac{3}{2}, 5)$

13.

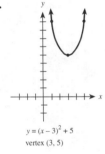

$y = (x - 3)^2 + 5$
vertex (3, 5)

15.

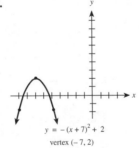

$y = -(x + 7)^2 + 2$
vertex (–7, 2)

17.

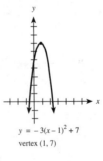

$y = -3(x - 1)^2 + 7$
vertex (1, 7)

19.

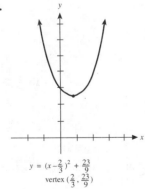

$y = (x - \frac{2}{3})^2 + \frac{23}{9}$
vertex $(\frac{2}{3}, \frac{23}{9})$

21.

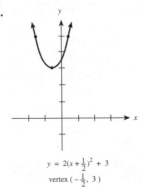

$y = 2(x + \frac{1}{2})^2 + 3$
vertex $(-\frac{1}{2}, 3)$

23.

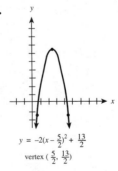

$y = -2(x - \frac{5}{2})^2 + \frac{13}{2}$
vertex $(\frac{5}{2}, \frac{13}{2})$

25. 90 tires, profit = \$2140   **27.** ± 4 **29.** ± 11 **31.** $\pm \frac{4}{3}i$ **33.** $\pm \frac{\sqrt{3}}{3}i$ **35.** $\frac{-1 \pm \sqrt{13}}{3}$ **37.** 1 or 5

39. $1 \pm \sqrt{2}$ **41.** $\frac{2 \pm 4i}{3}$ **43.** $\frac{-4 \pm \sqrt{7}}{3}$ **45.** $\frac{1 \pm \sqrt{3}}{2}$ **47.** $-\frac{1}{3}$ or 2   **49.** $\pm \frac{2\sqrt{5}}{5}$ **51.** $-\frac{9}{2}$ or 2

53. $-3 \pm \sqrt{2}$ **55.** $3 \pm i$ **57.** $-\frac{1}{3}$ or 4   **59.** $2 \pm 3i$ **61.** One rational solution   **63.** Two irrational solutions

65. $\frac{1}{2}$ and $\frac{3}{2}$ **67.** $60\sqrt{5} \doteq 134.16$ ft   **69.** $-5 + 2\sqrt{14} \doteq 2.5$ ft   **71.** $\frac{3 + \sqrt{29}}{2} \doteq 4.19$ sec   **73.** $\pm 3i, \pm 2$

75. $1, 9$ **77.** $-5, -\frac{5}{4}$ **79.** $-\frac{1}{5}, -\frac{1}{4}$ **81.** $-8, 125$ **83.** $\frac{1}{2}, \frac{5}{4}$ **85.** $\pm 3, -\frac{4}{3}$ **87.** $\pm \frac{1}{2}, -3$

89. $1, \frac{6}{7}, \frac{-1 \pm \sqrt{3}i}{2}$

91. $x \le -\frac{5}{2}$ or $x \ge 1$

93. All real numbers

95. $2 < x < 5$

97. $x < -1 - \sqrt{5}$ or $x > -1 + \sqrt{5}$

99. $5 - 3\sqrt{2} < x < 5 + 3\sqrt{2}$

101. $x = -\frac{4}{3}$

103. $x < -4$ or $x \ge \frac{3}{2}$

105. $-\frac{2}{3} < x < 2$

107. $-3 \le x \le 4$ or $x \ge 7$

CHAPTER 7   Test (p. 466)

1.
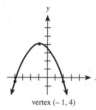
vertex $(-1, 4)$

3. $-\frac{11}{2}$ or $-\frac{3}{2}$ **5.** $\frac{-3 \pm 2\sqrt{2}}{2}$ **7.** $\pm \frac{3\sqrt{2}}{2}i, \pm 2\sqrt{2}$ **9.** $\pm \frac{1}{3}, -\frac{5}{3}$

11. $x \le -\frac{3}{2}$ or $x \ge 4$
$(-\infty, -\frac{3}{2}] \cup [4, \infty)$

13. $4 < x \le 5$
$(4, 5]$

15. $-50 + 10\sqrt{29} \doteq 3.85$ yd

CHAPTERS 1–7   Test Your Memory (p. 467)

1. $\frac{1}{4x^3y^{11}}$ **3.** $\frac{1}{3x^2(5x + 8)}$ **5.** $\frac{2x - 5}{x - 7}$ **7.** $-7\sqrt[3]{5}$ **9.** $\frac{x + 2\sqrt{xy} + y}{x - y}$

11.
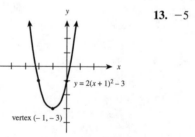
$y = 2(x + 1)^2 - 3$
vertex $(-1, -3)$

13. -5

15. 38 **17.** $y = -\frac{2}{5}x + \frac{2}{5}$ **19.** $y = -\frac{2}{3}x - \frac{20}{3}$ **21.** $\left\{-\frac{17}{3}\right\}$ **23.** $\{-5, 2\}$ **25.** $\varnothing$ **27.** $\{7\}$

29. $\left\{-\frac{3}{2}, -\frac{1}{3}\right\}$ **31.** $\left\{\pm 2, \pm \frac{1}{3}\right\}$ **33.** $\left\{-\frac{4}{3}, 2\right\}$ **35.** $\{2 \pm \sqrt{6}\}$ **37.** $x \le 4$
$(-\infty, 4]$

39. $-2 < x < 4$
$(-2, 4)$

41. $x < -6$ or $x > 2$
$(-\infty, -6) \cup (2, \infty)$

43.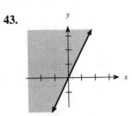

45. 12 nickels, 11 dimes, 6 quarters   **47.** 216 sq in.   **49.** $-5 + \sqrt{55} \doteq 2.4$ ft

CHAPTER 8   Exercises 8.1 (p. 477)

1.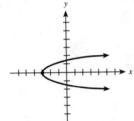

3.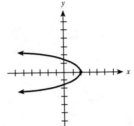

5.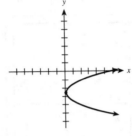

7.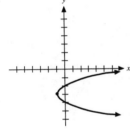

9.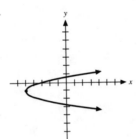

11.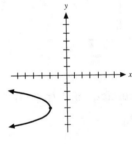

13.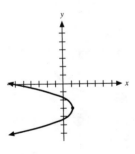

15.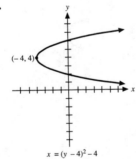

$(-4, 4)$

$x = (y - 4)^2 - 4$

17.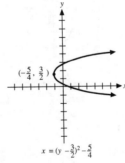

$\left(-\frac{5}{4}, \frac{3}{2}\right)$

$x = \left(y - \frac{3}{2}\right)^2 - \frac{5}{4}$

19.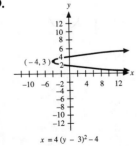

$(-4, 3)$

$x = 4(y - 3)^2 - 4$

21.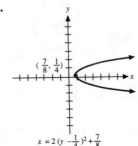

$\left(\frac{7}{8}, \frac{1}{4}\right)$

$x = 2\left(y - \frac{1}{4}\right)^2 + \frac{7}{8}$

23.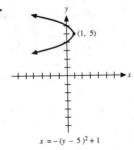

$(1, 5)$

$x = -(y - 5)^2 + 1$

25.

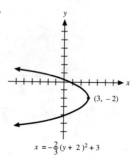

$x = -\frac{2}{3}(y + 2)^2 + 3$

$(3, -2)$

29.

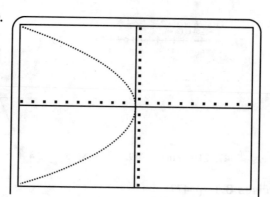

31.

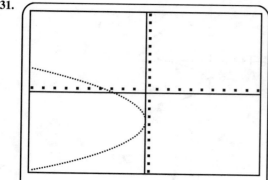

Exercises 8.2 (p. 484)

1.

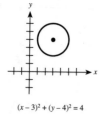

$(x - 3)^2 + (y - 4)^2 = 4$

3.

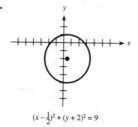

$(x - \frac{1}{2})^2 + (y + 2)^2 = 9$

5.

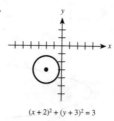

$(x + 2)^2 + (y + 3)^2 = 3$

7.

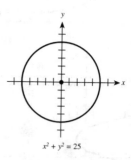

$x^2 + y^2 = 25$

9.

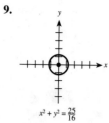

$x^2 + y^2 = \frac{25}{16}$

11.

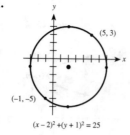

$(5, 3)$

$(-1, -5)$

$(x - 2)^2 + (y + 1)^2 = 25$

13.

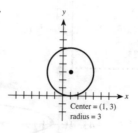

Center $= (1, 3)$
radius $= 3$

15.

Center $= (4, -2)$
radius $= 2$

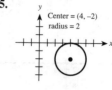

17.

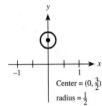

Center $= (0, \frac{3}{2})$
radius $= \frac{1}{2}$

19.

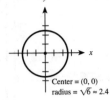

Center $= (0, 0)$
radius $= \sqrt{6} \approx 2.4$

21.

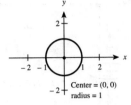

Center $= (0, 0)$
radius $= 1$

23.

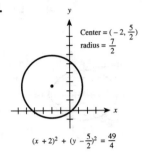

Center $= (-2, \frac{5}{2})$
radius $= \frac{7}{2}$

$(x + 2)^2 + (y - \frac{5}{2})^2 = \frac{49}{4}$

25.

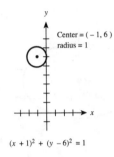

Center $= (-1, 6)$
radius $= 1$

$(x + 1)^2 + (y - 6)^2 = 1$

27.

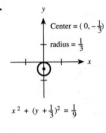

Center $= (0, -\frac{1}{3})$
radius $= \frac{1}{3}$

$x^2 + (y + \frac{1}{3})^2 = \frac{1}{9}$

29.

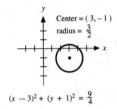

Center $= (3, -1)$
radius $= \frac{3}{2}$

$(x - 3)^2 + (y + 1)^2 = \frac{9}{4}$

33.

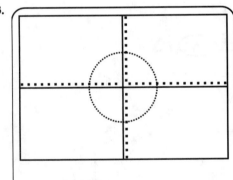

35.

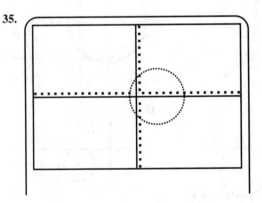

37. π **39.** $2\pi - 1$ **41.** $\dfrac{25\pi}{4} - 4$ **43.** $4\pi - 8$

ANSWERS TO ODD-NUMBERED PROBLEMS

Exercises 8.3 (p. 493)

1.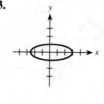

3.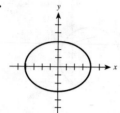

5.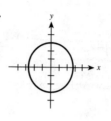

7.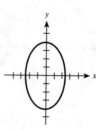

9.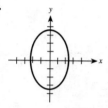

11.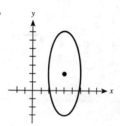

13.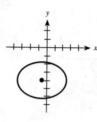

15.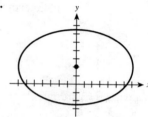

17.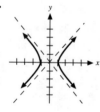

19.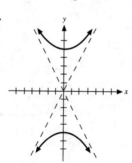

21.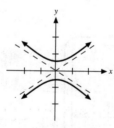

Exercises 8.4 (p. 501)

1.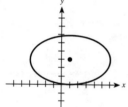

3.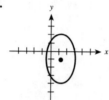

5.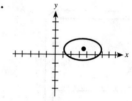

7.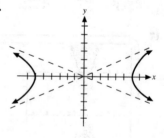

9.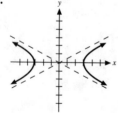

11.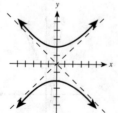

13.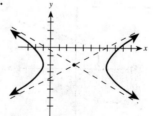

15.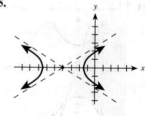

17.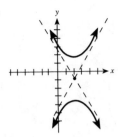

19.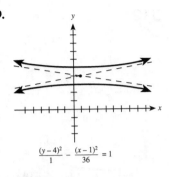

$$\frac{(y-4)^2}{1} - \frac{(x-1)^2}{36} = 1$$

21.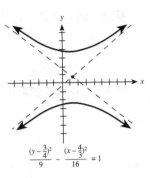

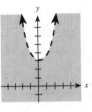

$$\frac{(y-\frac{3}{4})^2}{9} - \frac{(x-\frac{4}{3})^2}{16} = 1$$

Exercises 8.5 (p. 509)

1.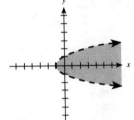

3.

5.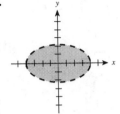

7.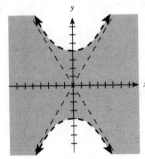

9.

11.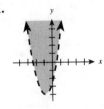

13.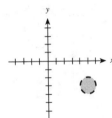

15.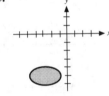

17.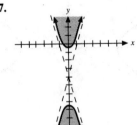

19.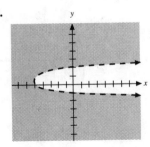

21.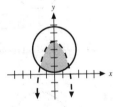

23.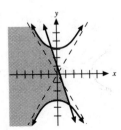

25.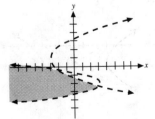

27.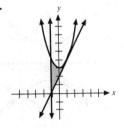

APPLIED ALGEBRA—YOUR TURN (p. 512)

1. $\dfrac{x^2}{(7.5)^2} + \dfrac{y^2}{(9.125)^2} = 1$; 79.5 in. = 6 ft 7.5 in.

CHAPTER 8   Review Exercises (p. 517)

1.

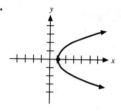

3.

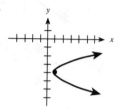

5.

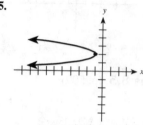

7.

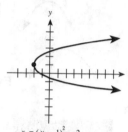

$x = (y - 1)^2 - 2$

9.

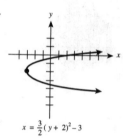

$x = \dfrac{3}{2}(y + 2)^2 - 3$

11.

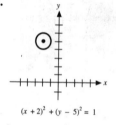

$(x + 2)^2 + (y - 5)^2 = 1$

13.

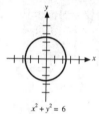

$x^2 + y^2 = 6$

15.

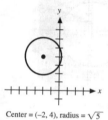

Center = (-2, 4), radius = $\sqrt{5}$

17.

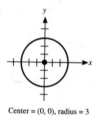

Center = (0, 0), radius = 3

19.

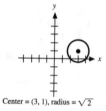

Center = (3, 1), radius = $\sqrt{2}$

21.

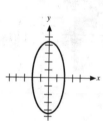

23.

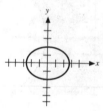

25.

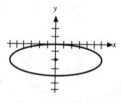

27.

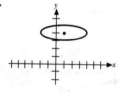

29.

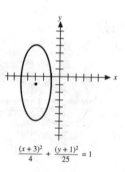

$\dfrac{(x + 3)^2}{4} + \dfrac{(y + 1)^2}{25} = 1$

31.

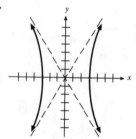

33.

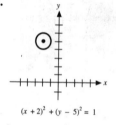

35.

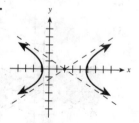

37.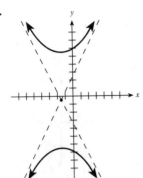

39.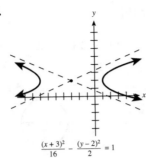

$$\frac{(x+3)^2}{16} - \frac{(y-2)^2}{2} = 1$$

41.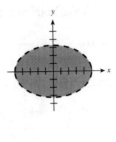

43.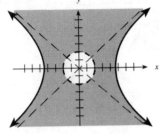

45.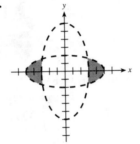

47.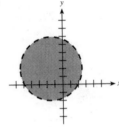

49.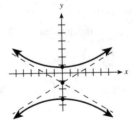

51.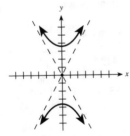

53.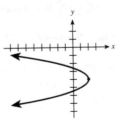

CHAPTER 8   Test (p. 518)

1.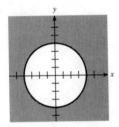

3.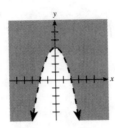

5.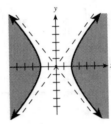

7.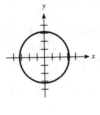

9. $(x + 3)^2 + (y + 1)^2 = 52$ **11.**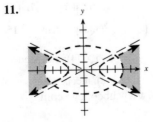

CHAPTERS 1–8 Test Your Memory (p. 519)

1. $\dfrac{27x^4}{y^2}$ **3.** 1 **5.** $-x\sqrt{3xy}$ **7.** $\dfrac{\sqrt{5xy}}{3y^2}$ **9.** $\dfrac{5}{2}$ **11.** 81 **13.** $y = 3x$ **15.** $y = -\dfrac{5}{2}x + \dfrac{1}{2}$

17. $(x - 2)^2 + (y + 3)^2 = 25$ **19.** 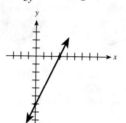 **21.** 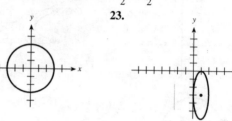 **23.**

25. $\{3 \pm 2i\}$ **27.** $\left\{-\dfrac{5}{2}, -\dfrac{3}{2}\right\}$ **29.** $\left\{-\dfrac{1}{10}\right\}$ **31.** $\left\{-\dfrac{12}{5}, \dfrac{14}{5}\right\}$ **33.** $\{2, -2, 4, -4\}$ **35.** $\{6\}$ **37.** $\{0\}$

39. 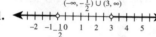 **41.** 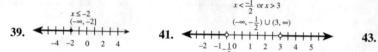 **43.**

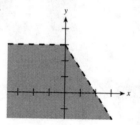

45. height is 4 in.; base is 11 in. **47.** 2 hr **49.** $3 + \sqrt{10} \doteq 6.2$ sec

CHAPTER 9 Exercises 9.1 (p. 530)

1. $(-1, 4)$; independent **3.** $(-2, 0)$; independent **5.** $\varnothing$; inconsistent **7.** $\{(x,y): 3x - y = 1\}$; dependent
9. $(-1, -3)$; independent **11.** $\left(\dfrac{1}{3}, \dfrac{7}{3}\right)$; independent **13.** $\varnothing$; inconsistent **15.** $\left(\dfrac{38}{5}, -\dfrac{17}{5}\right)$; independent
17. $\left(-\dfrac{13}{7}, -\dfrac{19}{7}\right)$; independent **19.** $(-13, -10)$; independent **21.** dependent **23.** $\left(-\dfrac{13}{9}, \dfrac{8}{3}\right)$; independent
25. $\left(\dfrac{5}{2}, \dfrac{9}{10}\right)$; independent **27.** $(-5, -3)$; independent **29.** $\varnothing$; inconsistent **31.** $\left(\dfrac{11}{16}, \dfrac{15}{8}\right)$; independent
33. $\left(-\dfrac{34}{29}, -\dfrac{2}{29}\right)$; independent **35.** dependent **37.** $(-2, 4)$; independent **39.** dependent **41.** $\left(\dfrac{21}{8}, -\dfrac{27}{2}\right)$;
independent **43.** $\left(\dfrac{1}{3}, 1\right)$ **45.** $(-3, -4)$ **47.** $\left(\dfrac{1}{23}, \dfrac{1}{16}\right)$ **51.** $(3, 1)$ **53.** $(-1.\overline{4}, 2.\overline{6})$ **55.** $(-3.186, 1.685)$

Exercises 9.2 (p. 537)

1. $(2, 1, 3)$; independent **3.** $(0, 1, 5)$; independent **5.** $(-1, -2, 4)$; independent **7.** $(-1, 4, 1)$; independent
9. $(0, 2, 3)$; independent **11.** $\left(-\dfrac{1}{2}, \dfrac{3}{2}, 1\right)$; independent **13.** $\left(\dfrac{1}{3}, \dfrac{2}{3}, \dfrac{1}{2}\right)$; independent **15.** $(0, 1, 5)$; independent
17. $\left(\dfrac{1}{2}, -\dfrac{2}{3}, \dfrac{4}{3}\right)$; independent **19.** $(-1, -2, -3)$; independent **21.** $(-1, 3, 4)$; independent **23.** Dependent
25. Dependent **27.** Inconsistent **29.** Inconsistent **31.** $(-1, 0, 3, 2)$; independent **37.** 6 days **39.** 2 hr
Internet Connection (p. 541) answers can be found at http://www.hbcollege.com

Exercises 9.3 (p. 548)

1. 4 and 21   **3.** 8 and -2 **5.** 4 and -6 **7.** 24 nickels, 16 dimes   **9.** 12 twenty-cent stamps, 9 thirty-two-cent stamps
11. 12 adult tickets, 25 children's tickets   **13.** 14 nickels, 8 dimes, 12 quarters   **15.** Pencil = 14¢, eraser = 6¢
17. Football = $9, basketball = $14   **19.** Saw = $22, hammer = $10, screwdriver = $4   **21.** 12 lb of Mr. Peanut, 8 lb of
King Peanut   **23.** $1000 in 9% account, $6000 in 11% account   **25.** 6.25 gal of Lone Star, 3.75 gal of Big Time
27. $A = 2, B = -1, C = 4$ **29.** 45   **31.** $x^2 + y^2 - 4x + 2y - 20 = 0$ **33.** 9 ft by 8 ft   **35.** 10 mi/hr and 4 mi/hr

Exercises 9.4 (p. 559)

1. $\{(0, 2), (2, 0)\}$ **3.** $\left\{(-2, 0), \left(\frac{6}{5}, \frac{8}{5}\right)\right\}$ **5.** $\{(2, 4)\}$ **7.** No real solution   **9.** $\{(4, 2)\}$ **11.** $\{(-2, -2), (-1, -4)\}$

13. $\left\{\left(-\frac{1}{3}, -3\right), \left(2, \frac{1}{2}\right)\right\}$ **15.** $\left\{\left(-4, -\frac{3}{5}\right), \left(3, \frac{4}{5}\right)\right\}$ **17.** No real solutions   **19.** $\{(7, 4), (-5, -2)\}$

21. $\left\{(0, -1), \left(\frac{3}{2}, \frac{5}{4}\right)\right\}$ **23.** $\{(2\sqrt{2}, 2), (2\sqrt{2}, -2), (-2\sqrt{2}, 2), (-2\sqrt{2}, -2)\}$ **25.** $\{(1, 1), (1, -1), (-1, 1), (-1, -1)\}$

27. No real solutions   **29.** $\left\{\left(-\frac{1}{2}, \frac{\sqrt{7}}{2}\right), \left(-\frac{1}{2}, -\frac{\sqrt{7}}{2}\right), (-1, 1), (-1, -1)\right\}$ **31.** $\{(0, 1)\}$ **33.** $\{(4, 2\sqrt{2}), (4, -2\sqrt{2})\}$

35. $\{(1, 2), (2, 1), (-1, -2), (-2, -1)\}$ **37.** $\left\{\left(\frac{1}{2}, 6\right), \left(-\frac{1}{2}, -6\right), (3, 1), (-3, -1)\right\}$ **39.** $\left\{\left(\frac{3}{2}, 2\right)\right\}$ **41.** 4 and 7

43. 4×12 ft   **45.** 3 and 4 ft   **47.** $\frac{x^2}{16} + \frac{y^2}{4} = 1$ **51.** $\{(-2.351, 5.526), (0.851, 0.724)\}$

53. $\{(-4.846, 3.949), (1.513, 1.829)\}$ **55.** $\pm 5\sqrt{2}$ **57.** $-\frac{1}{4}$ **59.** $\frac{17}{4}$

APPLIED ALGEBRA—YOUR TURN (p. 563)

1. $R(x) = 0.45x, C(x) = 0.20x + 5.00$ **3.** 20 lb   **5.** 180 lb

CHAPTER 9   Review Exercises (p. 565)

1. $(-1, -3)$; independent   **3.** $\varnothing$; inconsistent   **5.** $(3, 5)$, independent   **7.** $\{(x, y): x = 3y - 7\}$; dependent
9. $\left(-\frac{1}{2}, -\frac{5}{2}\right)$; independent   **11.** $(4, 0)$; independent   **13.** $\varnothing$; inconsistent   **15.** $(-1, -1)$; independent
17. $(2, -1, 3)$; independent   **19.** $\left(\frac{3}{2}, -\frac{5}{2}, \frac{1}{2}\right)$; independent   **21.** Dependent   **23.** $(3, -1, -1)$; independent
25. Dependent   **27.** 7 and 12   **29.** 22 adult tickets; 10 children's tickets   **31.** 25 qt of "Apple of Your Eye";
15 qt of "Generic"   **33.** 8 oz of Big D; 4 oz of Big G   **35.** Football = $18; basketball = $14; soccer ball = $15
37. $\{(2, -1), (-3, -6)\}$ **39.** $\left\{\left(-\frac{5}{2}, -8\right), (4, 5)\right\}$ **41.** $\left\{(0, -3), \left(-\frac{3}{4}, -\frac{15}{4}\right)\right\}$ **43.** $\{(\sqrt{7}, \sqrt{2}), (\sqrt{7}, -\sqrt{2}),$
$(-\sqrt{7}, \sqrt{2}), (-\sqrt{7}, -\sqrt{2})\}$ **45.** $\left\{(3, 1), (-3, 1), \left(\frac{\sqrt{29}}{2}, -\frac{3}{4}\right), \left(-\frac{\sqrt{29}}{2}, -\frac{3}{4}\right)\right\}$ **47.** $\{(4, 2), (-4, -2), (2, 4), (-2, -4)\}$
49. 6 ft $\times$ 12 ft

CHAPTER 9   Test (p. 567)

1. $\left(-\frac{11}{4}, -\frac{7}{2}\right)$; independent   **3.** $\varnothing$; inconsistent   **5.** $(-1, 2, 4)$; independent   **7.** $\left\{(-1, 0), \left(\frac{3}{5}, \frac{4}{5}\right)\right\}$
9. $\{(5, 2), (-5, -2), (2, 5), (-2, -5)\}$ **11.** 10 lb of "Let Your Garden Grow"; 20 lb of "This Might Work"

CHAPTERS 1–9   Test Your Memory (p. 568)

1. $\frac{25}{9x^{14}y^{12}}$ **3.** $-12x\sqrt{3x}$ **5.** $\frac{2x - 1}{(x + 4)(x - 2)}$ **7.** 0   **9.** $-\frac{27}{125}$ **11.** $y = -\frac{1}{3}x + \frac{13}{3}$ **13.** $x^2 + (y + 4)^2 = 36$

15.

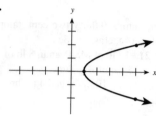

17.

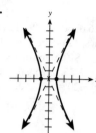

19.

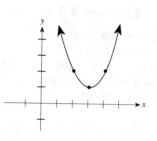

21. $\left\{2, -2, -\dfrac{1}{3}\right\}$

23. $\varnothing$ **25.** $\left\{-\dfrac{7}{2}\right\}$ **27.** $\left\{\dfrac{1 \pm 3i}{2}\right\}$ **29.** $\{-5\}$ **31.** $\left(\dfrac{3}{2}, -\dfrac{1}{2}\right)$; independent   **33.** $(-4, 1)$; independent

35. $(-2, 1, 4)$; independent   **37.** $\{(4, 1), (-4, -1), (1, 4), (-1, -4)\}$ **39.**

$$-1 < x < 5$$
$$(-1, 5)$$
◄─┼┼┼┼◇┼┼┼┼◇┼┼┼►
  −5−4−3−2−1 0 1 2 3 4 5 6 7 8 9

41.

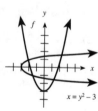

43. 7 quarters, 15 nickels   **45.** 90° F   **47.** 5 mi/hr

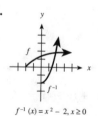

49. Ballpoint pen = 49¢, pencil = 12¢

Chapter 10   Exercises 10.1 (p. 584)

1.

$f^{-1}(x) = \dfrac{1}{3}x - \dfrac{1}{3}$

3.

$f^{-1}(x) = -\dfrac{1}{2}x + \dfrac{3}{2}$

5.

$f^{-1}(x) = 4x - 3$

7.

$f^{-1}(x) = \dfrac{3}{2}x + \dfrac{5}{2}$

9.

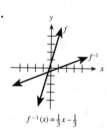
$x = y^2 - 3$

11.

$x = \dfrac{1}{2}y^2 + 2$

13.

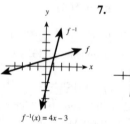
$x = 5 - 2y^2$

15.

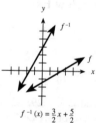
$x = \dfrac{6 - y^2}{4}$

17.

$f^{-1}(x) = x^2 - 2,\ x \geq 0$

19.

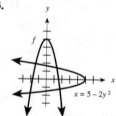
$f^{-1}(x) = x^2 + 1,\ x \geq 0$

21.

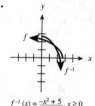
$f^{-1}(x) = \dfrac{-x^2 + 5}{2},\ x \geq 0$

23.

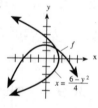
$f^{-1}(x) = (x - 1)^2,\ x \geq 1$

25.

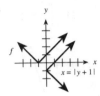

$x = |y + 1|$

27.

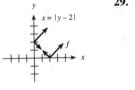

$x = |y - 2|$

29.

$x = |y| + 2$

31.

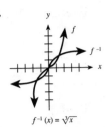

$f^{-1}(x) = \sqrt[3]{x}$

33.

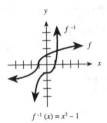

$f^{-1}(x) = x^3 - 1$

35. $f^{-1}(x) = \sqrt[5]{x}$ **37.** $f^{-1}(x) = \dfrac{3 - 2x}{x}$ **39.** $f^{-1}(x) = \dfrac{4x}{x - 1}$ **41.** Not 1 to 1

43. Not 1 to 1   **45.** Not 1 to 1   **47.** $f^{-1}(x) = x^2 - 2; x \geq 0$ **49.** $f^{-1}(x) = x^3 + 5$ **51. (a)** 3   **(b)** 13
(c) $7 - 2x$ **(d)** -4 **(e)** 4   **(f)** $2 - 2x$ **53. (a)** 10   **(b)** 65   **(c)** $x^2 - 10x + 26$ **(d)** 5   **(e)** -3
(f) $x^2 - 4$ **55. (a)** 4   **(b)** 4   **(c)** 4   **(d)** 8   **(e)** 8   **(f)** 8   **57.** Not inverses   **59.** Inverses

61. Inverses   **63.** Not inverses   **69.**

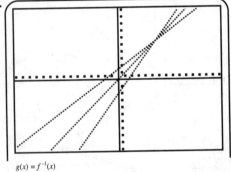

$g(x) = f^{-1}(x)$

71.

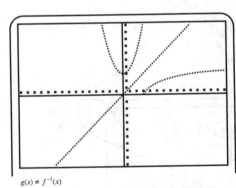

$g(x) \neq f^{-1}(x)$

73. -1 **75.** -2 **77.** -1

Exercises 10.2 (p. 594)

1.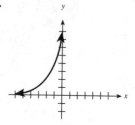

3.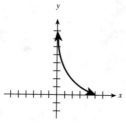

5.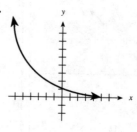

7.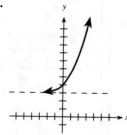

9.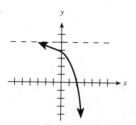

11.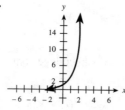

13.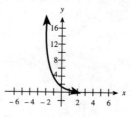

15.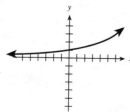

17.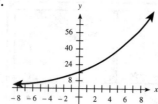

19.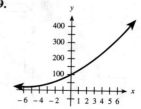

21. $\dfrac{1}{25}$ **23.** 9 **25.** $\dfrac{1}{125}$ **27.** 4 **29.** 27 **31.** 216 **33.** $\dfrac{1}{125}$ **35.** 8 **37.** 3 **39.** -2

41. 2 **43.** $-\dfrac{2}{3}$ **45.** -1 **47.** $\dfrac{9}{2}$ **49.** $-\dfrac{2}{3}$

51.

1940	6,400,000
1950	7,801,564
1960	9,510,063
1970	11,592,714
1980	14,131,454
2000	20,998,597

53. $5412.16, $7429.74, 9 yr

55. $1239.72, $7011.89 **57.** 57.8°, 65° **61.**

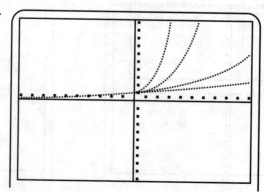

63.

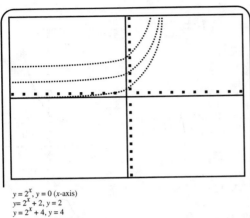

$y = 2^x, y = 0$ (x-axis)
$y = 2^x + 2, y = 2$
$y = 2^x + 4, y = 4$

65. $\dfrac{12}{5}$ **67.** -3 **69.** $\varnothing$ **71.** All real numbers

Internet Connection (p. 599) answers can be found at http://www.hbcollege.com

Exercises 10.3 (p. 606)

1. $\log_5 \dfrac{1}{125} = -3$ **3.** $\log 10,000 = 4$ **5.** $\log_6 6 = 1$ **7.** $x = \log_3 y$ **9.** $x + 2 = \log_3 y$ **11.** $2t = \ln P$

13. $12t = \log_{1.02} M$ **15.** $kt = \log_a \dfrac{Q}{Q_0}$ **17.** $8^2 = 64$ **19.** $9^{1/2} = 3$ **21.** $10^1 = 10$ **23.** $5^y = x$

25. $5^y = 2x + 1$ **27.** $27^x = 9$ **29.** $3^4 = 1 - 3x$ **31.** $x^2 = 16$ **33.** $10^4 = x + 1$ **35.** $e^y = x^2$

37. (a)   **(b)**   **39. (a)**   **(b)**

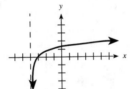

41.   **43.**   **45.**   **47. (a)** 12.603821

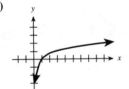
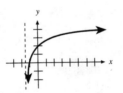
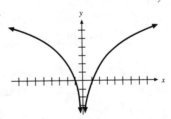
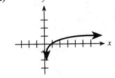
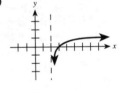

(b) $(4.6754)10^{61}$ **(c)** 1.0038574   **(d)** .00000351   **49.** 22; no

53.   **55.**

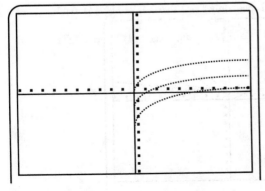
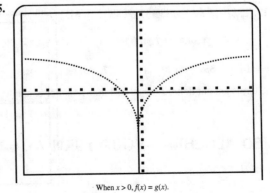

When $x > 0$, $f(x) = g(x)$.

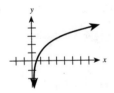

57. -0.5643 **59.** 0.9551 **61.** 0.9885 **63.** 0.9995

65. As x increases, $f(x)$ approaches 1 because $\log(20x - 17) - \log(2x + 9) = \log\left(\dfrac{20x - 17}{2x + 9}\right)$ and $\dfrac{20x - 17}{2x + 9}$ approaches 10, so $f(x)$ approaches $\log 10 = 1$.

Exercises 10.4 (p. 615)

1. $2\log_b x + 5\log_b y$ **3.** $3\log_b x + 3\log_b y - \dfrac{1}{2}\log_b z$ **5.** $5\log_b x - 2\log_b y - 4\log_b z$ **7.** $\dfrac{1}{2}\log_b x - \dfrac{3}{2}\log_b y$

9. $\dfrac{1}{2}\log_b x - \dfrac{1}{3}\log_b y - \dfrac{2}{3}\log_b z$ **11.** $\log_b x + 2t\log_b y$ **13.** $\log_b x^2 z\sqrt{y}$ **15.** $\log_b \dfrac{\sqrt[4]{xy^3}}{z}$ **17.** $\log_b \dfrac{\sqrt{x}}{y^2\sqrt[3]{z^2}}$

19. 1.9842775 **21.** -1.63032679 **23.** -1.2618595 **25.** 2 **27.** 125 **29.** 2 **31.** 3 **33.** $\dfrac{1}{3}$ **35.** 3

37. -2 **39.** $\dfrac{3}{2}$ **41.** $-\dfrac{3}{4}$ **43.** 3.4 **45.** 4.8 **47.** 4.8 **49.** 0.7 **51.** 0.4 **53.** 0.8 **55.** $\dfrac{1}{3}$

57. -1.85 **59.** Alaska, 8.5; San Francisco, 8.25 **61.** Let $\log_b N = n$ and rewrite in exponential form.
$$b^n = N$$
So, $N^r = (b^n)^r = b^{nr}$
Rewrite $N^r = b^{nr}$ in logarithmic form.
$$\log_b N^r = nr \text{ or } r \cdot n$$
$$= r \cdot \log_b N$$

63. $\ln \dfrac{(x + 3)^2\sqrt{2x - 1}}{|x^2 - x - 3|}$ **67.** $m = 2, n = \dfrac{1}{2}$ **69.** $m = -2, n = -\dfrac{1}{2}$ **71.** $m = 3, n = \dfrac{1}{3}$ **73.** $m = -3, n = -\dfrac{1}{3}$

75. $\log_a b \cdot \log_b a = 1$.
Proof: Let $\log_a b = m$. Then $a^m = b$.
Let $\log_b a = n$. Then $b^n = a$.
Hence $b = a^m = (b^n)^m = b^{mn}$.
Therefore, $mn = 1$ or $\log_a b \cdot \log_b a = 1$.

Exercises 10.5 (p. 623)

1. $\dfrac{\log 5}{\log 11} \doteq 0.6711877$ **3.** $\dfrac{\log 25}{\log 4} \doteq 2.3219281$ **5.** $\dfrac{\log 2.5}{\log 6} - 3 \doteq -2.4886084$ **7.** $-\dfrac{\log 4}{\log \dfrac{3.5}{32}} \doteq 0.6264398$

9. $\dfrac{3\log 2.54 - \log 49}{\log 7 + \log 2.54} \doteq -0.3805767$ **11.** $\dfrac{\log \dfrac{3}{4}}{\log 5.2} \doteq -0.1744946$ **13.** $\dfrac{\log \dfrac{8}{5}}{2\log 2.45} \doteq 0.2622530$ **15.** $\dfrac{1}{3}$ **17.** 32

19. -3 **21.** $\dfrac{4}{3}$ **23.** $\dfrac{1}{216}$ **25.** $100{,}000$ **27.** 3 **29.** 4 **31.** 8 **33.** 5 **35.** 3 **37.** 4

39. (a) $\dfrac{\log 1.5}{2\log 1.05} \doteq 4.15$ yr, which is about 4 yr, 2 mo. But since interest is paid only twice a year, it will be 4 yr, 6 mo.

(b) $\dfrac{\log 2}{2\log 1.05} \doteq 7.1$ yr, or after 7 yr, 6 mo **41.** The year 2009 **47.** $x \doteq 0.77$ **49.** $\{x : x > 4\}$ **51.** $\{x : x > -2\}$

53. $\{x : x > 6\}$ **55.** $\{x : x > -4\}$ **57.** All real numbers **59.** $\left\{\dfrac{4}{5}\right\}$

61. Some terms are not defined for all real numbers. For example, $\log(x - 4)$ is not defined when $x \le 4$.

APPLIED ALGEBRA—YOUR TURN (p. 625)

1. 97.5% of H_0

CHAPTER 10   Review Exercises (p. 627)

1. $f^{-1}(x) = \frac{1}{3}x + \frac{4}{3}$

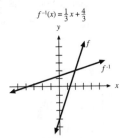

3. $f^{-1}(x) = \frac{3}{2}x - 2$

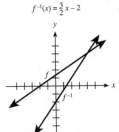

5. $x = 3 - \frac{1}{2}y^2$

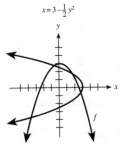

7. $f^{-1}(x) = 3 - x^2, x \geq 0$

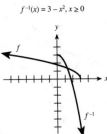

9. $f^{-1}(x) = \dfrac{2 - x}{x}$ **11.** Not 1 to 1   **13.** $f^{-1}(x) = \frac{1}{3}x^2 - \frac{1}{3}, x \geq 0$ **15. (a)** 5   **(b)** $4x + 13$ **(c)** 14

(d) $4x - 2$ **17.** Inverses   **19.** Inverses   **21.**

23.

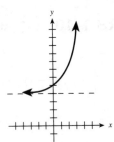

25.

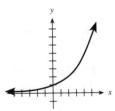

27. $\dfrac{1}{4}$ **29.** 256   **31.** -2 **33.** $-\dfrac{1}{2}$ **35.** 49,300   **37.** $\log_4 2 = \dfrac{1}{2}$

39. $\log_2 y = x - 5$ **41.** $\log_e B = kt$ **43.** $3^1 = 3$ **45.** $3^y = x + 2$ **47.** $7^{2t} = 100$

49.

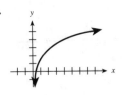

51.

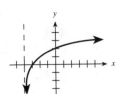

53. 14 days, 27 days   **55.** $\dfrac{3}{2}\log_b x - \dfrac{5}{2}\log_b y$

57. $\log_b D + rt \log_b e$ **59.** $\log_b \dfrac{\sqrt{x^3}}{z\sqrt{y}}$ **61.** 1.9746359   **63.** $\dfrac{1}{3}$ **65.** -3 **67.** 5.2   **69.** 1.2   **71.** $(3.2) \, 10^{-3}$,

$(4.0) \, 10^{-8}$, $(1.0) \, 10^{-8}$ **73.** $\dfrac{\log 7.5}{\log 3} + 4 \doteq 5.8340438$ **75.** $\dfrac{\log 2.5}{0.03 \log 2} \doteq 44.0642698$ **77.** -2 **79.** $\dfrac{1}{49}$ **81.** $\varnothing$

83. -1 **85.** $\dfrac{12}{25}$

CHAPTER 10 Test (p. 629)

1.
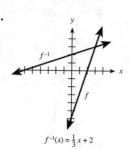

$f^{-1}(x) = \frac{1}{3}x + 2$

3.
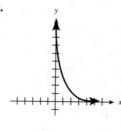

5. $\ln \dfrac{A}{A_0} = rt$ **7.** 9 **9.** $-\dfrac{2}{3}$ **11.** $\dfrac{1}{81}$

13. $3 - \dfrac{\log 1.5}{\log 7} \doteq 2.7916322$ **15.** 5

CHAPTERS 1–10 Test Your Memory (p. 631)

1. $\dfrac{27x^{36}}{8y^{12}}$ **3.** $5x\sqrt[3]{6}$ **5.** $\dfrac{x + 4}{a^2 - ab + b^2}$ **7.** -9 **9.** $f^{-1}(x) = \dfrac{x + 5}{2}$ **11.** $(g \circ f)(x) = 4x^2 - 20x + 29$

13. $y = -2x + 6$ **15.** $(x - 1)^2 + (y + 7)^2 = 25$ **17.**

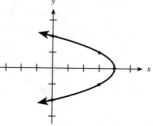

19.
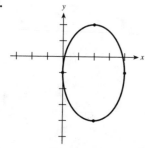

21.
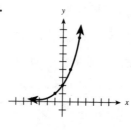

23. $\dfrac{1}{3}$ **25.** $\left\{ -\dfrac{2}{3}, \dfrac{1}{3} \right\}$ **27.** $\{3\}$

Harco rt W

29. $\left\{ 1, -1, \dfrac{5}{2}, -\dfrac{5}{2} \right\}$ **31.** $\left\{ \dfrac{3 \pm i}{5} \right\}$ **33.** $\left\{ -\dfrac{9}{2} \right\}$ **35.** $\{6\}$ **37.** $\{4\}$ **39.** $\left(-\dfrac{1}{5}, -\dfrac{1}{5} \right)$; independent

41. $(-3, -5)$; independent **43.** $\{(4, 3), (-4, 3), (3, -4), (-3, -4)\}$ **45.** 6 ft by 15 ft **47.** 9 mi/hr **49.** 6 lb of Jumbo Deluxe, 4 lb of Small Fry

CHAPTER 11 Exercises 11.1 (p. 638)

1. 5, 8, 11, 14, 17 **3.** 3, 7, 11, 15, 19 **5.** 3, 9, 27, 81, 243 **7.** $\dfrac{2}{3}, \dfrac{4}{9}, \dfrac{8}{27}, \dfrac{16}{81}, \dfrac{32}{243}$ **9.** $1, \dfrac{1}{2}, \dfrac{1}{3}, \dfrac{1}{4}, \dfrac{1}{5}$

11. $-3, 0, 5, 12, 21$ **13.** 3, 7, 13, 21, 26 **15.** $-3, 5, -7, 9, -11$ **17.** $1, -\dfrac{1}{2}, \dfrac{1}{3}, -\dfrac{1}{4}, \dfrac{1}{5}$ **19.** 37 **21.** $\dfrac{729}{125}$

23. $\dfrac{9}{365}$ **25.** $-\dfrac{115}{6}$ **27.** 55 **29.** 55 **31.** 363 **33.** $\dfrac{422}{243}$ **35.** $\dfrac{137}{60}$ **37.** 35 **39.** 75 **41.** -7

43. $\dfrac{47}{60}$ **45.** $-4 - 1 + 2 + 5 + 8 + 11 + 14 + 17 = 52$ **47.** $6 + 13 + 24 = 43$ **49.** $\dfrac{2}{3} + \dfrac{4}{9} + \dfrac{8}{27} + \dfrac{16}{81} = \dfrac{130}{81}$

51. $42 + 64 + 90 + 120 + 154 = 470$ **53.** $2 - 3 + 4 - 5 + 6 = 4$ **55.** $-\dfrac{1}{12} + \dfrac{1}{14} = -\dfrac{1}{84}$ **57.** $\displaystyle\sum_{i=1}^{7} i$

59. $\displaystyle\sum_{i=1}^{6} (2i + 5)$ **61.** $\displaystyle\sum_{i=1}^{5} 3^i$ **63.** $\displaystyle\sum_{i=1}^{4} 80\left(\dfrac{1}{2}\right)^{i-1}$ **65.** $\displaystyle\sum_{i=1}^{5} (-1)^{i+1} i^3$ **67.** $\displaystyle\sum_{i=1}^{\infty} \dfrac{1}{3^i}$ **69.** $\displaystyle\sum_{i=1}^{\infty} (3i - 1)$ **71.** \$256

Exercises 11.2 (p. 645)

1. 2 **3.** $-\dfrac{1}{2}$ **5.** 0 **7.** 64 **9.** 40 **11.** -14 **13.** 7 **15.** $2n + 5$ **17.** $-3n + 1$ **19.** $\dfrac{1}{2}n - 3$

21. $-4n + 5$ **23.** $-\dfrac{3}{2}n - \dfrac{1}{4}$ **25.** $a_1 = 4, d = 3$ **27.** $a_1 = -\dfrac{5}{2}, d = -2$ **29.** $a_1 = \dfrac{7}{6}, d = \dfrac{2}{3}$ **31.** 140

33. 722 **35.** $-\dfrac{800}{3}$ **37.** -299 **39.** 345 **41.** -2415 **43.** 552 **45.** $\dfrac{88}{15}$

Exercises 11.3 (p. 654)

1. 3 **3.** $\dfrac{2}{3}$ **5.** $-\dfrac{3}{2}$ **7.** 96 **9.** 972 **11.** $\dfrac{64}{81}$ **13.** $\dfrac{2187}{256}$ **15.** 2^n **17.** $\left(\dfrac{5}{2}\right)^{n+1}$ **19.** $-6(-3)^{n-1}$

21. $16\left(\dfrac{3}{4}\right)^{n-1}$ **23.** $\dfrac{7^n}{2^{n+2}}$ **25.** -189 **27.** $\dfrac{211}{2}$ **29.** $\dfrac{2255}{32}$ **31.** 27 **33.** $\dfrac{1}{4}$ **35.** Not possible **37.** 126

39. $\dfrac{422}{27}$ **41.** $\dfrac{3}{32}$ **43.** Not possible **45.** $\dfrac{4}{3}$ **47.** $-\dfrac{2}{3}$ **49.** $\dfrac{8}{9}$ **51.** $\dfrac{56}{99}$ **53.** $\dfrac{123}{999} = \dfrac{41}{333}$ **55.** $\dfrac{47}{90}$

57. $\dfrac{20}{11}$ **59.** $\dfrac{581}{198}$ **61.** 560 ft (up and down) **63.** \$31,886 **65.** \$2046 **69.** 211

71. (i) $S_n = \dfrac{1 - r^n}{1 - r}$ if $r \neq 1$

(ii) $S_n = \dfrac{1 - r^n}{r^{n-1}(1 - r)}$ if $r \neq 1$

(iii) The sum found in (i) is r^{n-1} times the sum found in (ii).

73. $\dfrac{11}{80}$

Exercises 11.4 (p. 660)

1. $x^6 + 12x^5y + 60x^4y^2 + 160x^3y^3 + 240x^2y^4 + 192xy^5 + 64y^6$ **3.** $x^5 - 15x^4 + 90x^3 - 270x^2 + 405x - 243$

5. $\dfrac{625}{16} - 125a + 150a^2 - 80a^3 + 16a^4$ **7.** 120 **9.** $\dfrac{1}{1680}$ **11.** 151,200 **13.** 4 **15.** $a^4 + 12a^3 + 54a^2$

$+ 108a + 81$ **17.** $81x^4 + 432x^3y + 864x^2y^2 + 768xy^3 + 256y^4$ **19.** $a^5 - \dfrac{15}{2}a^4b + \dfrac{45}{2}a^3b^2 - \dfrac{135}{4}a^2b^3 + \dfrac{405}{16}ab^4$

$- \dfrac{243}{32}b^5$ **21.** $x^7 - 14x^6y + 84x^5y^2 - 280x^4y^3 + 560x^3y^4 - 672x^2y^5 + 488xy^6 - 128y^7$ **23.** $\dfrac{1}{64}x^6 + \dfrac{3}{16}x^5y + \dfrac{15}{16}x^4y^2$

$+ \dfrac{5}{2}x^3y^3 + \dfrac{15}{4}x^2y^4 + 3xy^5 + y^6$ **25.** $x^5 + 10x^4y^2 + 40x^3y^4 + 80x^2y^6 + 80xy^8 + 32y^{10}$

APPLIED ALGEBRA—YOUR TURN (p. 662)

1. \$86,462,904 **3.** \$789,007

CHAPTER 11   Review Exercises (p. 664)

1. 6, 10, 14, 18, 22   **3.** $-1, 5, 15, 29, 47$ **5.** $-5, 8, -11, 14, -17$ **7.** 19   **9.** $\dfrac{81}{2}$ **11.** 121   **13.** $\dfrac{191}{32}$

15. $\dfrac{137}{120}$ **17.** $\dfrac{2}{5} + \dfrac{4}{25} + \dfrac{8}{125} = \dfrac{78}{125}$ **19.** $1 + 3 + 7 + 13 + 21 + 31 + 43 = 119$ **21.** $-3 + 5 - 7 + 9 = 4$

23. $\displaystyle\sum_{i=1}^{6} 2^i$ **25.** $\displaystyle\sum_{i=1}^{6} \dfrac{i+2}{i+1}$ **27.** 64 words/min   **29.** $-\dfrac{1}{2}$ **31.** -42 **33.** $3n + 2$ **35.** $\dfrac{1}{2}n + \dfrac{7}{2}$ **37.** $a_1 = 13$;

$d = 4$ **39.** -45 **41.** -24 **43.** $\dfrac{129}{2}$ **45.** 10,100   **47.** -3 **49.** 567   **51.** $-\dfrac{5}{16}$ **53.** $-15(-3)^{n-1}$

55. $\dfrac{1}{6}\left(\dfrac{3}{5}\right)^{n-1}$ **57.** $\dfrac{2101}{16}$ **59.** Not possible   **61.** $\dfrac{133}{3}$ **63.** 4   **65.** $\dfrac{11}{18}$ **67.** $\dfrac{95}{333}$ **69.** $\dfrac{500}{3}$ ft (up and down)

71. $729a^6 + 1458a^5b + 1215a^4b^2 + 540a^3b^3 + 135a^2b^4 + 18ab^5 + b^6$ **73.** 5040   **75.** $x^4 + 16x^3 + 96x^2 + 256x + 256$

77. $\dfrac{1}{32}x^5 - \dfrac{5}{24}x^4y + \dfrac{5}{9}x^3y^2 - \dfrac{20}{27}x^2y^3 + \dfrac{40}{81}xy^4 - \dfrac{32}{243}y^5$

CHAPTER 11   Test (p. 666)

1. $d = \dfrac{3}{2}, a_n = \dfrac{3}{2}n + \dfrac{5}{2}, s_{24} = 510$ **3.** 100   **5.** 910   **7.** Not possible   **9.** $\dfrac{56}{99}$ **11.** 1320   **13.** 127

CHAPTERS 1–11   Test Your Memory (p. 668)

1. $\dfrac{4y^{14/3}}{25x^4}$ **3.** $-7\sqrt{5x}$ **5.** $\dfrac{3x+4}{5x(x^2 + 2x + 4)}$ **7.** 17   **9.** $f^{-1}(x) = \dfrac{x-2}{3}$ **11.** $y = -\dfrac{2}{5}x + \dfrac{23}{5}$

13.   **15.**   **17.**

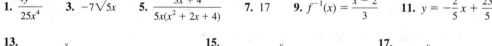

19.

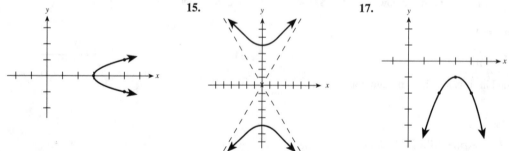

21. $-\dfrac{3}{4}$ **23.** 5040   **25.** 7644   **27.** $\left\{\dfrac{1 \pm 4i}{2}\right\}$ **29.** $\left\{-\dfrac{4}{3}, -\dfrac{1}{3}, \dfrac{1}{3}\right\}$

31. $\left\{-\dfrac{7}{5}, \dfrac{9}{5}\right\}$ **33.** $\left\{-\dfrac{7}{12}\right\}$ **35.** $\{-5\}$ **37.** $\{22\}$ **39.** $\left\{\dfrac{3}{2}\right\}$ **41.** $\{3\}$ **43.** $\{(-4, 3), (3, 4)\}$

45. 6 mm and 8 mm   **47.** 2 ft   **49.** 8 A's

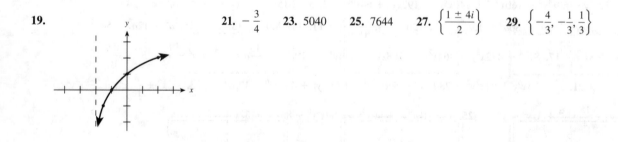

APPENDIX B   Exercises B.1 (p. 683)

1. 28   **3.** −6   **5.** 0   **7.** −10   **9.** 0   **11.** −2   **13.** 0   **15.** 126   **17.** −70   **19.** 0   **21.** 110
23. −110   **25.** 0   **27.** 84   **29.** −188   **31.** −76   **33.** 104   **35.** 42   **37.** 24   **43.** 208.08

Exercises B.2 (p. 688)

1. $(5, -1)$ **3.** $(3, 1)$ **5.** $(2, 0)$ **7.** $\left(-3, \dfrac{1}{2}\right)$ **9.** $\left(-\dfrac{3}{2}, -\dfrac{1}{3}\right)$ **11.** Inconsistent   **13.** Dependent

15. $(2, -1, 3)$ **17.** $(2, 0, -2)$ **19.** $\left(-1, \dfrac{1}{2}, 2\right)$ **21.** $\left(-2, 1, \dfrac{2}{3}\right)$ **23.** $\left(\dfrac{1}{2}, -1, -\dfrac{5}{2}\right)$ **25.** Dependent

27. Inconsistent   **29.** $(-1, 0, 3, 2)$ **31.** $(1.5, -3.2, -4.25, 0.6)$ **33.** $-\dfrac{24}{5}$ **35.** $\dfrac{2}{7}$ **37.** 2 or 5

39. Solve the equation $D = 0$ for a, where D is the determinant of the coefficient matrix.

APPENDIX C   Exercises C.1 (p. 697)

1. 32   **3.** 12   **5. (a)** 27   **(b)** 6   **(c)** 9   **(d)** 15   **7.** 144   **9.** n **11.** 2520   **13.** 4   **15.** 39,916,800
17. 362,880   **19.** 336   **21.** 720   **23.** 45,360   **25.** 280   **27.** 21   **29.** 4   **31.** 1   **33.** n
35. Permutations: *abc bac cab abd bad dab acd cad dac*
                           acb bca cba adb bda dba adc cda dca
                           bcd cbd dbc
                           bdc cdb dcb   **37.** 66   **39.** 4845   **41.** 20   **43.** 180
Combinations: *abc abd acd bcd*

45. (a) 8008   **(b)** 1260   **47.** ${}_nC_{n-r} = \dfrac{n!}{(n-r)![n-(n-r)]!} = \dfrac{n!}{(n-r)![r]!} = {}_nC_r$

Exercises C.2 (p. 708)

1. (a) $\dfrac{1}{6}$ **(b)** $\dfrac{1}{3}$ **(c)** $\dfrac{1}{2}$ **(d)** $\dfrac{1}{3}$ **3. (a)** $\dfrac{1}{52}$ **(b)** $\dfrac{1}{13}$ **(c)** $\dfrac{1}{4}$ **(d)** $\dfrac{4}{13}$ **5. (a)** $\dfrac{1}{12}$ **(b)** 0   **(c)** $\dfrac{1}{6}$

7. (a) $\dfrac{1}{6}$ **(b)** $\dfrac{5}{6}$ **9. (a)** $\dfrac{3}{8}$ **(b)** $\dfrac{1}{2}$ **11.** $\dfrac{2}{3}$ **13. (a)** $\dfrac{3}{8}$ **(b)** $\dfrac{11}{16}$ **15. (a)** 2,763,633,600

(b) $\dfrac{1}{2,763,633,600}$ **17.** 0.32513   **19.** 9.234×10^{-6} **21.** $\dfrac{15}{16}$ **23.** $\dfrac{1}{4}$ **25.** $\dfrac{1}{3}$ **27.** $\dfrac{1}{6}$ **29.** $\dfrac{3}{20}$ **31.** $\dfrac{20}{63}$

33. $\dfrac{25}{484}$ **35.** $\dfrac{9}{25}$

Appendix D (p. 714)

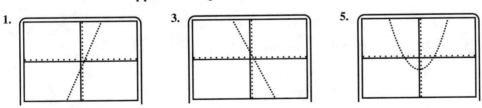

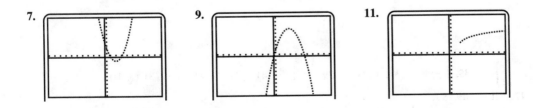

13.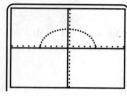

15.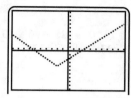

17.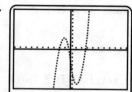

19.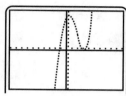

21. $x \doteq 1.67$ **23.** $x \doteq -2.83$ **25.** $x \doteq 3.65, x \doteq -1.65$ **27.** $x \doteq -0.45, x \doteq 4.45$

29. $x \doteq -1.84, x \doteq 5.41$ **31.** $x \doteq -3.00$ **33.** $x \doteq -2.61, x \doteq 3.58$ **35.** $x \doteq 2.84, x \doteq -0.44, x \doteq -2.40$

INDEX

Abscissa, 294
Absolute value, 7, 43
  equations, 89–94, 104
  inequalities, 96–99, 104
Absolute value equations, 89–94, 104
Absolute value inequalities, 96–99, 104
Addition
  of complex numbers, 277–278
  of fractions, 17–21
  of polynomials, 113–115
  of radicals, 261–262
  of rational expressions, 187–192
Additive inverse, 8, 35
  and rational expressions, 189
Additive property of equality, 60, 103
Additive property of inequality, 67, 103
Algebraic expressions
  and linear equations, 45
  and literal equations, 76–78
  and quotient law of exponents,
    168–169, 228
Applications
  and fractional equations, 219–224
  of linear equations, 80–87
  for linear systems of equations,
    543–548
  of literal equations, 76–78
  and quadratic equations, 151–155,
    429–435
Applied Algebra
  Benjamin Franklin Fund and
    compound interest, 633, 661–662

Break–even point, 521, 560–563
Expansion joints in a roadway, 57,
  101–102
Flow rate in a firehose, 237, 283–284
Friction loss in a firehose, 391,
  458–460
Greenhouse frame, 469–470,
  510–512
Heterozygosity of zoo populations,
  571–572, 624–625
Maximum horizontal distance of a
  stream of water, 293, 382–383
Maximum wind speed that windows
  can withstand, 109, 157
Trout population in a lake, 167,
  226–227
Wind–chill factor, 1, 48
Arithmetic sequences, 639–644, 662
  general term of, 640, 663
  sum of first n terms of, 643, 663
  terms of, 639–641
Arithmetic series, 642–643, 663
Array
  elements in square, 679
  for Pascal's triangle, 657
Associative property, 35
Asymptotes, 496, 516, 588, 602
Axes of ellipse, 490, 514–515
Axis
  of hyperbola, 498, 515
  of parabola, 392, 394–398, 461,
    470–471, 513

Base, 11
  of exponential function, 590, 591
  of logarithmic function, 602–603
Base e, 591, 604
  table for powers of, 673–674
Binomials, 110
  and difference of two squares
    formula, 122, 131–132, 158
  divisor as, 203–204, 207–211
  expansion of, 656–660, 664
  factoring of, 130–134, 158
  and Pascal's triangle, 657–658
  square of a binomial formula, 122
Binomial theorem, 659, 664
Boundary curves, 502
  equation of, 502
Boundary lines, 361
Boundary point
  for branch functions, 376
  decimal approximation of, 449
  as imaginary number, 450–451
  for quadratic inequalities, 445
  and rational inequalities, 452–454
Bounded segment of number line, 70
Boyle's law, 372–373
Branch or piecewise functions,
  375–380

Calculators
  and common logarithms, 604, 612,
    617–618

Calculators *(continued)*
  for exponential functions, 591
  graphing, 711–714
  and natural logarithms, 605, 612, 617
Cartesian coordinate system, 295
Center
  of circle, 478–479, 514
  of ellipse, 488, 490, 514
  of hyperbola, 494, 498, 500,
    515–516
Certain event, 701
Change of base property, 611–613, 626
Circle, 471, 478–484, 513–514
  circumference of, 28
  origin as center, 481, 514
Circumference of circle, 28
Closed interval, 72–73
Coefficients, 58
  and absolute value, 92
  and polynomials, 110
Columns of array, 679
Combination, 695–696
Combining like terms, 58, 113–114
Common denominators, 18, 189–190
Common difference, 639–640, 662
Common factor, 126–129
Common logarithms, 604
  table of, 674–677
Common ratio, 646, 663
Commutative property, 35
Complement of an event, 704
Completing the square
  to find center, 482–483, 491–493,
    500–501
  to find vertex, 403–406, 474
  solving quadratic equations by,
    414–415
  steps for, 406, 417, 461–462
Complex conjugates, 279
Complex fractions, 194–200
  with negative exponents, 199–200
Complex numbers, 275–281, 285
  addition of, 277–278
  arithmetic of, 285
  division of, 280–281
  multiplication of, 278–279
  subtraction of, 277–278
Composition, 580–583
Compound inequality, 71
Compound interest formula, 618, 626
Conditional probability, 704–705
Conic sections, 470–501, 512–516
  circles, 471, 478–484, 513–514
  ellipses, 471, 487–493, 514–515
  horizontal hyperbolas, 470–477,
    512–513
  hyperbolas, 471, 494–501, 515–516
Conjugate axis of parabola, 498, 515
Conjugates, 264
  complex, 279

Constants, 42
  degree of, 111
Constant term, 147
Coordinates, 294, 295–296
Cramer's rule, 684–688
Cubes, 11
  factoring sum and difference of two,
    132–134
  radicals and, 253

Decibel ratings, 614
Decimals, 21–23
  repeating, 21–22, 653–654
  terminating, 21
Degree
  of a polynomial, 111
  of a term, 111
Denominators, 14–23
  rationalizing the, 256–258, 265–266
Dependent linear system of equations,
  524
  and elimination method, 526
  and substitution method, 524
  and three variables, 533–534,
    536–537
Dependent variable, 321
Descartes, René, 295
Determinants, 679–683
  and Cramer's rule, 684–688
  defined, 679
  evaluation property of, 681–682
  higher order, 682
  minor of element in, 680
  third order, 680–682
Diagonal method for determinant,
  682–683
Diagonals, 27
Difference, 80
Difference of two cubes, 132–134, 158
Difference of two squares formula, 122,
  158
  and factoring binomials, 131–132,
    158
Direct proportion, 367
Direct variation, 367, 385
Discriminant, 422–423, 462
Distance equation, 76
  fractional equations for, 222–224
Distance formula, 352, 384
  and Pythagorean theorem, 350–352
Distance on number line, 89–91, 93,
  96, 97, 99
Distributive property, 35, 37
  and like terms, 58
  and polynomials, 112–113
Dividend, 12
Dividing out common factors, 177–180
Division, 12
  and complex numbers, 280–281

  of fractions, 17
  order of operations, 40–41
  of polynomials, 202–206
  of radicals, 255–256, 265–266
  of rational expressions, 184–185
  synthetic, 207–211
Divisor, 12
Domain, 303–306, 308

Elements
  of set, 2
  in square array, 679
Elimination method, 526–527, 533,
  564–565
  and nonlinear system of equations,
    554–557
  three variables, linear systems in,
    533, 564–565
Ellipse, 471, 487–493, 514–515
Empty set, 3
  solution set as, 63–64
Equalities, properties of, 34
Equations
  of boundary curve, 502
  of boundary line, 361
  of circle, 478–484, 513–514
  of ellipse, 487–493, 514–515
  exponential, 593–594, 617–619
  extraneous solution, 213, 268
  graph of, 298
  higher degree, 441–442
  of hyperbola, 494–501, 515–516
  logarithmic, 620–622
  systems of, 522–559
  two variables, solution in, 297–298
Equivalent
  equations, 59
  inequalities, 67
Evaluating algebraic expressions, 45
Evaluation property of third–order
  determinant, 681–682
Event, 700
  certain, 701
  complement of, 704
  impossible, 700
  independent, 707–708
  probability of, 700
Expansion
  of binomials, 656–660, 664
  of determinant by minors, 681–682
Experiment, 700
Exponential equation, 593–594,
  617–619
Exponential form, 602
Exponential functions, 586–594
  one–to–one property of, 593, 626
Exponential notation, 11
Exponents, 11, 170–172, 238–242
  integer, 168–175

laws of, 228, 238, 285
  power to a power law of, 119, 158, 228, 238, 285
  product law of, 118, 158, 228, 238, 285
  product to a power law of, 119, 158, 228, 238, 285
  quotient law of, 168–169, 175, 228, 238, 285
  quotient to a power law of, 169, 175, 228, 238, 285
Expression
  algebraic, 45
  rational, 177–211
Extraction of roots, 413–414, 462
Extraneous solutions, 213, 269

Factoring, 144–145, 158
  of binomials, 130–134
  by grouping, 128–129, 133–134, 140–141
  cubes, sum and difference of, 132–134
  difference of two squares, 130–134
  and equations in quadratic form, 437–441
  greatest common factor, 126–129
  of polynomials, 126–145
  and quadratic equations, 146–150
  of trinomials, 135–141
Factors and terms distinguished, 9, 60
Finite sequences, 634, 662
Finite set, 3
First coordinate, 294
Foci (Singular – Focus)
  of ellipse, 488
  of hyperbola, 494, 498
FOIL technique, 121–122
  and complex number multiplication, 278
  and radicals, 263–264
Fractional equations
  applications of, 219–224
  solutions for, 212–213
Fractions
  addition property, 238
  addition property of, 187
  common denominators, 18–19, 189–190
  complex, 194–200
  division property, 238
  division property of, 184
  lowest terms, 15
  multiplication property, 238
  multiplication property of, 181
  and negative exponents, 171–175
  reduction property, 228
  reduction property of, 14, 178
  simple, 194

simplified, 177
  subtraction property, 238
  subtraction property of, 188
Functional notation, 312–316, 384
Functions, 304–315
  combinations of, 314–315
  defined, 304
  defined by more than one equation, 375–380
  graph of, 305
  linear equations, 320–343
  one–to–one, 578–580
  vertical line test for, 306–307
Fundamental principle of counting, 692

General term, 634
  of arithmetic sequence, 640, 663
  of geometric sequence, 648, 663
Geometric sequences, 646–650, 663
  general term of, 648, 663
  sum of first n terms of, 649, 663
  sum of terms of an infinite, 651, 663
Geometric series, 650, 663
Golden ratio, 30–31
Graphing
  of branch functions, 375–380
  of circle, 479–484, 514
  of ellipse, 488–493, 514–515
  of equation, 298
  of exponential functions, 587–591
  of function, 305
  of hyperbola, 495–501, 516
  of inverse functions, 575–578
  of linear equations, 320–330
  of linear inequalities in two variables, 360–365
  of logarithmic functions, 603–605
  of nonlinear system of equations, 551–559
  of numbers, 32
  of ordered pairs, 296
  of parabola, 392–410, 471–476, 513
  of quadratic equations, 392–410, 470–501
  of second degree inequalities, 502–507
  of sets, 32, 43
  of slope of lines, 326–331
  of solution set, 67
  of system of equations, 522–524
Graphing calculator features
  GRAPH, 712
  RANGE, 712
  TRACE, 714
  ZOOM, 712
Graphing method, 522–524
  for three–dimensional solutions, 532
Greatest common factor, 126–129

Grouping, factoring by, 29–31, 133–134, 140–141
Grouping symbols, 41

Half–open interval, 72–73
Half planes, 361
Higher degree equations, 441–442
Hooke's law, 368–369
Horizontal line, 330, 384
Horizontal line test, 580
Horizontal parabolas, 470–477, 513
Hyperbola, 471, 494–501, 515–516

i, 275–276, 285
  simplifying to any integer power, 281
Identity property, 35
Imaginary numbers, 277
Imaginary part of complex number, 277
  and conjugate, 279
Imaginary solution, 423
Impossible event, 700
Inconsistent linear system of equations, 523
  and elimination method, 529, 536
  and substitution method, 524–526
  and three variables, 533–534
Independent events, 707–708
Independent linear system of equations, 522
  and three variables, 533
Independent variable, 321
Index of radical, 248
Index of summation, 636, 637
Inequalities
  interval notations, 72–73
  Linear, 66–74, 104
  polynomial, 455–456
  Properties of, 35, 67, 68
  Quadratic, 444–452, 462
  Rational, 452–455, 463
  Second–degree, 502–507
Inequality symbols, 33–34
Infinite geometric series, 651, 663
Infinite sequences, 634, 662
Infinite series, 635, 662
Infinite set, 3
Input and function, 312–313
Integer exponents, 168–175
Integers, 6–13
  as rational numbers, 14
Internet Connection
  average gasoline costs, 318–320
  energy supply and demand, 244–247
  predicting housing units in Hawaii, 347–349
  predicting smoking levels, 597–601
  U.S. trade balance, 538–542

Intersection of sets, 4
Interval notation, 72–73
Inverse functions, 572–584, 626
    and composition, 582–583
    graphing of, 575–578
    and property of logarithmic
        functions, 612–613
Inverse property, 35
Inverse variation, 370, 385
Irrational numbers, 26–29
Irrational solutions, 423
Is, is the same as, defined, 80

Joint variation, 371–372, 385

Latitude, 294
Leading coefficients, 112
Least common denominators (LCDs),
    18–19
    of complex fractions, 197
    and fractional equations, 212–215
    of rational expressions, 190
Like radicals, 261
Like terms, 58
    and polynomials, 113–114
Linear equations, 58–61, 104, 320–343
    forms of, 334–343
    slope of lines in graph, 326–331
    in three variables, 532
    in the variable x, 61
    vertical lines and, 330–331
Linear functions, 325
Linear inequalities in one variable,
    66–74, 104
    equivalent inequalities, 67
    reversing inequality symbol, 68–70
    solutions of, 66–67
    in the variable x, 66
Linear inequalities in two variables,
    360–365
    steps in graphing, 362
    system of inequalities, 363–365
Linear system of equations, 522–537
    applications of, 543–548
    and Cramer's rule, 684–688
    elimination method, 526–529,
        532–537, 564–565
    substitution method, 524–526, 564
    in three variables, 531–537
    in two variables, 522–529
Linear term, 147
Literal equations, 76–78
Logarithmic form, 602
Logarithmic functions, 602–615
    change of base property and,
        611–613, 626
    inverse functions and properties of,
        612–613

one–to–one property of, 621–622,
    626
    power property of, 609, 626
    product property of, 608, 626
    properties of, 608–615, 626
    quotient property of, 609, 626
    table of common logarithms,
        674–677
Longitude, 294
Lowest terms, fraction in, 15

Major axis of ellipse, 490
Members of set, 2
Midpoint formula, 356, 384
Minor axis of ellipse, 490
Minor of element
    defined, 680
    in determinant, 681–682
Monomials, 110
    divisor as, 202–203
Multiplication
    of complex numbers, 277–278
    of fractions, 15
    of polynomials, 118–124
    of radicals, 252, 263–265
    of rational expressions, 181–184
Multiplicative inverses, 35–36
Multiplicative property of equality, 60,
    103
Multiplicative property of inequality,
    68–69, 103

Natural logarithm function, 605
Natural numbers, 6–13
    powers and corresponding roots, 671
    radical equations raised to, 268
n factorial, 658, 663
Nonlinear system of equations,
    550–559
Notation
    functional, 312–316
    interval, 72–73
    set, 2, 49
    set builder, 42
    summation, 636–637, 662
Notation symbols, 49–50
Null set, 3
Numerators, 14–23

Ohm's law, 375
One–to–one function, 578–580
One–variable equations, 58
Open interval, 72–73
Ordered pairs, 294
    graphing of, 295–296
    for inverse functions, 573–574
    and logarithmic functions, 602

Ordered triples, 531
Order of operations, 40–41
Order of radical, 248
Order of real numbers, 33–34
Ordinate, 294
Origin, 295
    as center of circle, 482–483, 514
Outcome, 700
Output and function, 312–313

Parabola, 299, 301
Parabolas
    x– or y– axes, vertex not on,
        397–398, 402–409
    completing the square to find vertex,
        403–406, 474–476
    defined, 392
    graphing of, 392–410, 471–476, 513
Parallel lines, 340
Pascal's triangle, 657–658
Perfect square trinomials, 136, 158
    completing the square to form,
        400–401
Permutation, 693–694
Perpendicular lines, 340
Piecewise or branch functions,
    375–380
Pixels, 711–712
Plotting ordered pairs, 295–296
Points
    or ordered pairs, 298
    satisfying inequalities, 361–363,
        502–509
Point–slope form, 334–335, 384
    standard form, 336
Polynomials, 110–145
    addition of, 112–115
    defined, 110
    degree of, 111
    division of, 202–211
    examples of, 110, 111
    factoring of, 126–145
    inequalities, 455–456
    multiplication of, 118–124
    and quadratic inequalities, 444–452,
        462
    simplification of, 113–114
    standard form for, 112
    synthetic division of, 207–211
Population growth and exponential
    functions, 592–593
Power, 11
Power notation, 11
Power to a power law of exponents,
    119, 158, 228, 285
Power property of logarithms, 608, 626
Prime meridian, 294
Prime number, 14
Prime polynomials, 131

Probability
    of an event, 700
    Product rule for, 706
Product, 9, 80
Product law of exponents, 118, 158,
    228, 238, 285
Product to a power law of exponents,
    119, 158, 228, 238, 285
Pure imaginary number, 277
Pythagorean theorem, 27, 159
    and distance formula, 352

Quadrants, 296
Quadratic equations, 146–150,
    413–425
    applications of, 151–155, 429–435
    defined, 146
    discriminant used for, 422–423, 462
    equations in quadratic form,
        437–441
    extraction of roots theorem, 413
    graphing of, 392–410, 470–501
    and higher–degree equations,
        441–442
    minimum y values for, 397–398
    rational solutions to, 423, 463
    solving by completing the square,
        414–415
    summary, 427–428
    zero product theorem, 147, 159
Quadratic formula, 418–419, 462
    discriminant in, 422–423, 462
    standard form required for, 418–419
Quadratic inequalities, 444–452
    steps used to solve, 446, 451, 462
Quadratic term, 147
Quotient, 12, 80
Quotient law of exponents, 168–169,
    175, 228, 238, 285
Quotient to a power law of exponents,
    169, 175, 228, 285
    and radicals, 255
Quotient property of logarithms, 609,
    626

Radical conjugates, 264
Radical equations, 268–274, 285
Radicals, 248–266, 285
    division of, 265–266
    division property of, 255
    like radicals, 261
    multiplication property of, 252
    operations with, 261–266
    properties of, 285
    rational exponent, change to,
        249–250
    reduction of index of, 258–259
    simplification of, 253–259

Radical sign, 248
Radicand, 248
    and quadratic formula, 422–423
Radius of circle, 28–29, 478–479
Range, 303–306, 308
Rational exponents, 238–242
    defined, 238–240
    radical changed to, 249–250
Rational expressions, 177–211
    addition of, 187–192
    defined, 177
    division of, 184–185
    equations involving, 212–215
    least common denominators of, 190
    multiplication of, 181–184
    reduction to lowest terms, 177–180
    subtraction of, 188–191
Rational inequalities, 452–455, 463
Rationalizing the denominator,
    256–256, 265–266
Rational numbers, 14–15, 23, 177
Rational solution to quadratic
    equations, 423
Real number line, 32, 43–44
    and absolute value, 43–44
Real numbers, 31–46
    ordered pair of, 295
    order of, 33–34
    properties of, 35–36
    relations and, 303
Real part of complex number, 277
Reciprocals, 36
Rectangle, area of, 152, 159
Rectangular coordinate system, 295
Reduction property of fractions, 14,
    178
Reflexive property, 34
Relations, 303
Repeating decimals, 21–22
    and infinite geometric series,
        653–654
Rows of array, 679

Sample space, 700
Second coordinate, 294
Second–degree inequalities
    graphing of, 502–509
    system of, 507–509
Second–order determinants, 679
Sequences, 634, 653, 663
    arithmetic, 639–644, 663
    geometric, 646–650, 663
Series
    arithmetic, 642–643, 663
    geometric, 650, 663
    infinite, 635, 662
Set builder notation, 42
Set notation, 2
Sets, 2–5

equal sets, 4
graphing of, 32, 43
intersection of, 4
subsets, 2–3
union of, 4
Sign diagrams, 681–682
Sign of numbers, 7–13
    and absolute value, 43–44, 89
    of fractions, 16
    multiplicative inverses, 36
    and slope of line in graph, 329
    and third–order determinants,
        681–682
Simple fractions, 194
Simplified fractions, 177
Slope, 384
Slope–intercept form, 338, 384
Slope of lines, 326–331
Solutions, 59
    and additive property, 60
    to equation in three variables, 531
    to equation in two variables,
        297–298
    extraneous solutions, 213, 268, 621
    and multiplicative property, 60
    of system of equations, 522
Solution sets, 59
    and empty set, 63–64
    graphing of, 67
    linear inequalities and, 67
Solved for a variable, 76
Square
    of a binomial, 158
Square of a binomial formula, 122
Squares, 11
    difference of two squares factoring,
        131–132
    and irrational numbers, 26–27
    order of operations, 40–41
    radicals and, 253
    table of square roots, 672–673
Standard form
    formula requiring, 418–419
    for a polynomial in one variable,
        112
    for a quadratic equation, 146
    for a quadratic function, 400
    and quadratic inequalities, 445
Subsets, 2–3
Substitution method, 524–526, 564
    and nonlinear system of equations,
        550–554
Substitution property, 34
Subtraction
    of complex numbers, 277–278
    of fractions, 18
    of polynomials, 113–114
    of radicals, 261–262
    of rational expressions, 188–191
Sum, 9, 80

Sum of first n terms of arithmetic sequence, 643, 663
Sum of first n terms of geometric sequence, 649, 663
Summation notation, 636–637
Sum of terms of infinite geometric sequence, 651, 663
Sum of two cubes, 132–134
Sum of two cubes formula, 158
Symmetric property, 34
Synthetic division, 207–211
System of inequalities, 363–365
   for second–degree inequalities, 507–509
Systems of equations, 522–532
   nonlinear system of equations, 550–559

Tables
   e, power of, 673–674
   of powers and corresponding roots of natural numbers, 671
   of squares and square roots, 672–673
Terminating decimals, 21
Terms, 9
   of arithmetic sequence, 639–641
   degree, 111
   factors distinguished, 9, 62
   geometric sequence, general terms of, 648, 663
   like terms, 58, 113–114
   quadratic, 147
   of sequence, 634, 663
Third–order determinants, 679, 681–682
   evaluation property of, 681–682

minor of element in, 680
Three variables, linear equations in, 531–537
Transitive property
   of equality, 34
   of inequality, 35
Transverse axis, 498, 515
Tree diagram, 691
Triangle, area of, 159
Trichotomy property, 35
Trinomials, 110
   factoring of, 135–141
   grouping, factoring by, 140–141
   with leading coefficient not equal to one, factoring, 137–139
   with leading coefficient one, factoring, 135–136

Union of sets, 4

Variables, 42
   one–variable equations, 58
   and polynomial division, 202–211
   rational exponents on, 241–242
Variation, 366–373
   Boyle's law and, 372–373
   constant of, 367, 385
   Hooke's law and, 368–369
Vertex
   defined, 392, 471
   formula for finding, 407
   of parabola, 392, 394–398, 407, 471, 474, 513
Vertical format for polynomial multiplication, 123–124

Vertical line, 384
Vertical line test, 306–307
   and horizontal parabolas, 471, 477
Vertices
   of ellipse, 491
   of hyperbolas, 495, 498, 515–516

Work problems, 220–222
Word problems
   steps for solving, 87

X–axis, 295
X–coordinate, 296
X–intercept, 322
   of ellipse, 488–491
   of hyperbola, 494–495, 498

Y–axis, 295
Y–coordinate, 296
Y–intercept, 322
   of hyperbola, 494, 495, 498
   slope–intercept form, 338

Zero
   and absolute value, 43–44
   division by, 12–13
Zero product theorem, 147, 159
   extended to higher–degree equations, 442
   and quadratic equations, 415
   and quadratic inequalities, 446

DEFINITIONS AND LAWS OF EXPONENTS AND RADICALS

Let m and n be integers:

$x^m \cdot x^n = x^{m+n}$ Product law

$(x^m)^n = x^{mn}$ Power to a power law

$(xy)^n = x^n y^n$ Product to a power law

$\dfrac{x^m}{x^n} = x^{m-n}, x \neq 0$ Quotient law

$\left(\dfrac{x}{y}\right)^n = \dfrac{x^n}{y^n}, y \neq 0$ Quotient to a power law

$x^0 = 1, x \neq 0$ Zero exponent

$x^{-n} = \dfrac{1}{x^n}, x \neq 0$ Negative exponent

$x^{1/n} = \sqrt[n]{x}$ Rational exponent

$x^{m/n} = (x^{1/n})^m = \left(\sqrt[n]{x}\right)^m$

If a and b are not both negative,

$$\sqrt[n]{ab} = \sqrt[n]{a}\sqrt[n]{b} \qquad \sqrt[n]{\dfrac{a}{b}} = \dfrac{\sqrt[n]{a}}{\sqrt[n]{b}} \quad b \neq 0$$

SPECIAL PRODUCTS

$(a + b)(a - b) = a^2 - b^2$ Difference of two squares

$\left.\begin{aligned}(a + b)^2 = a^2 + 2ab + b^2\\(a - b)^2 = a^2 - 2ab + b^2\end{aligned}\right\}$ Square of a binomial

FACTORING FORMULAS

■ *Difference of two squares:*
$a^2 - b^2 = (a + b)(a - b)$

■ *Perfect square trinomials:*
$a^2 + 2ab + b^2 = (a + b)^2$
$a^2 - 2ab + b^2 = (a - b)^2$

■ *Sum of two cubes:*
$x^3 + y^3 = (x + y)(x^2 - xy + y^2)$

■ *Difference of two cubes:*
$x^3 - y^3 = (x - y)(x^2 + xy + y^2)$

FACTORING SUMMARY

Given a polynomial to factor:

1. Examine the terms to determine if there is a common factor. If so, factor out the greatest common factor.
2. Consider the number of terms.

Two terms	Difference of squares?
	Difference of cubes?
	Sum of cubes?
Three terms	Perfect square trinomial?
	Leading coefficient of 1?
	Leading coefficient not equal to 1?
Four or more terms	Factor by grouping

3. Examine all of the factors to determine if any factor can be factored further.

QUADRATIC EQUATIONS

■ *Extraction of roots property:* If $x^2 = p$, where p is any real number, then $x = \pm\sqrt{p}$.
■ *Zero product theorem:* Let a and b be real numbers. If $a \cdot b = 0$, then $a = 0$ or $b = 0$ or both.
■ *Quadratic formula:* If $ax^2 + bx + c = 0$, where a, b, and c are real numbers with $a \neq 0$, then the solutions of the equation are

$$x = \frac{-b \pm \sqrt{b^2 - 4ac}}{2a}$$

DISTANCE AND MIDPOINT FORMULAS

$d(A, B) = \sqrt{(x_2 - x_1)^2 + (y_2 - y_1)^2}$ $M = \left(\dfrac{x_1 + x_2}{2}, \dfrac{y_1 + y_2}{2}\right)$

MEDICAL
TECHNOLOGY

Design   Cooper–West   Consultant   Simon Nicholas,
Editor   Denny Robson   MA, MB, BChir, LRCP, MRCS,
Researcher   Cecilia Weston-Baker   Cambridge University and
Illustrators   Rob Shone and Cooper–West   London Hospital, UK.

© Aladdin Books Ltd

Designed and produced by
Aladdin Books Ltd
70 Old Compton Street
London W1

First Published in the
United States in 1986 by
Franklin Watts
387 Park Avenue South
New York NY 10016

ISBN 0 531 10199 1

Library of Congress Catalog
Card Number: 86 50032

Printed in Belgium

MODERN
TECHNOLOGY
MEDICAL
TECHNOLOGY

NICHOLAS WICKHAM

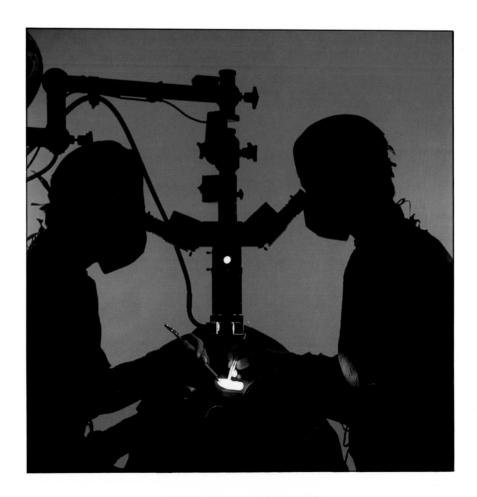

FRANKLIN WATTS
NEW YORK · LONDON · TORONTO · SYDNEY

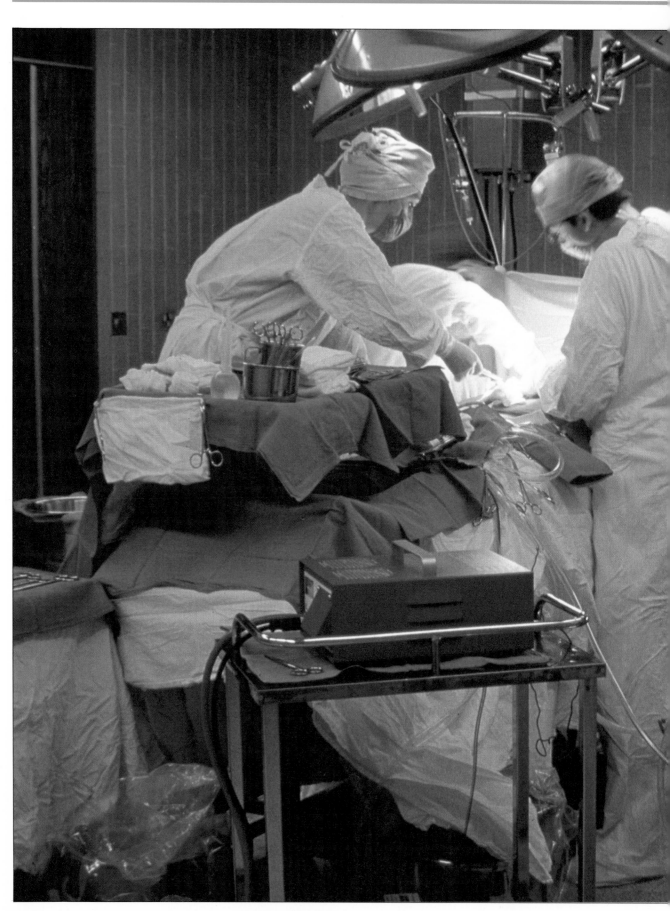

Open-heart surgery in a modern operating theater

Foreword

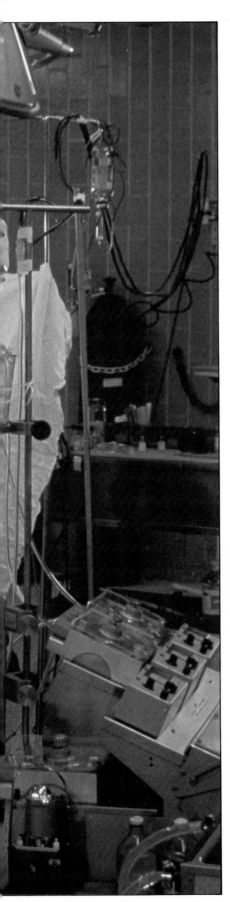

This century has seen a rapid increase in the application of science and technology to our everyday lives. Nowhere is this more apparent than in the field of medicine, as hospitals and their equipment become more sophisticated and complex. This use of medical technology can seem rather impersonal. But patients may now be investigated and treated much more easily and with less discomfort than in the past. Although operations like heart transplants are a result of medical progress, the use of some new techniques actually means that many major operations can be avoided. The medical technology responsible for some of these benefits is the subject of this book.

Contents

The modern hospital	6
Blood tests	8
Scanning technology	10
Nuclear medicine	12
Sound in medicine	14
Light in medicine	16
A world of detail	18
Operating theater	20
Artificial organs and transplants	22
Medicine at home	24
Medical emergencies	26
Medicine tomorrow	28
Datechart	30
Glossary	31
Index	32

The modern hospital

A large modern hospital contains several hundred patients and even more staff. All the different departments necessary for emergency as well as routine treatment are on one site. These departments house the equipment that helps doctors make an accurate diagnosis and prescribe effective treatment; for example, X rays and ultrasound machines, gamma cameras, scanners, fiber-optical instruments and lasers. Because so much of the hospital's service depends on electricity, it has its own generating plant that starts automatically in a power cut.

X rays, ultrasound, scanning
All these types of "body scanners" need to be linked to computers. The operator sits in an adjoining computer-control room. In the case of X-ray scanning, an image of the patient's body is generated by computer onto a screen.

Emergency bleeps
Bleeps, or radiopaging units, are carried by many hospital personnel. In an emergency, a team of doctors and nurses can be called instantly, which not only saves time, but also saves lives.

1. Main entrance
2. Emergency
3. Surgery theaters
4. Outpatients
5. Wards and intensive care
6. X rays, Ultrasound, Radiography
7. Admin, records, computers
8. Supplies
9. Research
10. Paramedics
11. Generator

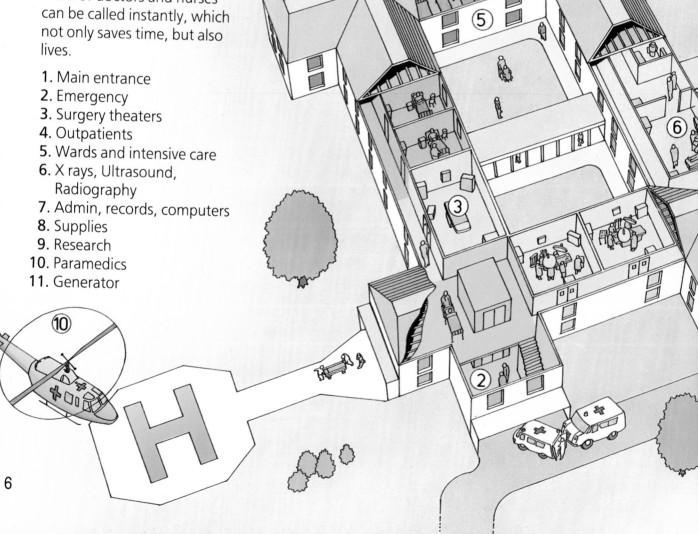

The operating theater

One important feature of a modern hospital is its operating theater. It is designed as a germ-free environment to reduce the risk of any infection and has three zones: the theater, a space for surgical preparations (pre-op) and post-op.

Intensive care

Every hospital has its own intensive care unit with life-support and monitoring systems. One member of staff can monitor all the patients in the unit from a central control panel.

Computer terminals

Television screens with keyboards linked to a computer are situated in laboratories and on the ward. Results of tests can be recalled directly from the computer. "Computer doctors" have been programmed to ask patients many of the routine questions that doctors ask.

Computer link-up

Today, computers help in many areas in a modern hospital. They book dates for admission, arrange operating times and follow up appointments. They can also monitor hospital services, such as the amount of drugs used. Many universities now have links with hospital computers. In this way, the information stored in the computer's memory bank, or "database," can be shared by doctors throughout the country.

Blood tests

Blood tests are often the first and the most important way of finding out what is wrong with a patient. Busy hospital laboratories handle hundreds of blood samples each day. Many different tests can be done on these blood samples by using automatic machines. There are three basic cell types; red cells which carry oxygen, white cells which fight infection and platelets which help prevent cuts from bleeding. Counting these different types of cells in a blood sample can show whether a patient has a blood disorder. Blood cells can be counted accurately and automatically in a "Coulter Counter."

▽ The diagram illustrates how the Coulter Counter works. Blood cells diluted by a salt solution (1) are sucked through a tiny hole (2) into a glass tube. As cells pass through, they interfere with an electric current (3) that is flowing across the hole. The machine detects this interference as a small electric pulse and shows it as a "blip" on a screen.

Coulter Counter

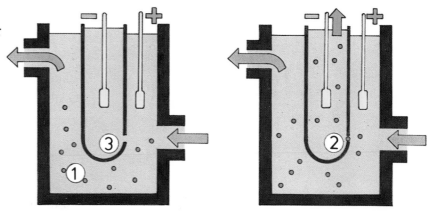

▽ This is a photograph of the Coulter Counter. The machine produces an electric signal or "blip" every time a cell is counted. The size of the blip varies with the size of cell. The machine counts the number of blips in a known volume of blood and works out the cell count. After one minute it displays the counts numerically and also as a graph (below), showing the size distribution of all three cell types.

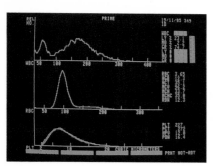

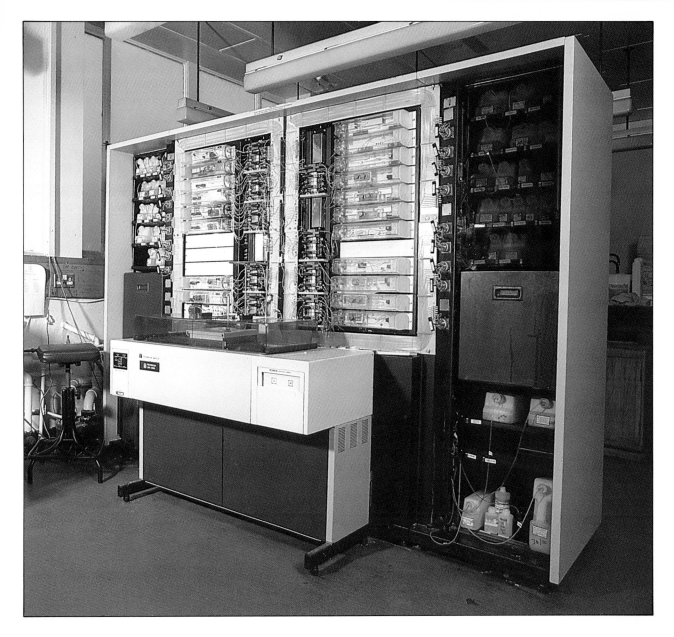

Changes in the blood

As well as cells, blood also contains proteins, fats, minerals and other chemicals. Blood circulates to every organ in the body. If an organ such as the kidney or liver is not working properly, there will be chemical changes in the blood. For example, in kidney failure, the levels of waste-products rise. In liver disease, the yellow appearance of the patient's skin, known as jaundice, is also the result of a build-up of a waste-product. There are several tests that check how the kidney and the liver are working. The machines that do these tests automatically are called "multi-channel analyzers." The results appear on a computer screen.

△ This photograph shows a "multi-channel analyzer," a machine that is able to carry out several different types of blood tests at the same time. In fact, up to 20 tests can be carried out at the same time by a machine of this kind. These tests together are known as a "blood profile."

9

Scanning technology

"Computerized axial tomography" is the full title of an X-ray technique, commonly called a CAT scan, that has changed the world of X-ray diagnosis in the past 15 years. The word tomography comes from the Greek, meaning to cut (*tomo*) and to write (*graphein*). The scanner moves around the patient and pictures are produced of cross sections through any part of the body, in great detail. Its great advantage is that you can see soft tissues inside the body very clearly.

▽ Beams of X rays (1) from an X-ray source (2) pass through a thin section of the body (3) while the scanner (4) rotates. Detectors (5) receive the information shown up by the X rays and pass it on to a computer for processing.

▽ In the photograph below, information from a patient's scan is being processed. The computer processes all the side views to construct a cross section on a TV screen. The insets show a body picture produced from the scanner (top) and a cross section through the body (bottom).

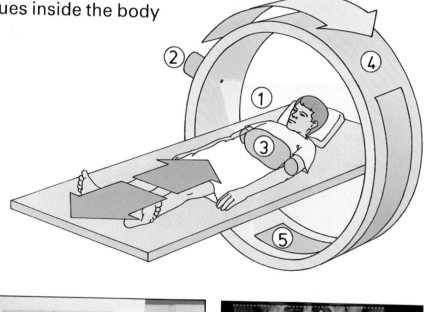

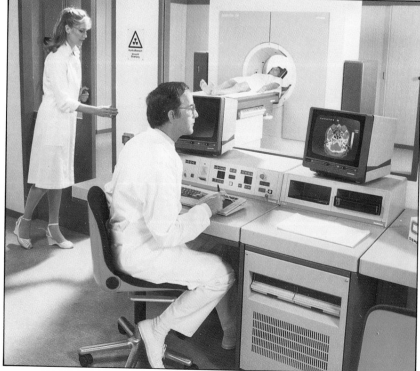

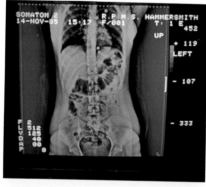

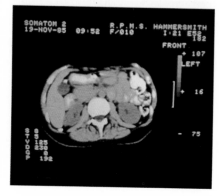

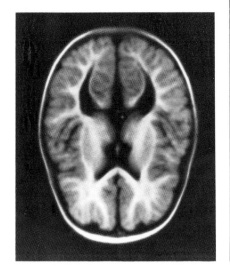

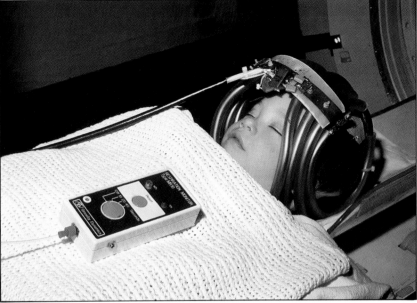

△ NMR scans are very good for looking at brain tissue. They show the white and gray matter very clearly, as can be seen in this baby's scan. No signal comes from bone, so the skull does not get in the way.

Scanning with nuclear magnetic resonance (NMR)

NMR does not use X rays, but uses radio waves. A whole body NMR scanner is very large and weighs 4.5 tons, because it contains powerful magnets. It detects minute changes that occur inside a body when the body is placed in a very strong magnetic field and exposed to pulses of radio waves. These are produced from metal coils that fit closely around the body. They also act as detectors and send their signals to a computer that builds up a picture like the CAT scanner. The scanning is done electronically.

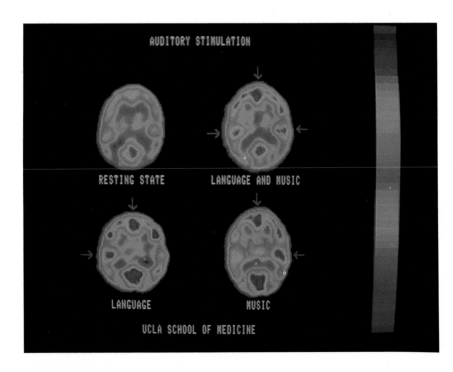

AUDITORY STIMULATION

RESTING STATE   LANGUAGE AND MUSIC

LANGUAGE   MUSIC

UCLA SCHOOL OF MEDICINE

◁ A research method called "positron emission tomography" (PET scan) has been used to study different parts of the body in action. A PET scanner works by picking up signals from a radioactive substance that has been injected into the patient's body. The photograph on the left shows how these signals are picked up and generated into a picture by the scanner's computer. Each of the four pictures represents a different activity of the brain.

11

Nuclear medicine

If radioactive chemicals are injected into the bloodstream, they will give off radiation. This allows doctors to trace their course through the body with radiation detectors and so look at how each organ and gland is working. Radioactive chemicals that produce "gamma rays" are commonly used, because these rays can penetrate the body well. The rays are detected by a "scintillation counter," which scans mechanically back and forth across the body. Many detectors used together form a gamma camera that can stay still and take pictures of whole organs. The pictures can be displayed on a computer screen, printed on paper, and color enhanced.

▽ In the diagram, a patient injected with a radioactive substance is being scanned by a gamma camera. The diagram on the right illustrates how the gamma camera works. When a gamma ray (1) hits a crystal (2) such as sodium iodide, it produces a small flash of light, or "scintillation" (3). This flash is detected by a light-sensitive emulsion (4) in a "photo-multiplier tube" (5). This produces a weak electric signal (6) that is strengthened up to a million times by the tube and fed to the computer (7) to make up the picture. In a gamma camera there may be over 50 crystals. To make sure that each one only sees gamma rays straight ahead, a lead screen with holes called a "collimator" (8) is used.

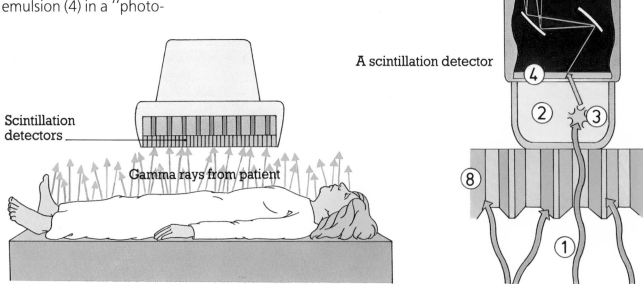

A scintillation detector

Scintillation detectors

Gamma rays from patient

Radioactive labels

Different tissues in the body use different chemicals. Therefore, when a radioactive substance, or "label," is attached to one of these chemicals, the organ that takes up and uses that substance can be studied. Doctors hope to develop a type of label that could be easily fixed to cancer cells. If these are labeled successfully, then the gamma camera will be able to detect cancer in its early stages.

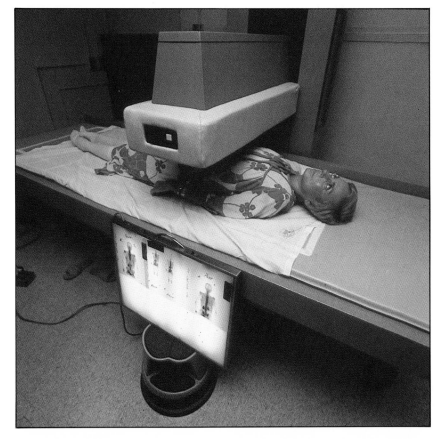

◁ This photograph shows a woman undergoing a gamma camera scan. This type of nuclear medicine is frequently used to determine whether a cancer has spread to other parts of the body.

▽ Below are two examples of the type of picture produced by a gamma camera. The different colors show how the radioactive label has been taken up by the organ. On the left is a scan of a thyroid gland; on the right is a scan of the liver.

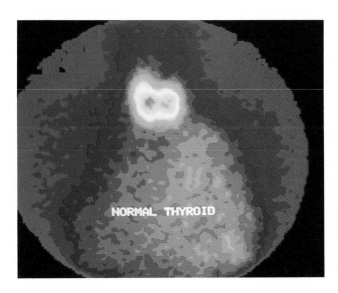

NORMAL THYROID

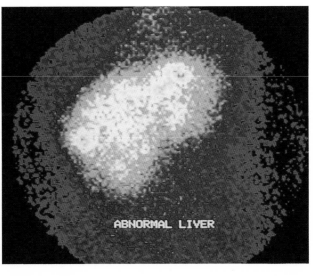

ABNORMAL LIVER

Sound in medicine

Ultrasound is high frequency sound that can be used to look inside the body. As the sound waves hit the different structures inside the body, they are bounced back as an echo. This produces a soundwave signal that a computer uses to make a picture on a screen. This is the same principle that bats use to navigate in the dark. The high frequency sound is produced by a "transducer," a machine that changes electrical energy to sound waves.

▷ The plastic transducer probe, illustrated in the diagram, contains a quartz crystal. It vibrates with the electric current to produce sound. In between pulses of sound, the transducer "listens" for the returning echoes. If the transducer is swept back and forth very quickly, then the computer can produce images from the returning echoes, which give very lifelike pictures.

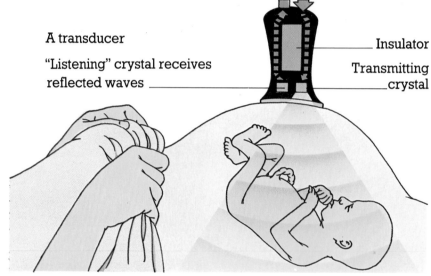

Electric signals to computer

Electricity

A transducer

"Listening" crystal receives reflected waves

Insulator

Transmitting crystal

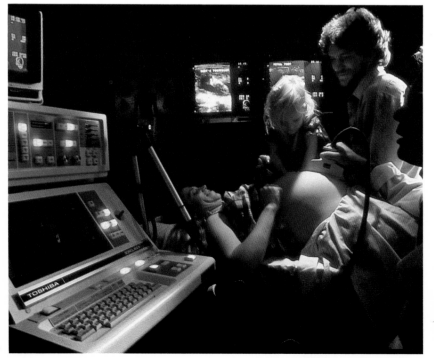

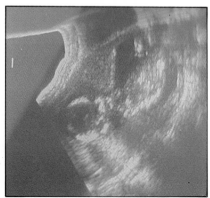

◁ The pregnant woman in the photograph is having an ultrasound scan. Ultrasound is a safe way of seeing babies in the womb. The photograph below shows an ultrasound image of a baby's skeleton.

▷ In the photograph, the patient has been hoisted into a waterbath. The shock waves are produced under water because they are transmitted through the human body better this way. The patient is anesthetized, as the shock waves are unpleasant and up to 2,000 may be given over one hour to smash big kidney stones.

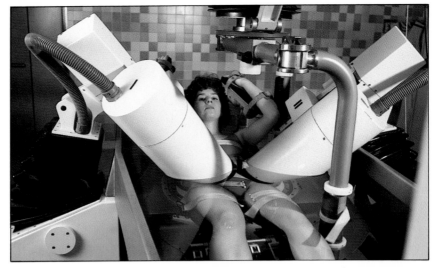

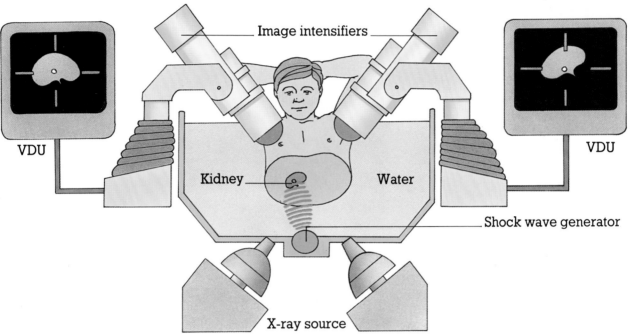

Image intensifiers

VDU

Kidney

Water

VDU

Shock wave generator

X-ray source

△ As the diagram illustrates, the patient is held in the waterbath in a cradle. This makes sure that the kidney stone is in the right place, as the shock wave is focused on a fixed point. The position is checked by X-ray cameras and screens called "image intensifiers," which are connected to video display units. Each shock wave only lasts one microsecond.

Shock wave treatment

If a soprano sings a high note, the sound waves may cause a glass to break. This is because they rattle the glass at a certain frequency, causing it to shatter. A similar principle has been used in a new machine – the "extracorporeal shock wave lithotripter," which destroys kidney stones by producing shock waves that are transmitted harmlessly through the body. It was developed following research conducted by an aircraft company in West Germany, to find out why canopies on jet airplanes shattered during flight. They fired rain drops at 8,000 km/h (5,000 mph) at the canopies and found that the sound waves produced caused the damage.

15

Light in medicine

It is sometimes very useful to be able to look directly inside different parts of the body. In the past, rigid tubes were used to look down the throat into the stomach, or down the windpipe. These are uncomfortable and cannot look around corners, except with mirrors. Nowadays, fiber-optical instruments, called "endoscopes," are used to look inside the body's various cavities. They are flexible and enable much more to be seen.

▽ Here, a doctor is using an endoscope. The tube is inserted through the patient's mouth, to reach into the stomach. A picture inside a stomach is seen in the photograph below.

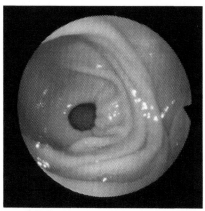

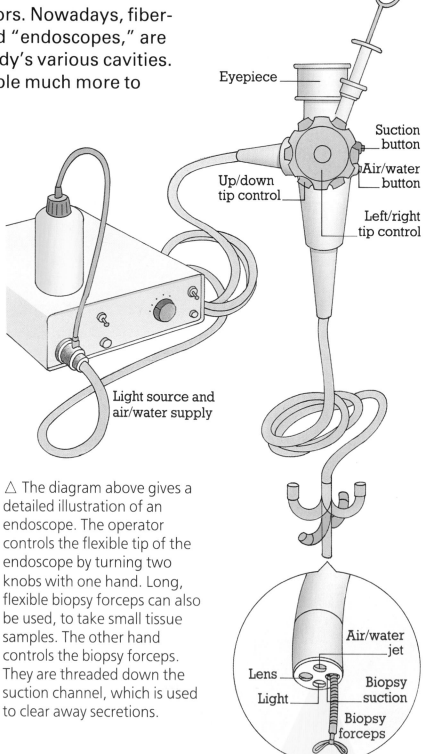

Biopsy forceps control

Eyepiece

Suction button

Air/water button

Up/down tip control

Left/right tip control

Light source and air/water supply

Air/water jet

Lens

Light

Biopsy suction

Biopsy forceps

△ The diagram above gives a detailed illustration of an endoscope. The operator controls the flexible tip of the endoscope by turning two knobs with one hand. Long, flexible biopsy forceps can also be used, to take small tissue samples. The other hand controls the biopsy forceps. They are threaded down the suction channel, which is used to clear away secretions.

How an endoscope works

A powerful light source is used to shine a light down one fiber-optic bundle. The picture seen in the eyepiece is carried by the fiber-optic viewing bundle, which is made up of thousands of tiny glass fibers. A beam of light passing along a glass fiber is reflected from the walls of the fiber like a mirror. The fibers stay in the same position in relation to each other and so an endoscope can be used to "see" around bends.

Laser treatment

Lasers are a very intense form of light. They have found many applications in medicine and surgery. Lasers have been used with endoscopes to stop stomach ulcers bleeding. The green-blue light of an argon laser, shone down an endoscope, is absorbed more by the red raw ulcer than the surrounding paler tissues. This produces localized heat that clots the blood and stops the bleeding.

▽ The patient in the photograph below is being treated to remove a birthmark. Birthmarks can be due to surface blood vessels. Treatment with an argon laser will stop the blood flow and allow the vessels to shrink without leaving a scar.

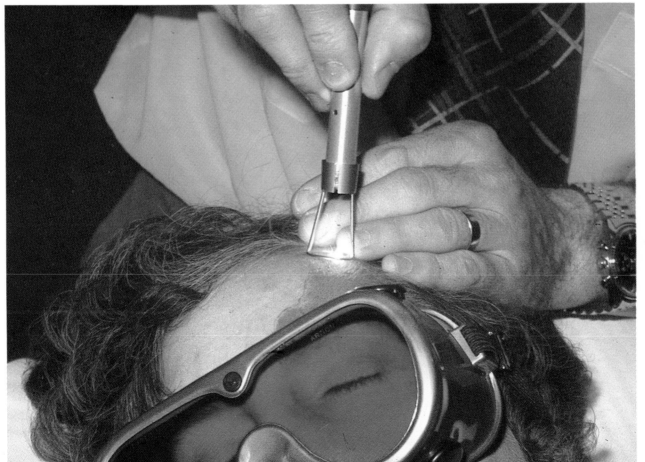

A world of detail

One of the most important tools in today's medical research is the electron microscope (EM). It can produce images of structures up to a million times smaller than the human eye can detect. Working with this size of magnification, scientists are able to study the detailed structure of viruses and cells, such as cancer cells, and information essential to the understanding of diseases is obtained.

Forensic medicine

Police scientists studying the causes of accidents or murders also use the EM, to match up different types of blood or hair to eliminate suspects. The EM can also show how a wound has been caused.

▷ These four electron micrographs show how a scanning EM reveals the surface details of structures as tiny as a single cell. Above left, hair is not smooth but scaly. The blobs are particles of dry shampoo. Above right are chains of spores of the fungus that produces penicillin. Below left, human fat cells are busy absorbing more fat. Below right, this picture also shows the behavior of a cell. These human cells from the lung act as scavengers. One cell is about to engulf a particle of debris.

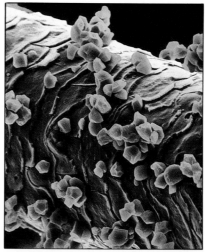
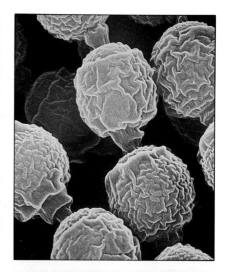

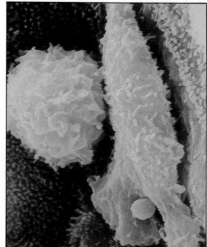

The electron image

Electron microscopes use beams of electrons to look at the very fine detail of cells. A beam of electrons is shone through a specimen such as a cell. The electrons that pass through the cell hit a fluorescent screen and this glows wherever the electrons strike. To form an accurate image on the screen, the cell has to stop some of the electrons passing through. This is done by staining the cell with metal salts. The images are therefore very clever shadow pictures that can be photographed. These are called "electron micrographs." Electron beams can also be reflected off the surface of a cell. This process, known as "scanning electron microscopy," gives very clear surface detail.

▽ A normal white blood cell and an abnormal one called a hairy cell. Studying EM pictures like these helps in understanding blood disorders.

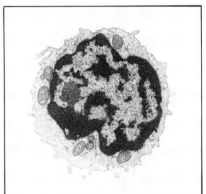

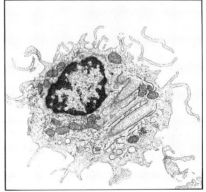

◁ In this electron microscope, the electron beams pass down the central column through the specimen to the screen below. The image is looked at through microscope eyepieces.

19

Operating theater

In the past, surgery was concerned mainly with the removal of diseased tissue. But today, with advances in surgical techniques and equipment, surgeons have much more scope in their work. It is now possible to do more to repair or replace damaged structures in the body.

Anesthesia

Modern surgery has been made possible by the development of safe anesthesia, which has allowed surgeons to perform longer and more complicated operations. The anesthetist monitors the patient's vital functions, all of which can be measured by electronic methods.

Robots

A very recent surgical development has been the use of robots. In California, a robot, "Neu-Robot Ole," is helping surgeons to perform brain surgery faster, and with greater precision.

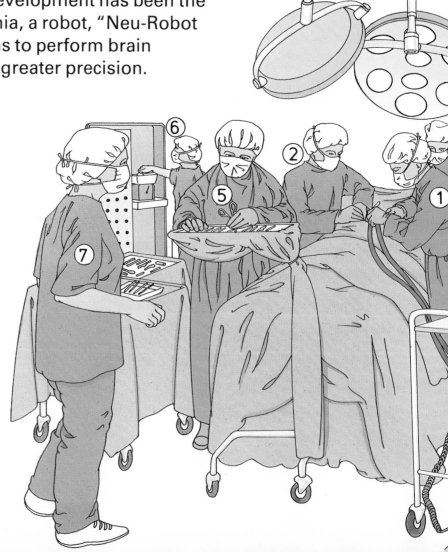

▷ A successful operation depends on teamwork. The surgeon (1) is helped by an assistant (2). As they operate, they check regularly with the anesthetist (3) that there are no problems. The anesthetist is assisted by a technician (4). A scrub nurse (5) assists the surgeon and another nurse (6) will supply the scrub nurse with sterile instruments as they are needed. The head nurse (7) is responsible for the efficient running of the theater and for the training of the nursing staff.

△ An electrocardiograph (right) monitors heart function, while above it are digital readouts for blood pressure, breathing and pulse rate. On the left is the anesthetic equipment — bottles, flowmeters and pressure devices.

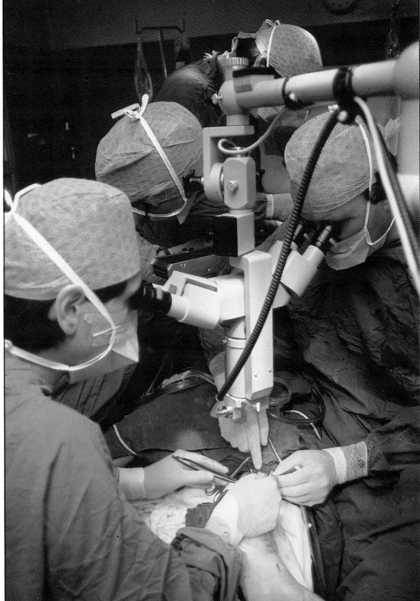

△ Operating microscopes, like the one in the photograph, are used for delicate surgery. They are often used in brain and eye operations. They are also used by surgeons who specialize in repairing very small blood vessels, illustrated in the photograph on the left. Improvements in this technique mean that severed fingers, hands and whole limbs can be reconnected with the nerves and blood supply and stitched back in place.

21

Artificial organs and transplants

Many organs can now be transplanted. Just as it is important to give blood of the right group, so it is important to match the tissues of donors with patients receiving organ transplants. This is called "tissue typing" and is much more complicated than blood grouping. Kidney, liver, pancreas, heart, lung and bone marrow transplants are now possible for certain patients, when no other treatment would save their lives.

Artificial organs

Machines can be constructed to take over the work of some of the body's organs or to replace them altogether. The heart-lung machine allows heart operations to take place. The kidney machine is used in kidney failure if no transplant is available. These are large machines and separate from the body. An artificial heart, the Jarvik 7, has been developed that can fit inside a human being and replace the natural heart.

▽ The kidney preserver (below, right) keeps donor kidneys alive. It cools them and soaks them with a special solution through their blood vessels. The photograph on the left illustrates the procedure involved in tissue-typing. Recent advances in this area have improved the chances of transplant survival.

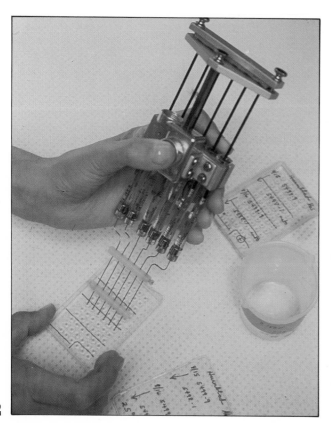

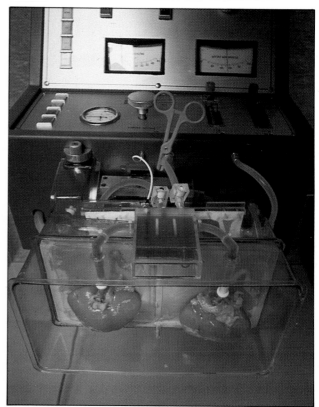

The Jarvik 7 artificial heart consists of two pumps side by side. One takes deoxygenated blood from the body (1) to the lungs (2) and the other takes oxygenated blood from the lungs (3) to the body (4). The outer case is made of polyurethane (5) on a rigid metal base (6).

Inside each pump there is a flexible silicone rubber diaphragm (7). When compressed air (8) is pumped between the rigid outer wall and the rubber diaphragm, blood is squeezed through valves (9) into the arteries. The valves ensure that the blood flows in the right direction.

The power to operate the heart comes from a compressor by the bedside, or from a portable unit. The air enters the Jarvik 7 through tubes that pass through the chest wall. An electric pump is being developed that could work inside the chest.

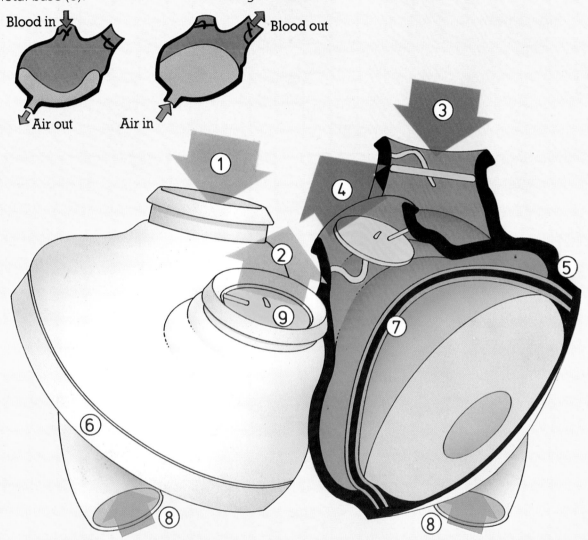

Blood in

Air out

Air in

Blood out

The artificial heart showing the rigid outer plastic case.

The main blood vessels are connected to the Jarvik 7.

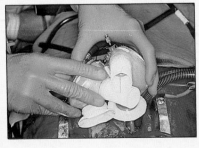

This miniature compressor allows patient mobility.

Medicine at home

Technology is helping to improve the quality of a patient's life at home, as well as in the hospital. The development of hundreds of new drugs has greatly improved treatment over the past 20 years. New ways of giving drugs have also improved the treatment of many patients. The use of ointments that are absorbed slowly through the skin, and the use of electrical pumps that can deliver a drug into the patient's blood vessels at a set rate, are just two examples. Inventions like these can give a patient the confidence to lead a more normal and active life.

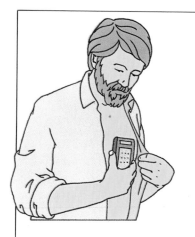

◁ This patient is using a cardiotrak ECG transmitter. It records the electrical activity in his heart. The recording is then transmitted as an audio signal over a normal telephone, to a cardiotrak receiver in the hospital.

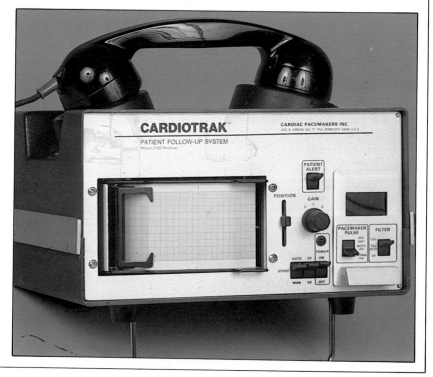

▷ This is a photograph of a cardiotrak receiver, which converts the incoming sound back into an electrical signal and prints out an ECG tracing. This can then be looked at and analyzed by the doctor, who can recommend treatment, or give the patient reassurance if needed.

Port-a-Cath drug delivery

This system consists of a built-in tube or "catheter" that is inserted in a neck vein under anesthesia. One end is attached to a metal reservoir with a plastic bung. This is fixed under the skin of the chest, so that needles may be inserted through the skin and the bung into the reservoir. Blood samples can then be taken and injections given. This system is very useful for some patients, as it avoids repeated injections into the arm veins.

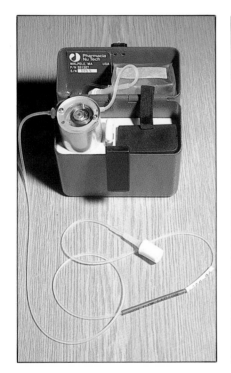

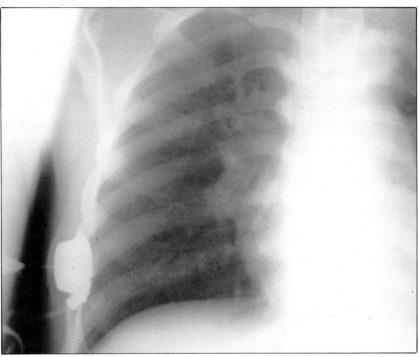

△ Continuous infusion pumps, like the one in the photograph, can be used with the Port-a-Cath. These are small, battery-powered pumps that deliver a drug at a set rate over 24 hours. This is a great improvement for those patients who have difficulty with regular injections.

△ The photograph above is an X ray of a patient with the Port-a-Cath. An X ray is always taken to check that the end of the catheter is in the correct position, just above the heart. The diagram on the left illustrates how the Port-a-Cath works.

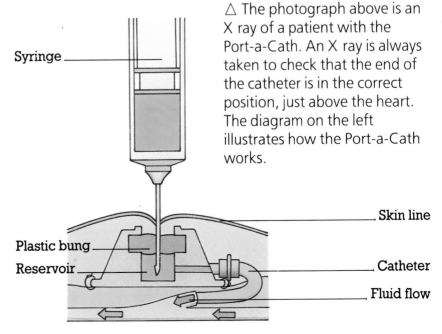

Syringe

Skin line

Plastic bung

Reservoir

Catheter

Fluid flow

Medical emergencies

Ambulance crews are now being trained in emergency medical procedures. These crews are called "paramedics" and their ambulances are equipped rather like a miniature accident and emergency room.

A paramedic ambulance

▽ In the photograph below, we see paramedics at the scene of an accident. They are equipped so that life-saving treatment can be started without delay. The patient in the airplane is being given an inflatable plastic splint, which can protect broken limbs from further damage.

In the ambulance, plasma solution is ready to be given directly into the bloodstream to replace blood loss. Oxygen can be given by a facemask or through a tube. Breathing can be assisted using the ventilator equipment, and there is an electro-cardiograph to record the heart beat. There is also an electric shock machine, to try and restart the heart if it stops beating.

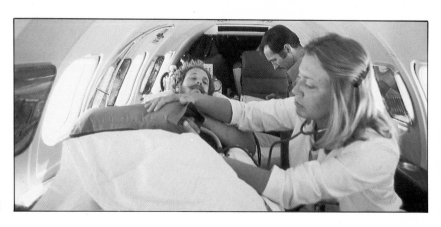

Disaster medicine

Helicopters and their crews play a vital role in reaching victims in inaccessible places in a disaster. After an earthquake, vibration detectors and sensitive listening devices help locate survivors under rubble. Infrared thermal detectors on the surface reveal warm spots produced by the body heat of people beneath. Space blankets, which look like aluminum foil and act as insulators, are used to cover victims.

Emergency health kits

In major disasters, the World Health Organization can supply emergency health kits. These contain enough drugs and basic medical equipment to treat hundreds of patients for a few days. The kits are packed in crates less than 3 cu.m (4 cu.yds) in size, weighing only 730 kg (1,600 lb), and are ready to be rushed to the scene of the emergency.

▽ A helicopter team arrives in an emergency. Equipped with winches and stretchers, they can lift people to safety and take them directly to the hospital, landing on rooftops if necessary.

Medicine tomorrow

Medical technology is at a very exciting stage of development. Research in biology and electronics has given rise to the study of "bioelectronics." Tiny electrical devices have been developed, which stimulate nerves so that muscles can be moved. A pacemaker is a bioelectronic device that sends electric signals to the heart causing it to beat. Another bioelectronic device stimulates nerves of the inner ear to improve hearing.

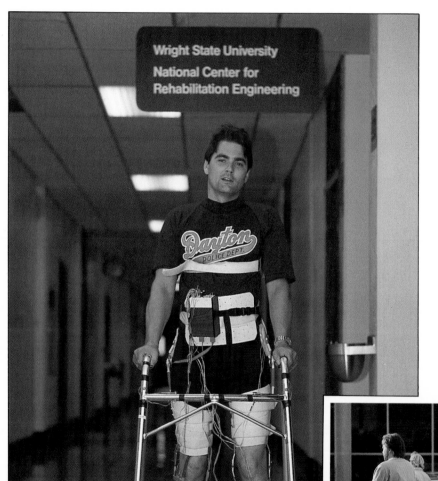

▽ The people in these photographs are walking with the aid of bioelectronics – tiny electrical devices that stimulate nerves. Research in this field means that people who previously had no hope of walking can now look forward to a better life. A future possibility is the production of completely artificial arms and legs that can be linked to and controlled by the brain.

Genetic engineering

Much of this book has been about machines. But the most fascinating machines of all are the tiny cells that make up our bodies. Cells are controlled by "genes:" all the information that is needed for the cell to work properly and to produce substances needed by the body is stored in its genes. Using some very clever techniques, scientists can now "cut out" an individual gene from human cells grown in the laboratory. Using these techniques, it has become possible to manufacture some of the substances that occur naturally within the body. This has been done in the production of human insulin and growth hormone, both of which are now used in treating patients.

Correcting hereditary diseases

Hereditary diseases are due to abnormal genes and many of these have been identified. The exciting prospect for the future is the possibility of inserting the correct gene into a patient's cells.

▽ Some patients suffer from diseases that are caused by a lack of certain substances. The scientists who work with producing these substances artifically from human cells grown in a laboratory are called "molecular biologists." They work in special laboratories like the one shown in the photograph below.

Datechart

1895

Wilhelm Konrad Roentgen discovered X rays. He won the Nobel prize in 1901 for his discovery. This was the first Nobel prize ever given for physics.

1939

The first commercial electron microscope was built by Ruska and Von Borries for Siemens in Germany.

1944

The first artificial kidney machine was built by Willem Johan Kolff who had worked on his discovery throughout World War II.

1958

Hal Oscar Anger, an American electronic engineer at Berkeley in California, developed the first gamma camera suitable for use in medicine.

1960

The first implantable heart pacemaker was inserted. Nowadays, thanks to microelectronics, they are no bigger than a matchbox. Earlier devices were very large and had to be pushed on a cart.

1964

The first practical laser was developed. The word LASER actually means Light Amplification by Stimulated Emission of Radiation.

1969

Dr Denton Cooley in Houston, Texas kept a patient alive with an artificial heart for 64 hours.

1971

The first CAT brain scanner was in use at Atkinson Morley Hospital, UK. Much of the development work was carried out by Gordon Hounsfield working for EMI in England. He later shared the Nobel prize in Medicine for this work in 1974.

1981

Artificial insulin, manufactured by Eli Lilly Company, was used in humans for the first time.

1982

First implantation of Jarvik 7 artificial heart into Dr Barney Clarke at Salt Lake City in Utah. He survived less than three months. Further operations have been carried out with increasing success. The longest survivors to date are nine months and 12 months.

1984

The gene for Factor VIII, a clotting factor in the blood, was isolated by Dr Edward Tuddenham at the Royal Free Hospital, London, England. Patients who have a deficiency of this factor will be treated in the future by pure human Factor VIII, produced by genetic engineering.

Glossary

Anesthesia During some operations, patients need to be unconscious for a period of time. This is achieved by giving the patient a mixture of gases while monitoring his vital functions such as breathing and heart beat. The loss of consciousness in a patient in this way during an operation is called "anesthesia." The word anesthesia means "without feeling."

Color enhancing This technique is used to color pictures on a computer screen (VDU screen). Each signal to the computer is shown up as a different color. Color enhancing is used in many different types of scans and makes computer-generated images easier to analyze.

Diagnosis Doctors are trained to work out what is wrong with our bodies. This is called a diagnosis. To make a diagnosis, they can use many different methods, e.g. blood tests.

Forensic medicine A branch of science concerned with producing information about how a crime has been committed. This medical evidence is often used in a court of law.

Organ A part of the body that has a special structure and is used for a special purpose such as eyes for seeing, stomach for digestion or lungs for breathing.

Scanning Different types of scanner are used to produce a picture of organs inside the body. They can show in detail the brain, heart or lungs. Often, the scan is taken from many angles – like a conventional X ray, and then analyzed by a computer.

Transplant The replacing of natural organs in the body with new organs, e.g. a heart.

Index

A

ambulance 26
artificial limbs 28
artificial organs 22

B

bioelectronics 28
blood 8-9

C

cancer 13, 18
CAT scan 10
cells 8, 18, 19, 29
computers 6, 7, 9, 10,
    11, 12
Coulter Counter 8

D

disasters 27
drugs 24, 25, 27

E

ECG 24, 26
electron microscope (EM)
    18-19
electronics 11, 28
endoscope 16-17

F

fiber-optics 6, 16, 17

G

gamma camera 6, 12, 13
genetic engineering 30

H

heart 22, 23, 25, 26
hereditary diseases 29

I

intensive care 7

J

Jarvik 7 22-3

K

kidney 9, 15, 22

L

laboratories 8, 29
lasers 6, 17
life-support systems 7
liver 9, 22

M

molecular biology 29

N

NMR scanning 11
nuclear medicine 12-13

O

operating theater 7

P

paramedics 26

R

research 28

S

scanners 6, 10, 11, 12, 13
shock wave treatment 15

T

tissue typing 22
transducer 14
transplants 22

U

ultrasound 6, 14

V

ventilators 26

W

World Health Organization 27

X

X rays 6, 10, 15

Acknowledgments
The publishers would like to thank the following organizations for their help in the preparation of this book:
Boehringer Corporation, BUPA Medical Center, Coulter Electronics, Daily Mail Newspaper, Dept of Health and Social Security – Euston Tower, Ferranti Systems, Hammersmith Hospital – CAT Scan dept and hematology dept, Key Med, Oxford Medical Systems, Siemens, St Martins Hospital and special thanks to Mrs Sarah Biggs.

Photographic Credits
Cover, contents page, title page and pages 11, 13, 14, 16, 17, 18, 20, 21, 22, 23, 26/27 and 29, Science Photo Library; pages 8, 9, 10 (insets) and 19, Alex Ramsay; page 10, Siemens; page 11, Jackie Pennock of Hammersmith Hospital; pages 13 and 19, Dept of Medical Illustration, Hammersmith Hospital; page 15, St Martins Hospitals Ltd; page 24, Dr Ward, St Bartholomew's Dept of Medical Illustration; page 25, Pharmacia; page 26, Frank Spooner; page 27, Zefa; page 28, Daily Mail.